Taxus

The genus *Taxus*

**Edited by
Hideji Itokawa
and Kuo-Hsiung Lee**

CRC Press
Taylor & Francis Group
Boca Raton London New York

CRC Press is an imprint of the
Taylor & Francis Group, an **informa** business
A TAYLOR & FRANCIS BOOK

T0175903

CRC Press
Taylor & Francis Group
6000 Broken Sound Parkway NW, Suite 300
Boca Raton, FL 33487-2742

First issued in paperback 2019

© 2003 by Taylor & Francis Group, LLC
CRC Press is an imprint of Taylor & Francis Group, an Informa business

Typeset in Garamond by
Newgen Imaging Systems (P) Ltd, Chennai, India

No claim to original U.S. Government works

ISBN-13: 978-0-415-29837-7 (hbk)
ISBN-13: 978-0-367-39546-9 (pbk)

This book contains information obtained from authentic and highly regarded sources. Reasonable efforts have been made to publish reliable data and information, but the author and publisher cannot assume responsibility for the validity of all materials or the consequences of their use. The authors and publishers have attempted to trace the copyright holders of all material reproduced in this publication and apologize to copyright holders if permission to publish in this form has not been obtained. If any copyright material has not been acknowledged please write and let us know so we may rectify in any future reprint.

Except as permitted under U.S. Copyright Law, no part of this book may be reprinted, reproduced, transmitted, or utilized in any form by any electronic, mechanical, or other means, now known or hereafter invented, including photocopying, microfilming, and recording, or in any information storage or retrieval system, without written permission from the publishers.

For permission to photocopy or use material electronically from this work, please access www.copyright.com (http://www.copyright.com/) or contact the Copyright Clearance Center, Inc. (CCC), 222 Rosewood Drive, Danvers, MA 01923, 978-750-8400. CCC is a not-for-profit organization that provides licenses and registration for a variety of users. For organizations that have been granted a photocopy license by the CCC, a separate system of payment has been arranged.

Trademark Notice: Product or corporate names may be trademarks or registered trademarks, and are used only for identification and explanation without intent to infringe.

British Library Cataloguing in Publication Data
A catalogue record for this book is available from the British Library

Library of Congress Cataloging in Publication Data
Taxus : the genus taxus / edited by Hideji Itokawa and Kuo-Hsiung Lee.
p. cm. – (Medicinal and aromatic plants – industrial profiles; v. 32)
Includes bibliographical references and index.
1. Paclitaxel. 2. Taxus. I. Itokawa, Hideji. II. Lee, Kuo-Hsiung. III. Series.

RC271.P27 T396 2003
615′.3252–dc21
2002151660

Visit the Taylor & Francis Web site at
http://www.taylorandfrancis.com

and the CRC Press Web site at
http://www.crcpress.com

Taxus

Medicinal and Aromatic Plants – Industrial Profiles

Individual volumes in this series provide both industry and academia with in-depth coverage of one major genus of industrial importance.
Edited by Dr Roland Hardman

The authors would like to dedicate this book in tribute to
Dr Monroe E. Wall, the discoverer of the novel anticancer agent Taxol

The authors would like to dedicate this book in tribute to
Dr Monroe E. Wall, the discoverer of the novel anticancer agent Taxol

Contents

List of contributors

David T. Brown
School of Pharmacy
University of Portsmouth
Hampshire, UK

Dairine Dempsey
Department of Pharmacognosy
School of Pharmacy
University of Dublin
Trinity College,
Dublin 2, Ireland

Ingrid Hook
Department of Pharmacognosy
School of Pharmacy
University of Dublin
Trinity College
Dublin 2, Ireland

Hideji Itokawa
Natural Products Laboratory
School of Pharmacy
University of North Carolina at Chapel Hill
Chapel Hill, NC 27599-7360, USA

Yoshinari Kikuchi
Gifu Academy of Forest Science and Culture
88 Sodai, Mino
Gifu, 501-3714, Japan

Mutsuo Kozuka
Natural Products Laboratory
School of Pharmacy
University of North Carolina at Chapel Hill
Chapel Hill, NC 27599-7360, USA

Kuo-Hsiung Lee
Natural Products Laboratory
School of Pharmacy
University of North Carolina at Chapel Hill
Chapel Hill, NC 27599-7360, USA

Susan Morris-Natschke
Natural Products Laboratory
School of Pharmacy
University of North Carolina at Chapel Hill
Chapel Hill, NC 27599-7360, USA

Hironori Ohtsu
Natural Products Laboratory
School of Pharmacy
University of North Carolina at Chapel Hill
Chapel Hill, NC 27599-7360, USA

Masayoshi Oyama
Natural Products Laboratory
School of Pharmacy
University of North Carolina at Chapel Hill
Chapel Hill, NC 27599-7360, USA

Ushio Sankawa
Toyama International Health Complex
Toyama Wellness Foundation International
  Research Center for Traditional Medicine
151 Tomosugi, Toyama 939-8224
Toyama Prefecture, Japan

Koichi Takeya
School of Pharmacy
Tokyo University of Pharmacy and
  Life Science
Horinouchi, Hachioji
Tokyo 192-03, Japan

Hui-Kang Wang
Natural Products Laboratory
School of Pharmacy
University of North Carolina at Chapel Hill
Chapel Hill, NC 27599-7360, USA

Xihong Wang
Natural Products Laboratory
School of Pharmacy
University of North Carolina at Chapel Hill
Chapel Hill, NC 27599-7360, USA

Zhiyan Xiao
Natural Products Laboratory
School of Pharmacy
University of North Carolina at Chapel Hill
Chapel Hill, NC 27599-7360, USA

Mitsuyoshi Yatagai
Graduate School of Agricultural and
  Life Sciences
The University of Tokyo1-1-1 Yayoi
Bunkyo-ku, Tokyo 113-8657
Japan

Preface to the series

There is increasing interest in industry, academia and the health sciences in medicinal and aromatic plants. In passing from plant production to the eventual product used by the public, many sciences are involved. This series brings together information which is currently scattered through an ever increasing number of journals. Each volume gives an in-depth look at one plant genus, about which an area specialist has assembled information ranging from the production of the plant to market trends and quality control.

Many industries are involved such as forestry, agriculture, chemical, food, flavour, beverage, pharmaceutical, cosmetic and fragrance. The plant raw materials are roots, rhizomes, bulbs, leaves, stems, barks, wood, flowers, fruits and seeds. These yield gums, resins, essential (volatile) oils, fixed oils, waxes, juices, extracts and spices for medicinal and aromatic purposes. All these commodities are traded worldwide. A dealer's market report for an item may say 'Drought in the country of origin has forced up prices'.

Natural products do not mean safe products and account of this has to be taken by the above industries, which are subject to regulation. For example, a number of plants which are approved for use in medicine must not be used in cosmetic products.

The assessment of safe to use starts with the harvested plant material which has to comply with an official monograph. This may require absence of, or prescribed limits of, radioactive material, heavy metals, aflatoxin, pesticide residue, as well as the required level of active principle. This analytical control is costly and tends to exclude small batches of plant material. Large-scale contracted mechanised cultivation with designated seed or plantlets is now preferable.

Today, plant selection is not only for the yield of active principle, but for the plant's ability to overcome disease, climatic stress and the hazards caused by mankind. Such methods as *in vitro* fertilization, meristem cultures and somatic embryogenesis are used. The transfer of sections of DNA is giving rise to controversy in the case of some end-uses of the plant material.

Some suppliers of plant raw material are now able to certify that they are supplying organically-farmed medicinal plants, herbs and spices. The Economic Union directive (CVO/EU No 2029/91) details the specifications for the *obligatory* quality controls to be carried out at all stages of production and processing of organic products.

Fascinating plant folklore and ethnopharmacology leads to medicinal potential. Examples are the muscle relaxants based on the arrow poison, curare, from species of *Chondrodendron*, and the anti-malarials derived from species of *Cinchona* and *Artemisia*. The methods of detection of pharmacological activity have become increasingly reliable and specific, frequently involving enzymes in bioassays and avoiding the use of laboratory animals. By using bioassay linked fractionation of crude plant juices or extracts, compounds can be specifically targeted which, for example, inhibit blood platelet aggregation, or have anti-tumour, or anti-viral, or any other required activity. With the assistance of robotic devices, all the members of a genus may be readily screened. However, the plant material must be *fully* authenticated by a specialist.

The medicinal traditions of ancient civilisations such as those of China and India have a large armamentaria of plants in their pharmacopoeias which are used throughout South-East Asia. A similar situation exists in Africa and South America. Thus, a very high percentage of the world's population relies on medicinal and aromatic plants for their medicine. Western medicine is also responding. Already in Germany all medical practitioners have to pass an examination in phytotherapy before being allowed to practise. It is noticeable that throughout Europe and the USA, medical, pharmacy and health related schools are increasingly offering training in phytotherapy.

Multinational pharmaceutical companies have become less enamoured of the single compound magic bullet cure. The high costs of such ventures and the endless competition from 'me too' compounds from rival companies often discourage the attempt. Independent phytomedicine companies have been very strong in Germany. However, by the end of 1995, eleven (almost all) had been acquired by the multinational pharmaceutical firms, acknowledging the lay public's growing demand for phytomedicines in the Western World.

The business of dietary supplements in the Western World has expanded from the health store to the pharmacy. Alternative medicine includes plant-based products. Appropriate measures to ensure the quality, safety and efficacy of these either already exist or are being answered by greater legislative control by such bodies as the Food and Drug Administration of the USA and the recently created European Agency for the Evaluation of Medicinal Products, based in London.

In the USA, the Dietary Supplement and Health Education Act of 1994 recognised the class of phytotherapeutic agents derived from medicinal and aromatic plants. Furthermore, under public pressure, the US Congress set up an Office of Alternative Medicine and this office in 1994 assisted the filing of several Investigational New Drug (IND) applications, required for clinical trials of some Chinese herbal preparations. The significance of these applications was that each Chinese preparation involved several plants and yet was handled as a *single* IND. A demonstration of the contribution of efficacy, of *each* ingredient of *each* plant, was not required. This was a major step forward towards more sensible regulations in regard to phytomedicines.

My thanks are due to the staffs of Harwood Academic Publishers and Taylor & Francis who have made this series possible and especially to the volume editors and their chapter contributors for the authoritative information.

Roland Hardman

Preface

The notable book "Taxol® Science and Application," was published in 1995. It was edited by the young, and sadly late, Dr Matthew Suffness of the National Cancer Institute, USA. The taxoid paclitaxel, trade name Taxol, was developed there as a medicinal cytotoxic agent and today it is used for ovarian, breast, lung and other cancers, which are in a metastatic state. Supplies of paclitaxel and now of related compounds are still dependent on Taxus raw material.

Since 1995, much work has been carried out worldwide on the numerous complex taxoids of the genus Taxus, resulting in a vast amount of information. Much of this literature, although not easily condensed and edited, has been put into this volume – an achievement of the chosen specialist authors. These authors have covered the taxoids of many species: botany and taxonomy, biosynthesis, plant tissue culture, detailed chemistry, physical methods of identification of the complex structures, structure–activity relationships, pharmacology, preclinical and clinical investigation and cultivation of selected plant species and varieties with attention to the levels of taxoids in various morphological parts by season or age and their extraction.

This volume is designed to be of value to all those who require a detailed perspective of taxoid science. At the same time it will meet the needs of those requiring selected information, for example, students and their teachers, horticulturalists, researchers in the pharmaceutical industry and many others.

We wish to thank all the authors for their important contributions. Moreover, we thank Dr Susan Morris-Natschke for her enthusiastic help in reviewing and making suggestions for the manuscript. Finally thanks are due to Dr Roland Hardman for his unfailing support and encouragement.

Editors

1   Introduction

Hideji Itokawa

Introduction

As mentioned in the book edited by Suffness (1995), "the approval of Taxol (paclitaxel) for marketing in December of 1992 was the culmination of 30 years of work that began with the collection of *Taxus brevifolia* in Washington state in 1962." Camptothecin was also discovered in the 1960s before taxol and was also commercially available only after 30 years of work. As is evident from these instances, a tremendous amount of work is necessary to develop new anticancer agents. Discovery and development constitute a complex story for each drug, especially for anticancer agents.

Taxol is now the best known and most studied member of the taxane diterpenoids, or taxoids. It is only one of over 350 members of this compound class. Because many taxoids can be modified chemically, a knowledge of other taxoid structures is important for the development of analogues. This book will cover the fundamental chemistry necessary to develop such analogues.

The taxane diterpenoids have been previously reviewed. Among the more comprehensive and accessible reviews are those by Lythgoe (1968), Morell (1976), Miller (1980), Suffness and Cordell (1985), Gueritte-Voegelein *et al.* (1987), Khan and Parveen (1987), Chen (1990), Blechert and Guenard (1990), Swindell (1991), Kingston (1991), Paquette (1992), Kingston *et al.* (1993), Suffness (1995), Das *et al.* (1995), Baloglu and Kingston (1999), and Parmer *et al.* (1999).

This review combines to some extent the scope of Blechert's, Kingston's, and Suffness's reviews cited above. The structures of the known naturally occuring taxane diterpenoids are described in Chapter 3. The chemistry of taxol and other taxanes are also described, followed by chapters describing approaches to the synthesis and pharmacology of key compounds. The literature has been covered as thoroughly as possible through April 2001.

Early history of anticancer development

The origin and history of taxol and other taxoids have been detailed previously by Wall (in the book by Suffness, 1995). The authors would like to paraphrase Wall's description herein.

The first key reports on chemotherapeutic agents used to treat human cancer were those of Goodman in 1946 on nitrogen mustards in leukemia, Farber in 1948 on aminopterin in childhood leukemia, and Burchenal in 1953 on 6-mercaptopurine in leukemias. At that time, the major anticancer discovery and development program in the USA was at the Sloan-Kettering Institute for Cancer Research. This program had developed out of a reorientation of the Second World War research on nitrogen mustards. More than 75% of all compounds evaluated in the USA were tested at Sloan-Kettering during the early 1950s. However, although these initially exciting results led to a strong interest in cancer research by a variety of chemists, biochemists, and biologists, the evaluative capacity was too low for the needs of the scientific community. In response, Congress directed the National Cancer Institute (NCI) to organize such a program. This directive led to the formation of the Cancer Chemotherapy National Service Center (CCNSC) in 1956 with the mission of supporting research in cancer drug discovery and treatment, and providing services to researchers nationwide, and later worldwide. Although a few plant extracts were evaluated from 1956 onward, the plant program began in earnest only in 1960 with contracts for collection, extraction, and screening of plant extracts (Zubrod *et al.*, 1966; Schepatz, 1976; Suffness, 1989).

The plant program at NCI was organized and nurtured for 15 years by the late J. Hartwell, who had made important contributions on podophyllotoxin and related lignans. First, it was decided to begin the collection program with native plants from the US, and an interagency agreement was negotiated with the US Department of Agriculture (USDA) to collect and identify plant materials for the screening program. This collaboration lasted from 1960 to 1981. In addition to collecting new plant samples, attempts were made to secure access to existing collections of plant extracts and one such collection was at the USDA Eastern Regional Research Laboratory under control of Monroe Wall. In 1958, the results of assays showed very strong activity for one plant in the USDA collection, the *Camptotheca acuminata*. This discovery led to the migration of Dr Wall to the Research Triangle Institute (RTI) in 1960, and the subsequent isolation by the RTI group of camptothecin, an exciting plant-derived antitumor agent (Wall *et al.*, 1966; Slichenmyer *et al.*, 1993).

Taxol discovery

Arthur Barclay of the USDA collection team was in Washington State in the Gifford Pinchot National Forest in August 1962 and collected samples of the Pacific yew (*T. brevifolia* Nutt.), family *Taxaceae*. The *Taxaceae* is a small family with one main genus, *Taxus*, which has an ancient reputation as a toxic and magical plant. The yew tree was used in ancient Greece and Rome to produce weapons and household implements. It was said that yew growing in Narvonia had such a power that those who sit or sleep under its shade suffered harm or, in many cases, even died.

Barclay collected two samples: PR-4959, stems and fruit, and PR-4960, stem and bark. Only the latter sample later proved active. Many samples were prepared at the Wisconsin Alumni Research Foundation and shipped to the NCI. After extracts were prepared, the first sample was tested against KB cells at Microbial Associates in Bethesda, Maryland in 1964. This extract was found to be cytotoxic, and the activity was confirmed in June 1964 at the University of Miami as the first demonstrated cytotoxic activity in Pacific yew. However, no toxicities were found in tests against L1210 and P-1798 cell lines. The decision to begin fractionation studies of *T. brevifolia* bark was made on the KB cytotoxicity alone. During his earlier studies on isolation of camptothecin, Wall noticed an excellent correlation between KB cytotoxicity and *in vivo* activity, and accordingly requested the assignment of other plants showing KB activity. The *Taxus* plant was assigned to Wall's laboratory for fractionation studies, and work began with the receipt of a 30 lb shipment of bark in late 1964. Subsequently, fractions from this extract were found to be active *in vivo* against P-1534 leukemia, Walker 256 carcinosarcoma, and P388 leukemia; all of these systems were used at various times to monitor the activity of fractions.

The separation methods available were much fewer in number and much less sophisticated than those available today. Thus, more separation steps were performed than would now be necessary. Further, the newer separation techniques of the late 1960s such as high-performance liquid chromatography (HPLC) were useful only in an analytical mode and could not be scaled up directly. The isolation of 0.5 g of pure taxol (1) from the *T. brevifolia* bark took about two years and the first pure sample was isolated in 1966. The structure was obviously complex with a molecular formula of $C_{47}H_{51}NO_{14}$. The molecule appeared to be in the taxane family, but at that time, not many members of the family had been elucidated structurally. Further, the extensive substitution and obviously large number of functionalities and asymmetric centers made solution of the structure by spectroscopic means and chemical derivatization or degradation to known compounds very difficult. At this time, none of the high resolution Nuclear Magnetic Resonance (NMR) experiments routinely used today, such as COSY, ROSEY, NOSY, COLOC, etc., were available. However, even with the most modern NMR techniques available today, the most unequivocal structural determinations of complex new natural products still rely on X-ray crystallography, and that approach was attempted for taxol. However, crystals of taxol and its halogenated derivatives were unsuitable for X-ray structure determination. The crystal structure of taxol itself has not yet appeared, although that of the related compound, taxotere (2), has been solved (Gueritte-Vogelein *et al.*, 1990). Solvolysis of taxol with methanol at 0°C gave three components: a tetraol (3), which was characterized by X-ray crystallography as its bisiodoacetate (4); a methyl ester derived from a side-chain (5), which was also characterized by X-ray crystallography as its p-bromobenzoate ester (6); and methyl acetate. From the structure of the tetraol (3), the main side-chain and the acetate side-chain could have been located at either positions 7, 10, or 13 in the parent molecule. Taxol was not oxidized by neutral manganese dioxide, indicating that both of the allylic positions C-10 and C-13 were esterified. Furthermore, basic manganese dioxide oxidation gave a product in which the main side-chain had been removed and C-13 had been oxidized to a ketone. These data, plus additional ^{1}H-NMR data indicating the acetate to be at C-10, resulted in secure assignments of the ester positions as shown in structure 1 (Figure 1.1).

The isolation of taxol was first presented at the American Chemical Society meeting in Miami Beach, FL, in 1967 (Wall and Wani, 1967). The main paper was published in 1971. The reported taxol yield was 0.02% from dried bark of *T. brevifolia,* and taxol was also reported to be present in other *Taxus* species, including *T. baccata* and *T. cuspidata.* The first structure activity data were also presented showing that both the taxol nucleus and the C-13 side-chain were essential for activity (Wani *et al.*, 1971).

An alkaloidal *Taxus* fraction, which was named "taxine" by Lucas (1856), is highly cardiotoxic and has been implicated in many stock poisonings and human poisonings by *Taxus* species (Lowe *et al.*, 1970; Alden *et al.*, 1977).

It was separated into taxines A and B by Graf *et al.* (1986). These alkaloids have common structural components of a multiple oxygenated taxane ring system and an alkaloidal side-chain at C-5 as seen in taxines A (7) and B (8) (Figure 1.2). The C-5 side-chain of the taxines is comparable to the C-13 side-chain in the antitumor taxanes. Taxine A (7) has a rearranged carbon skeleton, while the structure of taxine B was later revised to (8). The taxine B skeleton was identified as "taxane" in 1964 by Lythgoe, Nakanishi, and Ueo prior to the structural elucidation of taxol. Potier's group generated a biosynthetic hypothesis for taxol that involves the intermediacy of a C-5 to C-13 ester transfer (Gueritte-Vogelein *et al.*, 1987) and successfully demonstrated this hypothesis synthetically by converting taxine B to a baccatin V derivative (Ettouati *et al.*, 1991). Interestingly, the cardiotoxic alkaloidal fraction is relatively abundant in *T. baccata* (European yew), is even more prevalent in *T. cuspidata* (Japanese yew), but is almost absent in *T. brevifolia* (Tyler, 1960). Both the antitumor taxane fractions and the taxine alkaloidal fraction are soluble in ethanol, thus both would have been present in the primary extract. Therefore, if another species of *Taxus* had been investigated, first it is quite possible that subsequent *in vivo* experiments would have shown strong toxicity (toxic doses of the taxines are generally fatal within

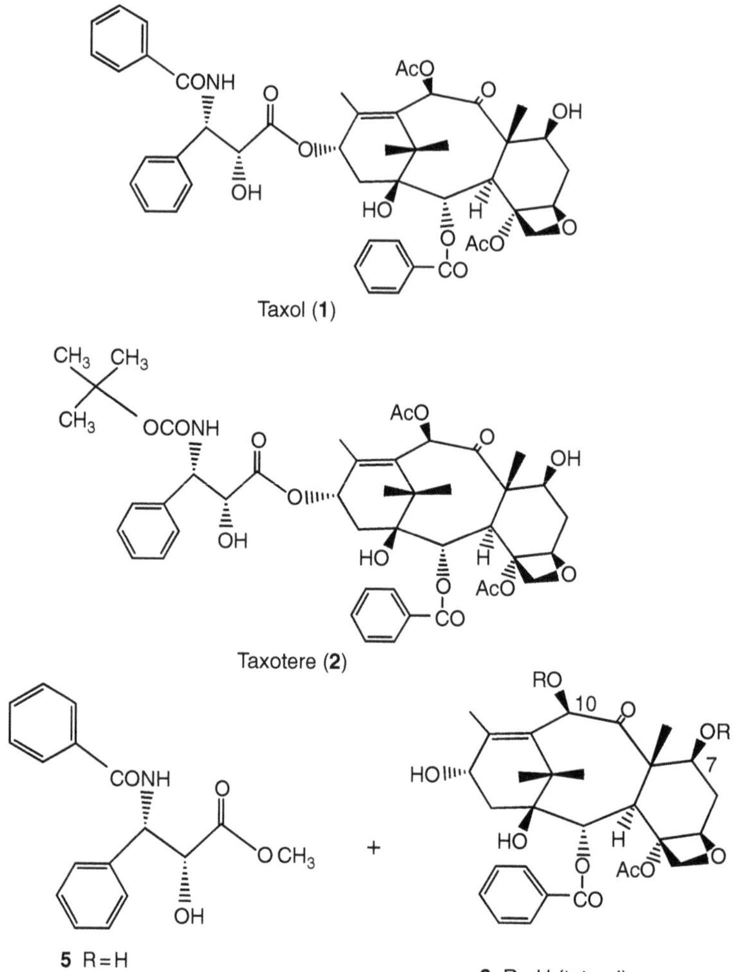

Taxol (**1**)

Taxotere (**2**)

5 R=H
6 R=CO-p-C$_6$H$_4$Br

3 R=H (tetraol)
4 R=COCH$_2$I

Figure 1.1 Secure assignments of the ester positions.

a day) and the project would never have been pursued. Or if fractionation had been pursued, the cytotoxicity-guided fractionation could have led to the toxic taxines rather than taxol.

The related baccatin III was isolated by Chan *et al.*, in 1966, but its structure was not eluci-dated at that time. A tentative structure, based on chemical and spectrospecific evidence, was published by De Marcano in 1970. Taylor isolated a substance that he called baccatin from the heartwood of *T. baccata* (1963). This compound was later renamed baccatin I (De Marcano and Halsall, 1970).

The structure of the related compound baccatin V (10) was determined by X-ray crystallogra-phy by Halsall in 1970 (De Marcano *et al.*, 1970). This finding led to a reconsideration of the structure of baccatin III, which was confirmed as (9) by comparison of its manganese dioxide oxidation product with that obtained from taxol (Figure 1.3). As mentioned before, solvolysis of taxol gave a tetraol (**3**) whose structure was established by X-ray crystallography of its

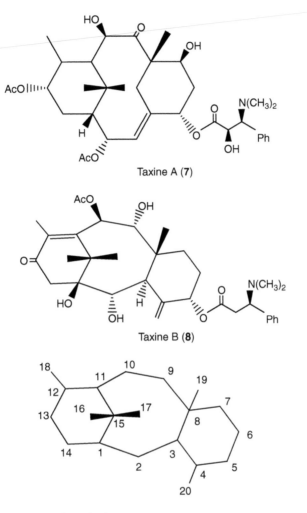

Taxine A (**7**)

Taxine B (**8**)

Figure 1.2 Taxane ring system and numbering.

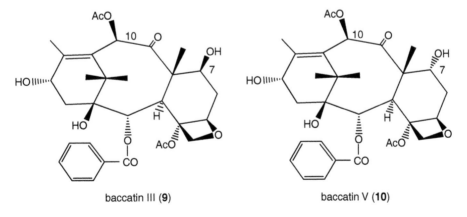

baccatin III (**9**)

baccatin V (**10**)

Figure 1.3 Structures of baccatin III and baccatin V.

Table 1.1 Chronology of Taxol discovery and development (by M. Wall)

Date	Event
June 1960	Beginning of NCI Plant Program contracts
August 1962	Collection of *T. brevifolia* in Washington
April 1964	Cytotoxicity of bark extract to KB cells
June 1964	Confirmation of KB cytotoxicity
October 1964	Shipment of 30 1b of bark to M.Wall at RTI
1965–67	Bioassay guided fractionation in KB, P1534, and Walker systems
October 1966	First isolation of pure Taxol (K172)
October 1966	Cytotoxic activity of pure Taxol to KB cells
?? 1967	Report of Taxol isolation to American Chemical Society national meeting
1965–67	Studies of activity of other Taxus species and variation of activity of *T. brevifolia* by location
1967	*In vivo* antitumor activity of Taxol in L1210 leukemmias and Walker 256 carcinoma
1968	*In vivo* antitumor activity of Taxol in P1534 and P388 leukemia
1968–69	Study of activity of *T. brevifolia* by plant part
May 1971	Chemical structure of Taxol published
April 1974	B16 melanoma activity observed
June 1975	B16 melanoma activity confirmed; Taxol meets development criteria
April 1977	Taxol accepted for development at Stage 2A
Fall 1977	Activity in B16 melanoma restricted to i.p.–i.p.
August 1978	Publication of Taxol as antimitotic drug
November 1978	Activity against MX-1 breast xenograft
December 1978	Confirmed activity against MX-1 breast xenograft
February 1979	Taxol published as promotor of microtuble assembly
June 1980	Cremophor formulation active against B16 melanoma; selected over PEG formulation
August 1980	Route and schedule dependency studies completed
October 1980	Taxol passes Decision Network stage 2B; approved for toxicology
June 1982	Toxicology completed; Cremophor a problem in dogs
November 1982	Decision Network stage 3 passed; Taxol approved for INDA filing
September 1983	Investigational New Drug Application filed
April 1984	Investigational New Drug Application approved
April 1984	Phase I clinical trials begin
April 1985	Decision Network stage 4 passed: approved for Phase II trials with use of 24-h continuous infusion and premedication regimen
1984–89	Phase II trials limited by drug supply
August 1989	Request for applications for cooperative Research and Development Agreement published in Federal Register
August 1989	Activity in advanced ovarian cancer published by Johns Hopkins group
November 1989	Selection of Bristol-Myers Squib as CRADA partner
June 1990	NCI Workshop on Taxol and Taxus
July 1990	Issuance of Request for Applications (RFA) for Taxol research
July–October 1991	13 research grants awarded from Taxol RFA
December 1991	Publication of activity of Taxol in metastatic breast cancer by M.D. Anderson group
1990–93	Large-scale production of Taxol by Hauser/BMS
July 1992	New Drug Application field with FDA
September 1992	Second NCI Workshop on Taxol and *Taxus*
December 1992	New Drug Application approved for refractory ovarian cancer
April 1994	Supplemental approval for metastatic breast cancer

7,10-bisiodoacetate (4). The skeletal configurations of this tetraol and baccatin III were found to be the same, but differed from that of baccatin V (10) only in the stereochemistry of the C-7 hydroxyl group. Therefore, the tetraol was identified as 10-deacetylbaccatin III.

Wall and Wani summarized about camptothecin and taxol in 1998 (Table 1.1).

Isolation and purification

A major issue in scale up was to determine which *Taxus* plant material possessed both high taxol content and adequate abundance. This investigation was conducted in three stages: (1) examination of various species and cultivars to look for the best species; (2) a geographic study of the best species to determine variation in taxol content by plant locality; and (3) examination of the best species from the best locations to determine which plant parts were best suited to production. In the first phase of the study, collections of numerous *Taxus* species were made by the USDA during 1965 and 1966, which included *T. baccata* L. (European yew), *T. cuspidata* Sieb. and Zucc. (Japanese yew), *T. globosa* Schlect. (Mexican yew), *T. floridiana* Nutt. (Florida yew), *T. canadensis* Marsh. (Canadian yew or ground yew), and examples of the main ornamental hybrids *T. x media* Rehd. (the cross of *T. baccata* and *T. cuspidata*), and *T. x hunnewelliana* Rehd. (the cross of *T. cuspidata* and *T. canadensis*). In the initial biological analyses, particularly cytotoxicity toward KB cells, none of these samples showed better activity than Pacific yew. Those extracts that had substantial activity were then processed by Wall's group at the RTI to determine taxol levels. For example, the level of taxol isolated from *T. cuspidata* was 0.0011% and from *T. baccata* was 0.0008%. These data led to the decision that Pacific yew was the best source of taxol. The next stage was to determine which parts of the tree had the greatest taxol content and the influence of geographic location on content.

Pacific yew trees can now be readily assayed by HPLC methods that can detect taxol in crude extracts and by immunoassays that are highly specific for taxol. However, although HPLC was available in the late 1960s, the reverse-phase columns that are now used to separate taxol and related compounds from other extract components were just becoming available and were extremely expensive and untried for taxol. Witherup *et al.* published (1989, 1990) the first paper identifying taxol in crude extracts by HPLC. They systematically examined the process working with six species of *Taxus*. They dried the biomass under vacuum, ground it in a Wiley mill, and then extracted it three times in a glass percolator for 16-h periods. They also used a second-stage partition between methylene chloride and water to improve the sample for analysis.

The first immunoassay was reported by Jaziri *et al.* (1991) and was not highly specific for taxol. Therefore, all analyses in the 1960s were done by bioassays, particularly cytotoxicity toward KB cells and *in vivo* antitumor activity in Walker 256 carcinosarcoma and P388 leukemia cell lines. The activity of *T. brevifolia* by plant part was as follows: stem bark > root > needles > wood > twigs.

An ideal extraction system for natural products is one that would completely remove only the product of interest, and leave behind all other components. Such a system rarely, if ever, can be developed and so various compromises are accepted. The objective of a primary extraction procedure is to obtain the maximum amount of product in a matrix that is as easy as possible to separate further into pure compound.

Xu examined the tissue from *Taxus chinensis* and sonicated the plant material in the presence of solvent for 5 min to help complete the extraction (Xu and Liu, 1991). Auriola also used sonication and further conducted the procedure at 4°C in order to minimize deterioration of the product (Auriola *et al.*, 1992). Recently, supercritical fluid extraction (SCFE) procedures have been reported and, while they are economical and environmentally desirable, the product still requires chromatography to separate the two major products, taxol and cephalomannine. Jennings *et al.* (1991) used liquid carbon dioxide with 3.8 mol% of ethanol as a modifier and found that they could achieve a yield of 0.008% taxol from the bark of *T. brevifolia*. This result is in good agreement with the generally reported yield of 0.01% obtained by conventional methods.

Using the needles as a biomass source requires that they first be extracted with a low-polarity solvent such as hexane to remove the cuticular lipids and fats. One team found that as much as 72% of interfering hexane-solubles can be removed from yew needles in this way (Witherup *et al.*, 1989). Castor was able to take advantage of the low polarity of unmodified liquid carbon dioxide to remove lipids and then increase polarity with alcohol modifier to obtain a yield of 0.044% taxol from needles of *T. media* and *T. brevifolia* (Castor, 1992).

Recently, Rao modified the extraction protocol by using a standard primary extraction with methanol followed by a clean-up partition with chloroform and water. Extract quality was significantly improved to the extent that subsequent chromatography steps were substantially less rigorous and overall yields of taxol were 0.02–0.04% (Rao, 1993).

No direct comparative study of these various extraction techniques has been published. Results from various laboratories are difficult to compare because the amount of taxol varies from plant to plant and the investigators were not using a standard, common sample of tissue. However, yields of taxol of approximately 0.01% from the bark and 0.025% from the needles of *T. brevifolia* are certainly reproducible and have been reported from many laboratories. One particularly rigorous investigator (Fang *et al.*, 1993) examining *T. cuspidata*, added measured amounts of diterpenoid standards to the dried, ground plant material in order to estimate the recovery of the various taxanes. Such a procedure adds significantly to the precision of the experiment but does not allow absolute amounts of taxanes to be measured. Taxanes can also be degraded between the time of harvest and the final extraction, and can be compartmentalized within tissue substructures. Numerous reports show that various species of *Taxus* show different ratios and quantities of taxanes. For a more extensive comparison of the various taxane congeners produced by the various species, seasonal and tissue variation in concentrations of *T. brevifolia* has been discussed. Seven taxanes (taxol, baccatin III, 10-deacetyltaxol, 10-acetylbaccatin III, 7-xylosyl-10-deacetyltaxol, cephalomannine, and brevifoliol) were analyzed in extracts from bark and foliage of Pacific yew (*T. brevifolia*) and a decreasing gradient concentration was found from stem base to branch tip. This corresponding decrease is attributed to generally higher taxane concentration in phloem tissue and the decrease in inner bark thickness from base to branch tip. Stem bark and needles were also sampled over a growing season. Typically, taxane concentrations in bark increased from May through August, whereas concentrations in needles changed little during that period. While most taxane concentrations were significantly lower in the needles than in the bark, two exceptions were baccatin III, which in the summer reached levels equivalent to bark, and brevifoliol, which increased from March to August, reaching levels in needles nine times greater than in bark (Vance *et al.*, 1994).

The taxol content of dried *Taxus* biomass was monitored monthly for 15 months. Intact and finely ground biomass was stored at room temperature (22–24°C) as well as under refrigeration (2–4°C). In addition, intact fresh clippings stored under refrigeration in sealed plastic bags for up to 10 weeks were evaluated for changes in taxol content. Analysis indicated that properly dried *Taxus* clippings could be stored either intact or powdered at room temperature or under refrigeration with no apparent loss of taxol content. The taxol content in fresh intact clippings was also stable for at least 10 weeks when stored under refrigeration (Elsohly *et al.*, 1994).

Clippings of *T. x media* "Hicksii" were stored for up to 27 days at 4°C and 22°C to determine the effect of storage conditions on the concentration of taxol and cephalomannine. Although the clippings stored at 22°C lost 12.7% more weight than when stored at 4°C for 27 days, the yield of neither taxol nor cephalomannine was affected by storage temperature. Taxol concentrations in clippings were 40% and 26% higher in plant material stored at 22°C and 4°C, respectively, for nine days when compared to concentrations at harvest. Thirty-five percent more cephalomannine was present after three and nine days of storage than at harvest time. After 27 days of storage, the yields for both taxol and cephalomannine were similar to the yield at harvest. These results indicate that the storage of fresh plant biomass for about nine days enhances the yield of both taxol and cephalomannine regardless of storage temperature (Schutzki *et al.*, 1994).

The concentrations of taxol and related compounds in the mature dried seeds of several *Taxus* species were determined by HPLC. In addition, the taxol level of *T. cuspidata* seeds at different stages of maturation was studied. The variation in the taxane content was dependent on individual trees within species, as well as among species. The average contents of taxol and its precursor, 10-deacetyl baccatin III (10-DAB III), in the mature seeds of different species were $67.3 \pm 83\,\mu\mathrm{gg}^{-1}$ dry wt, respectively. In mature seed parts of *T. cuspidata var. latifolia* obtained from tree U5, the embryo weight was *c.*0.2% of that of the dry weight of the whole seed and contained the highest level of taxol ($894\,\mu\mathrm{gg}^{-1}$) when compared with that obtained for the endosperm ($23\,\mu\mathrm{gg}^{-1}$) and testa ($376\,\mu\mathrm{gg}^{-1}$). However, the taxol content per seed was highest in the testa ($20.5\,\mu\mathrm{g}$), followed by the endosperm ($0.75\,\mu\mathrm{g}$) and embryo ($0.18\,\mu\mathrm{g}$). The taxol content of the fresh seed reached a maximum of $332\,\mu\mathrm{g}^{-1}$ and $20\,\mu\mathrm{g}^{-1}$ at the middle stage of seed maturation and then decreased with further maturation. The taxol identification in the seed was confirmed by electrospray mass spectrometry (Kwak *et al.*, 1995).

The concentrations of taxol and related compounds in the bark and needles of *T. cuspidata* grown on Mt Jiri, Mt Sobaek, and Cheju Island, and of *T. cuspidata var. latifolia* on Ullung Island in Korea were determined by HPLC. The taxane content varied significantly with the location and plant part. The taxol content in the bark of native yews from Mt Jiri and Mt Sobaek was high when compared to that reported for Pacific yew (*T. brevifolia*), whereas bark from trees on Cheju and Ullung Islands contained a much higher level of taxol than those of intermountain locations, on the basis of dry weight. The bark and needles of *T. cuspidata var. latifolia* on Ullung island also contained relatively high concentrations of 10-deacetylbaccatin III-0.0497% and 0.0545%, respectively-and indicated that environmental factors may affect the quantity. Taxol in the needles was confirmed by electrospray mass spectrometry. These results suggested that foliage from yew trees growing in their natural habitats on Cheju and Ullung Islands may provide a renewable source for taxol (Choi *et al.*, 1995). Evaluation of various methods of drying yew biomass was accomplished to test the effect on taxol content (Elsohly *et al.*, 1995). Cuttings from *T. x media* Hicksii were maintained in Gamborg's B5 liquid culture medium, where 10-deacetyl baccatin III, baccatin III, 10-deacetyl taxol, cephalomannine, 7-epi-10-deacetyl taxol and taxol accumulated in the incubation medium over time. Greater amounts of each taxane were recorded from plant material as compared with liquid medium. Medium was removed and replaced weekly, biweekly, or triweekly for up to nine weeks. The sum of all taxol recovered from the liquid that had been harvested from a yew cutting and extracted from the medium weekly over eight weeks was approximately equal to the amount of taxol that would be recovered from the fresh plant cutting, and the cutting still contained taxol. This suggests that *in vivo* cultures of yew cuttings may be a re-usable source of taxol and related taxanes (Hoffman *et al.*, 1996).

The average annual taxol content of shoots with dark green needles from Irish yew (*T. baccata var. fastiggiata*) was found to be 0.0075%. Maximum levels were recorded in April (0.010%) and minimum in February. Golden-leaved ("aurea") varieties contained only traces of taxol. Shoot growth took place mainly between May and July (Griffin and Hook, 1996).

The concentration of taxol declined in both the plant materials (30–40%) and their extracts (70–80%) upon storage for one year with no particular care. However, when kept in a freezer and not under direct sunlight, the taxol-rich fractions of the extracts were found to be quite stable (Das *et al.*, 1996). The roots of *Taxus wallichiana* gave nine known taxoids. They were paclitaxel, pentaacetoxytaxadiene, 1-hydroxybaccatin I, baccatin IV, baccatin III, taxusin, a C-oxygenated taxoid, a rearranged taxoid, and 7-xylosyl-10-deaceetyltaxol (Chattopadhyay *et al.*, 1998).

The recovery of taxol and related compounds was studied under different drying conditions, which included a tobacco drying barn, greenhouse, shadehouse, air-conditioned laboratory, oven, and freeze-drying. Based upon projections from analysis of fresh biomass, near total recoveries of the expected levels were observed for taxol and cephalomannine. For clippings dried under tobacco barn, greenhouse, oven, and freeze-drying conditions, however, only 75–85% of the expected values for 10-deacetyltaxol and 10-deacetylbaccatin III were found. When the length of drying was extended up to 10 and 15 days, as in the shadehouse and laboratoy conditions, the recovery of all taxanes was adversely affected (Elsohly *et al.*, 1997).

Recently, the needles of several yew species and cultivars were analyzed by high-pressure liquid chromatography for paclitaxel, 10-deacetylpaclitaxel, cepalomannine, baccatin III, 10-deacetylbaccatin III, and brevifoliol. About 750 samples were collected from five different locations in the Netherlands and the UK. The results showed a large variation in taxane content between the different species and cultivars. The content of paclitaxel and 10-deacetylbaccatin III varied from 0 to 500 μg^{-1} and 0 to 4800 μg^{-1} dried needles, respectively. Brevifoliol was found in a very high concentration in *T. brevifolia*. 10-Deacetylpaclitaxel, cephalomannine, and baccatin III were found in concentrations ranging from 0 to 500 μg^{-1} dried needles (Van Rozendaal *et al.*, 2000).

Isolation method

The original method for purifying taxol from *T. brevifolia* bark relied heavily on silica chromatography. Polysciences, Inc. developed this methodology into a large-scale preparative procedure, and although they subsequently produced the bulk of the early clinical supplies of taxol, the process was only partially described. It began with a methanol extraction and a partition between methylene chloride and water to remove non-taxane components. The procedure then utilized a series of chromatography steps. First was a rapid pass through Florisil™), followed by Sephadex LH-20™) and silica gel, with the final purification done by crystallization.

The current process in use by Hauser Chemical Research, Inc. to manufacture human-use taxol from *T. brevifolia* bark is proprietary, but still retains some chromatography steps. They have reported a significant improvement in the procedure, reducing the required *T. brevifolia* bark from 35,000 to 15,000 lb for the production of 1 kg of taxol (Stull and Jans, 1992).

Perhaps the most interesting new innovations are those that completely abandon adsorption chromatography. Rao *et al.* (1991) have reported the use of high-speed countercurrent chromatography to prepare taxol and other taxanes from *T. yunnanensis*. The group at the University of Brussels used the Ito multi-layer coil separator–extractor to process up to 300 mg per run of crude *Taxus* extract. From a single pass through this countercurrent extractor, they obtained 40% of the available taxol as 97% pure product and the remainder as an enriched fraction that was readily purified by preparative HPLC. Also reported is the isolation of new taxane diterpenoids from *T. canadensis* by a group at Abbott utilizing the Ito countercurrent system (Guanawardana *et al.*, 1992).

Polyclonal antibodies raised against 2′-succinyltaxol-bovine serum albumin (BSA) conjugate were used for the immunodetection of bioactive taxoids in chromatographic fractions of the stem bark extract of *T. baccata*. In addition to taxol, cephalomannine, and baccatin III, two new taxoids were isolated and their structures were elucidated as 4α, 7β-diacetoxy-2α, 9α-dibenzoxy-5β, 20-epoxy-10β, 13α, 15-trihydroxy-11(15-1)-abeo-tax-11-ene and taxol C using spectroscopic methods (Guo *et al.*, 1995).

The separation of a stem-bark extract of *T. baccata L. cv. Stricta* by high-speed countercurrent chromatography (HSCCC) resulted in several chromatographic fractions, which were analyzed by enzyme-linked immunosorbent assay (ELISA) using specific anti-10-deacetylbaccatin III antibodies. In addition to 10-deacetylbaccatin III and four known xylosyl taxoids, a new taxoid, 9-deacetyl-9-benzoyl-10-debenzoyltaxchinin A(I), was isolated. The cross-reactivity of I with anti-10-deacetylbaccatin III antibodies was examined (Guo *et al.*, 1996).

Microwave-assisted extraction (MAE) is a new method for extracting the taxane natural products, including paclitaxel, from *Taxus* needles. Various temperatures, times, and organic solvents were investigated to optimize the efficiency of the extraction. The effects of biomass to solvent ratio and of the system's water content on taxane recovery were also determined. Under appropriate MAE conditions using 95% ethanol, the extract of the needles was equivalent to that produced by a conventional extraction method (overnight shake of 5 g of needles in 100 ml of mathanol at ambient temperature). With optimized parameter settings, MAE was found to considerably reduce both extraction time and solvent consumption, while maintaning qualitative and quantitative taxane recovery relative to traditional solid/liquid extraction methods (Mattina *et al.*, 1999).

Taxol can be separated from cephalomannine by oxidation of a mixture of the two with OsO_4 and flash chromatography of the resulting products (Kingston *et al.*, 1992). Treatment of a mixture of paclitaxel and cephalomannine with bromine under mild conditions yields a readily separable mixture of paclitaxel and 2″,3″-dibromine. Cephalomannine can be regenerated by treating 2″,3″-dibromocephalomannine zinc in acetic acid (Rimoldi *et al.*, 1996).

Accelerated solvent extraction (ASE) of paclitaxel and related compounds from *T. cuspidata* (Japanese yew) bark has been investigated under various conditions. In ASE, pressure is applied to the sample extraction cell to maintain the heated solvent in a liquid state during the extraction. This method shortens the extraction time and increases recovery of target compounds. In this study, ASE produced greater amounts of paclitaxel, baccatin III, and 10-deacetylbaccatin III than those from ordinary solvent extraction at room temperature. The conditions providing the highest recovery of paclitaxel were as follows: solvent, $MeOH-H_2O$ (90 : 10); temperature, 150°C; and pressure, 10.13 Mpa (0.128% W/W recovery based on oven-dried sample powder). ASE does not require chlorinated solvents and can reduce solvent consumption because of its strong dissolving power. Moreover, with water alone, the recovery of paclitaxel and related compounds using ASE is much higher than with other extraction methods (Kawamura *et al.*, 1999).

An alternate stereochemical nomenclature for taxane skeleton

Since Wall, Wani, and McPhail first disclosed the structure of taxol in 1971 (Wani *et al.*, 1971), numerous reports about taxoids have appeared in the literature. The geminal dimethyl groups at C-15 were first numbered as C-19 and C-20 as seen in structure **1a** (Figure 1.4). Later, the numbering was changed to C-16 and C-17, which is the currently accepted numbering shown in the IUPAC recommendation for the unbridged taxane skeleton (**1b**) (IUPAC, 1999). Other structures have been seen and discussed in the literature[3–8] (Lythgoe, 1968; Miller, 1980; Kingston, 1991, 1993, 1995, 2000). Two different skeleton systems were proposed by Miller (**2a**) (Miller, 1980) and Lythgoe (**2b**) (Lythgoe, 1968). The former representation (**2a**) shows an accurate stereochemistry of the 16- and 17-methyl groups and has been used by Kingston in many

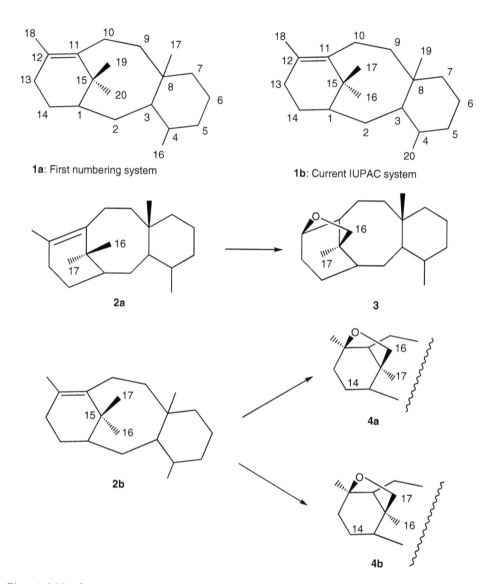

Figure 1.4 Numbering system of geminal dimethyl groups.

notable reviews of taxoid chemistry (Kingston, 1991, 1993, 1995, 2000). However, the latter system (**2b**) has generally been more popular, although it reverses the orientations of the 16-, 17-methyl groups from those expected (Lythgoe, 1968; Kingston, 1991). While the 16-methyl and 19-methyl groups are both β-configured, the 16-methyl must be shown with a dashed bond, and the 17-methyl, which is α to the 19-methyl, must be shown with a solid bond.

The nomenclature of the bridged oxido-type taxane is even more complex. From **2a**, formation of the –OCH₂– bridge would lead to the oxido structure **3** (Kingston, 1995). The α/β designations of the –O– and –CH₂-groups remain mutually consistent and the numbering agrees with the commonly used nomenclature of a C-12(16)oxido bridge. However, inconsistencies arise when the bridged oxido-type taxane is depicted from the more commonly used **2b** or **1b**.

With this skeleton system, the resultant oxido-bridged taxoid has been seen in the literature written in two ways, **4a** (Kingston, 2000) or **4b**. These two structural representations clearly reveal the ambiguities that occur with the β-configured oxido bridge; namely whether C-16 or C-17 is depicted graphically as β (solid wedged bond). Structure **4a** is consistent with the C-12(16)oxido designation, but is inconsistent with the α/β designation of **2b** and **1b**; namely C-16 is depicted as a solid wedged bond in **4a**, but with a dashed bond in **2b** and **1b**. In contrast, **4b** is consistent with the bond depictions present in **2b** and **1b**, but is inconsistent with the already established C-12(16)oxido designation. This designation is not in accordance with IUPAC nomenclature. **4b** (Shigemori *et al.*, 1999) is at present the only correct paper to be represented in accordance with the IUPAC nomenclature (http://www.chem.pmul.ac.uk/iupac/). However, the other papers could not understand the rule of IUPAC nomenclature correctly. It means that the nomenclature is somewhat difficult to understand and can cause confusions. That is the reason why the authors would propose the easily understanderble nomenclature.

The taxane skeleton was numbered by IUPAC about 30 years ago. At that time, determinations of complex stereochemistries were more difficult and less prevalent, and the nomenclature was decided according to a flat chemical concept as shown in Figure 1.5. The parent skeleton was numbered from carbon 1 to methyl 18 in a counterclockwise direction. Then the numbering proceeded clockwise from methyl 18 to methyl 20. Moreover, as the molecule was considered to be flat, the geminal dimethyls 16 and 17 must be below and above the molecular plane, respectively. The C-16 Me is oriented below the molecule and must have an α-configuration, while the C-17 Me is oriented in a β-configuration. This nomenclature is still maintained by IUPAC (Figure 1.4).

Presently, however, complex stereochemistry can be determined more exactly, and complete stereochemical concepts must be introduced into the nomenclature. The taxane skeleton is not

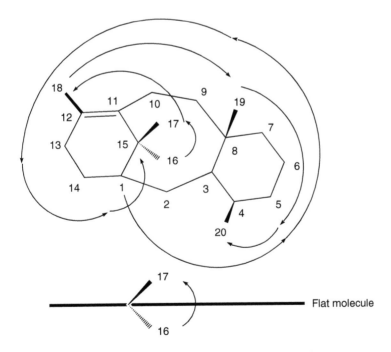

Figure 1.5 IUPAC numbering. Counterclockwise about taxane skeleton.

flat, but instead usually maintains a cage form, especially when a C-4(5)-oxetane ring is formed and rigidity increases (Boge *et al.*, 1999). As shown in (Figure 1.6), both of the geminal dimethyls C-16 and C-17 must be oriented outside of the cage in β-configurations. The numbering of these methyls would logically proceed clockwise from left to right. The resulting stereochemical representation and numbering would agree with the C-12(16)-oxido bridge nomenclature that is already accepted. Consequently, all problems and confusion would be eliminated.

Thus, to alleviate the present ambiguities, we propose herein a novel and convenient taxoid stereochemistry-based nomenclature. As shown in taxane 5 (Figure 1.7), the whole molecule would be taken into consideration when depicting the configuration of the C-16 and C-17 methyl groups, and both groups would be represented by solid wedged bonds. Because accepted numbering in the taxane skeleton proceeds from left to right, C-16 would be the geminal methyl group on the left side. Accordingly, the oxido bridge in 6 could have a definitive and consistent C-12(16) designation. Similar representations have appeared recently in the literature (Zamir, 1995, 1999; Wang, 1996; Zhang, 2000).

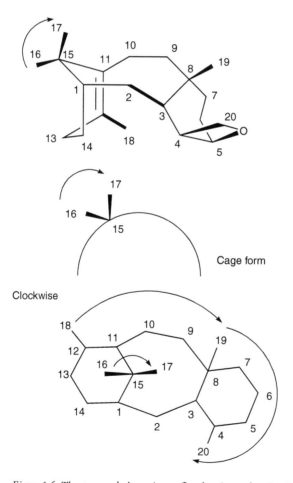

Figure 1.6 The taxane skeleton is not flat, but instead maintains a cage form.

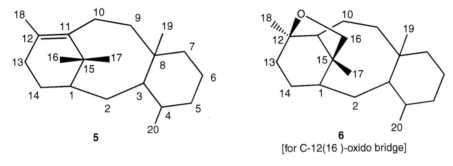

Figure 1.7 Taxoid stereochemistry-based nomenclature.

In conclusion, we propose in this volume the revised structure **5** for a general taxane and structure **6** for a C-12(16) oxido bridged taxoid. We suggest that these proposed designations would eliminate confusion in the stereochemistry of this compound class and accurately depict the physical orientation of the molecule (Figure 1.7).

References

Alden, C.L., Fosnaugh, C.J., Smith, J.B., and Mohan, R. (1977) Japanese yew poisoning of large domestic animals in the midwest. *J. Am. Vet. Med. Assn.*, 170, 314.

Auriola, S.O.K., Lepisto, A.-M., Naaranlajoki, S.P. (1992) Determination of taxol by high-performance liquid chromatography-thermospray mass spectrometry. *J. Chromatogr.*, 594, 153.

Baloglu, E. and Kingston, D.G.I. (1999) The taxane diterpenoids. *J. Nat. Prod.*, 62, 1448–1472.

Blechert, S. and Guenard, D. (1990) *Taxus* Alkaloids in *The Alkaloids*, Vol. 39, Chemistry and Pharmacology, ed. A. Brossi, p. 195. New York, Academic Press.

Boge, T.C., Hepperle, M., Vander Verde, D.G., Gunn, C.W., Grunewald, G.L., and Geog, G.I. (1999) The oxetane conformational lock of paclitaxel: structural analysis of D-secopaclitaxel. *Bioorg. & Med. Chem. Lett.*, 9, 3041–3046.

Castor, T.P. (1992) Improved isolation of taxol by supercritical fluid processing, 2nd NCI Workshop on taxol and *Taxus*, Arlington, VA.

Chan, W.R., Halsall, T.G., Hornby, G.M., and Oxford, A.W. (1966) Taxa-4(16),11-diene-5α,9α,10β,13α-tetraol, a new taxane derivative from the heartwood of yew (*T. baccata* L.): X-ray analysis of a p-bromobenzoate derivative. *Chem. Commun.*, 24, 923–925.

Chattopadhyay, S.K., Saha, G.C., Kulshrestha, M., Tripathi, Y., Sharma, R.P., and Metha, V.K. (1998) The taxoid constituents of the roots of *Taxus wallichiana*. *Planta Med.*, 64, 287–288.

Chen, W.M. (1990) Chemical constituents and biological activity of plants of *Taxus* genus. *Acta Pharm. Sin.* (*Yaoxue Xuebao*), 25, 227.

Choi, M.S., Kwak, S.-S., Liu, J.R., Park, Y.G., Lee, M.K., and An, N.-H. (1995) Taxol and related compounds in Korean native yews (*Taxus cuspidata*). *Planta Med.*, 61, 264–266.

Das, B., Rao, S.P., and Das, R. (1995) Naturally occurring rearranged taxoids. *Planta Med.*, 61, 393–397.

Das, B., Rao, S.P., and Kashinatham, A. (1996) Taxol content in the storage samples of the needles of Himalayan *Taxus baccata* and their extrats. *Planta Med.*, 64, 96.

De Marcano, D.P.D.C. and Halsall, T.G. (1970) The structures of baccatin III, a partially esterified octahydroxy-monoketo-taxane derivative lacking a double bond at C-4. *J. Chem. Soc., D (Chem. Commun.)*, 216.

De Marcano, D.P.D.C. and Halsall, T.G. (1970) The structure of the diterpenoid baccatin-I, the 4β,20-epoxide of 2α,5α,7β,9α,10β,13α-hexaacetoxytaxa-4(2),11-diene. *J. Chem. Soc. D*, 1381.

De Marcano, D.P.D.C. and Halsall, T.G. (1970) Crystallographic structure determination of the diterpenoid baccatin-V, a naturally occurring oxetan with a taxane skeleton. *J. Chem. Soc., D*, 1382.

Elsohly, H.N., Croom, E.M., El-Kashoury, E., Elsohly, M.A. (1994) Taxol content of stored fresh and dried *Taxus* clippings. *J. Nat. Prod.*, 57(7), 1025–1028.

Elsohly, H.N., El-Kashoury, E., Croom, E.M., Jr., Elsohly, M.A., Cochran, K.D., and McChesney, J.D. (1995) Effect of drying *Taxus* needles on their taxol content: the impact of drying intact clippings. *Planta Med.*, 61, 290–291.

Elsohly, H.N., Croom, E.M., Jr., Kashoury, E.A. El, Joshi, A.S., Kopycki, W.J., and McChensey, J.D. (1997) Efffect of drying conditions on the taxane content of the needles of ornamental *Taxus*. *Planta Med.*, 63, 83–85.

Ettouati, L., Ahond, A., Poupat, C., and Potier, P. (1991) Premiere hemisynthese d'un compose de type taxane porteur d'un grouppement oxetane en 4(20), 5, *Tetrahedron*, 47, 9823.

Fang, W., Wu, Y., Zhou, J., Chen, W., and Fang, Q. (1993) Qualitative and quantitative determination of taxol and related compounds in *Taxus cuspidata* Sieb et Zucc. *Phytochem. Anal.*, 4, 115.

Graf, E.S., Weinandy, S., Koch, B., and Breitmaier, E. (1986) 13C-NMR-Untersuchung von Taxin B aus *Taxus baccata* L. *Liebigs Ann. Chem.*, 1147.

Griffin, J. and Hook, I. (1996) Taxol content of Irish yews. *Planta Med.*, 62, 370–372.

Guanawardana, G.P., Premachandra, U., Burres, N.S., Whittern, D.N., Henry, R., Spanton, S., and McAlpine, J.B. (1992) Isolation of 9-dihydro-13-acetylbaccatin III from *Taxus canadensis*. *J. Nat. Prod.*, 55, 1686–1689.

Gueritte-Voegelein, F., Guenard, D., and Potier, P. (1987) Taxol and derivatives: a biogenetic hypothesis. *J. Nat. Prod.*, 50, 9–18.

Gueritte-Voegelein, F., Mangatal, L., Guenard, D., Potier, P., Guilhem, J., Cesairo, M., and Pascard, C. (1990) Structure of a synthetic taxol precursor: N-tert-butoxycarbonyl-10-deacetyl-N-debenzoyl taxol. *Acta Cryst.*, C46, 781.

Guo, Y., Vanhaelen-Fastre, R., Diallo, B., and Vanhaelen, M. (1995) Immunoenzymatic methods applied to the search for bioactive taxoids from *Taxus baccata*. *J. Nat. Prod.*, 58(7), 1015–1023.

Guo, Y., Diallo, B., Jaziri, M., Vanhaelen, R., and Vanhaelen, M. (1996) Immunological detection and isolation of a taxoid from the stem bark of *Taxus baccata*. *J. Nat. Prod.*, 59, 169–172.

Hoffman, A.M., Voelker, C.C.J., Franzen, A.T., Shiotani, K.S., and Sandhu, J.S. (1996) Taxanes exported from *Taxus x media* Hicksii cuttings into liquid medium over time. *Phytochemistry*, 43(1), 95–98.

IUPAC. *Pure Appl. Chem.*, 1999, 71, 587–643.

Jaziri, M., Diallo, B.M., Vanhaelen, M.H., Vanhaelen-Fastre, R.J., Zhiri, A., Becu, A.G., and Homes, J. (1991) Enzyme-linked immunosorbent assay for the detection and the semi-quantitative determination of taxane deterpenoids related to taxol in *Taxus* sp. and tissue cultures. *J. Pharm. Belg.*, 46, 93.

Jennings, D.W., Deutch, H.M., Zalkow, L.A., and Teja, S. (1991) Supercritical extraction of taxol from the bark of *T. brevifolia*, 2nd Int. conference on supercritical Fruids, Arlington, VA.

Kawamura, F., Kikuchi, Y., Ohira, T., and Yatagai, M. (1999) Accelerated solvent extraction of paclitaxel and related compounds from the bark of *Taxus cuspidata*. *J. Nat. Prod.*, 62, 244–247.

Khan, N.U.-D. and Parveen, N. (1987) The constituents of the Genus *Taxus*. *J. Sci. Ind. Res.*, 46, 512.

Kingston, D.G.I. (1991) The Chemistry of Taxol. *Pharmac. Ther.*, 52, 1.

Kingston, D.G.I. (1995) In *Taxol: Science and Applications*; Suffness, M., Ed.; CRC Press: Boca Raton, FL, pp. 288 and 291.

Kingston, D.G.I. (2000) *J. Nat. Prod.*, 63, 726–734.

Kingston, D.G.I., Gunatilaka, A.A.L., and Ivey, C.A. (1992) Modified taxols, 7. A method for the separation of taxol and cephalomannine. *J. Nat. Prod.*, 55(2), 259–261.

Kingston, D.G.I., Molinero, A.A., and Rimoldi, J.M. (1993) *Prog. Chem. Org. Nat. Prod.*, 61, 1–206.

Kingston, D.G.I., Molinero, A.A., and Rimoldi, J.M. (1993) The taxane diterpenoids. *Progress in the Chemistry of Organic Natural Products*, 61, 1–206.

Kwak, S.-S., Cho, M.-S., Park, Y.-G., Yoo, J.-S., and Liu, J.-R. (1995) Taxol content in the seeds of *Taxus* spp. *Phytochemistry*, 40(1), 29–32.

Lowe, J.E., Hintz, H.F., and Kingsbury, J.M. (1970) *Taxus cuspidata* (Japanese yew) poisoning in horses. *Cornell Vet.*, 60, 36.

Lucas, H. (1986, 1856) Uber ein in den blattern von *Taxus baccata* L. enthaltenes alkaloid (das taxin). *Arch. Pharm.*, 85, 145.

Lythgoe, B. (1968) The Taxus Alkaloids in *The Alkaloids*; Vol. 10, Chemistry and Physiology, ed. R.H.F. Manske, p. 597. New York, Academic Press.

Lythgoe, B., Nakanishi, K., and Ueo, S. (1964) Taxane. *Proc. Chem. Soc.*, 301.

Mattina, M.J.I., Berger, W.A.I., and Denson, C.L. (1999) Microwave-assisted extraction of taxanes from *Taxus* biomass. *J. Agric. Food Chem.*, 45(12), 4691–4696.

Miller, R.W. (1980) A brief survey of *Taxus* alkaloids and other taxane derivatives. *J. Nat. Prod.*, 43, 425.

Morell, I. (1976) Constituenti di *Taxus baccata* L. *Fitoterapia*, 47, 31.

Paquette, L.A. (1992) Studies directed toward the total synthesis of the Taxanes in *Studies in Natural Products Chemistry*, Vol. 11, ed. Atta-ur-Rahman, p. 3. Amsterdam, Elsevier.

Parmer, V.S., Jha, A., Bisht, K.S., Taneja, P., Singh, S.K., Kumar, A., Poonam, Jain, R., and Olsen, C.E. (1999) Constituents of the yew trees. *Phytochemistry*, 50, 1267–1304.

Rao, C., Liu, X., Zhang, P.L., Chen, W.M., and Fang, Q.C. (1991) Application of high-speed countercurrent chromatography for the isolation of natural products and preparative isolation of taxane deterpenoids and diterpene alkaloids. *Yaoxue xuebao*, 26, 510, (CA: 116, 55596t).

Rao, K.V. (1993) Taxol and related taxanes of *Taxus brevifolia* bark. *Pharm. Res.*, 10, 521.

Rimoldi, J.M., Molinero, A.A., Chordia, M.D., Gharpure, M.M., and Kingston, D.G.I. (1996) An improved method for the separation of paclitaxel and cephalomannune. *J. Nat. Prod.*, 59, 167–168.

Schepatz, S.A. (1976) History of the NCI and the plant screening program. *Cancer Treat. Rep.*, 60, 975.

Schutzki, R.E., Chandra, A., and Nair, N.G. (1994) The effect of postharvest storage on taxol and cephalomannine concentrations in *Taxus x media* 'Hicksit'. *Phytochemistry*, 37(2), 405–408.

Shigemori, H., Sakurai, C.A., Hosoyama, H., Kobayashi, A., Kajiyama, S., and Kobayashi, J. (1999) Taxezopidines J, K, and L, new taxoids from *Taxus cuspidata* inhibiting Ca^{2+}-induced depolymerization of microtubules. *Tetrahed.*, 55, 2553–2558.

Slichenmyer, W.J., Rowinsky, E.K., Donehower, R.C., and Kaufmann, S.H. (1993) The current status of camptothecin analogues as antitumor agents. *J. Nat. Cancer Inst.*, 85, 271.

Suffness, M. (1989) Development of antitumor natural products at the NCI, Gann Monogr. *Cancer Res.*, 36, 21.

Suffness M. (1995) *Taxol, Science and Applications*, Boca Raton, FL, CRC Press.

Suffness, M. and Cordell, G.A. (1985) *Taxus* Alkaloids in *The Alkaloids*, Chemistry and Pharmacology, Vol. 25, ed. A. Brossi, p. 6. New York, Academic Press.

Stull, D.P. and Jans, N.A. (September 1992) Current taxol production from yew bark and future production strategies, 2nd NCI Workshop on Taxol and *Taxus*, Arlington, VA.

Swindel, C.S. (1991) Taxane Diterpene Synthesis Strategies. A Review, *Org. Prep. Proced. Int.*, 23, 465.

Taylor, D.A. (1963) An extractive from *Taxus baccata*. West *Afr. J. Biol. Appl. Chem.* 7, 1.

Tyler, V.E., Jr. (1960) Note on the occurrence of taxine in *Taxus brevifolia. J. Am. Pharm. Assn. Sci. Ed.*, 49, 683.

Vance, N.C., Kelsey, R.G., and Sabin, E. (1994) Seasonal and tissue variaton in taxane concentrations of *Taxus brevifolia*. *Phytochemistry*, 36(5), 1241–1244.

Vance, N.C., Kelsey, R.G., and Sabin, E. (1994) Seasonal and tissue variaton in taxane concentrations of *Taxus brevifolia*. *Phytochemistry*, 36(5), 1241–1244.

Van Rozendaal, E.L.M., Lelyveld, G.P., and van Beek, T.A. (2000) Screening of the needles of different yew species and cultivars for paclitaxel and related taxoids. *Phytochemistry*, 53, 383–389.

Wall, J.L., Wani, M.C., Cook, C.E., Palmer, K.H., McPhail, A.T., and Sim, G.A. (1966) Plant antitumor agents. I. The isolation and structure of camptothecin, a novel alkaloidal leukemia and tumor inhibitor from *Camptotheca acuminata. J. Am. Chem. Soc.*, 88, 3888.

Wall, M.E. and Wani, M.C. (1967) Recent progress in plant anti-tumor agents, Paper M-006, 153rd National Meeting, American Chemical Society, Miami Beach, FL.

Wall, M.E. and Wani, M.C. (1998) Camptothecin and taxol. *The Alkaloids*, 50, 521–536.

Wang, X., Shigemori, H., and Kobayashi, J. (1996) *Tetrahedron*, 52, 12159–12164.

Wani, M.C., Taylor H.L., Wall, M.E., Coggon, P., and MacPhail, A.T. (1971) Plant antitumor agents. VI. The isolation and structure of taxol, a novel antileukemic and antitumor agent from *Taxus brevifolia*. *J. Am. Chem. Soc.*, 93, 2325–2327.

Witherup, K.M., Look, S.A., Stasko, T.G., McCloud, T.G., Isaaq, H.J., and Mushik, G.M. (1989) High performance liquid chromatographic separation of taxol and related compounds from *Taxus brevifolia*. *J. Liq. Chromatogr.*, 12, 2117.

Witherup, K.M., Look, S.A., Stasko, T.J., Ghiorzi, T.J., and Mushik, G.M. (1990) Taxus spp. Needles contain amounts of taxol comparable to the bark of *Taxus brevifolia*: analysis and isolation. *J. Nat. Prod.*, 53, 1249.

Xu, L.X. and Liu, A.R. (1989) Determination of taxol in *Taxus chinensis* by reversed phase HPLC. *Acta Pharm. Sin.*, 24, 552.

Xu, L.X. and Liu, A.R. (1991) Determination of taxol in *Taxus chinensis* by an HPLC method. *Acta Pharm. Sin.*, 26, 537.

Zamir, L.O., Nedea, M.E., Zou, Z.-H., Belair, S., Caron, G., Sauriol, F., Jacqmqin, E., Jean, F.I., Garneau, F.-X., and Mamor, O.J. (1995) *Taxus canadensis* taxanes: structures and stereochemistry. *Canad. J. Chem.*, 73, 655–665.

Zamir, L.O., Zhang, J., Wu, J., Sauriol, F., and Mamor, O.J. (1999) Five novel taxanes from *Taxus canadensis*. *J. Nat. Prod.*, 62, 1268–1273.

Zhang, Z. and Jia, Z. (1991) Taxanes from *Taxus chinensis*. *Phytochemistry*, 30, 2345–2348.

Zhang, J., Sauriol, F., Mamor, O., and Zamir, L.O. (2000) Taxoids from the needles of the Canadian yew. *Phytochemistry*, 54, 221–230.

Zubrod, C.G., Schepatz, S., Leiter, J., Endicott, K.M., Carrese, L.M., and Baker, C.G. (1966) The Chemotherapy program of the NCI: history, analysis, and plans. *Cancer chemother. Rep.*, 50, 349.

2 Biosynthesis of taxoids

Ushio Sankawa and Hideji Itokawa

Introduction

The natural product diterpenoid taxol is regarded as one of the most useful anticancer drugs of plant origin available today (Suffness, 1995). Taxol is effective in the treatment of ovarian and breast cancer, and shows promise for various other cancers, such as head and neck, lung, gastrointestinal, and bladder (Holmes *et al.*, 1995; Suffness, 1995). The fundamental anticancer mechanism of taxol is quite novel and unique. In contrast to Vinca alkaloids, colchicine, and other antitumor agents that act to prevent the polymerization of tubulin into microtubule, taxol promotes microtubule assembly and suppresses its depolymerization (Horwitz, 1992; Jordan and Wilson, 1995; Arbuck and Blaylock, 1995).

Taxol was originally obtained by extracting peeled bark of the Pacific yew (*Taxus brevifolia* Nutt.). The tree grows extremely slowly and big yew trees suitable for the extraction of taxol are several hundred years old. Thus, the harvesting of Pacific yew to produce taxol is destructive to the natural environment because the yields of taxol are too low to supply sufficient quantity for clinical use. Consequently, alternative methods of taxol supply have been extensively investigated. Although several groups (Cragg *et al.*, 1993; Holton *et al.*, 1994; Nicolau *et al.*, 1994, 1995; Masters *et al.*, 1995) achieved the total synthesis of taxol through elegant approaches, the high cost of this synthetic approach prohibits its commercial feasibility. Currently, taxol and its analogue, taxotere (docetaxel), are produced semi-synthetically by the acylation of 10-deacetyl-baccatin III and related compounds. These compounds are obtained from the needles of *Taxus brevifolia* as well as the European yew *Taxus baccata* which are propagated from cuttings (Guenard *et al.*, 1993; Georg *et al.*, 1994; Commercon *et al.*, 1995; Holton *et al.*, 1995).

Another approach would be to use cell suspension cultures of *T. brevifolia*. Several groups have succeeded in producing taxol or baccatin in reasonable yields from *Taxus* cell suspension cultures (Yukimine *et al.*, 1996; Hezari *et al.*, 1997), but this alternative production method is not commercially viable. Recently, taxol was isolated from the bark, leaves, branches, shells, and fruits of the hazelnut tree; however, the content is only about 10% of that in yew trees. Recently, a fungus, *Taxomyces andreanae*, was unexpectedly found to produce taxol and subsequent similar findings with other microorganisms confirmed its production (Stierle *et al.*, 1993; Strobel *et al.*, 1996). Because the fungus has been isolated from yew trees, it is said, "If a fungus could be coaxed into churning out the drug in vats, it would definitely have value" (Hoffman, 2000). This finding raised a question about the genes responsible for taxol's production (?). Comparative studies of these genes both in higher plants and fungi will clarify whether the genes were horizontally transferred from fungus to plant or vice versa. Similar cases are known; gibberellin and maytancin occur both in higher plants and microorganisms. The increasing demand of taxol for clinical use as an anticancer drug combined with limited supply from plant sources require a thorough understanding of its biosynthesis, in particular, the enzymatic reaction mechanism. Subsequently, cloning corresponding genes would allow genetic manipulation of plants or fungi to improve the yield of taxol production in these organisms.

Recently, excellent reviews on the biosynthesis of taxoids have been published (Floss and Mocek, 1995; Hezari and Ctoteau, 1997; Kobayashi and Shigemori, 1998; Shen *et al.*, 1999). In particular the review of Hezari and Croteau *et al.* (1997) covers all aspects of taxol skeleton formation including gene cloning of the enzymes involved. This chapter mainly describes the development of taxol biosynthesis according to their review.

The biosynthesis of the taxane skeleton (6/8/6-membered ring system) was previously believed to involve cyclization of geranylgeranyl diphosphate (GGPP) into taxa-4(20), 10(11)-diene (exotaxadiene). However, the enzyme that catalyzes the formation of taxadiene was cloned by Croteau *et al.* (1995) and the product of GGPP:taxadiene cyclase was (endo)taxadiene, not (exo)taxadiene. All the taxoids of *Taxus* origin are reasonably explained by cyclization and

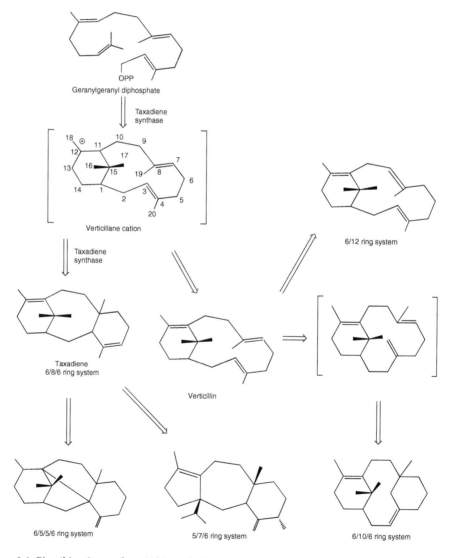

Figure 2.1 Plausible scheme of taxoid biosynthesis.

isomerization of verticillene and taxadiene. 3,11-cyclotaxanes (6/5/5/6-membered ring system) are derived from taxadiene (Kobayashi *et al.*, 1972) and 11(15 → 1)-*abeo*taxanes (5/7/6-membered ring system) are generated through an extended Wagner–Meerwein rearrangement in ring A of taxadiene (6/8/6-membered ring system) (Appendino *et al.*, 1993). On the other hand, the 2(3 → 20)-*abeo*taxanes (6/10/6-membered ring system) seem to be derived from verticilline through the cyclization of delta4(20),7-verticillene (Appendino *et al.*, 1994). The bicyclic taxane-related compounds (6/12-membered ring system) are derived from verticillene (Zamir *et al.*, 1995).

Taxadiene, the key intermediate of taxol biosynthesis, has been prepared enzymatically from isopentenyldiphosphate in cell-free extract of *Escherichia coli* by overexpressing genes encoding isopentenyl dophosphate isomerase, GGPP synthase and taxadiene synthase (Huang *et al.*, 2001) (Figure 2.1).

The scheme of taxol biosynthesis was summarized by Shen *et al.* (1999). The biosyntheses of the skeleton and side-chain of taxol proceed through independent pathways, after which the two moieties are combined to form taxol and related compounds as shown in Figure 2.2.

Earlier studies of taxol biosynthesis

The structure of taxol is quite complex and bears an unusual diterpene carbon skeleton, eight oxy-functional groups, and an assortment of appended acyl side-chains, for a total of 11 chiral centers (Wani *et al.*, 1971) (Figures 2.2(a,b)). As proposed first by Harrison *et al.* (1966), a proposal for taxol biosynthesis posited the initial formation of an olefinic taxane skeleton, followed by a series of oxygenation steps and, ultimately, acylation reactions (Zamir *et al.*, 1992; Eisenreich *et al.*, 1996). Based on precedents in the metabolism of other terpenoids (Gershenzon and Croteau, 1993), the pathway to taxol from the universal diterpenoid precursor GGPP must involve over a dozen individual enzymatic reactions. Deciphering the biosynthesis of taxol posed a considerable challenge because of its complex structure, low isolation yield, and experimental difficulty with *Taxus* species.

Early studies on the biosynthesis of taxol were carried out with *in vivo* feeding experiments of radiolabled precursors. The results indicated that acetate, mevalonate, and phenylalanine were the building blocks of taxol (Zamir *et al.*, 1992). Recently, however, Eisenreich *et al.* (1996) have shown that the taxane ring system is formed by a mevalonate-independent pathway found by Rohmer for isoprenoid biosynthesis (Schwender *et al.*, 1996).

In 1996, Eisenreich *et al.* reported the results of feeding experiments with ^{13}C labeled precursors in cell suspension cultures of *Taxus chinensis*. This methodology produced the diterpene taxuyunnanine C in a yield of 2.6% (dry weight basis). The incorporation of [U-$^{13}C_6$]glucose, [1-^{13}C]glucose, and [1,2-$^{13}C_2$]acetate into taxuyunnanine C was analyzed by NMR spectroscopy. The label from [1,2-$^{13}C_2$]acetate was incorporated into the four acetyl groups of taxuyunnanine C (Figure 2.3a), but not into the taxane moiety. [U-$^{13}C_6$]glucose was efficiently incorporated into both the taxane ring system and the acetyl groups (Figure 2.3b). The four isoprenoid units of taxuyunnanine C showed identical labeling patterns. The analysis of long-range ^{13}C-^{13}C couplings in taxuyunnanine C labeled by [U-$^{13}C_6$]glucose clearly demonstrated the involvement of an intramolecular rearrangement during the formation of isoprenyl diphosphate (IPP). The labeling patterns clearly demonstrated that taxuyunnanine C is biosynthesized through a mevalonate-independent pathway (Eisenrich *et al.*, 1996).

In exploring later stages of the taxol biosynthesis pathway, Floss and his colleagues carried out detailed feeding experiments with advanced metabolites to demonstrate the origin and timing of assembly of the N-benzoyl phenylisoserine ester (Fleming *et al.*, 1993, 1994; Schwender *et al.*, 1996). A description of this work and an overview of previous speculations on the biosynthetic pathway leading to taxol have been reviewed by Floss and Mocek (1995) (Figure 2.4).

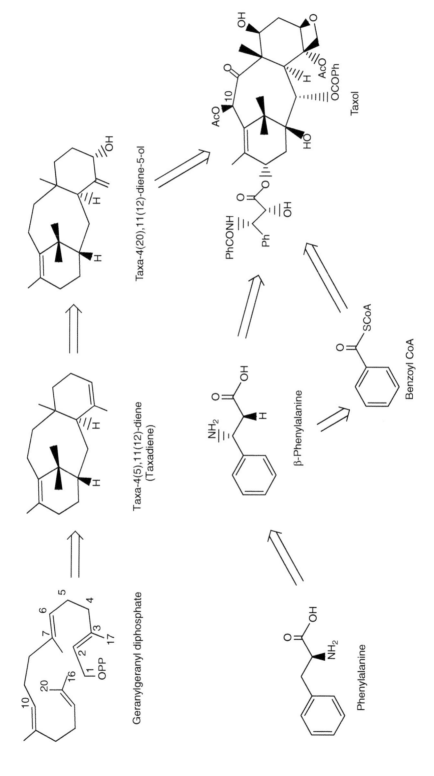

Geranylgeranyl diphosphate

Taxa-4(5),11(12)-diene
(Taxadiene)

Taxa-4(20),11(12)-diene-5-ol

Phenylalanine

β-Phenylalanine

Benzoyl CoA

Taxol

Figure 2.2a The scheme of taxol biosynthesis.

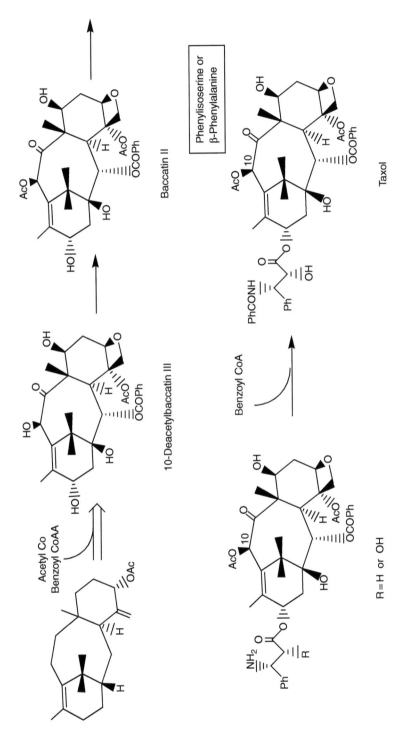

Baccatin II

10-Deacetylbaccatin III

Acetyl Co
Benzoyl CoAA

OAc

Phenylisoserine or
β-Phenylalanine

Benzoyl CoA

PhCONH

Ph

OH

Taxol

R=H or OH

NH₂

Ph

R

Figure 2.2b The scheme of taxol biosynthesis.

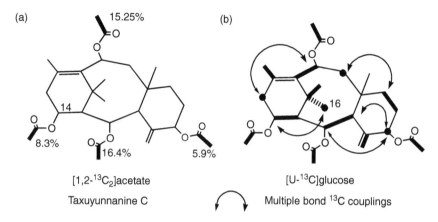

Figure 2.3 Biosynthetic pathway of taxuyunnanine C.

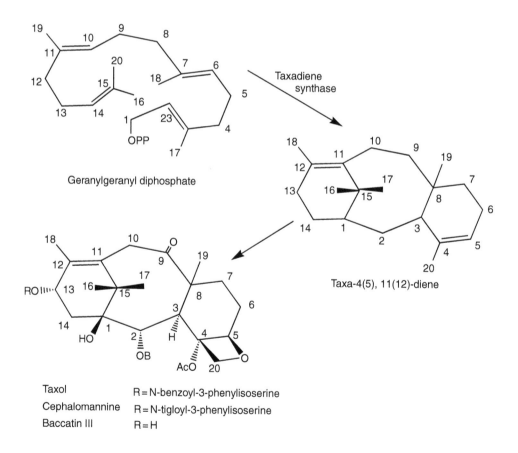

Taxol	R = N-benzoyl-3-phenylisoserine
Cephalomannine	R = N-tigloyl-3-phenylisoserine
Baccatin III	R = H

Figure 2.4 Biosynthetic pathway of taxol.

The first step of taxol biosynthesis; cyclization of GGPP with taxadiene synthase

The first committed step in the biosynthesis of taxol and related taxoids is the cyclization of the universal diterpenoid precursor GGPP to taxadiene [(taxa-4(5),11(12)-diene)] (Figure 2.1) (Koepp *et al.*, 1995), but not to an exocyclic isomer, taxa-4(20),11(12)-diene, as was proposed earlier based on the abundant occurrence of taxane metabolites with an exo-double bond structure (Gueritte-Voegelein *et al.*, 1987; Floss *et al.*, 1995). *In vivo* experiments with *T. brevifolia* stem sections, using labeled taxa-4(5),11(12)-diene, demonstrated the very efficient incorporation (30%) of the olefin into taxol and closely related taxoids. In addition, the endocyclic olefin was found in *T. brevifolia* bark at a low level (5–10 mg per kg dry wt.), whereas the exocyclic isomer was not detected (Koepp *et al.*, 1995). Both taxa-4(5),11(12)-diene and its 4(20)-isomer were subsequently synthesized (Rubenstein and Williams, 1995).

A cell-free system prepared from *T. brevifolia* stems catalyzed the divalent metal ion-dependent conversion of GGPP into taxadiene as essentially the only olefinic product. This enzyme activity was localized mainly in the bark and adjacent cambium cells. It was purified up to 600-fold by conventional chromatography and electrophoresis to give the pure protein. The combination of gel permeation chromatography and SDS-PAGE showed the cyclase to be a monomer of ~79 Da (Hezari *et al.*, 1995). This taxadiene synthase has been characterized with regard to optimum pH, kinetic constants, and possible active site residues. The cyclase requires divalent metal ion as a cofactor, and Mg^{2+} (Km ~0.16 mM) is more preferable to Mn^{2+} and other divalent metals.

The mechanism of the cyclization reaction has been investigated in detail and shown to proceed without detectable free intermediates or the preliminary formation of enzyme-bound taxa-4(20),11(12)-diene with subsequent isomerization. From these results, a stereochemical mechanism has been proposed for the taxadiene synthase reaction (Lin *et al.*, 1996). The initial cyclization of GPPP gives a transient verticillyl cation. Intramolecular transfer of the C-11 proton to C-7 initiates transannular B/C-ring closure to the taxenyl cation. Deprotonation at C-5 then yields the taxa-4(5),11(12)-diene product directly. Recently, proton migration was further studied with recombinant taxadiene synthase and the sequence above was firmly established. Proton transfer from C-11 to C-7 is facilitated by close proximity in the dimensional structure of the intermediate cation. Interestingly, analysis of reaction products revealed that taxadiene synthase yields principally taxa-4(5),11(12)-diene along with less amounts (*c.*5%) of the 4(2),11(12)-isomer, exo-taxadiene, resulting from the deprotonation of the C-20 methyl group, and the 3(4),11(12)-isomer arising from negligible proton loss from the bridgehead methine (C-3) (Williams *et al.*, 2000) (Figure 2.5).

The low levels of taxadiene synthase activity in *Taxus* stem tissue ($140 \, \text{pmol} \cdot \text{h}^{-1} \cdot \text{g}^{-1}$), the very low level of taxadiene in bark, and the efficient conversion of the olefin to taxoids *in vivo* suggested that the cyclization of GPPP to taxadiene was slow relative to later steps in the biosynthetic pathway to taxol. The potential role of the first committed step in the control of metabolic flux of biosynthesis pathway prompted a search for the cDNA encoding taxadiene synthase. Attempts to obtain the amino acid sequence information from the purified enzyme were unsuccessful, and led to the adoption of a PCR-based strategy using homologous sequence-based primers designed from the known terpenoid cyclases of plant origin. Of the more than 20 pairs of primers tested, only one set yielded a small DNA fragment possessing a "cyclase-like" sequence. This fragment was subsequently employed as a hybridization probe to screen a cDNA library constructed from *T. brevifolia* stem. Several clones of isolated cDNA were not full-length and a 5′ rapid amplification method (RACE) was applied to obtain full length cDNA. The full length cDNA of taxadiene synthase thus obtained was functionally expressed in *E. coli*, and the

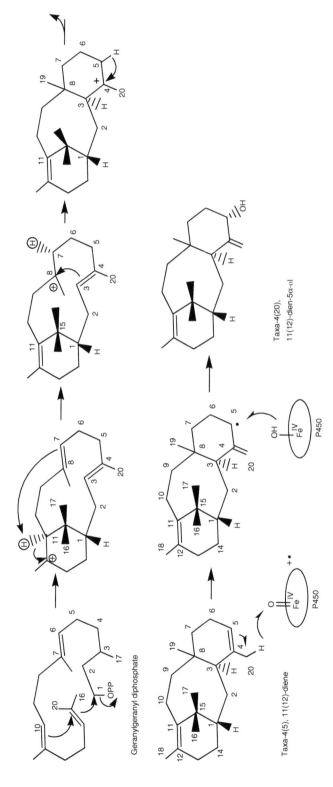

Geranylgeranyl diphosphate

Taxa-4(5), 11(12)-diene

Taxa-4(20), 11(12)-dien-5α-ol

Figure 2.5 Biosynthetic pathway of taxanes.

recombinant enzyme thereby confirmed the conversion of GGPP to taxadiene. The DNA sequence has an open reading frame of 2586 nucleotides, and the deduced polypeptide consists of 862 amino acid residues corresponding to a molecular weight of about 98,000. It contains a long plastidial targeting signal sequence at the N-terminal. Sequence comparison with other terpenoid synthases of higher plant origin showed a significant homology, especially with abietadiene synthase of gymnosperm origin, suggesting evolution from a common ancestry gene of this enzyme class (Wildung and Croteau, 1996).

The possible role of taxadiene synthase in controlling the metabolic flux of taxol's biosynthetic pathway could not be easily ascertained in intact *Taxus* stem tissue. Therefore, heterotrophic suspension cell cultures of the Canadian yew (*Taxus canadensis*) were employed as an experimental system. The enzyme from this source was isolated and shown to be chromatographically, electrophoretically, and kinetically identical to the taxadiene synthase of *T. brevifolia* stem. Analysis of enzyme activity levels during the time-course of taxol accumulation in developing cell cultures indicated that taxadiene synthase activity increased before taxol accumulation initiated and persisted to a stationary phase. Taxadiene synthase activity *in vitro* exceeds the maximum rate of taxol accumulation *in vivo* (Hezari and Croteau, 1997). These results suggest that, although taxadiene synthase represents a slow step, the rate-limiting step is further down the biosynthetic pathway.

Second step of taxol biosynthesis: taxadienol formation

No oxygenated taxoids bearing a 4(5)-double bond have been isolated from the plants, whereas taxoids with a 4(20)-ene-5-oxy functional group are common (Kingston *et al.*, 1993). These facts suggest that hydroxylation of taxa-4(5),11(12)-diene at C-5 is associated with migration of the double bond. This step may be next in the taxol biosynthetic sequence (Croteau *et al.*, 1995) (see Figure 2.2). Microsomal preparations from Taxus stem and cultured cells were shown to catalyze the NADPH dependent oxidative conversion of taxa-4(5),11(12)-diene into an alcohol. This compound was confirmed to be taxa-4(20),11(12)-diene-5α-ol by comparison to an authentic standard prepared by total synthesis. The stereochemistry at C-5 was rigorously and definitively established, as the β-epimer exhibits different chromatographic properties and a distinguishable mass spectrum. The hydroxylase, which is found mainly in the light membrane fraction, fulfilled all of the expected requirements of a cytochrome P450 monooxygenase (heme thiolate protein), such as inhibition by CO and blue light reversion (Hefner *et al.*, 1996). Many terpenoid hydroxylases are of the cytochrome P450 type (Mihaliak *et al.*, 1993). Whether the conversion of 4(5),11(12)-diene into 4(20),11(12)-diene-5α-ol involves an epoxide intermediate or abstraction of hydrogen radical from C-20 methyl followed by double bond migration and rebonding of hydroxyl radical at C-5 is not yet known. However, if the hydroxylation involves epoxidation followed by ring opening, it should require an enzyme other than P450. This mechanism is less likely, because the reaction is catalyzed only by P450. Nevertheless, the structure of the first oxygenated intermediate in taxol's biosynthesis constrains the mechanistic possibilities for subsequent elaboration of an oxetane moiety of taxol.

The role of taxa-4(20),11(12)-dien-5α-ol as a pathway intermediate was confirmed by *in vivo* conversion (in a range of 10–15% incorporation) of radioactive taxa-4(20),11(12)-dien-5α-ol into advanced taxoids, including 10-deacetylbaccatin III, cephalomannine, and taxol, in *T. brevifolia* stem disks (Hefner *et al.*, 1996). Radiochemically-guided fractionation of an extract of *T. brevifolia* dried bark additionally demonstrated that the alcohol (at 5–10 mg) and its esters (at 25–50–10 mg) were present as naturally occurring metabolites in the relevant tissue. Both of these observations were based on *in vivo* studies and suggest that this first oxygenation step in taxol biosynthesis is slow relative to subsequent metabolic transformations and, thus, is a worthy candidate for gene manipulation.

It has been established that taxa-4(20),11(12)-diene-5α-ol is the first oxygenated intermediate on the biosynthetic pathway to taxol. Recently, taxa-4(20),11(12)-diene-2α,5α-diol was synthesized and proved to be a potential candidate as the second oxygenated intermediate on the pathway (Vazquez and Williams, 2000).

The third step in taxol biosynthesis: acetylation of taxadienol and oxetane ring formation

Most naturally occurring taxoids are acetylated, or otherwise acylated, at C-5 (Kingston *et al.*, 1993; Baloglu and Kingston, 1999). This suggests that the Acylation of taxa-4(20),11(12)-dien-5α-ol could be the third step in taxol biosynthesis. Additionally, the 4(20)-ene-5α-acetate functional group should play a crucial role in the elaboration of the oxetane ring by a mechanism involving epoxidation of the 4(20)-double bond, followed by intramolecular acetoxy migration associated with epoxide ring expansion.

A possible mechanism for construction of the oxetane ring from the 4(20)-ene-5α-acetoxy functional grouping is epoxidation of the 4(20)-double bond, followed by intramolecular acetate migration and oxirane ring opening, shown in Figure 2.6 (Gueritte-Voegelein *et al.*, 1987; Floss and Mocek, 1995; Hefner *et al.*, 1996).

These considerations, as well as studies on subsequent oxygenation steps, prompted an investigation of the enzymatic acylation of taxa-4(20),11(12)-dien-5α-ol.

A soluble enzyme preparation from *T. canadensis* cell suspension cultures was shown to catalyze the acetyl coenzyme A dependent acylation of taxa-4(20),11(12)-diene-5α-ol. Membrane fractions were inactive in the acetylation reaction. Transacetylase in soluble fraction has been partially purified and characterized as a 50 kDa monomeric protein (with pI ~4.7) that accepts acetyl CoA, but not benzoyl CoA, as a substrate.

Subsequent oxygenation steps

Subsequent reactions in the taxol biosynthesis pathway include additional hydroxylations, acylations on the hydroxy groups, oxidation of hydroxyl into carbonyl, and the generation of the oxetane ring system. Numerous naturally occurring taxoids exhibit a broad range of oxygenation and acylation patterns, which provide no rigid guidance in predicting the order of functionalization beyond taxa-4(20).11(12)-dien-5α-yl acetate. Proposed sequence for the hydroxylation of taxa-4(5), 11(12)-diene to the level of a pentaol is based on the relative abundance of naturally

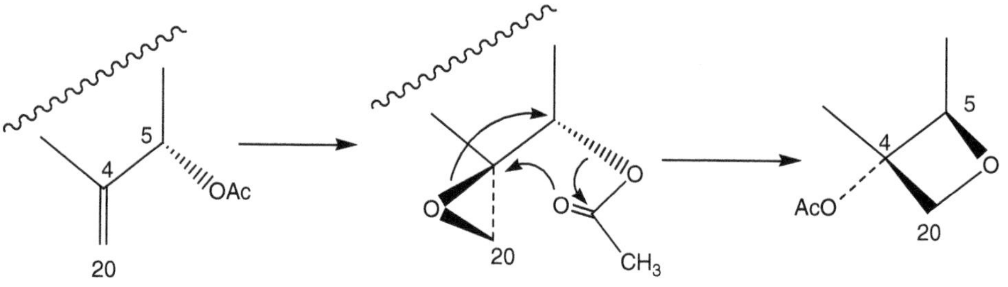

Figure 2.6

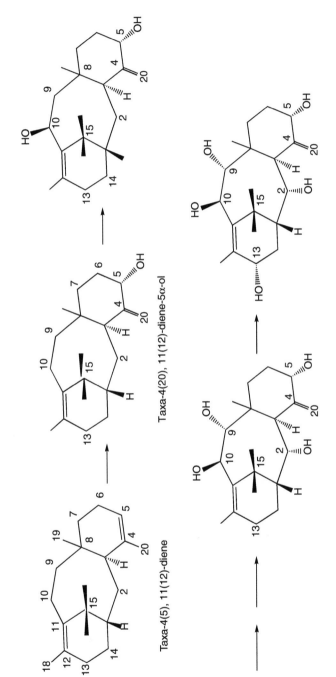

Taxa-4(5), 11(12)-diene

Taxa-4(20), 11(12)-diene-5α-ol

Figure 2.7

occuring taxoids. Thus, oxygenation likely starts from C-5 and is followed by C-10, C-2, C-9, and finally C-13 (Floss and Mocek, 1995; Croteau *et al.*, 1996). Taxoids bearing oxygen group at C-13 are abundant and this suggests a sequence of the tailoring of taxadiene to the level of a pentaol. The reactions are probably catalyzed by cytochrome P450 oxygenases. Acetylation at C-5 before further hydroxylation seems likely, but whether acylation of the various other hydroxy groups are required steps of the oxygenation sequence is unknown (Figure 2.7).

Oxygenations at C-7 and C-1 of the taxane nucleus are considered to be late steps, possibly after oxetane ring formation. However, oxetane formation itself seems to be a relatively late-stage transformation and presumably precedes acylation at C-13 and oxidation of the C-9 hydroxy to a carbonyl (cf. Figure 2.4) (Floss and Mocek, 1995; Croteau *et al.*, 1996).

When microsomal preparations of *T. canadensis* cell cultures were incubated with taxa-4(20),11(12)-dien-5α-ol in the presence of NADPH and O_2, the resulting product was more polar than the substrate. It showed chromatographic properties and mass spectrum (by HPLC-MS) consistent with a taxadiene diol. The reaction was catalyzed by a cytochrome P450 hydroxylase, suggesting that the other oxygenation steps in taxol biosynthesis may involve similar enzymes. By utilizing a microsomal preparation that had been carefully optimized to sustain cytochrome P450-type reactions (Hefner *et al.*, 1996), both taxa-4(5),11(12)-diene and taxa-4(20),11(12)-dien-5α-ol gave rise to products that were tentatively identified by HPLC-MS analysis as the corresponding diol, triol, and tetraol. The precise structures of these products are yet to be solved. Incubation of the optimized microsomes with taxa-4(20),11(12)-dien-5α-yl acetate yielded a single major product, identified by HPLC-MS analysis as a pentaol monoacetate. Taxa-4(20),11(12)-dien-5α-yl benzoate was not oxygenated by this system, indicating that the efficiency of the acetate ester as a substrate was not due simply to higher lipophilicity but rather it was recognized as a natural substrate by the set of microsomal hydroxylases that catalyze sequential hydroxylation. The successive series of hydroxylation reactions is now being systematically investigated in respect to substrates and the structural elucidation of the corresponding products, as well as the interpretation of additional acylation steps in the biosynthesis sequence.

Phenylalanine amino mutase (Walker and Floss, 1998)

Taxol possess a characteristic benzoylphenylisoserine acyl group. Phenylalanine amino mutase, which is responsible for the formation of β-phenylalanine, was investigated and identified for the first time in cell-free extracts of *T. brevifolia*. This enzyme has been shown to be a key intermediate in the biosynthesis of taxoid's phenylisoserine side-chain. β-Phenylalanine produced by the cell-free extract was found to have the same configuration as that present in the taxol side-chain. Studies with phenylalanines multiply labeled with ^2H and ^{15}N showed that the mutase reaction proceeds via intramolecular migration of the α-amino group to C-3 and replacement of the 3-proS hydrogen with retention of configuration (Figure 2.8).

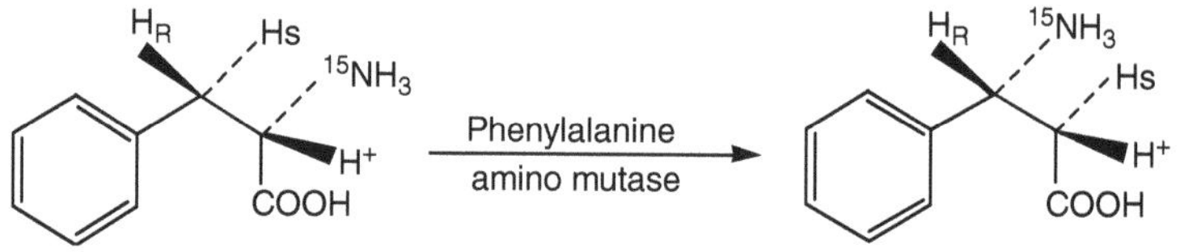

Figure 2.8 Intramolecular migration of the α-amino group.

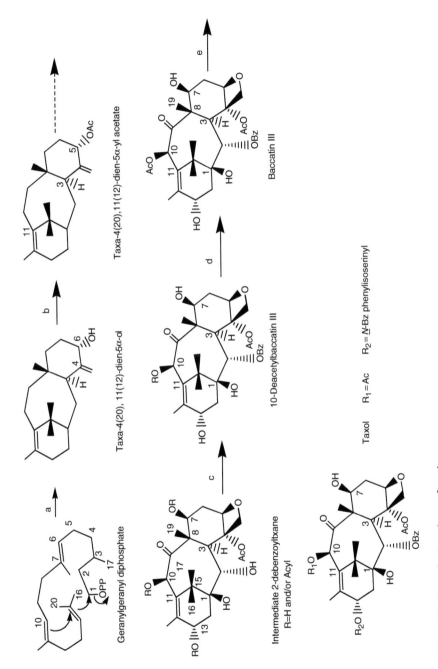

Figure 2.9 Biosynthetic pathway of taxol.

In contrast, microbial amino mutases catalyze intramolecular nitrogen migrations resulting in an inversion of configuration at the migration terminus. Thus, a different mechanism may be in operation; the 3-proS hydrogen migrates to the C-2 position in a process which is partly intermolecular in nature. This type of mutase has been known in other organisms and most require vitamin B_{12} as a cofactor. The amino mutase from *Taxus* apparently does not require vitamin B_{12}. It is unclear at present what cofactor is involved in the phenylalanine amino-2,3-mutase reaction (Walker and Floss, 1998).

Benzoyl transferase from *Taxus*

Recently, Walker and Croteau isolated a cDNA clone encoding a taxane 2α-0-benzoyltransferase from *Taxus cuspidata* (Walker and Croteau, 2000). The proposed outline of the taxol biosynthetic pathway about the cyclization of GGPP to taxadiene by taxadiene synthase and the hydroxylation to taxadien-5α-ol by taxadiene 5α-hydroxylase is shown in Figure 2.9: (a) the acetylation of taxadiene-5α-ol by taxa-4(20),11(12)-dien-5α-ol 0-acetyltransferase (TAT); (b) the conversion of a 2-debenzoyl "taxoid-type" intermediate to 10-deacetylbaccatin III by a taxane 2α-0-debenzoyltransferase (TBT); (c) the conversion of 10-deacetylbaccatin III to baccatin III by 10-deacetylbaccatin III 10-0-acetyltransferase (DBAT); (d) and the side-chain attachment to baccatin III to form taxol (e) are illustrated. The broken arrows indicate several as-yet-undefined steps (Walker and Croteau, 2000).

Conclusion

Improving the production of taxol or taxoid for drug semi-synthesis (Nicolau *et al.*, 1994) should be achieved by a thorough understanding of the complex biosynthesis pathway and its rate limiting steps. A systematic approach should reveal the sequence of oxygenation and acylation steps, the timing of the C-9 hydroxy dehydrogenation and the unique oxetane ring formation, and the coordination of these processes with the assembly of the N-benzoyl phenylisoserine side-chain (Fleming *et al.*, 1994; Hefner *et al.*, 1996).

The role of each enzymatic step in controlling the flux of the biosynthesis pathway can then be assessed by *in vivo* studies. The slow steps should be clarified, and suitable cloning strategies devised to acquire the corresponding genes. This approach will ultimately lead to engineering of Taxus or relevant microorganisms that overexpress key genes and to increased taxol production through the enhancement of corresponding enzymes (Walker and Croteau, 2001).

References

Appendino, G., Barboni, L., Gariboldi, P., Bombardelli, E., Gabetta, B., and Viterbo, D. (1993) *J. Chem. Soc., Chem. Commun.*, 1587.

Appendino, G., Cravotto, G., Enriu, R., Jakupovic, J., Gariboldi, P., Gabetta, B., and Bombardelli, E. (1994) *Phytochemistry*, 36, 407.

Arbuck, S.G. and Blaylock, B.A. (1995) Taxol: clinical results and current issues in development, in *Taxol: Science and Applications* (M. Suffness ed.), pp. 379–415, CRC Press, Boca Raton, FL.

Baloglu, E. and Kingston, D.G.I. (1999) The taxane diterpenoids. *J. Nat. Prod.*, 62, 1448–1472 .

Commercon, A., Bouzat, J.D., Didier, E., and Lavelle, F. (1995) *Taxane Anticancer Agents: Basic Science and Current Status* (G.I. Georg, T.T. Chen, I. Ojima, D.M. Vyas, eds), pp. 233–246, Amer. Chem. Soc., Washington, DC.

Cragg, G.M., Schepartz, S.A., Suffness, M., and Grever, M.R. (1993) *J. Nat. Prod.*, 56, 1657–1688.

Croteau, R., Hefner, J., Hezari, M., and Lewis, N.G. (1996) *Curr. Top. Plant Physiol.*, 15, 94–104.

Croteau, R., Hezari, M., Hefner, J., Koepp, A., and Lewis, N.G. (1995) in Taxane anticancer agents: Basic science and current status, (G.I. Georg, T.T. Chen, I. Ojima, D.M. Vyas, eds), pp. 72–80, Am. Chem. Soc., Washington, DC.

Eisenreich, W., Menhard, B., Zenk, P.J., and Bacher, A. (1996) *Proc. Natl. Acad. Sci.* USA, **93**, 6431–6436.

Fleming, P.E., Mocek, U., and Floss, H. (1993) *J. Am. Chem. Soc.*, **115**, 805–807.

Fleming, P.E., Knaggs, A.R., He, X.-G., Mocek, U., and Floss, H.G. (1994) *J. Am. Chem. Soc.*, **116**, 4137–4138.

Floss, H.G. and Mocek, U. (1995) in *Taxol: Science and Applications* (M. Suffness, ed.), pp. 191–208, CRC Press, Boca Raton, FL.

Georg, G.I., Ali, S., Zygmut, J., and Jayasinghe, L.R. (1994) *Exp. Opin. Ther. Patents*, **4**, 109–120.

Gershenzon, J. and Croteau, R. (1993) in *Lipid Metabolism in Plants* (T.S. Moore, Jr., ed.), pp. 339–388, CRC Press, Boca Raton, FL.

Guenard, D., Gueritte-Voegelein, F., and Potier, P. (1993) *Acc. Chem. Res.*, **26**, 160–167.

Gueritte-Voegelein, F., Guenard, D., and Potier, P. (1987) Taxol and derivatives: a biogenetic hypothesis. *J. Nat. Prod.*, **50**, 9–18.

Harrison, J.W., Scrowston, R.M., and Lythgoe, B. (1996) *J. Chem. Soc.* (C), 1933–1945.

Hefner, J. and Croteau, R. (1996) *Methods Enzymol.*, **272**, 243–250.

Hefner, J., Rubenstein, S.M., Ketchum, R.E.B., Gibson, D.M., Williams, R.M., and Croteau, R. (1996) *Chem. and Biol.*, **3**, 479–489.

Hezari, M. and Croteau, R. (1997) Taxol biosynthesis: An update. *Planta Medica*, **63**, 291–295 .

Hezari, M., Ketvhum, R.E.B., Gibson, D.M., and Croteau, R. (1997) *Arch. Biochem. Biophys.*, **337**, 185–190.

Hezari, M., Lewis, N.G., and Croteau, R. (1995) *Arch. Biochem. Biophys*, **322**, 437–444.

Hoffman, A. (2000) *Science*, vol. 288, 7 April, p. 27.

Holmes, F.A., Kudelka, A.P., Kavanagh, J.J., Huber, M.H., Ajani, J.A., and Valero, V. (1995) *Taxane Anticancer Agents: Basic Science and Current Status* (G.I. Geoge, T.T. Ojima, D.M. Vyas, eds), pp. 31–57, Amer. Chem. Soc., Washington, DC.

Holton, R.A., Kim, H.B., Somoza, C., Liang, F., Biediger, R.J., Boatman, P.D., Shindo, M., Smith, C.C., Kim, S., Nadizadeh, K., Gentile, L.N., and Liu, J.H. (1994) *J. Am. Chem. Soc.*, **116**, 1599–1600.

Holton, R.A., Biediger, R.J., and Boatman, P.D. (1995) *Taxol: Science and Applications* (M. Suffness, ed.), pp. 97–121, CRC Press, Boca Raton, FL.

Horwitz, S.B. (1992) *Trends Pharmacol. Sci.*, **13**, 134–136.

Huang, Q., Roessner, C.A., Croteau, R., and Scott, A.I. (2001) Engineering *Escherichia coli* for the synthesis of taxadiene, a key intermediate in the biosynthesis of taxol. *Bioorg. & Med. Chem.*, **9**, 2237–2242.

Jordan, M.A. and Wilson, L. (1995) Taxane anticancer agents: Basic science and current status (I.G. Georg, T.T. Chen, I. Ojima, D.M. Vyas, eds), pp. 138–153, Amer. Chem. Soc., Washington, DC.

Kingston, D.G.I., Molinero, A.A., and Rimoldi, J.M. (1993) *Prog. Chem. Org. Nat. Prod.*, **61**, 1–206.

Kobayashi, J. and Shigemori, H. (1998) Bioactive taxoids from Japanese yew *Taxus cuspidata* and taxol biosynthesis. *Heterocycles*, **47**(2), 1111–1133.

Kobayashi, T., Kurono, M., Sato, H., and Nakanishi, K. (1972) *J. Am. Chem. Soc.*, **94**, 2863.

Koepp, A.E., Hezari, M., Zajicek, J., Strofer Vogel, B., LaFever, R.E., Lewis, N.G., and Croteau, R. (1995) *Biol. Chem.*, **270**, 8686–8690.

Lin, X., Hezari, M., Koepp, A.E., Floss, H.E., and Croteau, R. (1996) *Biochem.*, **35**, 2968–2977.

Masters, J.J., Link, J.T., Snyder, L.B., Young, W.B., and Danishefsky, S.J. (1995) *Angew. Chem. Int. Ed. Engl.*, **34**, 1723–1726.

Mihaliak, C.A., Karp, F., and Croteau, R. (1993) *Methods Plant Biochem.*, **9**, 261–279.

Nicolau, K.C., Dai, W.-M., and Guy, R.K. (1994) *Angew. Chem. Int. Ed. Engl.*, **33**, 15–44.

Nicolau, K.C., Yang, Z., Liu, J.J., Ueno, H., Natermet, P.G., Guy, R.K., Claiborne, F.F., Renaud, J., Couladouros, E.A., Paullvannin, K., and Sorensen, E.J. (1994) *Nature*, **367**, 630–634.

Nicolau, K.C., Ueno, H., Liu, J.J., Natermet, P.G., Yang, Z., Renaud, J., Paulvannin, K., and Chadha, R. (1995) *J. Am. Chem. Soc.*, **117**, 653–659.

Rubenstein, S.M. and Williams, R.M. (1995) *J. Org. Chem.*, **60**, 7215–7223.

Schwender, J., Seeman, M., Lichtenthaler, H.K., and Rohmer, M. (1996) *Biochem. J.*, **316**, 73–80.

Shen, Y., Gan, F., Zhang, H., and Hao, X. (1999) Taxol biosynthesis. *Huaxue Yanjjiu Yu Yingyong*, 11(5), 460–464.

Stierle, A., Strobel, G., and Stierle, D. (1993) *Science*, 260, 214–216.

Strobel, G., Yang, X., Sears, J., Kramer, R., Sidhu, R.S., and Hess, W.M. (1996) *Microbiol.*, 142, 435–440.

Suffness, M. (1995) *Taxane Anticancer Agents: Basic Science and Current Status* (G.I. George, T.T. Chen, I. Ojima, D.M. Vyas, eds), pp. 1–17, Amer. Chem. Soc., Washington, DC.

Suffness, M. and Wall, M.E. (1995) in *Taxol: Science and Applications* (M. Suffness, ed.), pp. 3–25, CRC Press, Boca Raton, FL.

Vazquez, A. and Williams, R.M. (2000) Studies on the biosynthesis of taxol, synthesis of taxa-4(20),11(12)-diene-2α,5α-diol. *J. Org. Chem.*, 65, 7865–7869.

Walker, K.D. and Floss, H.G. (1998) Detection of a phenylalanine aminomutase in cell-free extracts of *Taxus brevifolia* and preliminary characterization of its reaction. *J. Am. Chem. Soc.*, 120, 5333–5334.

Walker, K.D. and Croteau, R. (2000) Taxol biosynthesis: Molecular cloning of a benzoyl-CoA: taxane 2α-O-benzoytransferase cDNA from Taxus and functional expression in Escherichia coli. *Proc. Natl. Acad. Sci. USA*, 97(25), 13591–13596.

Walker, K. and Croteau, R. (2001) Taxol biosynthesis genes. *Phytochemistry*, 58(1), 1–7.

Wani, M.C., Taylor, H.L., Wall, M.E., Coggon, P., and McPhail, A.T. (1971) *J. Am. Chem. Soc.*, 93, 2325–2327.

Wildung, M.R. and Croteau, R. (1996) *Biol. Chem.*, 271, 9201–9204.

Williams, D.C., Carrol, B.J., Jin, Q., Rithner, C.D., Lenger, S.R., Floss, H.G., Coates, R.M., Williams, R.M., and Croteau, R. (2000) Intramolecular proton transfer in the cyclization of geranylgeranyl dophosphate to the taxadiene precursor of taxol catalyzed by recombinant taxadiene synthase. *Chem. & Biol.*, 7(12), 969–977.

Yukimine, Y., Tabata, H., Higashi, Y., and Hara, Y. (1996) *Nature biotech.*, 14, 1129–1132.

Zamir, L.O., Nedea, M.E., and Garneau, F.X. (1992) *Tetrahed. Lett.*, 33, 5235–5236.

Zamir, L.O., Zhou, Z.H., Caron, G., Nedea, M.E., Sauriol, F., and Mamer, O. (1995) *J. Chem. Soc., Chem. Commun.*, 529.

3   Taxoids occurring in the genus *Taxus*

Hideji Itokawa

Structures of novel taxoids

Introduction

In the review of Blechert and Guenard (1990), taxoids were classified according to their structure.

In 1993, Kingston *et al.* (1993) edited a review entitled "The Taxane Diterpenoids" in "Progress in the Chemistry of Organic Natural Products." Moreover, the chemistry of taxoids was reviewed in the book "Taxol, Science and Applications" edited by Suffness in 1995. However, after that many related taxoids have been isolated from natural sources.

The first taxoids were isolated because of interest in the toxic constituents of yew. The English or European yew, *Taxus baccata*, has been known as a toxic plant for over 2000 years, and a reference to poisoning by yew occurs in Julius Caesar's history of the Gallic Wars. Chemical studies of the toxic constituents of yew date from 1856, when Lucas isolated a toxic material which he named taxine.

This toxic material later turned out to be a mixture of compounds, and subsequent workers elected to study degradation products which could be purified by the elimination of a basic group from a C-5 ester side-chain to yield a cinnamoyl group. The first taxoids whose structures were published were thus not natural products. The structure of taxinine was published by Kurono *et al.* (1963), and the structure of the related O-cinnamoyltaxicin-I triacetate was published by Lythgoe in a series of papers beginning in 1964 and culminating in a comprehensive review in 1968.

As mentioned in Chapter 1, De Marcano *et al.* (1970a) determined the structure of baccatin V by X-ray analysis and named as 5β, 20-epoxy-1β, 2α,4α,-7α,10β, 13α-hexahydroxytax-11-en-9-one-4α, 10β-diacetate 2α-benzoate. Tetraol (3 in Chapter 1), discovered by Wani *et al.* (1971), was found to have the same skeleton as baccatin V, except for the configuration of the hydroxyl group at C-7.

Until now more than 360 members of taxoids have been isolated from plants. First of all, these almost novel taxoids have the basic [9.3.1.0.^{3,8}] pentadecene ring system. This tricyclic system possesses a highly folded conformation, not obvious from its two-dimensional representation. The six-membered A ring, a distorted boat, is *cis*-fused to the eight-membered B ring which, in turn, has a boat–chair conformation. In the case of taxinine or baccatin derivatives, the remaining six-membered ring C has a distorted chair conformation and is *trans*-fused to ring B (Figure 3.1).

These first taxoids isolated as pure compounds have a C-4(20) double bond and an oxygen function at C-5. Compounds with these structural characteristics form the largest single group of natural taxoids, with over 120 examples of this class.

The other major subclasses are taxoids with a transannular bond, taxoids with a C-4(20) epoxide, and those with an oxetane ring (taxol is a member of this class). The rearranged taxoids were

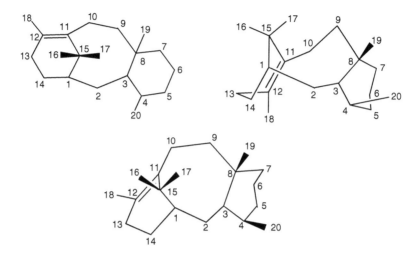

Figure 3.1 Taxene skeleton.

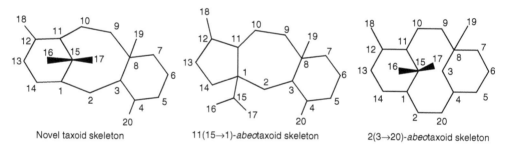

Figure 3.2 Taxoid skeleton.

reviewed by Rao (1995). The mother skeletons are divided in to three types shown in Figure 3.2. One is a novel taxoid, the second is 11(15 → 1)-abeotaxoid, and the third is 2(3 → 20)-abeotaxoid.

Recently, taxol was isolated from the plants outside of the taxaceae. One is found in the African fern pine, *Podocarpus gracilior* (*Podocarpaceae*) (Stahlhut *et al.*, 1999) and the other one is isolated from bark, leaves, branches, shells, and fruits of the hazelnut tree (Hoffman, 2000).

Taxoids with a C-4(20) exocyclic double bond

Taxoids with a hydroxyl group or acetate at C-5

A HYDROXYL GROUP OR ACETATE AT C-5

More than 130 taxoids have a C-4(20) exocyclic double bond and either a free hydroxyl group or an acetoxyl group at C-5, as shown in Figures 3.3–3.20. For example, taxusin, taxinine A, decinnamoyltaxinine J, 2α,9α,1β,5α,10β,13α-tetra hydroxy-taxa-4(20), 11-diene, etc.

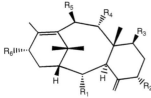

		R_1	R_2	R_3	R_4	R_5	R_6	Plant source	References
9α,10β,-Diacetoxy-5α,13α-dihydroxy-4(20),11-taxadiene	1a	H	OH	H	OAc	OAc	OH	*T. baccata* *T. mairei*	De Marcano *et al.* (1969) Liu *et al.* (1984)
5α,7β,9α,10β,13α-Pentaacetoxy-4(20),11-taxadiene	1b	H	OAc	OAc	OAc	OAc	OAc	*T.baccata* *T. mairei* *T. brevifolia*	De Marcano *et al.* (1969) Yeh *et al.* (1988) Rao *et al.* (1988)
2α,5α,9α,10β,13α-Pentaacetoxy-4(20),11taxadiene	1c	OAc	OAc	H	OAc	OAc	OAc	*T. baccata*	De Marcano *et al.* (1969)
2α,5α,7β,9α,10β,13α-Hexaacetoxy-4(20),11-taxadiene	1d	OAc	OAc	OAc	OAc	OAc	OAc	*T. baccata* *T. brevifolia*	De Marcano *et al.* (1969) Rao *et al.* (1988)
2α-(α-Methylbutyryl)oxy-5α-7β,10β-trihydroxy-4(20),11-taxadiene	1e	CO$_2$iBu	OAc	OAc	H	OAc	H	*T. baccata* *T. mairei*	De Marcano *et al.* (1969) Yeh *et al.* (1988)
5α-Hydroxy-2α-α-methylbutyryloxy-7β,9α,10β-trihydroxy-4(20),11-taxadiene	1f	CO$_2$iBu	OH	OAc	OAc	OAc	H	*T. baccata*	De Marcano *et al.* (1969)
2α-(α-Methylbutyryl) oxy-5α,7β,9α,10β-tetraacetoxy-4(20),11-taxadiene	1g	CO$_2$iBu	OAc	OAc	OAc	OAc	H	*T. baccata* *T. brevifolia*	De Marcano *et al.* (1969) Rao *et al.* (1998)
5α,9α,10β,13α-tatraacetoxy-4(20),11 taxadiene (Taxusin)	1h	H	OAc	H	OAc	OAc	OAc	*T. baccata* *T. mairei* *T. cuspidata* *T. brevifolia*	De Marcano *et al.* (1969) Ho *et al.* (1987) Miyazaki *et al.* (1968) Rao *et al.* (1998)
2α,5α-Dihydroxy-7β,9α10β,13α-tetra-acetoxy-4(20),11-taxadiene	1i	OH	OH	OAc	OAc	OAc	OAc	*Austrotaxus spicata*	Etouati *et al.* (1988)
7β,9α,10β-Triacetoxy-2α,5α,13α-trihydroxy-4(20),11-taxadiene	1j	OH	OH	OAc	OAc	OAc	OH	*Austrotaxus spicata*	Etouati *et al.* (1988)
5α-Hydroxy-2α,7β,9α,10β,13α-tetraacetoxy-4(20),11-taxadiene (Decinnamoyltaxinine J)	1k	OAc	OH	OAc	OAc	OAc	OAc	*T. brevifolia* *Austrotaxus spicata*	Kingston *et al.* (1982) Etouati *et al.* (1988)
5α,9α,10β,13α-Tetrahydroxy-4(20),11-taxadiene	1l	H	OH	H	OH	OH	OH	*T. baccata*	Chan *et al.* (1966)
2α,5α-trihydroxy-10β,13α-diacetoxy-taxa-4(20),11-dine	1m	OH	OH	H	OH	OAc	OAc	*T. chinensis*	Zhang *et al.* (1995)
2-Deacetoxy-5-decinnamoyl-taxinine J	1n	H	OH	OAc	OAc	OAc	OAc	*T.wallichiana*	Chattopadhay *et al.* (1995b)
Taxezopidine F	1o	OAc	OH	OAc	OAc	OAc	OH	*T. cuspidata (Japan)*	Wang *et al.* (1998)
Decinnamoyltaxine E	1p	OAc	OH	H	OAc	OAc	OAc	*T. chinensis*	Chen *et al.* (1999)
5α–Hydroxy-9α,10β,13α–triacetoxytaxa-4(20),11-diene	1q	H	OH	H	OAc	OAc	OAc	*T. mairei*	Yang *et al.* (1996)
2-Deacetyldecinnamoyltaxinine E	1r	OH	OH	H	OAc	OAc	OAc	*T. baccata*	Barboni *et al.* (1995b)
Compound 2	1s	OAc	OH	OH	OH	OAc	OAc	*T.canadensis*	Zamir *et al.* (1999)

Figure 3.3 A hydroxyl group or acetate at C-5.

Recently, many taxanes belonging to this class were isolated, so it was convenient to divide them into more subclasses, such as C-1 hydroxylated group and C-13 ketonized group and so on. Many derivatives of taxusin (1h) possessing various hydroxylated or acetylated functional groups have been extracted from leaves, barks, or woods. Many derivatives (1a–g) were found in the heart wood of *T. baccata* by De Marcano *et al.* (1969).

Kingston *et al.* (1982) added a compound (1k) from *T. brevifolia* with the other oxetane type taxoids. Ettouati *et al.* (1988, 1989) identified taxanes (1J) with other taxane-type alkaloids (2a–f, 3a–c) possessing a similar side-chain at position 5 from the leaves or bark of the genus *Austrotaxus*. A compound (1m) was isolated with a taxoid having an epoxide ring from *T. chinensis* (Zhang *et al.*, 1995). The stem bark of *T. wallichiana* gave the new taxoid 2-deacetoxydecinnamoyl taxine J (1n) (Chattopadhyay and Sharma, 1995b). Moreover, Taxazopidine F (1o) from the seeds and stems of Japanese yew was added to this class (Wang *et al.*, 1998). A total, then, of 19 compounds are in this class (Figure 3.3).

C-1 HYDROXYLATED GROUP

In this group six taxoids and three taxoids are classified respectively, as shown in Figure 3.4. Three taxoids (1-1e, 2h, and 2I) were isolated from needles of *T. baccata* (Barboni *et al.*, 1995b) whilst another (1-1f) was isolated from the root of *T. baccata* (Topcu *et al.*, 1994).

C-11 HYDROXYLATED OR C-10 KETONIZED GROUP

Taxezopidine K (1-1′b) was isolated from the seeds of the Japanese yew *T. cuspidata* (Shigemori *et al.*, 1999). Taxuspine P (1-1′a) (Kobayashi *et al.*, 1996) and Taxuspine D (1-1′c) (Kobayashi *et al.*, 1995a) were isolated from stems of the Japanese yew *T. cuspidata* (Figure 3.5).

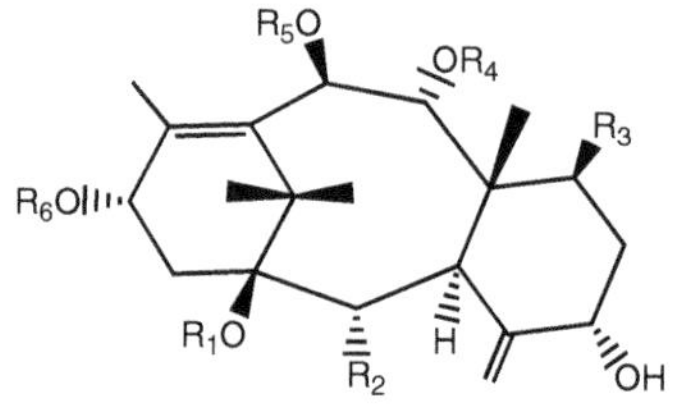

		R_1	R_2	R_3	R_4	R_5	R_6	Plant source	References
1β,2α,5α,9α,10β,13α-Hexahydroxy-4(20),11-taxadiene	1-1a	H	OH	H	H	H	H	*T. chinensis*	Jia *et al.* (1991)
2α-Benzoyloxy-9α,10β-diacetoxy-1β,5α,13α-trihydroxy-4(20),11-taxadiene	1-1b	H	OCOPh	H	Ac	Ac	H	*T. chinensis*	Jia *et al.* (1991)
2α-Benzoyloxy-10β,13α-diacetoxy-1β,5α,9α-trihydroxy--4(20),11-taxadiene	1-1c	H	OCOPh	H	H	Ac	Ac	*T. chinensis*	Jia *et al.* (1991)
1β,5α,7β,9α,10β,13α-hexaol-9α,10β-diacetate-7β-benzoate-taxa-4(20),11-diene	1-1d	H	H	OCOPh	Ac	Ac	H	*T. brevifolia*	Balza *et al.* (1991)
Decinnamoyl-1-hydroxytaxinine J	1-1e	H	OAc	Ac	Ac	Ac	Ac	*T. baccata*	Barboni *et al.* (1995b)
Taxa-4(20),11-diene-5αhydroxy-1β, 7β, 9α, 10β-tetraacetate	1-1f	Ac	H	Ac	Ac	Ac	H	*T. baccata*	Topcu *et al.* (1994)

Figure 3.4 C-1 Hydroxylated group.

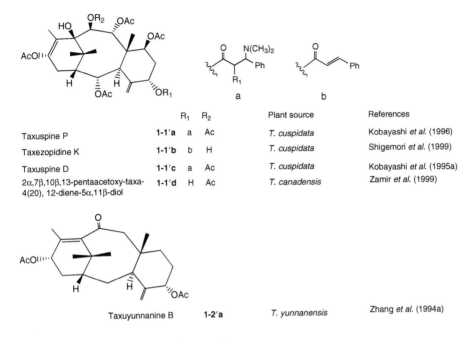

		R₁	R₂	Plant source	References
Taxuspine P	1-1′a	a	Ac	*T. cuspidata*	Kobayashi *et al.* (1996)
Taxezopidine K	1-1′b	b	H	*T. cuspidata*	Shigemori *et al.* (1999)
Taxuspine D	1-1′c	a	Ac	*T. cuspidata*	Kobayashi *et al.* (1995a)
2α,7β,10β,13-pentaacetoxy-taxa-4(20), 12-diene-5α,11β-diol	1-1′d	H	Ac	*T. canadensis*	Zamir *et al.* (1999)

Taxuyunnanine B	1-2′a	*T. yunnanensis*	Zhang *et al.* (1994a)

Figure 3.5 C-11 Hydroxylated or C-10 ketonized group.

C-13 KETONIZED GROUP

One more type of this group is ketonized at C-13 as shown in Figure 3.6. Taxinine A (1-2a) and taxinine H (1-2b) were isolated from *T. cuspidata* and *T. chinensis* (Chiang *et al.*, 1967). Taxuspine F (1-2c) and G (1-2d) were isolated from the Japanese yew *T. cuspidata* (Kobayashi *et al.*, 1995b). Taxezopine C (1-2f) and D (1-2g) were isolated from the Japanese yew (Wang *et al.*, 1998). Deacetyltaxinine A(1-2d) was isolated from the needles and stems of the Chinese yew *T. cuspidata* with two other rearranged taxoids by Tong *et al.* (1995). Taxuspinanane K(1-2g) was isolated from the stems of *T. cuspidata var. nana* by Morita *et al.* (1998a). Only one example, taxuyunanine B ketonized at C-10 (1-2′a), was isolated from *T. yunnanensis* (Zhang *et al.*, 1994a) (Figures 3.5 and 3.6).

OXETANE OR EPOXIDE RING OPENED GROUP

One (1-3a) of the six taxane diterpenoids is a new taxoid isolated from *T. chinensis var. mairei*, and is a rare example of a taxoid with a 4,11 double bond in this class (Shi *et al.*, 1999). Taxezopidine B (1-3b) was isolated from *T. cuspidata* by Wang *et al.* (1998). Two taxoids (1-4a, 1-4b) were isolated from *T. mairei* (Liang and Kingston, 1993). Taxchin A (1-4c) was isolated from *T. chinensis* by Tanaka *et al.* (1994). Taxumairol A (1-4d) was isolated from *T. mairei* (Shen *et al.*, 1996). Two novel taxoids, taxumairols N (1-4h) and O (1-4I), have been isolated from extracts of the roots of *T. mairei*. The structures of them were identified as 7β,9α,10β,13α-tetraacetoxy-2α,4α,5α,20-tetrahydroxy-11-ene   and   7β,9α,10β,13α-tetraacetoxy-1β,2α,4α,5α,20-pentahydroxytax-11-ene on the basis of 1D and 2D NMR techniques, including COSY, HMQC, HMBC, and NOESY experiments (Shen *et al.*, 2000). Taxichin B (1-5a and 1-5b) have

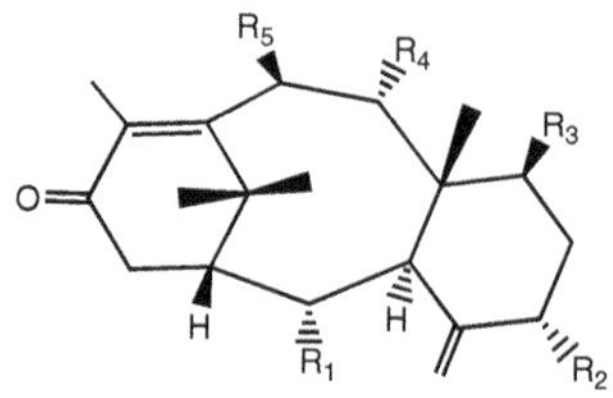

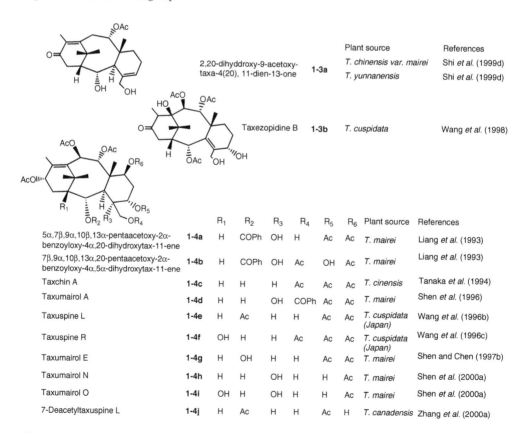

		R₁	R₂	R₃	R₄	R₅	Plant source	References
Taxinine A	1-2a	OAc	OH	H	OAc	OAc	T. cuspidata T. chinensis T. x media	Chiang et al. (1967) Chiang et al. (1975) Rao et al. (1996)
Taxinine H	1-2b	OAc	OAc	H	OAc	OA c	T. cuspidata T. x media	Chiang et al. (1967) Rao et al. (1996)
Taxuspine F	1-2c	OAc	OH	OAc	OAc	OAc	T. cuspidata (Japan)	Kobayashi et al. (1995b)
2-deacetyltaxinine A (Taxuspine G)	1-2d	OH	OH	H	OAc	OAc	T. cuspidata T. cuspidata (Japan)	Tong et al. (1995) Kobayashi et al. (1995b)
Taxezopidine C	1-2e	OH	OH	H	OAc	OH	T. cuspidata (Japan)	Wang et al. (1998)
Taxezopidine D	1-2f	OH	OH	H	OH	OAc	T. cuspidata (Japan)	Wang et al. (1998)
Taxuspinanane K(13-dehydro- 5,13-deacetyl-2-deacetoxy decinnamoyltaxinine J)	1-2g	H	OH	OAc	OAc	OAc	T. cuspidata var. nana	Morita et al. (1998a)
2,10-Di-O-acetyl-5- decinnamoyltaxicin I	1-2h	OAc	OH	H	OH	OAc	T. baccata	Barboni et al. (1995b)
Triacetyl-5- decinnamoyltaxicin I	1-2i	OAc	OH	H	OAc	OAc	T. baccata	Barboni et al. (1995b)

Figure 3.6 C-13 Ketonized group.

		Plant source	References
2,20-dihyddroxy-9-acetoxy- taxa-4(20), 11-dien-13-one	1-3a	T. chinensis var. mairei T. yunnanensis	Shi et al. (1999d) Shi et al. (1999d)
Taxezopidine B	1-3b	T. cuspidata	Wang et al. (1998)

		R₁	R₂	R₃	R₄	R₅	R₆	Plant source	References
5α,7β,9α,10β,13α-pentaacetoxy-2α- benzoyloxy-4α,20-dihydroxytax-11-ene	1-4a	H	COPh	OH	H	Ac	Ac	T. mairei	Liang et al. (1993)
7β,9α,10β,13α,20-pentaacetoxy-2α- benzoyloxy-4α,5α-dihydroxytax-11-ene	1-4b	H	COPh	OH	Ac	OH	Ac	T. mairei	Liang et al. (1993)
Taxchin A	1-4c	H	H	H	Ac	Ac	Ac	T. cinensis	Tanaka et al. (1994)
Taxumairol A	1-4d	H	H	OH	COPh	Ac	Ac	T. mairei	Shen et al. (1996)
Taxuspine L	1-4e	H	Ac	H	H	Ac	Ac	T. cuspidata (Japan)	Wang et al. (1996b)
Taxuspine R	1-4f	OH	H	H	Ac	Ac	Ac	T. cuspidata (Japan)	Wang et al. (1996c)
Taxumairol E	1-4g	H	OH	H	H	Ac	Ac	T. mairei	Shen and Chen (1997b)
Taxumairol N	1-4h	H	H	OH	H	H	Ac	T. mairei	Shen et al. (2000a)
Taxumairol O	1-4i	OH	H	OH	H	H	Ac	T. mairei	Shen et al. (2000a)
7-Deacetyltaxuspine L	1-4j	H	Ac	H	H	Ac	H	T. canadensis	Zhang et al. (2000a)

Figure 3.7 Oxetane or epoxide ring opened group.

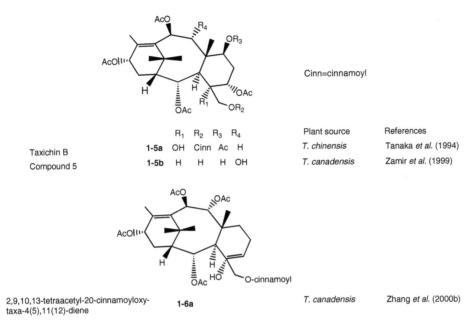

	R$_1$	R$_2$	R$_3$	R$_4$	Plant source	References
Taxichin B	**1-5a**	OH	Cinn	Ac	H	*T. chinensis*	Tanaka *et al.* (1994)
Compound 5	**1-5b**	H	H	H	OH	*T. canadensis*	Zamir *et al.* (1999)

Cinn=cinnamoyl

		Plant source	References
2,9,10,13-tetraacetyl-20-cinnamoyloxy-taxa-4(5),11(12)-diene	**1-6a**	*T. canadensis*	Zhang *et al.* (2000b)

Figure 3.8 Oxetane or epoxide ring opened group.

cinnamoyl group at C-5. Moreover, the taxoid (**1-6a**) also has a cinnamoyl group at C-5 and a double bond at C-4(5) (Figures 3.7 and 3.8).

Taxoids with an amino acyl group at C-5

AMINOACYL GROUP

Taxane-type alkaloids (**2a–f**, **3a–c**) possessing a similar side-chain at position 5 were identified from *Austrotaxus* by Ettouati *et al.* in 1988–89. Eighteen taxane-type diterpenes have been isolated from the trunk bark of *Austrotaxus spicata* thirteen of which are alkaloids. Eight compounds have an exocyclic-4(20) methylene group: austrospicatine (**2b**), 2′-desacetyl-austrospicatine (**2a**), 2′-desacetoxy austro-spicatine (**2f**), already found in the leaves of the same plant, comptonine, spicaledonine etc. 2′-Deacetoxyaustrospicatine (**2f**) from the stem bark of *T. baccata*. (Chauviere *et al.*, 1982). 1-Hydroxy-2-deacetoxytaxinine J (**6l**) and 7,2′-bisdeacetoxy austrospicatine (**2g**) were isolated from the barks of *T. wallichiana* together with 13 known compounds, including taxol. The structure of 7,2′-bisdeacetoxyaustrospicatine was further confirmed by X-ray crystallographic analysis by Zhang *et al.* (1995). Two taxoids, 13-deoxo-13-acetyloxy-7,9-diacetyl-1,2-dideoxytaxine B (**2h**) and 7-xylosyl-10-deacetyl taxol D (**11-1i**), were isolated from the stem bark of *Taxus baccata* cv. Stricta (Guo *et al.*, 1995b). Two taxoids (**3d,e**) were isolated from *T. baccata* (Appendino *et al.*, 1993b). Two new taxoids (**3-1a, b**) were isolated from *T. baccata* (Doss *et al.*, 1997). A taxoid (**3g**) was isolated from the seeds of *T. chinensis* (Chen *et al.*, 1999). Taxuspine Z (**3f**) was isolated from the stems of the Japanese yew *T. cuspidata* by Shigemori *et al.* (1997). A taxoid (**3-1d**) was obtained with two abeotaxanes (**13-1c** and **-1d**) from *T. yunnanensis* and *T. chinensis var. mairei* by Shi *et al.* (1999f). Two taxoids (**2I** and **3-1e**) were isolated from the Chinese yew, *T. chinensis var. mairei* by Shi *et al.* (1999b) (Figures 3.8–3.11).

C-13 KETONIZED GROUP

Comptonine (3-2a) and taxine B (3-2b) were isolated from *Austrospicata* and *T. baccata*, respectively (Ettouati *et al.*, 1989). Moreover, five taxoids belonging to this group were isolated

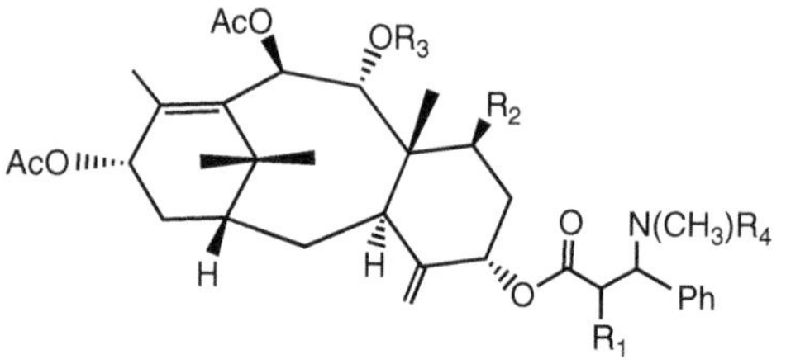

		R_1	R_2	R_3	R_4	Plant source	References
2'β-Deacetylaustrospicatine	2a	OH	OAc	Ac	Me	*Austrotaxus spicata*	Ettouati *et al.* (1989)
Austrospicatine	2b	Ac	OAc	Ac	Me	*Austrotaxus spicata*	Ettouati *et al.* (1989)
7β-Deacetylaustrospicatine	2c	Ac	OH	Ac	Me	*Austrotaxus spicata*	Ettouati *et al.* (1988)
7β,9α-Bisdeacetylaustrospicatine	2d	Ac	OH	H	Me	*Austrotaxus spicata*	Ettouati *et al.* (1988)
2'β,7β,9α-Trisdeacetylaustrospicatine	2e	OH	OH	H	Me	*Austrotaxus spicata*	Ettouati *et al.* (1988)
2'β-Deacetoxyaustrospicatine	2f	H	OAc	Ac	Me	*Austrotaxus spicata* *T. baccata*	Ettouati *et al.* (1989) Chauviere *et al.* (1982)
7,2'-bisdeacetoxyaustrospicatine	2g	H	H	Ac	Me	*T. wallichiana*	Zhang *et al.* (1995b)
13-Deoxo-13α-acetyloxy-7β,9α-diacetyl-1,2-dideoxytaxine B (=2f ?)	2h	H	OAc	Ac	Me	*T. baccata*	Guo *et al.* (1995b)
7,9,10,13-Tetraacetoxy-5(3'-methylamino-3'-phenyl)-propionyloxy-taxa-4(20),11-diene	2i	H	OAc	Ac	H	*T. mairei*	Shi *et al.* (1999b)

Figure 3.9 Amino acyl group at C-5.

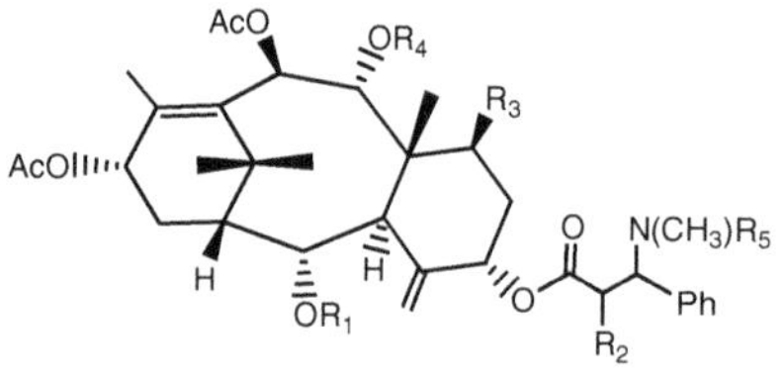

		R_1	R_2	R_3	R_4	R_5	Plant source	References
2α-Acetoxyaustrospicatine	3a	Ac	Ac	OAc	Ac	Me	*Austrotaxus spicata*	Ettouati *et al.* (1988)
2α-Acetoxy-2'β-deacetylaustrospicatine	3b	Ac	OH	OAc	Ac	Me	*Austrotaxus spicata*	Ettouati *et al.* (1988)
2α-Hydroxy-2'β-deacetoxyaustrospicatine	3c	H	H	OAc	Ac	Me	*Austrotaxus spicata*	Ettouati *et al.* (1988)
13-Deoxo-13α-acetyloxy-1-deoxy taxine B	3d	H	H	H	H	Me	*T. baccata*	Appendino *et al.* (1993b)
13-Deoxo-13α-acetyloxy-1-deoxy nor-taxine B	3e	H	H	H	H	H	*T. baccata*	Appendino *et al.* (1993b)
Taxuspine Z	3f	H	H	H	Ac	Me	*T. cuspidata*	Shigemori *et al.* (1997)
2α-Acetoxy-2',7-dideacetoxy-austrospicatine	3g	Ac	H	H	Ac	Me	*T. chinensis*	Chen *et al.* (1999)

Figure 3.10 Amino acyl group at C-5.

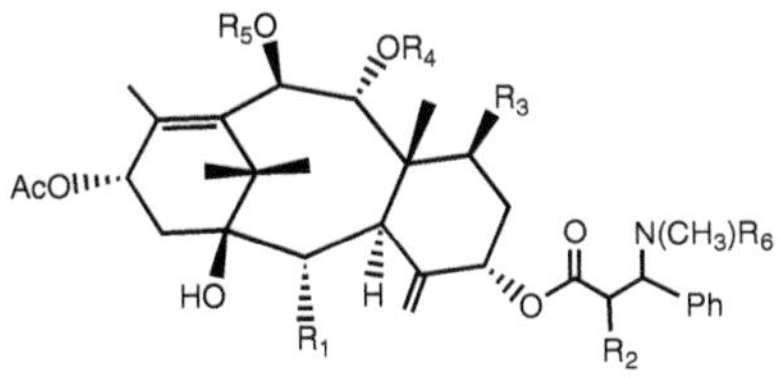

	R₁	R₂	R₃	R₄	R₅	R₆	Plant source	References

Structure table (R groups):

Compound name		R₁	R₂	R₃	R₄	R₅	R₆	Plant source	References
(+)-2-Acetoxy-2′,7-dideacetoxy-1-hydroxyaustrospicatine	**3-1a**	OAc	H₂	H	Ac	Ac	Me	*T. baccata*	Doss *et al.* (1997)
2,-Acetoxy-2′-deacetyl-1-hydroxyaustrospicatine	**3-1b**	OAc	OH	OAc	Ac	Ac	Me	*T. baccata*	Doss *et al.* (1997) Barboni *et al.* (1995)
13-Deoxo-13α-acetyloxy taxine B	**3-1c**	OH	H₂	H	H	Ac	Me	*T. baccata*	Appendino *et al.* (1993b)
1β,10β-dihydroxy-9α,13α-diacetoxy-5α-(3′-dimethylamino-3′-phenyl)-propionyloxytaxa-4(20),11-diene	**3-1d**	H	H₂	H	Ac	H	Me	*T. yunnanensis*	Shi *et al.* (1999f)
2α,9α,10β,13α-Tetraacetoxy-5α-(3′-methylamino-3′-phenyl)-propionyloxy-taxa-4(20),11-diene	**3-1e**	OAc	H₂	H	Ac	Ac	H	*T. mairei*	Shi *et al.* (1999b)

Figure 3.11 Amino acyl group at C-5.

from *T. baccata* by Jenniskens *et al.* (1996). A taxoid (**3-2h**) was isolated from the needles of *T. cuspidata* (Ando *et al.*, 1997).

Taxoids with a C-14 acyl group

Many taxoids with a functional group at C-14, as in taiwanxan (**4-2a**) (Ho *et al.*, 1987), were isolated from *T. chinensis var. mairei*. Four new taxoids were isolated from the roots of *T. yunnanensis*. Three (**4a–c**) of the compounds were analogues of taxuyunnanine C and the fourth was a 2-deacetoxytaxuyunnanine C derivative (Zhang *et al.*, 1995).

From the cell cultures of *T. chinensis var. mairei*, yunnanxane [2α, 5α, 10β-triacetoxy-14β-(2′-methyl-3′-hydroxyl)-butyryloxy-4(20), 11-taxadiene (**4i**), and four homologous esters, 2α, 5α, 10β, 14β-tetra-acetoxy-4(20), 11-taxadiene (**4d**), 2α, 5α, 10β-triacetoxy-14-propionyloxy-4(20), 11-taxxadiene (**4e**), 2α, 5α, 10β-triacetoxy-14β-isobutyloxy-4(20)m 11-taxadiene (**4f**), 2α, 5α, 10β-triacetoxy-14β-(2′-methyl)-butyryloxy-4(20), 11-taxadiene (**4g**) have been isolated (Ma *et al.*, 1994b).

The roots of *Taxus x media* gave two taxoids, the structures of which were established as 10-deacetyl-10-dehydro-7-acetyl taxol A (**11g**) and 10-deacetylyunnanxane (**4h**) by Gabetta *et al.* (1995b).

Three C-14 oxygenated taxanes were isolated from *T. yunnanensis* cell cultures (Cheng *et al.*, 1996), that is, 10-deacetyl yunnanxan (**4h**), 2α,10β,14β-Trihydroxy-5α–acetoxy-4(2),11-taxadiene (**4-1a**), and yunnanxan (**4-1b**). Many taxoids with a functional group at C-14, as in taiwanxan (**4-2a**), were isolated from *T. chinensis var. mairei* (Ho *et al.*, 1987). Two taxoids (**4-2c** and **-2d**) were isolated from the twigs of *T. mairei* by Yang *et al.* in 1996. It should be noted that compounds **5a–c** have a ketone at position 10 in contrast to that of ring B of baccatin III (at C-9). From the root of *T. baccata*, a taxoid (**5-1b**) was isolated with a taxoid (**1-1f**) (Topcu *et al.*, 1994) (Figures 3.12–3.15).

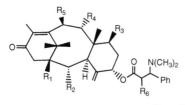

		R₁	R₂	R₃	R₄	R₅	R₆	Plant source	References
Comptonine	**3-2a**	H	H	OAc	OAc	OAc	OH	*Austrotaxus spicata*	Ettouati *et al.* (1989)
Taxine B	**3-2b**	OH	OH	H	OH	OAc	H	*T. baccata*	Etouati *et al.* (1991) Graf *et al.* (1957)
1-deoxytaxine B	**3-2c**	H	OH	H	OH	OAc	H	*T. baccata*	Jenniskens *et al.* (1996)
Isotaxine B (9-Acetyl-10deacetyltaxine B)	**3-2d**	OH	OH	H	OAc	OH	H	*T. baccata*	Jenniskens *et al.* (1996)
2-Acetyltaxine B	**3-2e**	OH	OAc	H	OH	OAc	H	*T. baccata*	Jenniskens *et al.* (1996)
2-Deacetoxy-9-acetoxy taxine B	**3-2f**	H	OH	H	OAc	OH	H	*T. baccata*	Jenniskens *et al.* (1996)
9-Acetoxytaxine B	**3-2g**	OH	OAc	H	OAc	OH	H	*T. baccata*	Jenniskens *et al.* (1996)
2'-Hydroxytaxine II	**3-2h**	H	OAc	H	OAc	OAc	OH	*T. cuspidata*	Ando *et al.* (1997)
1-deoxy-2-O-acetyltaxine B	**3-2i**	H	OAc	H	OH	OAc	H	*T. cuspidata*	Shi *et al.* (2000c)
1-deoxy-2,9-O-diacetyl-10-deacetyltaxine B	**3-2j**	H	OAc	H	OAc	OH	H	*T. cuspidata*	Shi *et al.* (2000c)
2-O-deacetyl taxine II	**3-2k**	H	OH	H	OH	OAc	H	*T. cuspidata*	Shi *et al.* (2000c)

Figure 3.12 C-13 Ketonized group.

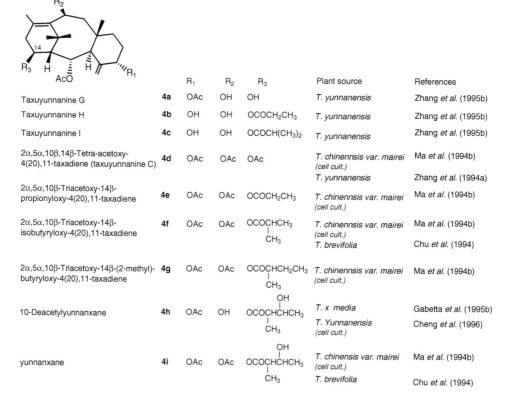

		R₁	R₂	R₃	Plant source	References
Taxuyunnanine G	**4a**	OAc	OH	OH	*T. yunnanensis*	Zhang *et al.* (1995b)
Taxuyunnanine H	**4b**	OH	OH	OCOCH₂CH₃	*T. yunnanensis*	Zhang *et al.* (1995b)
Taxuyunnanine I	**4c**	OH	OH	OCOCH(CH₃)₂	*T. yunnanensis*	Zhang *et al.* (1995b)
2α,5α,10β,14β-Tetra-acetoxy-4(20),11-taxadiene (taxuyunnanine C)	**4d**	OAc	OAc	OAc	*T. chinennsis var. mairei* (cell cult.) *T. yunnanensis*	Ma *et al.* (1994b) Zhang *et al.* (1994a)
2α,5α,10β-Triacetoxy-14β-propionyloxy-4(20),11-taxadiene	**4e**	OAc	OAc	OCOCH₂CH₃	*T. chinennsis var. mairei* (cell cult.)	Ma *et al.* (1994b)

2α,5α,10β-Triacetoxy-14β-isobutyryloxy-4(20),11-taxadiene	**4f**	OAc	OAc	OCOCHCH₃
				    |
				    CH₃
					T. chinennsis var. mairei (cell cult.)
					T. brevifolia
						Ma *et al.* (1994b)
						Chu *et al.* (1994)

2α,5α,10β-Triacetoxy-14β-(2-methyl)-butyryloxy-4(20),11-taxadiene	**4g**	OAc	OAc	OCOCHCH₂CH₃
				    |
				    CH₃
					T. chinennsis var. mairei (cell cult.)	Ma *et al.* (1994b)

10-Deacetylyunnanxane	**4h**	OAc	OH	OH
				    |
				OCOCHCHCH₃
				      |
				     CH₃
					T. x media
					T. Yunnanensis (cell cult.)
						Gabetta *et al.* (1995b)
						Cheng *et al.* (1996)

yunnanxane	**4i**	OAc	OAc	OH
				    |
				OCOCHCHCH₃
				      |
				     CH₃
					T. chinensis var. mairei (cell cult.)
					T. brevifolia
						Ma *et al.* (1994b)
						Chu *et al.* (1994)

Figure 3.13 Taxoids with C-14 acyl group.

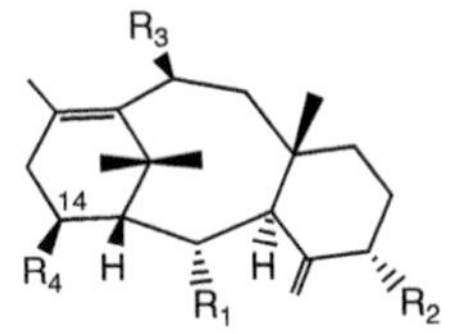

		R₁	R₂	R₃	R₄	Plant source	References
2α,10β,14β-Trihydroxy-5α-acetoxy-4(20),11-taxadiene	**4-1a**	H	OAc	OH	OH	*T. yunnnanensis (callus)*	Cheng *et al.* (1996)
yunnaxan	**4-1b**	OH	OAc	OAc	OAc	*T. yunnnanensis (callus)*	Cheng *et al.* (1996)
Taxuyunnanine J	**4-1c**	H	OAc	OH	OH	*T. yunnanensis*	Zhang *et al.* (1995b)

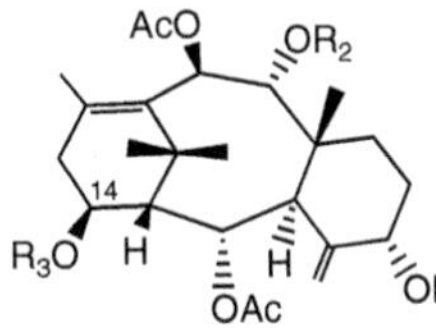

		R₁	R₂	R₃	Plant source	References
5α-hydroxy-14β-methylbutyloxy-2α,9α,10β-triacetoxy-4(20),11-taxadiene (taiwanxan)	**4-2a**	H	Ac	α-MεBυ	*T. x media*	Ho *et al.* (1987) Gabetta *et al.* (1995b)
2,5,9,10-tetraacetoxy-14b-methylbutyloxy-4(20),11-taxadiene (taxayunnanine B)	**4-2b**	Ac	Ac	α-MεBυ	*T. x media*	Gabetta *et al.* (1995b) Zhang *et al.* (1994a)
9α-Hydroxy14β-(2-methylbutyryl)oxy-2α,5α,10β-triacetoxytaxa-4(20),11-diene	**4-2c**	Ac	H	α-MεBυ	*T. mairei*	Yang *et al.* (1996)
2α,5α,9α,10β,14β-Pentaacetoxytaxa-4(20),11-diene	**4-2d**	Ac	Ac	Ac	*T. mairei*	Yang *et al.* (1996)
Taxuspine M	**4-2e**	Ac	Ac	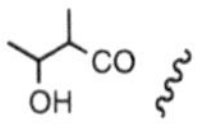 [structure]	*T. cuspidata var. nana*	Itokawa *et al.* (1999)

Figure 3.14 Taxoids with C-14 acyl group.

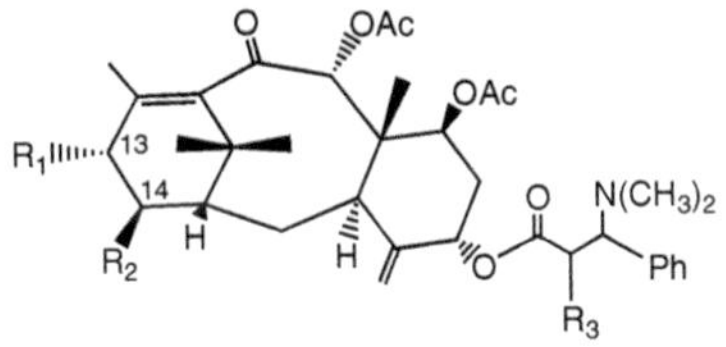

		R₁	R₂	R₃	Plant source	References
Austrotaxine	**5a**	OAc	OAc	OAc	*Austrotaxus spicata*	Etouati *et al.* (1988)
2'-Deacetylaustro-taxine	**5b**	OAc	OAc	OH	*Austrotaxus spicata*	Etouati *et al.* (1988)
2'-Deacetoxyaustrotaxine	**5c**	OAc	OAc	H	*Austrotaxus spicata*	Etouati *et al.* (1988)
2'-β,13α,14β-Trisdeacetylaustrotaxine	**5d**	OH	OH	OH	*Austrotaxus spicata*	Etouati *et al.* (1989)

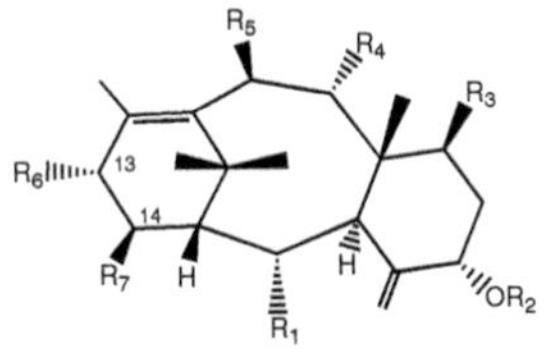

		R₁	R₂	R₃	R₄	R₅	R₆	R₇	Plant source	References
7β,9α-Diacetoxy-5α,13α,14β-trihydroxy-10-oxo-4(20),11-taxadiene	**5-1a**	H	H	OAc	=O	OAc	OH	OH	*Austrotaxus spicata*	Etouati *et al.* (1989)
Taxa-4-(20),11-diene-10β-methoxy-2α,5α-diacetoxy-14β, α-methylbutyrate	**5-1b**	OAc	Ac	H	H	OMe	H	OCOCH(CH₃)-C₂H₅	*T. baccata*	Topcu *et al.* (1994)

Figure 3.15 Taxoids with C-14 acyl group.

Taxoids with a cinnamoyl group at C-5

Recently, Itokawa *et al.* isolated 2α-deacetyltaxinine J (**6f**) from *T. cuspidata var. nana* (Morita *et al.*, 1998b). The bark of *T. chinensis* yielded 2-deacetoxy-7,9-dideacetyltaxinine J (**6k**) together with several known taxoids (Liang *et al.*, 1998). 2-Deacetoxytaxanine B (**6-1j**) was isolated from *T. wallichiana* by Shrestha *et al.* in 1997. Two taxane diterpenoids, 1-hydroxy-2-deacetoxytaxanine J (**6l**) and 7,2′-bisdeacetoxyaustrospicatine (**2g**), were isolated from the barks of *T. wallichiana* (Zhang *et al.*, 1995) (Figures 3.16 and 3.17).

Recently, dantaxusin B (**6p**) was isolated with dantaxusin A (**19a**) from *T. yunnanensis* by Shinozaki *et al.* (2001).

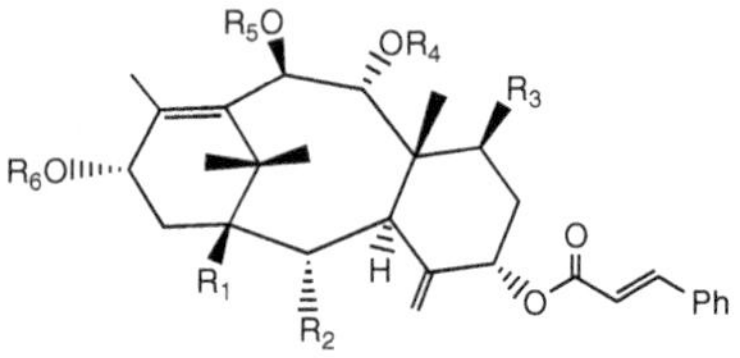

		R_1	R_2	R_3	R_4	R_5	R_6	Plant source	References
5α-Cinnamoyloxy-9α,10β,13α-triacetoxy-taxa-4(20),11-diene	**6a**	H	H	H	Ac	Ac	Ac	*T. mairei* T. chinensis	Yeh *et al.* (1988a) Zhang *et al.* (1991) Chen *et al.* (1999)
5α-Cinnamoyloxy-2α,13α,-dihydroxy-9α,10β-diacetoxy-4(20),11-diene	**6b**	H	OH	H	Ac	Ac	H	*T.chinensis*	Zhang and Jia (1991)
5α-Cinnamoyoxy-10β-hdroxy-2α,9α,13α-triacetoxy-taxa-4(20),11-diene	**6c**	H	OAc	H	Ac	H	Ac	*T. chinensis*	Zhang and Jia (1991)
5α-Cinnamoyloxy-2α,9α,10β,13α-tetraacetoxy-4(20),11-taxadiene(taxinine E)	**6d**	H	OAc	H	Ac	Ac	Ac	*T. mairei* T. cuspidata	Yeh *et al.* (1988b) Woods *et al.* (1968)
2α-Benzoyloxy-5α-cinnamoyloxy-9α,10β-diacetoxy-1β,13α-dihydroxy-4(20),11-taxadiene	**6e**	OH	OCOPh	H	Ac	Ac	H	*T. chinensis*	Jia, Zhang (1991) Zhang *et al.* (1989)
2α–Deacetyltaxinine J (taxuspinanane G)	**6f**	H	OH	OAc	Ac	Ac	Ac	*T. cuspidata var. nana*	Morita *et al.* (1998b)
Taxinine J	**6g**	H	OAc	OAc	Ac	Ac	Ac	*T. mairei* T.cuspidata T. brevifolia	Min *et al.* (1989) Woods *et al.* (1968) Rao and Juchum (1998)
2-Deacetoxytaxinine J	**6h**	H	H	OAc	Ac	Ac	Ac	*T. mairei* T. brevifolia T. baccata	Liang *et al.* (1988) Rao and Juchum (1998) Chauviere *et al.* (1982)
Taxezopidine G (taxuspinanane L)	**6i**	H	OH	H	Ac	Ac	Ac	*T. cuspidata* (Japan)	Wang *et al.* (1998) Morita *et al.* (1997c)
Taxezopidine H	**6j**	H	H	OAc	Ac	Ac	H	*T. cuspidata* (Japan)	Wang *et al.* (1998)
2-deacetoxy-7,9-dideacetyltaxinine J	**6K**	H	H	OH	H	Ac	Ac	*T. chinensis*	Liang *et al.* (1998)
1-Hydroxy-2-deaceetoxytaxinine J	**6l**	OH	H	H	Ac	Ac	Ac	*T. wallichiana*	Zhang *et al.* (1995a)
5α-(Cinnamoyl)oxy-7β–hydroxy-9α,10β,13α–triacetoxytaxa-4(20),11-diene	**6m**	H	H	OH	Ac	Ac	Ac	*T. mairei*	Yang *et al.* (1996)
9-Deacetyl-taxinine E	**6n**	H	OAc	H	H	Ac	Ac	*T. canadensis*	Zhang *et al.* (2000b)
9α,10β–Diacetoxy-5α–cinnamoyl 4920),11-dien-13α-ol	**6o**	H	H	H	Ac	Ac	H	*T. yunnanensi* T. cuspidata	Shi *et al.* (2000a)
Dantaxusin B	**6p**	H	H	H	H	Ac	Ac	*T. yunnanensis*	Shinozaki *et al.* (2001)

Figure 3.16 C-5 Cinnamoyl group at C-5.

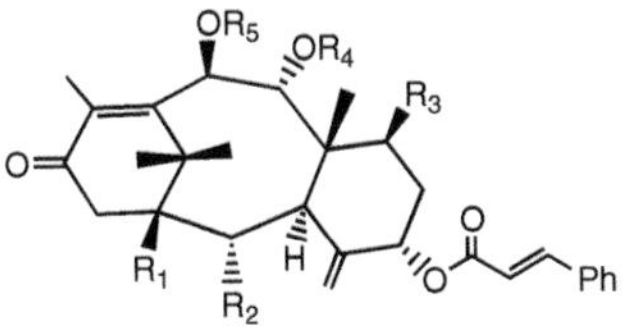

		R₁	R₂	R₃	R₄	R₅	Plant source	References
Taxinine	**6-1a**	H	OAc	H	Ac	Ac	*T. baccata*	Baxter *et al.* (1962)
								Dukes *et al.* (1967)
								Kurono *et al.* (1963)
							T. chinensis	Chiang (1975)
							T. cuspidata	Chiang *et al.* (1967)
								Woods *et al.* (1968)
							T. mairei	Liu *et al.* (1984)
Taxinine B	**6-1b**	H	OAc	OAc	Ac	Ac	*T. cuspidata*	Woods *et al.* (1968)
							T. mairei	Yeh *et al.* (1988)
5O-cinnamoyl-9-acetyltaxicin-II	**6-1c**	H	OH	H	Ac	H	*T. baccata*	Appendino *et al.* (1992b)
5O-cinnamoyl-10-acetyltaxicin-II	**6-1d**	H	OH	H	H	Ac	*T. baccata*	Appendino *et al.* (1992b)
2α-acetoxy-5α-cinnamoyl-9α,10β-dihydroxy-taxa-4(20),11-diene-13-one	**6-1e**	H	OAc	H	OH	OH	*T yunnanensis*	Shi *et al.* (1999i)
5O-cinnamoyl-taxicin I	**6-1e**	OH	OH	H	Ac	H	*T. baccata*	Baxter *et al.* (1962)
5O-cinnamoyl-taxicin I triacetate	**6-1f**	OH	OAc	H	Ac	Ac	*T. cuspidata*	Chiang *et al.* (1967)
							T. baccata	Baxter *et al.* (1962)
								Appendino *et al.* (1992)
5O-cinnamoyl-9-acetyltaxicin-I	**6-1g**	OH	OH	H	Ac	H	*T. baccata*	Appendino *et al.* (1992b)
5O-cinnamoyl-10 acetyltaxicin-I	**6-1h**	OH	OH	H	H	Ac	*T. baccata*	Appendino *et al.* (1992b)
Taxezopidine E	**6-1i**	H	OH	OH	H	Ac	*T. cuspidata*	Wang *et al.* (1998)
2-deacetoxytaxanine B	**6-1j**	H	H	OAc	Ac	Ac	*T. wallichiana*	Shrestha *et al.* (1997)
5-Cinnamoyl-9,10-diacetyltaxicin I	**6-1k**	OH	OH	H	Ac	Ac	*T. baccata*	Appendino *et al.* (1994)
10-Deacetyltaxinine B	**6-1l**	H	OAc	OAc	Ac	OH	*T. cuspidata*	Tong *et al.* (1994)
Taxinine NN-7	**6-1m**	H	OH	H	Ac	Ac	*T. cuspidata*	Kosugi *et al.* (2000)

Figure 3.17 C-13 Ketonized group with cinnamoyl group at C-5.

Taxoids with a transannular bond

TAXOIDS WITH A C-12(16)-OXIDO BRIDGE

These types of taxoids of 12 members were also obtained from natural sources. From *T. cuspidata* two taxoids (**7a**) and taxagifine (**7b**) previously extracted from *T. baccata*, have been isolated, each having a cinnamate group at C-5 and functionalized methyl group at positions 16 and 19, with formation of an tetrahydrofuran type bridge between C-12 and C-16 (Yoshizaki *et al.*, 1988).

Recently, 19-debenzoyl-19-acetyltaxinine M (**7g**) was isolated from *T. wallichiana* by Barboni *et al.* in 1995. Moreover, two taxoids added to this class, one is 19-acetoxytaxigifine (**7h**) isolated by Fukushima *et al.* in 1999 and by Shigemori *et al.* in 1999, independently, and another 5-O-acetyltaxinine M (**7i**) was isolated by Shi, Oritani *et al.* (1999d). The needles of *T. wallichiana* yielded 19-debenzoyl-19-acetyltaxinine M (**7g**) and two derivatives of brevifoliol (Barboni *et al.*, 1995a). Four taxoids, taxuspines Q – T (**14-2a, 1-4f, 7j,** and **7k**) were isolated from Japanese yew *T. cuspidata* (Wang *et al.*, 1996) (Figure 3.18).

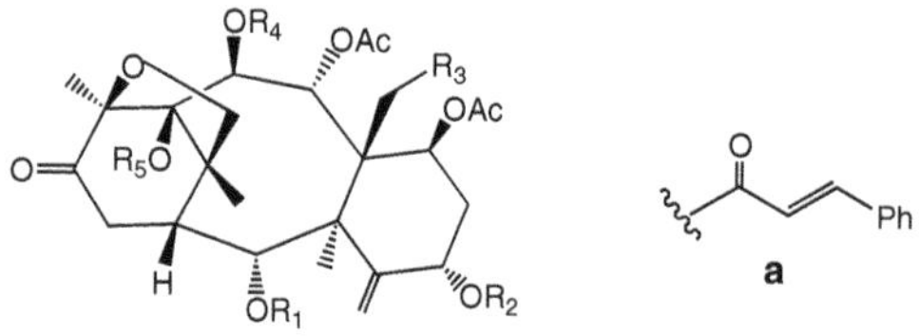

		R₁	R₂	R₃	R₄	R₅	Plant source	References
Taxacin	**7a**	Ac	a	OBz	Ac	H	*T. cuspidata*	Yoshizaki *et al.* (1988)
Taxagifine	**7b**	Ac	a	H	Ac	H	*T. cuspidata*	Yoshizaki *et al.* (1988)
							T. chinensis	Zhang *et al.* (1989)
								Zhang *et al.* (1990)
							T. baccata	Chauviere *et al.* (1982)
5,-Decinnamoyl-taxagifine	**7c**	Ac	H	H	Ac	H	*T. chinensis*	Chauviere *et al.* (1981)
Taxinine M	**7d**	Ac	H	OBz	Ac	H	*T. brevifolia*	Beutler *et al.* (1991)
5-Acetyl-5-decinnamoyl-taxagifine	**7e**	Ac	Ac	H	Ac	H	*T. chinensis*	Zhang *et al.* (1989)
								Zhang *et al.* (1990)
2-Deacetyl-5-decinnamoyltaxagifine	**7f**	H	H	H	Ac	H	*T. chinensis*	Zhang, Jia *et al.* (1991)
19-Debenzoyl-19-acetyltaxinine M	**7g**	Ac	H	OAc	Ac	H	*T. wallichiana*	Barboni *et al.* (1995)
19-acetoxytaxigifine(taxezopidine L)	**7h**	Ac	a	OAc	Ac	H	*T. chinensis*	Fukushima *et al.* (1999)
								Shigemori *et al.* (1999)
5-O-acetyltaxinine M (taxumairol R)	**7i**	Ac	Ac	OBz	Ac	H	*T. chinensis var. mairei*	Shi *et al.* (1999d)
							T. mairei	Shen *et al.* (2000b)
Taxuspine S	**7j**	Ac	a	OH	Ac	H	*T. cuspidata (Japan)*	Wang *et al.* (1996c)
Taxuspine T	**7k**	A c	a	OAc	H	H	*T. cuspidata (Japan)*	Wang *et al.* (1996c)
5-Decinnamoyl-11-acetyl-19-hydroxytaxagifine	**7l**	Ac	H	OH	Ac	Ac	*T. yunnanensis*	Chen *et al.* (1994)

Figure 3.18 C-12(16)-Oxido bridge.

TAXOIDS WITH A C-3(11) BRIDGE

There are 13 members of this class found from natural sources. 10-(-hydroxybutyryl)-10-deacetylbaccatin III (**10-3g**) and 5-cinnamoylphototaxicin II (**8a**) were isolated from the leaves of *T. baccata* (Appendino and Ozen, 1993a). Their structures were established by spectroscopic analysis (Soto and Castedo, 1998). Taxuspine C (**8-1i**) was isolated from Japanese yew *T. cuspidata* by Kobayashi *et al.* in 1994. Spicaledonine (**8-1a**) was isolated from *A. spicata* (Ettouati *et al.*, 1989). Although these types of compounds were isolated from natural sources, they have also been derived from normal taxoids by photosynthesis (Appendino *et al.*, 1992d). Three taxoids (**8-1f, -1g, -1h**) of this class were isolated from *T. canadensis* (Zamir *et al.*, 1998) (Figures 3.19 and 3.20).

Taxoids with a C4(20) epoxide

This class consists of 26 examples and is illustrated in Figures 3.21–3.24. Closely related members have an acetoxy group at C-2, but C-5 oxygen is variously substituted with hydrogen, acetyl, cinnamoyl, or basic side-chains. Most of these compounds also carry oxygenation at C-9, C-10, and C-13. These compounds shown in Figure 3.24 carry Winterstein-type side-chains (3-amino-3-phenyl propionic acid) at C-5, but an interesting group of four compounds from *A. spicata* were esterified at C-9 with nicotinic acid (Ettouati *et al.*, 1989).

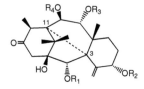

		R₁	R₂	R₃	R₄	Plant source	References
5-Cinnamoyl-9-acetylphototaxicin I	**8a**	H	cinnamoyl	H	Ac	*T. baccata L.*	Soto and Castedo (1998) Appendino and Ozen (1993)
5-Cinnamoyl-10-acetylphototaxicin I	**8b**	H	cinnamoyl	Ac	H	*T. baccata L.*	Appendino *et al.* (1992d)
3,11-Cyclotaxinine NN-2	**8c**	Ac	cinnamoyl	Ac	Ac	*T. cuspidata*	Kosugi *et al.* (2000)

Figure 3.19 C-3(11) bridge.

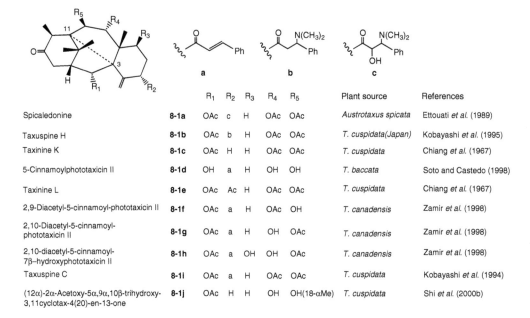

		R₁	R₂	R₃	R₄	R₅	Plant source	References
Spicaledonine	**8-1a**	OAc	c	H	OAc	OAc	*Austrotaxus spicata*	Ettouati *et al.* (1989)
Taxuspine H	**8-1b**	OAc	b	H	OAc	OAc	*T. cuspidata(Japan)*	Kobayashi *et al.* (1995)
Taxinine K	**8-1c**	OAc	H	H	OAc	OAc	*T. cuspidata*	Chiang *et al.* (1967)
5-Cinnamoylphototaxicin II	**8-1d**	OH	a	H	OH	OH	*T. baccata*	Soto and Castedo (1998)
Taxinine L	**8-1e**	OAc	Ac	H	OAc	OAc	*T. cuspidata*	Chiang *et al.* (1967)
2,9-Diacetyl-5-cinnamoyl-phototaxicin II	**8-1f**	OAc	a	H	OAc	OH	*T. canadensis*	Zamir *et al.* (1998)
2,10-Diacetyl-5-cinnamoyl-phototaxicin II	**8-1g**	OAc	a	H	OH	OAc	*T. canadensis*	Zamir *et al.* (1998)
2,10-diacetyl-5-cinnamoyl-7β–hydroxyphototaxicin II	**8-1h**	OAc	a	OH	OH	OAc	*T. canadensis*	Zamir *et al.* (1998)
Taxuspine C	**8-1i**	OAc	a	H	OAc	OAc	*T. cuspidata*	Kobayashi *et al.* (1994)
(12α)-2α-Acetoxy-5α,9α,10β-trihydroxy-3,11cyclotax-4(20)-en-13-one	**8-1j**	OAc	H	H	OH	OH(18-αMe)	*T. cuspidata*	Shi *et al.* (2000b)

Figure 3.20 Taxoids with a C-3(11) bridge.

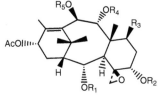

		R₁	R₂	R₃	R₄	R₅	Plant source	References
Baccatin I	**9a**	H	Ac	Ac	OAc	Ac	*T. baccata*	De Marcano and Halsall (1970b)
5α-Deacetylbaccatin I	**9b**	H	Ac	Ac	OAc	Ac	*T. baccata*	De Marcano and Halsall (1970b)
2α,5α,9α-trihydroxy-10β13α-diacetoxy-taxa-4β,20-epoxy-taxa-11-ene	**9c**	H	H	H	H	H	*T. chinensis*	Zhang *et al.* (1995)
1-Hydroxy-2,7,9-trideacetyl-baccatin I	**9d**	OH	OH	Ac	OH	H	*T. yunnanensis*	Chen *et al.* (1994)

Figure 3.21 A C-4(20) epoxide.

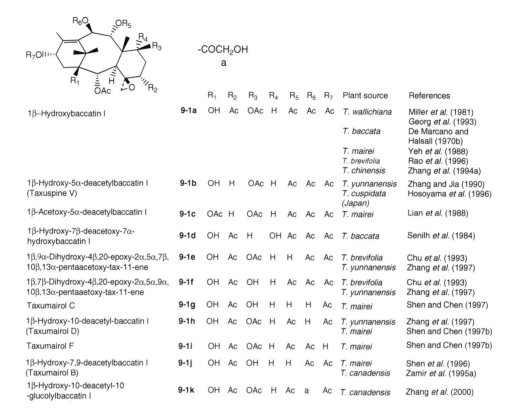

		R1	R2	R3	R4	R5	R6	R7	Plant source	References
1β–Hydroxybaccatin I	9-1a	OH	Ac	OAc	H	Ac	Ac	Ac	T. wallichiana	Miller et al. (1981) Georg et al. (1993)
									T. baccata	De Marcano and Halsall (1970b)
									T. mairei	Yeh et al. (1988)
									T. brevifolia	Rao et al. (1996)
									T. chinensis	Zhang et al. (1994a)
1β-Hydroxy-5α-deacetylbaccatin I (Taxuspine V)	9-1b	OH	H	OAc	H	Ac	Ac	Ac	T. yunnanensis T. cuspidata (Japan)	Zhang and Jia (1990) Hosoyama et al. (1996)
1β-Acetoxy-5α-deacetylbaccatin I	9-1c	OAc	H	OAc	H	Ac	Ac	Ac	T. mairei	Lian et al. (1988)
1β-Hydroxy-7β-deacetoxy-7α-hydroxybaccatin I	9-1d	OH	Ac	H	OH	Ac	Ac	Ac	T. baccata	Senilh et al. (1984)
1β,9α-Dihydroxy-4β,20-epoxy-2α,5α,7β,10β,13α-pentaacetoxy-tax-11-ene	9-1e	OH	Ac	OAc	H	H	Ac	Ac	T. brevifolia T. yunnanensis	Chu et al. (1993) Zhang et al. (1997)
1β,7β-Dihydroxy-4β,20-epoxy-2α,5α,9α,10β,13α-pentaaetoxy-tax-11-ene	9-1f	OH	Ac	OH	H	Ac	Ac	Ac	T. brevifolia T. yunnanensis	Chu et al. (1993) Zhang et al. (1997)
Taxumairol C	9-1g	OH	Ac	OH	H	H	H	Ac	T. mairei	Shen and Chen (1997)
1β-Hydroxy-10-deacetyl-baccatin I (Taxumairol D)	9-1h	OH	Ac	OAc	H	Ac	H	Ac	T. yunnanensis T. mairei	Zhang et al. (1997) Shen and Chen (1997b)
Taxumairol F	9-1i	OH	Ac	OAc	H	Ac	Ac	H	T. mairei	Shen and Chen (1997b)
1β-Hydroxy-7,9-deacetylbaccatin I (Taxumairol B)	9-1j	OH	Ac	OH	H	H	Ac	Ac	T. mairei T. canadensis	Shen et al. (1996) Zamir et al. (1995a)
1β-Hydroxy-10-deacetyl-10-glucolylbaccatin I	9-1k	OH	Ac	OAc	H	Ac	a	Ac	T. canadensis	Zhang et al. (2000)

Figure 3.22 A hydroxyl group or acetate at C-5.

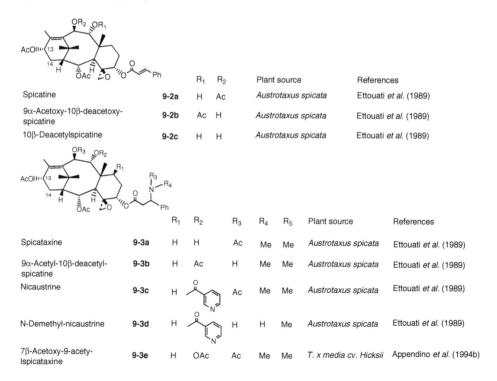

		R1	R2	Plant source	References
Spicatine	9-2a	H	Ac	Austrotaxus spicata	Ettouati et al. (1989)
9α-Acetoxy-10β-deacetoxy-spicatine	9-2b	Ac	H	Austrotaxus spicata	Ettouati et al. (1989)
10β-Deacetylspicatine	9-2c	H	H	Austrotaxus spicata	Ettouati et al. (1989)

		R1	R2	R3	R4	R5	Plant source	References
Spicataxine	9-3a	H	H	Ac	Me	Me	Austrotaxus spicata	Ettouati et al. (1989)
9α-Acetyl-10β-deacetyl-spicatine	9-3b	H	Ac	H	Me	Me	Austrotaxus spicata	Ettouati et al. (1989)
Nicaustrine	9-3c	H	[acyl]	Ac	Me	Me	Austrotaxus spicata	Ettouati et al. (1989)
N-Demethyl-nicaustrine	9-3d	H	[acyl]	H	H	Me	Austrotaxus spicata	Ettouati et al. (1989)
7β-Acetoxy-9-acetylspicataxine	9-3e	H	OAc	Ac	Me	Me	T. x media cv. Hicksii	Appendino et al. (1994b)

Figure 3.23 Cinnamoyl or aminoacyl group at C-5.

A hydroxyl group or acetyl group at C-5

Two taxoids were isolated from the needles of *T. chinensis* and their structures were established as 2α, 5α, 9α-trihydroxy-10β, 13α-diacetoxy-4β, 20-epoxy-taxa-11-ene (**9c**) and 2α, 5α, 9α -trihydroxy-10β,13α-diacetoxy-taxa-4(20),11-diene (**1m**) (Zhang, Z.P. *et al.*, 1995). An epoxide (**9-1a**) was isolated from *T. wallichiana* (De Marcano and Halsall, 1970). Spectroscopic reexamination of brevifoliol indicated its structure must be revised to a rearranged taxoid (**13a**) (Georg *et al.*, 1993). Two epoxides (**9-1e,f**) were isolated from the bark of *T. brevifolia* (Chu *et al.*, 1993). A new taxoid taxumairol F (**9-1i**) has been isolated with the other epoxides (**9-1g, h**) from the ethanolic extracts of the roots of *T. mairei*. Their structures were determined on the basis of spectral evidence and chemical correlation with 1-β-hydroxy-baccatin I (Shen and Chen, 1997b). Three taxoids (**9-1e, f**, and **h**) were isolated from the roots of *T. yunnanensis* by Zhang *et al.* in 1997 (Figures 3.21 and 3.22).

Taxoids with cinnamoyl or aminoacyl group

Five taxoids (**9-2a, 2b, 2c, 3a**, and **3b**) and the other taxoids (**9-3c** and **3d**) that esterified at C-9 with nicotinic acid were isolated from the trunk bark of *A. spicata* by Ettouati *et al.* in 1989. The roots of *T. x media* gave a taxane (**9-3e**) (Appendino *et al.*, 1994b) (Figure 3.23).

Taxoids with C-13 ketonized group

Three C-13 ketonized taxoids were isolated from the trunk of *A. spicata*. Two of them (**9-5a** and **5b**) were esterified at C-9 (Ettouati *et al.*, 1989) (Figure 3.24).

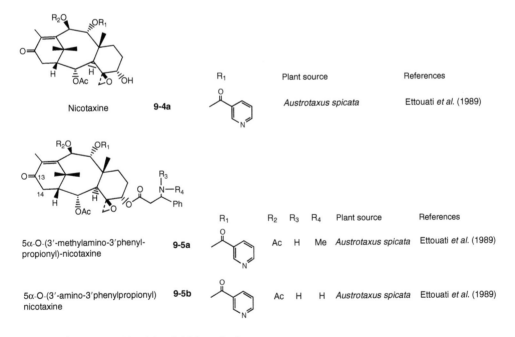

Figure 3.24 Epoxytaxoids with a C-13 ketonized group.

Taxoids with an oxetane ring

This group of taxoids is worthy of notice, because it includes taxol and its potential semisynthetic precursors baccatin III and 10-deacetylbaccatin III. Simple taxoids of this group are illustrated in Figures 3.25–3.35. All members isolated to date are highly oxygenated, with oxygen substituents at C-2, C-4, C-7, C-9, C-10, and C-13. Even the C-1 position, which is normally not oxygenated in the other series of compounds, is usually oxygenated in this series.

Taxoids with both an oxetane ring and a complex C-13 ester side-chain are illustrated in Figures 3.31–3.34. This group includes the antitumor agents taxol and cephalomannine, but it also includes other congeners and a series of xylosyl derivatives isolated from *T. baccata*. Until now many kinds of semisynthetic taxol side-chain analogues have been prepared; in this Chapter only naturally occurring taxoids have been discussed. Kingston discussed these compounds and their structure-activity relationship in his review (Kingston, 1991).

Simple taxoids with an oxetane ring

An oxetane type taxoid (10-2a) was isolated from *T. canadensis* (Zamir *et al.*, 1995a). A taxoid (10-3g) with hydroxybutyryl at C-10 was isolated from *T. baccata* (Soto and Castedo, 1998). Two taxoids (10-3h and 3i) were isolated from *T. chinensis* (Fuji *et al.*, 1993). A taxoid (10-1e) was isolated from the roots of *T. x media* (Appendino *et al.*, 1994). Four taxoids (10-2a, 2g, 2m, and 2n) were isolated from *T. canadensis* (Zamir *et al.*, 1995a). Two taxoids (10-2k and 10-2l) were isolated from the roots of *T. yunnanensis* (Zhang *et al.*, 1997). A taxoid (10-2l) was isolated from

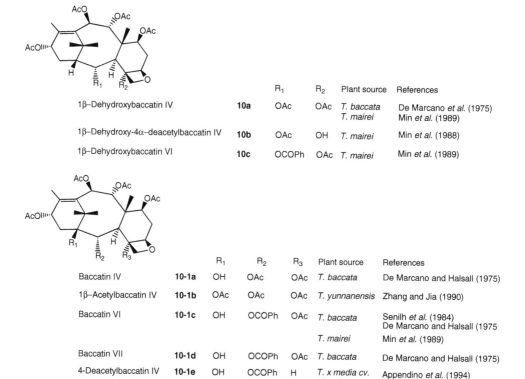

		R₁	R₂	Plant source	References
1β–Dehydroxybaccatin IV	**10a**	OAc	OAc	*T. baccata* *T. mairei*	De Marcano *et al.* (1975) Min *et al.* (1989)
1β–Dehydroxy-4α–deacetylbaccatin IV	**10b**	OAc	OH	*T. mairei*	Min *et al.* (1988)
1β–Dehydroxybaccatin VI	**10c**	OCOPh	OAc	*T. mairei*	Min *et al.* (1989)

		R₁	R₂	R₃	Plant source	References
Baccatin IV	**10-1a**	OH	OAc	OAc	*T. baccata*	De Marcano and Halsall (1975)
1β–Acetylbaccatin IV	**10-1b**	OAc	OAc	OAc	*T. yunnanensis*	Zhang and Jia (1990)
Baccatin VI	**10-1c**	OH	OCOPh	OAc	*T. baccata* *T. mairei*	Senilh *et al.* (1984) De Marcano and Halsall (1975 Min *et al.* (1989)
Baccatin VII	**10-1d**	OH	OCOPh	OAc	*T. baccata*	De Marcano and Halsall (1975)
4-Deacetylbaccatin IV	**10-1e**	OH	OCOPh	H	*T. x media cv.* *Hicksii*	Appendino *et al.* (1994)

Figure 3.25 Simple taxoids with an oxetane ring.

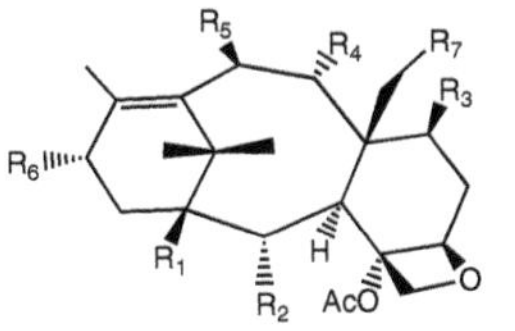

-COCH$_2$OH   -COCH$_2$OAc

a   b

		R$_1$	R$_2$	R$_3$	R$_4$	R$_5$	R$_6$	R$_7$	Plant source	References
7,9,10-Deacetyl-baccatin VI	10-2a	OH	OBz	OH	OH	OH	OAc	H	T. cuspidata var. nana	Morita et al. (1997b)
									T. canadensis	Zamir et al. (1995a)
Taxuspine E	10-2b	OH	OBz	OAc	OH	OH	OH	H	T. cuspidata (Japan)	Kobayashi et al. (1995)
7,9-Deacetyl baccatin VI	10-2c	OH	OBz	OH	OH	OAc	OAc	H	T. canadensis	Zamir et al. (1992)
									T. brevifolia	Rao et al. (1996b)
5β,20-Epoxy-1β-hydroxy-4α,7β,13α-triacetoxy-2α,9α,10β–tribenzoxy-tax-1-ene (?)	10-2d	OH	OBz	OAc	OBz	OBz	OAc	H	T. brevifolia	Chu et al. (1992)
2α,10β-Dibenzoxy-5β,20-epoxy-1β-hydroxy-4α,7β,9α,13α-tetraacetoxy-tax-11-ene (?)	10-2e	OH	OBz	OAc	OAc	OBz	OAc	H	T. brevifolia	Chu et al. (1992)
2α,7β-Dibenzoxy-5β20-epoxy-1β-hydroxy-4α,9α,10β,13α-teraacetoxy-tax-11-ene (?)	10-2f	OH	OBz	OBz	OAc	OAc	OAc	H	T. brevifolia	Chu et al. (1992)
7,9-Deacetylbaccatin IV	10-2g	OH	OAc	OH	OH	OAc	OAc	H	T. canadensis	Zamir et al. (1995a)
									T. brevifolia	Rao et al. (1996b)
9-Dihydro-13-acetylbaccatin III	10-2h	OH	OBz	OH	OH	OAc	OAc	H	T. canadensis	Gunawardana et al. (1992)
13-Deacetylbaccatin VI (?) (maybe revised to 14i)	10-2i	OH	OBz	OAc	OAc	OAc	OH	H	T. wallichiana	Barboni et al. (1993)
9(βH)-9-dihydro-19-acetoxy-10-deacetyl-baccatin III	10-2j	OH	OBz	OH	OH	OH	OH	OH	T. baccata	Appendino et al. (1992)
10-Deacetyl-baccatin VI	10-2k	OH	OBz	OAc	OAc	OH	OAc	H	T. yunnanensis	Zhang et al. (1997)
9-Deacetyl-baccatin VI	10-2l	OH	OBz	OAc	OH	OAc	OAc	H	T. yunnanensis	Zhang et al. (1997)
10-Hydroxyacetylbaccatin VI	10-2m	OH	OBz	OAc	OAc	a	OAc	H	T. canadensis	Zamir et al. (1995a)
7epi-9,10-Deacetylbaccatin VI	10-2n	OH	OBz	α-OH	OH	OH	OAc	H	T. canadensis	Zamir et al. (1995a)
7,9,13-Trideacetylbaccatin VI	10-2o	OH	OBz	OH	OH	OAc	OH	H	T. canadensis	Zhang et al. (2000a)
10-Deacetyl-10-glycolylbaccatin IV	10-2p	OH	OAc	OAc	OAc	a	OAc	H	T. canadensis	Zhang et al. (2000a)
10-acetylglycolylbaccatin VI	10-2q	OH	OBz	OAc	OAc	b	OAc	H	T. canadensis	Zhang et al. (2001)

Figure 3.26 Simple taxoids with an oxetane ring.

the needles of *T. wallichiana* (Barboni *et al.*, 1993). Mixed acetoxy-benzoxy taxane esters were added to this group (Figures 3.25 and 3.26). Three taxane derivatives (**10-2d, e** and **f**) were isolated from the bark of *T. brevifolia* (Chu *et al.*, 1992). From *T. brevifolia*, 7,9-deacetyl baccatin IV (**10-2g**), three other taxoids were isolated: baccatin IV (**10-1a**), 9-dihydro-13-acetylbaccatin III (**10-2h**) and 1β,7β-dihydroxy-4β,20-epoxy-2α,5α,9α,10β,13α-penta-acetoxy-tax-11-ene (**9-1f**) (Rao *et al.*, 1996b).

C-9 Ketonized group

The needles of the European yew, *T. baccata*, yielded two new analogues of 10-deacetylbaccatin III, identified as 13-*epi*-10-deacetylbaccatin III (**10-4a**) and 2-debenzoyl-2-tigloyl-10-deacetyl-baccatin III (**10-4b**) (Gabetta *et al.*, 1995a). The crystal structure of baccatin III, corresponding to the terpenoid core of paclitaxel, was determined by Gabetta *et al.* in 1995 (Figures 3.27 and 3.28). A taxoid (**10-4c**) was isolated from the stems of *T. cuspidata* by Morita *et al.* (1998a).

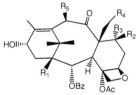

		R₁	R₂	R₃	R₄	R₅	Plant source	References
1-Dehydroxybaccatin III	10-3a	H	OH	H	H	OAc	*T. yunnanensis*	Zhang and Jia (1990)
Baccatin III	10-3b	OH	OH	H	H	OAc	*T. baccata*	Senilh *et al.* (1984)
10-Deacetylbaccatin III	10-3c	OH	OH	H	H	OH	*T. baccata* *T. yunnanensis* *T. wallichiana*	Chauviere *et al.* (1981) Zhang and Jia (1990) Appendino *et al.* (1992a)
19-Hydroxybaccatin III	10-3d	OH	OH	H	OH	OAc	*T. baccata* *T. yunnanensis*	Chauviere *et al.* (1981) Zhang and Jia (1990)
Baccatin V	10-3e	OH	H	OH	H	OAc	*T. baccata*	De Marcano *et al.* (1970)
1-Acetyl-10-deacetylbaccatin III	10-3f	OAc	OH	H	H	OH	*T. canadensis*	Zamir *et al.* (1992)
10-(β–hydroxybutyryl)-10-deacetylbaccatin III	10-3g	OH	OH	H	H	OCOCH₂CH(OH)CH₃	*T. baccata*	Soto and Castedo (1998)
7-epi-19-Hydroxybaccatin III	10-3h	OH	H	OH	OH	OAc	*T. chinensis*	Fuji *et al.* (1993)
10-Deacetyl -10-oxobaccatin V	10-3i	OH	H	OH	H	=O	*T. chinensis*	Fuji *et al.* (1993)

Figure 3.27 Taxoids with C-9 ketonized group.

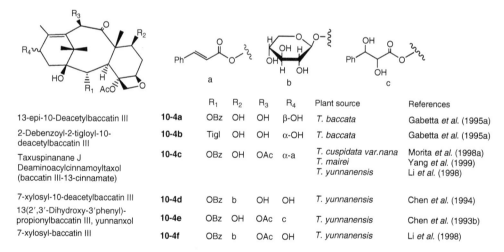

		R₁	R₂	R₃	R₄	Plant source	References
13-epi-10-Deacetylbaccatin III	10-4a	OBz	OH	OH	β-OH	*T. baccata*	Gabetta *et al.* (1995a)
2-Debenzoyl-2-tigloyl-10-deacetylbaccatin III	10-4b	Tigl	OH	OH	α-OH	*T. baccata*	Gabetta *et al.* (1995a)
Taxuspinanane J Deaminoacylcinnamoyltaxol (baccatin III-13-cinnamate)	10-4c	OBz	OH	OAc	α-a	*T. cuspidata var.nana* *T. mairei* *T. yunnanensis*	Morita *et al.* (1998a) Yang *et al.* (1999) Li *et al.* (1998)
7-xylosyl-10-deacetylbaccatin III	10-4d	OBz	b	OH	OH	*T. yunnanensis*	Chen *et al.* (1994)
13(2′,3′-Dihydroxy-3′phenyl)-propionylbaccatin III, yunnanxol	10-4e	OBz	OH	OAc	c	*T. yunnanensis*	Chen *et al.* (1993b)
7-xylosyl-baccatin III	10-4f	OBz	b	OAc	OH	*T. yunnanensis*	Li *et al.* (1998)

Figure 3.28 C-9 Ketonized and C-1 hydroxy group.

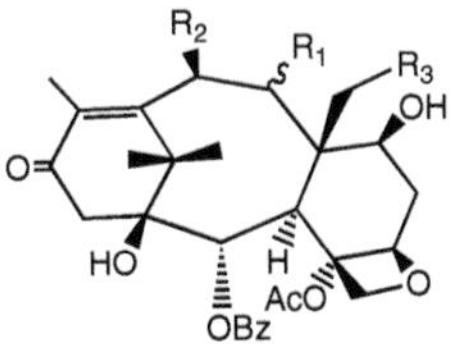

		R_1	R_2	R_3	Plant source	References
13-oxo-7,9-bisdeacetyl-baccatin VI (Taxuspinanane D)	**10-5a**	α-OH	OAc	H	*T. cuspidata var. nana*	Morita *et al.* (1998b)
Taxuspinanane C	**10-5b**	α-OH	OH	H	*T. cuspidata var. nana*	Morita *et al.* (1997a)
19-Hydroxy-13-oxobaccatin III	**10-5c**	=CO	OH	OH	*T. sumatrana*	Kitagawa *et al.* (1995)

Figure 3.29 C-13 ketonized group.

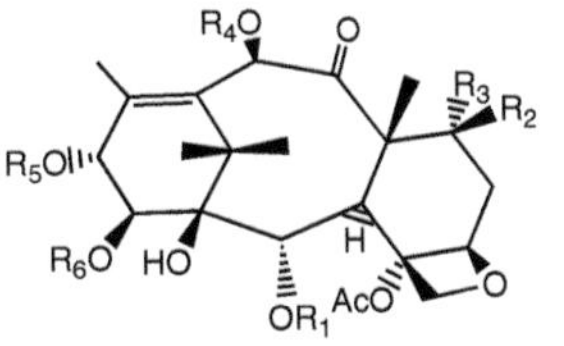

		R_1	R_2	R_3	R_4	R_5	R_6	Plant source	References
14β–Hydroxy-10-deacetylbaccatin III	**10-6a**	Bz	OH	H	H	H	H	*T. wallichiana*	Appendino *et al.* (1992a)
2-Debenzoyl-14β-benzoyloxy-10-deacetylbaccatin III	**10-6b**	OH	OH	H	H	H	OBz	*T. wallichiana*	Appendino *et al.* (1993e)

Figure 3.30 C-9 Ketonized group with hydroxyl group at C-14.

C-13 ketonized group or C-9 ketonized group with hydroxyl group at C-14

Morita *et al.* (1997a, 1998a,b) isolated five new taxoids (6o, 10-2a, 10-5a,b and 11-1i) from *T. cuspidata var. nana*. 10-5a and 5b are classified in this group. A taxoid (10-5c) was isolated from the needles of *T. sumatrana* (Kitagawa *et al.*, 1995). A taxoid (10-6a) was isolated from the needles of *T. wallichiana* (Appendino *et al.*, 1992a). The needles of *T. wallichiana* gave a taxoid (10-6b) (Appendino *et al.*, 1993e) (Figures 3.29 and 3.30).

Complex C-13 side-chain

In addition to taxol A, 10-deacetyltaxol A (11c), 10-deacetyl-7-*epi*-taxol A (11h), were isolated from the seeds of *T. mairei* (Shen *et al.*, 1999a). Three taxoids (11c, 11d, and 11-1g) were isolated from the needles of *T. x media* (Senilh *et al.*, 1984). The roots of *T. x media* yielded a taxoid (11g) (Gabetta *et al.*, 1995b). A taxoid (11i) was isolated from the stems of *T. cuspidata var. nana* by Morita *et al.* (1998). A new taxoid 2′-acetyl-7-epi-taxol (11j) was isolated from *T. canadensis*, together with taxoids (12l, 1-6a, 6p) by Zhang *et al.* (2000b).

A taxoid (11-1h) was isolated from the stem *of T. baccata* by Guo *et al.* (1995). Four new taxoids were isolated from cell cultures of *T. baccata*. Two were the aglycones corresponding to 7-O-xylosides of taxol C (11-2a) and 10-deacetyltaxol C (11-2b). The roots of *T. x media* gave two taxol analogues (11-2a, c) bearing a modified phenylisoserine side-chain (Barboni *et al.*, 1994b). The third (11-2c) had an *N*-methylated side-chain, while the fourth, named taxcultine (11-2d), contained an n-propyl group on the side-chain. All four compounds actively promoted tublin assembly (Ma *et al.*, 1994). A minor bioactive taxcultine (11-2d) was isolated from the needles of *T. canadensis* (Zamir *et al.*, 1996). New taxoid, taxuspinanane A (11-2e), showing potent cytotoxic activity, has been isolated from the stems of *T. cuspidata var. nana* (Morita *et al.*, 1997a).

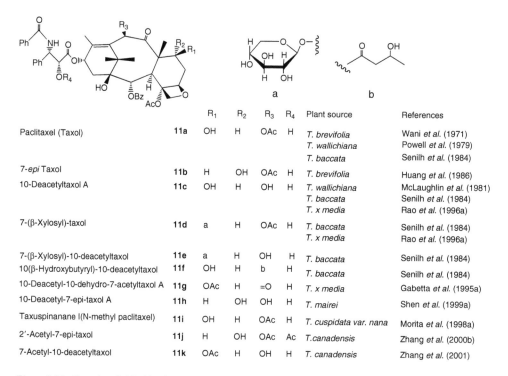

		R_1	R_2	R_3	R_4	Plant source	References
Paclitaxel (Taxol)	11a	OH	H	OAc	H	T. brevifolia T. wallichiana T. baccata	Wani et al. (1971) Powell et al. (1979) Senilh et al. (1984)
7-epi Taxol	11b	H	OH	OAc	H	T. brevifolia	Huang et al. (1986)
10-Deacetyltaxol A	11c	OH	H	OH	H	T. wallichiana T. baccata T. x media	McLaughlin et al. (1981) Senilh et al. (1984) Rao et al. (1996a)
7-(β-Xylosyl)-taxol	11d	a	H	OAc	H	T. baccata T. x media	Senilh et al. (1984) Rao et al. (1996a)
7-(β-Xylosyl)-10-deacetyltaxol	11e	a	H	OH	H	T. baccata	Senilh et al. (1984)
10(β-Hydroxybutyryl)-10-deacetyltaxol	11f	OH	H	b	H	T. baccata	Senilh et al. (1984)
10-Deacetyl-10-dehydro-7-acetyltaxol A	11g	OAc	H	=O	H	T. x media	Gabetta et al. (1995a)
10-Deacetyl-7-epi-taxol A	11h	H	OH	OH	H	T. mairei	Shen et al. (1999a)
Taxuspinanane I(N-methyl paclitaxel)	11i	OH	H	OAc	H	T. cuspidata var. nana	Morita et al. (1998a)
2′-Acetyl-7-epi-taxol	11j	H	OH	OAc	Ac	T.canadensis	Zhang et al. (2000b)
7-Acetyl-10-deacetyltaxol	11k	OAc	H	OH	H	T. canadensis	Zhang et al. (2001)

Figure 3.31 Complex C-13 side-chain.

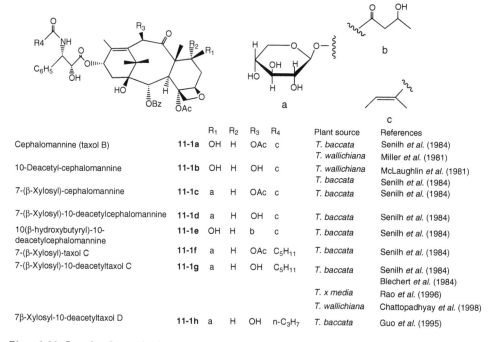

		R_1	R_2	R_3	R_4	Plant source	References
Cephalomannine (taxol B)	11-1a	OH	H	OAc	c	T. baccata T. wallichiana	Senilh et al. (1984) Miller et al. (1981)
10-Deacetyl-cephalomannine	11-1b	OH	H	OH	c	T. wallichiana T. baccata	McLaughlin et al. (1981) Senilh et al. (1984)
7-(β-Xylosyl)-cephalomannine	11-1c	a	H	OAc	c	T. baccata	Senilh et al. (1984)
7-(β-Xylosyl)-10-deacetylcephalomannine	11-1d	a	H	OH	c	T. baccata	Senilh et al. (1984)
10(β-hydroxybutyryl)-10-deacetylcephalomannine	11-1e	OH	H	b	c	T. baccata	Senilh et al. (1984)
7-(β-Xylosyl)-taxol C	11-1f	a	H	OAc	C_5H_{11}	T. baccata	Senilh et al. (1984)
7-(β-Xylosyl)-10-deacetyltaxol C	11-1g	a	H	OH	C_5H_{11}	T. baccata T. x media T. wallichiana	Senilh et al. (1984) Blechert et al. (1984) Rao et al. (1996) Chattopadhyay et al. (1998)
7β-Xylosyl-10-deacetyltaxol D	11-1h	a	H	OH	n-C_3H_7	T. baccata	Guo et al. (1995)

Figure 3.32 Complex C-13 side-chain.

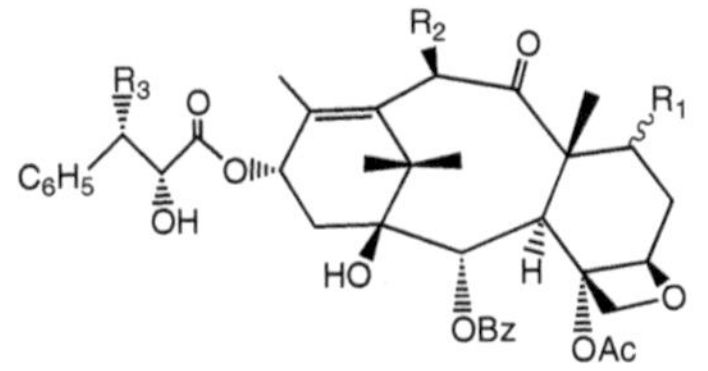

	R₁	R₂	R₃	Plant source	References
Taxol C **11-2a**	β–OH	OAc	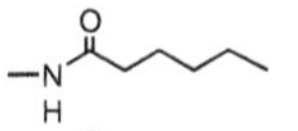	*T. x media* *T. baccata(cell cul.)*	Barboni *et al.* (1994) Ma *et al.* (1994a)
10-Deacetyltaxuyunnanine A (10-Deacetyltaxol C) **11-2b**	β–OH	OH	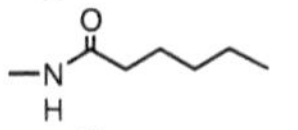	*T. yunnanensis* *T. baccata(cell cul.)*	H.Zhang *et al.* (1994b) Ma *et al.* (1994a)
N-Methyltaxol C **11-2c**	β–OH	OAc	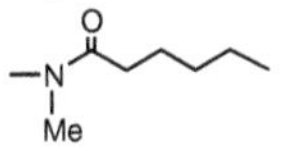	*T. x media* *T. baccata(cell cul.)*	Barboni *et al.* (1994b) Ma *et al.* (1994)
Taxcultine (taxol D) (N-debenzoyl-N-butanoyl-taxol) **11-2d**	β–OH	OAc	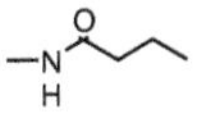	*T. baccata(cell cul.)* *T. canadensis*	Ma *et al.* (1994a) Zamir *et al.* (1996)
Taxuspinanane A **11-2e**	β–OH	OAc	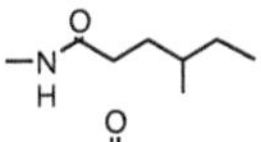	*T. x media cv. Hicksii* *T. cuspidata var.nana*	Appendino *et al.* (1994) Morita *et al.* (1997a)
N-Debenzoyl-N-butanoyl-10- deacetylpaclitaxel **11-2f**	β–OH	H	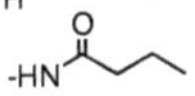	*T. baccata*	Gabetta *et al.* (1998)
N-Debenzoyl-N-prppanoyl-10- deacetylpaclitaxel **11-2g**	β–OH	H	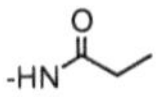	*T. baccata*	Gabetta *et al.* (1998)
N-Debenzoyl-N-(2-methylbutyryl)- paelitaxel **11-2h**	β–OH	OAc	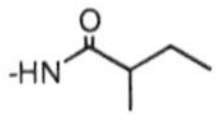	*T. x media cv. Hicksii*	Gabetta *et al.* (1999)
N-Debenzoyl-N-cinnamoyl paclitaxel **11-2i**	β–OH	OAc	-NH-Cinn	*T. x media cv. Hicksii*	Gabetta *et al.* (1999)
7-epi-Taxuyunnanine A **11-2j**	α–OH	OAc	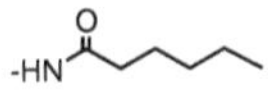	*T. yunnanensis*	Zhang *et al.* (1994b)
7-epi-10-Deacetyltaxuyunnanine A (taxuspinanane E) **11-2k**	αOH	OH	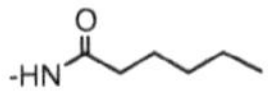	*T. yunnanensis* *T. cuspidata* *var. nana*	Zhang *et al.* (1994b) Morita *et al.* (1998b)
10-Deacetyl-10-oxo-7-epi- taxuyunnanine A **11-2l**	α–OH	=O	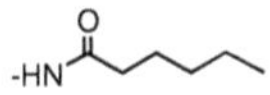	*T. yunnanensis*	Zhang *et al.* (1994b)
N-Acetyl-N-debenzoyltaxol **11-2m**	β-OH	OAc	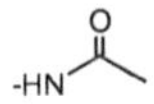	*T. canadensis*	Zhang *et al.* (2000a)
2'-acetyl-7-epi-cephalomannine **11-2n**	α-OH	β-OAc	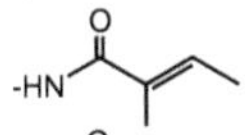	*T. canadensis*	Zhang *et al.* (2001)
10-deacetyl-10-oxo-7-epi- cephalomannine **11-2o**	α-OH	=O	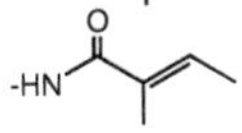	*T. canadensis*	Zhang *et al.* (2001)

Figure 3.33 Complex C-13 side-chain.

Four taxoids (11-2b, 2j, 2k, and 2l) were isolated from the roots of *T. yunnanensis* (Zhang *et al.*, 1994). *N*-Debenzoyl-*N*-butanoyl- and *N*-debenzoyl-*N*-propanoyl-10-deacetylpaclitaxel (11-2f and 2g) were identified as minor constituents of *T. baccata* by HPLC-MS (Gabetta *et al.*, 1998). The roots of *T. x media* Rehd. cv. Hicksii gave three novel analogues of paclitaxel modified at the N-acyl residue or at the ester group at C-2 (11-2h, 2i, 3c) (Gabetta *et al.*, 1999). 7-epi-Taxol (11b) and 10-deacetyl-10-oxo-7-epi-taxol (11-3a) were isolated from *T. brevifolia* (Huang *et al.*, 1986). Three taxoids taxuspine N–P (11-3b, 13-2b, 3-1d) have been isolated from stems of the Japanese yew *T. cuspidata* Sieb. Et Zucc. and the structures elucidated on the basis of spectro-scopic data. Taxuspine N is the first example of a taxane diterpene containing a 6/8/6-membered ring system with a 3-N,N-dimethylamino-3-phenylpropanoyl group (Winterstein's acid) at

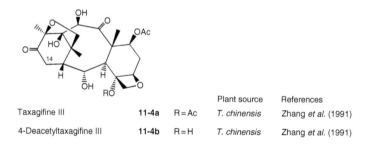

	Plant source	References
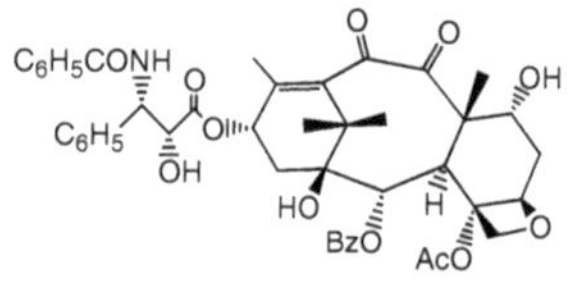 10-Deacetyl-10-oxo-7-epi-taxol (11-3a)	*T. brevifolia*	Huang *et al.* (1986)
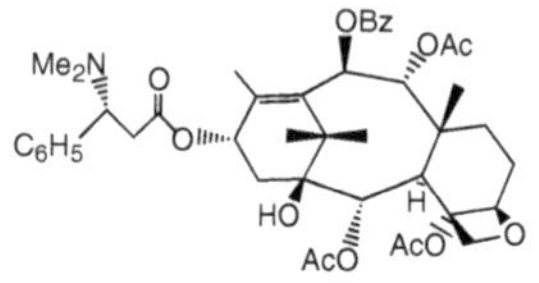 Taxuspine N (11-3b)	*T. cuspidata (Japan)*	Kobayashi *et al.* (1996)
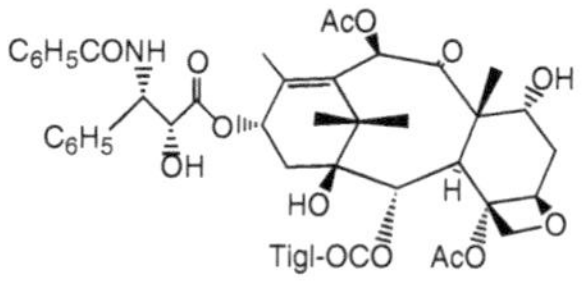 2-Debenzoyl-2-tigloyl paclitaxel (isocephalomannine) (11-3c)	*T. x media cv. Hicksii*	Gabetta *et al.* (1999)
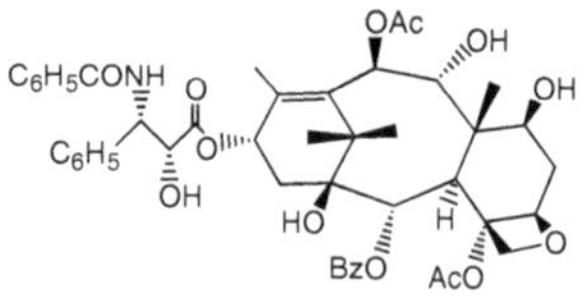 Dihydrotaxol (11-3d)	*T. yunnanensis* *T. wallichiana*	Chen *et al.* (1993b) Chattopadhyay and Saha (1996)

Figure 3.34 Complex C-13 side-chain.

			Plant source	References
Taxagifine III	**11-4a**	R = Ac	*T. chinensis*	Zhang *et al.* (1991)
4-Deacetyltaxagifine III	**11-4b**	R = H	*T. chinensis*	Zhang *et al.* (1991)

Figure 3.35 C-12(16)-Oxido bridge.

C-13 from yew tree. The absolute configuration of the 3-*N*,*N*-dimethylamino-3-phenylpropanoyl group was determined to be R by chiral HPLC analysis (Kobayashi *et al.*, 1996) (Figures 3.31–3.34).

C-12(16)-Oxido bridge

Five taxane diterpenes were isolated from the leaves and stems of *T. chinensis*. Their structures were established as 2-deacetyl-5-decinnamatetaxagifine (7f), taxagifine III (11-4a), 4-deacetyltaxagifine III (11-4b) (Zhang *et al.*, 1991) (Figure 3.35).

Taxoids with a C-3(8) cleaved ring (verticillene type)

Taxuspine U (12j) was isolated from Japanese yew *T. cuspidata* by Hosoyama *et al.* (1996).

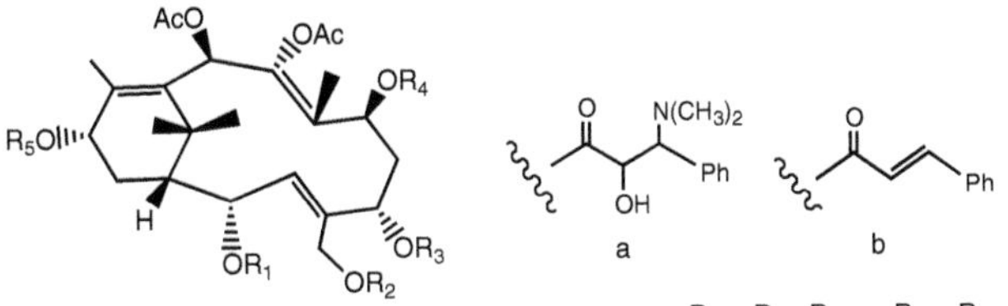

		R_1	R_2	R_3	R_4	R_5	Plant source	References
Taxachitriene A	12a	Ac	Ac	H	Ac	Ac	*T. chinensis*	Fang *et al.* (1996)
(2S,3R)-5-(N,N-Dimethyl-3'-phenylisoseryl) taxachitriene A	12b	Ac	Ac	a	Ac	Ac	*T. chinensis*	Shi *et al.* (1998b)
13-deacetyl canadensene	12c	Ac	H	H	Ac	H	*T. mairei*	Shi *et al.* (1999a)
7-deacetyl canadensene	12d	Ac	H	H	H	Ac	*T. mairei*	Shi *et al.* (1999a)
5-deacetyltaxachitriene B	12e	H	H	H	Ac	Ac	*T. chinensis*	Fang *et al.* (1996)
Taxachitriene B	12f	H	H	Ac	Ac	Ac	*T. chinensis*	Fang *et al.* (1996)
20-deacetyltaxachitriene A (canadensene)	12g	Ac	H	β–OH	Ac	Ac	*T. canadensis*	Zamir *et al.* (1995b)
5-epicanadense	12h	Ac	H	H	Ac	Ac	*T. canadensis*	Zamir *et al.* (1998)
Taxuspine X	12i	Ac	Ac	b	Ac	Ac	*T. cuspidata* (Japan)	Shigemori *et al.* (1997)
Taxuspine U	12j	H	H	Ac	H	Ac	*T. cuspidata* (Japan)	Hosoyama *et al.* (1996)
2-Deacetyltaxachitriene A	12k	H	Ac	Ac	Ac	Ac	*T. chinensis var. mairei*	Shi *et al.* (1998a)
5-epi-cinnamoylcanadensene	12l	Ac	H	b	Ac	Ac	*T. canadensis*	Zhang *et al.* (2000b)
7,9,10,13-tetraacetoxy-5-cinnamoyloxy-4-hydroxy-methyl-8,12,15-tetramethyl-bicyclo[9.3.1]pentadeca-3.8.11-trien-2-ol(2,20-dideacetyl tuxaspine X	12m	H	H	b	Ac	Ac	*T. cuspidata Sieb. et Zucc*	Shi *et al.* (1999h)
7,9,13,20-pentaacetoxy-5-cinnamoyloxy-8,12,15,15-tetramethyl bicyclo[9.3.1] pentadeca 3,8,11-trien-2-ol (2-deacetyl taxuspine X)	12n	H	Ac	b	Ac	aAc	*T. cuspidata Sieb. et Zucc*	Shi *et al.* (1999h)
9,10,13,20-tetraacetoxy-5-cinnamoyloxy-8,12,15,15-tetramethyl-bicyclo[9.3.1] pentadeca-3,8,11-trien-2,7-diol(2,7-dideacetyl taxuspine X)	12o	H	Ac	b	H	Ac	*T. cuspidata Sieb. et Zucc*	Shi *et al.* (1999h)

Figure 3.36 Verticillene type.

Taxuspines X (**12i**), Y (**13-2c**), Z (**3f**) have been isolated from stems of the Japanese yew *T. cuspidata* (Shigemori *et al.*, 1997). Taxumairenol M (**12-1a**) was isolated from the seeds of *T. mairei* by Shi *et al.* (1999). Three novel bicyclic taxane diterpenoids with a verticillene skeleton were isolated from the needles of Chinese yew, *T. chinensis var. mairei*. Their structures were established as **12a, b**, and **c** on the basis of 1D, 2D NMR and HR-MS spectroscopic analysis. The verticillene skeleton was considered as the biogenetic intermediate of taxane diterpenoids; isolation of bicyclic taxane diterpenoids with verticillene skeleton from this plant provides direct proof for this hypothesis (Shi *et al.*, 1999). Two bicyclic taxoids (**12a, 12b**) and two abeotaxanes (**13j, 13h**) were isolated from the bark and the needles of the Chinese yew (Shi *et al.*, 1998b). Three taxoids (**12a, 12e**, and **12f**) were isolated from the needles of *T. chinensis* (Fang *et al.*, 1996).

A taxoid (**12g**) was isolated from the needles of *T. canadensis* by Zamir *et al.* (1995b).

Three taxoids (**12k, 12-1d**, and **12-1e**) were isolated from the needles of the Chinese yew *T. chinensis var. mairei* (Shi *et al.*, 1998a) (Figures 3.36–3.38).

Miscellaneous ether linkage

Taxuspine K (**12-3a**) is the first example of a taxane diterpene from yew trees containing a 6/8/6-membered ring system with a tetrahydrofuran ring at C-2, C-3, C-4, and C-20 (Wang *et al.*,

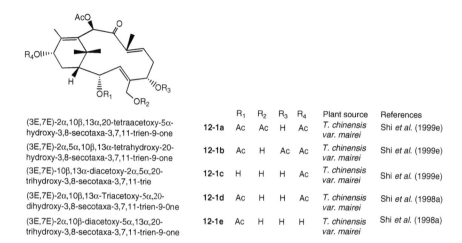

	R_1	R_2	R_3	R_4	Plant source	References
(3E,7E)-2α,10β,13α,20-tetraacetoxy-5α-hydroxy-3,8-secotaxa-3,7,11-trien-9-one **12-1a**	Ac	Ac	H	Ac	*T. chinensis var. mairei*	Shi *et al.* (1999e)
(3E,7E)-2α,5α,10β,13α-tetrahydroxy-20-hydroxy-3,8-secotaxa-3,7,11-trien-9-one **12-1b**	Ac	H	Ac	Ac	*T. chinensis var. mairei*	Shi *et al.* (1999e)
(3E,7E)-10β,13α-diacetoxy-2α,5α,20-trihydroxy-3,8-secotaxa-3,7,11-trie **12-1c**	H	H	H	Ac	*T. chinensis var. mairei*	Shi *et al.* (1999e)
(3E,7E)-2α,10β,13α-Triacetoxy-5α,20-dihydroxy-3,8-secotaxa-3,7,11-trien-9-One **12-1d**	Ac	H	H	Ac	*T. chinensis var. mairei*	Shi *et al.* (1998a)
(3E,7E)-2α,10β-diacetoxy-5α,13α,20-trihydroxy-3,8-secotaxa-3,7,11-trien-9-one **12-1e**	Ac	H	H	H	*T. chinensis var. mairei*	Shi *et al.* (1998a)

Figure 3.37 Verticillene type ketonized at C-9.

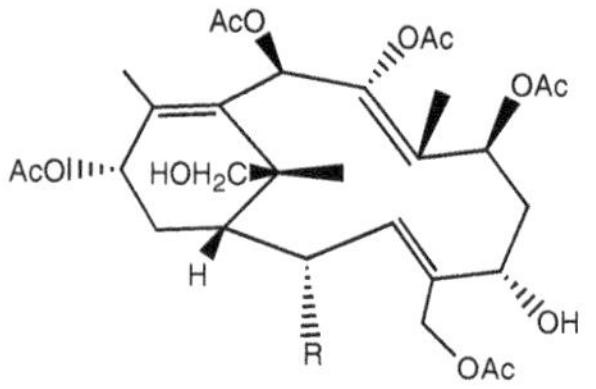

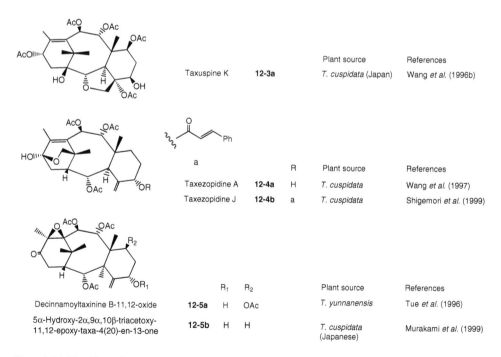

			Plant source	References
Taxumairol M	**12-2a**	R=OAc	*T. mairei*	Shen *et al.* (1999)

Figure 3.38 Verticillene type with C-16 hydroxylated group.

			Plant source	References
Taxuspine K	**12-3a**		*T. cuspidata* (Japan)	Wang *et al.* (1996b)

		R	Plant source	References
Taxezopidine A	**12-4a**	H	*T. cuspidata*	Wang *et al.* (1997)
Taxezopidine J	**12-4b**	a	*T. cuspidata*	Shigemori *et al.* (1999)

		R_1	R_2	Plant source	References
Decinnamoyltaxinine B-11,12-oxide	**12-5a**	H	OAc	*T. yunnanensis*	Tue *et al.* (1996)
5α-Hydroxy-2α,9α,10β-triacetoxy-11,12-epoxy-taxa-4(20)-en-13-one	**12-5b**	H	H	*T. cuspidata* (Japanese)	Murakami *et al.* (1999)

Figure 3.39 Taxoids with miscellaneous ether linkage.

1996b). Taxezopidine A (12-4a) has been isolated from the seeds of the Japanese yew *T. cuspidata* and is the first taxoid with hemiketal ring at C-11~C-13, C-15, and C-17 (Wang *et al.*, 1997). Three taxoids, taxezopidine J (12-4b) K (1-1f), and L (7h), were isolated from the seeds of the Japanese yew *T. cuspidata* Sieb. et Zucc (Shigemori *et al.*, 1999). Another type of transannular bond (C-13 → 16) was found in taxezopidine A (12-4a) and J (12-4b), as mentioned above.

Moreover, (C-11 → 12)epoxide, decinnamoyl-taxinine B-11, 12-oxide (12-5a), was isolated from the leaves and stems of *T. yunnanensis* by Yue *et al.* (1996). A taxoid (12-5b) was isolated from the needles of the Japanese yew, *T. cuspidata* Sieb et Zucc, together with 10 other taxoids (Murakami *et al.*, 1999) (Figure 3.39).

Structures of rearranged taxoids

Structures of 11(15 → 1)-abeotaxoids

C-4(20) Exocyclic double bond

A HYDROXYL GROUP OR ACETATE AT C-5

Although brevifoliol was previously assigned as a normal structure, it was reassigned as a member of this class by Appendino *et al.* (1993c,d). A brevifoliol type taxoid (13b) was isolated from *T. brevifolia* (Chen *et al.*, 1994). Two taxoids with a basic skeleton (13c, 13i) were isolated from *T. chinensis* (Fuji *et al.*, 1992). The structures of taxchinins D and G have been determined as 13g and 13l by Tanaka *et al.* (1994). Large-scale processing of the chloroform extract of the bark of *T. brevifolia* by reversed phase column chromatography on C-18 bonded silica using acetonitrile/ water/mixtures gave a taxane-rich fraction which emerged just after the elution of paclitaxel. Many taxoids belonging to the 11(15 → 1)-*abeo*taxane group, taxinine J group, brevifoliol group and those with an oxygenation pattern at 14 position were isolated. One of them was 13n (Rao and Juchum, 1998). Brevifoliol and its analogue, 9-deacetyl-9-benzoyl-10-debenzoyl-brevifoliol (13b) were isolated from *T. brevifolia* (Chen *et al.*, 1994).

A bevifoliol analogue 13m was isolated from *T. wallichiana* (Chattopadhay *et al.*, 1999). 9-Deacetyl-9-benzoyl-10-debenzoyltaxchinin A (13h) was isolated from the stem of bark of *T. baccata* along with 13b and 13c (Guo *et al.*, 1996). Teixidol (13-1b) was isolated from the leaves of *T. baccata*. Its structure, which features an 11(15 → 1)abeo-taxane skeleton consisting of a 5/7/6-membered ring system (Soto *et al.*, 1996). Two abeotaxanes (13-1c and 1d) with a taxoid (3-1d) were obtained from *T. yunnanensis* and *T. chinensis var. mairei* (Shi *et al.*, 1999). Taxaspine O (13-2b) was isolated from the stem of *T. cuspidata* (Kobayashi *et al.*, 1996). Taxuspinanane A (11-2e) and B (13-2d) were isolated from *T. cuspidata var. nana*. Their structures were elucidated by extensive 2D NMR and MS spectroscopic analysis (Morita *et al.*, 1997a) (Figures 3.40 and 3.41).

CINNAMOYL OR AMINOACYL GROUP AT C-5

Taxuspine A (13-3f) was isolated from the stems of Japanese yew *T. cuspidata* (Kobayashi *et al.*, 1994). The structures of taxchinin E (13-3g), H (13-3d), I (14-1h), J (14-2b), K (14m), taxchin A (1-4c) and taxchin B (1-6a), isolated from *T. chinensis*, have been elucidated by means of NMR spectroscopy. The first five possess a rearranged taxoid skeleton, and the rest have an ordinary taxane framework (Tanaka *et al.*, 1994). Taxuspine M (13-3a) was isolated from the stems of the Japanese yew *T. cuspidata* (Wang *et al.*, 1996b). Taxuspines E (10-2b), F (1-2c), G (1-2d),

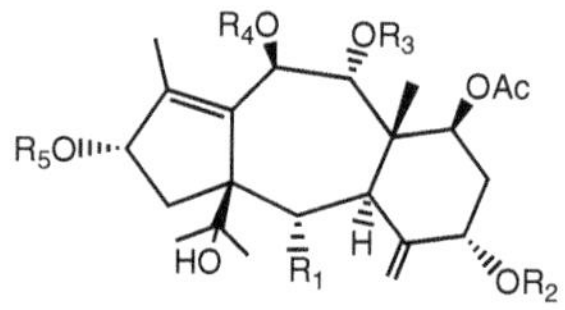

		R_1	R_2	R_3	R_4	R_5	Plant source	References
Breviforiol	13a	H	H	Ac	Bz	H	*T. wallichiana* *T. brevifolia*	Georg *et al.* (1993) Appendino *et al.* (1993) Georg *et al.* (1993)
9-Deacetyl-9-benzoyl-10-debenzoyl-brevifoliol	13b	H	H	Bz	H	H	*T. brevifolia*	Chen *et al.* (1994)
Taxchinin A	13c	OAc	H	Ac	Bz	H	*T. chinensis*	Fuji *et al.* (1992)
13-acetyl-2-deacetoxy-10-debenzoyltaxchinin A (Taxawallin F)	13d	H	H	Ac	H	Ac	*T. wallichiana*	Zhang *et al.* (1995b)
10-acetyl-2-deacetoxy-10-debenzoyltaxchinin A (Taxawallin H)	13e	H	H	Ac	Ac	H	*T. wallichiana*	Zhang *et al.* (1995b)
10-Debenzoyl-2α-acetoxy-brevifoliol	13f	OAc	H	Ac	H	H	*T. wallichiana*	Barbaroni *et al.* (1995a)
Taxchinin G	13g	OAc	H	Ac	H	Ac	*T. chinensis*	Li *et al.* (1993)
9-Deacetyl-9-benzoyl-10-debenzoyltaxchinin A	13h	OAc	H	Bz	H	H	*T. baccata*	Guo *et al.* (1996)
9-Benzoyl-2-deacetoxy-9-deacetyl-10-debenzoyl-10,13-diacetyltaxchinin A, Taxawallin D	13i	H	H	Bz	Ac	Ac	*T. wallichiana*	Fuji *et al.* (1992)
13-acetyl-9-deacetyl-9-benzoyl-10-debenzoyltaxchinin A	13j	OAc	H	Bz	H	Ac	*T. chinensis*	Shi *et al.* (1998b)
2α-Acetoxybrevifoliol	13k	OAc	H	Ac	Bz	H	*T. baccata*	Appendino *et al.* (1993)
Taxchinin D	13l	OAc	H	Ac	Bz	Ac	*T. chinensis*	Tanaka *et al.* (1993)
5,10,13-triacetyl-10-debenzoylbrevifoliol	13m	H	Ac	Ac	Ac	Ac	*T. wallichiana*	Chattopadhyay *et al.* (1999)
2-deacetyl-2α-benzoyl-5,13-diacetyltaxchinin A	13n	OBz	Ac	Ac	Bz	Ac	*T. brevifolia*	Rao *et al.* (1998)

Figure 3.40 A hydroxyl group or acetate at C-5.

H (**8-1b**), and J (**13-3b**), have been isolated together with known taxoids containing a series of taxol-related compounds from stems and leaves of the Japanese yew *T. cuspidata* sieb. et zucc. The structures and cytotoxicities of these taxoids (1–23) are described by Kobayashi *et al.* (1995). An abeo-taxane (**13-4a**) was isolated from the leaves of European yew *T. baccata* by Doss *et al.* (1997). Taxacustone (**13-2a**), 5α-*O*-β-D-gluco-pyranosyl)-10β-benzoyltaxacustone (**13-4b**) and an abeo-taxayuntin (**14j**) were isolated from the needles and stems of *T. cuspidata* by Tong *et al.* (1995). Three abeo-taxanes (**13-3f**, **13-4c**, and **13-4d**) were isolated from the needles of *T. brevifolia* (Chu *et al.*, 1994). The taxoid chinentaxunine (**13-3h**) has been isolated from the seeds of the Chinese yew *T. chinensis* (Shen *et al.*, 1999b). A new taxane diterpenoid from the seeds of *T. yunnanensis* (Shi *et al.*, 2000b) (Figure 3.42).

Taxoids with an oxetane ring

SIMPLE TAXOIDS WITH AN OXETANE RING

Taxumairol K (**14b**) was isolated from the roots of Formosan *T. mairei*. It exhibited mild cytotoxicity against Hela tumor cells (Shen *et al.*, 1998). Two taxoids were isolated from *T. brevifolia*.

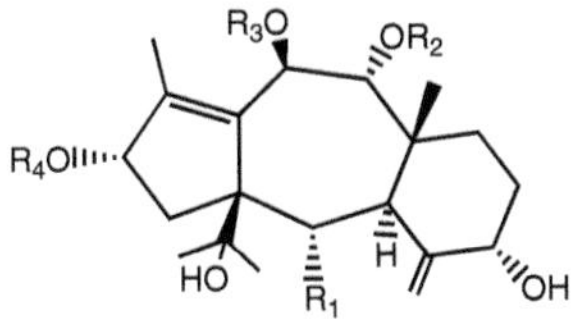

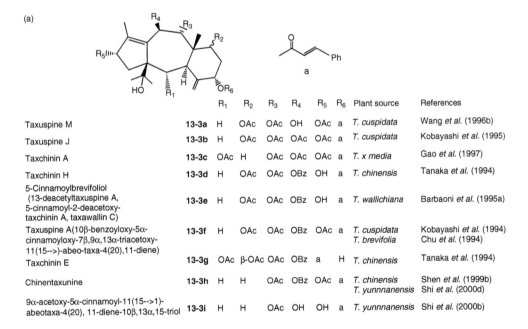

		R₁	R₂	R₃	R₄	Plant source	References
7-Debenzoyloxy-10-deacetylbrevifoliol	**13-1a**	H	Ac	H	H	*T. walichiana*	Appendino *et al.* (1993)
	13-1b	OAc	Ac	Ac	Ac	*T. baccata*	Soto *et al.* (1996)
Teixidol 5α-hydroxy-9α,10β,13α-triacetoxy-11(15-->1)abeotaxa-4(20),11-diene	**13-1c**	H	Ac	Ac	Ac	*T. yunnanensis* *T. chinensis var. mairei*	Shi *et al.* (1999f)
5α,13α-dihydroxy-9α,10β,-diacetoxy-11(15-->1)abeotaxa-4(20),11-diene	**13-1d**	H	Ac	Ac	H	*T. yunnanensis* *T. chinensis var. mairei*	Shi *et al.* (1999f)

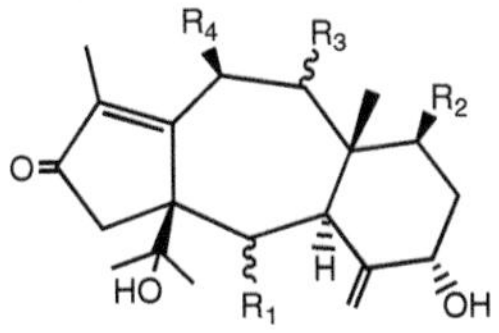

		R₁	R₂	R₃	R₄	Plant source	References
7-Debenzoyloxy-10-deacetylbrevifoliol(taxacustone)	**13-2a**	β-OAc	H	β-OAc	OH	*T. cuspidata*	Tong *et al.* (1995)
Taxuspine O	**13-2b**	α-OAc	OH	α-OAc	OAc	*T. cuspidata*	Kobayashi *et al.* (1996)
Taxuspine Y	**13-2c**	α-OAc	H	α-OAc	OBz	*T. cuspidata* (Japan)	Shigemori *et al.* (1997)
Taxuspinanane B	**13-2d**	α-OAc	OAc	α-OAc	OBz	*T. cuspidata*	Morita *et al.* (1997a)

Figure 3.41 Taxoids with a hydroxyl group or acetate at C-5.

(a)

		R₁	R₂	R₃	R₄	R₅	R₆	Plant source	References
Taxuspine M	**13-3a**	H	OAc	OAc	OH	OAc	a	*T. cuspidata*	Wang *et al.* (1996b)
Taxuspine J	**13-3b**	H	OAc	OAc	OAc	OAc	a	*T. cuspidata*	Kobayashi *et al.* (1995)
Taxchinin A	**13-3c**	OAc	H	OAc	OAc	OAc	a	*T. x media*	Gao *et al.* (1997)
Taxchinin H	**13-3d**	H	OAc	OAc	OBz	OH	a	*T. chinensis*	Tanaka *et al.* (1994)
5-Cinnamoylbrevifoliol (13-deacetyltaxuspine A, 5-cinnamoyl-2-deacetoxy-taxchinin A, taxawallin C)	**13-3e**	H	OAc	OAc	OBz	OH	a	*T. wallichiana*	Barbaoni *et al.* (1995a)
Taxuspine A(10β-benzoyloxy-5α-cinnamoyloxy-7β,9α,13α-triacetoxy-11(15-->)-abeo-taxa-4(20),11-diene)	**13-3f**	H	OAc	OAc	OBz	OAc	a	*T. cuspidata* *T. brevifolia*	Kobayashi *et al.* (1994) Chu *et al.* (1994)
Taxchinin E	**13-3g**	OAc	β-OAc	OAc	OBz	a	H	*T. chinensis*	Tanaka *et al.* (1994)
Chinentaxunine	**13-3h**	H	H	OAc	OBz	OAc	a	*T. chinensis* *T. yunnnanensis*	Shen *et al.* (1999b) Shi *et al.* (2000d)
9α-acetoxy-5α-cinnamoyl-11(15-->1)-abeotaxa-4(20), 11-diene-10β,13α,15-triol	**13-3i**	H	H	OAc	OH	OH	a	*T. yunnnanensis*	Shi *et al.* (2000b)

Figure 3.42a Cinnamoyl group at C-5.

(b)

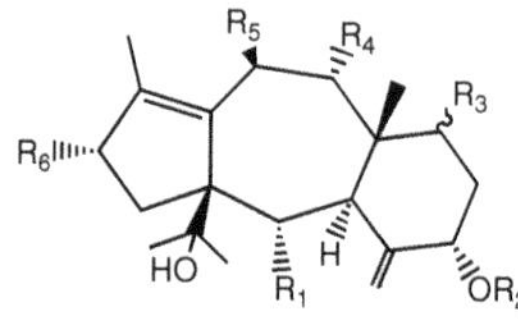

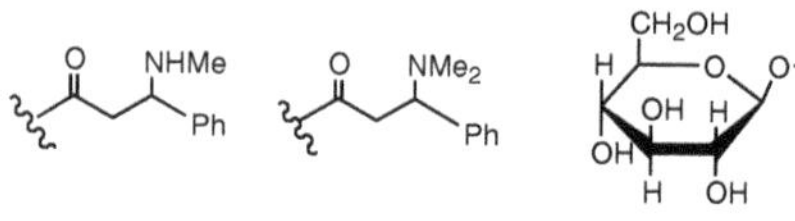

		R_1	R_2	R_3	R_4	R_5	R_6	Plant source	References
(–)-2α-acetoxy-2',7-dideacetoxy-1-hydroxy-11(15-->1)-abeo-austrospicatine	13-4a	α-OAc	a	H	α-OAc	OAc	OAc	*T. baccata*	Doss *et al.* (1997)
5α-O-(β–D-glucopyranosyl)-10β-β-benzoyltaxacustone	13-4b	β-OAc	c	H	β-OAc	OBz	=O	*T. cuspidata*	Tong *et al.* (1995)
10β-benzoyloxy-1β-hydroxy-5α-(3'-methylamino-3'phenyl)propanoxy-7β,9α,13α-triacetoxy-11(15-->1)-abeo-taxa-4(20), 11-diene	13-4c	H	b	OAc	α-OAc	OBz	OAc	*T. brevifolia*	Chu *et al.* (1994)
10β-benzyloxy-5α-(3'dimethylamino-3'-phenyl)propanoxy-1β-hydroxyl-7,9,13-triacetoxy-11(15-->1)-abeo-taxa-4(20),11-diene	13-4d	H	a	OAc	α-OAc	OBz	OAc	*T. brevifolia*	Chu *et al.* (1994)
13α-acetoxy-5α-(3'-dimethylamino-3'-phenyl)-propionyloxy-11(15-->1)-abeotaxa-4(20),11-diene-9α,10β-diol	13-4e	H	b	H	OH	OH	OAc	*T. yunnanensis*	Shi *et al.* (1999i)

Figure 3.42b Aminoacyl group at C-5.

They were 10β-benzoyloxy-2α,4α-diacetoxy-5β,20-epoxy-1β,7β,9α,13α-tetrahydroxy-11 (15 → 1)-abeotaxane (14c) and 7,9-deacetyl baccatin IV (10-2g), along with ponaterone A and three other taxanes: baccatin IV (10-1a), 9-dihydro-13-acetylbaccatin III (10-2h) and 1β,7β-dihydroxy-4β,20-epoxy-2α,5α,9α,10β,13α-penta-acetoxy-tax-11-ene (9-1f) (Rao *et al.*, 1996b). 13-Acetyl-13-decinnamoyltaxchinin B (14l) was isolated from the needles from the Himmalayan yew, *T. baccata* (Das *et al.*, 1995c). The needles of *T. wallichana* yielded 10,13-deacetylabeobaccatin IV (14a) (Chatopadhyay *et al.*, 1995a). Two taxane diterpenoids, taxayunnansin A (14i), were isolated from the roots of *T. yunnanensis*. Taxayunnanane E was identified as a unique taxane with a 2,20-oxetane ring (14-5a) (Zhang *et al.*, 1995a). The stem bark of *T. wallichiana* gave an *abeo*baccatin IV derivative (14g) (Suni *et al.*, 1996). The bark of *T. brevifolia* showed the presence of an *abeo*-taxoid (14c) and other taxoids (10-2g, 1a, and 2c) (Rao *et al.*, 1996b). Two *abeo*taxoids (14h and 14-1d) were isolated from the leaves and stems of *T. yunnanensis* (Yue *et al.*, 1995a).

The roots of *T. media* gave four *abeo*taxoids (14-1f, 1h, 1I, and 7a) (Barboni *et al.*, 1994a). Two *abeo*taxoids (14-1a and 13b) were isolated from *T. brevifolia* (Chen and Kingston, 1991). Two *abeo*taxoids (14-1q and 3c) were isolated from *T. chinensis* by Fuji *et al.* (1993). Four *abeo*taxoids (14m, 1h, 1q, and 3b) were isolated from *T. chinensis* by Tanaka *et al.* (1994). A 11(15 → 1)-*abeo*taxane diterpene, 7,9-dideacetyltaxayuntin (14-1n), and 11 known taxane diterpenes were isolated from the bark of *T. yunnanensis* (Zhong *et al.*, 1996). Nine known taxoids were isolated from the roots of *T. wallichiana*. One of them was 14-1h (Tanaka *et al.*, 1994). Taxayuntin (14j) was isolated from the needles and stems of *T. cuspidata* (Tong *et al.*, 1995). An abeo-taxoid (14-1n) was isolated from the bark of *T. yunnanensis* (Zhong *et al.*, 1996). Taxayuntin E (14-1d) and F (14h) were isolated from the leaves and stems of *T. yunnanensis* (Yue *et al.*, 1995). 7,13-dideacetyl-9,10-debenzoyltaxchinin C (14-1a) and 9-deacetyl-9-benzoyl-10-debenzoyl-brevifoliol (13b) were isolated from *T. brevifolia* (Chen and Kingston, 1991) (Figures 3.43 and 3.44).

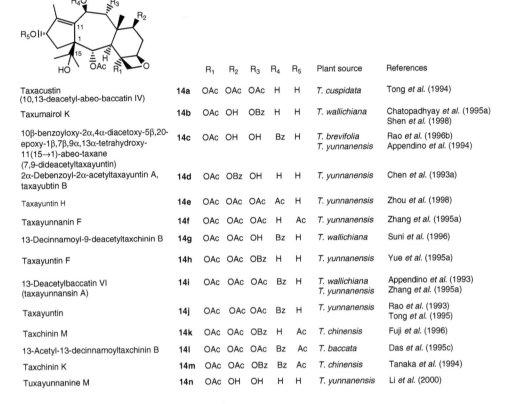

		R₁	R₂	R₃	R₄	R₅	Plant source	References
Taxacustin (10,13-deacetyl-*abeo*-baccatin IV)	**14a**	OAc	OAc	OAc	H	H	*T. cuspidata*	Tong *et al.* (1994)
Taxumairol K	**14b**	OAc	OH	OBz	H	H	*T. wallichiana*	Chatopadhyay *et al.* (1995a) Shen *et al.* (1998)
10β-benzoyloxy-2α,4α-diacetoxy-5β,20-epoxy-1β,7β,9α,13α-tetrahydroxy-11(15→1)-*abeo*-taxane (7,9-dideacetyltaxayuntin)	**14c**	OAc	OH	OH	Bz	H	*T. brevifolia* *T. yunnanensis*	Rao *et al.* (1996b) Appendino *et al.* (1994)
2α-Debenzoyl-2α-acetyltaxayuntin A, taxayubtin B	**14d**	OAc	OBz	OH	H	H	*T. yunnanensis*	Chen *et al.* (1993a)
Taxayuntin H	**14e**	OAc	OAc	OAc	Ac	H	*T. yunnanensis*	Zhou *et al.* (1998)
Taxayunnanin F	**14f**	OAc	OAc	OAc	H	Ac	*T. yunnanensis*	Zhang *et al.* (1995a)
13-Decinnamoyl-9-deacetyltaxchinin B	**14g**	OAc	OAc	OH	Bz	H	*T. wallichiana*	Suni *et al.* (1996)
Taxayuntin F	**14h**	OAc	OAc	OBz	H	H	*T. yunnanensis*	Yue *et al.* (1995a)
13-Deacetylbaccatin VI (taxayunnansin A)	**14i**	OAc	OAc	OAc	Bz	H	*T. wallichiana* *T. yunnanensis*	Appendino *et al.* (1993) Zhang *et al.* (1995a)
Taxayuntin	**14j**	OAc	OAc	OAc	Bz	H	*T. yunnanensis*	Rao *et al.* (1993) Tong *et al.* (1995)
Taxchinin M	**14k**	OAc	OAc	OBz	H	Ac	*T. chinensis*	Fuji *et al.* (1996)
13-Acetyl-13-decinnamoyltaxchinin B	**14l**	OAc	OAc	OAc	Bz	Ac	*T. baccata*	Das *et al.* (1995c)
Taxchinin K	**14m**	OAc	OAc	OBz	Bz	Ac	*T. chinensis*	Tanaka *et al.* (1994)
Tuxayunnanine M	**14n**	OAc	OH	OH	H	H	*T. yunnanensis*	Li *et al.* (2000)

Figure 3.43 Simple taxoids with an oxetane ring.

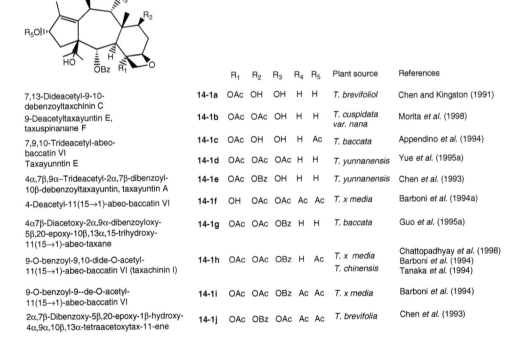

		R₁	R₂	R₃	R₄	R₅	Plant source	References
7,13-Dideacetyl-9-10-debenzoyltaxchinin C	**14-1a**	OAc	OH	OH	H	H	*T. brevifoliol*	Chen and Kingston (1991)
9-Deacetyltaxayuntin E, taxuspinanane F	**14-1b**	OAc	OAc	OH	H	H	*T. cuspidata var. nana*	Morita *et al.* (1998)
7,9,10-Trideacetyl-*abeo*-baccatin VI	**14-1c**	OAc	OH	OH	H	Ac	*T. baccata*	Appendino *et al.* (1994)
Taxayunntin E	**14-1d**	OAc	OAc	OAc	H	H	*T. yunnanensis*	Yue *et al.* (1995a)
4α,7β,9α–Trideacetyl-2α,7β-dibenzoyl-10β-debenzoyltaxayuntin, taxayuntin A	**14-1e**	OAc	OBz	OH	H	H	*T. yunnanensis*	Chen *et al.* (1993)
4-Deacetyl-11(15→1)-*abeo*-baccatin VI	**14-1f**	OH	OAc	OAc	Ac	Ac	*T. x media*	Barboni *et al.* (1994a)
4α7β-Diacetoxy-2α,9α-dibenzoyloxy-5β,20-epoxy-10β,13α,15-trihydroxy-11(15→1)-*abeo*-taxane	**14-1g**	OAc	OAc	OBz	H	H	*T. baccata*	Guo *et al.* (1995a)
9-O-benzoyl-9,10-dide-O-acetyl-11(15→1)-*abeo*-baccatin VI (taxachinin I)	**14-1h**	OAc	OAc	OBz	H	Ac	*T. x media* *T. chinensis*	Chattopadhyay *et al.* (1998) Barboni *et al.* (1994) Tanaka *et al.* (1994)
9-O-benzoyl-9--de-O-acetyl-11(15→1)-*abeo*-baccatin VI	**14-1i**	OAc	OAc	OBz	Ac	Ac	*T. x media*	Barboni *et al.* (1994)
2α,7β-Dibenzoxy-5β,20-epoxy-1β-hydroxy-4α,9α,10β,13α-tetraacetoxytax-11-ene	**14-1j**	OAc	OBz	OAc	Ac	Ac	*T. brevifolia*	Chen *et al.* (1993)

Figure 3.44 Simple taxoids with an oxetane ring.

		R_1	R_2	R_3	R_4	R_5	Plant source	References
2α-deacetyl-2α-benzoyl-13α-acetyltaxayuntin, taxayuntin C (2α,10β-dibenzoxy-5β,20-epoxy-1β-hydroxy-4α,7β,9α,13α-tetraacetoxytax-11-ene)	14-1k	OAc	OAc	OAc	Bz	Ac	*T. yunnanensis* *T. brevifolia*	Chen *et al.* (1993) Huang *et al.* (1998)
2,7-dideacetyl-2,7-dibenzoyl-taxayunnanine F taxayunnanine F	14-1l	OAc	OBz	OAc	H	Ac	*T. brevifolia*	Rao and Juchum (1998)
7-Deacetyltaxayuntin D	14-1m	OAc	OH	OBz	Bz	Ac	*T. brevifolia*	Rao and Juchum (1998)
7,9-Dideacetyltaxayuntin	14-1n	OAc	OH	OH	Bz	H	*T. yunnanensis*	Zhong *et al.* (1996)
7-Deacetyl-7-benzoyltaxchinin I	14-1o	OAc	OBz	OBz	H	Ac	*T. brevifolia*	Rao and Juchum (1998)
7-Deacetyl-7-benzoyl-taxayuntin C	14-1p	OAc	OBz	OAc	Bz	Ac	*T. brevifolia*	Rao and Juchum (1998)
9-Deacetyl-9-benzoyltaxayunntin C, taxayunntin D (Taxchinin C)	14-1q	OAc	OAc	OBz	Bz	Ac	*T. yunnanensis* *T. chinensis*	Chen *et al.* (1993) Fuji *et al.* (1993)

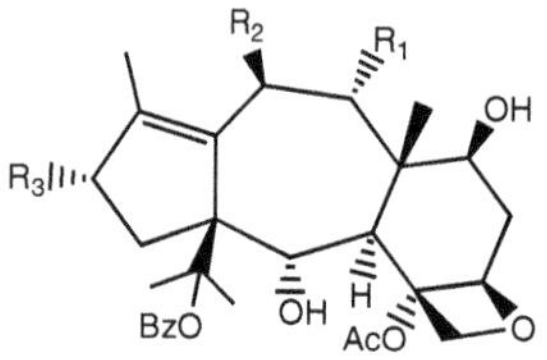

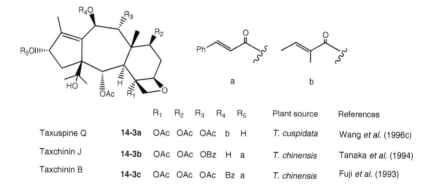

		R_1	R_2	R_3	Plant source	References
15-Benzoyl-10-deacetyl-2-debenzoyl-10-dehydro-abeo-baccatin III	14-2a	=O	=O	OH	*T. canadensis*	Zhang *et al.* (2000)
155-Benzoyl-2-debenzoyl-7,9-dideacetyl-abeo-baccatin VI	14-2b	OH	OAc	OAc	*T. canadensis*	Zhang *et al.* (2000)

Figure 3.44 (Continued).

		R_1	R_2	R_3	R_4	R_5	Plant source	References
Taxuspine Q	14-3a	OAc	OAc	OAc	b	H	*T. cuspidata*	Wang *et al.* (1996c)
Taxchinin J	14-3b	OAc	OAc	OBz	H	a	*T. chinensis*	Tanaka *et al.* (1994)
Taxchinin B	14-3c	OAc	OAc	OAc	Bz	a	*T. chinensis*	Fuji *et al.* (1993)

Figure 3.45 Taxoids with an oxetane ring and complex C-13 side-chain.

AN OXETANE RING AND A COMPLEX C-13 SIDE-CHAIN

Taxuspine Q (14-3a) was isolated from the Japanese yew *T. cuspidata* by Wang *et al.* (1996). Taxchinin B (14-3c) was isolated from the stems of *T. chinensis* by Fuji *et al.* (1993) (Figure 3.45).

MISCELLANEOUS ETHER LINKAGE GROUP

The needles of *T. wallichiana* gave the novel oxetane-type diterpenoids (10-6b and 14-4a) (Appendino *et al.*, 1993e). Two taxoid diterpenoids, taxayunnanine E (14-5a) and

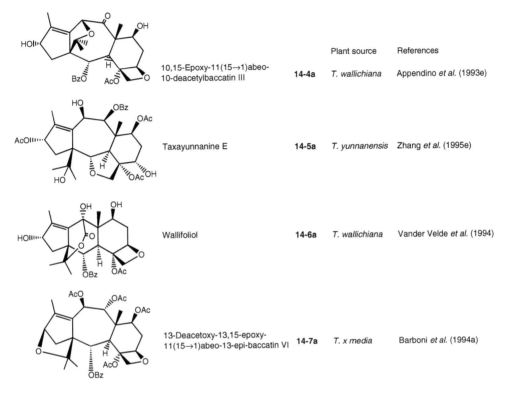

	Plant source	References
10,15-Epoxy-11(15→1)abeo-10-deacetylbaccatin III **14-4a**	*T. wallichiana*	Appendino *et al.* (1993e)
Taxayunnanine E **14-5a**	*T. yunnanensis*	Zhang *et al.* (1995e)
Wallifoliol **14-6a**	*T. wallichiana*	Vander Velde *et al.* (1994)
13-Deacetoxy-13,15-epoxy-11(15→1)abeo-13-epi-baccatin VI **14-7a**	*T. x media*	Barboni *et al.* (1994a)

Figure 3.46 Taxoids with miscellaneous ether linkage group.

taxayunnansin A (14i) were isolated from *T. yunnanensis* (Zhang *et al.*, 1995). Wallifoliol (14-6a) was isolated with known taxoids (taxol, cephalomannine, 10-deacetylbaccatin III, brevifoliol, 2-acetoxy-brevifoliol = taxchinin A) from the needles of Himalayan *T. wallichiana*. Wallichiana is the first diterpene to be found in nature with this particular 5/6/6/6/4 ring system (Vander Velde *et al.*, 1994). The roots of *T. x media* gave four taxoids of the *abeo* 11(15 → 1)-type (14-7a, 1f, 1h, and 1I). Taxoid 14-7a is conformationally fixed, whereas the latter three taxoids exist in solution as a mixture of rotamers (Barboni *et al.*, 1994a) (Figure 3.46).

AN OPENED OXETANE RING

Two 11(15 → 1) *abeo*taxoids (14e and 15b) were isolated from *T. yunnanensis* by Zhou *et al.* (1995). A taxoid (10-4c) and two *abeo*taxoids (15c and 15e) were isolated from the twigs of *T. mairei* (Yang *et al.*, 1999). Yunantaxusin A (15f), an 11(15 → 1)*abeo* taxane-type diterpene with an opened oxetane ring, has been isolated from the leaves and stems of *T. yunnanensis* (Zhang *et al.*, 1994) (Figure 3.47).

Taxoids with an C-4(20) epoxide

From *T. chinensis var. mairei* 16a was isolated by Shi *et al.* (1999). Taxuchinin A (16b) was isolated from the bark of *T. chinensis* by Zhang *et al.* (1994a,b) (Figure 3.48).

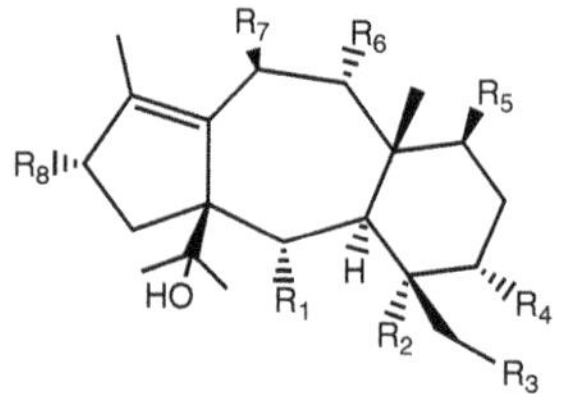

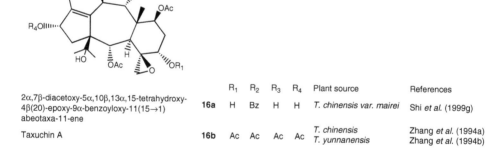

		R_1	R_2	R_3	R_4	R_5	R_6	R_7	R_8	Plant source	References
Taxayuntin G	**15a**	OAc	H	OH	OAc	H	OAc	OH	OH	*T. yunnanensis*	Yue *et al.* (1995b)
Taxayuntin J	**15b**	OH	H	OAc	OAc	H	OAc	OAc	OH	*T. yunnanensis*	Zhou *et al.* (1998)
Taxumain A	**15c**	OAc	H	OAc	OAc	H	OAc	OH	OH	*T. mairei*	Yang *et al.* (1999)
Taxuchin B	**15d**	OAc	OH	Cl	OH	H	OBz	OH	OH	*T. chinensis*	Fang *et al.* (1997)
Taxumain B	**15e**	OAc	H	OAc	OAc	H	OAc	OH	OAc	*T. mairei*	Yang *et al.* (1999)
Yunantaxusin A	**15f**	OH	OH	OAc	OAc	H	OAc	OBz	OH	*T. yunnanensis*	Zhang *et al.* (1994)
Tuxayunnanine K	**15g**	OH	OH	OBz	OAc	OH	OH	OH	OH	*T. yunnanensis*	Li *et al.* (2000)
Tuxayunnanine L	**15h**	OBz	OH	OAc	OH	OH	OH	OH	OH	*T. yunnanensis*	Li *et al.* (2000)
Tuxayunnanine N	**15i**	OH	H	OH	OAc	OAc	OAc	OH	OH	*T. yunnanensis*	Li *et al.* (2000)
Tuxayunnanine O	**15j**	OH	H	OH	OAc	OAc	OAc	OAc	OH	*T. yunnanensis*	Li *et al.* (2000)

Figure 3.47 Opened oxetane ring.

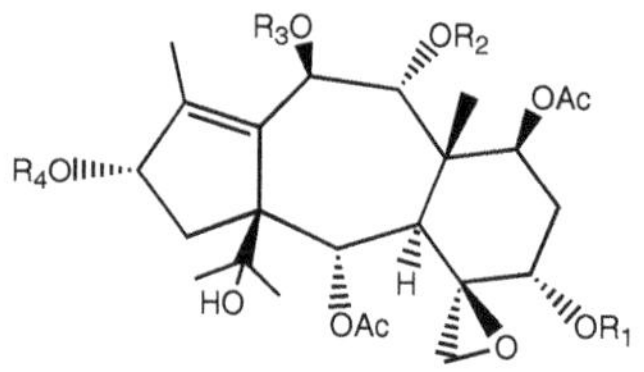

		R_1	R_2	R_3	R_4	Plant source	References
2α,7β-diacetoxy-5α,10β,13α,15-tetrahydroxy-4β(20)-epoxy-9α-benzoyloxy-11(15→1) abeotaxa-11-ene	**16a**	H	Bz	H	H	*T. chinensis var. mairei*	Shi *et al.* (1999g)
Taxuchin A	**16b**	Ac	Ac	Ac	Ac	*T. chinensis* *T. yunnanensis*	Zhang *et al.* (1994a) Zhang *et al.* (1994b)

Figure 3.48 Taxoids with an C-4(20) epoxide.

Structures of 2(3→20)-abeotaxoids with a C-4(20) double bond

Hydroxyl or acetate group at C-5

Two taxoids (**17a** and **17c**) were isolated from *T. x media* by Rao *et al.* (1996). Two rearranged taxoids, taxin B (**17d**), and 2-deacetyltaxin B (**17b**), were isolated from *T. yunnanensis* (Yue *et al.*, 1995c). Three taxoids were isolated from the bark of the Chinese yew, *T. mairei* (Shi *et al.*, 1999a) (Figure 3.49).

Ketonized at C-9 and an aminoacyl or cinnamoyl group at C-5

Taxuspinanane H (**17-1a**) was isolated from *T. cuspidata var. nana* by Morita *et al.* (1998a). Taxuspine B (**17-1b**) was isolated from the Japanese yew *T. cuspidata* by Kobayashi *et al.* (1994).

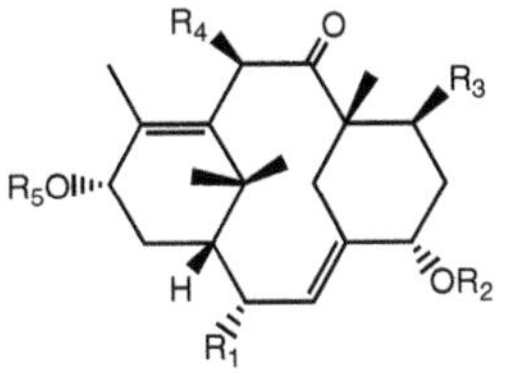

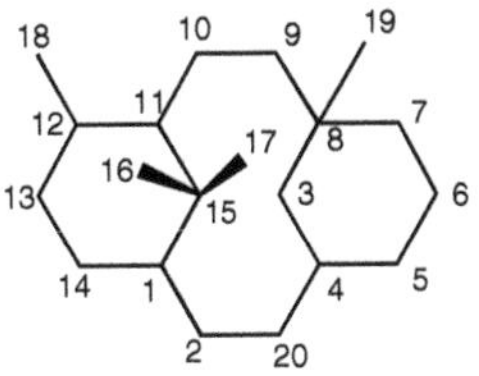

2(3→20)-abeotaxoid skeleton

	R_1	R_2	R_3	R_4	R_5	Plant source	References
Deaminoacyltaxine A	**17a**	OAc	H	OH	OH	Ac	*T. baccata*	Appendino *et al.* (1994)
2-Deacetyltaxine B	**17b**	OH	H	OAc	OAc	Ac	*T. yunnanensis*	Yue *et al.* (1995)
2α,7β,13α-triacetoxy-5α,10β-dihydroxy-9-keto-2(3→20)-abeo-taxane (taxaspine W)	**17c**	OAc	H	OAc	OH	Ac	*T. x media* *T. cuspidata*	Rao *et al.* (1996a) Hosoyama *et al.* (1996)
5α-Hydroxy-2α,7β,10β,13α-tetraacetoxy-2(3→20)abeotaxa-4(20),11-dien-9-one (Taxine B)	**17d**	OAc	H	OAc	OAc	Ac	*T. yunnanensis* *T. chinensis*	Yue *et al.* (1995c) Shi *et al.* (1999c)
2α5α,7β,13α-Tetraacetoxy-10β-hydroxyl-2(3→20) abeotaxane-9-one	**17e**	OAc	Ac	OAc	OH	Ac	*T. mairei*	Shi *et al.* (1999c)
7β,13α-Diacetoxy-2α,5α,10β-trihydroxyl-2(3→20) abeotaxane-9-one	**17f**	OH	H	OAc	OH	Ac	*T. mairei*	Shi *et al.* (1999c)
2α,7β-Diacetoxy-5α,10β,13α-trihydroxyl-2(3→20) abeotaxane-9-one	**17g**	OAc	H	OAc	OH	H	*T. mairei*	Shi *et al.* (1999c)

Figure 3.49 Taxoids with a hydroxyl or acetate group at C-5.

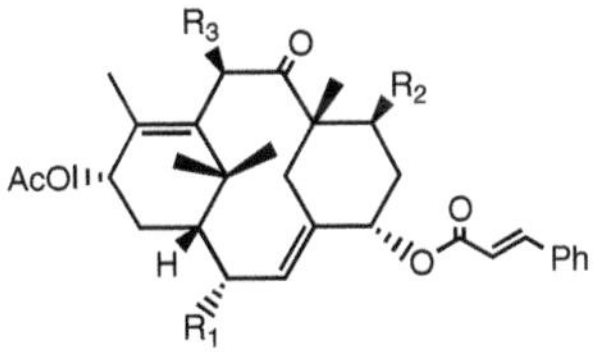

	R_1	R_2	R_3	Plant source	References
Taxuspinanane H, deaminoacyl-cinnamoyltaxine A	**17-1a**	OAc	OH	OH	*T. cuspidata*	Morita *et al.* (1998a)
Taxuspine B	**17-1b**	OAc	OAc	OH	*T. cuspidata (Japan)* *T. baccata*	Kobayashi *et al.* (1994) Barboni *et al.* (1995b)
5α-Cinnamoyl-7β,10β,13α-triacetoxy-2(3→20)-abeotaxa-2α-hydroxy-4(20),11-dien-9-one(2-deacetyl-10-O-acetyltaxuspine B	**17-1c**	OH	OAc	OAc	*T. chinensis var. mairei* *T. yunnanensis*	Shi *et al.* (1999d)
5α-Cinnamoyloxy-10β,13α-diacetoxy-2(3→0)abeotaxa-2α,7β-dihydroxy-4(20),11-diene-9-one	**17-1d**	OH	OH	OAc	*T. chinensis var. mairei* *T. yunnanensis* *T. baccata*	Shi *et al.* (1999d) Barboni *et al.* (1995b)
5α-Cinnamoyloxy-2α,7β,10β,13α-tetraacetoxy-2(3→20)abeotaxa-4(20),11-diene-9-one	**17-1e**	OAc	OAc	OAc	*T. chinensis var. mairei* *T. yunnanensis* *T. baccata*	Shi *et al.* (1999d) Barboni *et al.* (1995b)

Figure 3.50 Taxoids with cinnamoyl group at C-5.

A new basic taxoid, 2-deacetyltaxine A (**17-2a**), has been isolated from the leaves of *T. baccata* (Poupat *et al.*, 1994).

Six taxane diterpenoids and three known 2(3 → 20)*abeo*taxoids were isolated from the seeds of *T. chinensis var. mairei* and of *T. yunnanensis*. Their structures were established as 2α,20-dihydroxy-9α-acetoxytaxa-4(20),11-dien-13-one (**1-3a**), 5α-cinnamoyloxy-7β,10β,

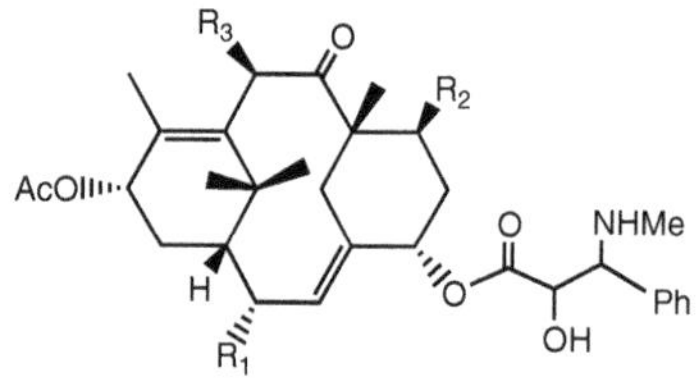

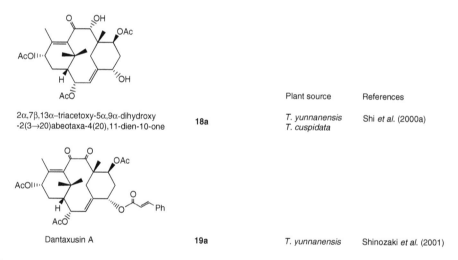

70   *Hideji Itokawa*

		R₁	R₂	R₃	Plant source	References

2-Deacetyltaxine A (taxine C)

17-2a OH OH OH *T. baccata* Poupat *et al.* (1994)

Taxine A

17-2b OAc OH OH *T. baccata* Graf and Berthold (1957) Graf *et al.* (1982)

7-O-Acetyltaxine A

17-2c OAc OAc OH *T. baccata* / *T. chinensis var. mairei* / *T. yunnanensis* Barboni *et al.* (1997) Shi *et al.* (1999d) Shi *et al.* (1999d)

2α,7β,10β,13α-tetraacetoxy-5α-phenylisoserinatoxy-2(3→20),11-dien-9-one (7,10-O-diacetyltaxine)

17-2d OAc OAc OAc *T. chinensis var. mairei* / *T. yunnanensis* Shi *et al.* (1999d)

Figure 3.51 Ketonized at C-9 with aminoacyl group at C-5.

		Plant source	References

2α,7β,13α–triacetoxy-5α,9α-dihydroxy-2(3→20)abeotaxa-4(20),11-dien-10-one   **18a**   *T. yunnanensis* / *T. cuspidata*   Shi *et al.* (2000a)

Dantaxusin A   **19a**   *T. yunnanensis*   Shinozaki *et al.* (2001)

Figure 3.52 Ketonized group at C-10 and C-9, 10.

13α-triacetoxy-2(3 → 20)*abeo*taxa-2α-ol-4(20),11-dien-9-one (**17-1c**), 5α-cinnamoyloxy-10β,13α-diacetoxy-2 (3 → 20) *abeo*-taxa-2α,7β-diol-4(20),11-dien-9-one (**17-1d**), 5α-cinnamoyloxy-2α,7β,10β,13α-tetra-acetoxy-2 (3 → 20) *abeo*taxa-4(20),11-dien-9-one (**17-1e**), 7-O-acetyltaxine A (**17-2c**), 2α, 7β, 10β, 13α-tetraacetoxy-5-phenylisoserinatoxy-2(3 → 20) *abeo*taxa-4(20),11-dien-9-one (**17-2d**), 5-O-acetyl- taxinine M (**7i**), 5α-hydroxy-2α,7β,10β, 13α-tetraacetoxy-2(3 → 20)*abeo*taxa-4(20),11-dien-9-one (**17d**), and 2-deacetyltaxine B (**17b**), on the basis of 1D and 2D NMR spectroscopic analysis. Compound 1, as a new one isolated from *T. chinensis var. mairei*, is a rare example of a taxoid with a 4,11-diene unit, which is considered to be an important intermediate in the biogenesis of taxoids. Compounds **17-1c, 1d, 1e, 2d,** and **7i** are new compounds, isolated from *T. yunnanensis* (Shi *et al.*, 1999d) (Figures 3.50 and 3.51).

Ketonized group at C-10

Recently, a taxoid with ketonized group at C-10 (**18a**) was isolated from *T. yunnanensis* and *T. cuspidata* (Shi *et al.*, 2000a). Moreover, dantaxusin A (**19a**) was isolated as a taxoid ketonized at C-9 and C-10 with dantaxusin B (**6p**) from *T. yunnanensis* by Shinozaki *et al.* (2001) (Figure 3.52).

Acknowledgement

The author would like to thank Dr H. Ohtsu for his valuable suggestions.

References

Ando, M., Sakai, J., Zhang, S., Watanabe, Y., Kosugi, K., Suzuki, T., and Hagiwara, H. (1997) A new basic taxoid from *Taxus cuspidata*. *J. Nat. Prod.*, **60**, 499–501.

Appendino, G. and Ozen, H.G. (1993) Four new taxanes from the needles of *Taxus baccata. Fitoterapia* **LXIV**(Suppl 1), 47–51.

Appendino, G., Gariboldi, P., Gabetta, B., Pace, R., Bombardelli, E., and Viterbo, D. (1992a) 14-Hydroxy-10-deacetylbaccatin III, a new taxane from Himalayan yew (*Taxus wallichiana* Zucc.). *J. Chem. Soc. Perkin.* 1, 2925–2929.

Appendino, G., Garibold, P., Pisetta, A., Bombardelli, E., and Gabetta, B. (1992b) Taxanes from *Taxus baccata*. *Phytochemistry*, **31**(120), 4253–4257.

Appendino, G., Ozen, H.C., Gariboldi, P., Gabetta, B., Bombaldeli, E. (1992c) *Fitoterapia* **64**(S1), 47–51.

Appendino, G., Lusso, P., Garibold, P., Bombardelli, E., and Gabetta, B. (1992d) A 3,11-Cyclotaxane from *Taxus baccata*. *Phytochemistry*, **31**(12), 4259–4262.

Appendino, G., Ozen, H.C., Fenoglio, I., Garibold, P., Gabetta, B., and Bombardelli, E. (1993a) Pseudoalkaloid taxanes from *Taxus baccata*. *Phytochemistry*, **33**(6), 1521–1523.

Appendino, G., Tagliapietra, S., Ozen, H.C., Gariboldi, P., Gabetta, B., Bombardeli, E. (1993b) Taxanes from the seeds of *Taxus baccata*. *J. Nat. Prod.*, **56**, 514–520.

Appendino, G., Barboni, L., Gariboldi, P., Gabetta, B., and Viterbo, D. (1993c) Revised structure of brevifoliol and some baccatin VI derivatives. *J. Chem. Soc. Chem. Commun.*, 1587–1589.

Appendino, G., Ozen, H.C., Gariboldi, P., Torregiani, E., Gabetta, B., Nizzola, R., and Bombardelli, E. (1993d) New oxetane-type taxanes from *Taxus wallichiana* Zucc. *J. Chem. Soc. Perkin Trans.*, 1, 1563–1566.

Appendino, G., Cravotto, G., Enriu, R., Jakupovic, Gariboldi, J.P., Gabetta, B., and Bombardelli, E. (1994a) Rearranged taxanes from *Taxus baccata*. *Phytochemistry*, **36**(2), 407–411.

Appendino, G., Cravotto, G., Enriu, R., Gariboldi, P., Barboni, L., Torregiani, E., Gabetta, B., Zini, G., and Bombardelli, E. (1994b) Taxoids from the roots of *Taxus x media* cv. Hicksii, *J. Nat. Prod.*, **57**(5), 607–613.

Baloglu, E. and Kingston, D.G.I. (1999) The taxane diterpenoids. *J. Nat. Prod.*, **62**, 1448–1472.

Barboni, L., Gariboldi, P., Torregiani, E., Appendino, G., Gabetta, B., Zini, G., and Bombardei, E. (1993) Taxanes from the needles of *Taxus wallichiana*. *Phytochemistry*, **33**(1), 145–150 (Structure should be revised to the corresponding 11(15 → 1)-abeo-taxoid, Appendino, G., Barboni, L., Gariboldi, P., Barboni, L., Gariboldi, P., Bombardelli, E., Gabetta, B., and Viterbo, D. (1993) *J. Chem. Soc. Chem. Comm.*, 1587–1589).

Barboni, L., Gariboldi, P., Torregiani, E., Appendino, G., Gravotto, G., Bombardelli, E., Gabetta, B., and Viterbo, D. (1994a) Chemistry and occurrence of taxane derivatives. Part 16. Rearranged taxoids from *Taxus x media* Rehd. Cv Hicksii. X-ray molecular structure of 9-O-benzoyl-9,10-dide-O-acetyl-11(15 → 1) abeo-baccatin VI. *J. Chem. Soc. Perkin Trans.* I, 3233–3238.

Barboni, L., Gariboldi, P., Torregiani, E., Appendino, G., Gabetta, B., and Bombadelli, E. (1994b) Taxol analogues from the roots of *Taxus x media*. *Phytochemistry*, **36**(4), 987–990.

Barboni, L., Gariboldi, P., Torregiani, E., Appendino, G., Varese, M., Gabetta, B., and Bombardelli, E. (1995a) Minor taxoids from *Taxus wallichiana*. *J. Nat. Prod.*, **58**(6), 934–939.

Barboni, L., Gariboldi, P., Appendino, G., Enriu, R., Gabetta, B., and Bombardelli, E. (1995b) New taxoids from *Taxus baccata* L. *Liebigs Ann.*, 345–349.

Begley, M.J., Frecknall, E.A., and Pattenden, G. (1984) *Acta Crystallogr.*, 40, 1745.

Beutler, J.A., Chmurny, G.M., Look, S.A., and Witherup, K.M. (1991) Taxinine M, A new tetracyclic taxane from *Taxus brevifolia*. *J. Nat. Prod.*, 54, 893.

Blechert, S. and Guenard, D. (1990) Taxus Alkaloids. In *The Alkaloids, Chemistry and Physiolosy*, vol. 39, edited by R.H.F. Manske, pp. 195–238. Academic Press, New York.

Blechert, S.S., Colin, M., Guenard, D., Picot, F., Potier, P., and Varenne, P. (1984) Mise en endence de nouveaux anlogues du taxol extraits de *Taxus baccata*. *J. Nat. Prod.*, 47(1), 131–137.

Bombaedelli, E., Gabetta, B., and Viterbo, D. (1993) *J. Chem. Soc. Chem. Comm.*, 1587–1589.

Chan, W.R., Halsall, T.G., Hornby, G.M., Oxford, A.W., Sabel, W., Bjamer, K., Fergson, G., and Robertson, J.M. (1966) Taxa-4(16),11-diene-5α,9α,10β,13α-tetraol, a new taxane deriivative from the heartwood of yew (*T. baccata* L.): X-ray analysis of a p-bromobenzoate derivative. *Chem. Commun.* 923.

Chattopadhyay, S.K. and Sharma, R.P. (1995) A taxane from the Himalayan yew, *Taxus wallichiana*. *Phytochemistry*, 39(4), 935–936.

Chattopadhyay, S.K., Shama, R.P., Appendino, G.A., and Garibold, P. (1995) A rearranged taxane from Himalayan yew. *Phytochemistry*, 39(4), 869–870.

Chattopadhyay, S.K., Saha, G.C., Kulshreshtha, M., Sharma, R.P., and Kumar, S. (1996) Studies on the Himalayan yew *Taxus wallichhiana*: Part IV – Isolation of dihydrotaxol and dibenzoyllated taxoids. *Indian J. Chem.*, 35B, 754–756.

Chattopadhyay, S.K., Saha, G.C., Kulshreshtha, M., Tripathi, V., Shama, R.P., and Mehta, V.K. (1998) The taxoid constituents of the roots of *Taxus wallichiana*. *Plant Med.*, 64, 287–288.

Chattopadhyay, S.K., Tripathi, V., Shamma, R.P., Shawl, A.S., Joshi, B.S., and Roy, R. (1999) A brevifoliol analogue from the Himalayan yew *Taxus wallichiana*. *Phytochemistry*, 50, 131–133.

Chauviere, G., Guenard, D., Picot, F., Senilh, V., and Potier, P. (1981) Analyse Structurale et Etude Biochimique de Produits Isoles de l'If: *Taxus baccata* L. (*Taxaceae*). *C.R. Acad. Sc. Paris, Serie II*, 293, 501.

Chauviere, G., Guenard, D., Pascard, C., Picot, F., Potier, P., and Prange, T. (1982) Taxagifine: New taxane derivative from *Taxus bacata* L. (*Taxaceae*). *J. Chem. Soc., Chem. Commun.*, 495.

Chen, W.M. (1990) Chemical constituents and biological activity of plants of *Taxus* genus. *Acta Pharm. Sin. (Yaoxue Xuebao)*, 25, 227.

Chen, R. and Kingston, D.G.I. (1994) Isolation and structure elucidation of new taxoids from *Taxus brevifolia*. *J. Nat. Prod.*, 57(7), 1017–1021.

Chen, Y.-C., Chen, C.-Y., and Shen, Y.-C. (1999) New taxoids from the seeds of *Taxus chinensis*. *J. Nat. Prod.*, 62, 149–151.

Chen, W.M., Zhou, J.Y., Zhang, P.L., and Fang, Q.C. (1993a) *Chin. Chem. Lett.*, 4, 695–698.

Chen, W.M., Zhou, J.Y., Zhang, P.L., and Fang, Q.C., (1993b) *Chin. Chem. Lett.*, 4, 699–702.

Chen, W.M., Zhang, P.L., and Zhou, J.Y. (1994) *Yaoxue Xuebao*, 29, 207–214.

Cheng, K., Fang, W., Yang, Y., Xu, H., Kong, M., He, W., and Fang, Q. (1996) C-14 Oxygenated taxanes from *Taxus yunnanensis* cell cultures. *Phytochemistry*, 42(1), 73–75.

Chiang, H.C. (1975) The constituents of *Taxus chinensis* Rehd. *Shi Ta Hsueh Pao*, 20, 147.

Chiang, H.C. Woods, M.C., Nakadaira, Y., Nakanishi, K. (1967) The structures of four new taxinine congeners, and a photochemical transannular reaction. *Chem. Commun.*, 1201.

Chu, A., Zajicek, J., Davin, L.B., Lewis, N.G., and Croteau, R.B. (1992) Mixed acetoxy-benzoxy taxane esters from *Taxus brevifolia*. *Phytochemistry*, 31(12), 4249–4252; Structure revision to the corresponding 11(15 → 1)-abeo-taxoid has been suggested, Huang, K., Liang, J., and Gunatilaka, A.A.L.(1998) *J. Chin. Pharm. University* (*Zhongguo Yaoke Daxue Xuebao*) 29, 259–266.

Chu, A., Davin, L.B., Zajicek, J., Lewist, N.G., and Croteau, R. (1993) Intramolecular acyl migrations in taxanes from *Taxus brevifolia*. *Phytochemistry*, 34(2), 473–476.

Chu, A., Furan, M., Davin, L.B., Zajicek, J., Towers, G.H.N., Soucy-Breau, C.M., Rettig, S.J., Croteau, R., and Lewis, N.G. (1994) Phenylbutanoid and taxane-like metabolites from needles of *Taxus brevifolia*. *Phytochemistry*, 36(4), 975–985.

Das, B., Rao, S.P., and Das, R. (1995a) Naturally occurring rearranged taxoids. *Planta Med.*, 61, 393–397.

Das, B., Rao, S.P., and Das, R. (1995b) Naturally occurring rearranged taxoids. *Planta Med.*, 61, 393–397.

Das, B., Rao, P., Srinivas, K.V.N.S., Yadav, J.S., and Das, R. (nee Chakrabarti): (1995c) A taxoid from needles of Himalayan *Taxus bacata*. *Phytochemistry*, 38(3), 671–674.

De Marcano, D.P. and Halsall, T.G. (1969) The Isolation of Seven New Taxane Derivatives from the Heartwood of Yew (*Taxus baccata* L.). *Chem.Comm.*, 1282.

De Marcano, D.P. and Halsall, T.G. (1970) The structure of the diterpenoid baccatin-I, the 4β,20-epoxide of 2α,5α,7β,9α,10β,13α-hexa-acetoxytaxa-4(20),11-diene. *J. Chem. Soc. D Chem. Commun.*, 1381.

De Marcano, D.P. and Halsall, T.G. (1975) The structure of some taxane diterpenoids, baccatin-III, -IV, and VI, and 1-dehydroxybaccatin-IV, possessing an oxetan ring. *J. Chem. Soc. Chem. Commun.*, 365.

De Marcano, D.P., Halsall, T.G., Castellano, E., and Hodder, O.J.R. (1970) Crystallographic structure determination of the diterpenoid baccatin-V, a naturally occurring oxetan with a taxane skeleton. *J. Chem. Soc. D Chem. Commun.*, 1382–1383.

Doss, R.P., Carney, J.R., Shanks, C.H., Jr., Williamsen, R.T., and Chamberlain, J.D. (1997) Two new taxoids from European yew (*Taxus baccata*) That act as pyrethroid insecticide synergists with the black vine weevil (*Otiorhnchus sulcatus*), *J. Nat. Prod.*, 60, 1130–1133.

Dukes, M., Eyre, D.H., Harrison, J.W., Scrowston, R.M., and Lythgoe, B. (1967) Taxine. Part V. The structure of taxicin II. *J. Chem. Soc. C.*, 448.

Ettouati, L., Ahond, A., Poupat, C., and Potier, P. (1989) Revision Structurale de la Taxine B, alcaloide. *Bull. Soc. Chem. Fr.*, 687.

Ettouati, L., Ahond, A., Convert, O., Poupat, C., and Potier, P. (1987) *Bull. Soc. Chim. Fr.*, 687.

Ettouati, L., Ahond, A., Convert, O., Laurent, D., Poupat, C., and Potier, P. (1988) Plantes de Nouvelle-Caledonie. 114. Taxanes isoles des feuilles d'*Austrotaxus spicata* Compton (*Taxacees*). *Bull. Soc. Chim. France*, 749.

Majoritaire des Feuilles de I'lf d'Europe, *Taxus baccata*. *J. Nat. Prod.* 54, 1455.

Fang, W.S., Fang, Q.C., and Liang, X.T. (1996) Bicyclic taxoids from needles of *Taxus chinensis*. *Planta Med.*, 62, 567–569.

Fang, W.S., Fang, Q.C., Liang, X.T., Lu, Y., Wu, N., and Zheng, Q.T. (1997) *Chin. Chem. Lett.*, 8, 231–232.

Fuji, K., Tanaka, K., and Li, B. (1992) Taxchinin A: a diterpenoid from *Taxus chinensis*. *Tetrahed. Lett.*, 33(51), 7915–7916.

Fuji, K., Yokoi, T., Shingu, T., Li, B., and Sun, H. (1996) *Chem. Pharm. Bull.*, 44, 1770–1774.

Fuji, K., Tanaka, K., Li, B., Shingu, T., Sun, H., and Taga, T. (1993) Novel diterpenoids from *Taxus chinensis*. *J. Nat. Prod.*, 56(9), 1520–1531.

Fukushima, M., Takeda, J., Fukamiya, N., Okano, M., Tagahara, K., Zhang, S.-X., Zhang, D.-C., and Lee, K.H. (1999) A new taxoid, 19-acetoxytaxagifine, from *Taxus chinensis*. *J. Nat. Prod.*, 62, 140–142.

Gabetta, B., Bellis, P., Pace, R., Appendino, G., and Viterbo, D. (1995a) 10-Deacetylbaccatin III analogue from *Taxus baccata*. *J. Nat. Prod.*, 58(10), 1508–1514.

Gabetta, B., Peterlongo, F., Zini, G., Barboni, L., Rafaini, G., Ranzuglia, Torregiani, E., Appendino,G., and Cravotto, G. (1995b) Taxanes from *Taxus x media*. *Phytochemistry*, 40(6), 1825–1828.

Gabetta, B., Orsini, P., Peterlongo, F., and Appendino, G. (1998) Paclitaxel analogues from *Taxus baccata*. *Phytochemistry*, 47(7), 1325–1329.

Gabetta, B., Fuzatti, N., Orsini, P., Peterlongo, F., Appendino, G., and Vander Velde, D.G. (1999) Paclitaxel analogues from *Taxus x media cv.* Hicksii, *J. Nat. Prod.*, 62, 219–223.

Gao, Y.-L., Zhou, J.-Y., Ding, Y., Fang, Q.-C. (1997) *Chin. Chem. Lett.*, 8, 1057–1058.

Georg, G., Gollapudi, S.R., Grunewald, G.L., Gunn, C.W., Himes, R.H., Rao, B.K., Liang, X.-Z., Mirhom, Y.W., and Mitscher, L.A. (1993) A reinvestigation of the taxol content of Himalayan *Taxus wallichiana* Zucc. and a revision of the structure of brevifoliol. *Bioorg. Med. Chem. Lett.*, 3(6), 1345–1348.

Graf, E. (1957) Taxin B, das amorphe Taxin und das Kristallisierte Taxinn A. *Pharmazeutische Zentralhalle* 96, 385.

Graf, E., Kirfel, A., Wolf, G.-J., and Breitmaier, E. (1982) Die Aufklarung von Taxin A aus *Taxus baccata* L. *Liebigs Ann. Chem.*, 376.

Grothaus, P.G., Bignami, G.S., O'Malley, S., Harada, K.E., Byrnes, J.B., Waller, D.F., Raybould, T.J.G., McGuire, M.T., and Alvarado, B.A. (1995) Taxane-specific monoclonal antibodies: Mesurement of taxol, bacatin III, and "Total taxanes" in *Taxus brevifolia* extracts by enzyme immunoassay. *J. Nat. Prod.*, 58(7), 1003–1014.

Gunawardana, G.P., Premachandran, U., Burres, N.S., Whittern, D.N., Henry, R., Spanton, S., and McAlpine, J.B. (1992) Isolation of 9-dihydro-13-acetylbaccatin III from *Taxus canadensis*. *J. Nat. Prod.*, 55(11), 1686–1689.

Guo, Y., Vanhaelen-Fastre, R., Diallo, B., Vanhaelen, M., Jaziri, M., Homes, J., and Ottinger, R. (1995a) Immunoenzymatic methods applied to the search for bioactive taxoids from *Taxus baccata. J. Nat. Prod.*, **58**, 1015–1023.

Guo, Y., Diallo, B., Jaziri, M., Vanhaelen-Fastre, R., and Vanhaelen, M. (1995b) Two new taxoids from the stem bark of *Taxus baccata. J. Nat. Prod.*, **58**(12), 1906–1912.

Guo, Y., .Diallo, Jaziri, M., Vanhaelen-Fastre, R., and Vanhaelen, M. (1996) Immunological detection and isolation of a new taxoid from the stem bark of *Taxus baccata. J. Nat. Prod.*, **59**, 169–172.

Ho, T.I., Lin, Y.C., Lee, G.H., Peng, S.M., Yeh, M.K., and Chen, F.C. (1987) Structure of Taiwanxan. *Acta Crystallogr. Sect.C: Cryst. Struct. Comm.*, **43**, 1380.

Hoffman, A. (2000) Science, **288**, 27–28.

Hosoyama, H., Innubushi, A., Katsui, T., Shigemori, H., and Kobayashi, J. (1996) Taxuspines U, V, and W, new taxane and related diterpenoids from the Japanese yew *Taxus cuspidata. Tetrahed.*, **52**(41), 13145–13150.

Huang, C.H.O., Kingston, D.G.I., Magri, N.F., and Samaranayake, G. (1986) New taxanes from *Taxus brevifolia. J. Nat. Prod.*, **49**(4), 665–669.

Huang, K., Liang, J., and Gunatilaka, A.A.L.(1998) Structure revision to the corresponding 11(15 → 1)-abeo-taxoid has been suggested. *J. Chin. Pharm. University (Zhongguo Yaoke Daxue Xuebao)*, **29**, 259–266.

Itokawa, H., Takeya, K., Hitotsuyanagi, Y., and Morita, H. (1999) Antitumor compounds isolated from higher plants. *Yakugaku Zasshi*, **119**(8), 529–583.

Jenniskens, L.H.D., Van Rozendaal, E.L.M., and Van Beek, T.A. (1996) Identification of six taxine alkaloids from *Taxus baccata* needles. *J. Nat. Prod.*, **59**, 117–123.

Jia, Z.J. and Zhang, Z.-P. (1991) Taxanes from *Taxus chinensis*. *Chin. Sci. Bull.*, **36**, 1174.

Kaczmarek, R. and Blechert, S. (1986) De-Mayo-Reaktionen mit Allen. Ein Kurzer Weg zum Bicyclo[5.3.1]undecansystem der Taxane. *Tetrahed. Lett.*, **27**, 2845.

Kingston, D.G.I. (1994) The Chemistry of Taxol, *Pharmac. Ther.*, **52**, 1.

Kingston, D.G.I., Hawkins, D.R., and Ovington, L. (1982) New taxanes from *Taxus brevifolia. J. Nat. Prod.*, **45**(4), 466–470.

Kingston, D.G.I., Molinero, A.A., and Rimoldi, J.M. (1993) The Taxane Diterpenoids. In *Progress in the Chemistry of Organic Natural Products*, vol. 61, 1–2206. Springer Verlag, New York.

Kitagawa, I., Mahmud, T., Kobayashi, M., Roemantyo, Shibuya, H., (1995) *Chem. Pharm. Bull.*, **43**, 365–367.

Kobayashi, J. and Shigemori, H. (1998) Bioactive taxoids from Japanese yew *Taxus cuspidata* and taxol biosynthesis. *Heterocycles*, **47**(2), 1111–1133.

Kobayashi, J., Ogiwara, A., Hosoyama, H., Shigemori, H., Yoshida, N., Sasaki, T., Li, Y., Iwasaki, S., Naito, M., and Tsuruo, T. (1994) Taxuspines A–C, new taxoids from Japanese Yew *Taxus* cuspidata inhibiting drug transport activity of P-glycoprotein in multidrug-resistant cells. *Tetrahedron*, **50**(25), 7401–7416.

Kobayashi, J., Hosoyama, H., Shigemori, H., Koiso, Y., and Iwasaki, S. (1995a) Taxuspine D, a new taxane diterpene from *Taxus cuspidata* with potent inhibitory activity against Ca2+ -induced depolymerization of microtubles. *Experientia*, **51**, 592.

Kobayashi, J., Inubushi, A., Hosoyama, H., Yoshida, N., Sasaki, T., and Shigemori, H. (1995b) Taxuspines E-H and J, new taxoids from the Japanese yew *Taxus cuspidata. Tetrahed.*, **51**(21), 5971–5978.

Kobayashi, J., Hosoyama, H., Katsui, T., Yoshida, N., and Shigemori, H. (1996) Taxuspines N, O, and P, new taxoids from Japanese yew *Taxus cuspidata. Tetrahedron*, **52**(15), 5391–5396.

Kosugi, K., Sakai, J., Zhang, S., Watanabe, Y., Sasaki, H., Suzuki, T., Hagiwara, H., Hirata, N., Hirose, K., Ando, M., Tomida, A., and Tsuruo, T. (2000) Neutral taxoids from *Taxus cuspidata* as modulators of multidrug-resistant tumor cells. *Phytochemistry*, **54**, 839–845.

Kurono, M., Nakadaira, Y., Onuma, S., Sasaki, K., and Nakanishi, K. (1963) Taxinine. *Tetrahed. Lett.*, 2153.

Li, B., Tanaka, K., Fuji, K., Sun, H., and Taga, T. (1993) Three new diterpenoids from *Taxus chinensis*. *Chem. Pharm. Bull.*, **41**(9), 1672–1673.

Li, S.H., Zhang, H.J., Yao, P., and Sun, H.D. (1998) Two new taxoids from *Taxus yunnanensis. Chin. Chem. Lett.*, **9**(11), 1017–1020.

Liang, J. and Kingston, D.G.I. (1993) Two new taxane diterpenoids from *Taxus mairei*. *J. Nat. Prod.*, 56(4), 594–599.

Liang, J.-Y., Huang, K.-S., and Guatilaka, A.A.L. (1998) A new 1,2-deoxotaxane diterpenoid from *Taxus chinensis*. *Planta Medica*, 64, 187–188.

Lucas, H. (1856) Uber ein in den blatten von *Taxus bacata* L. enthaltenes alkaloid (das Taxin). *Arch. Pharm.*, 85, 145.

Lythgoe, B. (1968) The Taxus Alkaloids. In *The Alkaloids*; vol. 10, *Chemistry and Physiology*, edited by R.H.F. Manske, pp. 597–627. New York, Academic Press.

Lythgoe, B., Nakanishi, K., and Ueo, S. (1964) Taxane. *Proc. Chem. Soc.*, 301.

Ma, W., Park, G.L., Gomez, G.A., Nieder, M.H., Adams, T.L., Aynsley, J.S., Sahai, O.P., Smith, J., Stahlhut, R.W., Hylands, P.J., Bitsch, F., and Shackleton, C. (1994a) New bioactive taxoids from cell cultures of *Taxus baccata*. *J. Nat. Prod.*, 57(1), 116–122.

Ma, W., Stahlhut, R.W., Adams, T.L., Park, G.L., Evans, W.A., Blumenthal, S.G., Gomez, G.A., Nieder, M.H., and Hylands, P.J. (1994b) Yunnanxane and its homologous esters from cell cultures of *Taxus chinensis var. mairei*. *J. Nat. Prod.*, 57, 1320–1324.

Morita, H., Gonda, A., Wei, L., Yamamura, Y., Takeya, K., and Itokawa, H. (1997a) Taxuspinananes A and B, new taxoids from *Taxus cuspidata var. nana*. *J. Nat. Prod.*, 60, 390–392.

Morita, H., Wei, L., Gonda, A., Takeya, K., and Itokawa, H. (1997b) A taxoid from *Taxus cuspidata var. nana*. *Phytochemistry*, 46(3), 583–586.

Morita, H., Gonda, A., Wei, L., Yamamura, Y., Wakabayashi, H., Shimizu, K. Takeya, K., Itokawa, H. (1997c) New Taxoids from *Taxus cuspidata var. nana*; Towards Nat. Med. Res. in the 21st Century (October 28–30, 1997, Kyoto, Japan), pp. 503–510.

Morita, H., Gonda, A., Wei, L., Yamamura, Y., Wakabayashi, H., Takeya, K., and Itokawa, H. (1998a) Taxoids from *Taxus cuspidata var. nana*. *Phytochemistry*, 48(5), 857–862.

Morita, H., Gonda, A., Wei, L., Yamamura, Y., Wakabayashi, H., Takeya, K., and Itokawa, H. (1998b) Four new taxoids from *Taxus cuspidata var. nana*. *Planta Med.*, 64, 183–186.

Murakami, R., Shi, Q.-W., and Oritani, T. (1999) A taxoid from the needles of the Japanese yew, *Taxus cuspidata*. *Phytochemistry*, 52, 1577–1580.

Parmer, V.S., Jha, A., Bisht, K.S., Tanja, P., Singh, S.K., Kumar, A., Poonam, Jain, R., and Olsen, C.E. (1999) Constituents of the yew trees. *Phytochemistry*, 50, 1267–1304.

Poupat, C., Ahond, A., and Potier, P. (1994) Nouveau taxoide basique isole des feuilies d'if, *Taxus baccata*: la 2-desacetyltaxine A. *J. Nat. Prod.*, 57(10), 1468–1469.

Rao, K.V. and Juchum, J. (1998) Taxanes from the bark of *Taxus brevifolia*. *Phytochemistry*, 47(7), 1315–1324.

Rao, C., Zhou, J.Y., Chen, W.M., Lu, Y., and Zheng, Q.T. (1993). *Chin. Chem. Lett.*, 4, 693–694.

Rao, K.V., Reddy, G.C., and Juchum, J. (1996a) Taxanes of the needles of *Taxus x media*. *Phytochemistry*, 43(2), 439–442.

Rao, K.V., Bhakun, R.S., Hanuman, J., Davies, R., and Johnson, J. (1996b) Taxanes from the bark of *Taxus brevifolia*. *Phytochemistry*, 41(3), 863–866.

Sako, M., Suzuki, H., Yamamoto, N., Hirota, K., and Maki, Y. (1998) Convenient method for regio/or chemo-selective O-deacylation of taxinine, a naturally occurring taxane diterpenoid. *J. Chem. Soc. Perkin Trace.* I, 417.

Senilh, V., Blechert, S., Colin, M., Guenard, D., Picot, F., Potier, P., and Varenne, er P. (1984) Mise en evidence de Nouveaux analogues du taxol extraits de *Taxus baccata*. *J. Nat. Prod.*, 47, 131–137.

Shen, Y.-C. and Chen, C.-Y. (1997a) Taxane diterpenes from *Taxus mairei*. *Planta Med.*, 63, 569–570.

Shen, Y.-C. and Chen, C.-Y. (1997b) Taxanes from the roots of *Taxus mairei*. *Phytochemistry*, 44(8), 1527–1533.

Shen, Y.-C., Tai, H.-R., and Chen, C.-Y. (1996) New taxane diterpenoids from the roots of *Taxus mairei*. *J. Nat. Prod.*, 59, 173–176.

Shen, Y.-C., Chen, C.-Y., and Kuo, Y.-H. (1998) A new taxane diterpenoid from *Taxus mairei*. *J. Nat. Prod.*, 61, 838–840.

Shen, Y.-C., Chen, C.-Y., and Chen, Y.-J. (1999a) Taxumairol M: a new bicyclic taxoid from seeds of *Taxus mairei*. *Planta Med.*, 65, 582–584.

Shen, Y.-C., Chen, Y.-J., and Chen, C.-Y. (1999b) Taxane diterpenoids from the seeds of Chinese yew *Taxus chinensis. Phytochemistry*, 52(8), 1565–1569.

Shen, Y.-C., Lo, K.-L., Chen, C.-Y., Kuo, Y.-H., and Hung, M.-C. (2000a) New taxanes with an opened oxetane ring from the roots of *Taxus mairei. J. Nat. Prod.*, 63, 720–722.

Shen, Y.-C., Prakash, C.V.S., and Hung, M.-C. (2000b) Taxane diterpenoids from the root bark of Taiwanese yew *Taxus mairei. J. Chin. Chem. Soc.*, 47, 1125–1130.

Shen, Y.-C., Prakash, C.V.S., , Chen, Y. J., Hwang, J.F., Kuo, Y.-H., and Chen, C.Y. (2001) Taxane diterpenoidsfrom the stem bark of *Taxus mairei. J. Nat. Prod.*, 64, 950–952.

Shi, Q.-W., Oritani, T., Sugiyama, T., and Kiyota, H. (1998a) Three novel bicyclic 3,8-secotaxane diterpenoids from the needles of the Chinese yew, *Taxus chinensis var. mairei. J. Nat. Prod.*, 61, 1437–1440.

Shi, Q.-W., Oritani, T., Sugiyama, T., and Kiyota, H. (1998b) Two new from *Taxus chinensis var. mairei. Planta Med.*, 64, 766–769.

Shi,Q.W., Oritani, T., and Sugiyama, T. (1999a) Two bicyclic taxane diterpenoid from the needles of *Taxus mairei. Phytochemistry*, 50, 633–636.

Shi, Q.-W., Oritani, T., Sugiyama, T., and Yamada, T. (1999b) Two novel pseudoalkaloid taxanes from the Chinese yew, *Taxus chinensis var. mairei. Phytochemistry*, 52, 1571–1575.

Shi, Q.-W., Oritani, T., and Sugiyama, T. (1999c) Three rearranged 2(3_20) abeotaxanes from the bark of *Taxus mairei. Phytochemistry*, 52, 1559–1563.

Shi, Q.-W., Oritani, T., Sugiyama, T., Murakami, R., and Wei, H.-Q. (1999d) Six new taxane diterpenoids from the seeds of *Taxus chinensis var. mairei* and *Taxus yunnanensis. J. Nat. Prod.*, 62, 1114–1118.

Shi, Q.W., Oritani, T., and Sugiyama, T. (1999e) Three novel bicyclic taxane diterpenoids with verticilline skeleton from the needles of Chinese yew, *Taxus chinensis var. mairei. Planta Med.*, 65, 356–359.

Shi, Q.-W., Oritani, T., Sugiyama, T., Zhao, D., and Murakami, R. (1999f) Three new taxane deterpenoids from seeds of the Chinese yew, *Taxus yunnanensis* and *T. chinensis var. mairei. Planta Med.*, 65, 767–770.

Shi, Q.-W., Oritani, T., Sugiyama, T., Kiyota, H., and Horiguchi, T. (1999g) Isolation from natural product *Taxus chinensis var, mairei. Heterocycles*, 51(4), 841–850.

Shi, Q.-W., Oritani, T., Sugiyama, T., Murakami, R., and Horiguchi, T. (1999h) Three new bicyclic taxane diterpenoids from the needles of Japanese yew, *Taxus cuspidata sieb. et Zucc. J. Asian Nat. Prod. Res.*, 2(1), 63–70.

Shi, Q.-W., Oritani, T., Sugiyama, T., Murakami, R., and Yamada, T. (1999i) Two new taxane diterpenenoids from the seeds of the Chinese yew, *Taxus yunnanensis. J. Asian Nat. Prod. Res.*, 2(1), 71–79.

Shi, Q.-W., Oritani, T., Kiyota, H., and Zhao, D. (2000a) Taxane diterpenoids from *Taxus yunnanensis* and *Taxus cuspida. Phytochemistry*, 54, 829–834.

Shi, Q.W., Oritani, T., and Zhao, D. (2000b) A new taxane diterpenoid from the seeds of *Taxus yunnanensis. Chin. Chem. Lett.*, 11(3), 235–238.

Shi, Q.-W., Oritani, T., Sugiyama, Oritani, T. (2000c) Three new taxane diterpenoids from the seeds of the Japanese yew, *Taxus cuspidata. Nat. Prod. Res.*, 14(4), 265–272.

Shi, Q.W., Oritani, T., and Zhao, D. (2000d) New taxane diterpenoid from seeds of the Chinese yew, *Taxus yunnanensis. Biosci. Biotechnol. Biochem.*, 64(4), 869–872.

Shigemori, H., Wang, X.-X., Yoshida, N., and Kobayashi, J. (1997) Taxuspines X–Z, new taxoids from Japanese yew *Taxus cuspidata. Chem. Pharm. Bull.*, 45(7), 1205–1208.

Shigemori, H., Sakurai, C.A., Hosoyama, H., Kobayashi, A., Kajiyama, S., and Kobayashi, J. (1999) Taxezopidines J, K, and L, new taxoids from Taxus cuspidata inhibiting Ca2 +-induced depolymerization of microtubles. *Tetrahedron*, 55, 2553–2558.

Shinozaki, Y., Fukamia, N., Fukushima, M., Okano, M., Nehira, Tagahara, K., Zhang, S.-X., Zhang, D.-C., and Lee, K.-H. (2001) Dantaxusin A and B, two new Taxoids from *Taxus yunnanensis. J. Nat. Prod.*, 64(8), 1073–1076.

Shrestha, T.B., Chetri, S.K.K., Banskota, A.H., and Manandhar, M.D. 2-Deacetoxytaxinine B: A new taxane from *Taxus wallichiana. J. Nat. Prod.*, 60, 820–821.

Soto, J. and Castedo, L. (1998) Taxoids from European yew, *Taxus baccata* L. *Phytochemistry*, 47(5), 817–819.

Soto, J., Fuentes, M., and L. Castedo, L. (1996) Teixidol, an abeo-taxane from European yew, *Taxus baccata. Phytochemistry*, 43(1), 313–314.

Stahlhut, R., Park, G., Petersen, R., Ma, W., and Hyland, P. (1999) The occurrence of the anti-cancer diterpene taxol in *Podocarpus gracillior pilger* (*Podocarpaceae*). *Biochem. System. Ecol.*, 27(6), 613–622.

Suffness, M., (Ed) (1995) "*Taxol, Science and Applications*", Washington D.C., CRS Press.

Suffness, M. and Cordell, G.A. (1985) Taxus alkaloids. In: *The Alkaloids, Chemistry and Pharmacology*, vol. 25, edited by A. Brossi, p. 6. New York, Academic Press.

Suni, K., Chattopadhyay, G.C., Gour, C.S., Saha, R.P., Ram, P.S. Susil, K., and Raya, R. (1996) A rearranged taxane from the Himalayan yew *Taxus wallichiana. Phytochemistry*, 42(3), 787–788.

Swindell, C.S. (1991) Taxane diterpene synthesis strategies. A review. *Org. Prep. Proced. Int.* 23, 465.

Tanaka, K., Fuji, K., Yokoi, T., Shingu, T., Li, B., and Sun, H. (1994) On the structures of six new diterpenoids, taxachinins E, H, I, J, K, and taxchinin B. *Chem. Pharm. Bull.*, 42(7), 1539–1541.

Tong, X.J., Fang, W.S., Zhou, J.Y., He, C.H., Chen, W.M., and Fang, Q.C. (1994) *Yaoxue Xuebao*, 29, 55–60.

Tong, X.J., Fang, W.S., Zhou, J.Y., He, C.H., Chen, W.M., and Fang, Q.C. (1993) *Chin. Chem. Lett.*, 4, 887–890.

Tong, X.-J., Fang, W.-S., Zhou, J.-Y., He, C.-H., Chen, W.-M., and Fang, Q.-C. (1995) Three new taxane diterpenoids from needles and stems of *Taxus cuspidata. J. Nat. Prod.*, 58(2), 233–238.

Topcu, G., Sultana, N., Akhtar, F., Rehman, H.-U., Hussain, T., Choudhary, M.I., and Atta-ur-Rahman, (1994) Taxane diterpenes from *Taxus bacata. Nat. Prod. Lett.*, 4(2), 93–100.

Vander Velde, D.G., Georg, G.I., Gollapudi, S.R., Jampani, H.B., Liang, X.-Z., Mitscher, L.A., and Ye, Q.-M. (1994) Wallifoliol, a taxol congener with a novel carbon skeleton, from Himalayan *Taxus wallichiana. J. Nat. Prod.*, 57(6), 862–867.

Wang, X.-X., Shigemori, H., and Kobayashi, J. (1996a) Taxezopindines B–H, new taxoids from Japanese yew *Taxus cuspidata. J. Nat. Prod.*, 61, 474–479.

Wang, X.-X., Shigemori, H., and Kobayashi, J. (1996b) Taxuspines K, L, and M, new taxoids from Japanese yew *Taxus cuspidata. Tetrahedron*, 52(7), 2337–2342.

Wang, X.-X., Shigemori, H., Kobayashi, J. (1996c) Taxuspines Q, R, S and T, new taxoids from Japanese yew *Taxus cuspidata. Tetrahedron*, 52, 12159–12164.

Wang, X.-X, Shigemori, H., and Kobayashi, J. (1997) Taxezopidine A, a novel taxoid from seeds of Japanese yew *Taxus cuspidata. Tetrahed. Lett.*, 38(43), 7587–7588.

Wang, X.-X, Shigemori, H., and Kobayashi, J. (1998) Taxezopidine B-H, new taxoids from Japanese yew *Taxus cuspidata. J. Nat. Prod.*, 61, 474–479.

Wani, M.C., Taylor, H.L., Wall, M.E.M., Coggon, P., and McPhail, A.T. (1971) Plant antitumor agents VI. The isolation and structure of taxol, a novel antileukemic and antitumor agent from *Taxus brevifolia. J. Am. Chem. Soc.*, 93, 2325–2327.

Woods, M.C., Chiang, H.-C., Nakadaira, Y., and Nakanishi, K. (1968) The nuclear overhauser effect, a unique method of defining the relative stereochemistry and conformation of taxane derivatives. *J. Am. Chem. Soc.*, 90, 522.

Yang, S.-J., Fang, J.-M., and Cheng, Y.S. (1996) Taxanes from *Taxus mairei. Phytochemistry*, 43(4), 839–842.

Yang, S.-J., Fang, J.-M., and Cheng, Y.-S. (1999) Abeo-taxanes from *Taxus mairei. Phytochemistry*, 50, 127–130.

Yeh, M.-K., Wang, J.-S., Liu, L.-P., and Chen, F.-C. (1988a) A new taxane derivative from the heartwood of *Taxus mairei. Phytochemistry*, 27, 1534.

Yeh, M.-K., Wang, J.-S., Liu, L.-P., and Chen, F.-C. (1988b) Some taxane derivatives from the heartwood of *Taxus mairei. J. Chin. Chem. Soc.*, 35, 309.

Yeh, M.K., Wan, J.S., Yang, W.L., Chen, F.C. (1988c) Proc. Natl. Sci. Counc. Repub. China Part A; *Phys. Sci. Eng.*, 12, 89–90.

Yoshizaki, F., Fukuda, M., Hisamichi, S., Ishida, T., and In, Y. (1988) Structures of taxane diterpenoids from the seeds of Japanese yew, *Taxus cuspidata. Chem. Pharm. Bull.*, 36, 2098.

Yue, Q., Fang, Q.-C., Liang, X.-T., and He, C.-H. (1995a) Taxayuntine and F: Two taxanes from leaves and stems of *Taxus yunnanensis. Phytochemistry*, 39(4), 871–873.

Yue, Q., Fang, Q.C., and Liang, X.T. (1995b) *Chin. Chem. Lett.*, 6, 225–228.

Yue, Q., Fang, Q.-C., Liang, X.-T., He, C.-H., and Jing, X.-L. (1995c) Rearranged taxoids from *Taxus yunnanensis. Planta Med.*, 61, 375–377.

Yue, Q., Fang, Q-C., and Liang, X.-T. (1996) A taxane-11,12-oxide from *Taxus yunnanensis. Phytochemistry*, 43(3), 639–642.

Zamir, L.O., Nedea, M.E., Belair, S., Sauriol, F., Mamer, O., Jacqmain, E., Jean, F.I., and Garneau, F.X. (1992) Taxanes isolated from *Taxus canadensis*. *Tetrahed. Lett.*, **33**(36), 5173–5176. Corrigendum, *ibid.*, 6548.

Zamir, L.O., Nedea, M.E., Zhou, Z.-H., Belair, S., Caron, G., Sauriol, F., Jacqmain, E., Jean, F.I., Garneau, F.-X., and Mamer, O. (1995a) *Taxus canadensis* taxanes: structures and stereochemistry. *Can. J. Chem.*, **73**, 655–665.

Zamir, L.O., Zhou, Z.H., Caron, G., Nedea, M.E., Sauriol, F., and Zamer, C. (1995b) Isolation of a putative biogenetic taxane precursor from *Taxus canadensis* needles. *J. Chem. Soc. Chem. Commun.*, 529–530.

Zamir, L.O., Nedea, M.E., Zhou, Z.-H., Caron, G., Sauriol, F., and Mamer, O. (1996) Isolation and semi-synthesis of a bioactive taxane from *Taxus canadensis*. *Phytochemistry*, **41**(3), 803–805.

Zamir, O.L., Zang, J., Kutterer, K., Sauriol, F., Mamer, O., Khiat, A., and Boulanger, Y. (1998) 5-Epi-canadesene and other novel metabolites of *Taxus canadensis*. *Tetrahedron*, **54**, 15845–15860.

Zamir, L.O., Zhang, J., Wu, J., Sauriol, F., and Mamer, O. (1999) Five novel taxanes from *Taxus canadensis*. *J. Nat. Prod.*, **62**, 1268–1273.

Zhang, H., Takeda, Y., Minami, Y., Yoshida, K., Matsumoto, T., Xiang, W., Mu, O., and Sun, H. (1994a) Three new taxanes from the roots of *Taxus yunnanensis*. *Chem. Lett.*, 957–960.

Zhang, H., Takeda, Y., Matsumoto, T., Minami, Y., Yoshida, K., Wei, X., Qing, M., and handong, S. (1994b) Taxol related diterpenes from the roots of *Taxus yunnanensis*. *Heterocycles*, **38**(5), 975–980.

Zhang, H., Tadeda, Y., and Sun, H. (1995a) Taxanes from *Taxus yunnanensis*. *Phytochemistry*, **39**(5), 1147–1151.

Zhang, H., Sun, H., and Takeda, Y. (1995b) Four new taxanes from the roots of *Taxus yunnanensis*. *J. Nat. Prod.*, **58**(8), 1153–1159.

Zhang, H., Mu, Q., Xiang, W., Yao, P., Sun, H., and Takeda, Y. (1997) Intramolecular transesterified taxanes from *Taxus yunnanensis*. *Phytochemistry*, **44**(5), 911–915.

Zhang, J.-Z., Fang, Q.-C., Liang, X.-T., He, C.-H., Kong, M., He, W.-Y., and Jin, X.-L. (1995a) Taxoids from the barks of *Taxus wallichiana*. *Phytochemistry*, **40**(3), 881–884.

Zhang, J.Z., Fang, Q.C., Liang, X.T., Kong, M., and Yi, H. (1995b) *Chin. Chem. Lett.*, **6**, 971–974.

Zhang, J., Sauriol, F., Mamer, O., and Zamir, L.O. (2000a) Taxoids from the needles of the Canadian yew. *Phytochemistry*, **54**, 221–230.

Zhang, J., Sauriol, F., Mamer, O., and Zamir, L.O. (2000b) New taxanes from needles of *Taxus canadensis*. *J. Nat. Prod.*, **63**, 929–933.

Zhang, J., Sauriol, F., Mamer, O., You, X.-L., Alaoui-Jamali, M.A., Batist, G., and Zamir, L.O. (2001) New taxane analogues from the needles of *Taxus canadensis*. *J. Nat. Prod.*, **64**, 450–455.

Zhang, S., Lee, C.T.-L., Chen, K., Kashiwada, Y., Zhang, D.-C., McPail, A.T., and Lee, K.H. (1994a) Structure and Stereochemistry of Taxuchin A, a new 11(15 → 1) Abeo-Taxane type diterpene from *Taxus chinensis*. *J. Chem. Soc. Chem. Commun.*, 1561–1562.

Zhang, S., Lee, C.T.-L., Kashiwada, Y., Chen, K., Zhang, D.-C., and Lee, K.H. (1994b) Yunantaxusin A, a new 11(15 → 1)-abeo-taxane from *Taxus yunnanensis*. *J. Nat. Prod.*, **57**(11), 1580–1583.

Zhang, Z., Jia, Z., Zhu, Z.-Q., Cui, Y.-X., Cheng, J.-L. and Wang, Q.-G. (1989) Studies on the chemical constituents of *Taxus*. *Chin. Sci. Bull.*, **21**, 1630.

Zhang, Z. and Jia, Z. (1990) Taxanes from *Taxus yunnanensis*. *Phytochemistry*, **29**, 3673.

Zhang, Z., Jia, Z., Zhu, Z.-Q., Cui, Y.-X., Cheng, J.-L., and Wang, Q.-G. (1990) New taxanes from *Taxus chinensis*. *Planta Med.*, **56**, 293.

Zhang, Z.P. and Jia, Z. (1991) Taxanes from *Taxus chinensis*. *Phytochemistry*, **30**, 2345–2348.

Zhang, Z.P., Wiedenfeld, H., and Roder, E. (1995) Taxane from *Taxus chinensis*. *Phytochemistry*, **38**(3), 667–670.

Zhong, S.-Z., Hua, Z.-X., and Fan, J.-S., (1996) A new taxane diterpene from *Taxus yunnanensis*. *J. Nat. Prod.*, **59**, 603–605.

Zhou, J.-Y., Zhang, P.-L., Chen, W.-M., and Fang, Q.-C. (1998) Taxayuntin H and J from *Taxus yunnanensis*. *Phytochemistry*, **48**(8), 1387–1389.

4 Physical methods for identification of the structures of taxoids

Masayoshi Oyama and Hideji Itokawa

UV, ORD/CD, and IR Spectroscopy

Although NMR spectroscopy is generally considered the most useful method for the structural elucidation of taxoids, other instrumental spectroscopic techniques add unique and confirmatory data. The UV spectra of taxoids containing only a ring A enone system display an anomalously long wavelength absorption band. Kingston *et al.* (1993) summarized about these spectroscopy. The overall ring strain of the taxane skeleton is, however, the major factor and this factor is apparent in the spectrum of the alcohol (1) which absorbs at 227 nm as opposed to 202 nm predicted for an unstrained alkene. The opening of ring C does not make a major difference in the spectrum, and the ring-opened product (2) has λ_{max} 276 nm. This is consistent with MM2 results mentioned by Swindell *et al.* (1985), because the A/B ring system alone accounts for the bulk of the strain energy of the taxane ring. However, as ketone (3) has a normal λ_{max} of 250 nm (Prelog *et al.*, 1949), ring strain alone cannot account for the whole effect, the gem-dimethyl groups at C-15 have been implicated (Lythgoe *et al.*, 1968).

The optical rotatory dispersion (ORD) and circular dichroism (CD) spectra of the taxane diterpenoids have provided valuable confirmatory evidence of their stereochemistry. Application of the benzoate sector rule to 5-benzoyl-isopropylidenetaxinol (4) confirmed its stereochemistry (Harada *et al.*, 1968), and the application of the dibenzoate chirality rule to the tetrahydrotaxinine dibenzoate (5) confirmed their stereochemistries (Harada and Nakanishi, 1969). Thus, the dibenzoate (5) showed negative benzoate chirality and to the stereochemistry shown (Figure 4.1). An independent investigation of the Cotton effects associated with the bridgehead olefin of several taxoids yielded similar conclusions (De Marcano *et al.*, 1970a). This study, which was carried out on taxoids lacking a C-13 carbonyl group, showed that these compounds all exhibited intense positive Cotton effects in the 214–230 nm region due primarily to the $\pi \rightarrow \pi^*$ olefinic absorption. These effects corresponded to those observed with (S)-(+)-*trans*-cyclodecene, in accordance with the previous stereochemical results.

The taxoids do not show any unusual infrared absorption properties. As shown in Figure 4.2, normal IR-absorption bands are present at 1730–1745 (ester C=O) and 1670–1685 (six-membered conjugated ketone).

Mass spectrometry

Relatively little information has been reported on the mass spectrometry of taxoids other than molecular ions of new compounds. The low volatility of taxoids has made them unsuitable for mass spectroscopic measurements, but recently modern instrumentation has, in principle, overcome this problem by using fast atom bombardment (FAB) or other soft ionization technique to produce molecular ions followed by collisional activation to produce daughter ion spectra

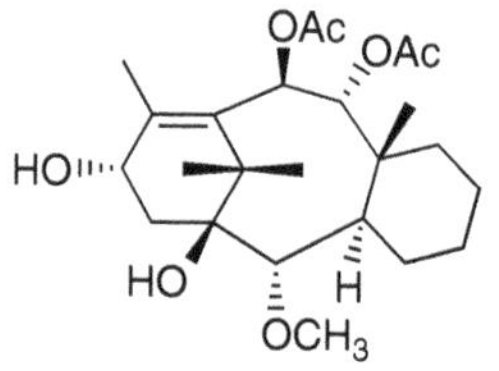

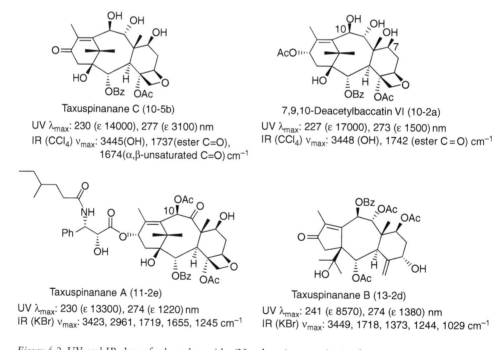

1

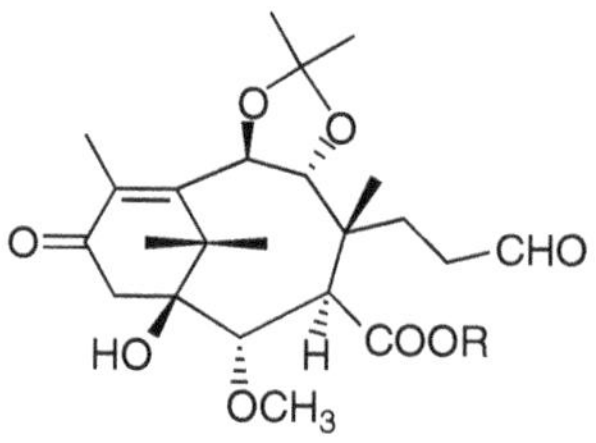

2

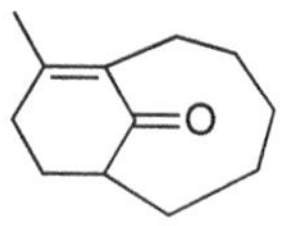

3

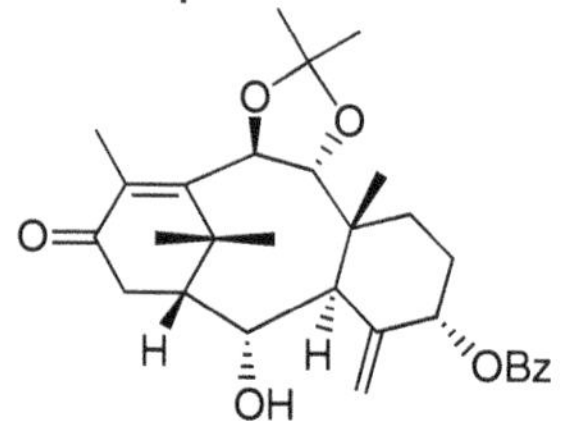

5-Benzoyl-isopropylydenetaxinol (**4**)

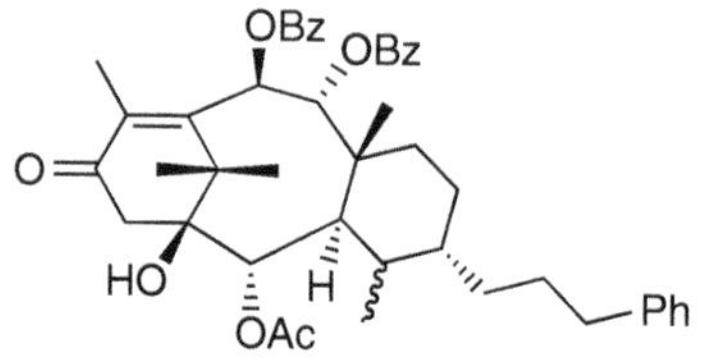

Tetrahydrotaxinine dibenzoate (**5**)

Figure 4.1

Taxuspinanane C (**10-5b**)

UV λ_{max}: 230 (ε 14000), 277 (ε 3100) nm
IR (CCl$_4$) ν_{max}: 3445(OH), 1737(ester C=O),
1674(α,β-unsaturated C=O) cm^{-1}

7,9,10-Deacetylbaccatin VI (**10-2a**)

UV λ_{max}: 227 (ε 17000), 273 (ε 1500) nm
IR (CCl$_4$) ν_{max}: 3448 (OH), 1742 (ester C=O) cm^{-1}

Taxuspinanane A (**11-2e**)

UV λ_{max}: 230 (ε 13300), 274 (ε 1220) nm
IR (KBr) ν_{max}: 3423, 2961, 1719, 1655, 1245 cm^{-1}

Taxuspinanane B (**13-2d**)

UV λ_{max}: 241 (ε 8570), 274 (ε 1380) nm
IR (KBr) ν_{max}: 3449, 1718, 1373, 1244, 1029 cm^{-1}

Figure 4.2 UV and IR data of selected taxoids. (Numbers in parenthesis after compound names are the numbers used in Chapter 3.)

(MS/MS) (Figure 4.3). Such an approach to the mass spectra of taxoids has been described in one abstract (McClure *et al.*, 1990). Similar studies were reported by Hoke *et al.* (1991). Some information on the fragmentation of the taxoids is scattered in the literature.

The original discoverers of taxol reported a molecular ion at *m/z* 853 and although not explicitly stated, this ion was presumably obtained under electron impact conditions (Wani *et al.*, 1971).

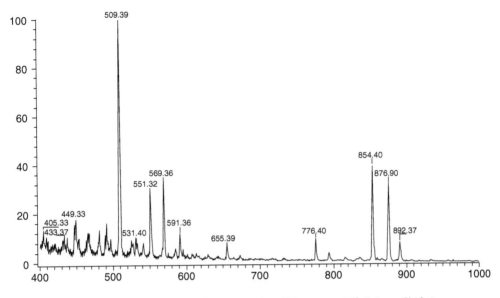

Figure 4.3 FAB-MS spectrum of taxol. (Tokyo University of Pharmacy & Life Science, Shida.)

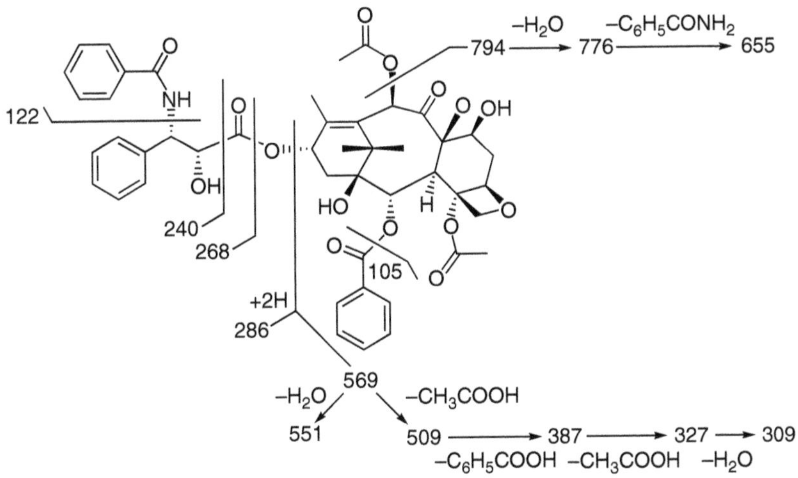

Figure 4.4 ESI-MS fragmentation of taxol.

Later studies report only fragment ions under EI conditions (Miller *et al.*, 1981), but the FAB mass spectrum of taxol shows intense molecular ions at MH^+, MNa^+, and MK^+. Fragment ions are observed corresponding to the loss of acetic acid and to the loss of the protonated C-13 side-chain (Magri, 1985).

Some assignments were reported by Kerns *et al.* (1994). In this chapter, the method integrates analytical HPLC, UV detection, UV spectroscopy, full-scan ionspray mass spectrometry, and tandem mass spectrometry on-line. The identification of taxane structure is based on comparing the mass spectrometric characteristics of the taxane with the taxol substructural template (Figure 4.4).

Ma *et al.* (1994) also reported some MS assignments of taxanes. High resolution mass spectral measurement (HRMS) of taxol C showed m/z 848.3857, consistent with the formula $C_{46}H_{58}NO_{14}$. Its electrospray ionization mass spectral (ESI-MS) fragmentations at m/z 770, 569, 551, and 509 indicated that the taxol skeleton was present. The ESI-MS spectrum of deacetyl-taxol C showed a protonated molecular ion at m/z 806 (HRMS measurement at m/z 806.3753 corresponding to a formula $C_{44}H_{56}NO_{13}$). The ESI-MS fragments of m/z 527, 509, and 280 indicated that this compound was a deacetyl derivative of taxol C. The HRMS measurement of N-methyltaxol C gave a protonated molecular ion at m/z 862.3753, corresponding to the formula, $C_{47}N_{60}NO_{14}$. ESI-MS fragment at m/z 294 suggested an additional methyl group in the side-chain of N-methyltaxol C (Figures 4.5 and 4.6).

As shown by Bitsch *et al.* (1993), the side-chain structure of taxoids can be partially eluci-dated by MS/MS of the C-13 side-chain fragments produced by electrospray ionization. Comparing the C-13 side-chain ion at m/z 294 of N-methyltaxol C with that at m/z 280 of taxol C, the daughter ions at m/z 196 and 130 clearly indicated that the methyl group is on the nitro-gen of the side-chain. The HRMS of taxcultine showed a protonated molecular ion at m/z 820.3559, which indicated the composition $C_{44}H_{54}NO_{14}$. The ESI-MS fragments at m/z 569, 551, and 509 indicated the presence of the taxol skeleton. The fragment at m/z 252 indicated a shorter aliphatic group on the C-13 side-chain than that of taxol C. By comparing the MS/MS spectra of the ion at m/z 252 with that of the ion at m/z 280 from taxol C, it is clear that a propyl group is attached at C-4'.

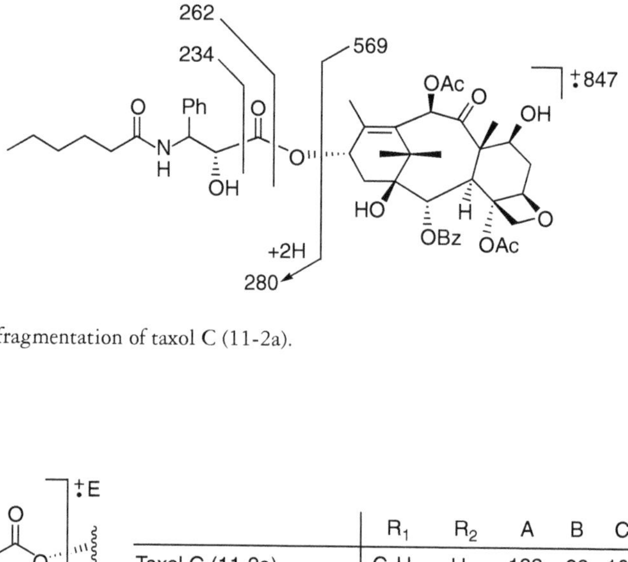

Figure 4.5 ESI-MS fragmentation of taxol C (11-2a).

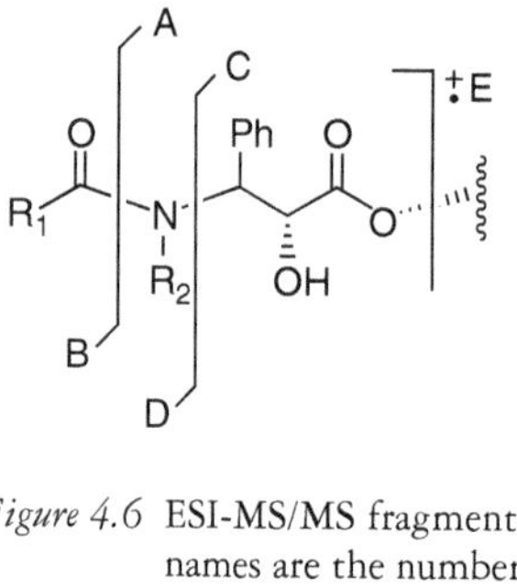

	R₁	R₂	A	B	C	D	E
Taxol C (11-2a)	C_5H_{11}	H	182	99	164	116	280
N-Methyltaxol C (11-2c)	C_5H_{11}	CH_3	196	99	164	130	294
Taxcultine (11-2d)	C_3H_7	H	182	71	164	88	252

Figure 4.6 ESI-MS/MS fragmentation of the C-13 side-chains. (Numbers in parenthesis after compound names are the numbers used in Chapter 3.)

Nuclear Magnetic Resonance spectroscopy

¹H-NMR spectroscopy

Nuclear Magnetic Resonance spectroscopy (NMR) is an important method for elucidating the structures of various compounds via instrumental analysis.

Studies of the ¹H-NMR spectra of taxoids were well reviewed by Miller (1980) and Kingston *et al.* (1993). As taxol has been the most representative taxoid, it must be discussed as a primary example. Detailed studies at 500 MHz have provided the most complete analysis of the spectrum (Chumurny *et al.*, 1992; Falzone *et al.*, 1992). Chumurny *et al.* (1992) mentioned ¹H- and ¹³C-NMR spectra of taxol and the related strategically important taxanes 7-epi-taxol and cephalomannine using 1D and 2D NMR techniques. Confirmation of the taxol assignments, in addition to preliminary conformational information for taxol, was obtained via NOE spectroscopy (Table 4.1). Falzone *et al.* (1992) and Chumurny *et al.* (1992) reported the following configurations at C-6, H-6a, or H-6α was shown with a dotted line and assigned as δ_H 2.54, while H-6b or H-6β was shown as a solid line and assigned as δ_H 1.88. On the other hand, Kingston *et al.* (1993) gave the reverse assignment of H-6α as δ_H 1.88 and H-6β as δ_H 2.54. The same assignment was observed at C-14. Hα must be expressed as a dotted line and Hβ as a solid line.

Falzone *et al.* (1992) also reported the ¹H- and ¹³C-NMR spectral data of taxol in methylene chloride using two dimensional methods. ¹³C chemical shift assignments were made based on inverse-detected two-dimensional NMR experiments. The NOESY data collected in this solvent suggested that the structure in solution is similar to the X-ray structure of the taxol analogue taxotere.

The ¹H NMR spectrum of taxol is shown in Figure 4.7 and its assignments are shown in Figure 4.8.

The ¹H-NMR spectral data of selected taxoids are presented in Tables 4.1–4.18. The data in these tables are taken directly from the cited literature references, and the assignments have been made with varying degrees of certainty. The structures of the cited compounds are shown in the following pages. Finally, the structural characteristic 5-oxotaxinine A (dimer) is added to the figures and tables, although this compound is not a natural product (Hosoyama *et al.*, 1998) (Figures 4.9–4.28) (Tables 4.19–4.37).

¹³C-NMR spectroscopy

The ¹³C-NMR spectrum of taxol is shown in Figure 4.4 and the assignments are shown in Figure 4.5. The ¹³C-NMR spectral data of selected taxoids are given in Tables 4.21–4.36. Particular ¹³C-NMR spectral assignments are also very useful for differentiating between normal taxanes and rearranged taxanes.

A 15-hydroxyl-11(15 → 1)abeo-taxane structure is strongly suggested by the low-field resonance (δ_C 60–70) of one of the aliphatic non-oxygenated quaternary carbons; this structural assignment can be confirmed by the analysis of the ROE effects associated to the tertiary hydroxyl proton. From such data, the structures of some previously assigned normal taxanes were revised to abeo taxanes (Appendino *et al.*, 1993b).

In another report, these relationships were also observed; δ_C 43.00 and 76.06 (C-8 and C-15) in taxuspinanane F (Morita *et al.*, 1998). Finally, the structural characteristic 5-oxotaxinine A (dimer) is added to the figures, although this compound is not natural product (Hosoyama *et al.*, 1998) (Figure 4.28) (Table 4.37).

Table 4.1 ^{1}H NMR spectra of Taxol (Paclitaxel) (11a)

Positions		COSY	TOCSY
2	5.67(d, 7.1)	H-3	H-3
3	3.79(d, 1.0, d, 7.0)	H-2, -20α	H-2, -20α
5	4.94(d, 0.8, d, 0.9, d, 2.3, d, 9.6)	H-6α,β, -20α,β	H-7, -6α,β
6α	2.54(d, 6.7, d, 9.7, d, 14.8)	H-5, -6β, -7	H-5, -6β, -7
6β	1.88(d, 2.3, d, 11.0, d, 14.7)	H-5, -6α, -7	H-5, -6α, -7
7	4.40(d, 6.7, d, 10.9, d, 4.3)	H-6α,β	H-5, -6α,β
10	6.27(s)		
13	6.23(q, 1.5, t, 15.4)	Me-14, H-14α,β	Me-18, H-14α,β
14α	2.35(d, 9.0, d, 15.4)	H-14β, -13	H-13, -14β
14β	2.28(d, 0.6, d, 9.0, d, 15.3)	H-14α, -13	H-13, -14α, Me-18
16	1.14(s)	H-13	H-13
17	1.24(s)		
18	1.79(d, 1.5)	H-13	H-13
19	1.68(s)		
20α	4.30(d, 1.1, d, 0.8, d, 8.4)	H-5, -20β, -3	H-20β, -3
20β	4.19(d, 1.0, d, 8.9)	H-5, -20α	H-20α
2′	4.78(d, 2.7, d, 5.4)	H-3′	NH-3′, H-2′
3′	5.78(d, 2.8, d, 8.9)	NH-3′, H-2′	NH-3′, H-2′
1-OH	1.98(br s)		
4-OAc	2.38(s)		
7-OH	2.48(br s)[d(4.4)]		
10-OAc	2.23(s)		
2′-OH	3.61(br s)[d(5.4)]		
3′-NH	7.01(d, 8.9)	H-3′	H-3′, -2′
o-Ph1	8.13(d, 1.3, d, 8.4)	p-Ph1, m-Ph1	p-Ph1, m-Ph1
m-Ph1	7.51(m)	o-Ph1, p-Ph1	o-Ph1, p-Ph1
p-Ph1	7.61(t, 1.4, t, 7.4)	o-Ph1, m-Ph1	o-Ph1, m-Ph1
o-Ph2	7.48(m)	p-Ph2, m-Ph2	p-Ph2, m-Ph2
m-Ph2	7.42(m)	o-Ph2, p-Ph2	o-Ph2, p-Ph2
p-Ph2	7.35(t, 1.6, t, 7.3)	o-Ph2, m-Ph2	o-Ph2, m-Ph2
o-Ph3	7.74(d, 1.2, d, 8.3)	p-Ph3, m-Ph3	p-Ph3, m-Ph3
m-Ph3	7.40(m)	o-Ph3, p-Ph3	o-Ph3, p-Ph3
p-Ph3	7.49(m)	o-Ph3, m-Ph3	o-Ph3, m-Ph3

Notes
Numbers in parenthesis after compounds names are the numbers used in Chapter 3.
Measured in $CDCl_3$ (500 MHz), δ values with TMS as internal standard.
The values in parenthesis are coupling constants (Hz).
Multiplicity: s = singlet, d = doublet, t = triplet, q = quartet, m = multiplet, br = broad.

X-ray crystallography

X-ray crystallographic analysis is the most important method for determining stereostructures. Several taxoids have been analyzed by this methodology. One of the first reported crystal structures was not a natural taxoid but rather the *p*-bromobenzoate derivative of the rearrangement product (6) (Chan *et al.*, 1966). The absolute configuration of taxinine (7) was established by X-ray analysis of the 14-bromo derivative of taxinol tetraacetate (9) (Shiro *et al.*, 1966, 1971). Other taxanes that have been analyzed by X-ray crystallography include baccatin V (10) (De Marcano *et al.*, 1970b; Castellano and Hodder, 1973), O-cinnamoyltaxicin-1-triacetate (8) (Begrey *et al.*, 1984), taxusin (11) (Ho *et al.*, 1987a), taxagifine (12) (Chaviere *et al.*, 1982), and taiwanxan (13) (Ho *et al.*, 1987b) (Figures 4.29–4.30).

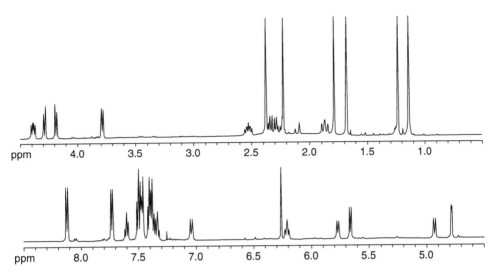

Figure 4.7 ¹H-NMR of taxol. (Tokyo University of Pharmacy & Life Science, Sakuma.)

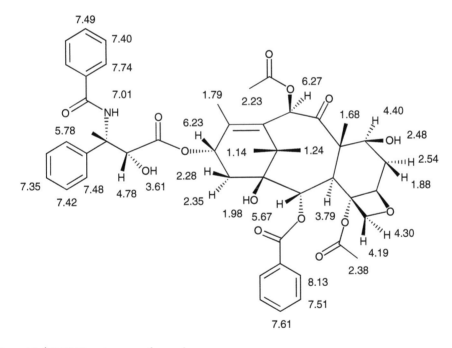

Figure 4.8 ¹H-NMR assignments for taxol.

Until now, the X-ray structure of taxol itself has never been determined. The original structure work was carried out on crystalline derivatives of a hydrolyzed product, as already mentioned in Chapter 1. Hydrolysis of taxol gave an N-benzoyl-β-phenylisoserine methyl ester, which was analyzed as its *p*-bromobenzoate derivative, and the tetraol 14, which was analyzed as

Table 4.2 ¹H NMR spectra of selected taxoids

Positions	5α-Hydroxy-9α,10β,13α-triacetoxytaxa-4(20),11-diene (1q)	2-Decinnamoyltaxinine E (1p)	¹H-¹H	Decinnamoyltaxinine E (1r)
1	1.76(m)	1.83(m)	H-2, 14	1.95(dd, 8.6, 2.1)
2		5.46(dd, 6.0, 7.5)	H-1, 3	4.19(dd, 6.2, 2.1)
3	3.26(d, 4.5)	3.56(d, 6.0)	H-2, 20	3.39(d, 6.2)
4	4.26(br s)			
5		4.24(br s)	H-6α, 6β	4.24(t, 2.7)
6α		1.60(m)	H-5	1.80(m)
6β		1.80(m)		1.59(m)
7α				1.89(m)
7β				1.60(m)
9	5.73(d, 10.2)	6.07(d, 10.5)	H-10	5.71(d, 10.4)
10	6.07(d, 10.2)	5.80(d, 10.5)	H-9	6.05(d, 10.4)
13	5.67(dd, 3.3, 7.8)	5.75(dd, 4.8, 10.5)	H-14, Me-18	5.76(ddq, 10.4, 5.0, 1.3)
14	1.12(dd, 4.5, 15.3)	1.52(dd-like, 5.0, 15.6)	H-13, 14β	1.40(dd, 15.7, 5.0)
	2.78(m)	2.64(ddd, 7.8, 10.5, 15.6)	H-1, 13, 14α	2.65(ddd, 15.7, 10.4, 8.6)
16	1.55(s)	1.03(s)		1.63(s)
17	0.97(s)	0.86(s)		1.04(s)
18	2.08(s)	2.15(s)		2.11(d, 1.3)
19	0.68(s)	1.71(s)		0.88(s)
20	4.71(br s)	5.24(br s)		5.30(d, 1.6)
	5.09(br s)	4.84(br s)		5.39(t, 1.6)
OAc		2.01(s)		2.08(s)
		2.05(s) ×2		2.04(s)
		2.11(s)		

Notes

Numbers in parenthesis after compounds names are the numbers used in Chapter 3.

Measured in CDCl₃, δ values with TMS as internal standard.

The values in parenthesis are coupling constants (Hz).

Multiplicity: s = singlet, d = doublet, t = triplet, q = quartet, m = multiplet, br = broad.

its 7,10-bisiodoacetate (Wani *et al.*, 1971). The full details of this work have never been published. However, the X-ray structure of the closely related semisynthetic compound taxotere (15) has been published (Gueritte-Voegelein *et al.*, 1990).

The side-chain ester of cephalomannine (16) was characterized by X-ray crystallography (Miller *et al.*, 1981) and the side-chain ester of taxol has recently been similarly characterized (Peterson *et al.*, 1991). The structure of the 2(3 → 20)-abeotaxoid, taxine A (17), has been determined by X-ray crystallography (Graf *et al.*, 1982).

Brevifoliol and some baccatin VI derivatives have been shown to have a rearranged 11(15 → 1) abeo taxane skeleton and not a taxane skeleton as originally reported, making abeo taxanes an emerging major structural type of taxoids (Appendino *et al.*, 1993a).

A diterpenoid with a novel basic skeleton has been isolated from *Taxus chinensis*, the structure of which was elucidated by spectral means and unambiguously determined by X-ray analysis (Fuji *et al.*, 1992).

X-ray crystallographic analysis of 1β-hydroxybaccatin I provided unambiguous characterization for the structures and relative stereochemistries of the other taxanes (Shen and Chen, 1997a) (Figure 4.31).

Table 4.3 ¹H NMR spectra of selected taxoids

Positions	Taxezopidine K (1-1'b)	Taxuspine D (1-1c)	Decinnamoyl-1-hydroxy-taxinine J (1-1e)
1	1.89(m)	1.89(m)	
2	5.76(d, 6.8)	5.80(d, 6.8)	5.55(d, 6.2)
3	3.61(d, 6.8)	3.69(d, 6.8)	3.66(d, 6.2)
5	5.24(m)	5.34(dd, 9.4, 6.2)	4.23(t, 2.5)
6α	2.03(m)	2.06(m)	1.95(m)
6β			1.59(m)
7	4.80(m)	4.84(m)	5.67(dd, 11.5, 5.4)
9	5.09(d, 4.4)	4.84(d, 4.8)	5.85(d, 10.9)
10	4.33(d, 4.4)	5.60(d, 4.8)	6.25(d, 10.9)
13		5.41(m)	5.95(ddq, 10.1, 5.3, 1.5)
14α	2.50(dd, 18.6, 8.0)	2.36(d, 17.8)	1.94(dd, 15.6, 5.3)
14β	2.30(d, 18.6)	2.57(dd, 17.8, 6.8)	2.34(dd, 15.6, 10.1)
16	1.52(s)	1.49(s)	1.63(s)
17	1.20(s)	1.17(s)	1.10(s)
18	1.50(s)	1.57(s)	2.20(d, 1.5)
19	1.57(s)	1.53(s)	0.96(s)
20α	5.20(s)	5.20(s)	5.29(s)
20β	5.04(s)	5.08(s)	4.73(s)
22	6.45(d, 16.0)	6.46(d, 16.0)	
23	7.70(d, 16.0)	7.70(d, 16.0)	
25,29	7.53(m)	25 7.54(m)	
26,28	7.39(m)	26 7.40(m)	
27	7.39(m)	7.40(m)	
2-OAc	1.96(s)	1.96(s)	Ac 2.09, 2.08, 2.03, 2.02, 1.98(s)
7-OAc	2.19(s)	2.19(s)	
9-OAc	2.02(s)	2.00(s)	
10-OAc		2.02(s)	
13-OAc	2.20(s)	2.20(s)	

Notes
Numbers in parenthesis after compounds names are the numbers used in Chapter 3.
Measured in CDCl₃, δ values with TMS as internal standard.
The values in parenthesis are coupling constants (Hz).
Multiplicity: s = singlet, d = doublet, t = triplet, q = quartet, m = multiplet, br = broad.

Crystal and solution state conformations of two taxoids, taxinine, and taxinine B, were analyzed by X-ray crystallographic analysis and ROE experiments. This study demonstrated that in solid state, both taxinine and taxinine B took similar cage conformations in which the A-ring and the cinnamoyl side-chain in the C-ring were spatially very close to each other. These conformations were also observed in their solution conformations deduced by ROE correlations in CDCl₃, Monte Carlo simulation using MM2 force field, and semiempirical molecular orbital calculation using the PM3 method (Morita *et al.*, 1997c) (Figure 4.32).

The stereochemistry of C-6,7-epoxy paclitaxel was determined by single crystal X-ray analysis of the baccatin derivative to be 6α, 7α (Altstadt *et al.*, 1998).

Table 4.4 ^{1}H NMR spectra of selected taxoids

Positions	2,20-Dihydroxy-9-acetoxy-taxa-4(20), 11-dien-13-one (1-3a)	^{1}H-^{1}H COSY	Taxezopidine B (1-3b)	Taxumairol (1-4d)	^{1}H-^{1}H COSY
1	2.52(dd, 7.4, 4.1)	H-2β, 14b	2.06(br s)	2.00(m)	
2	4.13(dd, 4.1, 9.1)	H-1β, 3α	6.30(br s)	4.24(dd, 11.7, 3.9)	H-1, 3
3	2.73(br d, 9.1)	H-2β, 5, 20		2.78(d, 3.9)	H-2
5	5.44(br s)	H-3α, 6, 20	4.10(br s)	5.26(br s)	H-6
6	2.10(m)	H-5, 7	6α: 1.75(m)	2.06(m)	H-5
			6β: 1.50(m)	1.87(m)	
7	1.55(m)	H-6	7α: 1.96(m)	5.58(dd, 12, 4.8)	H-6
			7β: 1.48(m)		
9	5.29(dd, 5.0, 8.5)	H-10α, 10β	6.11(d, 7.1)	5.82(d, 11.4)	H-10
10α	2.72(dd, 5.0, 14.3)	H-9β, 10β	10: 5.40(d, 7.1)	6.17(d, 11.4)	H-9
10β	2.90(dd, 8.5, 14.3)	H-9β, 10α			
12			2.91(q, 6.8)		
13				5.85(m)	H-14, Me-18
14α	2.45(d, 19.8)	H-14β	14: 2.86(m)	2.02(m)	
14β	2.75(dd, 7.4, 19.8)	H-1β, 14α		2.60(m)	
16	1.19(s)		1.71(s)	1.70(s)	
17	1.50(s)		1.00(s)	1.12(s)	
18	1.83(s)		1.52(d, 6.8)	2.27(s)	H-13
19	0.92(s)		1.21(s)	1.16(s)	
20α	4.18(br d, 2.0, 12.4)	H-3, 5, 20β	4.60(d, 12.0)	5.28(d, 11.8)	H-20
20β	4.32(br d, 1.6, 12.4)	H-3, 5, 20α	3.98(d, 12.0)	4.64(d, 11.8)	H-20
5-OH			2.86(br s)		
11-OH			3.01(br s)		
20-OH			1.75(br s)		
2-OAc			1.66(s)		
5-OAc				2.31(s)	
7-OAc				2.11(s)	
9-OAc	2.10(s)		1.65(s)	2.08(s)	
10-OAc			1.60(s)	2.05(s)	
13-OAc				1.98(s)	
OBz					
o				8.06(d, 7.5)	m-Bz
m				7.47(t, 7.5)	o,p-Bz
p				7.59(t, 7.5)	m-Bz
2αOH				3.80(d, 11.7)	H-2

Notes

Numbers in parenthesis after compounds names are the numbers used in Chapter 3.

Measured in CDCl$_3$, δ values with TMS as internal standard.

The values in parenthesis are coupling constants (Hz).

Multiplicity: s = singlet, d = doublet, t = triplet, q = quartet, m = multiplet, br = broad.

Table 4.5 ¹H NMR spectra of selected taxoids

Positions	Taxayunnanine G (*4a*)	Taxayunnanine H (*4b*)	10-Deacetylyunnanxane (*4b*)	¹H-¹H COSY
1	1.69(br d, 1.9)	1.78(d, 2.1)	1.88(d, 2.4)	H-2β
2	5.36(dd, 6.2, 2.0)	5.32(dd, 6.3, 2.2)	5.36(dd, 6.6, 2.4)	H-1β, 3α
3	2.82(d, 6.1)	3.19(d, 6.2)	2.92(d, 6.6)	H-2β
5	5.19(t, 2.7)	4.14(t, 2.8)	5.28(t, 2.9)	H-6α, 6β
6α	1.72(m)	1.66(m)	1.79(m)	H-5β, 7α, 7β
6β	1.72(m)	1.66(m)	1.79(m)	H-5β, 7α, 7β
7α	1.15(m)	1.05(m)	1.92(m)	H-6α, 6β, 7β
7β	1.86(m)	2.04(m)	1.25(m)	H-6α, 6β, 7α
9α	1.61(dd, 14.9, 5.6)	1.61(dd, 14.8, 5.6)	1.67(dd, 14.8, 5.6)	H-9β, 10α
9β	2.29(dd, 14.8, 11.8)	2.26(dd, 14.7, 11.7)	2.33(dd, 14.8, 11.7)	H-9α, 10α
10	5.05(dd, 11.8, 5.6)	5.12(dd, 11.7, 5.6)	5.06(dd, 11.7, 5.6)	H-9α, 9β
13α	2.56(dd, 18.5, 8.8)	2.74(dd, 18.8, 9.3)	2.82(dd, 18.9, 9.2)	H-13β, 14α
13β	2.46(dd, 18.6, 5.3)	2.29(dd, 18.4, 4.3)	2.37(dd, 18.9, 4.9)	H-13α, 14β
14	4.04(dd, 8.5, 5.5)	4.99(dd, 9.3, 4.7)	5.02(dd, 9.2, 4.9)	H-13α, 13β
16	1.18(s)	1.13(s)	1.69(s)	
17	1.66(s)	1.68(s)	1.18(s)	
18	1.94(s)	1.95(s)	1.97(s)	
19	0.78(s)	0.78(s)	0.84(s)	
20α	5.21(br s)	5.06(br s)	4.81(br s)	
20β	4.84(br s)	4.73(t, 1.4)	5.25(br s)	
2'			2.40(q, 7.2)	H-5'
3'			3.85(f, 6.3)	H-4'
Me-3'		1.08(t, 7.6)		
Me-4'			1.21(d, 6.3)	H-3'
Me-5'			1.15(d, 7.2)	H-2'
2-OAc	2.03(s)	2.00(s)	2.00(s)	
5-OAc	2.09(s)		2.16(s)	

Notes
Numbers in parenthesis after compounds names are the numbers used in Chapter 3.
Measured in CDCl₃, δ values with TMS as internal standard.
The values in parenthesis are coupling constants (Hz).
Multiplicity: s = singlet, d = doublet, t = triplet, q = quartet, m = multiplet, br = broad.

Table 4.6 ^{1}H NMR spectra of selected taxoids

Positions	2α-Deacetyltaxinine J Taxuspinanane G (6f)	5α-Cinnamoyloxy-2α,13α-dihydroxy-9α,10β-diacetoxy-4(20),11-diene (6b)	1-Hydroxy-2-deacetoxytaxinine (6i)	^{1}H-^{1}H COSY
1	2.16(1H, m)	1.94	2.00(m)	H-3
2	4.30(1H, m)	4.13(dd)	2.97(d, 5.5)	H-2
3	3.20(1H, d, 6.1)	3.19(d)	5.60(t, 2.8)	H-6
5	5.42(1H, m)	5.83(t)	1.97(m)	H-5, 7
6α	2.07(1H, m)	2.00(m)		
6β	1.90(dd, 5.1, 11.3)			
7	5.48(1H, m)	1.35(m)	5.67(dd, 11.4, 5.0)	H-6
9	5.70(1H, d, 9.4)	5.45(d)	5.95(d, 11.2)	H-10
10	6.19(1H, d, 9.4)	5.86(d)	6.31(d, 11.2)	H-9
13	5.87(1H, br t, 9.7)	4.25(br d)	5.96(br s)	H-14
14α	2.65(1H, m)	2.64(ddd)	2.58(dd, 14.8, 9.9)	H-13
14β	1.34(1H, m)	1.93(m)	1.53(dd, 14.8, 6.6)	
16	1.59(3H, s)	1.16(s)	1.18(s)	
17	1.15(3H, s)	1.14(s)	1.63(s)	
18	2.25(3H, s)	1.78(s)	2.33(s)	
19	1.14(3H, s)	1.58(s)	0.93(s)	
20α	5.58(1H, s)	5.46(s)	5.09(s)	
20β	5.49(1H, s)	5.55(s)	5.43(s)	
2′	6.66(1H, d, 16.0)	Cinn	7.77(d, 16.0)	
3′	7.78(1H, d, 16.0)	6.67(d)	6.55(d, 16.0)	
3′-Ph (q)		7.79(d)	Ph	
(o)	7.51(2H, m)	7.50(d)	7.49(m)	
(m)	7.40(2H, m)	7.40(d)	7.40(m)	
(p)	7.40(1H, m)			
7-OAc	1.81(3H, s)	Me	Me	
9-OAc	2.02(3H, s)	2.1(s)	1.73(s)	
10-OAc	2.04(3H, s)	2.28(s)	2.00(s)	
13-OAc	2.05(3H, s)		2.05(s)	
			2.07(s)	

Notes

Numbers in parenthesis after compounds names are the numbers used in Chapter 3.

Measured in CDCl$_3$, δ values with TMS as internal standard.

The values in parenthesis are coupling constants (Hz).

Multiplicity: s = singlet, d = doublet, t = triplet, q = quartet, m = multiplier, br = broad.

Table 4.7 ¹H NMR spectra of selected taxoids

Positions		19-Debenzoyl-19-acetyl-taxinine M (7g)	Taxezopidine L (7b) Shigemori (1999)	Taxezopidine L (7b) Fukushima (1999)
1		2.42(br d, 11.5)	2.40(dd, 11.2, 2.4)	2.53(m)
2		6.16(dd, 10.2, 2.0)	6.18(dd, 10.0, 2.4)	6.18(m)
3		3.63(d, 10.2)	3.46(d, 10.0)	3.47(d, 9.6)
5		4.38(br t, 2.5)	5.50(m)	5.37(m)
6α		2.16(m)	2.28(dd, 14.8, 6.0)	2.37(m)
6β		1.62(m)	1.73(m)	2.20(m)
7		5.44(dd, 10.5, 6.1)	5.52(m)	5.54(m)
9		5.48(d, 3.1)	5.48(d, 3.2)	5.54(m)
10		5.28(d, 3.1)	5.36(d, 3.2)	5.54(m)
14α		2.70(d, 19.2)	2.97(dd, 18.4, 11.2)	2.65(dd, 19.2)
14β		3.00(dd, 19.2, 11.5)	2.54(d, 18.4)	2.98(dd, 12.0, 19.2)
16		1.59(s)	1.59(s)	1.21(s)
16'				
17α		4.19(d, 8.0)	4.20(d, 8.0)	3.69(d, 8.0)
17β		3.69(d, 8.0)	3.69(d, 8.0)	4.20(d, 8.0)
18		1.17(s)	1.22(s)	1.25(s)
19α		4.40(d, 12.3)	4.44(d, 12.4)	4.34(d, 12.4)
19β		4.28(d, 12.3)	4.34(d, 12.4)	4.44(d, 12.4)
20α		5.31(s)	5.51(s)	4.59(s)
20β		4.44(s)	4.59(s)	5.54(m)
Me-17	22		6.80(d, 16.0)	6.80(d, 16.0)
Me-18	23		7.92(d, 16.0)	7.92(d, 16.0)
Me-19	25,29		7.81(m)	7.32–7.81(m)
	26,28		7.40(m)	
	27		7.40(m)	
OAc	2-OAc	2.18(s)	1.97(s)	1.97(s)
	7-OAc	2.11(s)	1.97(s)	1.97(s)
	9-OAc	2.10(s)	2.12(s)	2.20(s)
	10-OAc	1.99(s)	2.15(s)	2.12(s)
	19-OAc	1.96(s)	2.20(s)	2.15(s)

Notes

Taxezopidine L was assigned in two institutes independently.

Numbers in parenthesis after compounds names are the numbers used in Chapter 3.

Measured in CDCl₃, δ values with TMS as internal standard.

The values in parenthesis are coupling constants (Hz).

Multiplicity: s = singlet, d = doublet, t = triplet, q = quartet, m = multiplet, br = broad.

Table 4.8 ¹H NMR spectra of selected taxoids

Positions	5-Cinnamoylphototaxicin II (8-1d)		5-Cinnamoyl-10-acetyl-phototaxicin I (8b)		Taxuspine C (8-1i)
1	1.95(m)				2.18(m)
2	5.02(br s)		4.76(d)		6.14(d, 5.3)
5	5.57(t, 9)		5.57(br)		5.64(m)
6	2.20(t, 9)		2.16(m)		1.78(2H, m)
	1.60(m)		1.70(m)		
7	1.70(m)		1.9(m)		1.80(2H, m)
	1.30(m)		1.3(m)		
9	4.29(d, 9.5)		4.38(d)		5.70(d, 8.8)
10			5.4(d)		5.68(d, 8.8)
12	3.49(d, 7)		3.48(q)		3.55(q, 6.8)
14	2.80(d, 20)		3.03(d)		2.59(d, 20.5)
	2.50(dd, 20, 7)		2.35(d)		2.50(dd, 20.5, 6.8)
16	1.22(s)		1.1(s)		1.70(s)
17	1.45(s)		1.38(s)		1.22(s)
18	1.52(d, 7)		1.29(d)		1.29(d, 6.8)
19	1.43(s)		1.38(s)		1.32(s)
20	5.75(s)		5.84(br s)		5.85(s)
	5.61(s)		5.59(br s)		5.71(s)
Ph(o)	7.57(m)	2',6'	7.55(m)	22	6.39(d, 16.1)
Ph(m)	7.40(m)	3',4',5'	7.38(m)	23	7.67(d, 16.1)
Ph(p)	7.40(m)	7'	7.66(d)	25	7.56(2H, m)
		8'		26	7.38(2H, m)
2'	6.38(d, 16)	OAc	6.39(d)	27	7.38(m)
3'	7.67(d, 16)		2.15(s)	2-OAc	2.07(s)
				9-OAc	2.05(s)
				10-OAc	2.05(s)

Notes
Numbers in parenthesis after compounds names are the numbers used in Chapter 3.
Measured in CDCl₃, δ values with TMS as internal standard.
The values in parenthesis are coupling constants (Hz).
Multiplicity: s = singlet, d = doubler, t = triplet, q = quartet, m = multipler, br = broad.

Table 4.9 ^{1}H NMR spectra of selected taxoids

Positions	1β,9α-Dihydroxy-4β,20-epoxy-2α,5α,7β,10β,13α-pentaacetoxy-11-ene (9-1e)	Taxumairol C (9-1g)	1β-Hydroxy-7,9-deacetylbaccatin 1 (9-1j)
2	5.27(d, 3.8)	5.36(d, 3.3)	5.331(d, 3.7)
3	3.03(d, 3.8)	3.07(d, 3.3)	3.071(d, 3.7)
5	4.14(br d, 3.1)	4.21(br s)	4.202(t, 3.1)
6α	2.04(overlap)	2.05(m)	2.038(ddd, 3.7, 12.0, 15.1)
6β	1.78(m)	1.83(m)	1.887(ddd, 2.9, 4.4, 14.6)
7	5.32(dd, 11.5, 4.2)	4.21(m)	4.253(dd, 4.6, 11.7)
9	4.31(dd, 10.6, 8.8)	4.41(d, 10)	4.591(dd, 4.4, 10.7)
10	5.99(d, 10.6)	4.97(d, 10)	6.035(d, 10.7)
13	6.05(m)	6.08(m)	6.051(ddd, 1.5, 6.6, 9.7)
14α	1.80(dd, 14.8, 6.6)	1.83(m)	2.510(dd, 9.7, 14.9)
14β	2.44(dd, 14.8, 9.8)	2.52(m)	1.864(dd, 6.6, 14.7)
16	1.48(s)	1.59(s)	1.228(s)
17	1.18(s)	1.30(s)	1.527(s)
18	2.03(s)	2.00(s)	2.157(d, 1.5)
19	1.37(s)	1.39(s)	1.388(s)
20α	3.42(d, 5.2)	3.49(d, 5.1)	3.474(d, 5.4)
20β	2.26(d, 5.6)	2.31(d, 5.1)	2.322(d, 5.4)
OAc	2.15(s)	2.10(s)	2.184(s)
	2.05(s)	2.06(s)	2.134(s)
	2.03(s)	2.18(s)	2.099(s)
	1.97(s)		2.051(s)
	1.97(s)		
1-OH			1.944(s)
7-OH			4.725(s)
9-OH	2.78(d, 8.8)		3.574(d, 4.4)

Notes
Numbers in parenthesis after compounds names are the numbers used in Chapter 3.
Measured in CDCl$_3$, δ values with TMS as internal standard.
The values in parenthesis are coupling constants (Hz).
Multiplicity: s = singlet, d = doublet, t = triplet, q = quartet, m = multiplet, br = broad.

Table 4.10 ¹H NMR spectra of selected taxoids

Positions	Taxuspinanane C (**10-5b**)	Taxuspinanane D (**10-5a**)	14β-Hydroxy-10-deacetylbaccatin III (**10-6a**)
2	5.86(1H, d, 5.8)	5.87(1H, d, 5.9)	5.54(d, 7.1)
3	3.19(1H, d, 5.8)	3.19(1H, d, 5.9)	3.75(d, 7.1)
5	4.86(1H, br d, 9.2)	4.87(1H, d, 9.0)	4.90(br d, 9.1)
6α	2.51(1H, ddd, 7.5, 9.2, 14.5)	2.51(1H, m)	2.28(m)
6β	1.90(1H, m)	1.91(1H, m)	1.64(m)
7	4.38(1H, m)	4.44(1H, dd, 7.5, 17.6)	4.07(m)
9	4.30(1H, m)	4.44(1H, d, 10.6)	
10	5.00(1H, d, 10.2)	6.28(1H, d, 10.6)	7.13(d, 2.7)
13			4.41(br t, 5.5)
14α	2.87(1H, d, 19.6)	2.89(1H, d, 19.6)	3.79(t, 5.5)
14β	2.65(1H, d, 19.6)	2.64(1H, d, 19.6)	
16	1.73(3H, s)	1.70(3H, s)	0.92(s)
17	1.31(3H, s)	1.25(3H, s)	0.97(s)
18	2.00(3H, s)	2.10(3H, s)	1.91(s)
19	1.78(3H, s)	1.78(3H, s)	1.53(s)
20α	4.35(1H, d, 8.1)	4.35(1H, d, 8.5)	3.97(d, 8.2)
20β	4.16(1H, d, 8.1)	4.15(1H, d, 8.5)	3.99(d, 8.2)
4-OAc	2.18(3H, s)	2.18(3H, s)	2.24(s)
2-OBz	8.08(2H, dd, 1.4, 7.5)	8.08(2H, dd, 1.3, 7.5)	
	7.50(2H, t, 7.5)	7.50(2H, t, 7.5)	
	7.62(1H, t, 7.5)	7.63(1H, t, 7.5)	
10-OAc		2.19(3H, s)	
1-OH			4.43(s)
7-OH			5.01(d, 7.1)
10-OH			4.82(d, 2.7)
13-OH			5.53(d, 5.5)
14-OH			6.75(d, 5.5)

Notes
Numbers in parenthesis after compounds names are the numbers used in Chapter 3.
Measured in CDCl₃, δ values with TMS as internal standard.
The values in parenthesis are coupling constants (Hz).
Multiplicity: s = singlet, d = doublet, t = triplet, q = quartet, m = multiplet, br = broad.

Table 4.11 [italic] ^{1}H NMR spectra of selected taxoids

Positions	7-epi-10-Deacetyltaxol C (Taxuspinanane E) (11-2**k**)	Taxol C (11-2a)	Taxuspinanane A (11-2e)
2	5.74(1H, d, 7.4)	5.67(1H, d, 7.0)	5.95(1H, d, 7.0)
3	3.92(1H, d, 7.4)	3.78(1H, d, 6.9)	4.02(1H, d, 7.0)
5	4.89(1H, dd, 4.2, 8.6)	4.93(1H, br d, 9.5)	4.89(1H, dd, 1.8, 9.5)
6α	2.32(2H, m)	2.54(1H, m)	2.58(1H, m)
6β		1.88(1H, m)	2.11(1H, m)
7	3.67(1H, m)	4.40(1H, dd, 6.5, 10.7)	4.68(1H, m)
10	5.44(1H, s)	6.28(1H, s)	6.61(1H, s)
13	6.25(1H, m)	6.22(1H, m)	6.51(1H, br dd)
14α	2.34(1H, m)	2.30(1H, m)	2.58(1H, m)
14β	2.26(1H, m)	2.20(1H, m)	2.58(1H, m)
16	1.10(3H, s)	1.15(3H, s)	1.15(3H, s)
17	1.23(3H, s)	1.27(3H, s)	1.20(3H, s)
18	1.78(3H, s)	1.82(3H, s)	1.97(3H, s)
19	1.72(3H, s)	1.68(3H, s)	1.98(3H, s)
20α	4.41(1H, d, 8.8)	4.18(1H, d, 8.5)	4.35(1H, d, 8.6)
20β	4.38(1H, d, 8.8)	4.29(1H, d, 8.5)	4.36(1H, d, 8.6)
2'	4.68(1H, dd, 2.5, 2.4)	4.67(1H, d, 2.5)	4.63(1H, br dd)
3'	5.60(1H, dd, 2.5, 8.8)	5.57(1H, dd, 2.6, 8.9)	5.82(1H, dd, 2.0, 9.2)
5'	2.19(2H, t, 7.4)	2.20(2H, t, 7.1)	
6'	1.57(2H, m)	1.24(2H, m)	1.82(2H, m)
7'	1.23(2H, m)	1.24(2H, m)	1.27(1H, m)
			1.55(1H, m)
8'	1.23(2H, m)	1.24(2H, m)	1.14(1H, m)
9'	0.82(3H, t, 6.9)	0.84(3H, t, 6.7)	0.95(1H, m)
			1.12(1H, m)
10'			0.75(3H, t, 7.1)
11'			0.70(3H, d, 6.4)
7-OH	4.72(1H, d, 11.9)		2.97(1H, d, 4.2)
10-OH	4.12(1H, s)	2.24(1H, s)	
2'-OH	3.41(1H, d, 4.5)		3.76(1H, br d)
NH	6.47(1H, d, 8.8)	6.21(1H, d, 8.4)	
1-OH			2.04(1H, d, 4.2)
4-OAc	2.47(3H, s)	2.35(3H, s)	2.18(3H, s)
10-OAc			1.80(3H, s)
3'-Ph	7.39(4H, m)	7.40(4H, m)	7.39(2H, d, 7.3)
	7.35(1H, m)	7.34(m)	7.16(2H, m)
			7.11(1H, m)
2-OBz	8.16(2H, dd, 1.4, 7.3)	8.11(2H, dd, 1.2, 8.2)	8.33(2H, d, 1.3, 8.0)
	7.53(2H, t, 7.3)	7.61(2H, t, 7.3)	7.22(2H, m)
	7.62(1H, t, 7.3)	7.51(1H, t, 7.6)	7.18(1H, m)

Notes
Numbers in parenthesis after compounds names are the numbers used in Chapter 3.
Measured in $CDCl_3$, δ values with TMS as internal standard.
The values in parenthesis are coupling constants (Hz).
Multiplicity: s=singlet, d=doublet, t=triplet, q=quartet, m=multiplet, br=broad.

Table 4.12 ^{1}H NMR spectra of selected taxoids

Positions	Taxuspine X (12i)	Taxuspine U (12j)	2-Deacetyltaxacultriene A (12k)	^{1}H-^{1}H
1	1.81(m)	1.78(dd, 7.8, 4.2)	1.82(m)	H-14α, 2
2	5.80(dd, 11.1)	4.68(dd, 10.2, 4.2)	4.75(dd, 11.3, 4.1)	H-1, 3
3	5.99(d, 11.1)	5.59(d, 10.2)	6.51(br d, 11.3)	H-2, 5
5	5.75(m)	5.54(br s)	4.45(br s)	H-3, 6α, 6β, OH
6α	2.57(m)	2.43(ddd, 15.7, 8.4, 3.0)	2.03(m)	H-5, 6β, 7α
6β	2.14(m)	2.06(m)	2.61(ddd, 7.7, 12.2, 3.9)	H-5, 6α, 7β
7	5.50(d, 9.1)	4.40(d, 7.7)	5.05(br d, 7.7)	H-6α, 6β
10	7.27(s)	6.83(s)	6.91(d, 1.4)	H-19
13	5.28(d, 9.1)	5.42(d, 10.1)	5.32(br d, 8.8)	H-14β, 18
14α	2.55(m)	2.56(ddd, 16.4, 10.1, 8.4)	2.28(br d, 16.5)	H-1, 14β
14β	2.03(m)	2.11(m)	2.47(ddd, 16.5, 8.8, 7.1)	H-1, 13, 14α
16	1.12(s)	1.10(s)	1.18(s)	
17	1.30(s)	1.23(s)	1.12(s)	
18	2.28(s)	2.02(s)	1.91(s)	
19	1.65(s)	1.52(s)	1.64(d, 1.4)	H-10
20α	4.93(d, 12.9)	4.31(d, 12.3)	4.98(d, 12.6)	H-20β
20β	4.41(d, 12.9)	3.78(d, 12.3)	4.01(d, 12.6)	H-20α
22	6.60(d, 15.1)			
23	7.87(d, 15.1)			
25	7.53(m)			
26	7.40(m)			
27	7.40(m)			
2-OAc	1.98(s)			
5-OAc		2.21(s)	2.11(s)	
7-OAc	1.98(s)			
9-OAc	1.88(s)	2.21(s)		
10-OAc	2.06(s)	1.98(s)		
13-OAc	1.95(s)	2.12(s)		
20-OAc	2.20(s)			
OH			3.91(d, 4.1)	H-5

Notes

Numbers in parenthesis after compounds names are the numbers used in Chapter 3.

Measured in CDCl$_3$, δ values with TMS as internal standard.

The values in parenthesis are coupling constants (Hz).

Multiplicity: s = singlet, d = doublet, t = triplet, q = quartet, m = multiplet, br = broad.

Table 4.13 ¹H NMR spectra of selected taxoids

Positions	13-Deacetylcanadense (12c)	Taxamairol M (12-2a)	(3E,7E)-2α,10β,13α,20-Tetraacetoxy-5α-hydroxyl-3,8-seco-taxa-3,7,11-triene-9-one (12-1a)	¹H-¹H
1	1.75(br t)	2.15(m)	1.83(br t, 5.1)	H-14α, 2
2	5.72(dd)	5.73(dd, 4.5, 11.5)	5.72(dd, 11.0, 5.1)	H-1, 3
3	6.16(dd)	6.50(d, 11.5)	6.45(br d, 11.0)	H-2, 5
5	4.63(s)	4.47(br s)	4.70(br s)	H-3, 6α, 6β
6α	2.03(m)	1.90(m)	2.54(br d, 14.7)	H-5, 6β, 7α
6β	2.71(ddd)	2.60(m)	2.84(ddd, 3.6, 12.1, 14.7)	H-5, 6α, 7β
7	5.09(d)	5.10(d, 8.0)	6.41(br d, 12.1)	H-6α, 6β, 19
10	6.98(d)	6.93(s)	6.70(s)	
13	4.16(br d)	5.29(d, 9.0)	5.37(br d, 8.7)	H-14β, 18
14α	2.35(br d)	2.10(m)	2.19(br d, 16.5)	H-1, 14β
14β	2.45(ddd)	2.52(m)	2.52(m)	H-1, 13, 14α
16	1.21(s)	1.26(s)	0.90(s)	
17	1.10(s)	3.45(d, 11) 3.24(d, 11)	1.26(s)	
18	2.13(s)	1.98(s)	1.93(s)	
19	1.62(d)	1.65(s)	1.78(br s)	H-7
20α	4.56(d)	4.84(d, 12.7)	4.80(d, 12.9)	H-20β
20β	3.48(d)	4.38(d, 12.7)	4.24(d, 12.9)	H-20α
2-OAc	2.04(s)	OAc: 2.18	2.02(s)	
7-OAc	2.08(s)	OAc: 2.11		
9-OAc	2.21(s)	OAc: 2.03		
10-OAc	1.95(s)	OAc: 2.01	2.19(s)	
13-OAc		OAc: 2.22	2.16(s)	
20-OAc		OAc: 1.96(s)	2.05(s)	

Notes
Numbers in parenthesis after compounds names are the numbers used in Chapter 3.
Measured in CDCl₃, δ values with TMS as internal standard.
The values in parenthesis are coupling constants (Hz).
Multiplicity: s=singlet, d=doublet, t=triplet, q=quartet, m=multipler, br=broad.

Table 4.14 ^{1}H NMR spectra of selected taxoids

Positions	Taxuspine K (12-3a)	Taxezopidine J (12-4b)	Decinnamoyltaxinine B-11,12-oxide (12-5a)
1		2.18(m)	1.99(m)
2	4.26(d, 7.0)	5.48(dd, 5.6, 1.5)	5.78(dd, 5.9, 1.8)
3	2.81(d, 7.0)	3.13(d, 5.6)	3.27(d, 5.9)
5	4.36(m)	5.41(br s)	4.34(d, 3.0)
6α	2.05(m)	1.81(m)	2.00(m)
6β	1.76(m)	1.66(m)	1.70(m)
7α	5.11(dd, 11.0, 4.0)	1.87(m)	5.66(dd, 11.6, 5.4)
7β		1.83(m)	
9	5.18(d, 4.5)	5.71(d, 10.8)	5.43(d, 11.2)
10	6.05(d, 4.5)	6.09(d, 10.8)	6.02(d, 11.2)
13	6.02(t, 8.0)		
14α	2.40(m)	2.05(m)	2.30(d, 19.9)
14β	2.32(m)	1.73(d, 16.0)	2.64(dd, 19.9, 8.8)
16	1.51(s)	1.52(s)	1.83(s)
17α	1.18(s)	3.48(d, 8.1)	1.83(s)
17β		3.09(d, 8.1)	
18	1.80(s)	2.23(s)	2.07(s)
19	1.45(s)	0.92(s)	1.03(s)
20α	3.71(d, 11.7)	5.37(s)	5.35(s)
20β	4.34(d, 11.7)	4.93(s)	5.01(s)
21			
22		6.70(d, 16.0)	
23		7.70(d, 16.0)	
25,29		7.60(m)	
26,28		7.37(m)	
27		7.37(m)	
2-OAc	2.14(s)	2.06(s)	OAc: 2.02(s)
4-OAc	2.11(s)		2.02(s)
7-OAc	2.06(s)		2.02(s)
9-OAc	2.03(s)	2.04(s)	2.03(s)
10-OAc	2.10(s)	1.98(s)	
13-OAc			

Notes
Numbers in parenthesis after compounds names are the numbers used in Chapter 3.
Measured in CDCl$_3$, δ values with TMS as internal standard.
The values in parenthesis are coupling constants (Hz).
Multiplicity: s = singlet, d = doublet, t = triplet, q = quartet, m = multiplet, br = broad.

Table 4.15 ¹H NMR spectra of selected taxoids

Positions	Taxuspinanane B (13-2d)	Taxustone (13-2a)	Teixidol (13-1b)	5α-Hydroxy-9α,10β,13α-triacetoxy-11(15→1)abeotaxa-4(20),11-diene (13-1c)
2α	6.04(1H, d, 9.2)	5.67(br s)	5.90(d, 9.2)	2.17(dd, 14.2, 9.1)
2β				1.40(d, 14.2)
3	2.79(1H, d, 9.2)	3.17(br s)	3.42(d, 9.2)	2.84(br d, 9.1)
5	4.72(1H, br t)	4.23(br s)	1.6(br s)	4.29(br d)
6α	1.89(1H, m)			1.81(m)
6β	2.06(1H, m)			1.62(m)
7α	4.82(1H, t, 9.0)		1.73(br m)	2.07(m)
7β				1.60(m)
9	5.11(1H, d, 2.8)	5.94(d, 9.0)	5.73(d, 10)	5.76(d, 10.4)
10	6.30(1H, d, 2.8)	4.69(br d, 9.0)	6.15(d, 10)	6.14(d, 10.4)
13			5.5(t, 7.5)	5.45(br t, 7.4)
14α	2.75(1H, d, 18.6)	2.28(d, 18.5)	2.35(dd, 7.5, 15)	1.28(m)
14β	2.45(1H, d, 18.6)	2.63(d, 18.5)	1.9(dd, 7.4, 15)	2.48(dd, 13.7, 7.1)
16	1.14(3H, s)	1.05(s)	1.17(s)	0.74(s)
17	0.81(3H, s)	1.05(s)	1.15(s)	1.32(s)
18	1.95(3H, s)	2.03(s)	1.84(s)	1.80(d, 1.1)
19	1.83(3H, s)	1.32(s)	0.94(s)	1.13(s)
20α	4.78(1H, s)	4.39(s)	5.08(s)	5.05(br d)
20β	5.48(1H, s)	5.09(s)	4.44(s)	4.68(br d)
5-OH	1.89(1H, s)			
15-OH	1.17(1H, br d)			
2-OAc	1.97(3H, s)	OAc: 1.93(s)	OAc: 2.05(s)	
7-OAc	1.95(3H, s)	OAc: 2.13(s)	OAc: 2.0(s)	
9-OAc	1.99(3H, s)		OAc: 1.98(s)	2.03(s)
10-OAc			OAc: 1.97(s)	1.99(s)
13-OAc				2.01(s)
10-OBz	8.10(2H, d, 7.2) 7.47(2H, t, 7.2) 7.50(1H, t, 7.2)			

Notes
Numbers in parenthesis after compounds names are the numbers used in Chapter 3.
Measured in CDCl₃, δ values with TMS as internal standard.
The values in parenthesis are coupling constants (Hz).
Multiplicity: s = singlet, d = doublet, t = triplet, q = quartet, m = multiplet, br = broad.

Table 4.16 ^{1}H NMR spectra of selected taxoids

Positions	Taxuspinanane F (14-1b)		4-Deacetyl-11(15→1)-abeo-baccatin VI (14-1f)		Taxuyuntin E (14-1d)
2	6.10(1H, d, 7.4)		6.60(d, 8.0)		6.19(d, 7.6)
3	3.13(1H, d, 7.4)		2.38(d, 8.0)		3.13(d, 7.6)
5	4.89(1H, d, 7.8)		4.72(dd, 8.0, 2.7)		4.91(br d, 8.4)
6α	2.56(1H, m)		2.37(m)		2.35(m)
6β	1.89(1H, m)		1.97(m)		1.77(m)
7	5.35(1H, t, 8.3)		5.26(br s)		5.42(br t, 8.7)
9	4.28(1H, d, 9.8)		5.94(br s)		5.86(d, 10.2)
10	4.62(1H, d, 9.8)		6.18(d, 9.3)		4.63(d, 10.3)
13	4.57(1H, br t, 6.8)		5.60(t, 7.6)		4.50(br t, 7.1)
14α	1.75(1H, dd, 6.8, 14.6)		2.29(dd, 14.0, 7.0)		1.86(m)
14β	2.28(1H, dd, 6.8, 14.6)		2.49(dd, 14.0, 7.3)		2.16(m)
16-Me	1.04(3H, s)		1.13(s)		1.05(s)
17-Me	1.07(3H, s)		1.13(s)		0.97(s)
18-Me	1.97(3H, s)		1.85(s)		1.83(s)
19-Me	1.94(3H, s)		1.69(s)		1.61(s)
20α	4.13(1H, d, 7.7)		3.96(d, 7.8)		4.30(d, 7.5)
20β	4.50(1H, d, 7.7)		4.56(d, 7.8)		4.50(d, 7.6)
4-OAc	2.21(3H, s)	4-OH	2.70(s)	OAc	2.05(s)
7-OAc	2.05(3H, s)	15-OH	2.44(s)	OAc	2.09(s)
2-OBz	8.01(2H, d, 7.6)		7.95(d, 7.8)	OAc	2.16(s)
	7.46(2H, t, 7.6)		7.46(t, 7.8)		
	7.60(1H, t, 7.6)		7.59(t, 7.8)		

Notes

Numbers in parenthesis after compounds names are the numbers used in Chapter 3.

Measured in CDCl$_3$, δ values with TMS as internal standard.

The values in parenthesis are coupling constants (Hz).

Multiplicity: s=singlet, d=doublet, t=triplet, q=quartet, m=multiplet, br=broad.

Table 4.17 ¹H NMR spectra of selected taxoids

Positions	Taxin B (17***d***)	¹H-¹H	2-Deacetyltaxin B (17***b***)	¹H-¹H
1	1.71(1H, dd, 6.7, 2.0)	14β, 2β	1.70(1H, dd, 7.0, 2.5)	14β, 2β
2	5.76(1H, dd, 9.7, 1.7)	20, 1β	4.88(1H, dd, 9.5, 2.4)	20, 1β
3α	1.95(1H, d, 15.6)	3β	1.91(1H, d, 15.2)	3β
3β	2.67(1H, d, 15.7)	3α	2.56(1H, d, 15.3)	3α
5	4.47(1H, br d)	6α, 6β, 20	4.47(1H, br d)	6α, 6β, 20
6	2.10(2H, m)	7α, 5β	2.10(2H, br d)	7α, 5β
7	5.20(1H, dd, 10.6, 5.3)	6α, 6β	5.21(1H, t, 8.0)	6α, 6β
10	6.34(1H, s)		6.35(1H, s)	
13	5.35(1H, br d, 10.1)	14β	5.37(1H, br d, 10.2)	14β
14α	1.97(1H, d, 17.0)	14β	2.05(1H, d, 16.7)	14β
14β	2.69(1H, ddd, 17.0, 10.0, 7.8)	13β, 1β, 14α	2.66(1H, ddd, 16.7, 9.5, 7.6)	13β, 1β, 14α
16	1.12(3H, s)		1.14(3H, s)	
17	1.28(3H, s)		1.23(3H, s)	
18	1.96(3H, d, 2.4)		1.95(3H, br s)	
19	1.28(3H, s)		1.28(3H, s)	
20	5.66(1H, br d, 9.7)		5.71(1H, br d, 9.6)	
OAc	2.19(s)		2.17(s)	
	2.15(s)		2.16(s)	
	2.07(s)		2.06(s)	
	2.05(s)			

Notes
Numbers in parenthesis after compounds names are the numbers used in Chapter 3.
Measured in CDCl₃, δ values with TMS as internal standard.
The values in parenthesis are coupling constants (Hz).
Multiplicity: s = singlet, d = doublet, t = triplet, q = quartet, m = multiplet, br = broad.

Table 4.18 ¹H NMR spectra of selected taxoids

Positions	5α-Cinnamoyl-7β,10β,13α-triacetoxy-2(3 → 20)abeotaxa-2α-hydroxy-4(20),11-diene-9-one (17-1c)	¹H-¹H	5α-Cinnamoyl-10β,13α-diacetoxy-2(3 → 20)abeotaxa-2α,7β-dihydroxy-4(20),11-diene-9-one (17-1d)	¹H-¹H
1	1.70(dd, 8.2, 2.2)	H-2, 14β	1.70(dd, 8.4, 2.2)	H-2, 14β
2	4.64(br d, 9.6)	H-1, 20	4.62(br d, 11.3)	H-1, 20
3α	2.67(d, 15.4)	H-3β, 20	2.63(d, 14.8)	H-3β, 20
3β	1.92(d, 15.4)	H-3α	1.87(d, 14.8)	H-3α
5	5.73(br d, 6.3)	H-6α, 6β	5.60(br d, 6.6)	H-6α, 6β
6α	2.01(m)	H-5, 6β, 7	1.87(m)	H-5, 6β, 7
6β	2.24(m)	H-5, 6α, 7	2.36(m)	H-5, 6α, 7
7	5.38(dd, 11.8, 4.1)	H-6α, 6β	4.32(br d, 10.7)	H-6α, 6β
10	6.35(br s)	Me-18	6.44(br s)	Me-18
13	5.40(br d, 10.3)	H-14β, Me-18	5.43(br d, 11.8)	H-14β, Me-18
14α	1.93(m)	H-14β	1.92(m)	H-14β
14β	2.68(m)	H-1, 13, 14α	2.64(m)	H-1, 13, 14α
16	1.30(s)		1.25(s)	
17	1.21(s)		1.24(s)	
18	2.03(d, 1.2)	H-10, 13	1.93(d, 1.1)	H-10, 13
19	1.13(s)		1.13(s)	
20	5.46(br d, 9.6)	H-2, 3α	5.40(br d, 11.3)	H-2, 3α
2'	6.55(d, 16.2)	H-3'	6.50(d, 15.9)	H-3'
3'	7.85(d, 16.2)	H-2'	7.82(d, 15.9)	H-2'
5'	7.52(m)	H-6', 7'	7.50(m)	H-6', 7'
6'	7.40(m)	H-5', 7'	7.42(m)	H-5', 7'
7'	7.40(m)	H-5', 6'	7.42(m)	H-5', 6'
7-OAc	1.89(s)			
10-OAc	2.08(s)		1.96(s)	
13-OAc	2.15(s)		2.18(s)	

Notes

Numbers in parenthesis after compounds names are the numbers used in Chapter 3.

Measured in CDCl₃, δ values with TMS as internal standard.

The values in parenthesis are coupling constants (Hz).

Multiplicity: s = singlet, d = doublet, t = triplet, q = quartet, m = multiplet, br = broad.

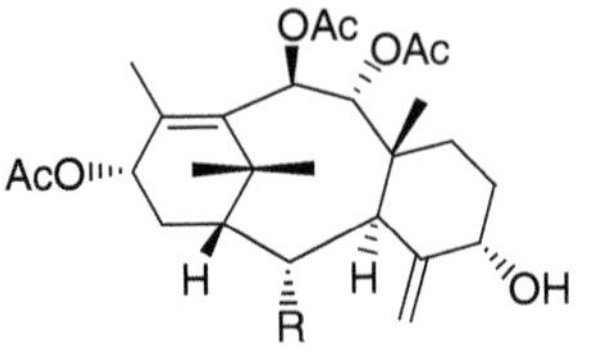

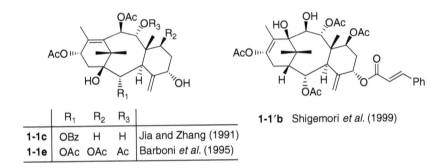

	R	
1p	OAc	Chen *et al.* (1999)
1q	H	Yang *et al.* (1996)
1r	OH	Barboni *et al.* (1995)

Figure 4.9 The structures of taxoids in Tables 4.2 and 4.20.

	R₁	R₂	R₃	
1-1c	OBz	H	H	Jia and Zhang (1991)
1-1e	OAc	OAc	Ac	Barboni *et al.* (1995)

1-1′b Shigemori *et al.* (1999)

Figure 4.10 The structures of taxoids in Tables 4.3 and 4.21.

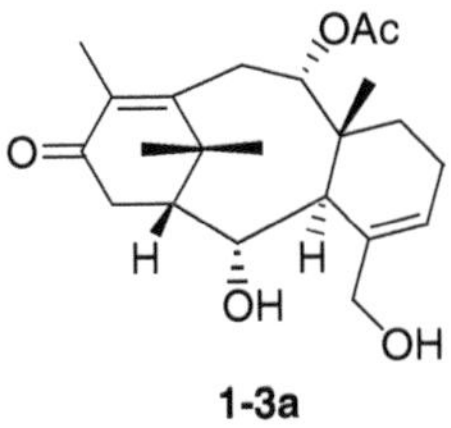

1-3a
Shi *et al.* (1999d)

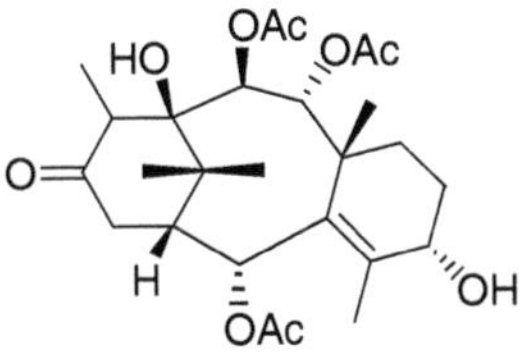

1-3b
Wang *et al.* (1998)

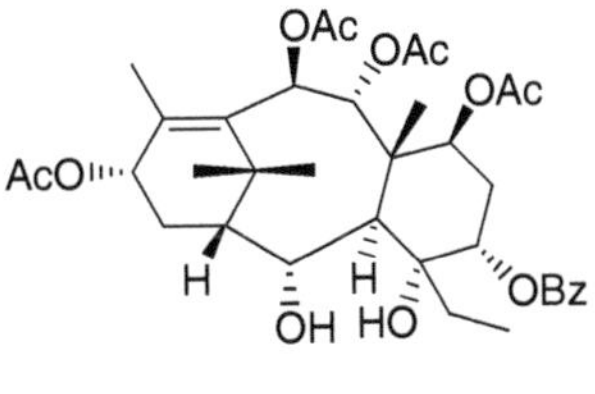

1-4d
Shen *et al.* (1996)

Figure 4.11 The structures of taxoids in Tables 4.4 and 4.22.

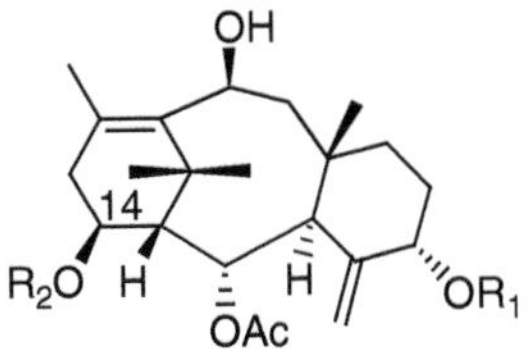

	R₁	R₂	
4a	Ac	H	Zhang *et al.* (1995)
4b	H	COCH₂CH₃	Zhang *et al.* (1995)
4h	Ac	COCHCHCH₃ with OH and CH₃	Cheng *et al.* (1996)

Figure 4.12 The structures of taxoids in Tables 4.5 and 4.23.

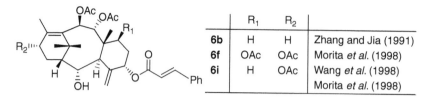

	R₁	R₂	
6b	H	H	Zhang and Jia (1991)
6f	OAc	OAc	Morita *et al.* (1998)
6i	H	OAc	Wang *et al.* (1998)
			Morita *et al.* (1998)

Figure 4.13 The structures of taxoids in Tables 4.6 and 4.24.

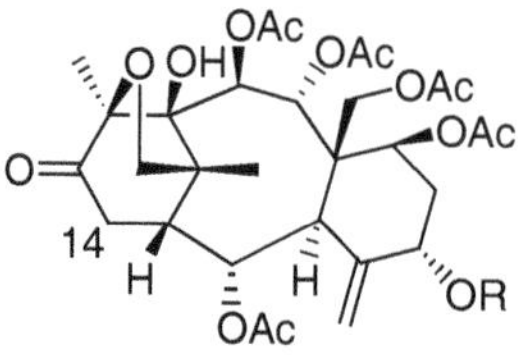

	R	
7g	H	Barboni *et al.* (1995)
7h	(acyl group, cinnamoyl)	Fukushima *et al.* (1999) Shigemori *et al.* (1999)

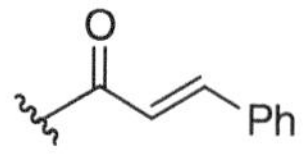

Figure 4.14 The structures of taxoids in Tables 4.7 and 4.25.

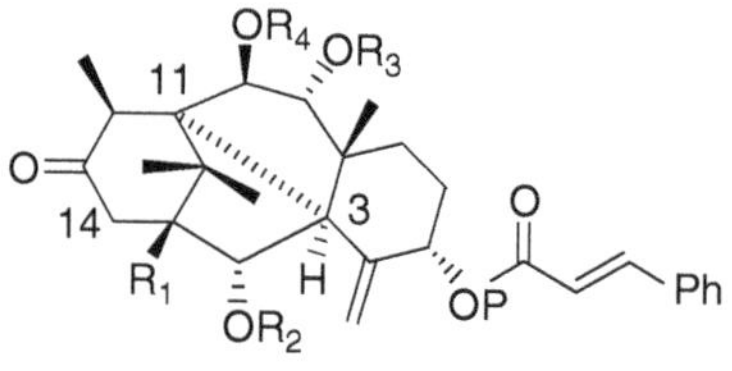

	R₁	R₂	R₃	R₄	
8b	OH	H	Ac	H	Appendino *et al.* (1992b)
8-1d	H	H	H	H	Soto and Castedo (1998)
8-1i	H	Ac	Ac	Ac	Kobayashi *et al.* (1994)

Figure 4.15 The structures of taxoids in Tables 4.8 and 4.26.

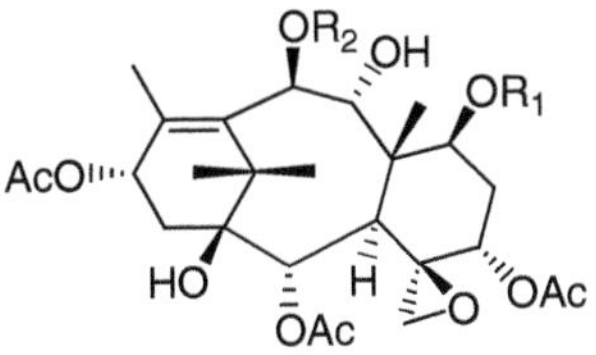

	R₁	R₂	
9-1e	Ac	Ac	Zhang *et al.* (1997)
9-1g	H	H	Shen and Chen (1997b)
9-1j	H	Ac	Zamir *et al.* (1995)

Figure 4.16 The structures of taxoids in Tables 4.9 and 4.27.

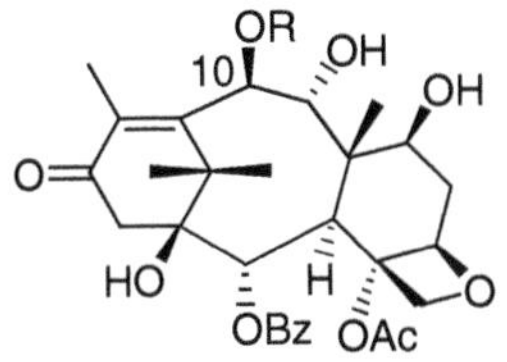

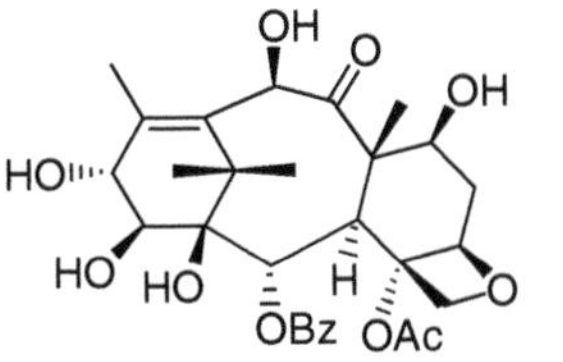

10-6a Appendino *et al.* (1992a)

	R	
10-5a	Ac	Morita *et al.* (1998)
10-5b	H	Morita *et al.* (1997b)

Figure 4.17 The structures of taxoids in Tables 4.10 and 4.28.

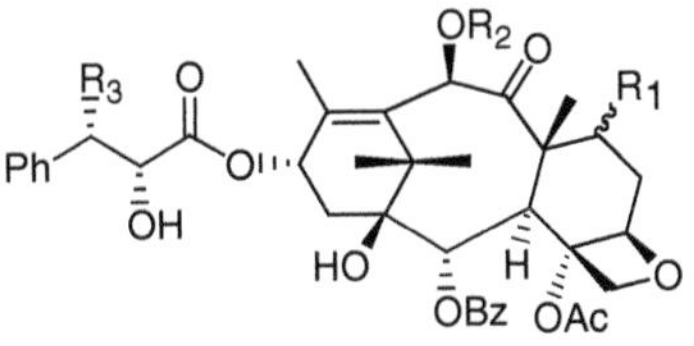

	R₁	R₂	R₃	
11-2a	β-OH	Ac	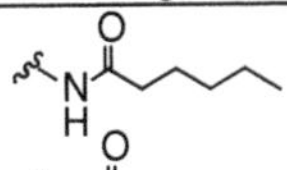	Ma *et al.* (1994)
11-2e	β-OH	Ac	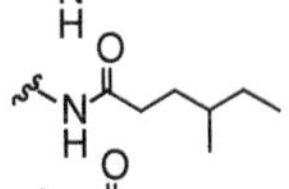	Morita *et al.* (1997a)
11-2k	α-OH	H	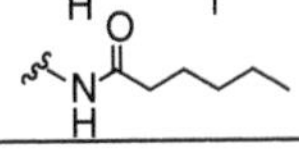	Morita *et al.* (1998)

Figure 4.18 The structures of taxoids in Tables 4.11 and 4.29.

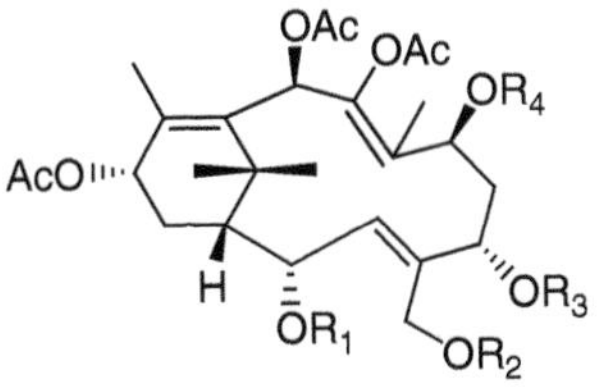

	R₁	R₂	R₃	R₄	
12i	Ac	Ac	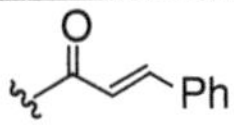	Ac	Shigemori *et al.* (1997)
12j	H	H	Ac	H	Hosoyama *et al.* (1996)
12k	H	Ac	Ac	Ac	Shi *et al.* (1998b)

Figure 4.19 The structures of taxoids in Tables 4.12 and 4.30.

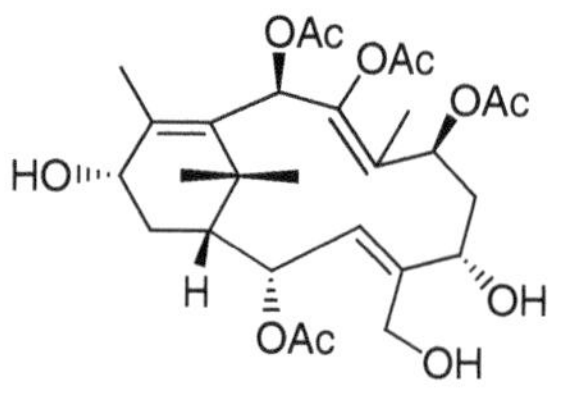

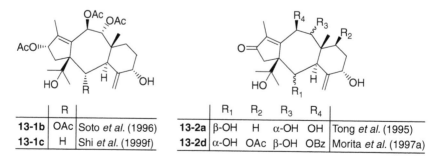

12c Shi *et al.* (1999a) 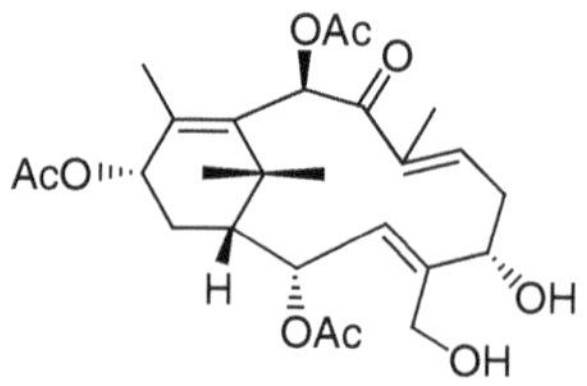   **12-1a** Shi *et al.* (1999e) 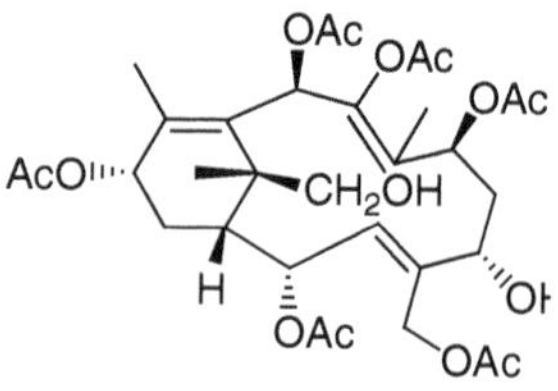   **12-2a** Shen *et al.* (1999a)

Figure 4.20 The structures of taxoids in Tables 4.13 and 4.31.

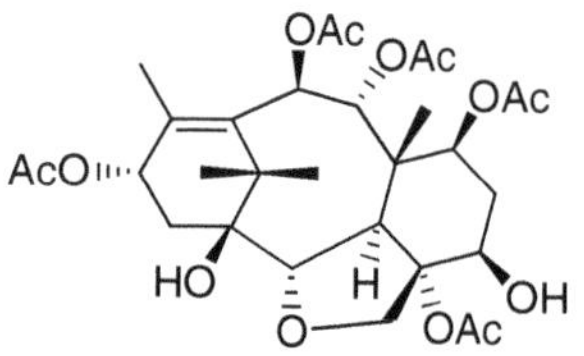

12-3a Wang *et al.* (1996) 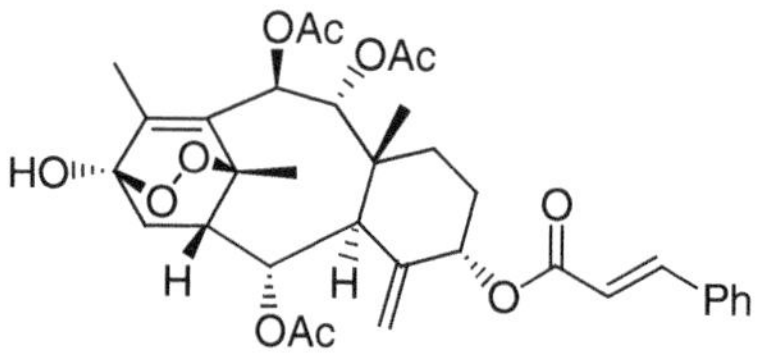   **12-4b** Shigemori *et al.* (1999) 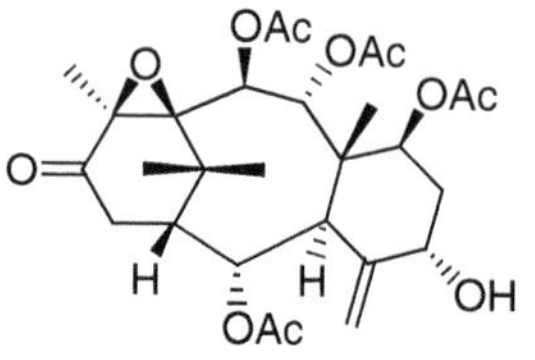   **12-5a** Yue *et al.* (1995b)

Figure 4.21 The structures of taxoids in Tables 4.14 and 4.32.

	R	
13-1b	OAc	Soto *et al.* (1996)
13-1c	H	Shi *et al.* (1999f)

	R₁	R₂	R₃	R₄	
13-2a	β-OH	H	α-OH	OH	Tong *et al.* (1995)
13-2d	α-OH	OAc	β-OH	OBz	Morita *et al.* (1997a)

Figure 4.22 The structures of taxoids in Tables 4.15 and 4.33.

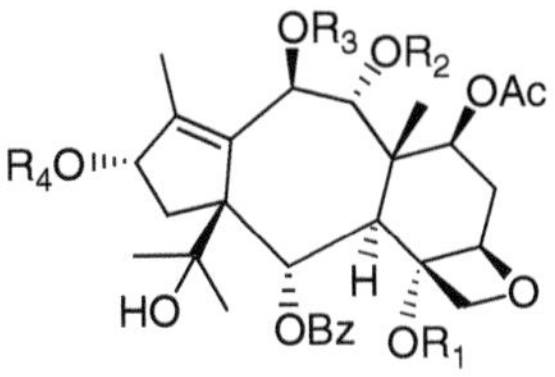

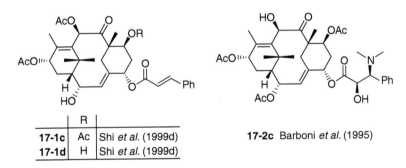

	R₁	R₂	R₃	R₄	
14-1b	Ac	H	H	H	Morita *et al.* (1998)
14-1d	Ac	Ac	H	H	Yue *et al.* (1995a)
14-1f	H	Ac	Ac	Ac	Barboni *et al.* (1994)

Figure 4.23 The structures of taxoids in Tables 4.16 and 4.34.

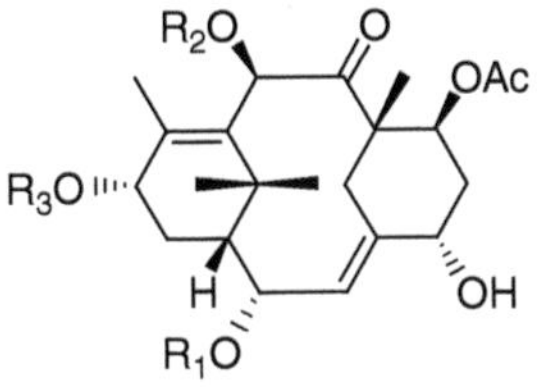

	R₁	R₂	R₃	
17b	H	Ac	Ac	Yue *et al.* (1995a)
17d	Ac	Ac	Ac	Yue *et al.* (1995b)
17g	Ac	H	H	Shi *et al.* (1999c)

Figure 4.24 The structures of taxoids in Tables 4.17 and 4.35.

	R	
17-1c	Ac	Shi *et al.* (1999d)
17-1d	H	Shi *et al.* (1999d)

17-2c Barboni *et al.* (1995)

Figure 4.25 The structures of taxoids in Tables 4.18 and 4.36.

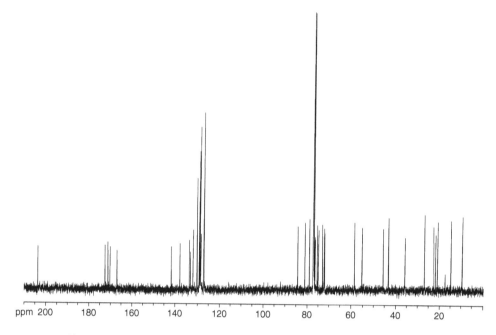

Figure 4.26 ^{13}C-NMR of taxol. (Tokyo University of Pharmacy & Life Science, Sakuma.)

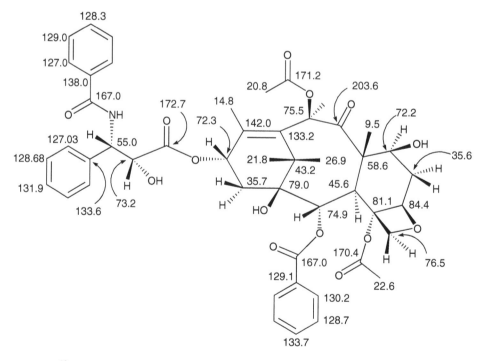

Figure 4.27 ^{13}C-NMR assignments for taxol.

Table 4.19 ^{13}C NMR spectra of taxol and cephalomannine

Positions		Taxol (Paclitaxel) (*11a*)		Cephalomannine (*11-1a*)
	1	79.0		79.0
	2	74.9		74.9
	3	45.6		45.5
	4	81.1		81.1
	5	84.4		84.4
	6	35.6		35.5
	7	72.2		72.1
	8	58.6		58.5
	9	203.6		203.8
	10	75.5		75.6
	11	133.2		133.1
	12	142.0		142.2
	13	72.3		72.2
	14	35.7		35.5
	15	43.2		43.1
	16	21.8		21.8
	17	26.9		26.8
	18	14.8		14.7
	19	9.5		9.5
	20	76.5		76.5
	1′	172.7		172.9
	2′	73.2		73.3
	3′	55.0		54.8
4-OAc	C=O	170.4		170.4
	Me	22.6		22.5
10-OAc	C=O	171.2		171.4
	Me	20.8		20.8
C=O	Ph1	167.0		167.1
q	Ph1	129.1		129.15
o	Ph1	130.2		130.3
m	Ph1	128.7		128.3
p	Ph1	133.7		133.8
q	Ph2	133.6		131.3
o	Ph2	127.03		127.0
m	Ph2	128.68		128.8
p	Ph2	131.9		129.0
C=O	Ph3	167.0	5′	169.1
q	Ph3	138.0	6′	138.2
o	Ph3	127.0	7′	132.0
m	Ph3	129.0	8′	13.9
p	Ph3	128.3	6′-Me	12.3

Notes
Numbers in parenthesis after compounds names are the numbers used in Chapter 3.
Measured in CDCl$_3$, δ values with TMS as internal standard.

Table 4.20 ^{13}C NMR spectra of selected taxoids

Positions	5α-Hydroxy-9α,10β13α-tri-acetoxytaxa-4(20),11-diene (1q)	2-Decinnamoyltaxinine E (1p)	Decinnamoyltaxinine E (1r)
1	39.5	47.7(d)	50.9(d)
2	32.3	72.7(d)	69.6(d)
3	36.1	41.5(d)	43.4(d)
4	153.3	147.5(s)	148.2(s)
5	77.4	69.7(d)	76.7(d)
6	26.4	30.2(t)	30.7(t)
7	29.0	26.4(t)	26.7(t)
8	43.4	44.7(s)	44.9(s)
9	74.3	76.4(d)	76.8(d)
10	72.8	71.5(d)	72.7(d)
11	136.1	134.2(s)	133.9(s)
12	137.2	137.9(s)	137.4(s)
13	70.2	77.2(d)	69.8(d)
14	26.0	28.6(t)	28.3(t)
15	38.7	37.1(s)	36.9(s)
16	27.4	32.2(q)	26.1(q)
17	32.3	25.8(q)	32.5(q)
18	15.8	15.8(q)	15.8(q)
19	17.1	17.3(q)	17.4(q)
20	111.1	115.2(t)	116.1(t)
OAc		170.6(s)	170.3(s)
		170.1(s)	170.0(s)
		170.0(s)	169.8(s)
		169.4(s)	
		20.7(q) x2	21.0(q)
		21.0(q)	20.9(q)
			20.8(q)

Notes
Numbers in parenthesis after compounds names are the numbers used in Chapter 3.
Measured in CDCl$_3$, δ values with TMS as internal standard.
Multiplicity: s = singlet, d = doublet, t = triplet, q = quartet.

Table 4.21 ^{13}C NMR spectra of selected taxoids

Positions	Taxezopidine K (1-1'**b**)	Taxuspine D (1-1c)	H coupled with C (HMBC)	Decinnamoyl-1-hydroxytaxinine J (1-1e)
1	50.60(d)	50.4(d)	H-3, 14α, 16, 17	78.4(s)
2	68.74(d)	68.3(d)	H-1, 14α, 14β	72.3(d)
3	41.06(d)	40.8(d)	H-1, 2, 7, 9, 10, 19	43.4(d)
4	142.96(s)	142.6(s)	H-3, 5, 2α	141.9(s)
5	67.58(d)	67.2(d)	H-3, 6, 2α	75.2(d)
6	32.70(t)	32.7(t)	H-5	37.1(t)
7	70.65(d)	70.4(d)	H-6, 9, 19	69.3(d)
8	43.03(s)	43.1(s)	H-2, 3, 6, 7, 9, 10, 19	47.9(s)
9	76.36(d)	76.4(d)	H-3, 10, 19	75.4(d)
10	78.55(d)	78.7(d)	H-9	71.3(d)
11	78.41(s)	77.3(s)	H-1, 9, 10, 16, 17, 18	134.5(s)
12	124.28(s)	124.3(s)	H-10, 14α, 14β, 18	140.3(s)
13	144.29(s)	144.2(s)	H-1, 14α, 14β, 18	70.6(d)
14	25.89(t)	25.8(t)	H-1, 2	37.0(t)
15	41.17(s)	41.3(s)	H-1, 2, 10, 14α, 16, 17	42.0(s)
16	24.46(q)	24.1(q)	H-17	20.8(q)
17	31.70(q)	31.3(q)	H-1, 16	29.0(q)
18	11.44(q)	11.6(q)		15.9(q)
19	14.80(q)	14.8(q)	H-3, 7, 9	13.0(q)
20	110.59(t)	110.9(t)	H-3, 5	116.2(t)
21	165.14(s)	165.2(s)	H-5, 22, 23	
22	117.84(d)	117.8(d)	H-23	
23	145.08(d)	145.1(d)	H-25	
24	134.29(s)	134.2(s)	H-22	
25,29	128.14(d)	128.1(d)	H-23, 26, 27	
26,28	128.95(d)	128.9(d)	H-25	
27	130.43(d)	130.4(d)	H-25	
				OAc
2-OAc	21.65(q)	21.4(q)		21.3(q)
	170.58(s)	170.6(s)	H-2	21.1(q)
7-OAc	21.31(q)	21.2(q)		20.9(q)
	170.31(s)	170.2(s)	H-7	20.9(q)
9-OAc	21.04(q)	21.7(q)		20.7(q)
	170.45(s)	169.3(s)	H-9	171.4(s)
10-OAc		21.2(q)		169.9(s)
		170.4(s)	H-10	169.8(s)
13-OAc	20.86(q)	20.2(q)		169.6(s)
	168.55(s)	168.5(s)		169.1(s)

Notes

Numbers in parenthesis after compounds names are the numbers used in Chapter 3.

Measured in CDCl$_3$, δ values with TMS as internal standard.

Multiplicity: s = singlet, d = doublet, t = triplet, q = quartet.

Table 4.22 ^{13}C NMR spectra of selected taxoids

Positions	2,20-Dihydroxy-9-acetoxytaxa-4(20),11-dien-13-one (1-3a)	Taxezopidine B (1-3b)	H coupled with C (HMBC)	Taxumairol (1-4d)
1	45.4	50.10(d)	H-2, 16, 17	51.1(d)
2	80.6	72.02(d)	H-1, 14	71.4(d)
3	44.3	137.08(s)	H-2, 19, 20β	45.6(d)
4	140.3	141.91(s)	H-2, 20β	77.2(s)
5	116.4	66.13(d)	OH-5	71.9(d)
6	22.7	20.22(t)		31.1(t)
7	31.6	23.79(t)	H-9, 19	68.7(d)
8	40.6	43.87(s)	H-2, 7α, 19	45.8(s)
9	78.6	75.82(d)	H-10, 19	75.3(d)
10	31.6	76.01(d)	H-9	71.0(d)
11	158.2	79.80(s)	H-1, 10, 16, 17, 18, OH-11	134.2(s)
12	136.0	50.79(d)	H-14, 18	137.3(s)
13	200.2	209.54(s)	H-14, 18	70.8(d)
14	35.8	38.03(t)	H-2	29.7(t)
15	37.4	44.50(s)	H-2, 10, 14, 16, 17	37.9(s)
16	25.8	23.81(q)	H-1, 17	26.9(q)
17	37.8	30.90(q)	H-16	31.7(q)
18	15.4	9.83(q)		15.0(q)
19	17.4	26.00(q)	H-9	14.7(q)
20	70.2	64.46(t)	H-5	66.7(t)
2-OAc		20.81(q)		
		168.00(s)	H-2	
5-OAc				172.0(s)
7-OAc				169.6(s)
9-OAc	21.6	20.83(q)		
	171.0	170.47(s)	H-9	170.2(s)
10-OAc		20.20(q)		
		171.48(s)	H-10	169.1(s)
13-OAc				170.2(s)
OBz				166.0(s)
I				129.6(s)
o				129.7(d)
m				128.6(d)
p				133.4(d)

Notes
Numbers in parenthesis after compounds names are the numbers used in Chapter 3.
Measured in CDCl$_3$, δ values with TMS as internal standard.
Multiplicity: s = singlet, d = doublet, t = triplet, q = quartet.

Table 4.23 ^{13}C NMR spectra of selected taxoids

Positions	Taxayunnanine G (4a)	Taxayunnanine H (4b)	10-Deacetylyunnanxane (4b)	^{13}C-^{1}H COSY
1	63.7	59.4	59.30	1.88(d)
2	71.4	71.2	70.60	5.36(dd)
3	41.9	39.8	41.95	2.92(d)
4	142.9	148.0	142.38	
5	78.8	76.4	78.32	5.28(t)
6	28.9	30.9	28.50	1.79(m)
7	33.8	33.2	33.95	1.25(m), 1.92(m)
8	39.7	40.0	39.66	
9	47.2	47.1	47.16	1.67(dd)
10	67.3	67.6	67.33	5.06(dd)
11	138.8	138.0	138.78	
12	132.9	133.6	132.20	
13	42.3	39.5	39.53	2.37(dd), 2.82(dd)
14	67.8	70.7	70.92	4.02(dd)
15	37.9	37.6	37.40	
16	31.8	32.1	25.34	1.69(s)
17	25.7	25.4	32.01	1.18(s)
18	21.1	21.0	21.04	1.97(s)
19	22.4	22.3	22.55	0.84(s)
20	116.7	113.4	116.62	4.81(br s), 5.25(br s)
1'		173.6	174.77	
2'		28.1	46.99	2.40(q)
3'		9.2	69.48	3.85(t)
4'			20.84	1.21(d)
5'			13.98	1.15(d)
			OAc	
2-OAc	21.5	21.4	21.87	
	169.6	169.9	21.38	
5-OAc	21.9		169.70	
	169.6		169.77	

Notes
Numbers in parenthesis after compounds names are the numbers used in Chapter 3.
Measured in CDCl$_3$, δ values with TMS as internal standard.
Multiplicity: s = singlet, d = doublet, t = triplet, q = quartet.

Table 4.24 ^{13}C NMR spectra of selected taxoids

Positions	2α-Deacetyltaxinine J Taxuspinanane G (6f)	5α-Cinnamoyloxy-2α,13α-dihydroxy-9α,10β-diacetoxy-4(20),11-diene (6b)	1-Hydroxy-2-deacetoxy-taxinine (6i)
1	51.61	51.08(d)	76.2
2	69.42	70.30(d)	37.0
3	44.45	45.71(d)	41.2
4	141.79	144.09(s)	145.6
5	5.42	70.58(d)	74.5
6	35.00	26.10(t)	34.6
7	70.27	28.34(t)	69.8
8	46.97	37.37(s)	46.6
9	76.36	75.93(d)	76.0
10	71.92	78.82(d)	71.3
11	134.20	136.99(s)	134.8
12	136.42	134.02(s)	139.2
13	70.63	75.93(d)	71.0
14	28.02	29.49(t)	41.6
15	37.56	44.97(s)	43.7
16	27.60	18.18(q)	27.3
17	32.11	27.18(q)	21.4
18	16.08	31.63(q)	15.3
19	13.90	15.29(q)	13.2
20	119.58	119.47(t)	116.1
		Cinn	Cinn
1'	166.47	166.41(s)	166.0
2'	118.23	118.78(d)	118.3
3'	146.16	145.43(d)	145.9
3'-Ph(q)	133.68	134.27(s)	138.1
(o)	128.23	128.05(d)	130.1
(m)	129.03	129.03(d)	130.7
(p)	130.69	130.55(d)	134.1
		OAc	OAc
7-OAc	20.85	21.02(q)	20.8
	169.94	21.32(q)	20.9
9-OAc	21.06	170.48(s)	21.0
	170.04	170.40(s)	21.8
10-OAc	21.06		169.0
	169.23		169.8
13-OAc	21.35		170.2
	170.68		170.5

Notes
Numbers in parenthesis after compounds names are the numbers used in Chapter 3.
Measured in CDCl$_3$, δ values with TMS as internal standard.
Multiplicity: s = singlet, d = doublet, t = triplet, q = quartet.

Table 4.25 ^{13}C NMR spectra of selected taxoids

Positions	19-Debenzoyl-19-acetyltaxinine M (7*b*)	Taxezopidine L (7*b*) Shigemori (1999)	Taxezopidine L (7*b*) Fukushima (1999)
1	48.7(d)	48.79(d)	48.7
2	70.4(d)	70.57(d)	70.5
3	38.4(d)	39.87(d)	39.9
4	144.6(s)	140.66(s)	140.8
5	72.7(d)	73.87(d)	73.8
6	39.2(t)	37.15(t)	37.1
7	68.7(d)	68.48(d)	69.2
8	49.7(s)	49.79(s)	49.1
9	69.5(d)	70.57(d)	68.4
10	64.1(d)	64.01(d)	65.3
11	80.3(s)	80.43(s)	82.3
12	91.4(s)	91.58(s)	91.7
13	204.5(s)	203.97(s)	204.4
14	33.9(t)	34.29(t)	32.3
15	49.1(s)	49.06(s)	49.7
16	15.1(q)	15.29(q)	15.1
17	82.2(t)	82.24(t)	80.4
18	12.1(q)	12.35(q)	12.2
19	61.6(t)	61.65(t)	61.6
20	113.0(t)	115.45(t)	115.4
21		165.91(s)	166.0
22		117.72(d)	117.7
23		146.24(d)	146.3
24		134.71(s)	134.7
25,29		128.77(d)	128.8
26,28		128.81(d)	127.9
27		130.27(d)	130.2
		2-OAc 21.50(q)	20.9
OAc	172.1(s), 170.5(s)	169.80(s)	168.9
	169.8(s), 168.5(s)	7-OAc 21.50(q)	20.6
	168.1(s)	172.14(s)	167.0
	21.4(q), 21.3(q)	9-OAc 21.97(q)	21.3
	20.8(q), 20.7(q)	168.45(s)	169.9
	20.7(q)	10-OAc 20.97(q)	20.9
		167.90(s)	170.8
		19-OAc 20.29(q)	21.3
		170.45(s)	172.2

Notes
Numbers in parenthesis after compounds names are the numbers used in Chapter 3.
Measured in CDCl$_3$, δ values with TMS as internal standard.
Multiplicity: s=singlet, d=doublet, t=triplet, q=quartet.

Table 4.26 ^{13}C NMR spectra of selected taxoids

Positions	5-Cinnamoyl-phototaxicin II (8-1d)		5-Cinnamoyl-10-acetylphototaxicin I (8b)		Taxuspine C (8-1i)	HMBC
1	50.9(d)		78.1(s)		47.8(d)	H-2, 14α, 14β, 16β, 17
2	75.8(d)		78.5(d)		76.5(d)	H-1, 14α, 14β
3	66.6(s)		61.9(s)		65.9(s)	H-2, 5, 7, 9, 12, 19
4	141.1(s)		142.1(s)		142.2(s)	H-2, 5, 20
5	76.2(d)		76.1(d)		77.2(m)	H-20
6	19.6(t)		25.9(t)		25.8(t)	H-5
7	25.8(t)		29.3(t)		31.2(m)	H-9, 19
8	42.3(s)		45.7(s)		44.5(s)	H-2, 7, 9, 19
9	83.9(d)		82.6(d)		79.5(d)	H-10, 19
10	82.0(d)		84.0(d)		82.3(d)	H-9
11	59.4(s)		56.4(s)		57.8(s)	H-1, 10, 12, 16α, 17
12	52.0(d)		51.0(d)		52.3(d)	H-10, 18
13	216.3(s)		214.2(s)		214.8(s)	H-1, 12, 14α, 14β, 18
14	38.3(t)		46.5(t)		38.8(t)	H-2
15	44.2(s)		45.0(s)		42.6(s)	H-1, 10, 12, 14α, 16α
16	26.8(q)		23.3(q)		28.8(q)	H-17
17	29.2(q)		22.9(q)		26.5(q)	H-16α
18	16.6(q)		15.8(q)		15.7(q)	H-12
19	25.2(q)		25.2(q)		26.7(q)	H-9
20	125.0(t)		126.7(t)		129.4(t)	H-5
1'	166.0(s)	1'	134.1(s)	21	165.7(s)	H-5, 22, 23
2'	117.8(d)	2',6'	128.7(d)	22	117.8(d)	H-23
3'	145.3(d)	3',5'	128.1(d)	23	145.3(d)	
		7'	145.4(d)	24	134.3(s)	H-22, 26
		8'	117.5(d)	25	128.2(d)	H-26
		9'	165.9(s)	26	128.8(d)	H-23, 25
		OAc	172.2(s)	27	130.3(d)	H-25
			21.2(q)	2-OAc	20.9(q)	
Ph(I)	134.3(s)				169.5(s)	H-2
(o)	128.1(d)			9-OAc	21.4(q)	
(m)	128.8(d)				170.9(s)	H-9
(p)	130.4(d)			10-OAc	21.1(q)	
					170.0(s)	H-10

Notes
Numbers in parenthesis after compounds names are the numbers used in Chapter 3.
Measured in CDCl$_3$, δ values with TMS as internal standard.
Multiplicity: s=singlet, d=doublet, t=triplet, q=quartet, m=multiplet.

Table 4.27 ¹³C NMR spectra of selected taxoids

Positions	1β,9α-Dihydroxy-4β,20-epoxy-2α,5α,7β,10β,13α-penta-acetoxy-11-ene (9-1e)	Taxumairol C (9-1g)	1β-Hydroxy-7,9-deacetyl-baccatin I (9-1j)
1	76.0	76.2(s)	76.01
2	72.0	71.3(d)	72.63
3	42.0	40.3(d)	40.21
4	58.5	58.3(s)	58.36
5	78.0	78.2(d)	78.19
6	30.6	33.2(t)	33.09
7	71.8	70.1(d)	69.60
8	45.8	45.9(s)	46.52
9	76.2	76.9(d)	78.36
10	72.1	71.0(d)	74.09
11	136.7	139.5(s)	136.13
12	139.1	137.2(s)	140.20
13	71.1	72.7(d)	71.16
14	38.6	38.5(t)	38.61
15	43.3	43.5(s)	43.39
16	22.1	21.8(q)	28.43
17	28.5	15.7(q)	22.05
18	15.5	28.8(q)	15.48
19	14.7	13.4(q)	13.54
20	49.5	49.8(t)	49.88
OAc-CO	170.4	169.9(s)	169.20
	170.1	169.1(s)	169.80
	169.3	170.1(s)	170.10
	169.3		169.80
	169.1		
OAc-Me	21.5	21.4(q)	21.70
	21.5	21.0(q)	21.21
	21.3	21.7(q)	21.38
	20.7		20.92
	20.7		

Notes

Numbers in parenthesis after compounds names are the numbers used in Chapter 3.

Measured in CDCl₃, δ values with TMS as internal standard.

Multiplicity: s=singlet, d=doublet, t=triplet, q=quartet.

Table 4.28 ^{13}C NMR spectra of selected taxoids

Positions	Taxuspinanane C (10-5b)	HMBC	Taxuspinanane D (10-5a)	14β-Hydroxy-10-deacetylbaccatin III (10-6a)
1	78.65	H-2, 3, 14α, 14β, 16, 17	78.39	75.24(s)
2	71.82	H-3, 14α, 14β	78.72	74.00(d)
3	45.75	H-2, 19	45.91	46.00(d)
4	81.43	H-3, 5, 20α	81.58	79.95(s)
5	83.77	H-6β, 20β	83.95	83.65(d)
6	38.11		37.97	36.58(t)
7	74.38	H-3, 5, 6α, 19	73.84	70.94(d)
8	44.48	H-2, 3, 6α, 19	45.02	57.26(s)
9	77.66	H-10, 19	76.31	210.04(s)
10	72.33		73.73	74.36(d)
11	158.21	H-16, 17, 18	154.36	135.27(s)
12	138.83	H-10, 18	140.11	139.53(s)
13	198.40	H-14α, 14β, 18	198.42	74.64(d)
14	42.92	H-2	42.97	72.23(d)
15	43.40	H-10, 14α, 16, 17	43.33	42.25(s)
16	20.23	H-17	20.53	26.55(q)
17	34.91	H-16	34.73	21.37(q)
18	13.88		13.96	14.89(q)
19	11.94	H-3	12.01	9.67(q)
20	76.46		76.38	75.37(t)
4-OAc	21.93	Me-4	21.93	22.31(q)
	170.16		170.11	169.81(s)
10-OAc			21.30	
			170.33	
2-OBz	133.91		134.00	165.40(s)
	128.77		128.81	129.72(s)
	130.09		130.11	129.80(d)
	128.99		129.37	128.68(d)
	166.93	H-2, H-2Ph(*p*)	166.95	133.29(s)

Notes

Numbers in parenthesis after compounds names are the numbers used in Chapter 3.

Measured in CDCl$_3$, δ values with TMS as internal standard.

Multiplicity: s = singlet, d = doublet, t = triplet, q = quartet.

Table 4.29 ^{13}C NMR spectra of selected taxoids

Positions	7-epi-10-Deacetyltaxol C (Taxuspinanane E) (11-2k)	Taxol C (11-2a)	Taxuspinanane A (11-2e)
1	79.34	78.89	79.45
2	75.54	74.88	75.69
3	40.34	45.52	46.26
4	82.15	81.04	81.64
5	82.70	84.32	84.55
6	36.70	35.55	36.19
7	77.93	72.12	72.71
8	57.36	58.54	59.15
9	215.11	203.49	203.52
10	75.99	75.50	76.03
11	135.77	133.59	133.96
12	137.93	141.93	142.23
13	72.48	72.37	72.71
14	35.40	35.60	36.33
15	43.64	43.18	43.68
16	22.56	21.89	22.11
17	26.03	26.80	26.92
18	14.44	14.84	14.82
19	16.76	9.56	10.07
20	77.81	76.42	76.54
1'	172.76	172.7	173.63
2'	73.20	73.05	73.10
3'	54.43	54.48	54.51
4'	172.93		
5'	36.41	36.58	172.99
6'	25.42	25.36	34.10
7'	31.35	31.28	32.51
8'	22.30	22.29	34.10
9'	13.85	13.85	29.46
10'			11.38
11'			18.84
4-OAc	20.69	22.61	22.55
	172.46	170.08	170.46
3'-Ph	138.18	137.90	127.60
	126.87	126.85	128.90
	128.81	128.50	128.15
	128.26	128.18	139.31
2-OBz	129.15	129.02	130.68
	130.32	130.10	128.89
	120.10	128.87	133.42
	133.70	133.03	130.35
	167.16	166.80	167.13

Notes
Numbers in parenthesis after compounds names are the numbers used in Chapter 3.
Measured in CDCl$_3$, δ values with TMS as internal standard.
Multiplicity: s=singlet, d=doublet, t=triplet, q=quartet.

Table 4.30 ^{13}C NMR spectra of selected taxoids

Positions	Taxuspine X (12i)	HMBC	Taxuspine U (12j)	2-Deacetyl Taxachitriene (12k)
1	46.47(d)	H-2, 13, 14α, 16, 17	48.0(d)	47.68
2	68.85(d)		66.7(d)	69.10
3	124.48(d)	H-5, 20α, 20β	127.5(d)	130.20
4	132.21(s)	H-5, 20α, 20β	135.4(s)	133.83
5	69.42(d)	H-3, 7, 20α	71.0(d)	69.10
6	35.11(t)	H-7	36.3(t)	37.16
7	66.74(d)	H-5, 6α, 19	63.7(d)	68.05
8	125.45(s)	H-6α, 19	128.8(s)	122.95
9	143.67(s)	H-7, 10, 19	139.9(s)	144.63
10	68.56(d)		68.3(d)	67.42
11	136.92(s)	H-10, 13, 16, 17, 18	136.5(s)	135.67
12	135.40(s)	H-10, 13, 14α, 18	136.0(s)	136.15
13	69.99(d)	H-18	69.8(d)	65.87
14	25.80(s)	H-2, 13	25.1(t)	24.76
15	36.19(s)	H-10, 16, 17	36.2(s)	35.89
16	33.20(q)	H-17	32.5(q)	34.03
17	25.17(q)	H-16	29.7(q)	24.13
18	17.01(q)		15.9(q)	16.85
19	12.23(q)	H-17	11.3(q)	11.99
20	59.67(s)	H-3	58.8(t)	59.38
21	165.65(s)	H-5, 22, 23		20.99
22	117.87(d)			
23	145.93(d)			
24	134.10(s)	H-22		
25	128.12(d)	H-27		
26	129.08(d)			
27	130.70(d)	H-25		
2-OAc	169.18(s)	H-2		
	20.84(q)			
5-OAc			169.8(s)	171.54
			20.8(q)	21.27
7-OAc	169.71(s)	H-7		171.12
	21.01(q)			21.59
9-OAc	167.83(s)		169.8(s)	168.05
	20.41(q)		21.5(q)	21.02
10-OAc	167.84(s)	H-10	169.4(s)	170.63
	21.05(q)		20.4(q)	21.27
13-OAc	170.58(s)	H-13	170.2(s)	171.12
	21.05(q)		21.5(q)	
20-OAc	170.63(s)	H-20		
	21.30(q)			

Notes
Numbers in parenthesis after compounds names are the numbers used in Chapter 3.
Measured in CDCl₃, δ values with TMS as internal standard.
Multiplicity: s=singlet, d=doublet, t=triplet, q=quartet.

Table 4.31 ^{13}C NMR spectra of selected taxoids

Positions	13-Deacetylcanadense (12c)	Taxumairol M (12-2a)	(3E,7E)-2α,10β,13α,20-Tetra-acetoxy-5α-hydroxy-3,8-seco-taxa-3,7,11-trien-9-one (12-1a)
1	47.10	39.9(d)	46.20
2	71.03	68.6(d)	68.31
3	122.74	125.6(d)	123.91
4	141.22	137.0(s)	140.61
5	68.23	68.2(d)	69.16
6	37.30	37.3(t)	33.52
7	68.48	67.3(t)	135.53
8	123.30	123.9(s)	139.40
9	144.18	144.0(s)	194.74
10	67.05	67.8(d)	76.16
11	133.96	133.0(s)	139.34
12	138.34	139.2(s)	134.12
13	68.75	69.0(d)	69.31
14	27.54	25.0(t)	23.99
15	35.98	41.7(s)	37.22
16	24.67	19.5(q)	34.11
17	34.01	71.9(t)	24.61
18	17.35	17.3(q)	17.39
19	12.69	12.9(q)	11.97
20	57.73	59.7(t)	59.76
2-OAc	171.86	OAc 172.5(s)	170.35
	20.49	170.7(s)	20.84
7-OAc	171.68	170.6(s)	
	21.41	169.8(s)	
9-OAc	168.14	168.0(s)	
	21.81	168.0(s)	
10-OAc	168.06	21.5(q)	169.68
	20.87	21.5(q)	20.92
13-OAc		21.4(q)	170.52
		21.2(q)	21.35
20-OAc		20.8(q)	169.73
		20.8(q)	21.37
		20.4(q)	

Notes
Numbers in parenthesis after compounds names are the numbers used in Chapter 3.
Measured in CDCl$_3$, δ values with TMS as internal standard.
Multiplicity: s=singlet, d=doublet, t=triplet, q=quartet.

Table 4.32 ^{13}C NMR spectra of selected taxoids

Positions	Taxuspine K (12-3a)	HMBC	Taxezopidine J (12-4b)	Decinnamoyl-taxinine B-11,12-oxide (12-5a)
1	76.27(s)	H-2, 3, 14α, 14β, 16, 17	48.30(d)	51.67(d)
2	85.04(d)	H-20β	70.00(d)	68.97(d)
3	50.33(d)	H-19, 20α, 20β	42.78(d)	40.30(d)
4	96.40(s)	H-3, 5, 20α, 20β	146.63(s)	143.30(s)
5	70.01(d)	H-6α	76.37(d)	75.24(d)
6	21.21(t)	H-7	28.45(t)	37.00(t)
7	70.65(d)	H-6α, 19	27.90(t)	69.09(d)
8	43.77(s)	H-2, 3, 7, 10, 19	43.74(s)	47.03(s)
9	77.57(d)	H-3, 10, 19	76.37(d)	
10	73.40(d)	H-9	70.00(d)	75.68(d)
11	134.44(s)	H-9, 10, 13, 16, 17, 18	128.96(s)	59.90(s)
12	136.67(s)	H-13, 14α, 14β, 18	141.40(s)	64.56(s)
13	71.43(d)	H-18	96.25(s)	208.35(s)
14	35.97(t)		35.35(t)	38.16(t)
15	42.60(s)	H-10, 14α, 16, 17	37.62(s)	38.55(s)
16	22.97(q)	H-17	17.36(q)	25.43(q)
17	27.58(q)	H-16	74.54(t)	29.17(q)
18	17.60(q)		11.90(q)	14.60(q)
19	15.61(q)	H-3	17.36(q)	13.24(q)
20	69.97(t)	H-5	118.03(t)	117.54(t)
21			166.34(s)	
22			118.08(d)	
23			145.65(d)	
24			134.45(s)	
25,29			128.30(d)	
26,28			128.95(d)	
27			130.41(d)	
2-OAc			169.51(s)	OAc 168.74(s)
			21.11(q)	168.11(s)
4-OAc	169.25(s)			169.53(s)
	22.69(q)			169.59(s)
7-OAc	168.99(s)	H-7		20.58(q)
	22.43(q)			20.72(q)
9-OAc	168.60(s)	H-9	169.97(s)	21.15(q)
	21.11(q)		20.91(q)	21.37(q)
10-OAc	168.47(s)	H-10	169.51(s)	
	21.00(q)		21.60(q)	
13-OAc	168.86(s)	H-13		
	21.17(q)			

Notes

Numbers in parenthesis after compounds names are the numbers used in Chapter 3.

Measured in CDCl$_3$, δ values with TMS as internal standard.

Multiplicity: s=singlet, d=doublet, t=triplet, q=quartet.

Table 4.33 ^{13}C NMR spectra of selected taxoids

Positions	Taxuspinanane B (13-2d)	5α-Hydroxy-9α,10β,13α-tri-acetoxy-11(15 → 1)abeotaxa-4(20),11-diene (13-1c)	Taxustone (13-2a)	Teixidol (13-1b)
1	62.66	62.94	65.07	68.74(s)
2	68.92	29.08	68.90	68.58(d)
3	45.40	39.19	45.18	43.65(d)
4	146.37	151.99	142.84	146.55(s)
5	65.67	73.07	80.29	75.50(d)
6	35.78	29.24	31.49	37.51(t)
7	70.57	27.2	31.49	30.38(d)
8	43.77	42.91	42.92	42.40(s)
9	74.34	78.01	68.90	76.89(d)
10	70.51	69.75	68.48	69.10(d)
11	148.11	137.41	150.24	146.83(s)
12	160.99	145.44	135.80	136.02(s)
13	206.87	79.68	207.57	79.23(d)
14	44.66	44.21	44.49	36.81(t)
15	75.60	75.68	75.59	75.37(s)
16	28.46	24.71	27.91	25.37(q)
17	27.56	27.04	27.91	27.65(q)
18	9.39	11.79	8.32	11.09(q)
19	14.45	16.80	26.55	17.05(q)
20	114.14	110.07	109.46	111.30(t)
2-OAc	170.26		OAc 170.75	OAc 171.11(s)
	21.67		21.03	170.93(s)
7-OAc	170.40		171.17	169.88(s)
	21.01		21.64	168.55(s)
9-OAc	169.53	170.21		21.73(q)
	20.65	21.20		21.11(q)
10-OAc		168.62		20.79(q)
		20.79		20.65(q)
13-OAc		170.99		
		20.93		
10-OBz	133.64			
	128.82			
	129.74			
	129.32			
	165.34			

Notes
Numbers in parenthesis after compounds names are the numbers used in Chapter 3.
Measured in CDCl$_3$, δ values with TMS as internal standard.
Multiplicity: s=singlet, d=doublet, t=triplet, q=quartet.

Table 4.34 ^{13}C NMR spectra of selected taxoids

Positions	Taxuspinanane F (14-1*b*)	4-Deacetyl-11(15 →1)-abeo-baccatin VI (14-1*f*)	Taxuyuntin E (14-1*d*)
1	68.05	67.9(s)	68.8(s)
2	68.75	69.8(d)	70.1(d)
3	44.56	49.4(d)	46.5(d)
4	80.14	73.7(s)	80.3(s)
5	85.10	86.4(d)	86.1(d)
6	34.81	34.0(t)	35.9(t)
7	71.61	70.8(d)	72.4(d)
8	43.00	43.3(s)	44.6(s)
9	78.78	76.2(d)	80.8(d)
10	68.20	68.3(d)	67.6(d)
11	137.24	136.5(s)	138.3(s)
12	147.46	146.3(s)	148.0(s)
13	77.84	79.7(d)	77.7(d)
14	39.92	36.8(t)	40.3(t)
15	76.06	75.4(s)	76.7(s)
16	27.69	26.1(q)	25.3(q)
17	24.73	28.0(q)	28.1(q)
18	11.28	11.9(q)	11.3(q)
19	13.14	12.8(q)	13.1(q)
20	74.98	78.8(t)	75.6(t)
2-OAc	171.14	OAc 170.9(s)	OAc 172.7(s)
	22.44	169.9(s)	172.1(s)
7-OAc	170.58	169.4(s)	171.9(s)
	21.74	167.9(s)	21.4(q)
		21.1(q)	21.8(q)
		21.1(q)	22.3(q)
		20.6(q)	
		20.6(q)	
2-OBz: 1'	129.67	129.7(s)	131.5(s)
2'	128.64	129.4(d)	130.8(d)
3'	133.55	128.1(d)	129.8(d)
4'	130.04	133.6(d)	134.8(d)
5'	133.55	128.1(d)	129.7(d)
6'	128.64	129.4(d)	134.4(d)
7'	166.09	165.5(s)	167.7(s)

Notes
Numbers in parenthesis after compounds names are the numbers used in Chapter 3.
Measured in CDCl$_3$, δ values with TMS as internal standard.
Multiplicity: s=singlet, d=doublet, t=triplet, q=quartet.

Table 4.35 ^{13}C NMR spectra of selected taxoids

Positions	Taxin B (17d)	2-Deacetyltaxin B (17b)	7-O-Acetyltaxine A (17-2c)
1	46.525	49.260	47.2(d)
2	76.280	67.284	70.8(d)
3	35.674	35.780	35.1(t)
4	131.683		131.4(s)
5	68.707	68.816	69.1(d)
6	35.092	35.184	32.1(t)
7	70.106	70.108	70.1(d)
8	52.848	52.890	53.2(s)
9	205.699		212.8(s)
10	77.845	77.967	76.6(d)
11	138.852	137.209	132.1(s)
12	136.939	136.956	135.7(s)
13	69.235	69.443	69.5(d)
14	26.426	26.456	27.4(t)
15	37.444	37.429	37.5(s)
16	34.897	35.125	24.5(q)
17	24.368	24.552	31.9(q)
18	18.647	18.619	16.5(q)
19	20.690	20.748	20.5(q)
20	123.950	127.885	124.8(d)
OAc	170.202	169.349 x3	170.9(s)
	169.919		169.9(s)
	169.340		169.5(s)
	21.338	21.004	21.3(q)
	21.030	20.832	21.0(q)
	20.831	20.748	20.8(q)
1'			172.7(s)
2'			69.6(d)
3'			71.4(d)
Ph-ipso			132.1(s)
(o)			129.9(d)
(m)			128.2(d)
(p)			128.1(d)
NMe$_2$			41.0(q)

Notes
Numbers in parenthesis after compounds names are the numbers used in Chapter 3.
Measured in CDCl$_3$, δ values with TMS as internal standard.
Multiplicity: s=singlet, d=doublet, t=triplet, q=quartet.

Table 4.36 ^{13}C NMR spectra of selected taxoids

Positions	5α-Cinnamoyloxy-7β,10β,13α-triacetoxy-2(3 → 20)abeotaxa-2α-hydroxy-4(20),11-dien-9-one (17-1c)	5α-Cinnamoyloxy-10β,13α-diacetoxy-2(3 → 20)abeotaxa-2α,7β-dihydroxy-4(20),11-dien-9-one (17-1d)
1	49.9	49.0
2	68.8	67.8
3	36.2	35.9
4	131.9	132.0
5	70.8	69.2
6	33.1	32.3
7	70.1	68.0
8	53.9	53.9
9	206.7	206.9
10	78.7	78.4
11	129.5	129.4
12	139.6	137.6
13	71.2	70.7
14	27.1	26.3
15	38.4	37.7
16	21.2	18.4
17	25.9	25.0
18	17.9	16.8
19	33.3	35.5
20	128.8	126.3
1'	167.3	166.3
2'	118.4	117.5
3'	147.1	146.3
4'	134.7	132.0
5'	129.8	129.1
6'	128.9	128.1
7'	131.5	130.8
7-OAc	171.2	
	21.7	
10-OAc	170.3	170.4
	21.9	21.3
13-OAc	171.3	170.0
	21.6	21.0

Notes
Numbers in parenthesis after compounds names are the numbers used in Chapter 3.
Measured in CDCl$_3$, δ values with TMS as internal standard.

Table 4.37 ^{1}H NMR and ^{13}C NMR spectra of selected taxoids (Dimer)

No.	^{1}H	^{13}C	5-Oxotaxinine A (dimer) HMBC (^{1}H)	No.	^{1}H	^{13}C	HMBC (^{1}H)
1	2.17(br dd)	47.8(d)	16,17	1'	2.12(m)	49.0(d)	14'α,16',17'
2	5.67(br dd)	72.0(d)	3	2'	5.71(dd, 4.4, 2.1)	71.2(d)	1',3',14'α,14'β
3	3.12(br d)	42.8(s)	1,2,9,19	3'	2.63(d, 4.4)	51.0(d)	1',2',19',20'
4		101.7(s)	20',3,6α,20α	4'		85.0(s)	3',6'β,20α
5		146.9(s)	3,6β	5'		206.4(s)	6'α,6'β,7'α,20'
6α	2.26(m)	25.5(t)		6'α	2.92(m)	33.3(t)	
6β	1.99(m)			6'β	2.21(m)		
7α	2.08(m)	29.0(t)	9,19	7'α	2.21(m)	32.9(t)	9',19'
7β	1.42(ddd,12.6,10.6,6.0)			7'β	1.62(m)		
8		42.8(s)	2,6α,9,19	8'		43.6(s)	2',9',19'
9	6.02(d, 10.7)	75.9(d)	19	9'	5.95(d, 10.5)	75.5(d)	19'
10	5.86(d, 10.7)	73.3(d)	10,16,17,18	10'	5.89(d, 10.5)	72.9(d)	16',17',18'
11		151.7(s)	14α,18	11'		150.4(s)	18'
12		138.5(s)		12'		137.9(s)	
13		199.3(s)	1,14α,18	13'		199.8(s)	14'α,14β,18'
14α	2.80(dd, 19.5, 6.3)	36.8(t)		14'α	3.24(d, 19.8)	36.0(t)	
14β	2.41(d, 19.5)			14'β	2.61(dd,19.8,6.8)		
15		40.0(s)	1,10,14α,16,17	15'		38.8(s)	10',14'α,14'β,16',17'
16	1.78(s)	25.4(q)	17	16'	1.74(s)	25.4(q)	17'
17	1.13(s)	37.2(q)	16	17'	1.10(s)	37.1(q)	16'
18	1.97(s)	13.6(q)		18'	1.90(s)	13.6(q)	
19	1.01(s)	18.5(q)	7β,9	19'	1.26(s)	18.9(q)	7'α
20α	2.75	26.3(t)	20'	20'	2.29(m)	22.1(t)	
20β	1.30						
OAc	2.11(s)	170.2(s)		OAc	2.11(s)	170.5(s)	
	2.03(s)	169.9(s)			2.02(s)	169.8(s)	
	2.00(s)	169.8(s)			2.00(s)	169.4(s)	
		21.5(q)				21.4(q)	
		20.7(q)				20.7(q)	
		20.6(q)				20.6(q)	

Notes

Measured in acetone-d_6, δ values with TMS as internal standard.

The values in parenthesis are coupling constants (Hz).

Multiplicity: s=singlet, d=doublet, t=triplet, q=quartet, m=multiplet, br=broad.

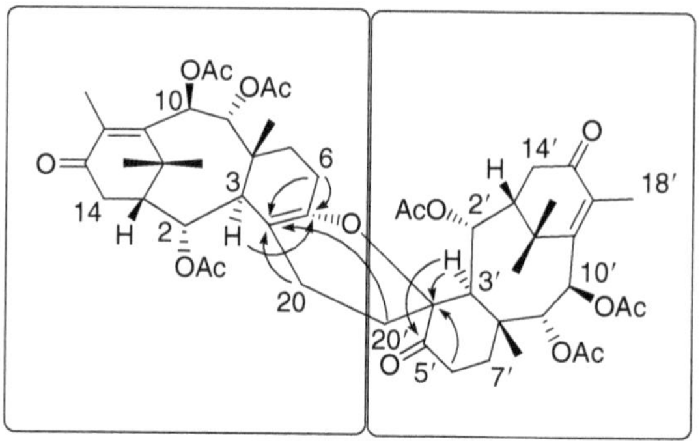

Figure 4.28 5-Oxotaxinine A.

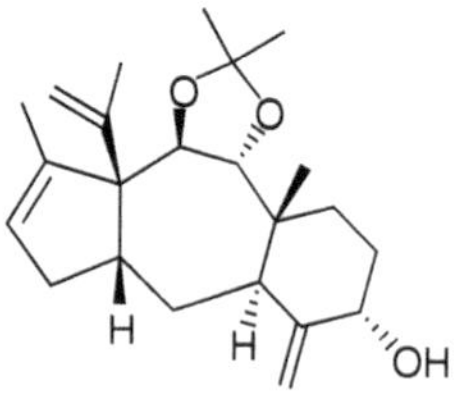

6

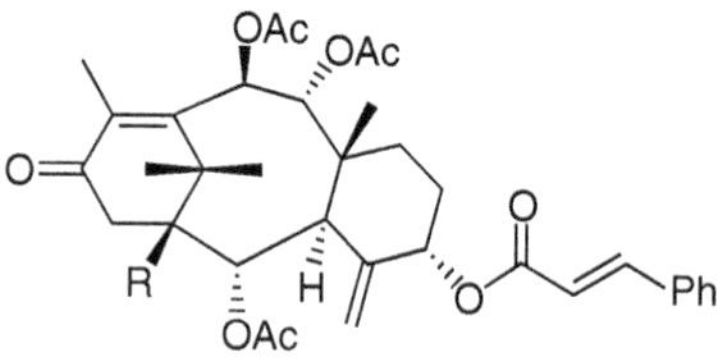

	R
7	H
8	OH

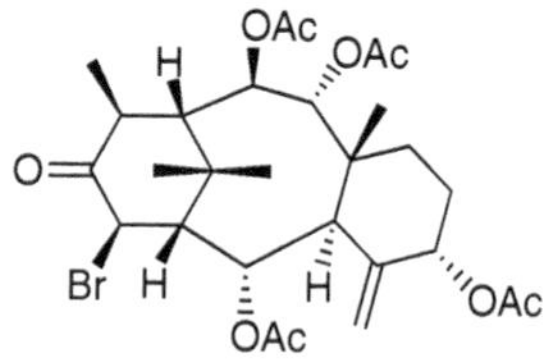

9

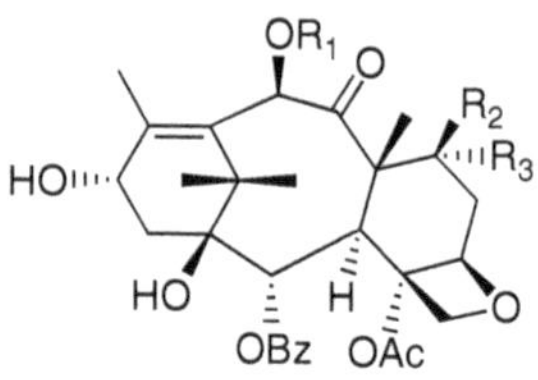

	R_1	R_2	R_3
10	Ac	H	OH
14	H	OH	H

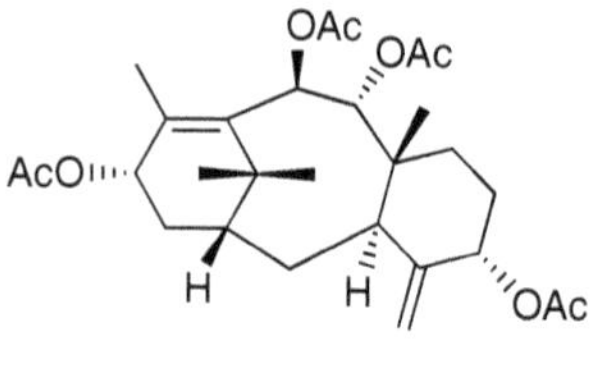

11

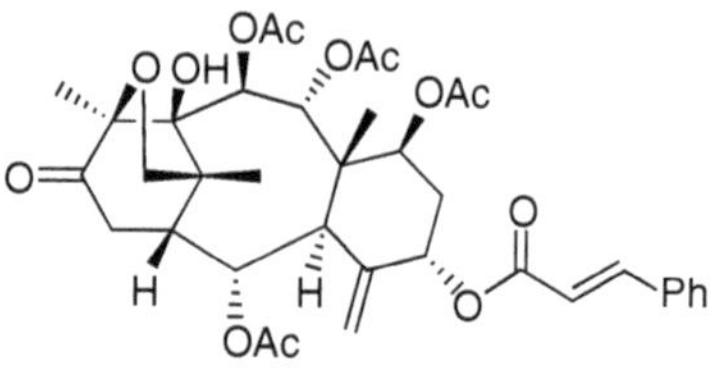

12

Figure 4.29

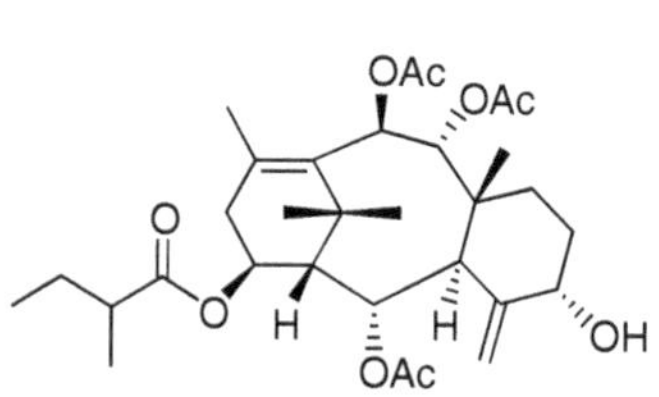

13

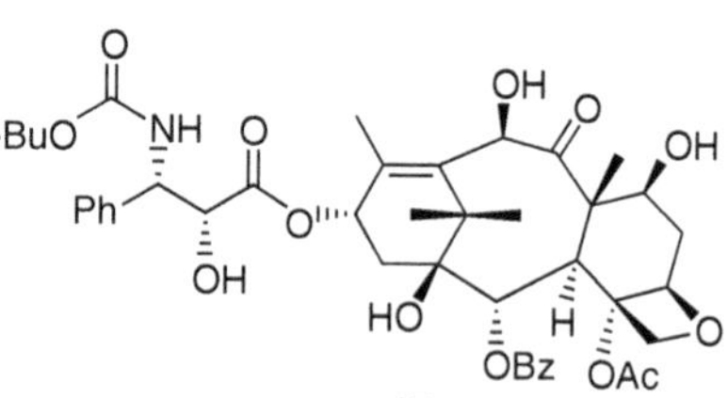

15

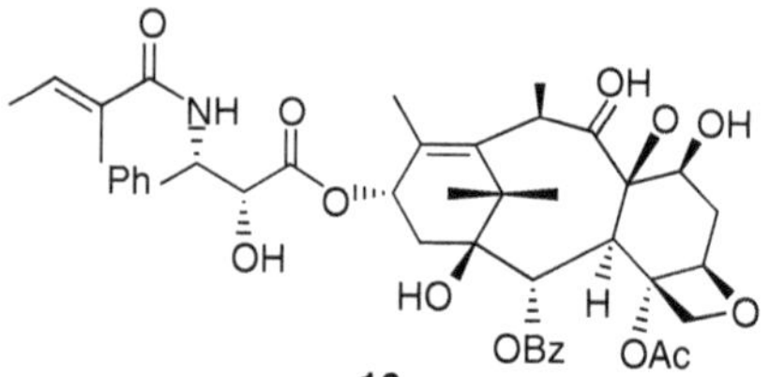

16

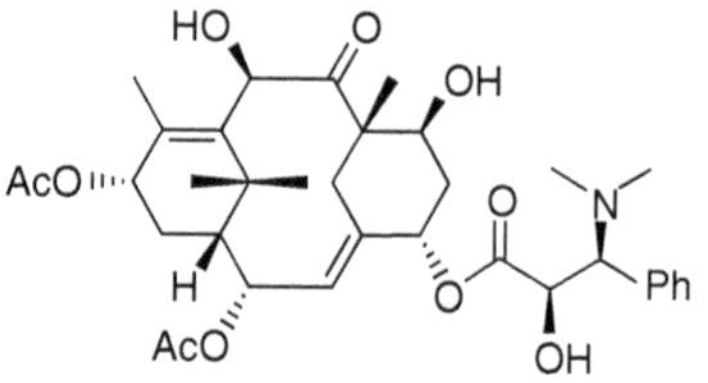

17

Figure 4.30

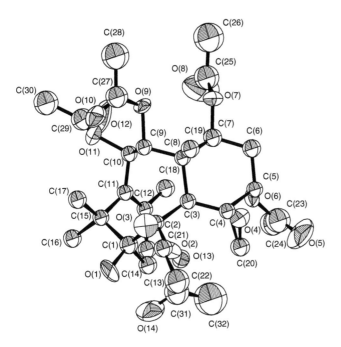

Figure 4.31 ORTEP diagram showing the crystallographic atom numbering scheme and solid state conformation of 1β-hydroxybaccatin I, the hydrogen atoms have been omitted for clarity. (Reprinted from Shen and Chen. Taxanes from the roots of *Taxu mairei*, *Phytochemistry*, 44 (8), 1531 (1997). Copyright 1997 with permission from Elsevier Science Ltd.)

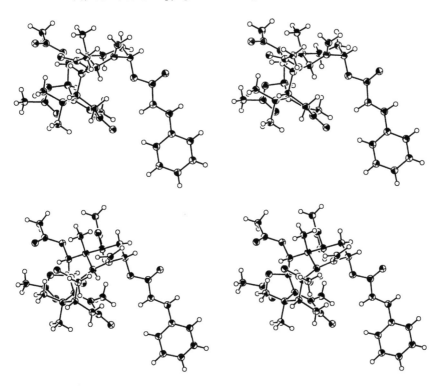

Figure 4.32 ORTEP drawing of stereoviews of taxinine and taxinine B, Above: taxinine; below: taxinine B. (Reprinted from Morita, H. *et al.*, Tetrahedron. 53 (13), 4623 (1997). Copyright 1997 with permission from Elsevier Science Ltd.)

Acknowledgement

The authors would like to thank Dr K. Hayashi for his valuable suggestions.

References

Altstadt, T.J., Gao, Q., Wittman, M.D., Kadow, J.F., and Vyas, D.M. (1998) Crystallographic determination of the stereochemistry of C-6,7 epoxy paclitaxel, *Tetrahed., Lett.*, **39** (28), 4965–4966.

Appendino, G., Gariboldi, P., Gabetta, B., Pace, R., Bombardelli, E., and Viterbo, D. (1992a) 14-Hydroxy-10-deacetylbaccatin III, a new taxane from Himalayan yew (*Taxus wallichiana* Zucc.). *J. Chem. Soc. Perkin Trans. 1*, 2925–2929.

Appendino, G., Lusso, P., Gariboldi, P., Bombardelli, E., and Gabetta, B. (1992b) A 3,11-Cyclotaxane from *Taxus baccata*. *Phytochemistry*, **31**(12), 4259–4262.

Appendino, G., Ozen, H.C., Gariboldi, P., Torregiani, E., Gabetta, B., Nizzola, R., and Bombardelli, E. (1993a) New oxetane-type taxanes from *Taxus wallichiana* Zucc. *J. Chem. Soc. Perkin Trans. 1*, 1563–1566.

Appendino, G., Barboni, L., Gariboldi, P., Gabetta, B., and Viterbo, D. (1993b) Revised structure of brevifoliol and some baccatin VI derivatives. *J. Chem. Soc., Chem. Commun.*, 1587–1589.

Barboni, L., Gariboldi, P., Torregiani, E., Appendino, G., Gabetta, B., and Bombadelli, E. (1994) Taxol analogues from the roots of *Taxus x media*. *Phytochemistry*, **36**(4), 987–990.

Barboni, L., Gariboldi, P., Appendino, G., Enriu, R., Gabetta, B., and Bombardelli, E. (1995) New taxoids from *Taxus baccata* L. *Liebigs Ann.*, 345–349.

Begrey, M.J., Frecknall, E.A., and Pattenden, G. (1984) X-ray structure of O-Cinnamoyltaxicin-I triacetate, C$_{35}$H$_{42}$O$_{10}$, from the yew, *Taxus baccata*. *Acta Cryst.*, **C40**, 1745.

Bitsch, F., Shackleton, C., Ma, W., Park, G., and Nieder, M. (1993) *Rapid. Comm. Mass Spectrum.*, 7, 891.

Castellano, E.E. and Hodder, O.J.R. (1973) The crystal and molecular structure of the diterpenoid baccatin V, a naturally occurring oxetane with a taxane skeleton. *Acta Cryst.*, **B29**, 2566.

Chan, W.R., Halsall, T.G., Hornby, G.M., Oxford, A.W., Sabel, W., Bjamer, K., Ferguson, G., and Robertson, J.M. (1966) Taxa-4(16),11-diene-5α,9α,10β,13α-tetraol, a new taxane derivative from the heartwood of yew (*T. baccata* L.): X-ray analysis of a *p*-bromobenzoate derivative. *Chem. Commun.*, 923.

Chaviere, G., Guenard, D., Pascard, C., Picot, F., Potier, P., and Prange, T. (1982) Taxagifine; New taxane derivative from *Taxus baccata* L. (Taxaceae). *J. Chem. Soc., Chem.Commun.* 495.

Chen, Y.-C., Chen, C.-Y., and Shen, Y.-C. (1999) New taxoids from the seeds of *Taxus chinensis*; *J. Nat. Prod.*, **62**, 149–151.

Cheng, K., Fang, W., Yang, Y., Xu, H., Kong, M., He, W., and Fang, Q. (1996) C-14 Oxygenated taxanes from *Taxus yunnanensis* cell cultures. *Phytochemistry*, **42**(1), 73–75.

Chumurny, G.N., Hilton, B.D., Brobst, S., Look, S.A., Witherup, K., and Beutler, J.A. (1992) ^{1}H- and ^{13}C-NMR assignments for taxol, 7-epi-taxol, and cephalomannine. *J. Nat. Prod.*, **55**(4), 414–423.

De Marcano, D.C., Halsall, T.G., Scott, A.I., and Wrixon, A.D. (1970a) Application of the olefin octant rule to some taxane derivatives: assignment of absolute configuration using the Cotton effect. *J. Chem. Soc. D (Chem. Commun.)*, 582.

De Marcano, D.C., Halsall, T.G., Castellano, E., and Hodder, O.J.R. (1970b) Crystallographic structure determination of the diterpenoid baccatin-V, a naturally occurring oxetane with a taxane skeleton. *J. Chem. Soc. D (Chem. Commun.)*, 1382.

Falzone, C.J., Benesi, A.J., and Lecomte, T.J. (1992) Characterization of taxol in methylene chloride by NMR spectroscopy. *Tetrahedron Lett.*, **33**(9), 1169–1172.

Fuji, K., Tanaka, K., and Li, B. (1992) Taxchinin A: A diterpenoid from *Taxus chinensis*. *Tetrahedron Lett.*, **33**(5), 7915–7916.

Fukushima, M., Takeda, J., Fukamiya, N., Okano, M., Yagahara, K., Zhang, S.-X., Zhang, D.-C., Lee, K.-H. (1999) A New Taxoid, 19-Acetoxytaxagifine, from *Taxus chinensis*. *J. Nat. Prod.*, **62**(1), 140–142.

Graf, E., Kirfel, A., Wolff, G.-J., and Breitmaier, E. (1982) Die Aufklarung von taxin A *aus Taxus baccata* L. *Liebigs Ann. Chem.*, 376.

Gueritte-Voegelein, F., Guenard, D., Mangatal, L., Potier, P., Guilhem, J., Cesario, M., and Pascard, C. (1990) Structure of synthetic taxol precursor: N-tert-butoxycarbonyl-10-deacetyl-N-debenzoyltaxol. *Acta Cryst.*, **C46**, 781.

Harada, N. and Nakanishi, K. (1969) A method for determining the chiralities of optically active glycols. *J. Am. Chem. Soc.*, **91**, 3989.

Harada, N., Ohashi, M., and Nakanishi, K. (1968) The benzoate sector rule, a method for determining the absolute configurations of cyclic secondary alcohols. *J. Am. Chem. Soc.*, **90**, 7349.

Ho, T.-I., Lee, G.H., Peng, S.-M., Yeh, M.-K., Chen, F.-C., and Yang, W.L. (1987a) Structure of taxusin. *Acta Cryst.*, **C43**, 1378.

Ho, T.-I., Lin, Y.-C., Lee, G.-H., Peng, S.-M., Yeh, M.-K., and Chen, F.-C. (1987b) Structure of taiwanxan. *Acta Cryst.*, **C43**, 1380.

Hoke, S.H.III, Wood, J.M., Cooks, R.G., Jayasuria, H., Heinstein, P., Chang, C.-J., Brobst, S.W., Wheeler, N., and Snader, K.M. (1991) Tandem Mass Spectrometric Analysis of Taxanes *from Taxus brevifolia*. *American Society Pharmacognosy 32nd annual Meeting, Chicago, IL, July, Abstract* p. 68.

Hosoyama, H., Inubushi, A., Katsui, T., Shigemori, H., and Kobayashi, J. (1996) Taxuspines U, V, and W, new taxane and related diterpenoids from the Japanese yew *Taxus cuspidata*. *Tetrahedron*, **52**(41), 13145–13150.

Hosoyama, H., Shigemori, H., Yasuko In, Ishida, T., and Kobayashi, J. (1998) Occurrence of a new dimeric compound of 5-oxotaxinine A through Diels-Alder cycloaddition. *Tetrahedron Lett.*, **39**, 2159–2162.

Jia, Z.J. and Zhang, Z.-P. (1991) Taxanes from *Taxus chinensis*. *Chin. Sci. Bull.*, **36**, 1174.

Kerns, E.H., Volk, K.J., and Hill, S.E. (1994) Profiling taxanes in *Taxus* extracts using LC/MS and LC/MS/MS techniques. *J. Nat. Prod.*, **57**(10), 1391–1403.

Kingston, D.G.I., Molinero, A.A., and Rimoldi, J.M. (1993) The taxane diterpenoids, *Progress in the Chem. of Org. Nat. Prod.*, **61**, 1–206.

Kobayashi, J. and Shigemori, H. (1998) Bioactive taxoids from Japanese yew *Taxus cuspidata* and taxol biosynthesis. *Heterocycles*, **47**(2), 1111–1133.

Kobayashi, J., Ogiwara, A., Hosoyama, H., Shigemori, H., Yoshida, N., Sasaki, T., Li, Y., Iwasaki, S., Naito, M., and Tsuruo, T. (1994) Taxuspines A–C, new taxoids from Japanese *Yew Taxus cuspidata* inhibiting drug transport activity of P-glycoprotein in multidrug-resistant cells. *Tetrahedron*, **50**(25), 7401–7416.

Kobayashi, J., Hosoyama, H., Shigemori, H., Koiso,Y., and Iwasaki, S. (1995) Taxuspine D, a new taxane diterpene from *Taxus cuspidata* with potent inhibitory activity against Ca^{2+}-induced depolymerization of microtubules. *Experientia*, **51**, 592.

Kobayashi, J., Inubushi, A., Hosoyama, H., Yoshida, N., Sasaki, T., and Shigemori, H. (1995) Taxuspines E–H and J, new taxoids from the Japanese yew *Taxus cuspidata*. *Tetrahedron*, **51**(21), 5971–5978.

Kobayashi, J., Hosoyama, H., Katsui, T., Yoshida, N., and Shigemori, H. (1996) Taxuspines N, O, and P, new taxoids from Japanese yew *Taxus cuspidata*. *Tetrahedron*, **52**(15), 5391–5396.

Li, B., Tanaka, K., Fuji, K., Sun, H., and Taga, T. (1993) Three new diterpenoids from *Taxus chinensis*. *Chem. Pharm. Bull.*, **41**(9), 1672–1673.

Li, S.H., Zhang, H.J., Yao, P., and Sun, H.D. (1998) Two new taxoids from *T. yunnanensis*. *Chin. Chem. Lett.*, **9**(11), 1017–1020.

Li, S.H., Zhang, H.J., Yao, P., Sun, H.D., and Fong, H.H.S. (2000) Rearranged taxanes from the bark of *Taxus yunnanensis*. *J. Nat. Prod.*, **63**, 1488–1491.

Liang, J. and Kingston, D.G.I. (1993) Two new taxane diterpenoids from *Taxus mairei*. *J. Nat. Prod.*, **56**(4), 594–599.

Liang, J. and Gunatilaka, A.A.L. (1998) *J. Chin. Pharm. University (Zhongguo Yaoke Daxue Xuebao)*, **29**, 259–266.

Lythgoe, B. (1968) The Taxus alkaloids in *The Alkaloids*, Vol. 10, *Chemistry and Physiology*, Ed.

Magri, N.F. (1985) Modified taxols as anticancer agents. Ph.D. Dissertation, Virginia Polytechnic Institute and State University, Blacksburg.

Ma, W., Park, G.L., Gomez, G.A., Nieder, M.H., Adams, T.L., Aynsley, J.S., Sahai, O.P., Smith, J., Stahlhut, R.W., Hylands, P.J., Bitsch, F., and Shackleton, C. (1994) New bioactive taxoids from cell cultures of *Taxus baccata*. *J. Nat. Prod.*, **57**(1), 116–122.

McClure, T.D., Reimer, M.L.J., and Schram, K.H. (1990) *The 38th ASMS Conference on Mass Spectrometry and Allied Topics*, WP 135.

Miller, R.W. (1980) A brief survey of Taxus alkaloids and other taxane derivatives. *J. Nat. Prod.*, **43**, 425.

Miller, R.W., Powell, R.G., Smith Jr., C.R., Arnold, E., and Clardy, J. (1981) Antileukemic alkaloids from *Taxus wallichiana* Zucc. *J. Org. Chem.*, 46, 1469.

Morita, H., Gonda, A., Wei, L., Yamamura, Y., Takeya, K., and Itokawa, H. (1997a) Taxuspinananes A and B, new taxoids from *Taxus cuspidata* var. *nana. J. Nat. Prod.*, 60, 390–392.

Morita, H., Wei, L., Gonda, A., Takeya, K., and Itokawa, H. (1997b) A taxoid from *Taxus cuspidata* var. *nana, Phytochemistry*, 46(3), 583–586.

Morita, H., Wei, L., Gonda, A., Takeya, K., Itokawa, H., Fukaya, H., Shigemori, H., and Kobayashi, J. (1997c) Crystal and Solution state conformations of two taxoids, taxinine and taxinine B. *Tetrahedron*, 53(13), 4621–4626.

Morita, H., Gonda, A., Wei, L., Yamamura, Y., Wakabayashi, H., Takeya, K., and Itokawa, H. (1998) Taxoids from *Taxus cuspidata* var. *nana. Phytochemistry*, 48(5), 857–862.

Morita, H., Gonda, A., Wei, L., Yamamura, Y., Wakabayashi, H., Takeya, K., and Itokawa, H. (1998) Four new taxoids from *Taxus cuspidata* var. *nana. Planta Med.*, 64, 183–186.

Murakami, R., Shi, Q.-W., and Oritani, T. (1999) A taxoid from the needles of the Japanese yew, *Taxus cuspidata. Phytochemistry*, 52, 1577–1580.

Parmer, V.S., Jha, A., Bisht, K.S., Taneja, P., Singh, S.K., Kumar, A., Poonam, Jain, R., and Olsen, C.E. (1999) Constituents of the yew trees. *Phytochemistry*, 50, 1267–1304.

Peterson, J.R., Do, H.D., and Rodgers, R.D. (1991) X-ray structure and crystal lattice interactions of the taxol side-chain methyl ester. *Pharm. Res.*, 8, 908.

Poupat, C., Ahond, A., and Potier, P. (1994) Nouveau taxoide basique isole des feuiles d'if, *Taxus baccata*: la 2-desacetyltaxine A. *J. Nat. Prod.*, 57(10), 1468–1469.

Prelog, V., Barman, P., and Zimmerman, M. (1949) Weitere Untersuchungen uber die Gultigkeitsgrenzen der Bredt'schen Rege. Eine Variante der Robinson's chem. Synthese von Cyclischen Ungesattigten Ketonen. *Helv. Chim Acta*, 32, 1284.

Rao, K.V. and Juchum, J. (1998) Taxanes from the bark of *Taxus brevifolia. Phytochemistry*, 47(7), 1315–1324.

Rao, C., Zhou, J.Y., Chen, W.M., Lu, Y., and Zheng, Q.T. (1993) *Chin. Chem. Lett.*, 4, 693–694.

Rao, K.V., Reddy, G.C., and Juchum, J. (1996) Taxanes of the needles of *Taxus x media. Phytochemistry*, 43(2), 439–442.

Rao, K.V., Bhakun, R.S., Hanuman, J., Davies, R., and Johnson, J. (1996) Taxanes from the bark of *Taxus brevifolia. Phytochemistry*, 41(3), 863–866.

Sako, M., Suzuki, H., Yamamoto, N., Hirota, K., and Maki, Y. (1998) Convenient method for regio- and/or chemo-selective O-deacylation of taxinine, a naturally occurring taxane diterpenoid. *J. Chem. Soc., Perkin Trans. I*, 417.

Shen, Y.-C. and Chen, C.-Y. (1997a) Taxane diterpenes from *Taxus mairei. Planta Med.*, 63, 569–570.

Shen, Y.-C. and Chen, C.-Y. (1997b) Taxanes from the roots of *Taxus mairei. Phytochemistry*, 44(8), 1527–1533.

Shen, Y.-C., Tai, H.-R., and Chen, C.-Y. (1996) New taxane diterpenoids from the roots *of Taxus mairei. J. Nat. Prod.*, 59, 173–176.

Shen, Y.-C., Chen, C.-Y., and Kuo, Y.-H. (1998) A new Taxane diterpenoid from *Taxus mairei. J. Nat. Prod.*, 61, 838–840.

Shen, Y.-C., Chen, C.-Y., and Chen, Y.-J. (1999a) Taxumairol M: a new bicyclic taxoid from seeds of *Taxus mairei. Planta Med.*, 65, 582–584.

Shen, Y.-C., Chen, Y.-J., and Chen, C.-Y. (1999b) Taxane diterpenoids from the seeds of Chinese yew *Taxus chinensis. Phytochemistry*, 52(8), 1565–1569.

Shen, Y.-C., Lo, K.-L., Chen, C.-Y., Kuo, Y.-H., and Hung, M.-C. (2000) New taxanes with an opened oxe- tane ring from the roots of *Taxus mairei. J. Nat. Prod.*, 63, 720–722.

Shi, Q.-W., Oritani, T., Sugiyama, T., and Kiyota, H. (1998a) Two new taxanes from *Taxus chinensis* var. *mairei. Planta Med.*, 64, 766–769.

Shi, Q.-W., Oritani, T., Sugiyama, T., and Kiyota, H. (1998b) Three novel bicyclic 3,8-secotaxane diter- penoids from the needles of the Chinese yew, *Taxus chinensis* var. *mairei. J. Nat. Prod.*, 61, 1437–1440.

Shi, Q.-W., Oritani, T., Sugiyama, T. (1999a) Two bicyclic taxane diterpenoid from the needles of *Taxus mairei*. *Phytochemistry*, 50, 633–636.

Shi, Q.-W., Oritani, T., Sugiyama, T., and Yamada, T. (1999b) Two novel pseudoalkaloid taxanes from the Chinese yew, *Taxus chinensis* var. *mairei*. *Phytochemistry*, 52, 1571–1575.

Shi, Q.-W., Oritani, T., Sugiyama, T. (1999c) Three rearranged 2(3 → 20) abeotaxanes from the bark of *Taxus mairei*. *Phytochemistry*, 52, 1559–1564.

Shi, Q.-W., Oritani, T., Sugiyama, T., Murakami, R., and Wei, H.-Q. (1999d) Six new taxane diterpenoids from the seeds of *Taxus chinensis* var. *mairei* and *Taxus yunnanensis*. *J. Nat. Prod.*, 62, 1114–1118.

Shi, Q.W., Oritani, T., and Sugiyama, T. (1999e) Three novel bicyclic taxane diterpenoids with verticilline skeleton from the needles of Chinese yew, *Taxus chinensis* var. *mairei*. *Planta Med.*, 65, 356–359.

Shi, Q.-W., Oritani, T., Sugiyama, T., Zhao, D., and Murakami, R. (1999f) Three new taxane diterpenoids from seeds of the Chinese yew, *Taxus yunnanensis* and *T. chinensis* var. *mairei*. *Planta Med.*, 65, 767–770.

Shigemori, H., Wang, X.-x., Yoshida, N., and Kobayashi, J. (1997) Taxuspines X–Z, new taxoids from Japanese yew *Taxus cuspidata*. *Chem. Pharm. Bull.*, 45(7), 1205–1208.

Shigemori, H., Sakurai, C.A., Hosoyama, H., Kobayashi, A., Kajiyama, S., and Kobayashi, J. (1999) Taxezopidines J, K, and L, new taxoids from *Taxus cuspidata* inhibiting Ca^{2+}-induced depolymerization of microtubles. *Tetrahedron*, 55, 2553–2558.

Shiro, M. and Koyama, H. (1971) Structure of taxinine: X-ray analysis of 2,5,9,10-tetra-0-acetyl-14-bromotaxinol. *J. Chem. Soc.*, 1342.

Shiro, M., Sato, T., and Koyama, H. (1966) The stereochemistry of taxinine: X-ray analysis of 2,5,9,10-tetra-0-acetyl-14-bromotaxinol. *Chem. Commun.*, 97.

Soto, J. and Castedo, L. (1998) Taxoids from European yew, *Taxus baccata* L. *Phytochemistry*, 47(5), 817–819.

Soto, J., Fuentes, M., and Castedo, L. (1996) Teixidol, an abeo-taxane from European yew, *Taxus baccata*. *Phytochemistry*, 43(1), 313–314.

Swindell, C.S., Isaacs, T.F., and Kanes, K.J. (1985) Bicyclo[5.3.1.]undec-1(10)-ene bridgehead olefin stability and the taxane bridgehead olefin. *Tetrahedron Lett.*, 26, 289.

Tong, X.-J., Fang, W.-S., Zhou, J.-Y., He, C.-H., Chen, W.-M., and Fang, Q.-C. (1995) Three new taxane diterpenoids from needles and stems of *Taxus cuspidata*. *J. Nat. Prod.*, 58(2), 233–238.

Wang, X.-X., Shigemori, H., and Kobayashi, J. (1996) Taxuspines K, L, and M, new taxoids from Japanese yew *Taxus cuspidata*. *Tetrahedron*, 52(7), 2337–2342.

Wang, X.-X., Shigemori, H., and Kobayashi, J. (1998) Taxezopidine B-H, new taxoids from Japanese yew *Taxus cuspidata*. *J. Nat. Prod.*, 61, 474–479.

Wani, M.C., Taylor, H.L., Wall, M.E., Coggon, P., and McPhail, A.T. (1971) Plant antitumor agents VI. The isolation and structure of taxol, a novel antileukemic and antitumor agent from *Taxus brevifolia*. *J. Am. Chem. Soc.*, 93, 2325.

Yang, S.-J., Fang, J.-M., and Cheng, Y.S. (1996) Taxanes from *Taxus mairei*. *Phytochemistry*, 43(4), 839–842.

Yue, Q., Fang, Q.-C., Liang, X.-T., and He, C.-H. (1995a) Taxayuntin E and F: Two taxanes from leaves and stems of *Taxus yunnanensis*. *Phytochemistry*, 39(4), 871–873.

Yue, Q., Fang, Q.-C., Liang, X.-T., He, C.-H., Jing, X.-L. (1995b) Rearranged taxoids *from Taxus yunnanensis*. *Planta Med.*, 61, 375–377.

Zamir, L.O., Nedea, M.E., Zhou, Z.-H., Belair, S., Caron, G., Sauriol, F., Jacqmain, E., Jean, F.I., Garneau, F.-X., and Mamer, O. (1995) *Taxus canadensis* taxanes: structures and stereochemistry. *Can. J. Chem.*, 73, 655–665.

Zhang, Z.P. and Jia, Z. (1991) Taxanes from *Taxus chinensis*. *Phytochemistry*, 30, 2345–2348.

Zhang, H., Sun, H., Takeda, Y. (1995) Four new taxanes from the roots of *Taxus yunnanensis*. *J. Nat. Prod.*, 58(8), 1153–1159.

Zhang, H., Mu, Q., Xiang, W., Yao, P., Sun, H., and Takeda, Y. (1997) Intramolecular transesterified taxanes from *Taxus yunnanensis*. *Phytochemistry*, 44(5), 911–915.

5   Plant tissue culture of taxoids

Koichi Takeya

Introduction

In the NCI-sponsored plant-screening program for natural antitumor substances, taxol (paclitaxel) was first isolated from the stem bark of the Pacific yew, *Taxus brevifolia* Nutt., which is a dioecious evergreen tree or shrub (Suffnes, 1995). Its chemical structure was established in 1971 (Wani *et al.*, 1971). Taxol is a potent cytotoxic diterpene amide that is active against B16 melanoma, MX-1 human mammary, and CX-1 colon xenografts (McGuire *et al.*, 1989). Taxol inhibits cell replication by blocking cell division in the G2/M phase of the cell cycle (Horowitz *et al.*, 1986) without significant effects on DNA, RNA, or protein syntheses (Horwitz, 1992). Taxol is currently the best-known drug approved for the treatment of ovarian and breast cancer; it also has potent antitumor activities against a variety of solid tumors. However, because *Taxus* trees grow very slowly and contain only a small amount of taxol, the supply of the drug from natural resources is limited. Additionally, its complex structure has proved challenging for complete chemical synthesis. Currently, semisynthetic taxol analogues are used as clinical drugs; however, their preparation is still dependent on a natural source for the taxane skeleton (Denis *et al.*, 1988; Holton *et al.*, 1992). More detailed reviews of taxol's chemistry, biological properties, pharmacology, and toxicology have been reported (Kingston, 1994; Nicolaou *et al.*, 1994; Suffnes, 1995).

Although taxol was initially isolated from *T. brevifolia*, it is also found in many *Taxus* species (Wani *et al.*, 1971; Vidensek *et al.*, 1990; Witherup *et al.*, 1990; Mattina and Paiva, 1992; Wheeler *et al.*, 1992; Choi *et al.*, 1994; Wickremesinhe and Arteca, 1994; Kwak *et al.*, 1995). Currently, *Taxus* trees are grown in commercial nurseries, and the needles and stems are allocated as an alternative or semisynthetic source of taxol (Joyce, 1993; Wheele and Hehnen, 1993). Thus, *T. brevifolia* may no longer be needed as a source of taxol (Cragg *et al.*, 1993). Plant cell and tissue culture is expected to be an useful method for the large-scale production of taxol and other natural taxanes in order to supply sufficient quantities of taxol for cancer chemotherapy.

Approaches to *Taxus* plant tissue culture

In the 1950s, La Rue (1953) and Tulecke (1959) initiated approaches to *Taxus* plant cell and tissue culture. At present, the production capability of taxol and related taxanes using cell and tissue culture has been established and the conditions suitable for fast-growing cultures to produce high levels of taxol have been studied in commercial production. Callus and cell suspension cultures have been established from *T. brevifolia* (Christen *et al.*, 1989, 1991; Wickremesinhe, 1992; Wickremesinhe and Arteca, 1993; Gibson *et al.*, 1993; Durzan and Ventimiglia, 1994; Ellis *et al.*, 1994; Chee, 1995; Ketchum *et al.*, 1995; Kim *et al.*, 1995), *T. baccata* (Jaziri *et al.*, 1991; Wickremesinhe, 1992; Wickremesinhe and Arteca, 1993; Guo *et al.*, 1994; Ma *et al.*, 1994;

Table 5.1 Studies on the tissue cultures of taxus species

Taxus species	Research summary	Reference
T. cuspidata	The effects of different organic solvents (paraffin, organic acid, alcohol and ester) and their volumetric fractions on the cell growth and taxol production were studied in two-liquid-phase and the two-stage culture.	Wu et al., 2000
T. yunnanensis	Viable protoplasts of Taxus yunnanensis were isolated from friable, light yellow callus. Protoplast yield was dependent on callus age, with a maximum from 20-day-old callus.	Luo et al., 2000
Taxus sp.	The adventitious root formation and taxol production of Taxus species were studied.	Dai et al., 2000
T. baccata	A high-yielding procedure was developed for the in vitro propagation of juvenile material of Taxus baccata involving a combination of seed handling and culture on WP culture medium.	Majada et al., 2000
Taxus sp.	Physiological and biochemical studies were conducted on a taxol-producing, fungi-resistant variant of Taxus grown in tissue culture.	Mei et al., 1999
T. yunnanensis	The effects of several precursors and chemical elicitors on the growth and taxol production in cell suspensions of Taxus yunnanensis were investigated.	Chen et al., 2000a
Penicillium sp.	10-Deacetylbaccatin III and other taxane derivatives were produced by Penicillium sp.	El Jaziri and Diallo, 2000
T. cuspidata	Bioreactor (3 mg/l of taxol and 74 mg/l total taxanes were obtained after 27 days of culture)	Son et al., 2000
Taxus sp.	Callus cultures (Generally, taxol contents in the callus strains of T. yunnanensis and T. chinensis were higher)	Chen et al., 2000b
T. cuspidata	Molecular cloning of a taxa-4(20),11(12)-dien-5α-ol-O-acetyl transferase cDNA from Taxus and functional expression in Escherichia coli.	Walker et al., 2000
T. chinensis T. yunnanensis T. baccata T. cuspidata	The growth and paclitaxel content in 4 Taxus cell suspensions.	Sun and Hu, 1999c
T. chinensis	The effect of mixing stress on the cell physiology and metabolites of Taxus chinensis in the bioreactor.	Pan et al., 1999
T. chinensis	Taxol production by immobilized cultures of Taxus chinensis suspense cell.	Wang et al., 2000b
T. cuspidata	Two modified media (SSS and RW medium) were used to induce callus formation from T. cuspidata and T. chinensis young stems in comparison to B5 medium.	Feng and Guo, 1999
T. chinensis	Optimization of plant cell suspension culture of Taxus chinensis var. mairei	Hu et al., 1999
T. chinensis var. mairei		
T. cuspidata	Molecular cloning of a 10-deacetylbaccatin III-10-O-acetyl transferase cDNA from Taxus and functional expression in Escherichia coli.	Walker et al., 2000
T. chinensis	The specific cDNA fragment (I) of taxadiene synthase from Taxus chinensis callus (82bp) was separated and analyzed by PCR.	Shi et al., 1999
T. media var. Hicksii	Hairy root cultures of Taxus media var. Hicksii as a new source of paclitaxel and 10-deacetylbaccatin III	Furmanowa and Syklowska-Baranek, 2000

(Continued)

Table 5.1 (Continued)

Taxus species	Research summary	Reference
T. chinensis	Taxus chinensis callus initiation, growth characterization and paclitaxel production were studied.	Sun and Hu, 2000
T. cuspidata	Two-phase cultures of T. cuspidata were performed using silicone cubes as a second phase in shake flasks for paclitaxel production.	Kim et al., 1999b
T. cuspidata	Production of taxol from cell culture of T. cuspidate.	Han et al., 1996
T. yunnanensis	The effects of the cultures of Penicillium citrinum, Absidia glauca, Mucor rouxianus and Botrytis cinerea on the growth and taxol production in suspension cells of Taxus yunnanensis were investigated.	Chen et al., 1999b
T. chinensis	A novel efficient method of Taxus chinensis zygote embryo-derived cell line initiation is reported which is suitable for large scale suspension culture to produce taxol.	Zhang et al., 2000
T. media	All stereoisomers of Me jasmonate (MJA) were prepared, and their effects on cell yield and promotion of paclitaxel (Taxol) and baccatin III production investigated in cell suspension cultures of T. media.	Yukimune et al., 2000
T. cuspidata	Endophytic fungi from Ginkgo biloba and Taxus cuspidata in Korea were screened for production of taxol.	Kim et al., 1999a
T. chinensis	The effects of both initial sucrose concentration and sucrose feeding on cell growth and accumulation of taxane diterpene (taxuyunnanine C) in suspension cultures of Taxus chinensis were investigated in detail.	Wang et al., 2000a
T. chinensis	The method of analysis of paclitaxel from callus of Taxus chinensis by HPLC was carried out.	Sun et al., 1999a
T. cuspidata	The suspension cells of Taxus cuspidata were transformed with Agrobacterium tumefaciens harboring binary vector pCAMBIA1302 encoding mgfp.	Kim et al., 2000
T. chinensis	The relationship between taxol production and inocula was investigated from different periods of the growth cycle of Taxus chinensis suspension cultures in the production medium (modified MS) with fungal elicitors derived from the bark of T. chinensis.	Zhang et al., 1999b
T. baccata	Geranylgeranyl diphosphate synthase, an enzyme that regulates taxane biosynthesis, was purified to homogeneity from cell cultures of T. baccata.	Laskaris et al., 2000
T. chinensis var. mairei	The effects of precursor feeding and mixed sugar supplements on the growth and taxol production of Taxus chinensis var. mairei in the biphasic-liquid culture were studied.	Wu et al., 1999
T. canadensis	Ethylene and methyl jasmonate interaction and binding models for elicited biosynthetic steps of paclitaxel in suspension cultures of Taxus canadensis.	Phisalaphong and Linden, 1999a
Taxus sp.	The invention provides a process for producing taxol by tissue culture of yew.	Sun et al., 1997
T. chinensis	The effect of Methyl-jasmonate (MJ) on taxol biosynthesis was investigated in detail.	Yu et al., 1999b
T. brevifolia	Leaves and callus of Kalanchoe daigremontiana and Taxus brevifolia were used to investigate nitric oxide-induced apoptosis in plant cells.	Pedroso et al., 2000
Taxus sp.	RP-HPLC analysis of paclitaxel and other taxoids was optimized. The analytical methodology was applied in the characterization and study of cell lines, providing interesting results concerning the biosynthesis of taxanes.	Theodoridis et al., 1999
T. cuspidata	Selection and proliferation of rapid growing cell lines from embryo derived cell cultures of yew tree.	Son et al., 1999

Species	Description	Reference
T. cuspidata	The yeast extract, coconut water, safflower seed oil, arachidonic acid, linolenic acid, jasmonic acid and Me jasmonate were added to Gamborg's B5 medium. The changes on productivity of baccatin III were estimated.	Shin *et al.*, 2000
T. cuspidata	Taxane diterpenes were manufactured with *Taxus cuspidata* and *T. cuspidata* var. *nana* by callus tissue culture.	Ando, 2000
T. cuspidata var. *nana*		
T. yunnanensis	Effects of L-glutamine and cinnamic acid on the growth of callus in cultures of Yunnan yew.	Chen *et al.*, 1999a
Taxus sp.	The diterpene cyclase taxadiene synthase from yew (*Taxus*) species transformed geranylgeranyl diphosphate to taxa-4(5),11(12)-diene as the first committed step in the biosynthesis of the anti-cancer drug taxol.	Williams *et al.*, 2000
T. cuspidata	Taxol and its derivatives were produced by inducing callus from the living cell of the leaves and stems of yew tree, and then by suspension culture of the callus cell in liquid medium.	Byun, 1997
T. baccata	A new process for the continuous production of baccatin III by enzymatic acetylation of 10-deacetylbaccatin III in an enzyme reactor was claimed.	Glund *et al.*, 1999
T. baccata	Small callus pieces excised from a selected 1-year-old stable callus line of *Taxus baccata* were grown on B5 solid medium. The callus growth and the production of taxol and baccatin III were tested.	Cusido *et al.*, 1999
T. chinensis	A method of manufacturing baccatin for use as an intermediate in the manufacture of taxol was described.	Bombardelli *et al.*, 1999
T. yunnanensis	10-L scale cell fermentation culture was studied under the optimum culture conditions of cell suspension culture from *Taxus yunnanensis*.	Gan *et al.*, 1997
T. cuspidata	Through the two-liquid-phase culture, effects of different organic solvents, their volume percentages, time of addition and phase toxicity on the growth of cells and yield of paclitaxel in cell culture of *T. cuspidata*.	Wu *et al.*, 1998b
T. cuspidata	Four diterpenoids, yunnaxane, taxuyunnanine C, yunnanxane and a novel compd., 7β-hydroxy-2α,5α,10β,14β-tetraacetoxy-4(20),11-taxadiene, have been isolated from the *T. cuspidata* callus culture.	Fedoreyev *et al.*, 1998
T. canadensis	Cell suspension cultures of *Taxus canadensis* and *Taxus cuspidate* rapidly produced paclitaxel (taxol) and other taxoids in response to elicitation with Me jasmonate.	Ketchum *et al.*, 1999b
T. cuspidata		
T. cuspidata	The effects of La, Ce on the growth of taxol biosynthesis and release of *Taxus cuspidata* cell suspension culture were studied.	Yuan *et al.*, 1998
Taxus sp.	An airlift bioreactor with 8 L for plant cell cultures was developed.	Lu *et al.*, 1998
T. chinensis	The effects of antioxidative chemicals on the prevention of discolorization of the callus cultures of *Taxus chinensis* were studied.	Huang *et al.*, 1999
T. chinensis var. *mairei*	Fresh weight biomass of callus originating from *Taxus chinensis* var. *mairei* young stem explants on different media was significantly different according to Duncan's multiple range test.	Zhang *et al.*, 1998
Taxus sp.	A method to increase the production level of paclitaxel by changing the temperature.	Choi *et al.*, 1999
T. canadensis	Suspension cultures of *Taxus canadensis* were elicited with Me jasmonate (MJ) under defined headspace ethylene concns.	Phisalaphong and Linden, 1999b
T. baccata	In a suspension culture of *Taxus baccata* cultivated in a liquid Medium according to Ericksson in a 6.2 L bioreactor paclitaxel and 8 of its analogs were identified.	Vanek *et al.*, 1999

(Continued)

Table 5.1 (Continued)

Taxus species	Research summary	Reference
T. cuspidata	Partially purified acetyl CoA:10–deacetylbaccatin–III–10–orthoxy–acetyltransferase from leaves and 3-year-old cell suspension cultures of T. cuspidata yielded baccatin-III from 10-deacetyl-baccatin-III in the presence of acetyl-CoA.	Pennington et al., 1998
T. camadensis	Cell suspension cultures of Taxus canadensis rapidly produced paclitaxel and other taxoids in response to elicitation with Me jasmonate.	Ketchum et al., 1999a
T. cuspidata	A simple method for the isolation of taxol and its analogs from the needles of Taxus cuspidata and production of taxoids from callus cultures.	Ando et al., 1998
Taxus sp.	The invention related to a method for producing taxane-type diterpenes by culturing cells of a taxane-type diterpene-producing plant.	Yukimune et al., 1998
T. chinensis	The isoenzymes of peroxidase, polyphenol oxidase, esterase, and superoxide dismutase, amylase and cytochrome oxidase in sinenxans-high-yield cell line Ts19 derived from Taxus chinensis were studied using PAGE.	Zhao et al., 1998
T. chinensis	The method comprised inoculating Taxus chinensis cell B5 in shake medium, culturing at 24–26° in dark room while adding inducer at logarithmic phase.	Yuan et al., 1997
T. chinensis	The paper studied high density suspension cultures of Taxus chinensis cells in both a 250 mL shake flask and a 1.5 L stirred bioreactor.	Wang et al., 1999
T. baccata	The results presented in this paper described for the first time the presence of an inducible enzyme in the biosynthetic pathway leading to the anticancer drug paclitaxel.	Laskaris et al., 1999a
T. chinensis	The effects of chitosan on the improvement of taxol biosynthesis and phenylalanine ammonia-lyase (PAL) activity in Taxus chinensis suspension cultures were studied.	Zhang et al., 1999a
T. media	Root proliferating callus occurred within 6–8 week on the basis of seedling hypocotyls of T. media cultivated on DCR medium containing 1 mg/L NAA and gelled with phytagar.	Furmanova et al., 1998
T. yunnanensis	The oligosaccharide could increase the growth rate of Taxus yunnanensis cultured cells by 36.7% and promote the taxol yield nearly 2-fold.	Gan et al., 1999
Taxus sp.	The invention provided new methods for production and isolation of taxanes and other metabolites from taxane-producing tissue culture of Taxus species.	Durzan and Ventimiglia, 1999
T. cuspidata	The fermentation broth of hyphomycete TCH68, an endophytic fungus associated with the inner bark of Taxus cuspidata Sieb. et Zucc., was found to induce mitotic arrest in vitro.	Jiang et al., 1998
T. cuspidata	The effects of addition of adsorbents on the production of taxol and taxane compounds were investigated in Taxus cuspidata cell culture.	Kwon et al., 1998
T. baccata	A Taxus baccata cell line had been characterized for growth, nutrient uptake, taxane production and for GGPP synthase activity.	Laskaris et al., 1999b
T. media	Tissue derived from callus of Taxus media was cultured with L-ascorbic acid and Me jasmonate for 14 days to produce more taxol than the control tissue cultured in the presence of Me jasmonate alone.	Kodama et al., 1999

Species	Description	Reference
T. cuspidata	Maleic acid hydrazide, N-phosphomethyl glycine, succinic acid 2.2-di-Me hydrazide, coconut milk, and yeast ext. were administered in the cell suspension culture system of T. cuspidata, and the production of baccatin III was analyzed.	Shin and Lim, 1998
T. cuspidata	A rotating wall vessel, designed for growth of mammalian cells under microgravity, was used to study shear effects on Taxus cuspidata plant suspension cell cultures.	Sun et al., 1999b
Taxus sp.	The correlation among primary metabolic pathway, MVA pathway, and taxol biosynthesis in Taxus cells had been studied by supplementing with some metabolic regulators.	Yu et al., 1999a
Taxus sp.	Disclosed were novel crystal complexes of baccatin III with imidazole, 2-methylimidazole or isopropanol, which were useful for isolating baccatin III from plant tissue cell culture and plant extracts containing Baccatin III.	Gibson et al., 1999
Taxus sp.	Antitumor taxanes were manufactured by culturing taxanes-producing cells from Taxus sp. in culture media.	Morita and Suita, 1998c
Taxus sp.	This invention provided methods whereby taxol, baccatin III, and other taxol-like compounds, or taxanes, could be produced in very high yield from all known Taxus species.	Bringi et al., 1997
T. brevifolia	The establishment of an efficient transformation system for Taxus brevifolia was studied using Agrobacterium rhizogenes A4(ATCC31798).	Huang et al., 1997
Taxus sp.	High-yield sinenxans cell lines of Taxus spp. distributed in China were selected and cultured for providing enough precursors for semi-synthesis of taxol-like anticancer drugs.	Wu et al., 1998a
T. wallichiana	Cell culture of Taxus wallichiana had been established and taxol-producing cell lines selected by cell line cloning.	Jha et al., 1998
T. cuspidata	Taxanes were manufactured by cultivation of Taxus sp. cell in carbohydrate-containing media.	Morita and Suita, 1998a
T. cuspidata	Taxanes were manufactured by cultivation of Taxus sp. cell in indolebutyric acid (IBA)-containing media.	Morita and Suita, 1998b
T. cuspidata	The suspension culture of self-immobilized aggregates of Taxus cuspidata, consisting of spherical, compact aggregates displaying some level of cellular/tissue differentiation was established.	Xu et al., 1998
T. yunnanensis	Taxol was manufactured by callus tissue culture of Taxus such as T. yunnanensis.	Zheng et al., 1997
T. yunnanensis	Callus culture system was established from Taxus yunnanensis, using different explants including young stems and needles on MS and B2 medium.	Luo et al., 1997
T. media	The effects of phenylalanine, sucrose, and mannitol on the growth and production of baccatin III, 10-deacetyl-baccatin III, and taxol in suspension cells of Taxus media were studied.	Chen et al., 1998
T. cuspidata var. nana	Cut leaves or stems of T. cuspidata var. nana were cultured in a B5 growth medium at 25° under dark for callus induction.	Tachibana et al., 1998
T. chinensis	The work studied the kinetics of suspension cultures of Taxus chinensis as our first step of engineering approach to the bioprocess scale-up and optimization.	Wang et al., 1997b
T. chinensis	A cell culture of Taxus chinensis was found to produce taxoids in 4% yield (dry wt.) after stimulation with Me jasmonate. sixteen taxoids were isolated from the culture medium and cell mass.	Menhard et al., 1998
T. chinensis	By studying the conditions of entrapping Taxus chinensis suspense cells with sodium alginate, the suitable immobilizing operation conditions were established.	Yu et al., 1998
T. brevifolia	Small cell clusters with high taxol contents were selected.	Kawamura et al., 1998
T. cuspidata		

(Continued)

Table 5.1 (Continued)

Taxus species	Research summary	Reference
T. baccata	Me jasmonate increased taxane production in suspension cultures of *Taxus baccata*.	Moon et al., 1998
Taxus sp.	A fluorescence polarization immunoassay (FPIA) was developed for the quantitative determination of paclitaxel in serum, crude *Taxus* extracts and *Erwinia taxi* culture medium.	Bicamumpaka and Page, 1998
Taxus sp.	The study presented a protocol for analyzing taxol and five related taxanes from tissue culture samples of *Taxus* spp. by HPLC using a reverse phase C18 column.	Wu and Zhu, 1997a
T. chinensis	The biosynthesis of taxoids was studied conveniently in a cell culture of *Taxus chinensis*, which produced taxoids in 4% (dry wt.) yield.	Eisenreich et al., 1998
T. wallachiana	*Pestalotiopsis microspora* was an endophyte associated with many plants including *Taxus wallachiana*.	Li et al., 1998
Taxus sp.	Explants showed that 2,4-D 0.1, 6-BA 2, or GA3 2 mg/L alone could enhance the germination of yew embryo.	Mei et al., 1998
Taxus sp.	The invention related to a method for mass production of taxol by semi-continuous culture of *Taxus* genus plant with a high yield.	Choi et al., 1996
T. media	The effect of some culture conditions and media components on callus growth rate and production of taxanes in callus of *Taxus media* var. *Hatfieldii* was analyzed.	Furmanowa et al., 1997
T. baccata T. cuspidata	Taxanes were manufactured by culturing *Taxus* in the presence of amino acids selected from asparagine, alanine, proline, and peptone 0.0–4.0 wt./vol.%.	Takami and Takigawa, 1997
T. chinensis T. baccata	Two *Taxus* cell suspension cultures were used as a model system to demonstrate the similarities of biomass accumulation and secondary metabolite (taxane) production.	Srinivasan et al., 1997
Taxus sp.	Taxol, baccatin III, and their analogs were manufactured by tissue culture of *Taxus* sp. in a medium containing 0.1–10 mmol I-/L.	Takami and Takigawa, 1996
Taxus sp.	Taxol is manufactured by two-step callus tissue culture of five species of *Taxus*.	Wang et al., 1994
Taxus sp.	Five separate cell lines, three of *Taxus canadensis* and two of *Taxus cuspidata*, were used to test the effect of carbohydrates and plant growth regulators on the growth of cells and production of paclitaxel in culture.	Ketchum and Gibson, 1996
T. cuspidata	The effects of N, P, and cell immobilization on growth and taxol production were investigated using cell culture of *Taxus cuspidata*.	Park et al., 1995b
T. mairei	A new HPLC method for the determination of taxol in tissue culture of *Taxus mairei* was described.	Zhao et al., 1997
Taxus sp.	The metabolic kinetics of the *Taxus* cell suspension culture in a bioreactor including the cell growth curve, the sugar consumption curve, the PO43-consumption curve, the pH curve and protein contents curve, and the taxol contents in cells curve were studied.	Mei et al., 1997
T. brevifolia	A method for regeneration of *Taxus brevifolia* from immature zygotic embryos via somatic embryogenesis was described.	Chee, 1996
Taxus sp.	Taxane diterpenes (I) were manufactured with I-producing taxane in the presence of compounds that promoted increase of intercellular Ca concentration and/or activation of protein kinase C.	Tabata and Hara, 1995

Species	Description	Reference
T. brevifolia	Growth characteristics and the production of taxol were studied in *Taxus brevifolia* cell suspension cultures.	Yun *et al.*, 1995
Taxus sp.	An antitumor compound NSC-LSC2 (I) was manufactured by culturing cells originate from *Taxus* sp., recovery of the cultured cells, and collection of I from the cells.	Takigawa *et al.*, 1996
T. cuspidata	The effects of light and culturing temperature on production of baccatin III, the precursor of taxol, were studied with the undifferentiated callus induced from *Taxus cuspidata*.	Shin and Kim, 1996
T. cuspidata	The production characteristics for Taxol (paclitaxel) using free and immobilized cells of *Taxus cuspidata* were investigated in a perfusion culture bioreactor.	Seki *et al.*, 1997
Taxus sp.	Taxane diterpenes (I) are manufactured with I-producing *Taxus* in the presence of jasmonic acid and derivatives.	Hara and Yukimune, 1997
T. brevifolia	The cyclization of geranylgeranyl diphosphate to taxa-4(5),11(12)-diene represented the 1st committed, and a slow, step in the complex biosynthetic pathway leading to the anticancer drug taxol.	Hezari *et al.*, 1997
T. baccata	Stable growing callus cultures were recovered from stem segments of *Taxus baccata*.	Brukhin *et al.*, 1995
T. baccata	The effect of six nutrient media as well as the concentration ratio of the growth regulators auxin and cytokinin on callus induction in different organs of *Taxus baccata* were investigated.	Filonova *et al.*, 1996
T. baccata	A procedure for obtaining taxanes, esp. taxol, was accomplished by cultivating *Taxus baccata* in liquid medium under conditions that disproportionately stimulated the initiation of growth of substance-rich roots.	Derdulla *et al.*, 1997
T. brevifolia	A study on the cultivation of *Taxus brevifolia* suspension cells in polyurethane foam matrix was performed to investigate the effects of immobilization on cell growth and the production of taxol.	Park *et al.*, 1995a
T. chinensis	Inoculum size (1.5–6.0 g dry wt./L) significantly affected cell growth and accumulation of intracellular and extracellular taxol in *T. chinensis*.	Wang *et al.*, 1997a
Taxus sp.	A microgravity liquid-suspension culture method was developed for growing aggregates of plant cells.	Soon-Shiong, 1997
Taxus sp.	*T. baccata* and several other plants had a UDP-glucose: leucinostatin A glucosyl transferase that catalyzed the production of leucinostatin A β di-O-glucoside from leucinostatin A. This glucoside, also made by the fungus, had a lower bioactivity against plants, fungi and a breast cancer cell line, BT-20, than leucinostatin A.	Strobel and Hess, 1997
Taxus sp.	A novel process for extraction, isolation and separation of taxanes, particularly paclitaxel, from natural sources such as bark, needles, and twigs from *Taxus* species, tissue cultures, and fungi was described.	Pandey and Yankov, 1997
T. brevifolia	Two strains of *Pestalotiopsis* spp. (JCM 9685 and JCM 9686), endophytic fungi of *T. brevifolia*, produced several new compounds when grown in liquid culture.	Pulici *et al.*, 1997
Taxus sp.	A process for producing phytochemicals from plant cell cultures was provided.	Dutta and Pandey, 1997
Taxus sp.	Various plant cell cultures were screened for the selection of biotransformation capability.	Kim *et al.*, 1995a
Taxus sp.	The invention provided new sources of taxanes and other metabolites from members of the order Coniferales that were not in the genus *Taxus*.	Durzan and Ventimiglia, 1997

Srinivasan *et al.*, 1995; Zhiri *et al.*, 1995), *T. canadensis* (Fett-Neto *et al.*, 1992), *T. cuspidata* (Fett-Neto *et al.*, 1992, 1993, 1995; Wickremesinhe, 1992; Wickremesinhe and Arteca, 1993; DiCosmo *et al.*, 1993; Mirjalili and Linden, 1995, 1996), *T. floridana* (Salandy *et al.*, 1993) and *T. media* (Wickremesinhe, 1992; Wickremesinhe and Arteca, 1993, 1994). Many excellent references have reviewed prior research on establishing callus, suspension, embryo, root, hairy root, shoot and nodule cultures, the taxoid content achieved, and the medium conditions and elicitors for the optimizing of taxoids production (Wickremesinhe and Arteca, 1998; Fett-Neto *et al.*, 1996; Gibson *et al.*, 1995; Seki and Furusaki, 1999; Roberts and Shuler, 1998; Han *et al.*, 2000; Wang *et al.*, 2000; Ramawat, 1999; Erdemoglu and Sener, 1999; Das *et al.*, 1999; Sun and Linder, 1999; Qui and Han, 1998; Wu and Zhu, 1997; Chen and Zhu, 1997; Ciddi and Kokate, 1997; Jaziri *et al.*, 1996). Therefore, this chapter will cover the more recent research references reported from 1997 to 2000. Table 5.1 summarizes the various tissue cultures established from *Taxus* species. The main research emphasis has been to establish the capacity of taxol and the related taxoids in the tissue cultures and also to establish fast-growing cultures in high taxol content strains. The eventual goal is to establish the conditions suitable for scale up and commercial production.

Conclusion and prospective

Generally, because many plant products are accumulated in small amounts in the intact plants and cultured tissues, these products cannot be mass produced efficiently by plantation or tissue culture methods. In order to increase the productivity of taxol and related taxoids by tissue cultures, various strategies, including optimization of culture conditions, selection of high-producing cell lines, the use of elicitors, the addition of precursors, and other approaches have been examined by many researchers. The mass production of taxol and related taxoids using *Taxus* plant tissue cultures should be commercially feasible. Phyton Catalytic and ESCA Genetics in the USA are now demonstrating the commercial production of taxol using cell cultures. However, the details of the actual commercial production are not offered by these companies.

References

Ando, M., Sakai, J., Zhang, S., Kosugi, K., Watanabe, Y., Minato, H., Fujisawa, H., Sasaki, H., Suzuki, T., Hagiwara, H., Hirata, N. and Hirose, K. (1998) A simple method for the isolation of taxol and its analogs from the needles of *Taxus cuspidata* and production of taxoids from callus cultures, *Tennen Yuki Kagobutsu Toronkai Koen Yoshishu*, **40**, 353–358.

Ando, S. (2000) Production method for taxane diterpenes by plant tissue culture. Jpn. Kokai Tokkyo Koho JP 2000106893 A2 18 Apr 2000, 14 pp.

Bicamumpaka, C. and Page, M. (1998) Development of a fluorescence polarization immunoassay (FPIA) for the quantitative determination of paclitaxel. *J. Immunol. Meth.*, **212**, 1–7.

Bombardelli, E., Menhard, B. and Zenk, M. H. (1999) Enzymic preparation of baccatin from 10-deacylbaccatin with an acetyltransferase of *Taxus chinensis*, Ger. DE 19804815 C1 15 Apr 1999, 16 pp.

Bringi, V., Kakrade, P., Prince, C. L. and Roach, B. L. (1997) Enhanced production of taxanes by cell cultures of *Taxus* species. PCT Int. Appl. WO 9744476 A1 27 Nov 1997, 106 pp.

Brukhin, V. B., Moleva, I. R., Filonova, L. H., Grakhov, V. P., Blume, Y. B. and Bozhkov, P. V. (1996) Proliferative activity of callus cultures of *Taxus baccata* L. in relation to anticancer diterpenoid taxol biosynthesis, *Biotechnol. Lett.*, **18**, 1309–1314.

Byun, S.-Y. (1997) Process for the high yield production of taxol by cell culture. Repub. Korea KR 9709157 B1 7 Jun 1997.

Chee, P. P. (1995) Organogenesis in *Taxus brevifolia* tissue cultures. *Plant Cell Rep.*, **14**, 560–565.

Chee, P. P. (1996) Plant regeneration from somatic embryos of *Taxus brevifolia*. *Plant Cell Rep.*, **16**, 184–187.

Chen, Y. and Zhu, W. (1997) Tissue cell and embryo culture of *Taxus*. *Zhiwu Shenglixue Tongxun*, 33, 213–219.

Chen, Y., Wu, W., Hu, Q. and Zhu, W. (1998) Effects of phenylalanine, sucrose and mannitol on the growth and production of taxol, baccatin III and 10-deacetylbaccatin III in suspension cells of *Taxus media*. *Yaoxue Xuebao*, 33, 132–137.

Chen, Y., Zhu, W., Wu, Y. and Hu, Q. (1999a) Effects of L-glutamine and cinnamic acid on the growth of callus in cultures of Yunnan yew. *Zhongcaoyao*, 30, 542–544.

Chen, Y., Zhu, W., Wu, Y. and Hu, Q. (1999b) Effects of fungus elicitors on taxol production in suspension cells of *Taxus yunnanensis*. *Shengwu Gongcheng Xuebao*, 15, 522–524.

Chen, Y., Zhu, W., Wu, Y. and Hu, Q. (2000a) Effect of precursors and chemical elicitors on the growth and taxol production in cell suspensions of *taxus yunnanensis*. *Hubei Daxue Xuebao, Ziran Kexueban*, 22, 91–94.

Chen, Y., Wu, Y., Hu, Q. and Zhu, W. (2000b) Differences in callus induction and taxol contents of several yew (*Taxus* L.) species. *Zhongcaoyao*, 31, 216–218.

Choi, M.-S., Kwak, S.-S., Liu, J.-R., Park Y.-G., Lee, M.-K. and An, N.-H. (1994) Taxol and related compounds in Korean native yew (*Taxus cuspidate*). *Planta Med.*, 61, 264–266.

Choi, H.-K., Adams, T. L., Stahlhut, R. W., Kim, S.-I., Yun, J.-H., Song, B.-K., Kim, J.-h., Song, J.-S., Hong, S.-S. and Lee, H.-S. (1996) A method for mass production of taxol by semi-continuous culture. PCT Int. Appl. WO 9634110 A1 31 Oct 1996, 23 pp.

Choi, H. J., Choi, H. K., Kim, S. I., Yun, J. H., Son, J. S., Chang, M. S., Choi, E. Kim, H. R., Hong, S. S. and Lee, H. S. (1999) Mass production of paclitaxel by changing the temperature of the medium during the plant cell culture. PCT Int. Appl. WO 9900513 A1 7 Jan 1999, 26 pp.

Christen, A. A., Bland, J. and Gibson, D. M. (1989) Cell culture as a means to produce taxol. *Proc. Am. Assoc. Cancer Res.*, 30, 566.

Christen, A. A., Gibson, D. M. and Bland, J. (1991) Production of taxol or taxol-like compounds in cell culture. *US Patent*, 5, 019, 504.

Ciddi, V. and Kokate, C. (1997) Taxol: a novel anticancer drug from cell cultures of *Taxus* species. *Indian Drugs*, 34, 354–359.

Cragg, G. M., Schepartz, S. A., Suffness, M. and Grever, M. R. (1993) The taxol supply crisis: new NCI policies for handling the large-scale production of novel natural product anticancer and antiviral agents. *J. Nat. Prod.*, 56, 1657–1668.

Cusido, R. M., Palazo, J., Navia-Osorio, A., Mallol, A., Bonfill, M., Morales, C. and Pinol, M. T. (1999) Production of taxol and baccatin III by a selected *Taxus baccata* callus line and its derived cell suspension culture. *Plant Sci.* (Shannon, Irel.), 146, 101–107.

Dai, J., Chen, Y. and Zhu, W. (2000) Adventitious root formation and taxol production of *Taxus* species. *Zhiwu Shenglixue Tongxun*, 36, 14–18.

Das, B., Kashinatham, A. and Anjani, G. (1999) Natural taxols. *Indian J. Chem. B: Org. Chem. Incl. Med. Chem.*, 38B, 1018–1024.

Denis, J.-N., Greene, A. E., Guenard, D., Gueritte-Voegelein, F., Mangatal, L. and Poiter, P. (1988) A highly efficient, pratical approach to natural taxol. *J. Am. Chem. Soc.*, 110, 5917–5919.

Derdulla, H.-J., Becker, K., Glund, K., Haeupke, K., Ewald, D., Roensch, H., Zocher, R., Hacker, C. and Weckwerth, W. (1997) Production of taxanes. Ger. DE 19623338 C1 4 Sep 1997, 7 pp.

DiCosmo, F., Fett-Neto, A. and Malanson, S. (1993) Organization of taxol production and growth in cell cultures of *Taxus cuspidata*. In *International Yew Resources Conference*, Berkeley, CA, March 12–13, p. 10 (Abst.).

Durzan, D. J. and Ventimiglia, F. (1994) Free taxanes and the release of bound compounds having taxane antibody reactivity by xylanase in female, haploid-derived cell suspension cultures of *Taxus brevifolia*. *In Vitro Cell Dev. Biol.*, 30P, 219–227.

Durzan, D. J. and Ventimiglia, F. F. (1997) Recovery of taxanes from conifers. PCT Int. Appl. WO 9730352 A1 21 Aug 1997, 21 pp.

Durzan, D. J. and Ventimiglia, F. F. (1999) Recovery of taxanes from plant material. PCT Int. Appl. WO 9925705 A1 27 May 1999, 26 pp.

Dutta, A. and Pandey, R. C. (1997) Production and screening process for phytochemical production by plant cell cultures. PCT Int. Appl. WO 9721801 A2 19 Jun 1997, 24 pp.

Eisenreich, W., Menhard, B., Lee, M. S., Zenk, M. H. and Bacher, A. (1998) Multiple oxygenase reactions in the biosynthesis of taxoids. *J. Am. Chem. Soc.*, 120, 9694–9695.

El Jaziri, M. and Diallo, B. (2000) Taxane and taxine derivatives from *Penicillium* sp. PCT Int. Appl. WO 2000018883 A1 6 Apr 2000, 19 pp.

Ellis, D. D., Zeldin, E. L., Broghagen, M., Russin, W. A., Mocek, U., Flemming, P. E., Floss, H. G. and McCown, B. H. (1994) Taxol production, localization, and biotransformation of precursors in *Taxus* shoot and nodule cultures. *In Vitro Cell Dev. Biol.*, 30P, 229.

Erdemoglu, N. and Sener, B. (1999) The biosynthesis of taxol and derivatives. *Ankara Univ. Eczacilik Fak. Derg.*, 28, 99–116.

Fedoreyev, S. A., Vasilevskaya, N. A., Veselova, M. V., Denisenko, V. A., Dmitrenok, P. S., Ozhigova, I. T., Muzarok, T. I., Zhuravlev and Yu. N. (1998) A new C-14 oxygenated taxane from *Taxus cuspidata* cell culture. *Fitoterapia*, 69, 430–432.

Feng, Y. and Guo, Z. (1999) Low ammonium medium used in *Taxus* spp. callus inducement and taxol biosynthesis. *Tianran Chanwu Yanjiu Yu Kaifa*, 11, 7–13.

Fett-Neto, A. G. and Dicosmo, F. (1996) Production of paclitaxel and related taxoids in cell cultures of *Taxus cuspidata*: perspectives for industrial applications. In *Plant Cell Culture Secondary Metabolism*: Toward industrial application, edited by F. Dicosmo and M. Misawa, pp. 139–166. Boca Raton, New York: CRC Press.

Fett-Neto, A. G., DiCosmo, F., Reynolds, W. F. and Sakata, K. (1992) Cell culture of *Taxus* as a source of the antineoplastic drug taxol and related taxanes. *Bio/Technology*, 10, 1572–1575.

Fett-Neto, A. G., Melanson, S. J., Sakata, K. and DiCosmo, F. (1993) Improved growth and taxol yield in developing calli of *Taxus cuspidata* by medium composition modification. *Bio/Technology*, 11, 731–734.

Fett-Neto, A. G., Pennington, J. J. and DiCosmo, F. (1995) Effect of white light on taxol and baccatin III accumulation in cell cultures of *Taxus cuspidata* Sieb. and Zucc. *J. Plant Physiol.*, 146, 584–590.

Filonova, L. G., Malysheva, L. V., Grakhov, V. P., Blyum and Ya. B. (1996) The peculiarities of regulation of callus genesis and biosynthesis of taxol in common yew. *Biotekhnologiya*, 8, 38–44.

Furmanowa, M., Glowniak, K., Syklowska-Baranek, K., Zgorka, G. and Jozefczyk, A. (1997) Effect of picloram and methyl jasmonate on growth and taxane accumulation in callus culture of *Taxus media* var. *Hatfieldii*. *Plant Cell, Tissue Organ Cult.*, 49, 75–79.

Furmanowa, M., Syklowska-Baranek, K., Jaziri, M. and Glowniak, K. (1998) Production of paclitaxel and 10-deacetylbaccatin III in differentiated callus and adventitious root of *Taxus media* var. *Hicksii*. *Herba Pol.*, 44, 252–257.

Furmanowa, M. and Syklowska-Baranek, K. (2000) Hairy root cultures of *Taxus media* var. *Hicksii* Rehd. as a new source of paclitaxel and 10-deacetylbaccatin III. *Biotechnol. Lett.*, 22, 683–686.

Gan, F.-Y., Deng, B. and Shen, Y.-M. (1999) The structure of an oligosaccharide and its effect on cultured plant cells. *Stud. Plant Sci.*, 6, 409–414.

Gan, F., Zheng, G., Peng, L., Luo, J. and Deng, B. (1997) Study on cell fermentation culture of *Taxus yunnanensis*. *Tianran Chanwu Yanjiu Yu Kaifa*, 9, 97–101.

Gibson, D. M., Ketchum, R. E., Hirasuna, T. J. and Shuler, M. L. (1993) Cell cultures of *Taxus brevifolia* (Pacific yew) and other *Taxus* sp. For taxol production. *Int. Yew Resources Conf.*, Berkeley, California, March 12–13, p. 15 (Abstr.).

Gibson, D. M., Ketchum, R. E. B., Hirasuna, T. J. and Shuler, M. L. (1995) Potential of plant cell culture for taxane production. In *Taxol: Science and Application*, edited by M. Suffness, pp. 71–95. Boca Raton, New York: CRC Press.

Gibson, F. S., Wei, J., Dillon, John L., Jr. and Vemishetti, P. (1999) Novel crystalline complexes of baccatin III with imidazole, 2-methylimidazole or isopropanol for isolation and purification of baccatin III. PCT Int. Appl. WO 9936413 A1 22 Jul 1999, 25 pp.

Glund, K., Hoffmann, M., Weckwerth, W. and Zocher, R. (1999) Process for the preparation of baccatin III by enzymatic synthesis. Eur. Pat. Appl. EP 959138 A1 24 Nov 1999, 7 pp.

Guo, Y., Jaziri, M., Diallo, B., C-Faster, R. J., Zhiri, A., Vanhaelen, M., Homes, J. and Bombardelli, E. (1994) Immunological detection quantitation of 10-deacetylbaccatin III in *Taxus* sp. plant and tissue cultures. *Biol. Chem. Hoppe-Seyler*, 375, 281–287.

Han, K., Park, E., Park, D., Park, S., Park, J. and Kim, S. (1996) Production of taxol from cell culture of *Taxus cuspidate*. Repub. Korea KR 9610562 B1 2 Aug 1966.

Han, K.-H., Gordon, M. P. and Floss, H. G. (2000) Genetic transformation of *Taxus* (yew) to improve production of taxol. In *Biotechnology in Agriculture Forestry*, 44 (Transgenic Trees), edited by Y. P. S. Bajaj, pp. 291–306. Berlin: Springer-Verlag.

Hara, Y. and Yukimune, T. (1997) Commercial manufacture of taxane diterpenes with *Taxus*. Jpn. Kokai Tokkyo Koho JP 09056387 A2 4 Mar 1997 Heisei, 18 pp.

Hezari, M., Ketchum, R. E. B., Gibson, D. M. and Croteau, R. (1997) Taxol production and taxadiene synthase activity in *Taxus Canadensis* cell suspension cultures. *Arch. Biochem. Biophys.*, 337, 185–190.

Holton, R. A., Liu, J. H., Gentile, L. N. and Beidiger, R. J. (1992) Semi-synthesis of taxol. In *Proc. 2nd Natl. Cancer Institute Workshop on Taxol and Taxus*.

Horwitz, S. B., Lothsteia, L., Manfredi, J. J., Mellado, W., Parness, J., Roy, S. N., Schiff, P. B., Sorbara, L. and Zeheb, R. (1986) Taxol: mechanism of action and resistance. *Ann. NY Acad. Sci.*, 466, 733–743.

Horwitz, S. B. (1992) Mechanism of action of taxol. *Trends Pharmacol. Sci.*, 13, 134–136.

Hu, P., Li, S., Yuan, Y., Sun, A. and Hu, C. (1999) Optimization of plant cell suspension culture of *Taxus chinensis* var. *mairei*. *Tianran Chanwu Yanjiu Yu Kaifa*, 11, 30–35.

Huang, Z., Mu, Y., Zhou, Y., Chen, W., Xu, K., Yu, Z., Bian, Y. and Yang, G. (1997) Transformation of *Taxus brevifolia* by *Agrobacterium rhizogenes* and taxol production in hairy roots culture. *Yunnan Zhiwu Yanjiu*, 19, 292–296.

Huang, H., Lu, M. and Mei, X. (1999) Selection of antioxidant in the cell cultures of *Taxus chinensis*. *Huazhong Ligong Daxue Xuebao*, 27, 107–109.

Jaziri, M., Diallo, B.M., Vanhaelen, M., Vanhaelen-Faster, R. J., Zhiri, A., Becu, A. G. and Homes, J. (1991) Enzyme-liked immunosorbent assay for the detection and semi-quantitative determination of taxane diterpenoids related to taxol in *Taxus* sp. and tissue cultures. *J. Pharm. Belg.*, 46, 93–99.

Jaziri, M., Zhiri, A., Guo, Y.-W. Dupont, J.-P., Shimomura, K., Hamada, H., Vanhaelen, M. and Homes, J. (1996) Taxus sp. cell, tissue and organ cultures as alternative sources for taxoids production: a literature survey. *Plant Cell, Tissue Organ Cult.*, 46, 59–75.

Jha, S., Sanyal, D., Ghosh, B. and Jha, T. B. (1998) Improved taxol yield in cell suspension culture of *Taxus wallichiana* (Himalayan yew). *Planta Med.*, 64, 270–272.

Jiang, W., Jin, W., Zhang, Y. and Wei, Y. (1998) Chemical studies on metabolites of an endophytic fungus associated with *Taxus cuspidate*. *Zhongguo Kangshengsu Zazhi*, 23, 263–266.

Joyce, C. (1993) Taxol: search for a cancer drug. *BioScience*, 43, 133–136.

Kawamura, M., Shigeoka, T., Tahara, M., Takami, M., Ohashi, H., Akita, M., Kobayashi, Y. and Sakamoto, T. (1998) Efficient selection of cells with high taxol content from heterogeneous *Taxus* cell suspensions by magnetic or fluorescent antibodies. *Seibutsu Kogaku Kaishi*, 76, 3–7.

Ketchum, R. E. B. and Gibson, D. M. (1996) Paclitaxel production in suspension cell cultures of *Taxus*. *Plant Cell, Tissue Organ Cult.*, 46, 9–16.

Ketchum, R. E. B., Gibson, D. M. and Gallo, L. G. (1995) Media optimization for maximum biomass production in cell cultures of Pacific yew. *Plant Cell Tissue Organ Cult.*, 42, 185–193.

Ketchum, R. E. B., Tandon, M., Gibson, D., Begley, T. and Shuler, M. L. (1999a) Isolation of labeled 9-dihydrobaccatin III and related taxoids from cell cultures of *Taxus canadensis* elicited with methyl jasmonate. *J. Nat. Prod.*, 62, 1395–1398.

Ketchum, R. E. B., Gibson, D. M., Croteau, R. B. and Shuler, M. L. (1999b) The kinetics of taxoid accumulation in cell suspension cultures of *Taxus* following elicitation with methyl jasmonate. *Biotechnol. Bioeng.*, 62, 97–105.

Kim, H.-K., Park, Y.-S. and Kim, D.-I. (1995a) Screening of plant cell cultures with biotransformation capability. *Nonmunjip – Sanop Kwahak Kisul Yonguso* (Inha Taehakkyo), 23-1, 589–595.

Kim, J.-H., Yun, J.-H., Hwang, Y.-S., Byun, S. Y. and Kim, D.-I. (1995b) Production taxol and related taxanes in *Taxus brevifolia* cell cultures: effect of sugars. *Biotechnol. Lett.*, 17, 101–106.

Kim, S., Strobel, G., Ford, E. (1999a) Screening of taxol-producing endophytic fungi from *Ginkgo biloba* and *Taxus cuspidata* in Korea. *Agric. Chem. Biotechnol.* (Engl. Ed.), 42, 97–99.

Kim, C., Hong, S., Son, S. and Chung, I. (1999b) Improved paclitaxel production in a two-phase suspension culture of *Taxus cuspidata* using silicone cubes. *Biotechnol. Bioprocess Eng.*, 4, 273–276.

Kim, C. H., Kim, K. I. and Chung, I. S. (2000) Expression of modified green fluorescent protein in suspension culture of *Taxus cuspidate. J. Microbiol. Biotechnol.*, 10, 91–94.

Kingston, D. G. (1994) Taxol: the chemistry and structure–activity relationships of a novel anticancer agent. *Trends Biotechnol.*, 12, 222–227.

Kodama, O., Tamokami, S., Hara, Y., Tabata, T. and Matsubara, K. (1999) Promotion of plant second metabolite production by jasmonic acids or coronatines and radical generation promoters and manufactured of the metabolites. Jpn. Kokai Tokkyo Koho JP 11046783 A2 23 Feb 1999 Heisei, 13 pp.

Kwak, S.-S., Choi, M.-S., Park, Y.-G., Yoo, J.-S. and Liu, J.-R. (1995) Taxol content in the seeds of *Taxus* spp. *Phytochemistry*, 40, 29–32.

Kwon, I. C., Yoo, Y. J., Lee, J. H. and Hyun, J. O. (1998) Enhancement of taxol production by in situ recovery of product. *Process Biochem.* (Oxford), 33, 701–707.

La Rue, C. D. (1953) Studies on growth and regeneration in gametophytes and sporophytes of gymnosperms. In *Abnormal and pathological plant growth*, Report Symp. Brookhaven National Laboratory, J Upton, New York, pp. 187–208.

Laskaris, G., Bounkhay, M., Theodoridis, G., van der Heijden, R., Verpoorte, R. and Jaziri, M. (1999a) Induction of geranylgeranyl diphosphate synthase activity and taxane accumulation in *Taxus baccata* cell cultures after elicitation by methyl jasmonate. *Plant Sci.* (Shannon, Irel.), 147, 1–8.

Laskaris, G., De Jong, C. F., Jaziri, M., Van Der Heijden, R., Theodoridis, G. and Verpoorte, R. (1999b) Geranylgeranyl diphosphate synthase activity and taxane production in *Taxus baccata* cells. *Phytochemistry*, 50, 939–946.

Laskaris, G., van der Heijden, R. and Verpoorte, R. (2000) Purification and partial characterization of geranylgeranyl diphosphate synthase, from *Taxus baccata* cell cultures. An enzyme that regulates taxane biosynthesis. *Plant Sci.*, 153, 97–105.

Li, J.-Y., Sidhu, R. S., Bollon, A. and Strobel, G. A. (1998) Stimulation of taxol production in liquid cultures of *Pestalotiopsis microspora. Mycol. Res.*, 102, 461–464.

Lu, M., Yu, F., Hong, Q. and Lin, L. (1998) Research on *Taxus* cell suspension culture and metabolic kinetics in airlift bioreactor. *Huazhong Ligong Daxue Xuebao*, 26, 110–112.

Luo, J., Niu, B., Jia, J. and Zheng, G. (1997) Establishment of tissue culture from *Taxus yunnanensis. Shengwu Gongcheng Xuebao*, 13, 326–330.

Luo, J.-P., Mu, Q. and Gu, Y.-H. (2000) Protoplast culture and paclitaxel production by *Taxus yunnanensis, Plant Cell Tissue Organ Cult.*, 59, 25–29.

Ma, W., Park, G. L., Gomez, G. A., Nieder, N. H., Adams, T. L., Aynsley, J. S., Sahai, O. P., Smith, R. J., Stahlhut, R. W., Hylands, P. J., Bitsvh, F. and Shackleton, C. (1994) New bioactive taxoids from cell cultures of *Taxus baccata. J. Nat. Prod.*, 57, 116–122.

Majada, J. P., Sierra, M. I. and Sanchez-Tames, R. (2000) One step more towards taxane production through enhanced *Taxus* propagation. *Plant Cell Rep.*, 19, 825–830.

Mattina, M. J. I. and Paiva, A. A. (1992) Taxol concentration in Taxus cultivars. *J. Environ. Hortic*, 10, 187–191.

McGuire, W. P., Rowinsky, F. K., Rosenshein, N. B., Grumbine, F. C., Ettinger, D. S., Armstrong, D. K. and Donehower, R. C. (1989) A unique antineoplastic agent with significant activity in advanced ovarian epithelial neoplasms. *Ann. Intern. Med.*, 111, 273–279.

Mei, X., Gong, W., Su, X. and Ke, T. (1999) Screening of fungi-resistant variant of *Taxus* and a study of its physiological and biochemical properties. *Huazhong Ligong Daxue Xuebao*, 27, 107–109.

Mei, X., Lu, M., Yu, L. and Dai, Q. (1997) *Taxus* cell suspension culture and metabolic kinetics. *Huazhong Ligong Daxue Xuebao*, 25, 104–106.

Mei, X., Wen, C., Liu, L. and Ke, T. (1998) Studies on the seedling culture of Yew embryo in vitro. *Huazhong Ligong Daxue Xuebao*, 26, 5–7.

Menhard, B., Eisenreich, W., Hylands, P. J., Bacher, A. and Zenk, M. H. (1998) Taxoids from cell cultures of *Taxus chinensis. Phytochemistry*, 49, 113–125.

Mirjalili, N. and Linden, J. C. (1995) Gas phase composition effects on suspension cultures of *Taxus cuspidata. Biotechnol. Bioeng.*, 48, 123–132.

Mirjalili, N. and Linden, J. C. (1996) Methyl jasmonate induced production of taxol in suspension cultures of *Taxus cuspidata*: ethylene interaction and induction models. *Biotechnol. Prog.*, 12, 110–118.

Moon, W. J., Yoo, B. S., Kim, D.-I. and Byun, S. Y. (1998) Elicitation kinetics of taxane production in suspension cultures of *Taxus baccata* Pendula. *Biotechnol. Tech.*, 12, 79–81.

Morita, E. and Suita, Y. (1998a) Manufacture of taxanes by tissue culture of *Taxus* plant. Jpn. Kokai Tokkyo Koho JP 10052294 A2 24 Feb 1998 Heisei, 5 pp.

Morita, E. and Suita, Y. (1998b) Manufacture of taxanes by tissue culture of *Taxus* plant. Jpn. Kokai Tokkyo Koho JP 10052295 A2 24 Feb 1998 Heisei, 5 pp.

Morita, E. and Suita, Y. (1998c) Manufacture of taxanes by cell culture of *Taxus* sp. Jpn. Kokai Tokkyo Koho JP 10052296 A2 24 Feb 1998 Heisei, 7 pp.

Nicolaou, K. C., Dai, W.-M. and Guy, R. K. (1994) Chemistry and biology of taxol. *Angew. Chem. Int. Ed. Engl.*, 33, 15–44.

Pan, Z., Zhong, J., Wu, J., Takagi, M. and Yoshida, T. (1999) Fluid mixing and oxygen transfer in cell suspensions of *Taxus chinensis* in a novel stirred bioreactor. *Biotechnol. Bioprocess Eng.*, 4, 269–272.

Pandey, R. C. and Yankov, L. K. (1997) Isolation and purification of paclitaxel and cephalomannine. PCT Int. Appl. WO 9721696 A1 19 Jun 1997, 64 pp.

Park, Y.-S., Hwang, Y.-S. and Kim, D.-I. (1995a) Taxol production by immobilized plant cell cultures. *Nonmunjip – Sanop Kwahak Kisul Yonguso* (Inha Taehakkyo), 23(1), 583–588.

Park, J. H., Kim, C. H., Chung, I. S. and Lee, S. W. (1995b) Production of taxol from in vitro cell culture of *Taxus cuspidate*. *Acta Hortic.*, 390 (International Symposium on Medicinal and Aromatic Plants, 1994), 127–132.

Pedroso, M. C., Magalhaes, J. R. and Durzan, D. (2000) A nitric oxide burst precedes apoptosis in angiosperm and gymnosperm callus cells and foliar tissues. *J. Exp. Bot.*, 51, 1027–1036.

Pennington, J. J., Fett-Neto, A. G., Nicholson, S. A., Kingston, D. G. I. and Dicosmo, F. (1998) Acetyl CoA: 10-deacetylbaccatin-III-10-O-acetyltransferase activity in leaves and cell suspension cultures of *Taxus cuspidate*. *Phytochemistry*, 49, 2261–2266.

Phisalaphong, M. and Linden, J. C. (1999a) Ethylene and methyl jasmonate interaction and binding models for elicited biosynthetic steps of paclitaxel in suspension cultures of *Taxus canadensis*. In *Biol. Biotechnol. Plant Horm. Ethylene II* [Proc. EU-TMR-Euroconf. Symp.], edited by A. K. Kanellis, pp. 85–94. Dordrecht: Kluwer Academic Publishers.

Phisalaphong, M. and Linden, J. C. (1999b) Kinetic studies of paclitaxel production by *Taxus canadensis* cultures in batch and semicontinuous with total cell recycle. *Biotechnol. Prog.*, 15, 1072–1077.

Pulici, M., Sugawara, F., Koshino, H., Okada, G., Esumi, Y., Uzawa, J. and Yoshida, S. (1997) Metabolites of Pestalotiopsis spp., endophytic fungi of *Taxus brevifolia*. *Phytochemistry*, 46, 313–319.

Qiu, D. and Han, Y. (1998) Recent advances in the research on *Taxus* cell, tissue, organ cultures and genetic engineering. *Tianran Chanwu Yanjiu Yu Kaifa*, 10, 90–98.

Ramawat, K. G. (1999) Production of antitumour compounds. In *Biotechnology*, edited by K. G. Ramawat and J. M. Merillon, pp. 67–88. Enfield: Science Publishers.

Roberts, S. and Shuler, M. L. (1998) Strategies for bioproduct optimization in plant cell tissue cultures. In *BioHydrogen* [Proc. Int. Conf. Biol. Hydrogen Prod.], Meeting Date 1997, edited by O. R. Zaborsky, pp. 483–491. New York: Plenum Publishing Corp.

Salandy, A., Grafton, L., Uddin, M. R. and Shafi, M. I. (1993) Establishing an embryogenic cell suspension culture system in Florida yew (*Taxus floridana*). *In Vitro Cell Dev. Biol.*, 29, 75A.

Seki, M. and Furusaki, S. (1999) Medium recycling as an operational strategy to increase plant secondary metabolite formation: continuous taxol production. In *Plant Cell Tissue Cult. Prod. Food Ingredients* [Proc. Symp.], Meeting Date 1997, 157–163, edited by T.-J. Fu, G. Singh and W. R. K. Curtis, New York: Academic/Plenum Publishers.

Seki, M., Ohzora, C., Takeda, M. and Furusaki, S. (1997) Taxol (Paclitaxel) production using free and immobilized cells of *Taxus cuspidata*. *Biotechnol. Bioeng.*, 53, 214–219.

Shi, Q., Li, J., Cheng, K. and Zhu, P. (1999) Separation of specific cDNA fragment of taxadiene synthase from *Taxus chinensis* callus. *Zhiwu Shenglixue Tongxun*, 35, 285–287.

Shin, S.-W. and Kim, Y.-S. (1996) Production of taxane derivatives by cell culture of Korean *Taxus* species (I). *Saengyak Hakhoechi*, 27, 262–266.

Shin, S.-W. and Lim, S. (1998) Effects of several elicitors and amino acids on production of taxane derivatives in cultured cells. *Saengyak Hakhoechi*, 29, 360–364.

Shin, S.-W., Kim, Y. S. and Lim, S. (2000) Elicitors for the regulation of baccatin III biosynthesis in plant cell culture system. *Yakhak Hoechi*, 44, 60–65.

Son, S. H., Choi, S. M., Choi, K. B., Lee, Y. H., Lee, D. S., Choi, M. S. and Park, Y. G. (1999) Selection and proliferation of rapid growing cell lines from embryo derived cell cultures of yew tree (*Taxus cuspidata* Sieb. et Zucc). *Biotechnol. Bioprocess Eng.*, 4, 112–118.

Son, S. H., Choi, S. M., Lee, Y. H., Choi, K. B., Yun, S. R., Kim, J. K., Park, H. J., Kwon, O. W., Noh, E. W., Seon, J. H. and Park, Y. G. (2000) Large-scale growth and taxane production in cell cultures of *Taxus cuspidata* (Japanese yew) using a novel bioreactor. *Plant Cell Rep.*, 19, 628–633.

Soon-Shiong, P. (1997) Method for growing plant cells in liquid suspension cultures and chemotherapeutic agents derived from plants. PCT Int. Appl. WO 9731100 A1 28 Aug 1997, 14 pp.

Srinivasan, V., Pestchanker, L., Moser, S., Hirasuna, T. J., Taticek, R. A. and Shulet, M. L. (1995) Taxol production in bioreactors: kinetics of biomass accumulation, nutrient uptake, and taxol production by cell suspensions of *Taxus baccata*. *Biotechnol. Bioeng.*, 47, 666–676.

Srinivasan, V., Roberts, S. C. and Shuler, M. L. (1997) Combined use of six-well polystyrene plates and thin layer chromatography for rapid development of optimal plant cell culture processes. Application to taxane production by *Taxus* sp. *Plant Cell Rep.*, 16, 600–604.

Strobel, G. A. and Hess, W. M. (1997) Glucosylation of the peptide leucinostatin A, produced by an endophytic fungus of European yew, may protect the host from leucinostatin toxicity. *Chem. Biol.*, 4, 529–536.

Suffness, M. (1995) *Taxol: Science and Applications*, Boca Raton: CRC Press.

Sun, J. and Hu, Z. (2000) Callus initiation, growth characterization and paclitaxel production from *Taxus chinensis*, *Xibei Daxue Xuebao, Ziran Kexueban*, 30, 55–59.

Sun, X. and Linden, J. C. (1999) Shear stress effects on plant cell suspension cultures in a rotating wall vessel bioreactor. *J. Ind. Microbiol. Biotechnol.*, 22, 44–47.

Sun, J., Han, K., Park, D. and Park, E. (1997) Process for producing taxol by cell culture of *Taxus*. Repub. Korea KR 9710602 B1 28 Jun 1997.

Sun, J., Hu, Z. and Cheng, X. (1999a) Analysis of paclitaxel from callus of *Taxus chinensis* by HPLC. *Xibei Daxue Xuebao, Ziran Kexueban*, 29, 159–161.

Sun, B., Zhang, G., Liu, D. and Hu, Z. (1999b) Cell culture of *Taxus* and paclitaxel produce. *Zhiwu Shenglixue Tongxun*, 35, 135–140.

Sun, B., Zhang, G., Liu, D. and Hu, Z. (1999c) Comparison of growth and paclitaxel content in 4 Taxus cell suspensions. *Zhiwu Shenglixue Tongxun*, 35, 279–280.

Tabata, H. and Hara, Y. (1995) Commercial manufacture of taxane diterpenes with *Taxus*. Jpn. Kokai Tokkyo Koho JP 09028392 A2 4 Feb 1997 Heisei, 7 pp.

Tachibana, A., Yoshida, M., Ito, K., Oki, T., Azuma, A. and Kubota, M. (1998) Manufacture of taxol by plant tissue culture. Jpn. Kokai Tokkyo Koho JP 10042888 A2 17 Feb 1998 Heisei, 8 pp.

Takami, M. and Takigawa, K. (1996) Manufacture of taxanes by tissue culture of *Taxus* species. Jpn. Kokai Tokkyo Koho JP 08298993 A2 19 Nov 1996 Heisei, 5 pp.

Takami, M. and Takigawa, K. (1997) *Taxane* manufacture with *Taxus* by plant tissue culture. Jpn. Kokai Tokkyo Koho JP 09023894 A2 28 Jan 1997 Heisei, 7 pp.

Takigawa, K., Takami, M., Asari, T. and Fukumoto, K. (1996) Antitumor compound NSC-LSC2 and its manufacture by cell culture of *Taxus*. Jpn. Kokai Tokkyo Koho JP 08291062 A2 5 Nov 1996 Heisei, 4 pp.

Theodoridis, G., Laskaris, G. and Verpoorte, R. (1999) HPLC analysis of taxoids in plant and plant cell tissue cultures. *Am. Biotechnol. Lab.*, 17, 40, 42, 44.

Tulecke, W. (1959) The pollen cultures of C. D. LaRue: A tissue from the pollen *Taxus*. Bull. Torrey Bot. Club, 86, 283–289.

Vanek, T., Mala, J., Saman, D., Silhava, I. (1999) Production of taxanes in a bioreactor by *Taxus baccata* cells. *Planta Med.*, 65, 275–277.

Vidensek, N., Lim, P., Campbell, A. and Carlson, C. (1990) Taxol content in bark, wood, root, leaf, twig and seedling from several *Taxus* species. *J. Nat. Prod.*, 53, 1609–1610.

Walker, K. and Croteau, R. (2000) Molecular cloning of a 10-deacetylbaccatin III-10-O-acetyl transferase cDNA from *Taxus* and functional expression in *Escherichia coli. Proc. Natl. Acad. Sci. USA*, 97, 583–587.

Walker, K., Schoendorf, A. and Croteau, R. (2000) Molecular cloning of a Taxa-4(20),11(12)-dien-5a-ol-O-Acetyl Transferase cDNA from *Taxus* and functional expression in *Escherichia coli. Arch. Biochem. Biophys.*, 374, 371–380.

Wang, J., Mei, X. and Yu, L. (1994) Method for producing taxol by cell culture. Faming Zhuanli Shenqing Gongkai Shuomingshu CN 1096820 A 28 Dec 1994, 11 pp.

Wang, H.-Q., Zhong, J.-J. and Yu, J.-T. (1997a) Enhanced production of taxol in suspension cultures of *Taxus chinensis* by controlling inoculum size. *Biotechnol. Lett.*, 19, 353–355.

Wang, H., Zhong, J., Chen, X. and Yu, J. (1997b) Kinetic study on suspension cultures of *Taxus chinensis* for production of anticancer drug taxol. *Huadong Ligong Daxue Xuebao*, 23, 679–683.

Wang, H., Zhang, Y., Chen, X., Yu, J. and Zhong, J. (1999) High density suspension cultivation of *Taxus chinensis* cells. *Huadong Ligong Daxue Xuebao*, 25, 55–58.

Wang, H. Q., Yu, J. T. and Zhong, J. J. (2000a) Significant improvement of taxane production in suspension cultures of *Taxus chinensis* by sucrose feeding strategy. *Process Biochem.* (Oxford), 35, 479–483.

Wang, J., Lu, H. and Shu, W. (2000b) Studies on taxol-producing endophytic fungus. *Weishengwuxue Tongbao*, 27, 58–60.

Wang, K. (2000) Taxol production by immobilized cultures of *Taxus chinensis* suspense cell. *Yantai Daxue Xuebao, Ziran Kexue Yu Gongchengban*, 13, 25–30.

Wani, M. C., Taylor, H. L., Wall, M. E., Coggan, P. and McPhail, A. T. (1971) Plant antitumor agents. VI. The isolation and structure of taxol, a novel antileukemic and antitumor agent from *Taxus brevifolia. J. Am. Chem. Soc.*, 93, 2325–2327.

Wheeler, N. C. and Hehnen, M. T. (1993) Taxol, a study in technology commercialization. *J. For.*, 91, 15–18.

Wheeler, N. C., Jech, K., Masters, S., Brobst, S. W., Alvarado, A. B., Hoover, A. J. and Snader, K. M. (1992) Effects of genetic, epigenetic, and environmental factors on taxol content in *Taxus brevifolia* and related species. *J. Nat. Prod.*, 55, 432–440.

Wickremesinhe, E. R. M. (1992) Callus and cell suspension cultures of Taxus as a source of taxol and related taxanes. *PhD Thesis*, The Pennsylvania State University, University Park, Pennsylvania.

Wickremesinhe, E. R. M. and Arteca, R. N. (1993) Establishment of *Taxus* callus cultures, optimization of growth and confirmation of taxol production. *Plant Cell Tissue Organ Cult.*, 35, 181–193.

Wickremesinhe, E. R. M. and Arteca, R. N. (1994) Roots of hydroponically grown *Taxus* plants as a source of taxol and related taxanes. *Plant Sci.*, 101, 125–135.

Wickremesinhe, E. R. M. and Arteca, R. N. (1998) XXI *Taxus* species (yew): in vitro culture, and the production of taxol and other secondary metabolites. In *Biotechnology in Agriculture and Forestry*, Vol. 41: *Medicinal and Aromatic Plants X*, edited by Y. P. S. Bajaj, pp. 415–442. Berlin, Heidelberg: Springer-Verlag.

Williams, D. C., Wildung, M. R., Jin, A. Q., Dalal, D., Oliver, J. S., Coates, R. M. and Croteau, R. (2000) Heterologous expression and characterization of a "pseudomature" form of taxadiene synthase involved in paclitaxel (taxol) biosynthesis and evaluation of a potential intermediate and inhibitors of the multistep diterpene cyclization reaction. *Arch. Biochem. Biophys.*, 379, 137–146.

Witherup, K. M., Look, S. A., Stasko, W. M., Ghiorzi, J. T. and Muschik, G. M. (1990) *Taxus* spp. needles contain amounts of taxol comparable to the bark of *Taxus brevifolia*: analysis and isolation. *J. Nat. Prod.*, 53, 1249–1255.

Wu, Y. and Zhu, W. (1997a) High performance liquid chromatographic determination of taxol and related taxanes from *Taxus* callus cultures. *J. Liq. Chromatogr. Relat. Technol.*, 20, 3147–3154.

Wu, Y. and Zhu, W. (1997b) Research and development of taxol and analogs. *Tianran Chanwu Yanjiu Yu Kaifa*, 9, 98–104.

Wu, Y., Zhu, W., Lu, J., Hu, Q. and Li, X. (1998a) Selection and culture of high-yield sinenxans cell lines of *Taxus* spp. distributed in China. *Zhongguo Yaoxue Zazhi* (Beijing), 33, 15–18.

Wu, Z., Yuan, Y., Hu, Z. (1998b) Studies on the two-liquid-phase culture of Japanese yew (*Taxus cuspidata*) for the production of paclitaxel. *Zhongcaoyao*, 29, 589–592.

Wu, Z.-L., Yuan, Y.-J., Liu, J.-X., Xuan, H.-Y., Hu, Z.-D.,Sun, A.-C. and Hu, C.-X. (1999) Study on enhanced production of taxol from *Taxus chinensis* var. *mairei* in biphasic-liquid culture. *Zhiwu Xuebao*, 41, 1108–1113.

Wu, Z.-L., Yuan, Y.-J., Ma, Z.-H. and Hu, Z.-D. (2000) Kinetics of two-liquid-phase *Taxus cuspidata* cell culture for production of Taxol. *Biochem. Eng. J.*, 5, 137–142.

Xu, J. F., Yin, P. Q., Wei, X. G. and Su, Z. G. (1998) Self-immobilized aggregate culture of *Taxus cuspidata* for improved taxol production. *Biotechnol. Tech.*, 12, 241–244.

Yu, L., Li, W., Liu, X. and Zhang, C. (1998) Taxol production by immobilized cultures of *Taxus chinensis* suspension cells. *Huazhong Ligong Daxue Xuebao*, 26, 1–4.

Yu, L., Li, W., Mei, X. and Wu, Y. (1999a) Improving the effect of metabolic pathway regulation on taxol biosynthesis. *Huazhong Ligong Daxue Xuebao*, 27, 100–103.

Yu, L., Zhu, M., Zhou, Y., Ke, T. and Mei, X. (1999b) The induction effect of methyl-jasmonate on taxol biosynthesis. *Tianran Chanwu Yanjiu Yu Kaifa*, 11, 1–7.

Yun, J.-H., Kim, J.-H. and Kim, D.-I. (1995) A study on the growth characteristics and taxol production in *Taxus brevifolia* cell cultures. *Nonmunjip – Sanop Kwahak Kisul Yonguso* (Inha Taehakkyo), 23(1), 575–582.

Yuan, Y., Hu, G., Na, P., Wang, C., Zhou, Y. and Shen, P. (1997) Method for improving content and release of taxol in *Taxus chinensis* cell. Faming Zhuanli Shenqing Gongkai Shuomingshu CN 1158356 A 3 Sep 1997, 5 pp.

Yuan, Y., Hu, G., Wang, C., Jing, Y., Zhou, Y. and Shen, P. (1998) Effect of rare earth compounds on the growths, taxol biosynthesis and release of *Taxus cuspidata* cell culture. *Zhongguo Xitu Xuebao*, 16, 56–60.

Yukimune, Y., Matsubara, K. and Hara, Y. (1998) Methods for producing taxane-type diterpenes. Eur. Pat. Appl. EP 887419 A2 30 Dec 1998, 15 pp.

Yukimune, Y., Hara, Y., Nomura, E., Seto, H. and Yoshida, S. (2000) The configuration of methyl jasmonate affects paclitaxel and baccatin III production in *Taxus* cells. *Phytochemistry*, 54, 13–17.

Zhang, Z., Yang, J. and Wu, Y. (1998) Taxol production in tissue culture of *Taxus chinensis* var. *Mairei*. *Xibei Zhiwu Xuebao*, 18, 488–492.

Zhang, C., Yu, L., Mei, X. and Li, Y. (1999a) Effect of chitosan on the improvement of taxol biosynthesis and phenylalanine ammonia-lyase activity in *Taxus chinensis* suspension cultures. *Huazhong Ligong Daxue Xuebao*, 27, 104–106.

Zhang, C., Mei, X. and Yu, L. (1999b), Relationship between taxol production and inocula from different periods of the growth cycle in *Taxus chinensis* suspension cultures. *Huazhong Ligong Daxue Xuebao*, 27(Suppl. 1), 64–66.

Zhang, C., Mei, X., Yu, L. and Liu, X. (2000) Initiation and growth of *Taxus chinensis* embryo-derived cell line and production of taxol. *Huazhong Ligong Daxue Xuebao*, 28, 82–84.

Zhao, J., Zhu, W. and Wu, Y. (1998) Kinetic study on isoenzymes and activities of several enzymes in sinenxans-high-yield cell line Ts19 in tissue cultures of *Taxus chinensis*. *Xibei Zhiwu Xuebao*, 18, 339–347.

Zhao, F., Geng, Z., Liu, Q., Wan, X. and Chen, Z. (1997) Determination of taxol in tissue culture of *Taxus mairei* by reversed-phase high-performance liquid chromatography. *Fenxi Huaxue*, 25, 941–943.

Zheng, G., Gan, F. and Luo, J. (1997) Production technology for taxol with *Taxus*. Faming Zhuanli Shenqing Gongkai Shuomingshu CN 1146488 A 2 Apr 1997, 23 pp.

Zhiri, A., Maciejewska, K., Jazari, M., Homes, J. and Vanhaelen, M. (1995a) Establishment of *Taxus baccata* callus cultures and evaluation of taxoid production. *Med. Fac. Landbouww Uni. Gent.*, 60/4a, 2111–2114.

Zhiri, A., Jazari, Guo, Y., Vanhaelen-Faster, R. J., M., Homes, J., Yoshimatsu, K. and Shimomura, K. (1995b) Tissue cultures of *Taxus baccata* as a source of 10-deacetylbaccatin III, a precursor for the hemisynthesis of taxol. *Biol. Chem. Hoope-Seyler*, 376, 583–586.

6 The commercial cultivation of *Taxus* species and production of taxoids

Yoshinari Kikuchi and Mitsuyoshi Yatagai

Introduction

Among higher plants, trees contain the most diverse and unique components, although many have not been identified or used (Morita and Yatagai, 1994; Tachibana, 1995). It is highly possible that tree extracts contain unknown but useful physiologically active substances (Akiyama, 1996). The pharmaceutical values of temperate and polar tree species are being reinvestigated (Stringer, 1994).

Taxus brevifolia Nutt is attracting special attention for its taxol (Wani *et al.*, 1971) content (see Figure 6.1) (Willam *et al.*, 1996). Although the demand for this carcinostatic taxoid is likely to increase due to its strong tumor-inhibitory activity (Kobayashi and Shigemori, 1995), the sources and production methods are limited (Denis and Greene, 1994). However, *Taxus* trees grow very slowly, and the extraction of taxol from the bark should be avoided to protect forests. Because *Taxus* species contain only small amounts of taxol and similar compounds, mass production methods are being actively investigated (Witherup *et al.*, 1989; Kingston *et al.*, 1990; Arthur *et al.*, 1992; Weishuo *et al.*, 1993; Appendino *et al.*, 1994; Appendino, 1995; Guo *et al.*, 1996; John *et al.*, 1996).

In 1994, two research groups in the United States accomplished the total synthesis of taxol (Horton *et al.*, 1994; Nicolau *et al.*, 1994). In Japan, Mukaiyama *et al.* also succeeded in this goal (Mukaiyama, 1997). Their methods use complicated reaction pathways, and simpler methods for industrial production are being researched and developed (Kuwajima, 1997).

10-Deacetylbaccatin III, which is a homolog of taxol, can be extracted at high yields from the needles of *Taxus baccata* and is considered the most appropriate precursor for synthesizing taxol derivatives (Ojima *et al.*, 1991, 1992). The supply of 10-deacetylbaccatin III depends on natural production.

Tissue culture is also being widely studied (Tachibana *et al.*, 1994; Ellis *et al.*, 1996; Pezzuto, 1996; Yukimine *et al.*, 1996), but tissues in the form of calli turn brown and stop growing within six to twelve months. Accordingly, various problems must be solved to develop industrially useful production methods for producing taxol that are entirely artificial or that use tissue cultures. Chemical synthesis and biotechnology approaches usually need advanced technologies and large-scale investment and are difficult to use in mountainous forestry areas (Kikuchi *et al.*, 1997). Industries that use biological resources will be important in the twenty-first century, and technologies should be developed and introduced for the effective utilization of forest resources.

This chapter describes methods of producing taxol using forest resources (*Taxus* trees), which can be transferred to forestry areas, and establishing self-sustaining and stable systems for supplying pharmaceutical materials.

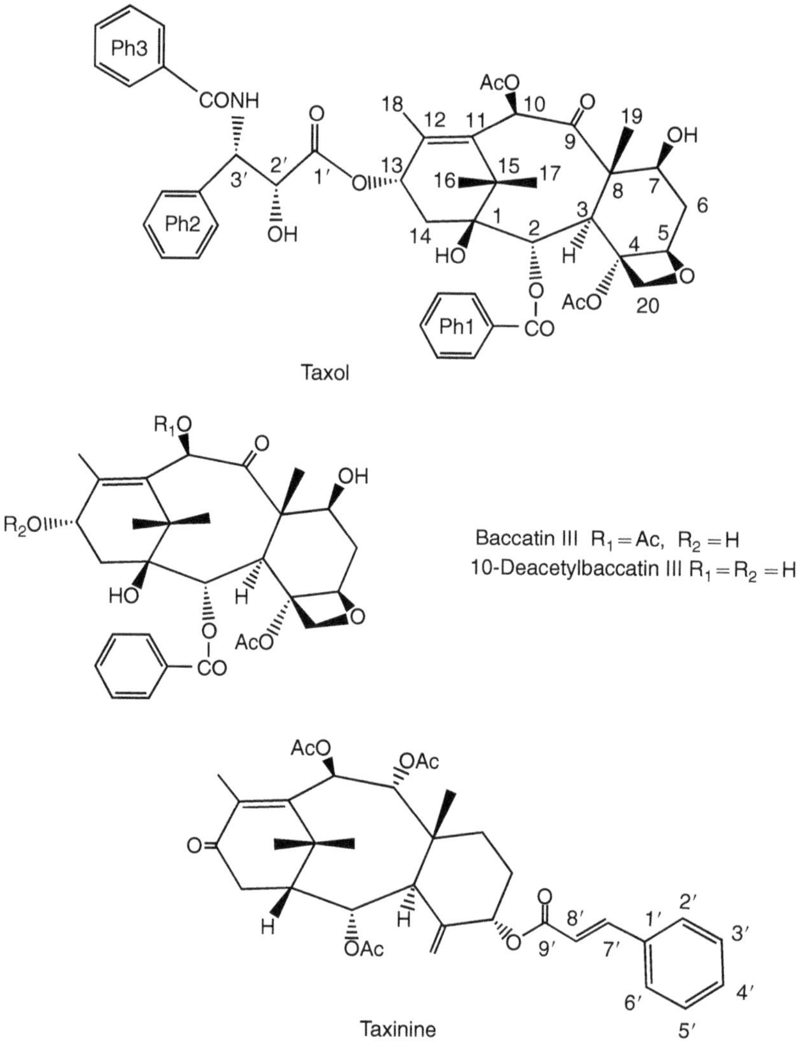

Figure 6.1 Chemical structure of taxol, baccatin III, 10-deacetylbaccatin III, and taxinine.

Characteristics of *Taxus* species

The classification of *Taxus* tree species by Krüssmann (1983) is shown in Table 6.1.

Taxol is contained in *Taxus* species other than *Taxus brevifolia* Nutt (Pacific yew), such as *Taxus cuspidata* Sieb. et Zucc. (Ohira *et al.*, 1996) and *Taxus cuspidata* var. nana *Rehd* (Tachibana *et al.*, 1995), which also grow in Japan.

Taxus cuspidata, or Japanese yew, is an evergreen arbor and grows in temperate and subarctic forests in Hokkaido, Honshu, Shikoku, and Kyushu, and in polar forests in Sakhalin and the Kuril Islands (Kurata, 1971). The tree has various names, such as *Araragi* and *Onko*, in Hokkaido, and grows in gardens and roadsides as well as in natural forests. The Japanese name, *Ichii*, comes from the title of an old Japanese official who used maces made from this tree.

Table 6.1 Classification of the genus *Taxus* (Krüssmann)

Trivial name	Krüssmann classification
European yew	*T. baccata* L.
Himalayan yew	*T. wallichiana* Zucc.
Chinese yew	*T. celebica* Li.
Japanese yew	*T. cuspidata* Sieb et Zucc.
Pacific yew	*T. brevifolia* Nutt.
Mexican yew	*T. globosa* Schlechtd.
Florida yew	*T. floridana* Nutt.
Canadian yew	*T. canadaessis* Marsh.

The genera name *Taxus* means "bow" in Greek, which is also the origin of the word "toxic" (Ichinohe, 1994).

The needles are straight, 1.5–3 cm long, and 2–4 mm wide, and are attached spirally to the dwarf shoots and in two rows to the stems. The extract is a diuretic and causes hypotension, and is used as a Chinese medicinal herb (Izawa, 1995). The plant is dioecious. The male flower (or "cone") is brownish yellow, spherical, and approximately 3 mm in diameter. The female flower (or "cone") is light green or brownish light green, oval, 2.5 mm long, and blooms in April to May. The fruit (or "aril") is oval and approximately 8 mm in diameter. The aril matures red, or yellow in some varieties such as *Taxus cuspidata* var. *luteo-baccata et Tatewaki*, in September to October. The bark is reddish brown and is covered by shallow longitudinal scars. The wood is fine, hard, lustrous, and beautiful (Saitho, 1990) and is used for houses and art crafts.

Characteristics of extract components

Studies on the components of yew extracts began in 1856 when Lucas (Lucas, 1856) extracted an alkaloid from yew trees and named it taxine. Ueda (1913, 1914) reported the first study on the components of extracts from Japanese yew needles. This study was followed by those of Kondo and Amano (1992) and Takahashi and Kondo (1925), who succeeded in isolating taxinine (Takahashi, 1931). Compounds with the taxol skeleton are specific to *Taxus* species, including Japanese yew, and are being studied worldwide (Appendino, 1995).

The rest of this section describes constituents found in *Taxus* species and the recent trends in chemical studies.

Taxinines A, H, K, and L were isolated from *T. cuspidata* Sieb. et Zucc. (Tachibana *et al.*, 1994), Ginkgetin, isorhamnetin, quercetin, tasinine, and taxinine B were isolated from the needles of *T. cuspidata* var. nana *Rehd* (Tachibana *et al.*, 1994). 10-Dacetylbaccatin III is isolated in high yields from fresh needles of *T. baccata* and is used as an important precursor for synthesizing taxol (Ojima *et al.*, 1991). Cephalomannine (John *et al.*, 1996), which is used to treat leukemia, was isolated from *Taxus wallichiana* Zucc. (Pezzuto, 1996). The pith of yew trees also contains the pigment rhodoxanthin, which is used as a coloring for foods and cosmetics (Czeczuga, 1986; US Patent, 1969, 1971).

The needles contain various carotenoids, such as xanthophyll, zeaxanthin, violaxantin, neoxanthin, and eschsoltzaxanthone (Tachibana *et al.*, 1994), and flavonoids, such as amentoflavone (Parveen *et al.*, 1985), sequoiaflavone (Khan *et al.*, 1976), sotetsuflavone (Modica, 1962), ginkgetin (Das *et al.*, 1994), and taxifolin (Chuang *et al.*, 1989, 1990).

Phenols are species dependent. *T. brevifolia* contains large quantities of betuligneol (Chu *et al.*, 1994) and *T. wallichiana* contains much betuloside (Kuhn and Brockmann, 1933).

Taxoid components of yew trees have the following characteristics:

1   Taxoids are contained only in *Taxus* species.
2   Most taxoids are difficult to synthesize industrially.
3   Taxol and related compounds are medically useful.

Taxol is used in the United States and other nations as a cancer chemotherapy agent against ovarian and mammary cancers, and is also being clinically tested against many other cancers (Turuo and Nagahiro, 1995). Many taxoids are cytotoxic, but no compound has been found to be a better anticancer agent than taxol (Suffness and Cordell, 1985). It is highly possible that yew trees contain other unknown but useful compounds that are physiologically active (Kikuchi *et al.*, 1998).

Accordingly, taxoids are being actively researched throughout the world in order to discover new compounds that have similar tumor-inhibiting effects as taxol but different physiological activity. For example, taxuspines B and C which are found in yew stems, inhibit the ability of p-glycoprotein to make cells resistant against cancer chemotherapy agents (Kobayashi *et al.*, 1994; Fuji, 1996; Kobayashi, 1996). Also, components of *Taxus celebica* Li. are being investigated as potential new anticancer agents, and taxamarins A, B, and C from the tree bark have carcinostatic activity (Liang *et al.*, 1987).

Although many components of yew trees are being investigated for their medicinal properties, bioengineering or chemical methods for synthesizing the compounds have not been established. Therefore, taxoids are appropriate targets of production systems that utilize forest resources.

Methods for obtaining materials containing taxoids

Quantitative analysis of taxoids in each part of T. cuspidata

To establish a method of producing taxol using yew forests, the amounts of taxol, baccatin III, and 10-deacetylbaccatin III in living tissues must be fully understood.

We quantitatively analyzed taxol, baccatin III, and 10-deacetylbaccatin III in each tissue of Japanese yew trees (16 years old) growing in an artificial forest in Engaru-cho, Monbetsu-gun, Hokkaido. In addition, we examined the seasonal change in the content of these compounds in the yew tree needles (current and previous years needles).

Figure 6.2 shows the taxoid contents in each tissue of sampled yew trees (*T. cuspidata*) as dry weight percentages.

The highest percentage of taxol in the 16-year-old yew trees was in the bark, followed by the roots, branches, needles, and wood. In a joint study with Ohira and Yatagai in September 1995, 40-year-old naturally grown yew trees in the same region showed the following taxol profile: bark, needles, roots, branches, seeds, and wood (Ohira *et al.*, 1996). Compared to our results for the 16-year-old trees, the 40-year-old trees contained less taxol in the bark and similar amounts in the other tissues. Although the trees differed in age, year of sampling, and extraction method, our research showed that artificially grown yew trees are as useful for the production of taxol as naturally grown trees.

The amounts of taxol and 10-deacetylbaccatin III in the needles of the 16-year-old trees were equivalent to the amounts that were contained in the needles of yew trees grown in Korea and sampled in November 1994 (taxol: 0.022–0.0173%, 10-deacetylbaccatin III: 0.0497–0.0545% (Choi *et al.*, 1995)). 10-Deacetylbaccatin III content was highest in the roots, followed by

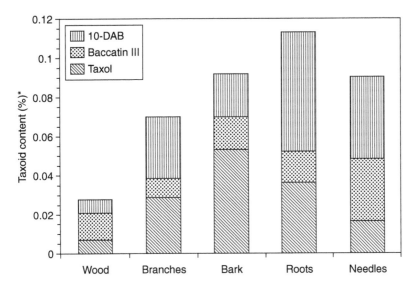

Figure 6.2 Taxoid content in 16-year-old trees of *T. cuspidata*.

Note
* Average percentages of 3 samples based on oven-dried material. 10-DAB: 10-deacetylbaccatin III.

needles, branches, bark, and wood. Baccatin III was maximal in the needles, followed by the bark, roots, wood, and branches. The amounts of taxol, baccatin III, and 10-deacetylbaccatin III varied greatly depending on the tissue. The bark contained approximately three times more taxol than the needles. The amount of 10-deacetylbaccatin III in needles was almost double the amount in the bark; however, content was maximal in the roots.

Although the specimens contained maximum amounts of taxol in the bark and 10-deacetylbaccatin III in the roots, sampling the bark or roots may seriously damage the trees, while needles may be used as a sustainable source for extracting taxol and other compounds. Because of the relatively large taxoid content in the bark and roots, we also investigated the use of these tissues.

The seasonal changes in taxol, baccatin III, and 10-deacetylbaccatin III content in needles as dry weight percentage are shown in Figure 6.3.

The needles sampled in June of the current year contained less taxol than the corresponding needles of the previous year. However, in current needles, the taxol content increased in December, and was almost identical to that from needles of the previous year. For needles of the previous year, those collected in September contained slightly more taxol than those collected in June or December, but the difference was not significant.

In needles of the current year, those collected in June contained the largest amount of baccatin III and those collected in September contained the least amount.

For 10-deactylbaccatin III, the needles of the previous year did not show much difference among the sampling times, while in the current year, the amount was least in June and most in December samples.

Seasonal changes in the components of Japanese yew needles have been reported for taxinine (Yoshizaki, 1988), but this is the first report of seasonal changes in the content of taxol, baccatin III, and 10-deacetylbaccatin III in the needles of current and previous years (Kikuchi, 2001).

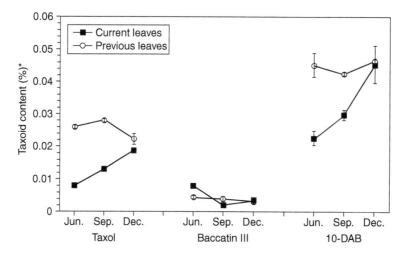

Figure 6.3 Seasonal variation of taxoids in the needles of *T. cuspidata*.

Note
* Percentages based on oven-dried material. Bars show mean ± SD. 10-DAB: 10-deacetylbaccatin III.

The seasonal changes were especially notable for taxol and 10-deacetylbaccatin III content in current leaves, suggesting that these compounds are biosynthesized and accumulated as the trees grow. Correspondingly, we also observed that baccatin III in current needles decreases from June to September.

As baccatin III was abundant in current leaves in the growing period and taxol was abundant after the growing period, needles could be collected at times appropriate for each purpose. The needles collected in June of the current year were yellowish green. They turned deep green as they grew. The amounts of taxol and 10-deacetylbaccatin III are likely to increase as the needles grow.

Our investigation revealed that needles of the previous year show fewer seasonal changes than current needles in the content of taxol, baccatin III, and 10-deacetylbaccatin III, which were almost uniform throughout the year.

Japanese yew trees are planted in gardens and along streets, and are also widely cultivated in Hokkaido. These artificially grown yew trees are periodically trimmed, usually at the end of June, after the growth of needles and branches in the spring or in early to mid-August after the summer growth (Saitho, 1986). Considering the seasonal change in taxol content, needles should be collected when those of the current year have grown, or in September, which is almost the traditional pruning time. The trimmed needles and branches may be collected and used as the source to extract useful taxoids.

Quantitative analysis of taxoids in the needles and seedlings of Japanese yew trees

To understand the overall trends in taxoid content, we quantified taxol, baccatin III, and 10-deacetylbaccatin III in the needles of naturally grown and planted yew trees. We also investigated the taxoids in the seedlings of Japanese yew.

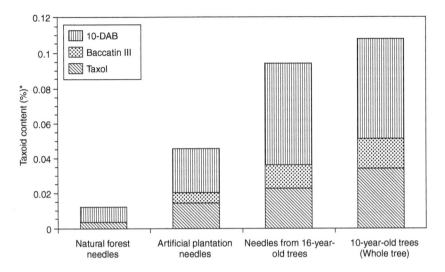

Figure 6.4 Taxoid content in the young trees and the needles of *T. cuspidata*.

Note
* Average percentages of 10 needles and 5 trees based on oven-dried material. 10-DAB: 10-deacetylbaccatin III.

Figure 6.4 shows the taxoid content in the needles and seedlings. The dry weight percentages of taxol, baccatin III, and 10-deacetylbaccatin III were higher in the needles of 16-year-old yew trees that were nursery cultivated than the trees growing in natural forests or parks. Even in the same region, the taxoid contents differed greatly depending on the sites where the trees grew. The yew trees from the natural forest were 6–50 cm (breast height) in diameter and 3–10 m high, and were almost of the same size as those growing in the nursery or parks. Because the yew trees in the natural forest were shaded by taller trees, photosynthesis in these yew trees could have been low and the trees could accumulate only small amounts of taxoids. On the other hand, the yew trees that were growing in the park and the nursery received sufficient light (Hatano and Sasaki, 1993) and produced large quantities of taxoids.

Environmental adaptation and physiological and ecological reactions of Japanese yew trees have not been well studied or understood. However, it is likely that the growing conditions, environmental factors, and stresses (Kikuchi *et al.*, 1997) greatly affect the taxoid content in yew needles.

We then determined the dry weight percentages of taxoids in a whole yew plant (10-year-old seedling), because taxoids are also contained in tissues other than needles. A 10-year-old yew seedling contained more taxol, baccatin III, and 10-deacetylbaccatin III than the needles of 16-year-old yew trees. Compared to the needles of the yew trees growing in the natural forest, the 10-year-old seedling contained 11 times more taxol, 34 times more baccatin III, and 10 times more 10-deacetylbaccatin III.

These results show that the needles of planted yew trees contain a higher taxoid content than the needles of natural forests and the whole plants of 10-year-old seedlings contain more taxoids than the needles alone.

Therefore, it is possible to produce more taxol by using whole yew seedlings rather than only the needles, and it is more efficient to produce yew seedlings than to collect trimmed needles.

Growth of seedlings

Intensive nursery cultivation of yew seedlings can be mechanized, and thus, is easier than planting and growing yew trees in mountain forests. The costs of planting, growing, and harvesting are also cheaper in a nursery than in a forest. As described in the previous section, 10-year-old seedlings contain large amounts of taxoids. Therefore, sufficient material for extracting taxol and other useful compounds can likely be produced by planting seedlings in large nurseries.

We measured the growth of nursery seedlings at different ages and investigated the relationship between the period of cultivation and the use of seedlings. The mean height, the mean dry weight of each tissue, and the mean annual growth (dry weight) for 4-year-old, 8-year-old, and 10-year-old seedlings are shown in Figures 6.5–6.7.

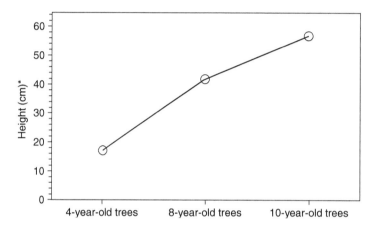

Figure 6.5 Relation between height and age of *T. cuspidata*.

Note
* Average of 5 trees.

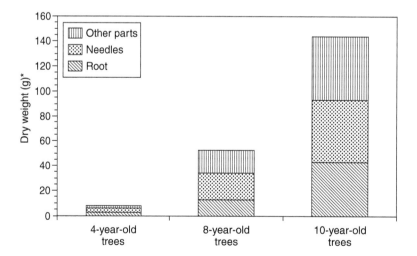

Figure 6.6 Dry weight of parts of young trees of *T. cuspidata*.

Note
* Average of 5 trees based on oven-dried material.

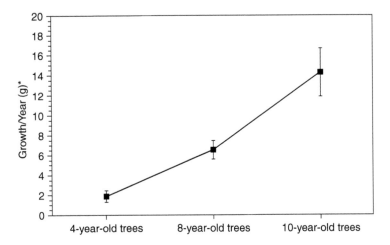

Figure 6.7 Growth/Year of young trees from *T. cuspidata*.

Note
* Based on oven-dried material. Bar show mean ± SD.

The mean dry weight percentage of the needles compared to the whole plant was 48% in the 4-year-old seedlings, 34% in the 8-year-old seedlings, and 35% in the 10-year-old seedlings. The total weight of needles increased as trees grew older, and the weight of needles of a 10-year-old plant was approximately double that of an 8-year-old seedling and approximately 15 times larger than that of a 4-year-old seedling. The mean dry weight percentage of the roots to whole plant was 35% in the 4-year-old seedlings, 24% in the 8-year-old seedlings, and 30% in the 10-year-old seedlings. Therefore, it is probably most effective to use the whole plants, including the roots, to extract taxol.

The biomass – the product of the dry weight per seedling and number of seedlings (Masuzawa, 1997) – was 0.15 tons per hectare for the 4-year-old seedlings, 1.06 tons per hectare for the 8-year-old seedlings, and 2.87 tons per hectare for the 10-year-old seedlings. The annual growth in biomass was 0.04 tons per hectare for the 4-year-old seedlings, 0.13 tons per hectare for the 8-year-old seedlings, and 0.29 tons per hectare for the 10-year-old seedlings.

From the biomass and the taxol content, it is estimated that 10-year-old seedlings produce approximately 984 g per hectare of taxol. To treat one ovarian cancer patient in the treatment period, approximately 1–2 g of taxol are needed (Akiyama, 1996). Assuming that taxol is used only for ovarian cancer, one hectare of 10-year-old seedlings would produce sufficient chemotherapy agents for 50–100 patients. Because 50,000 ovarian cancer patients are reported per year in the United States (Akiyama, 1996), at least 50–100 kg/year of taxol should be produced, which would require 50–100 hectares of 10-year-old seedlings (20,000 seedlings per hectare).

Conclusion

Taxoid concentrations in yew plants are very low, so an enormous amount of plant material is needed to produce taxol industrially. Intensive cultivation of Japanese yew trees, in which large numbers of seedlings are grown for short periods of time (short-rotation intensive culture), is likely to be effective for systematic and stable production of whole plants or needles (Sasaki, 1989).

Our experiments showed that: (1) seedlings contain relatively large amounts of taxoids; and (2) 8- to 10-year-old seedlings are most efficient in terms of biomass. We also found that taxol can be extracted from: (1) trimmed needles of yew trees planted in gardens, parks, and roadsides; (2) needles of seedlings, and (3) the whole plant (the latter two methods are quite promising). We further identified the basic conditions for practical and industrial utilization of yew tree components. These methods should be verified by conducting large-scale experiments in mountain forests and nurseries. It should be relatively easy to introduce the system into forestry areas by utilizing the characteristics of the tree species, regional environmental conditions, and existing production facilities.

Industrial production of taxol using chemical synthesis or tissue culture will require significant time until the processes can be established. Thus, at present, extraction of taxol by cultivating yew seedlings is the most promising and fastest method for producing taxol in large amounts.

Investigation of taxol production systems

Construction of a test site

Industrial production of taxol requires large quantities of materials that should be of sufficient quality to meet the demands of production systems. Cultivation methods should be improved and stable production systems should be established to grow yew trees as a medicinal crop. As described in the previous section, the production and utilization of seedlings are likely to be promising.

For industrial and practical utilization of extract components of Japanese yew, large-scale verification tests should be conducted. For example, studies are being conducted on the cultivation of medicinal plants, such as *Polygala tennifolia* (Fujino *et al.*, 1998), *Glenia littoralis* (Ishizuka *et al.*, 1998), *Paeonia lactiflora* (Hatakeyama *et al.*, 1988), and *Bupleurum falcatum* (Minami *et al.*, 1995), and on the application of fertilizers and cultivation periods (Sakai *et al.*, 1996a), the relationship between shading conditions and growth (Sakai *et al.*, 1996b), and germination and raising of seedlings (Isoda and Shoji, 1994).

As described in the section on "Growth of seedlings," we isolated taxol from Japanese yew seedlings (whole plant), and estimated the production of taxol. We also developed a taxol production system that utilizes forest resources (seedlings) and investigated the possibility of technology transfer to forestry areas. We then constructed a test site at Engaru-cho, Monbetsu-gun, Hokkaido, in June 1997. We measured the content of taxoids in each tissue of the seedlings that were cultivated at the test site.

The dry weight percentages of taxoids for each tissue in a 9-year-old seedling are shown in Figure 6.8.

Taxol was found most intensively in the roots of the seedling, which was twice that of the needles. Needles that were collected in August from a 16-year-old yew tree in another site of the same region contained 0.0231% taxol, 0.0133% baccatin III, and 0.0580% 10-deacetylbaccatin III. These values were similar to those of the 9-year-old seedling. Baccatin III was found in almost the same amounts in the needles and roots, and 10-deacetylbaccatin III was abundant in the roots.

The dry weight percentages of taxol for each tissue of yew trees of various ages are shown in Figure 6.9.

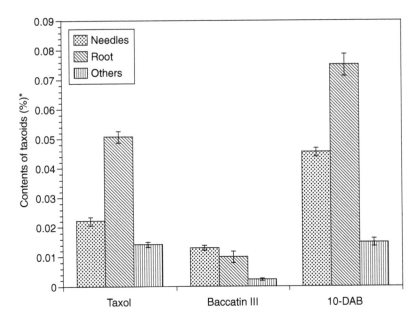

Figure 6.8 Taxoid content in 9-year-old seedling from *T. cuspidata*.

Note
* Percentages based on oven-dried material. Bars show mean ± SD. 10-DAB: 10-deacetylbaccatin III.

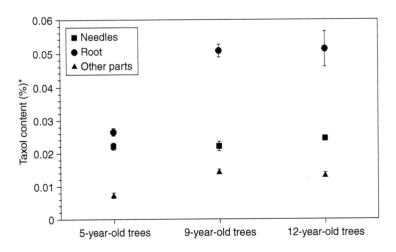

Figure 6.9 Taxol contents in 5-year-old trees, 9-year-old trees, and 12-year-old trees from *T. cuspidata*.

Note
* Percentages based on oven-dried material. Bars show means ± SD.

The roots contained the highest percentage of taxol, followed by needles, then by other tissues. Almost equal amounts of taxol were found in needles collected in June from yew trees of various ages, and the difference by age was small.

In tissues other than needles, less taxol was found in a 5-year-old tree than in 9- or 12-year-old yew trees. The difference was especially notable in the roots. The tissues of the 9- and 12-year-old trees contained almost the same amounts of taxol.

Conclusion

Thus we confirmed that taxol was also contained in high percentages in tissues other than the needles, and reconfirmed that the use of whole plants was most efficient. This analysis also supported the effectiveness of the method we propose, which is to plant, cultivate, and harvest yew seedlings in short cycles.

Overall, our study identified the histological distributions of taxol, baccatin III, and 10-deacetylbaccatin III in Japanese yew trees. The compounds are likely to be produced as the trees grow and undergo various biological activities. Because there is a mutual correlation between the growth rate and the amounts of components in *Chamaecyparis obtusa* (Ohotani *et al.*, 1996), we should also analyze the relationship between growth and taxoids in Japanese yew trees.

Methods for preparing materials for extraction

Most of the physiologically active substances contained in trees are chemically and thermally unstable. Thus, the conditions used for preparing specimens can affect the amounts of components (Anetai *et al.*, 1998). The most widely and consistently used method is to dry plant parts (Higashi *et al.*, 1997). It is also safe, reliable, and inexpensive.

We found that the seedlings and needles of Japanese yew trees are likely to be reliable and stable materials for producing taxol. Harvested and collected yew plants or needles should be preserved inexpensively without deterioration of taxol until they are used for compound extraction. Various methods and conditions should be examined, including preparation methods (Kikuchi *et al.*, 2000).

Stability of taxol with various drying times

As yew needles contain large amounts of water, we investigated drying and preserving methods to improve extraction efficiency and to stabilize quality.

We pulverized the needles of Japanese yew, either dried them in the dark at room temperature or froze them, and investigated the changes in taxol content during the process. The same processes were also applied to the bark. The quantification results are shown in Figures 6.10–6.12 for each preservation method.

Dried needles gave lower taxoid yields than fresh needles. The longer the drying period, the lower the yield. Needles that had been dried for 48 h contained 2/3 of the taxoid content of fresh needles, and those dried for one year contained only half of the fresh amount. The bark also showed a similar tendency; bark that had been dried for 48 h contained approximately 7/10 and that dried for one year contained approximately 6/10 of the amount of the fresh bark (Figure 6.10).

Needles that had been dried for 48 h produced the maximum amount of taxol. The yield was 0.020%, which is almost twice the yield in fresh needles. The amount of taxol extracted from bark that had been dried for one year was almost equivalent to that extracted from fresh bark. Taxol content in bark peaked at a drying time of 192 h (Figure 6.11). Thus the efficiency of taxol extraction is likely to be improved by reducing the moisture content in the needles and the bark.

Under all drying conditions, the yield of taxol per unit dry weight of needles was almost constant, and fluctuated only slightly without a sharp reduction in taxol content.

Before the experiment, we had been concerned about taxol deterioration caused by enzymes or oxygen during the drying process. However, because the plant parts were dried by controlling the humidity rather than by heating, little deterioration occurred.

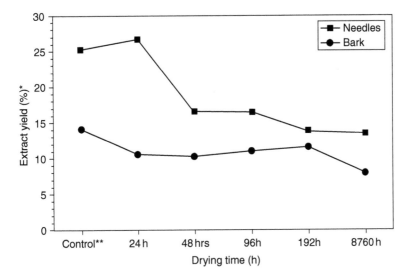

Figure 6.10 Change on extract yield in the needles and bark from *T. cuspidata* during drying time.

Notes
* Average percentages of 5 samples based on oven-dried material.
** Control, fresh samples.

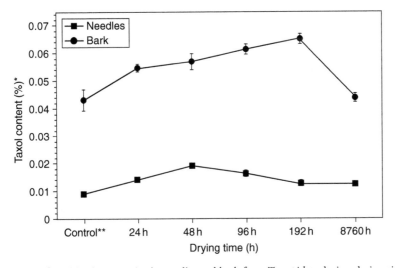

Figure 6.11 Taxol content in the needles and bark from *T. cuspidata* during drying time.

Notes
* Percentages based on oven-dried material. Bars show means ± SD.
** Control, fresh samples.

The moisture content of the needles was 50.89% at 24 h, 23.99% at 48 h, and 10.59% at 192 h. Because we had pulverized the specimens, the water content dropped sharply at first. Reduction in water content decreased the total weight of the specimen and improved the efficiency of the extraction and fractionation processes. In our experiments, the dried needles were

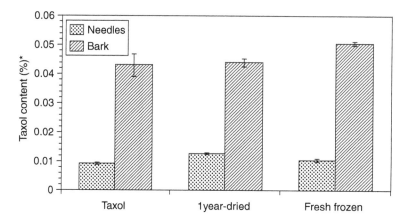

Figure 6.12 Taxol content in the needles and bark of *T. cuspidata*.

Note
* Percentages based on oven-dried material. Bars show means ± SD.

not contaminated by fungi during one year of storage. It has been reported that fungi grow on the leaves of *Crotow sublyratus* unless their water content is reduced below 10% (Ichikawa, 1996). Accordingly, reduction of water content is essential for preserving plant materials.

CONCLUSION

Our results showed that needles dried at room temperature under ordinary pressure for one year did not show notable reduction in taxol content. The efficiency of taxol extraction was improved by drying the needles and the bark. The needles that had been dried for 48 h produced the maximum amount of taxol. Frozen needles were almost equivalent to the dried needles in taxol content and yield per unit dry weight, and showed no sharp deterioration of taxol. Therefore, it is also possible to freeze-preserve yew materials.

Stability of taxol at various drying temperatures

We then investigated the effects of drying temperature on taxol content by pulverizing, freezing, and drying the fresh needles and bark at various temperatures. We first tested simple air drying methods (in sun and in shade) of frozen needles and bark. The results of the quantification analysis are shown in Figures 6.13 and 6.14.

The needles showed lower extract contents at higher drying temperatures. The amount of extract from a specimen that had been dried at 100°C was only 25% of the amount of the specimen dried at 0°C (Figure 6.13).

The yield of taxol from the needles was maximum when the specimens were dried at 25°C, which was 1.6 times more than the yield at 0°C. The needles dried at 100°C produced the least amount of taxol, only 35% of the yield at 25°C (Figure 6.14).

The bark also showed a similar trend. The bark specimen that had been dried at 25°C produced the maximum amount of taxol, and that dried at 100°C produced the least. Our experiments showed that drying specimens at high temperature seriously reduced the yield of taxol.

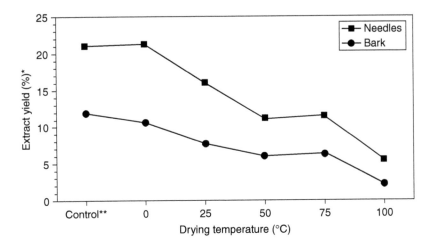

Figure 6.13 Effect of drying temperature on the extract yield from the needles and bark of *T. cuspidata*.

Notes
 * Average percentages of 5 samples based on the oven-dried material.
** Control, fresh frozen samples.

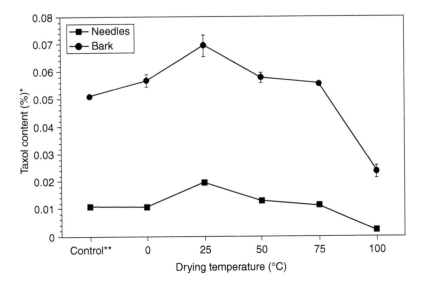

Figure 6.14 Effect of drying temperature on taxol content in the needles and bark from *T. cuspidata*.

Notes
 * Percentages based on oven-dried material. Bars show means ± SD.
** Control, fresh frozen samples.

The taxol measurements from specimens that had been air dried after freezing are shown in Figure 6.15.

From both the needle and bark specimens, taxol was most efficiently extracted when the specimens were air dried in the shade. This result agrees well with a report that the air drying of

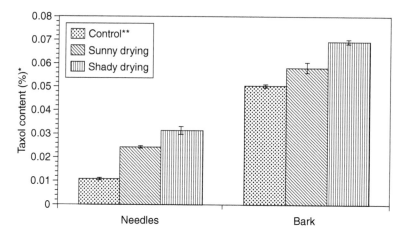

Figure 6.15 Taxol content in the needles and bark from *T. cuspidata* dried in sun and shade.

Notes
* Percentages based on oven-dried material. Bars show means ± SD.
** Control, fresh frozen samples.

the roots of *Astragalus mongholicus* caused less reduction in component contents than forced drying with hot air (80°C) (Anetai *et al.*, 1996). The moisture content was 19.93% for the needles that had been artificially dried at 25°C and 3.76% for the specimens dried at 50°C. Natural air drying in the shade also reduced the water content to 16.74%, suggesting the possibility of using this simple preparation method rather than more intense methods.

CONCLUSION

In summary, this experiment showed that the yield of taxol was highest when the frozen needles were dried at 25°C and was lowest when the specimens were dried at 100°C. It also showed that drying frozen specimens at high temperature seriously decreases the yield of taxol. Taxol was most efficiently extracted when specimens were air dried in the shade. The bark specimens also showed similar trends. The amount of taxol in a yew tree is very small. To efficiently extract this limited resource, the preparation conditions are quite important (Robert *et al.*, 1992). Appropriate preparation methods will enable a prolonged storage and sustainable supply of yew materials.

Development of efficient extraction methods

Trees usually contain small amounts of physiologically active substances, which are chemically and thermally unstable. Obtaining these compounds from plants requires several processes, such as extraction, condensation, isolation, and purification. Because substances are lost at each stage, a simple production method should be developed (Chun *et al.*, 1994; John *et al.*, 1996; Kawamura *et al.*, 1999).

We investigated the efficiency of three widely used extraction methods to separate taxol, baccatin III, and 10-deacetylbaccatin III from needles of Japanese yew: ordinary solvent extraction (OSE), supercritical fluid extraction (SFE), and accelerated solvent extraction (ASE) methods.

In the ordinary solvent extraction method, the most widely used solvent is a mixture of methanol and dichloromethane. However, because taxoids are extracted for medicinal purposes, we should identify solvents that extract large quantities of taxol but have no medical impact. In addition, deterioration rate and decreased extraction efficiency varies with solvents. Our experiment showed that acetone and methanol efficiently extract taxol (see Figures 6.16 and 6.17).

We then investigated supercritical fluid extraction under low pressures. An outline of the SFE method is shown in Figure 6.18.

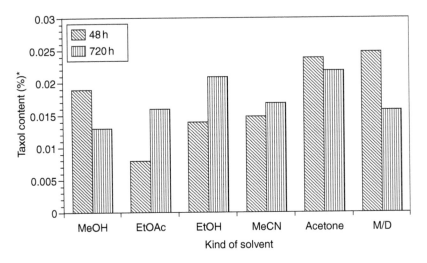

Figure 6.16 Taxol content in the needles of *T. cuspidata* extracted with various solvents.

Note
* Average percentages of 3 samples based on oven-dried material. MeOH, methanol; EtOAc, ethylacetate; EtOH, ethanol; MeCN, acetonitril; M/D, MeOH : dichloromethane (1:1, v/v).

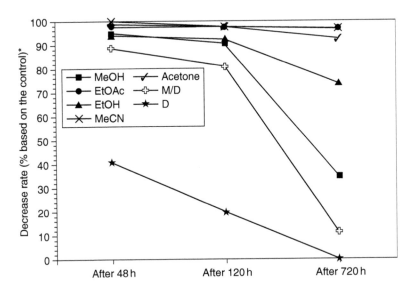

Figure 6.17 Decrease of taxol with time in various solvents.

Note
* Percentages of 3 samples based on oven-dried material. MeOH, methanol; EtOAc, ethylacetate; EtOH, ethanol; MeCN, acetonitril; M/D, MeOH: dichloromethane (1:1, v/v); D, dichloromethane.

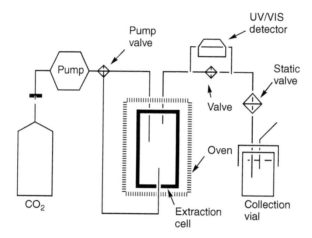

Figure 6.18 Schematic diagram of Supercritical Fluid Extraction (SFE).

We used a supercritical fluid extractor (Thermo Instrument System, supercritical gas extraction screening system X-10-05), carbon dioxide, and extraction pressures of 50, 200, and 300 kgf/cm^2, extraction temperature of 40°C, and a 2-h extraction period. Methanol (25 ml/hr) was added as a modifier to increase the solubility of the supercritical fluid. The extract was condensed in an evaporator at 50°C. This method yielded only little taxol even at the highest pressure of 300 kgf/cm^2. Addition of the methanol modifier improved the extraction efficiency and enabled selective extraction of taxol. Addition of methanol also enabled the extraction of taxol by the liquid carbon dioxide extraction method, although the extraction temperature had to be controlled (Kikuchi and Kawamura, 1999).

Using an accelerated solvent extractor (Dionex, ASE-200), we extracted components from each tissue specimen (extraction pressure: 200 kgf/cm^2; amount of solvent for each specimen: 400 ml; extraction temperature: 50°C; extraction period: 15 min). The extracts were condensed in evaporators at 50°C. The mechanism of the accelerated solvent extractor is shown in Figure 6.19.

A specimen in the extraction cell is incubated in the oven, into which the solvent is pumped. The cell is then kept at the given pressure for the given time. The pressure inside the cell is then released by opening the static valve, and the extract is collected in a collection vial at atmospheric pressure by washing the pipe with nitrogen gas.

Using the accelerated solvent extraction method, we succeeded for the first time by this method to extract taxol from Japanese yew (Kikuchi *et al.*, 1997a). The time required to extract taxol was only 15 min. The accelerated solvent extraction method (extraction pressure: 200 kgf/cm^2; extraction temperature: 50°C; extraction time: 15 min; solvent: methanol) efficiently extracted taxol from the bark. We investigated the optimum conditions for extracting taxol from the needles (solvent: methanol), and found that extraction conditions of 100 kgf/cm^2, 100°C, and 15 min can efficiently and selectively extract taxol (Figure 6.20). We also found that we could extract taxol using water as solvent (Kikuchi and Kawamura, 1999). Addition of ethanol to the water improved the extraction coefficient of taxol (Kikuchi, 2001).

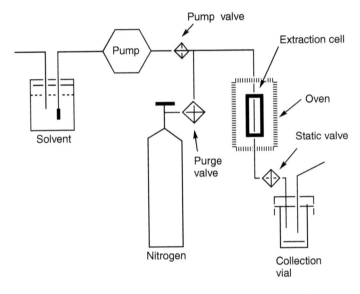

Figure 6.19 Schematic diagram of Accelerated Solvent Extraction (ASE).

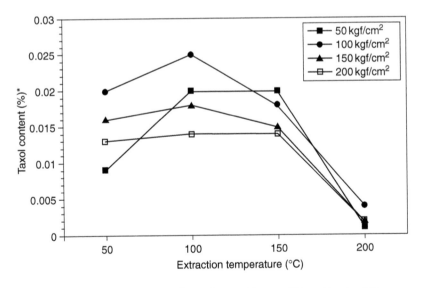

Figure 6.20 Taxol content in the needles from 16-year-old trees of *T. cuspidata* by ASE.

Note
* Average percentage of 3 samples based on oven-dried material.

Table 6.2 Effects of using various extraction methods on the needles of *T. cuspidata*

Extraction methods	OSE	SFE	ASE
Extraction pressure	$1\,kgf/cm^2$	$300\,kgf/cm^2$	$100\,kgf/cm^2$
Extraction temperature	Room temperature	40°C	100°C
Extraction time	48 h	2 h	15 min
Extraction solvent	MeOH	CO_2 + MeOH	MeOH
Taxol content in the extract*	0.030%	0.130%	0.090%
Taxol content in the needles*	0.016%	0.006%	0.025%
Manipulation	Simple	Difficult	Automation
Instrument price	Cheap	High	High

Note
* Percentage based on oven-dried material.

Conclusion

Table 6.2 summarizes the optimal conditions for taxol extraction using the three extraction methods, and compares the characteristics of the three devices.

These data indicate that the accelerated solvent extraction method can extract more taxol than the other two methods and is much faster. The supercritical fluid extraction method also has higher taxol selectivity.

Of the extracting instruments, the batch extractor is the cheapest and is easy to design and operate. The supercritical fluid extractor is very expensive. The experimental device cost more than 10 million yen ($\doteqdot$ US $935,000), and machines for industrial use would cost more. Supercritical fluid extractors are also difficult and complicated to use because they must be operated at high pressures. The accelerated solvent extractor is automatic and relatively easy to use. However, the device has only just been developed, is expensive, and is difficult to maintain.

These results suggest that at present the ordinary solvent extraction method is the best technique for extracting taxol. However, because various extractors are being improved in terms of design, cost, and performance, extraction methods should be further investigated (Kikuchi, 2001).

Isolation and purification of taxol

We have shown that using whole yew plants is the most efficient means to extract taxol. To verify its effectiveness, we isolated taxol from the extracts of 9-year-old seedlings of Japanese yew, estimated the yield of taxol, and attempted to create a basic model for industrial production of taxol.

Experimental methods

SPECIMENS

In June 1997, we randomly sampled five 9-year-old seedlings from a nursery in Engaru-cho, Monbetsu-gun, Hokkaido. The seedlings were rinsed with water, separated into needles, roots, and other parts, and pulverized. The total fresh weight of the five seedlings were 923.87 g composed of needles: 383.76 g (water content: 61.71%); roots: 199.21 g (water content: 51.42%); and other parts: 340.90 g (water content: 43.80%). The dry weight was 435.31 g.

The pulverized specimens of the needles, roots, and other parts were mixed together, air dried in the shade for 48 h, and used to extract taxol (see Figure 6.21).

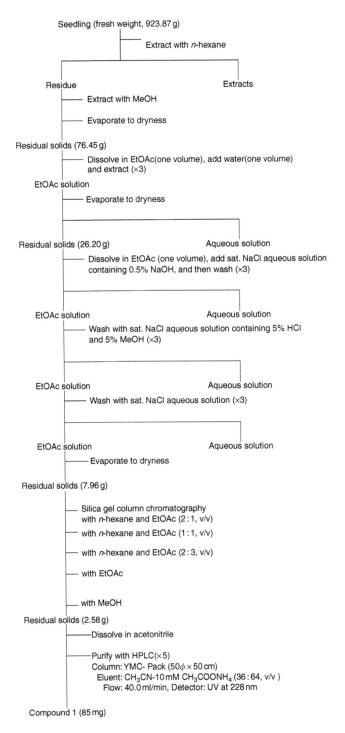

Seedling (fresh weight, 923.87 g)
├── Extract with n-hexane

Residue   Extracts
├── Extract with MeOH
├── Evaporate to dryness

Residual solids (76.45 g)
├── Dissolve in EtOAc(one volume), add water(one volume)
│   and extract (×3)

EtOAc solution
├── Evaporate to dryness

Residual solids (26.20 g)   Aqueous solution
├── Dissolve in EtOAc (one volume), add sat. NaCl aqueous solution
│   containing 0.5% NaOH, and then wash (×3)

EtOAc solution   Aqueous solution
├── Wash with sat. NaCl aqueous solution containing 5% HCl
│   and 5% MeOH (×3)

EtOAc solution   Aqueous solution
├── Wash with sat. NaCl aqueous solution (×3)

EtOAc solution   Aqueous solution
├── Evaporate to dryness

Residual solids (7.96 g)
├── Silica gel column chromatography
│   with n-hexane and EtOAc (2 : 1, v/v)
├── with n-hexane and EtOAc (1 : 1, v/v)
├── with n-hexane and EtOAc (2 : 3, v/v)
├── with EtOAc
├── with MeOH

Residual solids (2.58 g)
├── Dissolve in acetonitrile
├── Purify with HPLC(×5)
│   Column: YMC- Pack (50φ × 50 cm)
│   Eluent: CH₃CN-10 mM CH₃COONH₄ (36 : 64, v/v)
│   Flow: 40.0 ml/min, Detector: UV at 228 nm

Compound 1 (85 mg)

Figure 6.21 Extraction and separation of Compound 1 from the seedlings of *T. cuspidata*.

EXTRACTION

Taxoids were extracted from the specimen (whole seedlings) with five volumes of n-hexane for 24 h. The solvent was removed and the insoluble residue was then extracted with ten volumes of methanol for 48 h. The solution was condensed at 50°C, and 76.45 g of the methanol extract was obtained.

The methanol extract was further separated with three volumes of ethyl acetate and three volumes of distilled water. The ethyl acetate soluble fraction was condensed at 50°C, and 26.20 g of ethyl acetate extract was obtained.

The ethyl acetate extract was extracted three times with 300 ml of ethyl acetate and 300 ml of saturated sodium chloride solution that contained 0.5% NaOH, and then rinsed three times with 300 ml of saturated sodium chloride solution that contained 5% HCl and 5% MeOH. The product was then rinsed three times with 300 ml of saturated sodium chloride solution. The solvent was evaporated at 50°C, and 7.96 g of ethyl acetate extract was obtained.

PURIFICATION

The extract was then separated using column chromatography (column: 4 cm diameter and 50 cm long, packing: 300 g of wakogel c-300). The extract was eluted with successive mixtures of hexane and ethyl acetate, 2:1 (v/v), 1:1, and 2:3. The residue was eluted with ethyl acetate and then with methanol. Each fraction was examined for taxol using TLC, and the ethyl acetate fraction contained taxol. After evaporation, the residue (2.58 g) was dissolved in acetonitrile (2 ml per 100 mg of extract), and subjected to fractionation chromatography using a Shimazu LC-8A HPLC under the following conditions:

Column:	YMC, YMC-Pack (50 mm φ × 50 cm);
Detection:	UV 228 nm;
Eluent:	10 mM ammonium acetate water solution:acetonitrile = 36:64 (v/v); and
Speed of elution:	40 ml/min.

The peak fraction at an elution time of 28–42.5 min was collected and condensed using an evaporator. Ethyl acetate was added to separate the water fraction. The ethyl acetate soluble fraction was condensed and dried in an evaporator, and 250 mg of pale yellow solid was obtained. This solid was again subjected to fractionation chromatography, and 85 mg of white solid (Compound 1) was obtained. Compound 1 was identified by various analytical and spectroscopic techniques and comparison with known preparations, and was quantified using HPLC.

TLC analysis used Merck Silica gel 60 F254 as the carrier. The plate was developed with chloroform:methanol (9:1, v/v), and the compound was observed under UV (254 nm). The Rf was 0.86.

The HPLC retention time was 20.2 min under the following analytical conditions:

Column:	Senshu Kagaku, PEGASIL C8 (4.6 φ × 250 mm);
Detection:	UV 228 nm;
Eluent:	10 mM ammonium acetate water solution:acetonitrile = 55:45 (v/v);
Column temperature:	40°C; and
Amount injected:	5 μl.

The compound was identified by comparing its HPLC retention time with that of a known taxol sample (20.2 min), obtained from Sigma. The amount of taxol was determined using HPLC workstation and calibration curves that were prepared using reference standards

(LC10 SPD-MA)(Shimazu-co, 1996). The melting point was determined as 254°C using a Yamato Science A-9200 apparatus.

^{1}H-NMR and ^{13}C-NMR measurements were conducted using tetramethyl silane (TMS) as the internal standard, chloroform deuteride (CDCl$_3$) as the solvent, and an FT-NMR gauge JEOL-LA400 (400 MHz).

The spectral data of the isolated Compound 1 were:

^{1}H-NMR in CDCl$_3$: δ (Chemical shift) 1.14 (3^{1}H, s(singlet), Me-16), 1.24 (3H, s, Me-17), 1.68 (3H, s, Me-19), 1.80 (3H, d, *J*(coupling constant) = 1.0 Hz, Me-18), 1.89 (1H, ddd (double double doublet), *J* = 2.44, 10.99, 14.65 Hz, H-6b), 2.24 (3H, s, 10-OAc), 2.30 (1H, dd (double doublet), *J* = 8.85, 1.57 Hz, H-14β), 2.35 (1H, dd, *J* = 8.85, 15.57 Hz, H-14a), 2.39 (3H,s, 4-OAc), 2.55 (1H, m, H-6a), 3.80 (1H, d (doublet), *J* = 6.71 Hz, H-3), 4.20 (1H, dd, *J* = 1.0, 8.55 Hz, d, *J* = 5.2 Hz, H-20b), 4.31 (1H, d, *J* = 8.85, H-20a), 4.40 (1H, ddd, *J* = 4.4, 6.6, 11.0 Hz, H-7), 4.80 (1H, dd, *J* = 5.4, 2.7 Hz, H-2'), 4.95 (1H, bd(broad doublet), *J* = 7.32 Hz, H-5), 5.68 (1H, d, *J* = 6.71 Hz, H-2), 5.80 (1H, dd, *J* = 2.44, 8.55 Hz, H-3'), 6.24 (1H, s, H-13), 6.27 (1H, s, H-10), 7.02 (1H, d, *J* = 8.9 Hz, 3'-NH), 7.35 (1H, m, *p*-ph2), 7.40(1H, m, *m*-ph3), 7.42 (2H, m, *m*-ph2), 7.48 (2H, m, 0-ph2), 7.50 (1H, m, *p*-ph3), 7.51 (2H, m, *p*-ph1) ,7.62 (1H, m, *p*-ph1), 7.75 (2H, dd, *J* = 1.0, 8.35 Hz, 0-ph3), 8.14 (2H, dd, *J* = 1.22, 8.54 Hz, 0-ph1), ^{13}C-NMR (in CDCl$_3$): δ 9.55 (C-19), 14.84 (C-18), 20.85 (10-OAc, Me), 21.82 (C-16), 22.62 (4-OAc, Me), 26.85 (C-17), 35.60 (C-6), 35.68 (C-14), 43.17(C-15), 45.61(C-3), 55.03 (C-3'), 58.60 (C-8), 72.16 (C-7), 72.38 (C-13), 73.25, (C-2'), 75.03 (C-2), 75.59 (C-10), 76.54 (C-20), 79.12 (C-1), 81.23 (C-4), 84.44 (C-5), 127.04 (0-ph2, 0-ph3), 128.36 (*p*-ph3), 128.70 (*m*-ph2), 128.72 (*m*-ph1), 129.02 (*m*-ph3), 129.15 (q-ph1), 130.20 (0-ph1), 131.97 (*p*-ph2), 133.18 (C-1), 133.61(q-ph2), 133.71(*p*-ph1), 138.00 (q-ph3), 142.00 (C-12), 167.01 (C=O ph1, ph3), 170.36, (4-OAc, C=O), 171.26 (10-OAc, C=O), 172.73 (C-1'), and 203.62 (C-9)

Results and discussion

The HPLC and ^{1}H-NMR and ^{13}C-NMR measurements of Compound 1 agreed with those of the known taxol sample, and Compound 1 was identified as taxol.

In the experiment, we isolated 85 mg of taxol (purity: 97%) from five seedlings of Japanese yew (fresh weight: 923.87 g) (taxol yield: 0.0195%). Approximately 17 mg of taxol was produced from each seedling.

Taxol was extracted using the ordinary solvent extraction method. To treat the large volume of plant tissues, we needed large quantities of organic solvents. The batch extractor operates at room temperature and under atmospheric pressure and is easy to use. However, using supercritical fluid extraction and accelerated solvent extraction methods, we selectively extracted taxol, improved the extraction efficiency, and reduced the amount of solvent. Therefore, it is possible to improve the productivity of taxol by utilizing and improving the advantages of these methods.

We efficiently isolated and purified taxol by using solvents that have little impact on taxol. Our methods may be applied to produce taxol, industrially, from forest resources. Our study also provided basic data to develop low-cost, efficient industrial methods for extracting and purifying baccatin III and 10-deacetylbaccatin III, which are precursors of taxol (Maki and Sako, 1993).

Japan has also started to use taxol to treat ovarian and mammary cancers, and chemotherapeutic methods are actively being investigated (Turuo and Nagahiro, 1995). The demand for taxol is likely to rise, and taxol will attract increasing attention as an anticancer agent.

Table 6.3 The production cost of seedlings
T. *cuspidata* and taxol

Classifications	Cost (thousand yen per hectare)
Seedling production	7,648
Preparation	315
Extraction	4,688
Purification	2,423
Total	15,074

System plan

The processes and economy of taxol production should be understood to transfer the technologies for taxol production to forestry areas (Kikuchi, 2001).

Based on the results of our experiments, we drew up system plans for taxol production using Japanese yew seedlings. We divided the production processes into four broad classifications (seedling production, preparation, extraction, purification) and estimated the time and costs for producing crude taxol from 9-year-old seedlings (1 ha) (Table 6.3).

To produce 15,000 Japanese yew seedlings, 387 man-days and 7,648 thousand yen are needed. For the price of a seedling, we used the value given in a normal price list (tree-plantation, Hokkaido, fiscal 1999). The price includes the production costs and benefits. Taxoids can be extracted using existing facilities, so new facilities are not needed.

For preparing extraction materials, 39 man-days and 315 thousand yen are needed. Pulverizers should be newly purchased.

The process of material production takes as long as 13 years, from collection of seeds to delivery of products, which are extracted from 9-year-old seedlings. Therefore, trimmed needles of yew trees growing in parks and gardens should be used as well as seedlings grown in nurseries. As yew trees contain very little taxol when they are young, methods should be investigated and developed to reduce the seedling stage. Such possibilities include cutting (Saitho, 1986), mass propagation of superior clones, optimum fertilizer application methods, bio-breeding (Shibata, 1994), and technologies for accelerating germination. Intensive plantation of yew seedlings in nurseries would enable easy and mechanical control.

In the extraction process, 247 man-days and 4,688 thousand yen are needed, and in the purification process, 130 man-days and 2,423 thousand yen are needed. The costs vary depending on the present systems and facilities, and the above values were based mainly on solvent prices.

Because expensive production facilities and chemical technologies are needed, to purify taxol, production would likely be done by chemical manufacturers. We used our experimental values for the yields of extraction and purification; therefore the total taxol yields should be reinvestigated for industrial production.

In our estimation, we assumed production of 17 mg taxol per seedling. With 15000 seedlings produced per hectare, each hectare can produce 255 g of taxol. As approximately 1–2 g of taxol are needed to treat an ovarian cancer patient in the treatment period, this amount is sufficient for 125–250 patients.

In monetary terms, 15,074 thousand yen per hectare was estimated: 7,648 thousand yen to produce yew materials, 315 thousand yen to prepare the materials, 4,688 thousand yen to extract taxoids, and 2,423 thousand yen to purify taxol. Thus, approximately 59 thousand yen is needed to produce one gram of taxol. The estimated price of taxol is 410,040 thousand yen for 255 g or

1,608 thousand yen per gram. Therefore, the price of taxol (410,040 thousand yen) is approximately 27 times the production cost (15,074 thousand yen), and significant profit is expected.

Conclusion

To use taxol as a medicine, crude taxol must be further purified and formulated. It is possible to develop a new industry in forestry areas by growing seedlings in these areas and entrusting chemical companies to extract, purify, and sell taxol.

Introduction of production systems that utilize forest resources should be considered for those chemotherapy agents that are medically understood but for which bioengineering or chemical production methods are not established, such as taxol.

For industrial production of taxol, verification tests should be conducted using large fields, and purification and other production methods should be further researched and developed. We believe that taxol can be produced in forestry areas using forest resources.

Conclusion

This chapter described a method of producing taxol using forest resources (Japanese yew), which can be transferred into forestry areas and provide a stable supply of medicinal materials.

Taxol is noted as a natural anticancer agent that is effective against ovarian and lung cancers. Taxol is contained in *Taxus* species, including *T. cuspidata*, which grows in Japan. Due to its tumor-inhibiting properties, the demand for taxol is likely to increase, although the supply is limited.

Taxol can be produced in forestry areas using forest resources. Intensively cultivated and relatively young seedlings of Japanese yew are promising sources from which taxol can be extracted. Our studies showed efficient methods for extracting taxol that are appropriate for the characteristics of taxol components and tissues of Japanese yew. We also suggested methods for the commercial cultivation of *Taxus* species and production of taxoids.

The components of *Taxus* species are being actively studied throughout the world. However, forest resources contain many physiologically active substances of unknown possibilities. It is highly possible that continued studies will lead to the discovery of new physiological activities and new physiologically active substances. Therefore, we will continue to search for new ways of using forest resources, besides the conventional production of wood and trees, and to propose new forestry industries (such as forest factories) that provide employment opportunities for people in forestry areas.

Industries, including forestry, that use biological resources and bio-engineering are strategically important, and agricultural crops are a new focus. New forestry industries may be created by conducting research and development together with other fields of study, such as botany, bio-engineering, organic chemistry, and pharmacy.

References

Akiyama, T. (1996) Studies on the development of medicinal agents from forest resources, *Ringyo-gijutsu*, 651, 11–14.

Anetai, M., Masuda, T. and Takasugi, M. (1996) Preparation and furanocomarin composition of Glehnia root produced in Hokkaido, *Natural Medicines*, 51(4), 331–334.

Anetai, M. *et al.* (1998) Preparation and chemical evaluation of Astragali Radix produced in Hokkaido, *Natural Medicines*, 52(1), 10–13.

Appendino, G., Cravotto, G., Gariboldi, P., Gabetta, B. and Bombardelli, E. (1994) The chemistry and occurrence of taxane derivatives. X, *Gazzetta Chimica Italiano*, 124, 1–4.

Appendino, G. (1995) The phytochemistry of the tree, *Nat. Prod. Rep.*, 349–360.

Arthur, G. Fett Neto. and Frank, di Cosmo. (1992) Distribution and amounts of taxol in different shoot parts of *Taxus cuspidata, Plant Med.*, 58, 464–466.

Choi, M., Kwak, S., Liu, J., Park, Y., Lee, M. and An, N. (1995) Taxol and related compounds in Korean native yews (*Taxus cuspidata*), *Plant Med.*, 61, 264–266.

Chu, A. *et al.* (1994) *Phytochemistry*, 36, 975.

Chuang, L.C., Chen, K.J., Lin, Y.S. and Chen, F.C. (1989) *Taiwan ko Hsueh.*, 42, 29.

Chuang, L.C. *et al.* (1990) *Huaxue Xuebao.*, 48, 275.

Chun, M., Shin, H. and Lee, H. (1994) Supercritical fluid extraction of taxol and baccatin III from needles of *Taxus cuspidata*, *Biotechnology Techniques*, 8, 547–550.

Czeczuga, B. (1986) *Biochem. Syst. Ecol.*, 14, 13.

Das, B. *et al.* (1994) *Fitoterapia.*, 65, 189.

Denis, J.N. and Greene, A.E. (1994) *J. Am. Chem. Soc.*, 110, 5917–5919.

Ellis, D.D. *et al.* (1996) Taxol production in nodule cultures of *Taxus,* 59, 246–250.

Fuji, K. (1996) Development of combination chemotherapy including paclitaxel and docetaxel, *Workshop on Taxoids*, p. 7.

Fujino, H., Suzuki, S., Yoshizaki, M., Satake, M. and Konda, H. (1998) Studies on cultivation of polygala tenuifolia WILLD. I. Seed germination and storage. *Natural Medicines*, 52(2), 97–102.

Gordon, M. *et al.* (1993) The Taxol supply Crisis, *J. Nat. Prod.*, 56(10), 1657–1668.

Guo, Y., Diallo, B., Jazir, M., Vanhaelen-Fastre, R. and Vanhaelen, M. (1996) Immunological detection and isolation of a new taxoid from the stem bark of *Taxus baccata*, *J. Nat. Prod.*, 59, 169–172.

Hatakeyama, Y., Kumagai, T., Katsuki, S., Honma, N., Ishizaki, S. *et al.* (1998) Studies on culivation and breeding of paeonia lactiflora PALLAS(2) Line variation of characteristics on growth and constituent, *Natural Medicines*, 52(2), 109–115.

Hatano, K. and Sasaki, S. (1993) *Jumoku-no seicho to kankyo*, pp. 133–159, Tokyo: Youken-dou.

Higashi, A. *et al.* (1997) Studies on the effect of preparation methods of platycodon root on its quality, *Natural Medicines*, 51(1), 56–62.

Horton, R.A., Somoza, C., Kim, H.-B., Liang, F., Biediger, R. *et al.* (1994) *J. Am. Chem. Soc.*, 116, 1597–1598.

Ichikawa, K. (1996) *Farumashia*, 32(6).

Ichinohe, Y. (1994) Dokuso-no-Saigiki, pp. 147–155, Tokyo: kensei-sha.

Ishizuka, Y., Hayashi, K. and Morita, A. (1998) Studies on the cultivation of Saposhnikovia divaricata (Trucz) (II). Seasonal variation of root growth, methanol extract and constituent content, *Natural Medicines*, 52(2), 151–155.

Isoda, S. and Shoji, J. (1994) Studies on the cultivation of eleuthrrococcus senticosus MAXI.II. On the germination and raising of seeding, *Natural Medicines*, 48(1), 75–81.

Izawa, K. (1995) *The Herbal Colour Encyclopedia 2*, pp. 217, Tokyo: Shufu-no-to.

Jaziri, M., *et al.* (1997) A new concept for the Isolation of bioactive taxoids, *Current Topics in Phytochemistry*, 1, 67–76.

Jennings, D.W., Howard, M.D., Leon, H.Z. and Amyn, S.T. (1992) Supercritical extraction of taxol from the bark of *Taxus Brevifolia*, *The Journal of Supercritical Fluids*, 5 , 1–6.

John, M. *et al.* (1996) An improved method for the separation of paclitaxel and cephalomannine, *J. Nat. Prod.*, 59, 167–168.

Kawamura, F., Kikuchi, Y., Ohira, T. and Yatagai, M. (1999) Accelerated solvent extraction of paclitaxel and related compounds from the bark of *Taxus cuspidata.*, *J. Nat. Prod.*, 62(2), 244–247.

Khan, M.S.Y., Kumar, I. *et al.* (1976) *Plant Med.*, 30, 82.

Kikuchi, Y., Ohira, T. and Yatagai, M. (1997a) Accelerated solvent extraction on taxane compounds from *Taxus cuspidata.*, *Journal of Wood Science*, 43(11), 971–974.

Kikuchi, Y., Kawamura, F., Ohira, T. and Yatagai, M. (1997b) Approach to supply of taxanes from natural sources, *Journal of the Japanese Forestry Society*, 79(4), 239–241.

Kikuchi, Y. *et al.* (1988) *Mokuzai-hiyatuka*, pp. 160–161, Akita: Akita Prefectual University.

Kikuchi, Y. and Kawamura, F. (1999) Liquid CO_2 extraction of taxanes from needles of *Taxus cuspidata, Holz als Roh-und Werkstoff.*, 57(2), 12.

Kikuchi, Y., Kawamura, F., Ohira, T. and Yatagai, M. (2000) Studies on production of taxol from natural sources I.-Preparation and taxol composition of the needles from *Taxus cuspidata*, *Natural Medicines*, 54(1), 14–17.

Kikuchi, Y. (2001) *Studies on production of taxol (anticancer agent) from natural sources*, Grant-in-Aid for Scientific research (No. 09556033) pp. 1–142. Tokyo: Ministry of Education, Science, and Culture of Japan.

Kingston, D.G.I. *et al.* (1990) The chemistry of taxol, a clinically useful anticancer agent, *J. Nat. Prod*, 53(1), 1–12.

Kobayashi, J. *et al.* (1994) *Tetrahedron.*, 50, 7401.

Kobayashi, J. and Shigemori, H. (1995) *Kagaku to Seibutsu*, 33(8), 538–545.

Kobayashi, J. (1996) *Workshop on Taxoids*, p. 8.

Kondo, H. and Amano, U. (1922) *Yakugakuzashi*, 42, 1075.

Kuhn, R. and Brockmann, B.D.H. (1933) *Chem. Ges.*, 66, 828.

Krüssmann, G. (1983) *Handbuch der Nadelgehölze*, pp. 317–333, Berlin, Hamburg: Paul Parey.

Kurata, S. (1971) *Illustrated Important Forest Trees of Japan* V, p. 1–2, Tokyo: Chiku Shuppan.

Kuwajima, I. (1997) *Chemistry.*, 52, 30–33.

Liang, J.Y. *et al.* (1987) *Chem. Pharm. Bull.*, 35, 2613.

Lucas, H. (1856) *Arsh, Pharm.*, 85, 145.

Maki, Y. and Sako, M. (1993) The chemistry of taxane diterpenoids: the present status of development of taxol as a new type of antitumor agent, *Youkigousei-kagaku-kyokai-shi*, 51, 298–306.

Masuzawa, T. (1997) *Ecology of the Alpine Plants*, pp. 68–74, Tokyo: The University of Tokyo-Press.

Minami, M., Sugino, M., Hata, K., Hasegawa, C. and Ogawa, K. (1995) Physiological response and improvement of tolerance to environmental stress in *Bupleurum falcatun* L. (I) Effects of dry stress on growth and saikosaponins content of one-year-old, Rosette Plant (I), *Natural Medicines*, 49(2), 137–147.

Modica, G.Di. (1962) *Atti Accad Nazl Lincei Rend Classe Sci Fis Mat e Nat.*, 32, 87.

Morita, S. and Yatagai, M. (1994) Antimite components the hexane extractives from Domaiboku of Yakusugi (*Cryptomeria japonica*), *Journal of Wood Science*, 40(9), 996–1002.

Mukaiyama, T. (1997) *Chemistry Today*, 10, 56–61.

Nicolaou, K.C., Yang, Z., Liu, J.J., Ueno, H., Nantermet, P.G., Guy, R.K., Claiborne, C.F., Renaud, J., Couladouros, E.A., Paulvannan, K., Sorensen, E.J. (1994) *Nature*, 367(17), 630–634.

Ohira, T., Kikuchi, Y. and Yatagai, M. (1996) Extractives of *Taxus Cuspidata* I - Content of taxol and its related compounds, *Journal of Wood Science*, 42, 1234–1242.

Ohtani, Y., Hazama, M. and Sameshima, K. (1996) Crucial chemical factors for termiticidal activity of hinoki wood (*Chamaecyparis obtusa*) II. Variation in termiticidal activities among five individual samples of hinoki wood, *Journal of Wood Science*, 42(12), 1228–1233.

Ojima, I., Habus, I. and Zhao, M. (1991) Ibid., 56, 1681.

Ojima, I., Habus, I., Zhao, M., Zucco, M., Park, Y.H., Sun, C.M. and Brigaud, T. (1992) *Tetrahedron*, 48(34), 6985–7012.

Parveen, N., Taufeeq, H.M. and Ud-Din. K.N. (1985) *J. Nat. Prod.*, 48, 994.

Pezzuto, J. (1996) Taxol production in plant cell culture comes of age. *Nature Biotechnology*, 14, September, 1083.

Robert, C.H. *et al.* (1992) Harvesting, drying and storage of cultivated taxus clipping: an exploratory study, Workshop on Taxus, Taxol, Taxoteve. pp. 23–24

Saitho, S. (1986) *Onko-The Nature of Hokkaido*, pp. 24–46, Hokkaido: Hokkaido-newpaper-press.

Saitho, S. (1990) *Tree & Shrubs of Hokkaido*, pp. 63–64, Tokyo: Arinishi.

Sakai, E., Iida, O., Saito, Y., Oono, A. and Satake, M. (1996a) Studies on the cultivation of lithospermum erythrorhizon SIEB.et Zucc.-The relation between fertilization and cultivation period, *Natural Medicines*, 50(1), 41–44.

Sakai, E., Shibata, T., Kawamura, T., Hisata, Y. *et al.* (1996b) Pharmacognostical studies of Houttuyniae Herba (2), growth and flavonoid glycoside contents of *Houttuynia cordata* THUNB. cultivated under shade condition., *Natural Medicines*, 50(1), 45–48.

Sasaki, S. (1989) Research development on short rotation intensive culture of trees for biomass usage, *Journal of Wood Science*, 35, 865–874.

Shibata, M. (1994) Selection breeding in forest trees, *Journal of Wood Science*, 40(7), 681–686.

Shimazu-co. (1996) *The Handbook of CLASSLC10/M10A*, Kyoto, Shimazu-Co.

Stringer, R. (1994) *The State of Food and Agriculture-Forest Development and Policy Dilemmas.* p. 12, Tokyo: kokusai-shokuryu-nougyo-kyokai.

Suffnes, M. and Cordell, G.A. (1985) The Alkaloids', ed. by R.H.E.Manke, *Academic Press.*, **25**, 1.

Tachibana, S., Itho, K., Ohkubo, K. and Towers, G.H.N. (1994a) Leaf of flavonoids of *Taxus brevifolia*, *Journal of Wood Science*, 40(12), 1294–1397.

Tachibana, S., Matsuo, A., Itho, K., and Oki, T. (1994b) Extractives in the leaves and bark of *Taxus cuspidata* Sieb. et. Zucc.*var.nana* Reder callus cultures. *Journal of Wood Science*, 40(9), 1008–1013.

Tachibana, S., Watanabe, E., Itho, K. and Oki, T. (1994c) Formation of taxol in *Taxus cuspidata* Sieb. et. Zucc.*var.nana* Reder callus cultures, *Journal of Wood Science*, 40(11), 1254–1258.

Tachibana, S. (1995) Utilization of biologically active substances in trees, *Journal of Wood Science*, 41, 967–977.

Takahashi, T. and Kondo, H. (1925) *Yakugakuzashi*, 45, 861.

Takahashi, T. (1931) *Yakugakuzashi*, 51, 401.

Turuo, T. and Nagahiro, S. (1995) *Current Review 3 Konnichi-no Gan-kakagaku-ryohou*, pp. 24, 55–59, 127–132, Tokyo: Cyugai-igaku.

Ueda, R. (1913) *Yakugakuzashi*, 33, 923; (1914) *Yakugakuzashi*, 34, 631. US Patent, 1969, 3466331; US Patent, 1971, 3624105.

Wani, M.C., Taylor, H.L. and Wall, M.E. (1971) Plant antitumor agents VI. The isolation and struture of taxol, a novel antileukemic and antitumor agent from *Taxus brevifolia. J. Am. Chem. Soc.*, 93, 2325–2327.

Weishuo, F. *et al.* (1993) Qualitative and quantitative determination of taxol and related compounds in *Taxus cuspidata Sieb* et zucc, *Phytochemical Analysis*, 4, 115–119.

Willam P. Mcguire. *et al.* (1996) Cyclophoshamide and cisplatin compared with paclitaxel and cisplatin in patients with stage III and stage IV ovarian cancer, *The New England Journal of Medicine*, 334(1), 1–6.

Witherup, K.M. *et al.* (1989) High performance liquid chromatographic separation of taxol and related compounds from *Taxus brevifolia*, *Journal of Liquid Chromatography*, 12(11), 2117–2132.

Yoshizaki, F., Yanagihashi, R. and Hisamichi, S. (1988) Determination of taxinine and seasonal variation of its content in the leaf of Japanese yew (*Taxus cuspidata*), *Natural Medicines*, 42(2), 151–152.

Yukimune, Y., Tabata, H., Higashi, Y. and Hara, Y. (1996) Methyljasmonate-induced overproduction of paclitaxel and bacctin III in Taxus cell suspension culture, *Nat. Biotech.*, 14, 1129–1132.

7 Analytical aspects of taxoids

Mutsuo Kozuka, Susan Morris-Natschke,
and Kuo-Hsiung Lee

Introduction

Taxol (Paclitaxel, 1)* was first isolated from the bark of *Taxus brevifolia* Nutt. (Pacific yew) by
Wani and Wall, *et al.* (1971). It has unique antitumor activity and currently is an important
anticancer drug, which is highly effective for the treatment of ovarian, breast, and other cancers.
However, due to the slow growth and low abundance of this tree and low taxol content in the
bark, its limited natural resource has prompted numerous studies on alternative supply routes,
including synthesis and semi-synthesis, cell and tissue culture methods, search for taxoid-
producing microorganisms, and isolation from various *Taxus* species including *T. baccata* L.
(European yew; Sénilh *et al.*, 1984), *T. cuspidata* Sieb. et Zucc. (Japanese yew; Fang *et al.*, 1993),
T. canadensis Marshall (Canadian yew; Zamir *et al.*, 1992), *T. chinensis.* Rehder (Chinese yew; Fuji
et al., 1993), *T. globosa* Schltdl. (Mexican yew; Nicholson *et al.*, 1992), *T. floridana* (Florida yew;
Rao and Johnson, 1998), *T. wallichiana* Zucc. (Himalayan yew, Miller *et al.*, 1981), *T. yunnanensis*
C.Y. Cheng, W.C. Cheng and L.K. Fu (Zhang *et al.*, 1994), *T. sumatrana* (Miq.) de Haub.
(Kitagawa *et al.*, 1995), their variations, and their hybrids such as *T. x media* Hicksii (Witherup
et al., 1990). In addition, needles, stems, wood, roots, stem clippings, and foliage have been
tested for their taxoid content (Vidensek *et al.*, 1990; Witherup *et al.*, 1990). The low concentrations of taxol in these plant materials, together with the complexity of the matrix, make
extraction and purification of taxoids quite difficult. For example, from the view point of natural
resources preservation, needles or clippings rather than bark, are more reproducible and better
sources of taxol; however, purification of taxoids from these sources is more difficult due to the
presence of waxes, chlorophyll, and many other endogenous compounds in these materials
(Schutzki *et al.*, 1994). Such difficulties have challenged many researchers to devise modified,
effective, and new isolation procedures for taxol and other taxoids from natural sources, including cell and tissue cultures. Comprehensive and excellent reviews of taxol by Wall and Wani
(1998), Suffness and Wall (1995), and Jaziri and Vanhaelen (2001) were published previously.

Extraction, separation, and purification procedures

Extraction, separation, and purification procedures for taxoids have been discussed in numerous
reports, including comprehensive excellent reviews by Snader (1995) and Theodoridis and
Verpoorte (1996). Examples of these procedures will be given briefly below. For more details in
this research area, refer to the reviews indicated above and literature cited therein.

---

* Taxol® is a registered trademark of Bristol-Myers Squibb Company and the generic name of the drug is paclitaxel.
However, in this chapter, we will use the name "taxol" for the compound as originally named by Wall *et al.* (1971).

Taxol (paclitaxel) (**1**) R₁ = β-OH, R₂ = Ac
10-Deacetyl-taxol (**2**) R₁ = β-OH, R₂ = H
7-*epi*-Taxol (**3**) R₁ = α-OH, R₂ = Ac
7-*epi*-10-Deacetyl-taxol (**4**) R₁ = α-OH, R₂ = H

Cephalomannine (**5**) R₁ = β-OH, R₂ = Ac
10-Deacetyl-cephalomannine (**6**) R₁ = β-OH, R₂ = H

Docetaxel (Taxotere®) (**7**)

Baccatin III (**8**) R₁ = β-OH, R₂ = Ac
10-Deacetyl-Baccatin III (**9**) R₁ = β-OH, R₂ = H

Conventional solvent extraction

Taxol (1) was first isolated from *T. brevifolia* (Pacific yew) by the following procedures (Wani *et al.*, 1971). An alcohol extract of the stem bark was concentrated and partitioned between water and chloroform. The residue from the chloroform extract was subjected to bioassay-guided fractionation using three successive chromatographic procedures on Florisil, Sephadex LH-20, and silica gel. The isolated active principle was crystallized from aqueous methanol to give taxol as needles.

Thereafter, many taxoids were isolated from various *Taxus* species (Baloglu and Kingston, 1999). Most of the isolation methods used methanol (Vidensek, 1990) at room or reflux temperature (Glowniak *et al.*, 1996), 95% ethanol (Kopycki *et al.*, 1994), ethanol (Powell *et al.*, 1979), ethanolic percolation (McLaughlin *et al.*, 1981), ethanol/water/acetic acid (80 : 19 : 1; van Rozendaal *et al.*, 1997), methanol/chloroform (1 : 1; Fang *et al.*, 1993), or methanol/methylene chloride (1 : 1; Witherup *et al.*, 1990) as an initial extraction solvent and followed by liquid–liquid partitioning between water and chloroform (Wani *et al.*, 1971) or methylene chloride (Huang *et al.*, 1986) or toluene (Kobayashi *et al.*, 1994) or ethyl acetate (Kitagawa *et al.*, 1995) to remove hydrophilic components (Vidensek, 1990). Subsequently, taxoids were separated and purified from the above organic solvents by various chromatographic methods, including traditional chromatographies (open column; dry column: Zamir *et al.*, 1998; TLC: Jenniskens *et al.*, 1996) using silica gel or high performance reversed-phase (RP) liquid chromatography (HPLC) followed by crystallization. Rao (1993) published a simple and practical scheme of conventional solvent extraction. The method involved an initial extraction with methanol, partition between water and chloroform, single column chromatography using silica gel, Florisil, or a RP C$_{18}$-bounded silica, and crystallization. All three adsorbents gave comparable yields of the key components, taxol (1), 10-deacetylbaccatin-III (9) and 10-deacetyltaxol-7-xyloside, but the RP column was more convenient and the listed taxoids could be purified by direct crystallization. The author emphasized that the yields of taxol (0.02–0.04%) and of the related taxoids were high because this method involved fewer steps than other methods. When needles were used as the plant material, pre-extraction with hexane was performed to remove non-polar substances such as waxes (Wheeler *et al.*, 1992; Fett Neto and diCosmo, 1992), because such materials would interfere with RP HPLC.

However, conventional extraction methods generally give relatively low yields of taxol, are quite time-consuming, and use large amounts of expensive and environmentally hazardous solvents, such as chlorinated hydrocarbons. Thus, several modified and new extraction and separation methods have been developed to achieve the following goals: (1) improve extraction efficiency; (2) reduce extraction time; (3) reduce the amount and costs of solvent used for extraction and separation; (4) protect our environment from hazardous waste solvents; and (5) scale up extraction quantities (Mattina *et al.*, 1997).

Alcohol-based extraction methods could eliminate the use of chlorinated hydrocarbon solvent systems, and degradation of the taxoids during the drying process could be avoided by using fresh plant material. Accordingly, Ketchum *et al.* (1999) compared various solvent systems for the extraction of taxoids and of contaminating residue from leaves of *T. brevifolia* and *T. x media*. Taxol recovery paralleled solvent polarity. However, although a more polar solvent such as methanol removed more taxol than a less polar solvent such as acetone, it also resulted in more total plant residue. A general extraction method with 40% ethanol-water followed by C$_{18}$ solid phase extraction (SPE) gave the best overall results with 500-fold taxol enrichment. Using these methods, taxol constituted 0.01–0.02% of the dry weight of the leaf tissue from *T. x media* and 0.04% of dried leaves from *T. brevifolia*. MacEachern-Keith *et al.* (1997) investigated taxol stability in several solvent systems at various temperatures and pressures. Kinetics experiments indicated that the apparent activation energy barrier for taxol degradation, including 7-epimerization, is highly dependent on experimental conditions and taxol is most stable in aqueous SH medium, less stable in DMSO and least stable in methanol. Kubota and Abe (1999) further compared taxol stability in various solvent systems at 25°C for 48 h and 120 h. Taxol was stable in acetone, methanol, acetonitrile, methyl acetate, diethyl ether, and water, while it decomposed in ethanol, acetone-silica gel, hexane-acetone, tetrahydrofuran, isopropyl alcohol, diisopropyl ether, benzene, and especially, in methylene chloride, and chloroform. Decomposition may be caused by the presence of trace amounts of halogens and peroxides.

Jenniskens *et al.* (1996) used acid extraction to readily obtain the taxane alkaloid taxine in crude form. For further purification, the crude alkaloids were subjected to various chromatographic methods including column chromatography over neutral alumina, preparative TLC on silica gel, flash chromatography over silica gel and HPLC using a C$_{18}$ column. Six taxine alkaloids were isolated from the needles of *T. baccata*; four were new compounds.

Other methods

Supercritical fluid extraction and supercritical fluid chromatography

Supercritical fluid extraction (SFE) was first used by industries and was later introduced into laboratory work with the development of supercritical fluid chromatography (SFC) and instruments for laboratory-scale SFE. Excellent review articles (for SFE: Vandana and Teja, 1995 and

Jarvis and Morgan, 1997; for SFC: Berger, 1997) have been published on this research area. Although SFE has been applied mainly to non-polar natural products and relatively little information has been found on the extraction of polar compounds, this method has real advantages over conventional solvent extraction. For instance, because taxoids do not decompose under supercritical conditions, they can be obtained in higher yields and as cleaner products than by using organic solvents. In addition, supercritical fluids are mild solvents and their use resolves some environmental problems.

Jennings *et al.* (1992) used carbon dioxide with 3.8 mol% of ethanol as a co-solvent for taxoid extraction from the bark of *T. brevifolia*. Chun *et al.* (1994) extracted taxol (1) and baccatin III (8) from the needles and seeds of *T. cuspidata* using 3% ethanol as a co-solvent in carbon dioxide at 40°C and 30.4 Mpa. They modified extraction conditions, varied temperature and pressure, and compared ethyl acetate, methanol, dichloromethane, and diethyl ether as the co-solvent with carbon dioxide. They also found that waxy substances were co-extracted with taxoids under these conditions, but these contaminants could be removed by pre-extraction with hexane (Chun *et al.*, 1996).

Vandana *et al.* (1996) extracted taxol from the bark of *T. brevifolia* using the following conditions: 320 K and 331 K, pressure range of 10.3 to 38.1 Mpa using supercritical nitrous oxide and nitrous oxide plus ethanol mixture. The latter mixtures could extract most of the taxol present in the bark and the extractions were more efficient than those using carbon dioxide plus ethanol.

Vandana and Teja (1997) studied the solubility of taxol in supercritical carbon dioxide and nitrous oxide, and found that taxol was more soluble in the latter medium. However, the solubility in both cases was rather low, and co-solvents would be needed for the practical extractions of taxol from plant materials. Also, because of the safety hazard posed by nitrous oxide, carbon dioxide with a co-solvent would be the preferred supercritical fluid (SCF) in a practical extraction.

Recently, Nalesnik *et al.* (1998) studied the solubility of pure taxol in supercritical carbon dioxide and found that taxol can show differing degrees of solubility, depending on whether it is as a pure solute or is extracted from a complex matrix found in nature. The natural plant matrix contains numerous compounds including natural acids that may alter the solubility of taxol in SCF CO_2 from the actual solubility of pure taxol. Measuring the solubility of pure taxol in SCF CO_2 can elucidate these matrix effects and aid in the development of mass-transfer models.

Kikuchi and Kawamura (1999) used low pressure extraction using liquid carbon dioxide or a liquid carbon dioxide and methanol mixture to extract taxol and taxoids from the needles of *T. cuspidata*. The former extractions were processed at 50 kgf/cm^2 pressure and 40°C temperature, and the latter extractions were done at 40–60°C with a pressure of 50 kgf/cm^2. Both extraction methods were found to be more effective than the conventional liquid methanol extraction procedure. The addition of co-solvent increased the rate of taxol extraction and removed a significant amount of the taxol present in the bark.

Supercritical fluid chromatography is a complimentary technique to HPLC and GC, and its advantages include the possibility of analysis for thermally labile compounds and use of both HPLC and GC detectors such as UV-Vis and flame-ionization. Instruments for both capillary and packed column SFC are commercially available.

The first study on using SFC to analyze taxol-related compounds was reported by Heaton *et al.* (1993). They analyzed taxicin I and II from English yew by SFC using a cyano column with a carbon dioxide and methanol gradient. Both capillary and packed column separations were compared with the packed columns giving a better quantitative analysis of the taxoids. However, this work was only preliminary, and was not suitable for purity/impurity analysis. Bristol-Myers Squibb and Hewlett-Packard groups (Jagota *et al.*, 1996) later reported analysis of taxol and related compounds using SFC. The separation of taxol and five related taxoids was compared on two SFC instruments. The typical analysis time was 20 min on system I and 4 min on system II. Therefore, system II was selected for further studies and successfully separated taxol from sixteen impurities or degradation products in about 35 min. The separations were achieved on a diol column with a mobile phase of carbon dioxide with a methanol gradient and detection by UV at 227 nm. Impurities or taxol degradation products behaved comparably on HPLC and SFC, and the latter method gave results in half the run time of HPLC. Thus, SFC can be used for purposes of impurity/purity profiling and stability-indication.

Microwave-assisted extraction

Mattina *et al*. (1997) found that microwave-assisted extraction (MAE) can reduce both extraction time and solvent consumption. With *Taxus* needles as the plant material and using the MES-1000 microwave solvent extraction system from CEM Corp. with 95% ethanol rather than the methanol conventionally used, solvent consumption was reduced considerably. The most favorable time and temperature combinations were 85°C and 9–10 min. Using the above parameters and after partial purification of the taxoid extract with SPE, MAE is quantitatively and qualitatively equivalent to the conventional extraction methods consisting of overnight shaking of 5 g of needles in 100 mL of methanol at room temperature.

Accelerated solvent extraction

Accelerated solvent extration (ASE) is a new analytical method that can shorten extraction time and increase the yields of target compounds. In ASE, pressure is applied to the sample extraction cell to maintain the heated solvent in a liquid state during the extraction. Kikuchi *et al*. (1997) and Kawamura *et al*. (1999) investigated the extraction of taxol (1) and related taxoids from the bark of *T. cuspidata* using a Dionex Accelerated Solvent Extractor ASE-200 under various conditions. Using experimental conditions of 90 : 10 MeOH/H$_2$O, 150°C, and 10.13 Mpa, the yields of taxol, baccatin III (8) and 10-deacetylbaccatin III (9) were higher than those using conventional solvent extraction of the same material at room temperature. ASE does not require chlorinated solvents and, thus, can avoid taxol degradation. Solvent consumption is reduced because of increased dissolving power. Moreover, even with water alone, the yields of taxol and taxoids using ASE were much higher than with other extraction methods. Extraction with water would help protect the environment and reduce extraction costs. Based on these results, ASE will certainly be expanded to industrial applications for extraction of taxoids from *Taxus* plant resources.

Fast pyrolysis

Although taxoids are not volatile, they can be obtained by fast pyrolysis. Recently, Legge *et al*. (Cass *et al*., 2001) investigated a new unique method to extract taxoids from needles, twigs, and whole clippings of *T. canadensis* using a Waterloo Fast Pyrolysis Process (WFPP) apparatus. With this method, no more than 20% of the taxol (1) present in the *T. canadensis* biomass was recovered and other taxoids present in the biomass could not be analyzed because of the phenolic interference. Although numerous separation techniques were attempted to purify the taxoids from the pyrolysis oil, one disadvantage of this method remains the presence of much higher quantities of phenolics than of taxoids in the oil, which interfere with the isolation of the desired taxoids. Thus, for this method to be practical, pyrolysis conditions, including temperature and separation procedures, must be further examined.

Large-scale process

In 1986, Virginia Polytechnic Institute and Polysciences, Inc. groups (Huang *et al.*, 1986) reported an early large-scale procedure essentially based on the method of Wani *et al.* (1971). The method included extraction of *T. brevifolia* bark (8061 lbs) with methanol, partition of the wet extract (2400 lbs) between methylene chloride and water, chromatography of the methylene chloride extract (95 kg) using Florisil and silica gel columns [taxol (1) rich fraction: 13 kg], and finally crystallization (yield of crude taxol: 911 g). Fuji *et al.* (1993) reported later a large-scale extraction of taxoids from the needles and stems of *T. chinensis*. The methanol extract (3.3 kg) of dried materials (60 kg) was partitioned between water and methylene chloride and yielded 850 g of methylene chloride extract. The extract was dissolved in diethyl ether and washed successively with H_2O, 5% $NaHCO_3$, 1N-HCl, and H_2O. The residue (321 g, neutral materials) of the ether solution was subjected to column chromatography (cc) on silica gel, then repeated chromatography including preparative recycle HPLC with polystyrene column, preparative TLC (silica gel), cc (LH-20), and recrystallization. Seventeen pure compounds resulted: five new and twelve known diterpenoids including taxol (yield: $1.1 \times 10^{-4}\%$). Other early stage large-scale preparative procedures were reviewed by Snader (1995).

Many recently published methods have concentrated on a preparative or semi-preparative separation of taxoids. Reversed-phase chromatography provides a rapid and efficient method to separate and purify taxol. However, it is very important to scale-up the isolation process in order to obtain a significant amount of taxol.

Rao *et al.* (1995) developed a new effective large-scale process based on a single reversed-phase column chromatography for taxol and related taxoids from *T. brevifolia*. This process is a modification of their laboratory scale method (Rao, 1993). It was successfully run on a pilot-plant scale for the production of pure taxol (yield: 0.04%) and related taxoids including 10-deacetyl baccatin III (9, 0.02%), 10-deacetyl taxol-7-xyloside (0.1%), 10-deacetyl taxol-C-7-xyloside (0.04%), 10-deacetyl-cephalomannine-7-xyloside (0.006%), taxol-7-xyloside (0.008%), 10-deacetyl taxol (2, 0.008%), and cephalomannine (5, 0.004%) from the bark; brevifoliol (0.17%) from the needles; and 10-deacetyl taxol-C-7-xyloside (0.01%) and 10-deacetyl taxol-C (0.006%) from the wood. This method includes the following three steps: (1) methanol extraction and concentration of the extract under reduced pressure at <30°C; (2) liquid–liquid partition of the methanol extract between chloroform and water; and (3) C_{18}–bonded silica gel column chromatography of the chloroform extractable fraction in 25% acetonitrile-water, eluting with a step gradient of 30–50% acetonitrile-water. Eight different taxoids, including taxol, were crystallized from different fractions. Taxol and four taxoids were purified by recrystallization, and the other three taxoids required a short silica gel column chromatography. Taxol was completely freed from cephalomannine by selective ozonolysis. The entire process including single reversed-phase column chromatography and direct crystallization is simpler, more efficient (it provides taxol and other important taxoids simultaneously), and more productive (it gives increased yields of taxoids, especially, of taxol).

Subsequently, the same group (Rao *et al.*, 1996) presented a pilot-plant scale chromatographic process (50–200 lbs of plant material) for the isolation of taxol and several other taxoids from the needles of *T. x media* Hicksii and *T. floridana*. Using similar processes, the former species gave taxol (0.012–0.015%), brevifoliol, and another four taxoids from the dry needles and the latter species gave taxol (0.01%), 10-deacetyl baccatin III (9, 0.06%), and two other taxoids from the fresh needles. Preparative separation and purification of taxoids from the crude extracts of *T. canadensis* using the Zorbax-based bonded material "Zorbax SW-Taxane" in isocratic elution was reported by Wu, D.R. *et al.* (1995). The method provides high selectivity and recoveries of five compounds from the taxoid extract in either reversed or normal phase. The process works at high alcohol concentrations and the increased solubility of the taxoids at these concentrations leads to a greater output. "Zorbax SW-Taxane" packed column offers a large surface area and, therefore, a high taxoid saturation capacity.

Yan *et al.* (1996) developed a low-cost method for the isolation and purification of taxol from *T. yunnanensis*. The bark or needles were extracted with 95% ethanol, and the extract was partitioned between water and methylene chloride. The methylene chloride extract was chromatographed on Celite 545-flash column, eluted with hexane and methylene chloride. Then, the latter fraction was chromatographed twice on a silica gel 60 medium-pressure flash column. Taxol was separated from cephalomannine by RP C_{18} preparative HPLC. The purity of product was ≥98.8%, and the yield of taxol was 0.028% of the bark and 0.0088% of the needles.

Wu, D.J. *et al.* (1997) developed an efficient and economical low-pressure liquid chromatography process for recovery and purification of taxol from plant-tissue culture (PTC) broth. The PTC broth was diluted with ethanol to dissolve taxol and then passed through a column packed with a high-capacity polystyrene divinyl-benzene copolymer resin (Dow Chemical Co.). A stepwise increase in ethanol concentration in the mobile phase (ethanol/water) was used to concentrate and compress the taxoid bands to as high as 29-fold of influent concentrations (about 1 mg/mL). A recycling analysis was then applied to separate the concentrated taxol band from other taxoid bands, achieving higher purity (95%) with higher yield (>90%) than those of conventional solvent extraction methods. The use of ethanol/water as a mobile phase provided additional advantages of being cost effective and environmentally better compared to typical solvents in the solvent extraction. In this process, the same low-pressure columns were used to capture, concentrate, and purify taxol. Their theoretical predictions agreed with the step-wise elution and recycled chromatography data. After validation, simulations were used to explore various design and operating alternatives. Analysis of the alternatives showed that the solvent consumption, that is, process, cost could be further reduced by using higher feed concentration, larger loading volume, smaller particle size, and optimal gradient and recycle strategies.

Nair (1997) also developed a new inexpensive process for the isolation and purification of taxanes, particularly taxol, cephalomannine, baccatin III, and deacetylbaccatin III, from ornamental yew tissue. A specific solvent mixture of water and 50–95% methanol, ethanol, or acetone was used as an initial extraction solvent, and the resulting extract was treated with activated carbon. The taxanes were separated from the crude extract by a normal phase chromatographic step through vacuum and then medium pressure column chromatographic separation using silica gel as an adsorbent. Heating in a furnace above 500°C removed adsorbed organic materials and regenerated the silica gel.

Yang *et al.* (1998) developed an efficient preparative method to purify taxol from a crude chloroform extract of *T. yunnanensis* using industrial preparative liquid chromatography (IPLC) system on a polymeric stationary phase (D956 resin: macroporous methacrylate-divinylbenzene copolymer). With 4.8 kg of crude extract using a single-system apparatus through three chromatographic runs, this system yielded 49.7 g of taxol (purity >99%) within 155 h without any organic solvent waste. The taxol recovery was over 80%. In comparative studies, the D956 resin showed greater selectivity and capacity than silica gel and C_{18} phases. The IPLC system could be used repeatedly over 50 times without degradation and consumed only 3.6 L of acetone in the mobile phase for 1 g of pure taxol. This method was first to accomplish truly industrial scale production of taxol from a crude extract. Xue *et al.* (2000 and 2001) investigated a large scale separation of taxol from semi-purified extract of *T. yunnanensis* bark. One difficulty in purification of taxol is separation from the closely related analogues, 7-*epi*-10-deacetyltaxol (4) and cephalo-mannine. Although ozonolysis can be used to eliminate cephalomannine from taxol, the separation of taxol and 7-*epi*-10-deacetyltaxol is complicated by the fact that *T. yunnanensis* bark contains a much higher content of 7-*epi*-10-deacetyltaxol than in other *Taxus* species.

Accordingly, the authors studied the chromatographic behavior of these three taxoids on a reverse phase C_{18} bonded column with different mobile phases. The process was carried out in four steps. In the first step, the raw bark extract was purified by normal phase chromatographic system, which gave a semi-purified bark extract. In the second step, ozonolysis of cephaloman-nine in the extract was performed. As the third step, a second purification of the ozonized semi-purified bark extract was conducted using normal-phase liquid chromatography. The last step was the most important, purification by preparative reverse-phase HPLC using a C_{18} Nova-pak™ column in a Waters PrepLC™ 4000 preparative HPLC system, methanol/water (60 : 40) as mobile phase. In a 60 min run, 1.9 g of taxol was isolated from 2.0 g at a 99.5+ % level of purity; 95+ % taxol was obtained in a yield more than 98%. Park *et al.* (1999 and 2000) puri-fied taxol from yew tree extracts using solvent extraction and normal phase HPLC in isocratic mode. Impurities were removed in the pretreatment step with an open silica gel column. In sub-sequent analytical chromatography, solvent components, as well as flow rate and injection amounts, were varied in order to determine a correlation equation between optimum resolution (cephalomannine/taxol and taxol/10-deacetyltaxol) and the mobile phase components (cf. Park *et al.*, 1998). The analytical experimental conditions were extended to preparative liquid chro-matography. Zhang, Z.Q. *et al.* (2001) proposed a rate model for non-linear liquid chromatogra-phy under normal pressure from studies of the basic transport phenomena between solid phase and liquid phase. They applied this method to taxol separation on C_{18}-silica gel and confirmed that their mathematical model could be useful for scaling up chromatographic operation.

Separation and purification

Improved separation processes are needed to effectively remove taxoids from their natural sources. Pure taxoids are mainly separated and purified from extracts by using HPLC with reversed-phase columns. However, several additional chromatographic techniques have been applied to taxoid separation and purification, such as flash chromatography (Kingston *et al.*, 1992), thin-layer chromatography (TLC; Stasko *et al.*, 1989), counter-current chromatography, including high speed counter-current chromatography (HSCCC; Vanhaelen-Fastre *et al.*, 1992) and centrifuged counter-current chromatography (Kobayashi *et al.*, 1995), simulated moving-bed chromatography (SMB Chromatography; Wu, D.J. *et al.*, 1999), capillary ion electrophoresis (Issaq *et al.*, 1995) and capillary electrochromatography (Chen *et al.*, 2001).

Separation and purification by high performance liquid chromatography

Separation and purification procedures for taxol and related taxoids are complicated and trouble-some because taxol content in natural materials is so small and in addition, many endogenous compounds interfere or co-elute, especially cephalomannine and 7-*epi*-taxol. Therefore, in general, only small laboratory scale procedures have been available and selectivity, recovery, and yield of taxol have been low. The following efforts are being made to overcome the above obstacles.

Witherup *et al.* (1989) first reported that cyano and phenyl bonded silica gel columns oper-ated in the reverse-phase mode offer the needed resolution for HPLC analysis of taxol (1) and four related taxoids, cephalomannine (5), 10-deacetylcephalomannine (6), baccatin III (8), and 10-deacetyl baccatin III (9), in sample mixtures obtained from *T. brevifolia* bark. The authors examined both isocratic and gradient elution, and found both columns more suitable for taxoid analysis than C_{18} columns. Forgács (1994) used porous graphitized carbon (PGC) columns for taxol analysis from *T. baccata* bark and foliage. PGC has the advantage of being practically inert to acidic and alkaline eluents and, thus, it can be used at any pH. PGC also shows different retention characteristics, with both adsorption and reversed phase supports. The eluents for PGC generally are water and an organic modifier miscible with water, which are typical reversed-phase eluents. The taxol content in the bark and foliage was determined using a mixture of dioxane/water (46 : 54) as the mobile phase. Dioxane was chosen in order to elute the highly hydrophobic taxol from the column because methanol and acetonitrile did not achieve an adequate elution of this compound from the graphite surface. In addition, it was presumed that any taxol analogues co-eluting with taxol could be detected with peak purity tests. Németh-Kiss *et al.* (1996a,b) used PGC columns to determine taxol content of *Taxus* species in Hungary by HPLC. The method was reliable and could be used for the separation and quantitative determination of taxol in both the bark and foliage.

Solid phase extraction

Solid phase extraction is a powerful method for sample pre-treatment and is widely used to remove interfering substances in the matrix and to concentrate the analyte in order to increase the sensitivity of analysis. It is a more subtle and selective method than conventional solvent extraction, which requires several pre-treatment processes, including liquid–liquid extraction (LLE), is not very selective, and often requires hazardous solvents. Because many types of solid phase, that is, polar, non-polar, and ion exchange phases, are available, different interactions between the analyte and the solid phase are possible in SPE, whereas LLE is limited to only partition equilibrium in the liquid–liquid phase. In addition, SPE has advantages of time saving and low solvent consumption.

In most cases, two general separation procedures are possible: analytes can either (1) be absorbed on the SPE packing material or (2) directly flow through while the interfering substances are retained. In the first method, the sample is drawn through the solid phase, and the analyte molecules are enriched on the adsorbent. Most interfering components and solvent molecules are not held, and any retained interfering components must be washed from the adsorbent with a suitable washing solution. Finally, the desired analyte is removed from the adsorbent by elution with a suitable solvent. In general, polar compounds are easily adsorbed to a polar adsorbent from a non-polar environment and are eluted with a polar solvent. The opposite holds true for non-polar compounds. They are easily adsorbed from a polar environment onto non-polar surface. Elution is achieved by less polar solvents. In some cases, interfering components may remain on the adsorbent, which offers another possibility for the pre-purification of difficult matrices, such as waste oils or sludge. For if only the interfering components are retained, the solid phase can be used to simply filter the sample.

Mattina and co-workers (Mattina and Paiva, 1992; Elmer *et al.*, 1994; Mattina and MacEachern, 1994) proposed the use of SPE on C_{18} cartridges to extract taxol (1) from needles of 14 *Taxus* cultivars. The crude methanol extracts were pre-treated by C_{18} SPE, permitting collection of a fraction in which the taxoids eluted quantitatively. The procedure was successfully upscaled by using large SPE cartridges. Celite (Kopycki *et al.*, 1994) and silica (Lauren *et al.*, 1995) columns were also used for pretreating an extract of *T. baccata* foliage. Verpoorte's group (Theodoridis *et al.*, 1998a) studied SPE for sample pretreatment prior to HPLC analysis of *Taxus* cell suspension cultures. Various types of SPE materials were tested with taxoid standards and samples of various origin. Comparison between the different cartridges and the different elution solvents were discussed in terms of extraction recovery and sample cleanup. Zhang *et al.* (1998) reported that taxol, cephalomannine (5), and N-debenzoyl-N-phenyl acetyl taxol were successfully removed from a sample of *T. yunnanensis* by using an γ-Al$_2$O$_3$ SPE column. They reported distinct advantages with this method over liquid–liquid partitioning as the taxol-rich fractions did not contain any hydrophilic peaks in the HPLC, thus saving separation time and reducing solvent usage. The same authors (Zhang *et al.*, 2000a,b) subsequently reported an improved method that used a single alkaline Al$_2$O$_3$ chromatography for separating taxol from extracts of *T. cuspidata* callus cultures, and, at the same time, converts other taxoids to taxol through catalysis by the Al$_2$O$_3$ column. Under optimized operating conditions, the taxol yield was 170% of that in the crude extracts and taxol purity was raised to 29% from 0.6% in the initial callus extracts. Hata *et al.* (1999) reported that taxol was adsorbed on FSM-type mesoporous silicas with pore sizes larger than 1.8 nm, while it was not adsorbed with pore size less than 1.6 nm. The adsorption behavior of taxol was also solvent-dependent; taxol was not adsorbed into mesoporous silicas when methanol or acetone was used while it was adsorbed when dichloromethane or toluene was used. The adsorption–desorption of an extract from yew needles was performed using the mesoporous silica and several elution solvents. In the adsorption, many components including taxol were adsorbed into mesoporous silica. When the desorption of the extract components was performed with a mixture of methanol and water with varied ratios, taxol and a few other components were separated from the adsorbed mixture. Compared with the conventional method, the most noteworthy point of this process was that taxol was concentrated with a limited number of the other components from the extract. This finding remarkably simplified the subsequent column chromatographic separation and reduced the amount of solvent.

More recently, the same group (Zhang *et al.*, 2000c, 2001) reported a simplified and sensitive method for taxol separation and purification using reversed phase chromatography at normal pressure. They compared Sep-Pak C$_{18}$-silica gel (SPE), silica gel SPE, Al$_2$O$_3$ SPE and liquid–liquid partition for the initial separation of taxol from *T. yunnanensis* extract. Among the SPE methods, Al$_2$O$_3$ SPE gave the best results, and taxol content improved to 10% from 0.65%. In order to prepare taxol in large scale and reduce production expense, C$_{18}$-silica gel RP chromatography at normal pressure was used after the initial separation step. The final purity of taxol was more than 98% with a recovery of 85% without HPLC purification.

Separation of taxol and cephalomannine

Cephalomannine differs from taxol only at the 3′-N-alkyl group in C-13 ester side-chain. In the latter compound, this group is a benzoyl moiety and, in the former compound, it is a trigloyl group. Because of the structural similarity and because cephalomannine is present in most plant sources in about half the quantity of taxol, separating taxol (1) from cephalomannine (5) is an extremely difficult process, which has attracted much research attention.

Cardellina (1991) achieved an efficient separation of taxol and cephalomannine from *T. brevifolia* by normal phase HPLC on a cyano bonded phase column. Subsequently, Wickremesinhe and Arteca (1993) published an improved protocol for the extraction and purification of taxol and cephalomannine from plant tissue culture samples. Their protocol allowed large numbers of samples to be processed with a minimal use of solvents and could be completed in a relatively short period of time. The procedure involved a methanol extraction followed by the use of C$_{18}$ Sep-pak cartridges, before analysis by isocratic HPLC using a reverse-phase analytical phenyl column. Their purification protocol consisted of a 65% methanol/chloroform partitioning step followed by C$_{18}$ bonded silica column chromatography and finally, C$_{18}$ preparative HPLC. The same authors described similar analytical procedures and confirmation of HPLC peaks for taxol,

10-deacetyl-taxol, and cephalomannine by mass spectrometry (Wickremesinhe and Arteca, 1994 and 1996).

Jung *et al.* (1997, 1998) reported separation of taxol from a mixture of cephalomannine, taxol, and 10-deacetyltaxol by various mobile phase compositions using isocratic RP-HPLC. The most desirable resolution was obtained with a binary system of 60/40 vol% water/ acetonitrile and a ternary system of 53/39/5 vol% water/acetonitrile/methanol. At the above conditions, the separation times were 32 min and 40 min for the ternary and binary systems, respectively.

Chiou *et al.* (1997) used countercurrent chromatography (Model CCC-1000, Pharma Tech) to separate taxol from cephalomannine on a preparative scale. In their experiments, near baseline resolution of the two components was achieved and recovery of taxol was 90% with purity above 98%.

Ito *et al.* (Du *et al.*, 1998) applied recycling high-speed countercurrent chromatography for separation of taxol and cephalomannine from 50 mg of a mixture (taxol: 40%; cephalomannine: 51%). The two components were recycled twice, and yields were 25 mg of taxol and 19.2 mg of cephalomannine.

Park *et al.* (1998, 2000) applied isocratic normal-phase HPLC with mobile phases of hexane, 1-propanol, and methanol to separate taxol from cephalomannine and 10-deacetyltaxol. The resolutions of cephalomannine/taxol and 10-deacetyltaxol/taxol were correlated mathematically in terms of mobile phase composition. The best correlation equation contained linear and interaction terms of the mobile phase composition. With hexane/1-propanol/methanol (96 : 2 : 2) as the mobile phase, the calculated resolution agreed well with the experimental data.

As described above, various chromatographic methods, including HPLC, have been used to separate taxoids from mixtures; however, on a large-scale basis, the cost of taxol becomes very expensive. Thus, an alternative separation method that does not require rigorous and costly chromatography would be valuable. Kingston *et al.* (1992) reported that taxol can be separated in excellent yield from cephalomannine by oxidation of a mixture of these two compounds with osmium tetraoxide and flash chromatography of the resulting products over a column of silica gel. Under these reaction conditions, cephalomannine formed diol derivatives while taxol was unchanged. The reaction could also be run by a catalytic method, to reduce the cost and toxicity problems associated with osmium tetraoxide. This method was simple and efficient for the separation of taxol in pure form from mixtures of taxol and cephalomannine. The same group (Rimoldi *et al.*, 1996) also reported that treating a mixture of taxol and cephalomannine with bromine under mild conditions yields a readily separable mixture of taxol and 2′, 3′-dibromo cephalomannine. Treating the dibromo-derivative with zinc in acetic acid regenerated cephalomannine. A similar technique was applied by Chen and Luo (1997) to purify taxol from the stems of *T. chinensis*. The purity was 99.19% and the recovery rate from plant material was 70%.

High-speed countercurrent chromatography (HSCCC)

In the early stage studies on taxol (1) and taxoids, Craig countercurrent distributions were used (Wall and Wani, 1967). This procedure was extremely tedious and required extensive glassware cleanup. Now, more progressive and sophisticated instruments based on this technique allow for high-speed countercurrent chromatography (HSCCC). This new instrumental method could be adapted to the scale-up separation of natural products. Vanhaelen-Fastre *et al.* (1992) used an Ito multi-layer coil separator–extractor and applied 300 mg of pre-purified sample from the stem bark of *T. baccata* per run. They obtained one fraction containing 40% of the total amount of recovered taxol (97% pure form, 3% cephalomannine), together with fractions containing cephalomannine (90% pure form, 5), and a taxol and cephalomannine mixture.

Guanawardana *et al.* (1992) used high-speed planetary coil countercurrent chromatography (pccc) with *T. canadensis* extracts. They isolated 1.27% of taxol and a new taxoid, 9-dihydro-13-acetylbaccatin III, from needles and twigs. Ito's group reported semi-preparative separation and purification of taxol and analogs from the bark of *T. yunnannensis* (Cao *et al.*, 1998a) and 10-deacetylbaccatin-III (>98% purity) from the needles of *T. chinensis* (Cao *et al.*, 1998b) by HSCCC (Model GS10A2 multilayer coil planet centrifuge CCC). Centrifuged countercurrent chromatography (Model LLB-M, Sanki Laboratories) has also been used to separate taxuspines from *T. cuspidata* by Kobayashi *et al.* (1995, 1996; Hosoyama *et al.*, 1996).

Other methods

SIMULATED MOVING-BED CHROMATOGRAPHY

Simulated moving-bed (SMB) chromatography has been widely used for various industrial applications, for example, sugar and hydrocarbon purification. All industrial-scale SMBs are low-pressure systems and use large adsorbent particles (600–1000 μm in diameter). These low-pressure SMBs have proven to be economical for commodity chemicals. SMB chromatography has the same advantages as recycle chromatography, namely, high product purity, high yield, and low desorbent requirements. Wu D.J. *et al.* (1999) studied both design and optimization issues of low-pressure SMBs for the purification of taxol. Li *et al.* (2000) purified taxol from *T. cuspidata* using polymer resin column MN and SMB chromatography. A high purity product was obtained.

CAPILLARY ION ELECTROPHORESIS AND CAPILLARY ELECTROCHROMATOGRAPHY

Capillary electrophoresis (CE) is a rapid separation technique that operates on the basis of the different migration of charged and uncharged species in an electric field. CE can also be combined with other processes to provide high separation efficiency and selectivity with low solvent consumption and operation costs. Capillary ion electrophoresis (CIE) combines CE with an osmotic flow modifier and indirect UV detection and capillary electrochromatography (CEC) combines CE with HPLC. Nair *et al.* (1993) studied the application of CIE to bulk drug purification, including taxol, and recently, Chen *et al.* (2001) applied semipreparative scale CEC to separate a mixture of taxol and baccatin III.

Analysis of taxoids

HPLC analysis of taxoids, a survey

At present, taxoids are mainly analyzed by reversed-phase (RP) HPLC, as excellently reviewed in *Taxol Analysis by HPLC* (Theodoridis and Verpoorte, 1996). Only a few reports have discussed analysis of taxoids by normal-phase high performance liquid chromatography (NP HPLC). Analysis of taxoids in plant materials and biological fluids encounters the following difficulties. Firstly, taxol content is very low in *Taxus* plants; second, plant extracts contain various endogenous compounds, and third, the structures of each taxoid differ only slightly. Accordingly, HPLC analysis of taxoids often results in poorly resolved peaks and requires a long analysis time. Thus, other techniques including modified TLC, micellar electrokinetic chromatography (MEKC), capillary electrophoresis, biochemical procedures based on tubulin-formation, and immuno-assays such as enzyme linked immunosorbent assay (ELISA), radioimmuno-assay (RIA) and competitive inhibition enzyme immuno-assay (CIEIA) have also developed as reviewed by

Theodoridis and Verpoorte (1996), Jaziri *et al.* (1996), and Jaziri and Vanhaelen (2001). However, this section will focus on HPLC alone and combined with other techniques.

HPLC stationary phases, mobile phases, and detectors

Various stationary phases are used for RP-HPLC analysis of taxoids. Plant material is often analyzed on phenyl, diphenyl, and pentafluorophenyl (PFP) columns but C_{18}, C_8, and cyano phases are also used. In many cases, C_{18} columns are preferred for analysis of biological samples. Recently, new specialized "taxane" phases, such as Phenomenex Taxol (Curosil, RP on silica), Whatman TAC 1, Metachem Taxil, Supercosil LC-F® (all three are pentafluorophenyl on silica, RP), and Zorbax SW-Taxane (bonded to silica) have been developed for taxoid analysis. In addition, a Fluofix® (branched polyfluorinated alkyl, bonded to silica) RP column was used for determination of taxol (1) and five commonly known taxoids (Aboul-Enein and Serignese, 1996). A non-commercially available D956 resin, composed of macroporous methacrylate-divinyl-benzene copolymers, was also developed for preparative LC (Yang *et al.*, 1998).

Recently, Tian *et al.* (2001) developed four phenyl-silica gels with reaction on γ-aminopropyl-triethoxy silane (APTS) silica gel. The synthesized gels were used as the stationary phase of normal pressure chromatography to separate and purify taxol. The gel synthesized by gas-phase reaction gave the best results. The purity of taxol was 77.4% and the recovery was about 90%. Additionally, a porous graphitized column has been used for taxol analysis (Forgacs *et al.*, 1994). For taxoid analysis, the HPLC mobile phase usually consists of mixtures of methanol, acetonitrile, and water, or mostly, ammonium acetate buffer.

Analysis of taxoids requires sensitive and selective detection. Ultraviolet (UV) detection is the usual detection method in HPLC of taxoids; however, taxol and related compounds such as cephalomannine (5) and baccatin III (8) exhibit similar UV spectra with a minimum wavelength at 210–215 nm and two maxima at 225–232 nm and 270–280 nm in methanol. [taxol λmax nm (ε): 230 (29,000), 274 (1670), 282 (1175); cephalomannine λmax nm (ε): 221 (27,300), 229 (26,300), 275 (1250), 282 (1100); baccatin III λmax nm (ε): 230nm (13900), 274 (1000), 282 (850) (Sénilh *et al.*, 1984)]. Therefore, selectivity is rather low. Although most HPLC analysis is performed at 227 nm or 228 nm, wavelengths up to 230 nm are also used. Detection at lower wavelengths, for example, 210 nm (Schutzki *et al.*, 1994) and 200 nm (Glowniak *et al.*, 1996), is rare, because in this region, increased noise and decreased sensitivity and selectivity are observed. At higher wavelengths, for example, 254 nm (Harvey *et al.*, 1991) and 270 nm (Candellina II, 1991), the molecular coefficient is rather low, but dual or multiple detection uses both lower and upper UV regions, e.g. 227 and 273 nm (Kopycki *et al.*, 1994), 228 and 280 nm (Castor and Tyler, 1993), and 230 nm (for quantitation) and 280 nm (for LC traced) (Elmer *et al.*, 1994; Mattina and MacEachern, 1994). Finally, detection has been performed simultaneously at three wavelengths (227, 254, and 270 nm) (Vanhaelen-Fastre *et al.*, 1992).

HPLC combined with other techniques

HPLC coupled to mass spectrometry, that is, HPLC-MS and HPLC-MS/MS (Hoke *et al.*, 1992), has also been used to determine taxol fragmentation and identify taxoids. Various ionization techniques such as electron impact, fast atom bombardment (McClure *et al.*, 1992), thermospray (Auriola *et al.*, 1992), electrospray (Bitsch *et al.*, 1993), chemical ionization (Chmurny *et al.*, 1993), ion spray (Kerns *et al.*, 1994), and matrix-assisted laser desorption (Gimon *et al.*, 1994) have been used. Selective ion monitoring increases sensitivity and selectivity, and MS detection raises detection limits to the 100–200 pg level.

Coupling HPLC with NMR (HPLC-NMR) is one of the most powerful techniques for the separation and structural elucidation of unknown compounds in mixtures. Hostettmann *et al.* (Wolfender *et al.*, 2001) presented an excellent review on the application of LC-NMR in phytochemical analysis. The utility of HPLC-NMR for taxane analysis was exemplified by analysis of taxanes from three *Taxus* species (Schneider *et al.*, 1998).

Examples of taxoid analysis are described briefly below; refer to the review of Theodoridis and Verpoorte (1996) for more details in this field.

HPLC analyses of plant materials and cell cultures

There are numerous literature reports on analyses of plant materials and cell cultures (Baloglu and Kingston, 1999). Jaziri *et al.* (1996) published a comprehensive literature survey entitled *Analytical Methods for Taxoid Production in Taxus* sp. *Cultures.*

Preparation of cell culture extracts

Taxus cell cultures contain much lower quantities of pigments, waxy substances, and non-polar lipids than do needles or bark extracts. Thus, extraction of taxoids from callus or cell culture materials is simplified. Typically, the dried callus or cell sample is macerated in methanol, methylene chloride, or a methanol–methylene chloride (1 : 1) mixture. After solvent partition with an aqueous phase, the non-polar extract is dried and redissolved in an appropriate solvent or buffer prior to taxol (1) analysis.

HPLC analysis using C_{18} and C_8 columns

Vidensek *et al.* (1990) analyzed the taxol content in various parts of several *Taxus* species, including *T. brevifolia*, *T. baccata*, *T. cuspidata* and *T. media* Rehder, by HPLC using a C_{18} column and methanol–water (68 : 32) as a mobile phase. Harvey *et al.* (1991) reported high-resolution microcolumn HPLC on octadecyl silica for the separation of taxol in crude extracts of twig and needle foliage of *T. brevifolia*. In this technique, baseline resolution between taxol and cephalomannine were readily achieved using isocratic elution (acetonitrile–methanol–water = 4 : 3 : 3).

Kelsey and Vance (1992) measured taxol and cephalomannine (5) concentrations in the bark and foliage of *T. brevifolia* growing in the shade and at a site exposed to sunlight by HPLC on a RP-C_{18} column with a mobile phase of methanol–water (65 : 35). Mattina and Paiva (1992) analyzed taxol content in the needles of 14 cultivars of *Taxus* sp. by HPLC on a C_{18} column with acetonitrile–methanol–water (35.2 : 20 : 44.8). Schutzki *et al.* (1994) investigated the effect of storage conditions on the concentrations of taxol and cephalomannine in clippings of *T. x media* "Hicksii." The analysis was conducted by HPLC using a C_{18} column and acetonitrile–water (1 : 1) as a mobile phase.

Fang *et al.* (1993) studied qualitative and quantitative determination of taxol and three related taxanes, 10-deacetyltaxol (2), cephalomannine (5), and 10-deacetyl cephalomannine (6), in various parts of *T. cuspidata*. The four compounds were well separated on a RP-C_{18} column with a mobile phase composed of water–acetonitrile–tetrahydrofuran (55 : 35 : 10). Bitsch *et al.* (1993) analyzed five authentic taxoids, including taxol, and extracts from cell cultures derived from various yew tree species by microbore RP-18 column HPLC-electrospray MS, and obtained excellent spectra on 5–10 ng of separated compounds. In the HPLC-ESMS mode, only 10% of the eluent was mass-analyzed, 90% would be available for recovery through fraction collecting.

Glowniak *et al.* (1996) compared combinations of two preparation methods, SPE and preparative TLC, with HPLC for determination of taxol (1) and cephalomannine (5) in twigs and needles of various *Taxus* species. In both cases, the quantitative analysis of taxol (1) and cephalomannine (5) was carried out with RP-18 HPLC using 50% acetonitrile as a mobile phase in <10 min. While both preparation methods can be recommended for the screening analysis of taxol and cephalomannine in various *Taxus* organs, purification by SPE seemed superior to preparative TLC in the precision of the quantitative analysis.

Van Rozendaal *et al.* (1997) developed an analytical method for the large-scale qualitative and quantitative screening of six related taxoids, including taxol, in needles of different yew species and cultivars. The needles were extracted with ethanol–water–acetic acid (80 : 19 : 1) and the extract was purified from non-polar substances, including cinnamoyl taxanes that co-eluted with taxol, by small-scale partition chromatography on Extrelelut®. The taxoids were determined by HPLC on standard C_{18} column with an elution gradient of acetonitrile-water adjusted to pH 2.5 with orthophosphoric acid and ultraviolet detection at 227 and 280 nm. The method provided efficient extraction, high sensitivity, and reproducible data. Furthermore, C_{18} RP-HPLC is economic, operation is simple, and one sample requires only 20 min for analysis.

Wu and Zhu (1997) presented a protocol for analyzing taxol and five related taxoids from tissue culture samples of *Taxus* species by RP-HPLC on a C_{18} column. Sep-Pak C_{18} cartridges were used for semi-purification of the crude samples prior to analyses. Taxol (1), 10-deacetyltaxol (2), cephalomannine (5), 10-deacetyl-cephalomannine (6), baccatin-III (8), and 10-deacetyl baccatin-III (9) were well separated using a mobile phase of methanol–acetonitrile–water (25 : 35 : 45).

Adeline *et al.* (1997) reported extraction and analysis by HPLC using silica and C_{18} column of taxine A, isotaxine B, and 10-deacetylbaccatin III (both basic and neutral taxoids) from *Taxus* needles. Schneider *et al.* (1998) carried out reversed-phase (C_{18}) HPLC–NMR analyses of taxane diterpenoids from three *Taxus* species, *T. chinensis* var. *mairei*, *T. canadaensis*, and *T. media*, employing a stopped-flow technique. Several taxoids were identified from 500 mg needle samples without prior isolation. This technique allowed rapid identification of various types of taxoids in partially purified plant extracts by SPE (C_{18}). HPLC–NMR technique will undoubtedly be employed extensively not only in the phytochemical screening of taxoid-containing plant materials but also for other classes of natural products.

More recently, Theodoridis *et al.* (2001) first reported HPLC-PDA and HPLC-MS analysis of taxines and cinnamates from a semi-purified *T. baccata* needles extract. With RP-HPLC (C_{18} columns), more than 18 taxines and cinnamates were detected by photodiode array detection and LC/MS; 10 were positively identified. Furthermore, 10-deacetyl baccatin III and other taxanes were also found in the extract. Taxines were also detected in numerous plant and cell culture extracts and identified based on correlation of retention time and spectral data. LC-electrospray MS verified the identification of the known taxines in *T. baccata* seeds, needles, and pollen.

Because taxol is susceptible to hydrolysis and epimerization (formation of 7-*epi*-taxol as the major product) in animal culture media, its concentration decreases with time and analysis speed is crucial. Recently, Nguyen *et al.* (2001) developed a quick analysis of taxoids from large numbers of plant cell suspension samples. The method was optimized to analyze a range of taxoids of differing polarity, and allowed separation and identification of a standard mixture of taxol and 12 related taxoids within 15 min, using a Microsorb-MV C_8 column with a acetonitrile–water gradient. Test samples of suspension cultures of *T. cuspidata* cv *Densiformiis* were also analyzed by this method.

HPLC using phenyl columns

Phenyl columns have also been used for determination of taxanes in both bark and needle extracts. These columns are superior in separating naturally occurring taxanes from one another and are better for the separation of polar taxanes from endogenous compounds in extracts of *Taxus* biomass.

Castor and Tyler (1993) developed an analytical method using methanol extraction, methylene chloride partitioning, and silica mini-column clean up to remove interfering compounds. Then, taxol (1), 10-deacetyl taxol (2), 7-*epi*-10-deacetyltaxol (4), cephalomannine (5), and baccatin-III (8) in *T. x media* needles were separated on a phenyl column eluted with a mixture of acetonitrile–methanol–water (12 : 53 : 33). The influence of co-eluting compounds on taxol purity was quantified from absorbance data at 228 nm and 280 nm. Hoke *et al.* (1992, 1994) reported tandem MS with desorption chemical ionization used either as the main analysis method or combined with HPLC for determining taxol (1), cephalomannine (5), and baccatin III (8) in crude bark and needle extracts of *T. brevifolia*. A parent ion scan was used to determine the weight percentages by standard addition. In an alternative experiment, the concentration of taxol in the same samples was determined by MS/MS using trideuterated 10-deacetyltaxol as an internal standard. HPLC using a phenyl column, isocratic elution with methanol–acetonitrile–aqueous ammonium acetate buffer (pH 4.4) (20 : 32 : 48), and external standard was also used to determine the weight percentages of the three compounds in the same extracts. The HPLC method provided a good means of taxane analysis at high picomole to low nanomole levels but took 40 min per sample. The MS/MS method was the best method of analysis examined in this study. By MS/MS, taxol was quantitated in the extracts at low picomole level or better for all samples analyzed with an analysis time of less than 5 min per sample.

Liu *et al.* (1997) analyzed taxol, related taxoids, and yew tree bark extracts using an HPLC system with small-bore and micro-bore phenyl columns. Both diode array detector and ionspray mass spectrometry were incorporated into this system, providing additional spectral and structural information for identification of unknown samples. Three chromatographic columns (4.6, 2, and 1 mm i.d.) were evaluated using a standard mixture of taxol and three analogues. The experimental results correlated well with theoretical calculations with respect to the sensitivity enhancement. The combination of miniaturized HPLC with ionspray MS remarkably improved MS performance as compared to conventional LC/MS. Therefore, this miniaturized HPLC system provides fast analysis and sensitive detection for taxoids and other natural products.

HPLC using CN columns

As with phenyl columns, cyano columns are more appropriate than C_{18} columns for taxol analysis in bark extracts. The columns can often use isocratic elution.

Auriola *et al.* (1992) determined taxol in bark and needle of *T. cuspidata* by HPLC-thermospray-MS. After pretreatment of the sample by SPE (Bond-Elut C_{18}), isocratic HPLC of taxol on a Zorbax CN column was conducted. Eluted taxol was quantitated by selected-ion recording of the protonated molecule.

HPLC using other columns

Tailor-made columns are usually developed to solve special problems, and the chemical and physical properties of the packing materials are designed to have the appropriate characteristics for a specific analyte or a group of analytes. PRP columns are suitable for separation of taxol from closely related taxanes.

Richheimer *et al.* (1992) reported a RP HPLC method for assay of taxol from bulk drug or process samples. The PFP columns used are more selective than phenyl columns for taxol and related taxoids. However, a high carbon loaded diphenyl column was superior to the PFP method for separating polar taxoids such as baccatin III and 10-deacetylbaccatin III from other polar material occurring in methanol extracts of *Taxus*. These methods were found suitable for assaying both taxol bulk drug and taxol in crude *Taxus* extracts.

Ketchum and Gibson (1993) detected taxanes from crude extracts of *T. brevifolia* callus cells using two new specialty columns, Phenomenex Taxol™, and Metachem Taxsil™, developed specifically for taxane analysis. Using an isocratic system (acetonitrile–sodium acetate buffer), taxol (1), cephalomannine (5), and 7-*epi*-10-deacetyltaxol (4) were eluted with baseline separation. The isocratic HPLC method permitted rapid analysis of large numbers of crude extracts, generally showed excellent resolution of the taxanes, and enabled the detection of 7-*epi*-taxol (3) in 20 min. There were differences between the columns. The biggest problem with the Metachem Taxsil™ column was the difficulty in resolving baccatin III (8) and 7-xylosyl-10-deacetyltaxol. These two compounds co-eluted on the Metachem column, but were clearly resolved on the Phenomenex column.

Ketchum and Gibson (1995) later reported a novel method to isolate a mixture of taxanes from cell suspension cultures of *T. cuspidata*. The aqueous suspension medium was pre-filtered and centrifuged to remove cellular debris, and then passed through either nylon or PVDF membranes. Contaminants were washed from the membranes, while the taxanes were retained by the filters. Elution with methanol or ethanol then produced a relatively clean mixture of taxanes. This technique eliminates the need for a solvent partitioning step to extract the crude taxane mixture, and also avoids the use of chlorinated hydrocarbons that are typically used in these extractions. This method provides a rapid, efficient, and inexpensive means of extracting taxanes from cell suspension medium, as well as reducing the total volume of solvent used and the associated environmental pollution.

In this paper, a cell suspension medium of *T. cuspidata* was analyzed using a Phenomenex Taxol G (Curosil G) column eluted with acetonitrile–water (52.5 : 47.5). Eleven taxanes were separated within 30 min and seven were identified. Kerns *et al.* (1994) used HPLC/MS and HPLC/MS/MS to analyze taxanes in extracts from *T. brevifolia* and *T. baccata*. The instrumental components included HPLC (RP PFP or RP phenyl column), UV detector, ionspray MS, and tandem MS on-line. The data obtained for 18 taxanes from natural sources provided a taxane profile database useful for the rapid identification of taxanes in mixtures and samples of limited quantity.

Kopycki *et al.* (1994) reported more efficient separation of seven taxoids in *Taxus* plant extracts using Curosil G column and three gradient elution systems rather than pentafluorophenyl (Taxil) or diphenyl bonded silica (Supercosil LC-DP) columns. The method was linear for the taxoids within the tested concentration range of 0.1–2.4 µg injected. The purging and regeneration procedures allowed more than 600 injections to be made onto the same column without any back pressure problems. Later, these authors used the same method to analyze six taxoids in the needles of 57 *Taxus* cultivars (ElSohly *et al.*, 1995).

Theodoridis *et al.* (1998) determined taxol and related compounds in plant and cell culture extracts by RP HPLC in isocratic and gradient modes. They compared three general purpose columns (one phenyl and two C_{18}) and one specialized column (Zorbax SW Taxane). With a gradient system, 13 taxoids in a standard mixture were clearly separated on a phenyl column within 30 min, while in isocratic mode, 11 of the 13 taxoids were detected by LC-MS with an electrospray interface. The system was extended to plant extracts (*T. baccata* bark), and 11 taxoids of the standard mixture and two unknown taxoids were detected. The specialized taxane column also achieved base line resolution of 13 taxoids in isocratic and gradient modes; however, it took 55 min and had a short column lifetime.

Cass *et al.* (1999) determined taxane concentrations in clippings of *T. canadensis* using HPLC (using Supelcosil LC-F column) with n-octylbenzamide as an internal standard. The addition of the internal standard reduced the standard deviation of the calculated taxol content by one-third. The presence of taxol was confirmed by photodiode array spectra. Impurities co-eluted with other taxanes, and one compound was tentatively identified as 9-dihydro-13-acetylbaccatin III, the major taxane in *T. canadensis* (Zamir *et al.*, 1995), based on retention time and UV spectral data.

Analysis of taxol from fungi

Stierle *et al.* (1993, 1995) discovered a novel endophytic taxol-producing fungus, *Taxomyces andreanae*, in the inner bark of *T. brevifolia*. This report was not only the first isolation of a taxol-producing microorganism, but also the first example of an organism other than *Taxus* species to produce taxol. The fungal extract was purified with repeated silica gel column chromatography, prep. TLC (silica gel) and HPLC (silica gel). The presence of taxoids was confirmed by mass spectrometry, chromatographic behavior, monoclonal antibody analysis, and 9KB cytotoxicity studies. However, the amounts of taxoids produced by this fungus were very low. As estimated by two methods, electrospray MS and quantitative competitive inhibition enzyme immunoassay (CIEIA) technique, only 24–50 ng of taxol were produced per litre of culture broth. The same authors (Strobel *et al.*, 1996) isolated another endophytic fungus, *Pestalotiopsis microspora*, from the inner bark of a small limb of *T. wallichiana*. It was shown to produce taxol in mycelial culture at the level of 60–70 μg per litre. Pagé and Landry (1996) discovered a new taxol-producing bacteria, *Erwinia taxi*, that produces taxol and other taxanes. Noh *et al.* (1999) isolated a novel microorganism, *Pestalotia heterocornis*, that produces taxol. The fungus was isolated from soil collected in a yew forest and was shown to produce taxol in semi-synthetic liquid media. After incubation, culture fluid was extracted with methylene chloride, and the extract residue chromatographed twice on silica gel and by reversed-phase C_{18}-HPLC. The presence of taxol in the fungal extract was confirmed by FAB mass spectrometry and NMR spectroscopy. The maximum yield of taxol was 31 μg per liter. Thus, *P. heterocornis* is an excellent candidate for producing taxol by fermentation technology.

Analyses of drugs and biological and clinical samples

In general, analysis of biological and clinical samples is mainly aimed at the determination of taxol (1), docetaxel (7), and their metabolites in blood plasma, serum, bile, and urine. The complexity of these matrices and the interference of endogenous constituents are less than that in plant materials; therefore, sample pretreatment is simple and chromatographic high resolution is not essential. Various methods were applied to samples from clinical studies of taxol in cancer patients and are useful for pharmacokinetic studies of taxol.

Analysis of taxol in biological samples has been performed using HPLC with UV detection, HPLC with mass spectrometry, and immunoassay with lower detection limits of 10, 0.2, and 0.3 ng/mL, respectively. Although the HPLC-MS method gives greater sensitivity and higher specificity than HPLC-UV methods, its utility for routine taxol analysis is limited partly because this highly specialized instrumentation is not generally available. The immunoassay also provides high sensitivity but lacks the specificity of HPLC methods. For these reasons, taxol analysis is still most commonly performed using the HPLC-UV method, especially with C_{18} column phases. In order to decrease the interference from biological matrices such as plasma, most studies used SPE alone (Verginol *et al.*, 1992), SPE coupled with LLE (Longnecker *et al.*, 1987;

Rizzo *et al.*, 1990), SPE with protein precipitation (Grem *et al.*, 1987) or protein precipitation (Wiernik *et al.*, 1987). For more details of pretreatment of biological samples, refer to the review of Theodoridis and Verpoorte (1996).

Monsarrat *et al.* (1990) investigated taxol metabolism in rat bile and urine. The bile was extracted with ethyl acetate and the extract was evaporated to dryness. The residue was dissolved in a mixture of methanol–water (65 : 35) and purified by a semi-preparative RP C_{18} HPLC column with a mobile phase of methanol–water (65 : 35). The fractions containing metabolites were combined, methanol was evaporated, the remaining solution was lyophilized and the final residue was dissolved in methanol. HPLC analysis of metabolites was performed with an ODS column using a mobile phase of methanol–water (60 : 40). The metabolites were identified using both desorption chemical ionization and FAB MS and ^{1}H-NMR. As in humans, no taxol metabolites were detected by HPLC in rat urine. However, significant hepatic metabolism of taxol was seen with nine taxol metabolites detected by HPLC in the rat bile. The C13 side-chain had been removed in only one minor metabolite, baccatin III. The two hydroxylated metabolites obtained as major products were as active as taxol in preventing cold microtubule disassembly.

Vergniol *et al.* (1992) developed a rapid, selective, and reproducible semi-automated HPLC method for determining docetaxel in biological fluids. Plasma pretreatment involved a SPE step (C_2 micro-columns), followed by RP C_{18} column HPLC analysis with methanol-0.3% ortho-phosphoric acid (67.5 : 32.5) as a mobile phase. The validated quantitation range of this method was 10–2500 ng/mL in plasma with a coefficient of variation 11%. Burris *et al.* (1993) used a similar method to determine docetaxel in patient plasma and urine during a phase I clinical trial of the drug.

Willey *et al.* (1993) developed an isocratic HPLC method for quantitative determination of taxol in human plasma. The analysis required 0.5 mL of plasma, and detected taxol by UV absorbance at 227 nm, following extraction and concentration. Taxol in the plasma was extracted with ammonium acetate buffer (pH : 5.0) and the extract was purified with semi-automated SPE by Cyano Bond Elut columns. HPLC analyses of the samples were conducted on a RP C_8 column. This method allowed the taxol quantitation at concentrations of 10–1000 ng/mL in human plasma. The observed recovery for taxol was 83%. 7-*Epi*-taxol, a biologically active stereoisomer, and baccatin III, a degradation product, were also separated chromatographically from taxol by this assay. Its simplicity allows the analysis of at least 100 samples per day. In addition, semi-automation of the SPE procedure facilitates greater efficiency of the analysis.

Mase *et al.* (1994) developed a simpler and reproducible reversed phase HPLC method using an internal standard for the determination of taxol in biological fluids. The sample preparation involved only a SPE process on Sep-pak C_{18}, and HPLC was performed on a C_{18} column using acetonitrile-2 mM phosphoric acid (45 : 55) as a mobile phase with detection at 227 nm. The intra- and inter-day coefficients of validation were very low (7%). This method can be applied to human, dog, and rat plasma as well as urine.

Leslie *et al.* (1993) reported stability problems with taxol during an HPLC analysis in mouse plasma. The assay method included a simple protein precipitation step by centrifugation to prepare the samples for HPLC analysis using C_{18} column with acetonitrile/water (40 : 60) as a mobile phase. The taxol peak occurred at 8.88 min and cephalomannine at 7.36 min. The plasma samples obtained from mice were stored at $-20°C$ during the three-week period prior to analysis and no evidence of sample instability was observed during storage.

Song and Au (1995) analyzed taxol in human plasma, cell culture medium, and dog bladder tissue by isocratic HPLC with automated column switching. Biological fluids were extracted with ethyl acetate, and dog bladder tissue was homogenized in acetonitrile or ethyl acetate. Ethyl acetate was ultimately selected for the tissue extraction as its extract contained fewer endogenous substances than that with acetonitrile. The organic solvent extracts were centrifuged, the supernatants were evaporated to dryness, and the residue was subjected to HPLC analysis using a C_8-clean up column and a C_{18}-analytical column. Samples were injected onto the clean-up column and eluted with the clean-up mobile phase (37.5% aqueous acetonitrile). Concurrently, the analytical mobile phase (49% aqueous acetonitrile) was directed through the analytical column. A second sample could be loaded onto the clean-up column while the first sample was eluting from the analytical column, thus reducing the HPLC analysis time to about 15 min per sample. This method required LLE to recover the free and protein-bound taxol and avoids the often-used second extraction of taxol from biological matrices by off-line SPE and the use of gradient elution. The lower detection limit (5 ng/ml) and accuracy of this assay are better or equal to other HPLC-UV methods that require more extensive sample preparation. Thus, column-switching techniques for HPLC analysis of drugs in biological samples can be used for on-line sample clean up and sample pre-concentration, resulting in significant advantages such as increased separation power, reduced analysis time, and reduced cost. The authors also compared the extraction efficiency of LLE with ethyl acetate to that of protein precipitation with trichloroacetic acid followed by LLE with ethyl acetate. With the latter technique, taxol and cephalomannine co-precipitated with the plasma proteins. Because it gave high recovery in a one step process, direct LLE was the method of choice.

In a study of taxol metabolism and pharmacokinetics in human patients, Gianni *et al.* (1995) purified taxol from human plasma with Bond Elut C_{18} (SPE) and HPLC conducted with C_{18} column and isocratic elution monitored at 230 nm. Recovery of taxol was 95% and of cephalomannine was 90%. Forgács and Cserháti (1995) investigated hydrophobicity and hydophobic surface area of 21 commercial anticancer drugs, including taxol, on a C_{18} column using methanol/0.025 M potassium dihydrogen phosphate mobile phases with methanol concentrations varying from 0–90%. The aim of this study was to determine a correlation between the lipophilicity and specific surface area of a non-homologous series of anticancer drugs by RP HPLC for future quantitative structure–activity relationship studies in drug design.

Huizing *et al.* (1993, 1995a) developed a reversed-phase HPLC using an APEX C_8 column eluted with acetonitrile–methanol–0.02 M ammonium acetate buffer pH 5.0 for the quantitative determination of taxol in human urine and plasma. The authors also compared SPE and LLE as sample pretreatment. The SPE procedure involved extraction on Cyano Bond Elut columns and LLE used n-butyl chloride as an organic extraction fluid. The recoveries of taxol in human urine were 79% and 75% for SPE and LLE, respectively. The lower limit of quantitation is 0.01 μg/mL for SPE and 0.25 μg/mL for LLE. The same group also reported the HPLC analysis of taxol and three major metabolites in human plasma (Huizing *et al.*, 1995b) and in mouse plasma, tissue, urine, and feces (Sparreboom *et al.*, 1995).

Sparreboom *et al.* also reported a sensitive quantitative determination of docetaxel (Loos *et al.*, 1997) and taxol (Sparreboom *et al.*, 1998) in human plasma by similar procedures involving LLE (to avoid using SPE) and isocratic RP HPLC using C_{18} column.

Shao and Locke (1997) reported base line separation and quantitative determination of taxol and 14 related taxoids in bulk drug and injectable dosage forms. The analysis was achieved on reversed-phase HPLC using PFP columns and aqueous acetonitrile gradients as mobile phases. The elution order was related to molecular size, the number of acetylated hydroxyl groups, and the substitution of a xylosyl group at the 7-position. The methods developed for the injectable drug form allowed good resolution of taxoids from the excipient Cremophor EL, a polyethoxylated castor oil used with ethanol to solubilize taxol. These methods were fully validated excellent analytical methods for bulk drugs.

Huizing *et al.* (1998) observed that although RP HPLC methods with SPE as sample pretreatment procedure are frequently used to quantify taxol in human plasma, recovery problems often occur. The major problems were a large batch-to-batch difference in performance of the SPE (Cyano bond elut) columns and the effects of the pharmaceutical vehicle, such as Cremophor EL, on the performance of SPE. At concentrations exceeding 1.0%, Cremophor EL greatly influenced the recovery of taxol from human plasma using the SPE procedure. Also, recoveries decreased about 10–40% depending on the quality of the SPE columns. To avoid these problems, 2′-methyltaxol was used as an internal HPLC standard.

Volk *et al.* (1997) presented profiling and substructural analysis of taxol degradants by LC-MS and LC-MS/MS. This single instrumental approach integrated analytical HPLC (PFP RP column), UV detection, full scan electron spray MS, and MS/MS to rapidly and accurately elucidate structures of impurities and degradants of taxol. In these studies, degradants were profiled using molecular structures, chromatographic behavior, molecular mass, and MS/MS substructural information. Using this methodology, detailed structural information was obtained for potential degradants in less than one day. This library provides a foundation for future work involving the analysis of new taxol degradation products.

Sottani *et al.* (1997) studied mass separation and structural characterization of taxol metabolites directly in rat bile by high performance liquid chromatography/ion spray mass spectrometry (HPLC/ISP-MS) and HPLC/ISP tandem mass spectrometry (HPLC/ISP-MS/MS) without previous isolation. HPLC/ISP-MS yielded information on the molecular weights of several hydroxylated derivatives, while HPLC/ISP-MS/MS allowed the on-line structure characterization of all metabolites present in different ratios in rat bile. Using this method, nine metabolites, including three dihydroxy- and four monohydroxy-taxols, were detected, distinct from other endogenous contaminants. This study provided a highly sensitive analytical tool based on the selectivity of tandem mass spectrometry combined with chromatographic separation to detect new taxol derivatives directly in biological fluids in order to study the effect of metabolism on the clinical activity of taxoids.

The same authors (Sottani *et al.*, 1998) later presented a new analytical method that had some advantages over the prior assay. The newer assay was based on HPLC/ISP-MS/MS, and a fully automated SPE sample preparation method using a CN Sep-pak cartridge was essential. It had a lower quantification limit (5 ng/mL) than that of previous methods. The coefficient validation was <10%, and intra- and inter-day precision were within 12%. A run time of only 6 min made this assay fast and suitable for clinical pharmacokinetic studies requiring rapid analysis of many samples. The assay is highly specific, and endogenous plasma constituents and common concomitants of medications do not interfere or compromise the quantitative results. This assay is applicable to clinical pharmacokinetic studies of taxol in humans. Taxol is a cytotoxic agent; therefore, careless handling could lead to a potential exposure risk for personnel involved in preparation and/or administration duties. In this regard, several exposure routes must be taken into account, including inhalation of the aerosolised drug (e.g. during insertion and withdrawal of the needles to and from the vials) and skin absorption (e.g. during vial spillage or contamination with urine of patients under therapy or accidental ingestion). In order to assess exposure levels and identify situations with potential exposure to taxol, the compound must be measured in environmental matrices such as filters, pads, gloves, and wipe samples.

Recently, Sottani *et al.* (2000) developed a new method for the quantitative analysis of trace levels of taxol in environmental samples. The method is based on analysis by HPLC-MS/MS, and is suitably precise (CV 20%) and sensitive (lower limit of quantification, LLQ, 10 and 2 ng for wipe samples and air filter samples) to be used in cancer risk assessment studies. Aqueous extraction, a single LLE with ethyl acetate, and solvent exchange into an HPLC mobile phase were used as sample pretreatment.

The use of LLE is based on the observation by Huizing *et al.* (1998) that Cremophor EL, a taxol-formulation vehicle, has negative effects on the recovery of taxol during SPE. Accordingly, the authors chose HPLC-MS/MS to avoid the SPE process. This method seems to be suitable for monitoring trace levels of taxol in environmental matrices.

Martin *et al.* (1998) validated a rapid, isocratic RP HPLC assay for taxol in human plasma and urine. Taxol was extracted from these sources by LLE using diethyl ether (extraction efficiency: 90%). The assay was linear in the range 25–1000 ng and required a sample volume of 50 μL to 1 mL of plasma or urine. With UV detection at 227 nm, the quantification limits were 25 (plasma) and 40 ng/mL (urine). Its simplicity allowed the analysis of at least 30 samples/day. The method has been applied to a dose-dependence pharmacokinetic study in children involved in a Phase I trial.

Rouini *et al.* (1998) developed a rapid, simple, and sensitive isocratic HPLC method to measure the concentration of docetaxel (7, taxotere®) in plasma samples with UV detection at 227 nm. The method used a column switching technique with a C_{18} column as clean up column, and a C_8 column as the analytical column. A phosphate buffer–acetonitrile mixture was used as the mobile phase and taxol was used as an internal standard. The plasma samples were extracted using a SPE method and recovery was 92% for docetaxel. With this system, the retention times of docetaxel and taxol were 7.2 and 8.5 min, respectively. The detection limit of docetaxel in plasma was 2.5 ng per mL. In conclusion, this method is a fast simple, reproducible assay for docetaxel and is applicable in clinical and pharmcokinetic studies.

Coudoré *et al.* (1999) published a simple isocratic RP HPLC assay for taxol in rat serum using a rapid single step LLE. The HPLC used a C_{18} column and a methanol–water (70 : 30) mobile phase following a single-step extraction from the serum with dichloromethane. This method required short analysis time per sample with a detection limit acceptable for toxicokinetic studies.

Garg and Ackland (2000) developed a new HPLC method for the determination of docetaxel in human plasma or urine. The method was sensitive and simpler than previous methods, and applicable to use in clinical pharmacokinetic analysis. Plasma samples were extracted with SPE method using a cyanopropyl end-capped column followed by isocratic RP HPLC with UV detection at 227 nm. Using this method, the retention times for docetaxel and taxol (internal standard) were 8.5 min and 10.5 min, respectively, with good resolution and without any interference from endogenous plasma constituents or docetaxel metabolites. The total run time was only 13 min. The LLQ was 5 ng/mL using 1 mL of plasma. This method is also suitable for determining docetaxel in urine samples under the same conditions.

Fraier *et al.* (1998) developed a sensitive and selective HPLC assay for the determination of taxol, 7-*epi*-taxol, and PNU 166945, a new polymer bound taxol derivative, in dog plasma and urine. The method involves SPE of taxol and 7-*epi*-taxol, a possible degradation product of taxol, from plasma and urine on cyanopropyl columns. HPLC was performed on a RP-octyl column with UV detection at 229 nm. Determination of total taxol (free plus polymer bound) was achieved after release of taxol from the polymeric carrier by chemical hydrolysis. Concentration of PNU 166945 was determined by subtraction of free taxol from total taxol. The limit of quantitation of this method was 5 ng/mL for taxol and 7-*epi*-taxol and 20 ng/mL for PNU 166945 (as taxol equivalent). The method was validated in mouse and rat plasma, and could also be used for the same compounds in human plasma and urine.

Lee *et al.* (1999) reported a rapid, simple, and sensitive isocratic HPLC with diode array UV detection for micro-sample analysis of taxol in mouse plasma. The analysis used a Capcell-pak octadecyl analytical column and acetonitrile −0.1% phosphoric acid in deionized water (55 : 45) as a mobile phase. Taxol and *n*-hexyl *p*-hydroxybenzoic acid (internal standard) were extracted from plasma by one-step extraction with *tert*.-butyl methyl ether. Taxol and the internal standard were eluted at 3.4 min and 5.4 min, respectively. No interfering peaks were observed and the total run time was 10 min. The extraction recovery was >90% for both taxol and the internal standard. The LOD (limit of detection) and LOQ were 5 and 10 ng/mL, respectively, for taxol using a 100 μL plasma sample volume. This assay method provided excellent extraction recovery, sensitivity, accuracy, and precision, with short run time. This method was applied to a taxol pharmacokinetic study in mice where limited volumes were available.

Malingré *et al.* (2001) reported HPLC assays of taxol and docetaxel, which were based on the methods of Wiley *et al.* (1993) and Huizing *et al.* (1995). SPE with Cyano Bond Elut columns were used for sample pretreatment, 2′-methyl-taxol was the internal standard, and HPLC on a RP-C$_8$ column with acetonitrile–methanol–water containing 0.01 M ammonium acetate pH 5.0 (4 : 1 : 5) as the mobile phase and detection by UV at 227 nm was the analytical method. Their validated analytical assays showed very good performance over a prolonged period of time in a routine hospital laboratory setting.

Other methods

Micellar electrokinetic chromatography/capillary zone electrophoresis

Irregardless of the many reports of HPLC analysis of taxoids, baseline separation of taxol (1), cephalomannine (5), and baccatin III (8) from needle extracts is complicated by co-elution of interfering compounds in the extracts, especially in a single chromatographic run (Chmurny *et al.*, 1993). Micellar Electrokinetic Chromatography (MEKC) can resolve mixtures of neutral and charged molecules, and is quite efficient at separating a standard mixture of taxol, cephalomannine, baccatin III, and their deacetyl derivatives. The method is simpler, has less solvent waste compared to HPLC, and requires a smaller sample. MEKC is also applicable to the determination of taxol in plant tissue culture, the resolution of products from semi-organic synthesis, and the study of taxol metabolites in biological fluids. Capillary zone electrophoresis (CZE) is also a powerful microanalytical method because it can quickly resolve both large biomolecules as well as small ions and requires only a few nL of sample and a few μL of electrolyte. Compared to HPLC, capillary electrophoresis is simpler and needs less organic solvent.

Chan *et al.* (1994) applied MEKC (with capillary electrophoresis units and photodiode array detection) for the separation of taxol (1), cephalomannine (5), and baccatin III (8) from the crude extracts of needles and bark of *Taxus* species. According to their data, MEKC gave a baseline separation of taxol from other co-eluting compounds and analogs, which were difficult to separate by HPLC. Thus, this method may be a more suitable technique than HPLC for the analysis of taxol and related taxanes. The same group (Issaq *et al.*, 1995) used both CZE and MEKC in cancer research. Shao and Locke (1998) separated taxol and 14 related taxanes used this method within 11.5 min. An aqueous acetonitrile buffer containing sodium dodecyl sulfate (SDS) surfactant allowed resolution of the 15 taxoids from each other and from the principal matrix ingredient in the injectable dosage form of the drug, Cremophor EL (polyethoxylated castor oil). Hempel *et al.* (1996) developed a simple and sensitive method for the determination of taxol in plasma and urine using MEKC with SDS as additive in the run buffer. The samples were extracted and pre-concentrated with *tert.*-butyl methyl ether. The limit of detection for taxol was 20 ng/mL for plasma and 125 ng/mL for urine samples.

Biochemical and immunological methods

Rapid, sensitive, and accurate screening methods are need to analyze large numbers of samples from *Taxus* cultivars and to optimize taxoid production by *Taxus* tissue cultures and microorganisms.

However, although HPLC methods are the most frequently used techniques for the determination of taxoids, some are time consuming while others are relatively insensitive. Therefore, simpler and more sensitive methods, such as biochemical and immunological assays have been developed. Among them, an assay based on tubulin-formation (Hamel *et al.*, 1982) was highly sensitive for taxol but could not detect inactive analogs such as baccatin III (8) and 10-deacetyl-baccatin III (9). Immunological methods such as enzyme linked immunosorbent assay (ELISA, Jaziri *et al.*, 1991), radioimmunoassay (RIA, Leu *et al.*, 1993), competitive inhibition enzyme immunoassay (CIEIA, Grothaus *et al.*, 1993), and fluorescence polarization immunoassay (FPIA, Bicamumpaka and Pagé, 1998) were also developed and used for the determination of taxanes in human plasma and tissue of *Taxus* species. However, these methods have a serious disadvantage: a cross-reaction of taxol (1) and cephalomannine (5). Jaziri and Vanhaelen (2001) described recent advances in taxoid analysis by immunological and other biological methods in their review.

CIEIA has been developed for quantifying taxanes in extracts of *T. brevifolia* tissue and in human plasma. The method can detect taxol (1) and cephalomannine (5) at concentrations as low as 0.3 ng/mL, but does not detect baccatin III (8) and 10-deacetyl-baccatin III (9) at concentrations below 1000 ng/mL (Grothaus *et al.*, 1993).

Guo *et al.* (1995, 1996) reported a combination of chromatographic procedures and ELISA in the search for new taxoids in *T. baccata* stem bark. They found that the polyclonal antibodies-based immunoassay was a powerful tool for the detection of new taxoids. Russin *et al.* (1995) used immunocytochemical techniques with polyclonal antitaxol antiserum to localize taxol in the stem of *T. cuspidata*.

Jaziri *et al.* (1997) developed a new detection method for novel bioactive taxoids from natural sources. The method used two complementary biological tools: antibodies (anti-taxol and anti-10-deacetylbaccatin III) and naturally occurring receptor(s) for taxoids (tubulin–microtubule complex). The combination of the immunoassay (ELISA) with microtubule bioassay was successfully applied to bioassay-guided fractionation of crude stem bark extracts from *T. baccata*. Several known and new taxoids were detected and isolated. This screening assay could also be effective for the discovery of new microtubule-stabilizing agents based on molecular features of taxol. Both assays, when associated with chromatographic methods, are powerful tools for the screening of alternative sources of bioactive taxoids, and new compounds showing the same biological activities as taxol.

Svojanovsky *et al.* (1999) developed a new, sensitive, and highly specific indirect competitive ELISA for determination of low taxol concentrations in physiological fluids. The working range of the assay is pg/mL to 100 ng/mL. Taxol was determined in plasma, plasma ultrafiltrate, and saliva, thus, demonstrating the applicability of the newly developed ELISA method to pharmacokinetic studies for free, biologically active taxol species in cancer patients.

Guillemard *et al.* (1999a) developed a rapid and sensitive RIA for the measurement of taxol and related taxanes in biological specimens using commercially available reagents. Because taxol competitively inhibits the binding of radioactive taxol to polyclonal anti-taxane antibody or anti-taxane monoclonal antibody, this method using polyclonal antibody can detect taxol in concentrations as low as 0.87 nM. This RIA is useful for the measurement of taxol and taxanes in biological specimens, including culture medium of taxol producing microorganisms, and may be useful for the measurement of taxol and related taxanes in other potential natural sources.

The same group (Guillemard *et al.*, 1999b) also developed a rapid and very sensitive enzyme immunoassay to measure taxol and related taxanes in crude extracts of *Taxus* sp., in human serum, and in culture medium of taxol-producing microorganisms. For the ELISA, taxol was chemically modified by the introduction of an amine capable of coupling with biotin.

The presence of taxol or related taxanes competitively inhibited the binding of taxol-biotin to anti-taxane monoclonal antibody. This assay detected taxol in concentrations as low as 33 pM, and the affinity of the antibody was higher for taxol than for cephalomannine, baccatin III, and 10-deacetyl baccatin III. The sensitivity of this assay makes it useful for estimating the taxol and taxane content of *Taxus* sp. extracts and for monitoring the levels of taxol in the serum and other biological fluids of treated patients.

Shrestha *et al.* (2001) used an immunoassay with monoclonal antibodies, TLC, and HPLC-ESI-MS/MS to confirm taxol production from three endophytic fungi, *Sporormia minima*, *Trichothecium* sp., and an unidentified fungus, which were obtained from the stem of *T. wallichiana* collected in Nepal Himalaya. These fungi were new taxol producers and were shown to produce taxol in a culture medium.

Bicamumpaka and Pagé (1998) developed a fluorescence polarization immunoassay (FPIA) for the quantitative determination of taxol in serum, crude *Taxus* extracts and culture medium of *Erwinia taxi*, a taxol producing microorganism. The assay was performed using antitaxol monoclonal antibody and FITC-labeled taxol, which was formed by introducing an amino function into taxol enabled coupling with fluorescein isothiocyanate. Taxol competitively inhibited the binding of the monoclonal antibody with FITC-labeled taxol causing decreased polarization. This method detected taxol at a concentration as low as 2 nM and was quite simple. Therefore, FPIA is useful for estimating the taxol content in *Taxus* crude extracts and culture medium of taxol producing microorganisms and as an assay for monitoring taxol levels in treated patients.

Thin layer chromatography

In many reports, thin layer chromatography (TLC) is still used in conjunction with other chromatographic methods for the detection and separation of taxol and taxoids. Several reports using TLC in taxoid analysis have been published recently and are described below.

Stasko *et al.* (1989) reported the separation of taxol (1) and cephalomannine (5) from other compounds found in the bark of *T. brevifolia* by two multimodal two-dimensional TLC methods. Both cyano modified and diphenyl modified silica gel plates were used. The first dimension was developed in an organic mobile phase and the second in an aqueous organic mobile phase. In both procedures, taxol was resolved from cephalomannine and at least 20 additional compounds. However, although this two-dimensional TLC system did separate taxol from other taxanes, no quantitative results could be obtained.

Matysik *et al.* (1995) used a TLC technique to determine the taxol content of extracts from the needles of three *Taxus* species, *T. baccata*, *T. cuspidata*, and *T. media*. The development on thin layers of silica Si 60 included two stages: first stage; 60% heptane solution of a mixture of 5% methanol and 95% chloroform; second stage; 70% heptane solution of the same solution of methanol in chloroform. Taxol was satisfactorily separated from the neighboring peaks to permit its densitometric determination.

Srinivasan *et al.* (1997) combined six-well plate technology and TLC (silica gel/CHCl$_3$-CH$_3$CN) for rapid taxane analysis to establish conditions for an optimal *Taxus* (*T. chinensis* and *T. baccata*) cell suspension culture processes for taxane production.

Glowniak *et al.* (1999) applied zonal TLC to determine contents of taxol and 10-deacetyl baccatin III (9) in some *Taxus* species. The method included RP SPE for purification of crude methanolic extracts of different yew species, followed by zonal TLC of the purified extracts on silica gel with a multicomponent mobile phase. Taxoid fractions were separated from the polar compounds and partially separated fractions containing taxol and 10-deacetyl baccatin III were obtained. RP HPLC of these isolated fractions then quantified the two taxoids in different yew species and varieties.

The same group (Glowaniak and Mroczek, 1999) investigated many mobile phases including solvent systems and gradient elutions to select the best TLC conditions (silica gel F_{254} plates) for micro-preparative isolation of 10-deacetylbaccatin III, baccatin III, taxol, and cephalomannine from methanolic extracts of fresh and dried needles and stems of *T. baccata*. The best mobile phase, benzene–chloroform–acetone–methanol (20 : 92.5 : 15 : 7.5), was used to isolate the analyzed taxoids by preparative (Prep.) TLC.

Acknowledgments

The authors are grateful to Dr Y. Kikuchi who kindly provided us with many reports of his work on taxol. We also thank many researchers who provided reprints of their work.

References

Aboul-Enein, H.Y. and Serignese, V. (1996) Liquid chromatographic determination of taxol and related derivatives using a new polyfluorinated reversed-phase column. *Anal. Chim. Acta*, 319, 187–190.

Adeline, M.-T., Wang, X.-P., Poupat, C., Ahond, A. and Potier, P. (1997) Evaluation of taxoids from *Taxus* sp. crude extracts by high performance liquid chromatography. *J. Liq. Chromatogr. & Rel. Technol.*, 20, 3135–3145.

Auriola, S.O.K., Lepistö, A.-M. and Naaranlahti, T. (1992) Determination of taxol by high performance liquid chromatography-thermospray mass spectrometry. *J. Chromatogr.*, 594, 153–158.

Baloglu, E. and Kingston, D.G.I. (1999) The taxane diterpenoids. *J. Nat. Prod.*, 62, 1448–1472.

Berger, T.A. (1997) Separation of polar solutes by packed column supercritical fluid chromatography. *J. Chromatogr. A*, 785, 3–33.

Bicamumpaka, C. and Pagé, M. (1998) Development of a fluorescence polarization immunoassay (FPIA) for the quantitative determination of paclitaxel. *J. Immunol. Methods*, 212, 1–7.

Bitsch, F., Ma, W., Macdonald, F., Nieder, M. and Shackleton, C.H. (1993) Analysis of taxol and related diterpenoids from cell cultures by liquid chromatography-electrospray mass spectrometry. *J. Chromatogr.*, 615, 273–280.

Burris, H., Irvin, R., Kuhn, J., Kalter, S., Smith, L., Schaffer, D., Fields, S., Weiss, G., Eckardt, J., Rodriguez, G., Rinaldi, D., Wall, J., Cook, G., Smith, S., Vreeland, F., Bayssas, M., LeBail, N. and Von Hoff, D. (1993) Phase I clinical trial of taxotere administered as either a 2-hour or 6-hour intravenous infusion. *J. Clin. Oncol.*, 11, 950–958.

Cao, X., Tian, Y., Zhang, T.Y. and Ito, Y. (1998a) Semi-preparative separation and purification of taxol analogs by high-speed countercurrent chromatography. *Prep. Biochem. & Biotechnol.*, 28, 79–87.

Cao, X., Tian, Y., Zhang, T.Y. and Ito, Y. (1998b) Separation and purification of 10-deacetylbaccatin III by high-speed countercurrent chromatography. *J. Chromatogr. A*, 813, 397–401.

Cardellina II, J.H. (1991) HPLC separation of taxol and cephalomannine. *J. Liq. Chromatogr.*, 14, 659–665.

Cass, B.J., Scott, D.S. and Legge, R.L. (1999) Determination of taxane concentrations in *Taxus canadensis* clippings using high performance liquid chromatographic analysis with an internal standard. *Phytochemical Analysis*, 10, 88–92.

Cass, B.J., Piskorz, J., Scott, D.S. and Legge, R.L. (2001) Challenges in the isolation of taxanes from *Taxus canadensis* by fast pyrolysis. *J. Anal. Appl. Pyrolysis*, 57, 275–285.

Castor, T.P. and Tyler, T.A. (1993) Determination of taxol in *Taxus media* needles in the presence of interfering components. *J. Liq. Chromatogr.*, 16, 723–731.

Chan, K.C., Alvadaro, A.B., McGuire, M.T., Muschik, G.M., Issaq, H.J. and Snader, K.M. (1994) High-performance liquid chromatography and micellar electrokinetic chromatography of taxol and related taxanes from bark and needle extracts of *Taxus* species. *J. Chromatogr. B*, 657, 301–306.

Chen, C. and Luo, S. (1997) New isolation technology of taxol from *Taxus chinensis*. *Zongguo Yoyao Gongye Zazhi*, 28, 344–347.

Chen, J.-R., Zare, R.N., Peters, E.C., Svec, F. and Frechét, J.J. (2001) Semipreparative capillary electrochromatography. *Anal. Chem.*, 73, 1987–1992.

Chiou, F.-Y., Kan, P., Chu, I.-M. and Lee, C.-J. (1997) Separation of taxol and cephalomannine by countercurrent chromatography. *J. Liq. Chrom. & Rel. Technol.*, 20, 57–61.

Chmurny, G.N., Paukstelis, J.V., Alvarado, A.B., McGuire, M.T., Snader, K.M., Muschik, G.M. and Hilton, B.D. (1993) NMR structure determination and intramolecular transesterification of four diacetyl taxinines which co-elute with taxol obtained from *Taxus* x. *media* Hicksii. *Phytochemistry*, 34, 477–483.

Chun, M.K., Shin, H.W. and Lee, H. (1996) Supercritical-fluid extraction of paclitaxol and baccatin-III from needles of *Taxus cuspidata. J. Supercrit. Fluids*, 9, 192–298.

Chun, M.K., Shin, H.W., Lee, H. and Liu J.R. (1994). Supercritical-fluid extraction of taxol and baccatin-III from needles of *Taxus cuspidata. Biotechnol. Tech.* 8, 547–550 (1994).

Coudoré, F., Authier, N., Guillaume, D., Béal, A., Duroux, E. and Fialip, J. (1999) High-performance liquid chromatographic determination of paclitaxel in rat serum: application to a toxicokinetic study. *J. Chromatogr. B*, 721, 317–320.

Du, Q.-Z., Ke, C.-Q. and Ito, Y. (1998) Recycling high-speed countercurrent chromatography for separation of taxol and cephalomannine. *J. Liq. Chrom. & Rel. Technol.*, 21, 157–162.

Elmer, W.H., Mattina, M.J.I. and MacEachern, G.J. (1994) Sensitivity of plant pathogenic fungi to taxane extracts from ornamental yew. *Phytopathology*, 84, 1179–1185.

ElSohly, H.N., Croom, Jr., E.M., Kopycki, W.J., Joshi, A.S., ElSohly, M.A. and McChesney, J.D. (1995) Concentrations of taxol and related taxanes in the needles of different *Taxus* cultivars. *Phytochem. Anal.*, 6, 149–156.

Fang, W.-S., Wu, Y., Zhou, J.-Y., Chen, W.-M. and Fang, Q.-C. (1993) Qualitative and quantitative determination of taxol and related taxanes in *Taxus cuspidata* Sieb. et Zucc. *Phytochem. Anal.*, 4, 115–119.

Fett Neto, A.G. and DiCosmo, F. (1992) Distribution and amounts of taxol in different shoot parts of *Taxus cuspidata. Planta Med.*, 58, 464–466.

Forgács, E. (1994) Use of porous graphitized carbon column for determining taxol in *Taxus baccata. Chromatographia*, 39, 740–742.

Forgács, E. and Cserháti, T. (1995) Determination of hydrophobicity of non-homologous series of anticancer drugs by reversed-phase high-performance liquid chromatography. *J. Chromatogr. B*, 664, 277–285.

Fraier, D., Cenacchi, V. and Frigerio, E. (1998) Determination of a new polymer-bound paclitaxel derivative (PNU 166945), free paclitaxel and 7-epipaclitaxel in dog plasma and urine by reversed-phase high-performance liquid chromatography with UV detection. *J. Chromatogr., A*, 797, 295–303.

Fuji, K., Tanaka, K., Li, B., Shingu, T., Sun, H. and Taga, T. (1993) Novel diterpenoids from *Taxus chinensis. J. Nat. Prod.*, 56, 1520–1531.

Garg, M.B. and Ackland, S.P. (2000) Simple and sensitive high-performance liquid chromatography method for the determination of docetaxel in human plasma or urine. *J. Chromatogr. B*, 748, 383–388.

Gianni, L., Kearns, C.M., Giani, A., Capri, G., Viganó, L., Locatelli, A., Bonadonna, G. and Egorin, M.J. (1995) Nonlinear pharmacokinetics and metabolism of paclitaxel and its pharmacokinetic/pharmacodynamic relationships in humans. *J. Clin. Oncol.*, 13, 180–190.

Gimon, M.E., Kinsel, G.R., Edmondson, R.D., Russel, D.H., Prout, T.R. and Ewald, H.A. (1994) Matrix-assisted laser desorption/ionization time-of-flight mass spectrometry of paclitaxel and related taxanes. *J. Nat. Prod.*, 57, 1404–1410.

Glowniak, K., Zgórka, G., Józefczyk, A. and Furmanowa, M. (1996) Sample preparation for taxol and cephalomannine determination in various organ of *Taxus* sp. *J. Pharm. & Biomed. Anal.*, 14, 1215–1220.

Glowniak, K., Wawrzynowicz, T., Hajnos, M. and Mroczek, T. (1999a) The application of zonal thin-layer chromatography to the determination of paclitaxel and 10-deacetyl-baccatin III in some *Taxus* species. *JPC-J. Planar Chromatogr.-Mod. TLC*, 12, 328–335.

Glowniak, K. and Mroczek, T. (1999b) Investigation on preparative thin-layer chromatographic separation of Taxoids from *Taxus baccata* L. *J. Liq. Chrom. & Rel. Technol.*, 22, 2483–2502.

Grem, J. L., Tutsch, K.D., Simon, K.J., Alberti, D.G., Wilson, J.K.V., Tormey, Swaminathan, and Trump, D. (1987) Phase I study of taxol administered as a short i.v. infusion daily for 5 days. *Cancer Treat. Rep.*, 71, 1179–1184.

Grothaus, P.G., Raybould, T.J.G., Bignami, G.S., Lazo, C.B. and Byrnes, J.B. (1993) An enzyme immunoassay for the determination of taxol and taxanes in *Taxus* sp. tissues and human plasma. *J. Immunol. Methods*, 158, 5–15.

Guillemard, V., Landry, N. and Pagé, M. (1999a) A radioimmunoassay for the measurement of paclitaxel and taxanes in biological specimens using commercially available reagents. *Oncol. Rep.*, 6, 1257–1259.

Guillemard, V., Bicamumpaka, C., Boucher, N. and Pagé, M. (1999b) Development of a very sensitive luminescence assay for the measurement of paclitaxel and related taxanes. *Anticancer Res.*, 19, 5127–5130.

Guanawardana, G.P., Premachandran, U., Burres, N.S., Whittern, D.N., Henry, R., Spanton, S. and McAlpine, B. (1992) Isolation of 9-dihydro-13-acetylbaccatin III from *Taxus canadensis. J. Nat. Prod.*, 55, 1686–1689.

Guo, Y., VanHaelen-Fastré, R., Diallo, B., Vanhaelen, M., Jaziri, M., Homes, J. and Ottinger, R. (1995) Immunoenzymatic methods applied to the search for bioactive taxoids from *Taxus baccata. J. Nat. Prod.*, 58, 1015–1023.

Guo, Y., Diallo, B., Jaziri, M., VanHaelen-Fastré, R. and Vanhaelen, M. (1996) Immunological detection and isolation of a new taxoids from the stem bark of *Taxus baccata. J. Nat. Prod.*, 59, 169–172.

Hamel, E., Lin, .M. and Johns, D.G. (1982) Tublin-dependent biochemical assay for the antineoplastic agent taxol and application to measurement of the drug in serum. *Cancer Treat. Rep.*, 66, 1381–1386.

Harvey, S.D., Campbell, J.A., Kelsey, R.G. and Vance, N.C. (1991) Separation of taxol from related taxanes in *Taxus brevifolia* extracts by isocratic elution reversed-phase microcolumn high-performance liquid chromatography. *J. Chromatogr.*, 587, 300–305.

Hata, H., Saeki, S., Kimura, T., Sugahara, Y. and Kuroda, K. (1999) Adsorption of taxol into ordered mesoporous silicas with various pore diameters. *Chem. Mater.*, 11, 1110–1119.

Heaton, D.M., Bartle, K.D., Rayner, C.M. and Clifford, A.A. (1993) Application of super-critical fluid chromatography to the production of taxanes as anti-cancer drugs. *HRCJ. High Resol. Chromatogr. Commun.*, 16, 666–670.

Hempel, G., Lehmkuhl, D., Krümpelmann, S., Blaschke, G. and Boos, J.(1996) Determination of paclitaxel in biological fluids by micellar electrokinetic chromatography.*J. Chromatogr.*, 745, 173–179.

Hoke II, S.H., Wood, J.M., Cooks, R.G., Li, X.-H. and Chang, C.-J. (1992) Rapid screening for taxanes by tandem mass spectrometry. *Anal. Chem.*, 64, 2313–2315.

Hoke II, S.H., Cooks, R.G., Chang, C-J., Kelly, R.C., Qualls, S.J., Alvadaro, B., McGuire, M.T. and Snader, K.M. (1994) Determination of taxane in *T. brevifolia* extracts by tandem mass spectrometry and high-performance liquid chromatography.*J. Nat. Prod.*, 57, 277–286.

Hosoyama, H., Inubushi, A., Yoshida, N., Katsui, T., Shigemori, H. and Kobayashi, J. (1996) Taxuspines U,V, and W, new taxane and related diterpenoids from the Japanese Yew *Taxus cuspidata. Tetrahedron*, 52, 13145–13150.

Huang, C.H.O., Kingston, D.G.O., Magri, N.F., Samaranayake, G. and Boettner, F.E. (1986) New taxanes from *Taxus brevifolia. J. Nat. Prod.*, 49, 665–669.

Huizing, M.T., Rosing, H., Koopmans, F.P. and Beijnen, J.H. (1998) Influence of Cremophor EL on the quantification of paclitaxel in plasma using high-performance liquid chromatography with solid-phase extraction as sample pretreatment.*J. Chromatogr. B*, 709, 161–165.

Huizing, M.T., Rosing, H., Koopman, F., Keung, A.C.F., Pinedo, H.M. and Beijnen, J.H. (1995a) High-performance liquid chromatographic procedures for the quantitative determination of paclitaxel (Taxol) in human urine.*J. Chromatogr. B*, 664, 373–382.

Huizing, M.T., Sparreboom, A., Rosing, van Tellingen, O., Pinedo, H.M. and Beijnen, J.H. (1995b) Quantification of paclitaxel metabolites in human plasma by high-performance liquid chromatography. *J. Chromatogr. B*, 674, 261–268.

Huizing, M.T., Keung, A.C.F., Rosing, H., van der Kuij, V., Ten Bokkel Huinik, W.W., Mandjes, I.M., Dubbelman, A.C., Pinedo, H.M. and Beijnen, J.H. (1993) Pharmaco-kinetics of paclitaxel and its metabolites in a randomized comparative study in platinum-pre-treated ovarian cancer patients.*J. Clin. Oncol.*, 11, 2127–2135.

Issaq, H.J., Chan, K.C., Muschik, G.M. and Jaini, G.M. (1995) Application of capillary zone electrophoresis and micellar electrokinetic chromatography in cancer research.*J. Liq. Chromatogr.*, 18, 1273–1288.

Jagota, N.K., Nair, J.B., Frazer, R., Klee, M. and Wang, M.Z. (1996) Supercritical fluid chromatography of paclitaxel. *J. Chromatogr. A*, 721, 315–322.

Jarvis, A.P. and Morgan, E.D. (1997) Isolation of plant products by supercritical-fluid extraction. *Phytochemical Analysis*, 8, 217–222.

Jaziri, M. and Vanhaelen, M. (2001) Bioactive Taxoid Production from Natural Sources. In *Bioactive Compounds from Natural Sources, Isolation, Characterisation and Biological Properties*, edited by C. Tringali, pp. 283–303. New York: Taylor and Francis.

Jaziri, M., Guo, Y.-W., Vanhaelen-Fastre, R. and Vanhaelen, M. (1997) A new concept for the isolation of bioactive taxoids. *Curr. Top. Phytochem.*, 1, 67–76.

Jaziri, M., Diallo, B.M., Vanhaelen, M.H., Vanhaelen-Fastre, R.J., Zhiri, A., Becu, A.G. and Holmes, J. (1991) Enzyme-linked immunosorbent assay for the detection and the semi-quantitative determination of taxane diterpenoids related to taxol in *Taxus* sp. and tissue culture. *J. Pharm. Belg.*, 46, 93–99.

Jaziri, M., Zhiri, A., Guo, Y.-W., Dupont, J.-P., Shimomura, K., Hamada, H., Vanhaelen, M. and Holmes, J. (1996) *Taxus* sp. cell, tissue and organ cultures as alternative sources for taxoids production: A literature survey. *Plant Cell, Tissue and Organ Cult.*, 46, 59–75.

Jennings, D.W., Deutsch, H.M., Zalkow, L.H. and Teja, S. (1992) Supercritical extraction of taxol from the bark of *T. brevifolia*. *J. Supercrit. Fluids.*, 5, 1.

Jenniskens, L.H.D., van Rozendaal, E.L.M., van Beek, T.A., Wiegerinck, P.H.G. and Scheeren, H.W. (1996) Identification of six taxine alkaloids from *Taxus baccata* needles. *J. Nat. Prod.*, 59, 117–123.

Jung, Y.-A., Chung, S.-T. and Row, K.-H. (1997) Resolution of taxol in terms of mobile phase composition in reversed-phase HPLC. *Hwahak Konghak*, 35, 712–716.

Jung, Y.-A., Chung, T. and Row, K.-H. (1998) Resolution of taxol by mobile phase composition in isocratic RP-HPLC. *Nonmunjip-sanop Kwahak Kisui Yonguso* (Inha Taehakkyo), 26, 139–145.

Kawamura, F., Kikuchi, Y., Ohira, T. and Yatagai, M. (1999) Accelerated solvent extraction of paclitaxel and related compounds from the bark of *Taxus cuspidata*. *J. Nat. Prod.*, 62, 244–247.

Kelsey, R.G. and Vance, N.C. (1992) Taxol and cephalomannine concentrations in the foliage and bark of shade-grown and sun-exposed *Taxus brevifolia* trees. *J. Nat. Prod.*, 55, 912–917.

Kerns, E.H., Volk, K.J., Hill, S.E. and Lee, M.S. (1994) Profiling taxanes in *Taxus* extracts using LC/MS and LC/MS/MS techniques. *J. Nat. Prod.*, 57, 1391–1403.

Ketchum, R.E.B. and Gibson, D.M. (1993) Rapid isocratic reversed phase HPLC of taxanes on new columns developed specifically for taxol analysis. *J. Liq. Chromatogr.*, 16, 2519–2530.

Ketchum, R.E.B. and Gibson, D.M. (1995) A novel method of isolating taxanes from cell suspension cultures of yew (*Taxus* spp.) *J. Liq. Chromatogr.*, 18, 1093–1111.

Ketchum, R.E.B., Luong, J.V. and Gibson, D.M. (1999) Efficient extraction of paclitaxel and related taxoids from leaf tissue of *Taxus* using a potable solvent system. *J. Liq. Chrom. & Rel. Technol.*, 22, 1715–1732.

Kikuchi, Y. and Kawamura, F. (1999) Liquid carbon dioxide extraction of taxol and its related compounds from the needles of *Taxus cuspidata*. *Holz als Roh- und Werkstoff*, 57, 12.

Kikuchi, Y., Kawamura, F., Ohira, T. and Yatagai, M. (1997) Accelerated solvent extraction of taxane compounds from *Taxus cuspidata*. *Mokuzai Gakkaishi.*, 43, 971–974.

Kingston, D.G.I., Gunatilaka, A.A.L. and Ivey, C.A. (1992) Modified taxols, 7. A method for the separation of taxol and cephalomannine. *J. Nat. Prod.*, 55, 259–261.

Kitagawa, I., Mahmud, T., Kobayashi, M., Roemantyo and Shibuya, H. (1995) Taxol and its related taxoids from the needles of *Taxus sumatrana*. *Chem. Pharm. Bull.*, 43, 365–367.

Kobayashi, J., Hosoyama, H., Katsui, T., Yoshida, N. and Shigemori, H. (1996) Taxuspines N, O, and P, new taxoids from Japanese Yew *Taxus cuspidata*. *Tetrahedron*, 52, 5391–5396.

Kobayashi, J., Inubushi, A., Hosoyama, H., Yoshida, N., Sasaki, T. and Shigemori, H. (1995) Taxuspines E~H and J, new taxoids from Japanese Yew *Taxus cuspidata*. *Tetrahedron*, 51, 5971–5978.

Kobayashi, J., Ogiwara, A., Hosoyama, H., Shigemori, H., Yoshida, N., Sasaki, T., Li, Y., Iwassaki, S., Naito, M. and Tsuruo, T. (1994) Taxuspines A~C, New taxoids from Japanese Yew *Taxus cuspidata* inhibiting drug transport activity of P-glyco-protein in multidrug-resistant cells. *Tetrahedron*, 50, 7401–7416.

Kopycki, W.J., ElSohly, H.N. and McChesney, J.D. (1994) HPLC determination of taxol and related compounds in *Taxus* plant extracts. *J. Liq. Chromatogr.*, 17, 2569–2591.

Kubota, M. and Abe, T. (1999) *Jyumoku Seiri-kinousei Bussitsu Kenkyu hokokusyo*, pp. 221–222. Tokyo: Jyumoku Seiri-kinousei Bussitsu Gijyutsu Kenkyukumiai.

Lauren, D.R., Jensen, D.J. and Douglas, J.A. (1995) Analysis of taxol, 10-deacetylbaccatin III and related compounds in *Taxus baccata. J. Chromatogr., A*, 712, 303–309.

Lee, S.-H., Yoo, S.D. and Lee, K.-H. (1999) Rapid and sensitive determination of paclitaxel in mouse plasma by high-performance liquid chromatography. *J. Chromatogr. B*, 724, 357–363.

Leslie, J., Kujawa, J.M., Eddington, N., Egorin, M. and Eiseman, J. (1993) Stability problems with taxol in mouse plasma during analysis by liquid chromatography. *J. Pharm. Biomed. Anal.*, 11, 1349–1352.

Leu, J.-G., Chen, B.-X., Schiff, P.-B. and Erlanger, B.F. (1993) Characterization of polyclonal and mono-clonal anti-taxol antibodies and measurements of taxol in serum. *Cancer Res.*, 53, 1388–1391.

Li, B., Xiao, G.-Y., Lin, B.-C. and Ma. Z.-D. (2000) Separation of taxol by using simulated moving bed chromatography. *Anshan Gangtie Xueyuan Xuebao*, 23, 244–248.

Liu, J., Volk, K.J., Mata, M.J., Kerns, E.H. and Lee, M.S. (1997) Miniaturized HPLC and ionspray mass spectrometry applied to the analysis of paclitaxel and taxanes. *J. Pharm. Biomed. Anal.*, 15, 1729–1739.

Longnecker, S.M., Donehower, R.C., Cates, A.E., Chen, T.-L., Brundett, R.B., Grochow, L.B., Ettinger, D.S. and Colvin, M. (1987) *Cancer Treat. Rep.*, 71, 53–59.

Loos, W.J., Verweij, K., Nooter, K., Stoter, G. and Sparreboom, A. (1997) Sensitive determination of docetaxel in human plasma by liquid–liquid extraction and reversed-phase high-performance liquid chromatography. *J. Chromatogr., B*, 693, 437–441.

McClure, T.D., Schram, K.H. and Reimer, M.L.J. (1992) The mass spectrometry of taxol. *J. Am. Soc. Mass Spectrom.*, 3, 672–679.

MacEachern-Keith, G.J., Wagner Butterfield, L.J. and Mattina, M.J.I. (1997) Paclitaxel stability in solu-tion. *Anal. Chem.*, 69, 72–77.

McLaughlin, J.M., Miller, R.W., Powell, R.G. and Smith C.R. Jr. (1981) 19-Hydroxy baccatin III, 10-deacetylcephalomannine, and 10-deacetyltaxol: New anti-tumor taxanes from *Taxus wallichiana. J. Nat. Prod.*, 44, 312–319.

Malingré, M.M., Rosing, H., Koopman, F.J., Schellens, J.H.M. and Beijnen, J.H. (2001) Performance of the analytical assays of paclitaxel, docetaxel, and cyclosporin A in a routine hospital laboratory setting. *J. Liq. Chromatogr. & Rel. Technol.*, 24, 1697–2717.

Martin, N., Catalin, J., Blachon, M.F. and Durand, A. (1998) Assay of paclitaxel (Taxol) in plasma and urine by high performance liquid chromatography. *J. Chromatogr. B*, 709, 281–288.

Mase, H., Hiraoka, M., Suzuki, A. and Nakanomyo, H. (1994) Determination of new anti-cancer drug, paclitaxel, in biological fluids by high performance liquid chromatography. *Yakugaku Zasshi*, 114, 351–355.

Mattina, M.J.I. and Paiva, A.A. (1992) Taxol concentration in *Taxus* cultivars. *J. Environ. Hort.*, 10, 187–191.

Mattina, M.J.I. and MacEachern, G.J. (1994) Extraction, purification by solid-phase extraction and high-performance liquid chromatographic analysis of taxanes from ornamental *Taxus* needles. *J. Chromatogr. A*, 679, 269–275.

Mattina, M.J.I., Berger, W.A.I. and Denson, C.L. (1997) Microwave-assisted extraction of taxanes from *Taxus* biomass. *J. Agric. Food Chem.*, 45, 4691–4696.

Matysik, G,. Glowniak, K., Józefczyk, A. and Furmanowa, M. (1995) Stepwise gradient thin-layer chro-matography and densitometric determination of taxol in extracts from various species of *Taxus*. *Chromatographia*, 41, 485–487.

Miller, R.W., Powell, R.G., Smith, C.R., Jr., Arnold, E. and Clardy, J. (1981) Anti-leukemic alkaloids from *Taxus wallichiana* Zucc. *J. Org. Chem.*, 46, 1469.

Monsarrat , B., Mariel, E., Cros, S., Garès, M., Guénard, D., Guéritte-Voegelein, F. and Wright, M. (1990) Taxol metabolism. Isolation and identification of three major metabolites of taxol in rat bile. *Drug Metab. Dispos.*, 18, 895–901.

Nair, M.G. (1997) "Process for isolation and purification of taxol and taxanes from *Taxus* spp." PCT Int. Appl. PIXXD2 WO 9709443 A1 19970313; *Chem. Abs.*, 126: 255480.

Nair, J.B. and Izzo, C.G. (1993) Anion screening for drugs and intermediates by capillary ion electrophoresis. *J. Chromatogr.*, 640, 445–461.

Nalesnik, C.A., Hansen, B.N. and Hsu, J.T. (1998) Solubility of pure taxol in supercritical carbon dioxide. *Fluid Phase Equilibria*, 146, 315–323.

Németh-Kiss, V., Forgács, E., Cserháti, T. and Schmidt, G. (1996a) Determination of taxol in *Taxus* species grown in Hungary by high-performance liquid chromatography-diode array detection, effect of vegetative period. *J. Chromatogr. A*, 750, 253–256.

Németh-Kiss, V., Forgács, E., Cserháti, T. and Schmidt, G. (1996b) Taxol content of various *Taxus* species in Hungary. *J. Pharm. Biomed. Anal.*, 14, 997–1001.

Nguyen, T., Eshraghi, J., Gonyea, G., Ream, R. and Smith, R. (2001) Studies on factors influencing stability and recovery of paclitaxel from suspension media and cultures of *Taxus cuspidata* cv *Densiformis* by high-performance liquid chromatography. *J. Chromatogr. A*, 911, 55–61.

Nicholson, R., Schemluck, M., Estrada, E. and Brobst, S. (1992) Taxol levels of *Taxus globosa*, the Mexican yew, Second NCI workshop on taxol and Taxus, NCI, Alexandria, VA Sept. 23–24, 1992, poster A-4.

Noh, M.-J., Yang, J.-G., Kim, K.-S., Yoon, Y.-M., Kang, K.A., Han, H.-Y., Shim, S.-B. and Park, H.J. (1999) Isolation of a novel microorganism, *Pestalotia heterocornis*, producing paclitaxel. *Biotech. & Bioeng.*, 64, 620–623.

Pagé, M. and Landry, N. (1996) "Bacterial mass production of taxanes with *Erwinia*." US P5561055; *Chem. Abs.*, 125: P245821.

Park, Y.K., Chung, S.T. and Row, K.H. (1998) Separation of taxol by empirical correlation of mobile phase composition and resolution. *Nonmunjip-Sanop Kwahak kisui Yonguso (Inha Taehakkyo)*, 26, 147–152, *Chem. Abs.*, 130:227812.

Park, Y.K., Chung, S.T. and Row, K.H. (1999) Preparative chromatographic separation of taxol from yew. *J. Liq. Chrom. & Rel. Technol.*, 22, 2755–2761.

Park, Y.K., Row, K.H. and Chung, S.T. (2000) Adsorption characteristics and separation of taxol from yew tree by NP-HPLC. *Sep. Purif. Technol.*, 19, 27–37.

Powell, R.G., Miller, R.W. and Smith, C.R. Jr. (1979) Cephalomannine; a new antitumor alkaloid from *Cephalotaxus manii. J. Chem. Soc. Chem. Commun.*, 102–104.

Rao, K.V. (1993) Taxol and related taxanes. I. Taxanes of *Taxus brevifolia* bark. *Pharmaceut. Res.* 10, 521–524.

Rao, K.V. and Johnson, J.H. (1998) Occurrence of 2,6-dimethoxy cinnamaldehyde in *Taxus floridanna* and structural revision of taxiflorine to taxchinin M. *Phytochem.*, 49, 1361–1364.

Rao, K.V., Bhakuni, R.S., Juchum, J. and Davies, R.M. (1996) A large-scale process for paclitaxel and other taxanes from the needles of *Taxus x media Hicksii* and *Taxus floridana* using reverse phase column chromatography. *J. Liq. Chrom. & Rel. Technol.*, 19, 427–447.

Rao, K.V., Hanuman, J.B., Alvarez, C., Stoy, M., Juchum, J., Davies, R.M. and Baxley, R. (1995) A new large-scale process for taxol and related taxanes from *Taxus brevifolia. Pharmaceut. Res.*, 12, 1003–1010.

Richheimer, S.L., Tinnermeier, D.M. and Timmons, D.W. (1992) High-performance liquid chromatographic assay of taxol. *Anal. Chem.*, 64, 2323–2326.

Rimoldi, J.M., Molinero, A.A., Chordia, M.D., Gharpure, M.M., and Kingston, D.G.I. (1996) An improved method for the separation of paclitaxel and cephalomannine. *J. Nat. Prod.*, 59, 167–168.

Rizzo, J., Riley, C., Von Hoff, D., Kuhn, J., Phillips, J. and Brown, T. (1990) Analysis of anticancer drugs in biological fluids: determination of taxol with application to clinical pharmacokinetics. *J. Pharm. Biomed. Anal.*, 8, 159–164.

Rouini, M.R., Lotfolahi, A., Stewart, D.J., Molepo, J.M., Shirazi, F.H., Verginiol, J.C., Tomiak, E., Delorme, F., Vernillet, L., Giguere, M. and Goel, R. (1998) A rapid reversed phase high performance liquid chromatographic method for the determination of docetaxel (Taxotere®) in human plasma using a column switching technique. *Pharm. Biomed. Anal.*, 17, 1243–1247.

Russin, W.A., Ellis, D.D., Gottwald, J.R., Zeldin, E.L., Brodhagen, M. and Evert, R.F. (1995) Immunocytochemical localization of taxol in *Taxus cuspidata. J. Plant Sci.*, 156, 668–678.

Schutzki, R., Chandra, A. and Nair, M.G. (1994) The effect of post harvest storage on taxol and cephalomannine concentrations in *Taxus X media* 'Hicksii' Rehd. *Phytochem.*, 37, 405–408.

Schneider, B., Zaho, Y., Blitzke, T., Schmitt, B., Noolandeh, A., Sun, X. and Stöckigt, J. (1998). Taxane analysis by high performance liquid chromatography-nuclear magnetic resonance spectroscopy of *Taxus* species. *Phytochemical Analysis*, 9, 237–244.

Sénilh, V., Blechert, S., Colin, M., Guénardo, D., Picot, F., Potier, P. and Varenne, P. (1984) Mise evidence de nouveaux analogues du taxol extraitis de *Taxus baccata. J. Nat. Prod.*, 47, 131–137.

Shao, L.K. and Locke, D.C. (1997) Determination of paclitaxel and related taxanes in bulk drug and injectable dosage forms by reversed phase liquid chromatography. *Anal. Chem.*, 69, 2008–2016.

Shao, L.K. and Locke, D.C. (1998) Separation of paclitaxel and related taxanes by micellar elektrokinetic capillary chromatography. *Anal. Chem.*, 70, 897–906.

Shrestha, K., Strobel. G.A., Shrivastava, S.P. and Gewali (2001) Evidence for paclitaxel from three new endophytic fungi of Himalayan yew of Nepal. *Planta Med.*, 67, 374–376.

Snader, K.M. (1995) Detection and isolation. In Taxol®, *Science and Applications*, edited by M. Suffness, pp. 277–286. Boca Raton: CRC Press.

Song, D. and Au, J.L.-S. (1995) Isocratic high-performance liquid chromatographic assay of taxol in biological fluids and tissues using automated column switching. *J. Chromatogr. B.* 663, 337–344.

Sottani, C., Turci, R., Micoli, G., Fiorentino, M.L. and Minoia, C. (2000) Rapid and sensitive determination of paclitaxel (Taxol®) in environmental samples by high-performance liquid chromatography tandem mass spectrometry. *Rapid Comm. Mass Spectrometry*, 14, 930–935.

Sottani, C., Minoia, C., D'Incalci, M., Paganini, M. and Zucchetti, M. (1998) High performance liquid chromatography tandem mass spectrometry procedure with automated solid phase extraction sample preparation for the quantitative determination of paclitaxel (Taxol®) in human plasma. *Rapid Comm. Mass Spectrometry*, 12, 251–255.

Sottani, C., Minoia, C., Colombo, A., Zucchetti, M., D'Incalci, M. and Fanelli, R. (1997) Structural characterization of mono- and dihydroxylated metabolites of paclitaxel in rat bile using liquid chromatography/ion spray tandem mass spectrometry. *Rapid Comm. Mass Spectrom.*, 11, 1025–1032.

Sparreboom, A., van Tellingen, O., Nooijen, W.J. and Beijnen, J.H. (1995) Determination of paclitaxel and metabolites in mouse plasma, tissue, urine, and faeces by semi-automated reversed-phase high performance liquid chromatography. *J. Chromatogr. B*, 664, 383–391.

Sparreboom, A., de Bruijn, P., Nooter, K., Loos, W.J., Stoter, G. and Verweij, J. (1998) Determination of paclitaxel in human plasma using single solvent extraction prior to isocratic reversed-phase high performance liquid chromatography with ultraviolet detection. *J. Chromatogr. B*, 705, 159–164.

Srinivasan, V., Roberts. S.C. and Shuler, M.L. (1997) Combined use of six-well polystyrene plates and thin layer chromatography for rapid development of optimal plant cell culture processes: application to taxane production by *Taxus* sp. *Plant Cell Reports*, 16, 600–604.

Stasko, M.W., Witherup, K.M., Ghiorzi, T.J., McCloud, T.G., Look, S., Muschik, G.M. and Issaq, H.J. (1989) Mutimodal thin layer chromatographic separation of taxol and related compounds from *Taxus brevifoifolia. J. Liq. Chromatogr.*, 12, 2133–2143.

Stierle, A., Strobel, G. and Stierle, D. (1993) Taxol and taxane production by *Taxomyces andreanae*, an endophytic fungus of the pacific yew. *Science*, 260, 214.

Stierle, A., Strobel, G., Stierle, D., Grothaus, P. and Bignami, G. (1995) The search for a taxol-producing microorganism among the endophytic fungi of the pacific yew, *Taxus brevifolia. J. Nat. Prod.*, 58, 1315–1324.

Strobel, G., Yang, X., Sears, J., Kramer, R., Sidhu, R.S. and Hess, W.H. (1996) Taxol from *Pestalotiopsis microspora*, an endophytic fungus of *Taxus wallachianna. Microbiology*, 142, 435–440.

Suffness, M. and Wall, M.E. (1995) Discovery and development of taxol. In Taxol®, *Science and Applications*, edited by M. Suffness, pp. 3–25. Boca Raton: CRC Press.

Svojanovsky, S.R., Egodage, K.L., Wu, J., Slavik, M. and Wilson, G.S. (1999) High sensitivity ELISA determination of taxol in various human biological fluids. *J. Pharm. Biomed. Anal.*, 20, 549–555.

Theodoridis, G. and Verpoorte, R. (1996) Taxol analysis by high performance liquid chromatography: a review. *Phytochemical Analysis*, 7, 169–184.

Theodoridis, G., de Jong, C.F., Laskaris, G. and Verpoorte, R. (1998a) Application of SPE for the HPLC analysis of taxanes from *Taxus* cell cultures. *Chromatographia*, 47, 25–34.

Theodoridis, G., Laskaris, G., de Jong, C.F., Hofte, A.J.P. and Verpoorte, R. (1998b) Determination of paclitaxel and related diterpenoids in plant extracts by high-performance liquid chromatography with

UV detection in high-performance liquid chromatography-mass spectrometry. *J. Chromatogr. A*, **802**, 297–305.

Theodoridis, G., Laskaris, G., Van Rozendaal, E.L.M. and Verpoorte, R. (2001) Analysis of taxines in *Taxus* plant material and cell cultures by HPLC photodiode array and HPLC-electrospray mass spectrometry. *J. Liq. Chromtogr. & Rel. Technol.*, **24**, 2267–2282.

Tian, G.-L., Zhang, Z.-Q and Su, Z.-G. (2001) Synthesis of phenyl-silica gel and its application in purification of taxol. *Sepu*, **19**, 47–50.

Vandana, V. and Teja, A.S. (1995) Supercritical extraction of paclitaxel using CO_2 and CO_2-ethanol mixture. In *Innovations in Supercritical Fluids*, edited by K.W. Hutchenson and N.R. Foster, pp. 429–443. Washington: American Chemical Society (ACS Symposium Series 608).

Vandana, V. and Teja, A.S. (1997) The solubility of paclitaxel in supercritical CO_2 and N_2O. *Fluid Phase Equilibria*, **135**, 83–87.

Vandana, V., Teja, A.S. and Zalkow, L.H. (1996) Supercritical extraction and HPLC analysis of taxol from *Taxus brevifolia* using nitrous oxide and nitrous oxide + ethanol mixtures. *Fluid Phase Equilibria*, **116**, 162–169.

Vanhaelen-Fastre, R., Diallo, B., Jaziri, M., Faes, M.-L., Homes, J. and Vanhaelen, M. (1992) High-speed countercurrent chromatography of taxol and related diterpenoids from *Taxus baccata*. *J. Liq. Chromatogr.*, **15**, 697–706.

Van Rozendaal, E.L.M., Lelyveld, G.P. and van Beek, T.A. (1997) A simplified method for the determination of taxanes in yew needles by reversed-phase (C_{18}) high pressure liquid chromatography. *Phytochemical Analysis*, **8**, 286–293.

Vergniol, J.C., Bruno, R., Montay, G. and Frydman, A. (1992) Determination of taxotere in human plasma by a semi-automated high-performance liquid chromatographic method. *J. Chromatogr.*, **582**, 273–278.

Vidensek, N., Lim., P. Campbell, A. and Carlson, C. (1990) Taxol content in bark, wood, root, leaf, twig, and seedling from several *Taxus* species. *J. Nat. Prod.*, **53**, 1609–1610.

Volk, K.J., Hill, S.E., Kerns, E.H. and Lee, M.S. (1997) Profiling degradants of paclitaxel using liquid chromatography-mass spectrometry and liquid chromatography-tandem mass spectrometry substructural techniques. *J. Chromatogr. B*, **696**, 99–115.

Wall, M.E. and Wani, M. C. (1967) Recent progress in plant anti-tumor agents, Abstracts of Papers, 153rd National Meeting, American Chemical Society, Miami Beach , FL, No. M-006.

Wall, M.E. and Wani, M. C. (1998) History and future prospects of camptothecin and taxol. In *The Alkaloids, Chemistry and Biology*, Vol. 50, edited by G.A. Cordell, pp. 509–536. San Diego: Academic Press.

Wani, M.C., Taylor, H.L., Wall, M.E., Coggon, P. and McPhail, A.T. (1971) The isolation of taxol, a novel antileukemic and antitumor agent from *Taxus brevifolia*. *J. Am. Chem. Soc.*, **93**, 2325–2326.

Wheeler, N.C., Jech, K., Masters, S., Brobst, S.W., Alvarado, A.B., and Hoover, A.J. and Snader, K.M. (1992) Effects of genetic, epigenetic, and environmental factors on taxol content in *Taxus brevifolia* and related species. *J. Nat. Prod.*, **55**, 432–440.

Wickremesinhe, E.R.M. and Arteca, R.N. (1993) Methodology for the identification and purification of taxol and cephalomannine from *Taxus* callus cultures. *J. Liq. Chromatogr.*, **16**, 3263–3274.

Wickremesinhe, E.R.M. and Arteca, R.N. (1994) Roots of hydroponically grown *Taxus* plants as a source of taxol and related taxanes. *Plant Sci.* **101**, 125–135.

Wickremesinhe, E.R.M. and Arteca, R.N. (1996) Effects of plant growth regulators applied to the roots of hydroponically grown *Taxus x media* on the production of taxol and related taxanes. *Plant Sci.* **121**, 29–38.

Wiernik, P.H., Schwartz, E.L., Einzig, A., Strauman, J.J., Lipton, R.B. and Dutcher, J.P. (1987) Phase I trial of taxol given as a 24-hour infusion every 21 days: Responses observed in metastatic melanoma. *J. Clin. Oncol.*, **5**, 1232–1239.

Willey, T.A., Bekos, E.J., Gaver, R.C., Duncan, G.F., Tay, L.K., Beijnen, J.H. and Farmen, R.H. (1993) High performance liquid chromatographic procedure for the quantitative determination of paclitaxel (Taxol®) in human plasma. *J. Chromatogr.*, **621**, 231–238.

Witherup, K.M., Look, S.A., Stasko, M.W., McCloud, T.G., Issaq, H.J. and Muschik, G.M. (1989) High performance liquid chromatographic separation of taxol and related compounds from *Taxus brevifolia. J. Liq. Chromatogr.*, 12, 2117–2132.

Witherup, K.M., Look, S.A., Stasko, M.W., Ghiorzi, T.J., Muschik, G.M. and Cragg, G. M. (1990) *Taxus* spp. needles contain amounts of taxol comparable to the bark of *Taxus brevifolia*: analysis and isolation. *J. Nat. Prod.*, 53, 1249–1255.

Wolfender, J.-L., Ndjoko, K. and Hostettmann, K. (2001) The potential of LC-NMR in phytochemical analysis. *Phytochem. Anal.*, 12, 2–22.

Wu, Y.-G. and Zhu, W.-H. (1997) High performance liquid chromatographic determination of taxol and related taxanes from *Taxus* callus cultures. *J. Liq. Chrom. & Rel. Technol.*, 20, 3147–3154.

Wu, D.-R., Lohse, K. and Greenblatt, H.C. (1995) Preparative separation of taxol in normal- and reversed-phase operations. *J. Chromatogr. A*, 702, 233–241.

Wu, D.-J., Ma, Z. and Wang, N.-H.L. (1999) Optimization of throughput and desorbent consumption in simulated moving-bed chromatography for paclitaxel purification. *J. Chromatogr. A.*, 855, 71–89.

Wu, D.-J., Ma, Z., Au, B.W. and Wang, N.-H.L. (1997) Recovery and purification of paclitaxel using low-pressure liquid chromatography. *AIChE. J.*, 43, 232–242.

Xue, J., Chen, J.-M., Zhang, L. and Bu, H.-S. (2000) Large-scale process for high purity Taxol from bark extract of *Taxus yunnanesis. J. Liq. Chrom. & Rel. Technol.*, 23, 2499–2512.

Xue, J., Cao, C.-Y., Chen, J.-M., Bu, H.-S. and Wu, H.-M. (2001) A large-scale separation of taxanes from the bark extract of *Taxus yunnanesis* and ^{1}H- and ^{13}C-NMR assignments for 7-*epi*-10-deacetyltaxol. *Chinese J. Chem.*, 19, 82–90.

Yan, J., Fan, J. and Wang, J. (1996) Improved purification of taxol. *Zhongguo Yiyao Gongye Zazhi.*, 27, 531–534.

Yang, X., Liu, K. and Xie, M. (1998) Purification of taxol by industrial preparative liquid chromatography. *J. Chromatogr. A.*, 813, 201–204.

Zamir, L.O., Zhang, J.-Z., Kutterer, K., Sauriol, F., Mamer, O., Khiat, A. and Boulanger, Y. (1998) 5-Epi-canadensene and other novel metabolites of *Taxus canadensis. Tetrahedron*, 54, 15845–15860.

Zamir, L.O., Nedea, M.E., Belair, S., Sauriol, F., Mamer, O., Jacqmain, E., Jean, F.I., Jean, F.I. and Garneau, F.X. (1992) Taxanes isolated from *Taxus canadensis. Tetrahedron Lett.*, 33, 5173–5176.

Zamir, L.O., Nedea, M.E., Zhou, Z.-H., Belair, S., Caron, G., Sauriol, F., Jacqmain, E., Jean, F.I., Garneau, F.X. and Mamer, O. (1995) *Taxus canadensis* taxanes: structures and stereochemistry. *Can. J. Chem.*, 73, 655–665.

Zhang, Z.-Q. and Su, Z.-G. (2000a) Single step Al_2O_3 chromatography for improved separation and recovery of taxol. *Biotechnology Letters*, 22, 1657–1660.

Zhang, Z.-Q. and Su, Z.-G. (2000b) Recovery of taxol from the extract of *Taxus cuspidatae* callus cultures with Al_2O_3 chromatography, *J. Liq. Chrom. & Rel. Technol.*, 23, 2683–2693.

Zhang, Z.-Q., Wang, Y.-S., Tian, G.-L. and Su, Z.-G. (2000c) Separation and purification of taxol with solid-phase extraction and reversed phase chromatography. *Yaowu Shengwu Jishu*, 7, 157–160.

Zhang, Z.-Q., Wang, Y.-S. and Su, Z.-G. (2001) Purification of taxol with reversed phase chromatography at normal pressure. *Gaoxiao Huaxe Gongcheng Xuebao*, 15, 56–60.

Zhang, Z.-Q., Wei, X.-G., Tian, G., Xu, J.-F. and Su, Z.-G. (1998) Improved HPLC method for taxol determination with Al_2O_3 solid-phase extraction. *Biotechnology Techniques*, 12, 633–636.

Zhang, S., Lee, C.T.-L., Kashiwada, Y., Chen, K., Zhang, D.-C. and Lee, K.H. (1994) Yunantaxusin A, a new 11(15 → 1)ABEO-taxane from *Taxus yunnanensis. J. Nat. Prod.*, 57, 1580–1583.

8 Chemistry of taxol and related taxoids

Hui-Kang Wang, Hironori Ohtsu, and Hideji Itokawa

Introduction

Taxol (1) isolated from the Pacific yew, *Taxus brevifolia*, has attracted considerable attention in many research areas due to its potent anticancer activities and complex structure. Consequently, numerous chemical reactions have been reported, and Kingston has reviewed the chemistry of both taxol (2000), and the taxane diterpenoids (Kingston *et al.*, 1993).

In this chapter, we will describe the different kinds of reactions, interconversions, and semisyntheses of taxol and related taxoids.

Acylation and other protective group chemistry

10-Deacetylbaccatin III(10-DAB) (2) and 7-TES-10-DAB were selectively acylated using Pseudomonas cepacia lipase as the catalyst and dichloroacetic anhydride or trichloroacetic

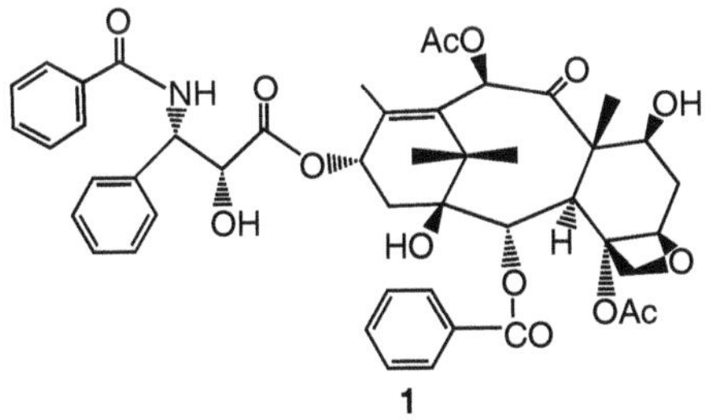

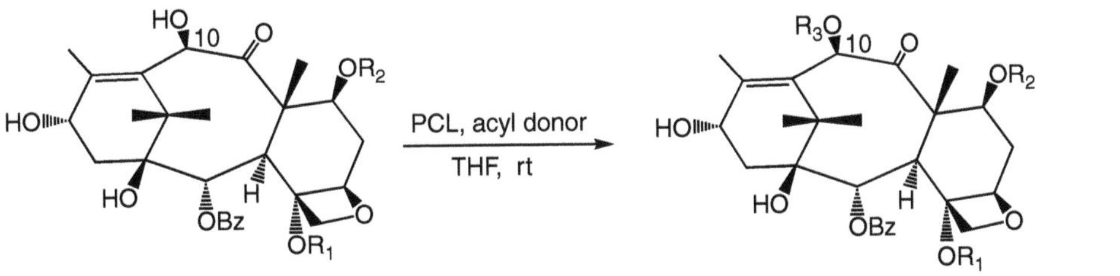

2 R_1 = Ac; R_2 = H
3 R_1 = Ac; R_2 = TES
4 R_1 = H; R_2 = TES

5 R_1 = Ac; R_2 = R_3 = DCA; R_4 = H
6 R_1 = Ac; R_2 = TCA; R_3 = R_4 = H
7 R_1 = Ac; R_2 = TES; R_3 = H; R_4 = DCA
8 R_1 = Ac; R_2 = TES; R_3 = H; R_4 = TCA
9 R_1 = H; R_2 = TES; R_3 = H; R_4 = DCA

PCL = Pseudomonas cepacia lipase
TES = triethylsilyl; DCA = dichloroacetyl; DCAA = dichloroacetic anhydride;
TCAA = trichloroacetic anhydride

Figure 8.1 Selective enzymatic acylation of 10-deacetylbaccatin III.

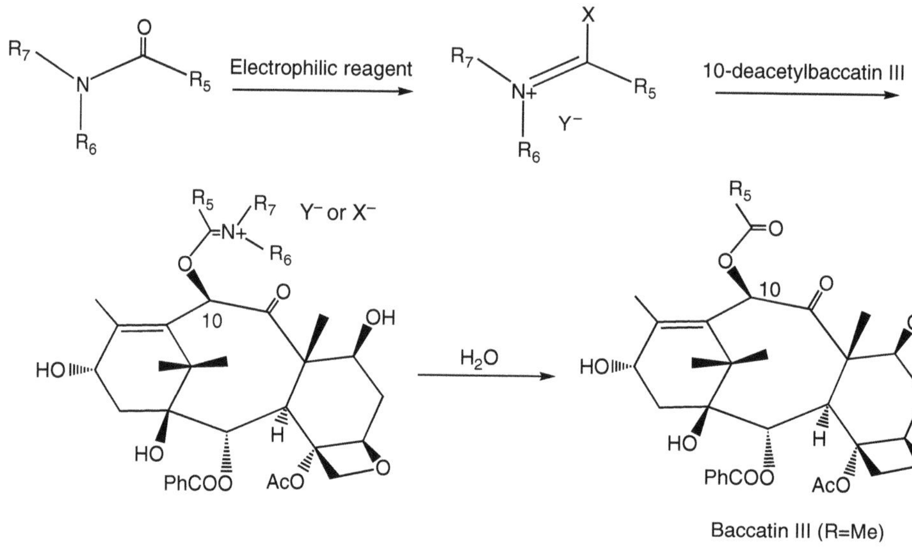

Figure 8.2 Selective acylation at C-10.

anhydride as the acylating agent. In a typical experiment, the heterogeneous solution containing substrate (2–4), acylating agent (9 equiv.), triethylamine (0–12 equiv.), and enzyme (1 mg/mg substrate) in THF was stirred at r.t. and the reaction progress was followed by TLC analysis. When the reaction was complete, the products (5–9) were isolated by silica gel chromatography (Lee *et al.*, 1998a) (Figure 8.1). The selective C-10 acylation of 10-deacetylbaccatin III to baccatin III and derivatives is efficiently catalyzed by lanthanide trifluoromethanesulfonates. Baccatin III, now readily available through this procedure, is an important precursor for an economically viable semisynthesis of taxol and its derivatives (Damen *et al.*, 1998).

New methyleniminium salts were investigated as acylating agents of 10-deacetylbaccatin III. Thus, baccatin III was synthesized in one step with 79% isolated yield and 98% selectivity using the iminium salt prepared from mesyl chloride and N-ethylacetamide (Cravalle *et al.*, 1998) (Figure 8.2).

Taxol can be readily nitrated with nitric acid in acetic anhydride. The reaction was also reported with three natural analogues of taxol, 10-deacetyl baccatin III (10), 10-deacetyl paclitaxel (11), and 10-deacetyl paclitaxel-7-xyloside (12). In each case, the reaction yielded the corresponding tri-13, tri-14, and penta-15 nitrate esters, respectively. After workup, all of these compounds were crystallized readily from diethyl ether to give near quantitative yields (>95%) without chromatographic separation. No case nitration was observed at the sterically hindered 1-hydroxyl. Proton and carbon NMR values of the completely nitrated compounds are given in this report (Rao and Johnson, 1998). Under milder conditions, partially esterified products can be prepared. The regioselectivity of these partial nitrations was studied (Figure 8.3).

Oxetane ring-opening and formation

Analysis of the ^{1}H NMR data of taxol (16), in comparison with its oxetane ring-opened analogue D-secotaxol (17), suggests that the oxetane moiety (D-ring) of taxol serves as a conformational lock for the diterpene moiety and the C-13 side-chain (Boge *et al.*, 1999) (Figure 8.4).

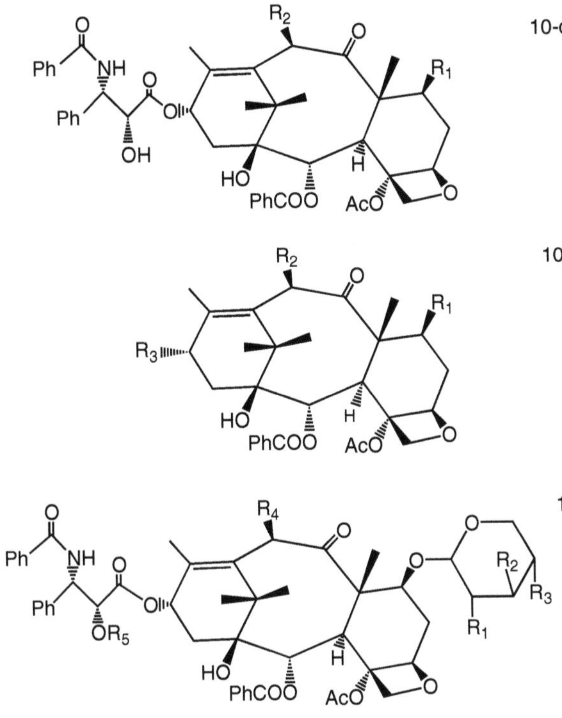

Figure 8.3 Nitration of taxoids.

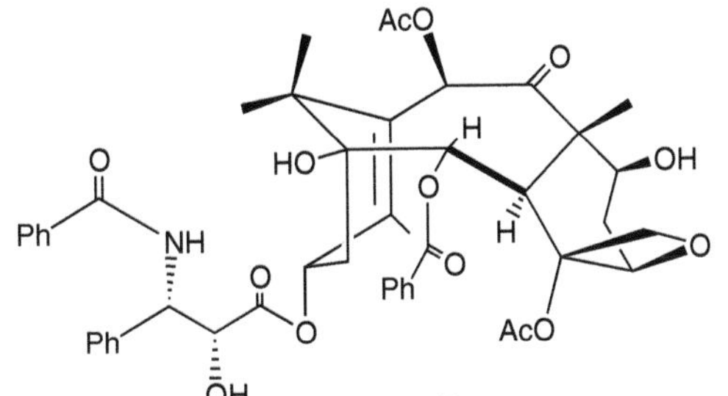

10-deacetyl paclitaxel

11 $R_1=OH$, $R_2=OH$

Nitration

14 $R_1=ONO_2$, $R_2=ONO_2$, $R_3=ONO_2$

10-deacetyl baccatin III

10 $R_1=OH$, $R_2=OH$, $R_3=OH$

Nitration

13 $R_1=ONO_2$, $R_2=ONO_2$, $R_3=ONO_2$

10-deacetyl paclitaxel-7-xyloside

12 $R_1=OH$, $R_2=OH$, $R_3=OH$,
$R_4=OH$, $R_5=OH$

Nitration

15 $R_1=ONO_2$, $R_2=ONO_2$, $R_3=ONO_2$,
$R_4=ONO_2$, $R_5=ONO_2$

16 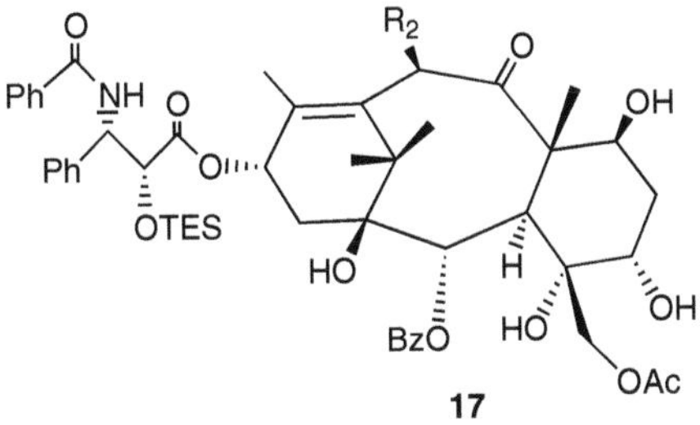 **17**

Figure 8.4 Idealized stereostructure of taxol (16).

Reaction of taxol with Meerwein's reagent led to a product (18) with an opened oxetane ring. Ethylation of taxol's oxetane oxygen would give 19 followed by participation of the neighboring acetoxy group as proposed by Raber (Raber and Guida, 1974) to yield the cation 20. Intramolecular trapping of 20 by the ether oxygen would lead to 21, which could then form the ortho ester 22 by reaction with a nucleophile or on aqueous workup. The ortho ester 22 would hydrolyze to the stable final product 18 under the acidic workup conditions (Samaranayake *et al.*, 1991) (Figure 8.5).

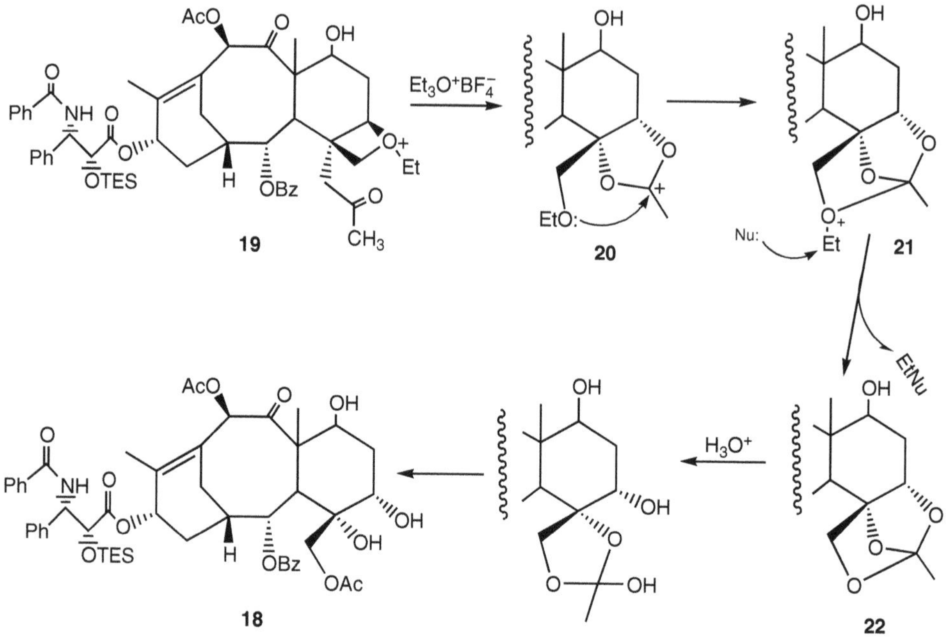

Figure 8.5 Oxetane ring opening reaction.

Epoxide ring-opening and formation

Compound 23 was converted to 24 via a hydroxyl-directed epoxidation, which led exclusively to the C4(20)-α-epoxide in 94% isolated yield. Acetylation of the latter at C-2 using standard procedures, followed by crotonization with DBU then furnished the desired epoxy dienone 25. Treatment with H_2O_2 and NaOH in methanol converted the dienone 25 to the bis-epoxy-enone intermediate 26. Treating 25 under the same conditions but for longer times gave the *tris*-epoxy intermediate 27 as the major component of the reaction mixture along with 26 (Hernando *et al.*, 2000a) (Figure 8.6).

These epoxide products contained functional groups suitable for further elaboration to various taxoids, thus demonstrating the validity of this four-step synthesis of the taxoid ABC framework by the transmetallation (C10−C11)-aldol (C1−C2) route (Hernando *et al.*, 2000b).

Epimerization of 6α-hydroxy-7-*epi*paclitaxel, which was prepared from taxol in four steps in high yield, gave 6α-hydroxypaclitaxel (28), the major human metabolite of taxol. Various epimerization conditions were investigated, and the optimum conditions using DBU in xylenes afforded 80–88% isolated yield based on unrecovered starting material, along with 4-deacetyl analogs as minor products. The stereochemistry of taxol 6,7-epoxide (29) was revised in the course of this work (Yuan and Kingston, 1998) (Figure 8.7).

Rearrangement reactions

Taxanes are a type of terpene, and various rearrangements occur in such molecules. The chemical reactivities of the carbonyl groups of triketone 30 are widely disparate. The carbonyl group in the central ring is most susceptible to transannular aldolization. Compound 30 undergoes O-silylation initially in ring A and subsequently in ring C.

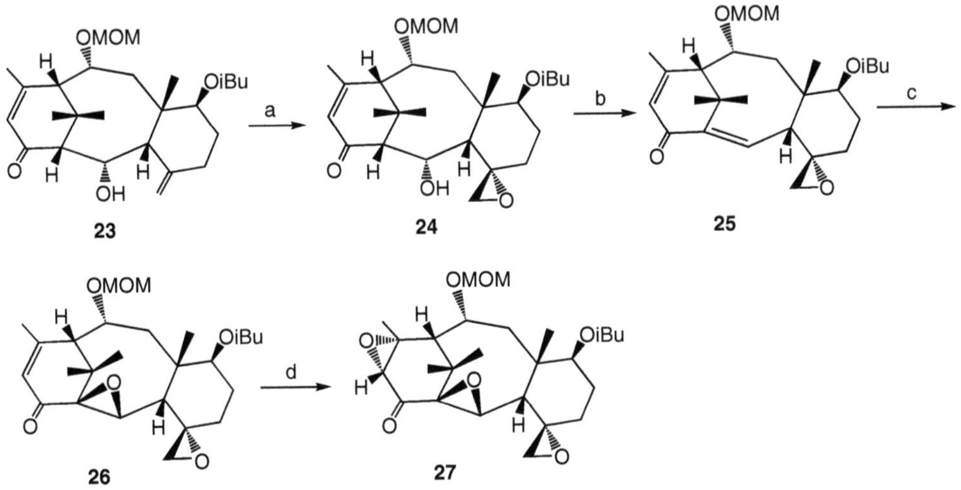

Figure 8.6 (a) Vo(acav)$_2$, *t*-BuOOH in dec., PhH, Δ; (b) Ac$_2$O, py., DMAP, then DBU; (c) 30% H$_2$O$_2$, 6N NaOH, MeOH, 0°C to rt.

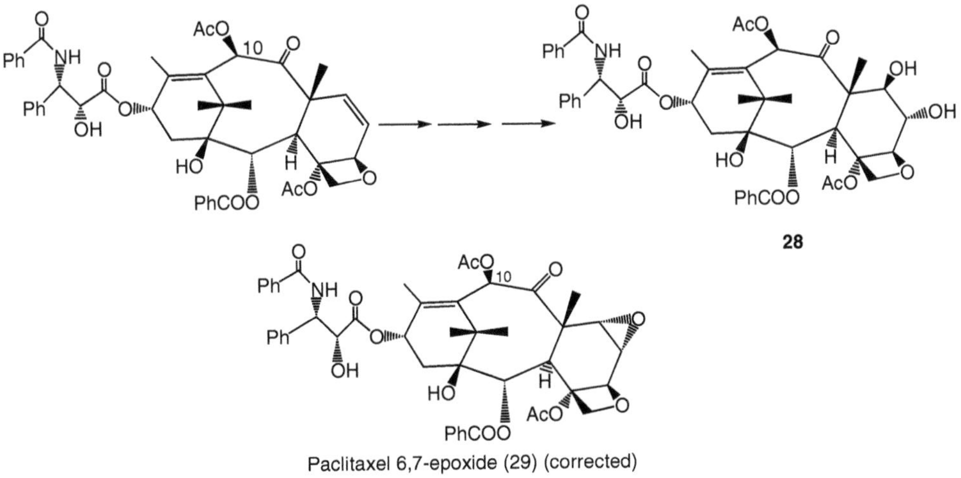

Paclitaxel 6,7-epoxide (29) (corrected)

Figure 8.7 Synthesis of 6-hydroxylpaclitaxel.

Under swamping silylation conditions, an unusual intramolecular hydride shift occurs to afford **31** (Paquette *et al.*, 1992) (Figure 8.8).

Taxicin I (**32**), an intermediate for the synthesis of 7-deoxypaclitaxel, was converted to the bisacetal, 1,15-secotaxane (**33**), by methanolysis. A vinylogous retroaldol reaction is followed by acetalization and a transannular hydride shift. The former reaction opens ring A, giving a 1,15-secotaxane with two keto groups. The closure of a 10,13-acetal bridge brings H-9 in close proximity to the C-1 carbonyl, setting the stage for the hydride shift from C-9 to C-1, the transannular version of an α-ketol rearrangement to **34**. Formation of a C-1, C-9 acetal bridge then terminates the reaction (Appendino *et al.*, 1997) (Figure 8.9).

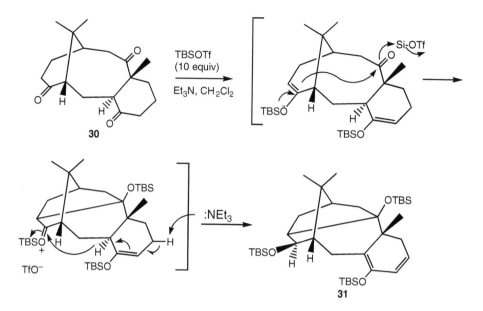

Figure 8.8 An unprecedented silyl triflate-promoted hydride shift within a taxane derivative.

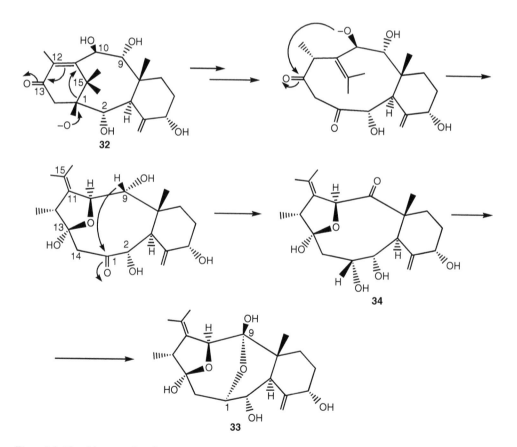

Figure 8.9 Novel base-catalyzed rearrangement of the taxane skeleton.

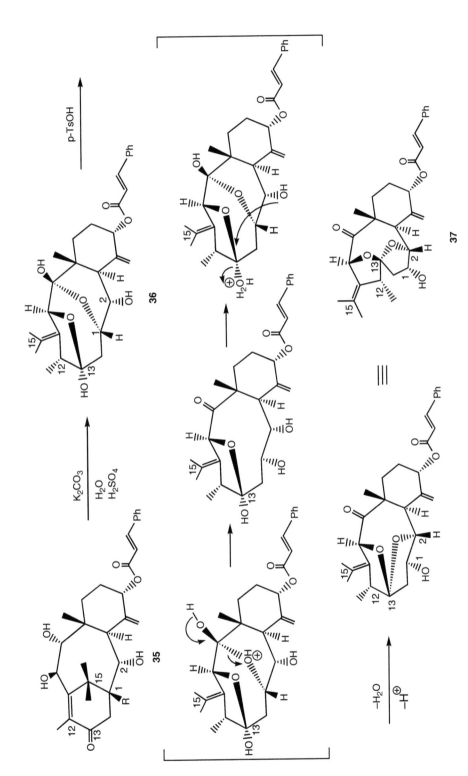

Figure 8.10 Rearrangement of *O*-cinnamoyltaxicin to a C-13 spiro-taxane.

During the large-scale synthesis of an *O*-cinnamoyltaxicin (35), an intermediate in the semisynthesis of 7-deoxypaclitaxel derivatives, a side-product (36) was formed via a vinylogous retro-aldol reaction and a long-range hydride shift from *O*-cinnamoyltaxicin(I) under alkaline reaction conditions. Compound 36 has two hemi-acetal bridges at C-1, C-9 and C-10, C-13. Compound 37 was formed from side-product 36 under acidic reaction conditions and is the first C-13 spiro-taxane described in the literature. This spiro-taxane has two acetal bridges between C-1, C-13 and C-10, C-13 (Van Rosendaal *et al.*, 2000) (Figure 8.10).

Photochemistry

The transannular photocyclization of taxicins was discovered by Nakanishi (Chiang *et al.*, 1967). The reaction presumably takes place via a C-3, C-11 diradical, which is formed by hydrogen transfer from C-3 to C-12, the enone a-carbon (Wolff *et al.*, 1972).

Irradiation (low pressure mercury lamp) of some taxicin derivatives (38–40) gave their corresponding 3,11-cycloderivatives (41–43) in quantitative yield. In all cases, the phototaxicins were obtained as mixtures of *E*- and *Z*-isomers at the cinnamoyl residue. But, if this irradiation had been carried out in the presence of iodine, the mixture would have contained mainly the *Z*-isomer (Appendino *et al.*, 1992) (Figure 8.11).

Photolysis of taxol with 254 nm UV light is therefore an effective method to prepare a novel taxol analog featuring a Carbon–Carbon bond between C3 and C11. As in the first step of the oxadi-π-methane rearrangement, the reaction most likely proceeds through the T_1 (π, π^*) of the C-9 carbonyl group, and leads to diradicaloid species 46. A favorable geometry permits intramolecular hydrogen transfer from C-3 to C-12, as observed in the related rearrangement of taxinine. Finally, transannular bond formation in 47 leads to 44 (Chen *et al.*, 1992) (Figure 8.12).

A novel and non-radioactive bifunctional photoaffinity probe (BPP) (48) was designed and synthesized in order to study of the paclitaxel-microtubules interactions. This new BPP-taxoid bears a 3-nitro-5-trifluoromethyl-diazirinyl-phenoxyacetyl group at C3'N (for photoaffinity/photocleavage) and a biotin subunit at C-7 (for affinity chromatography) by Nakanishi and Fang (Fang *et al.*, 1998; Lin *et al.*, 2000) (Figure 8.13).

Side-chain formation

Radiofrequency encoded combinatorial (REC) chemistry is a recently developed nonchemical encoding strategy in library synthesis. Encoded chemical libraries of complex molecular

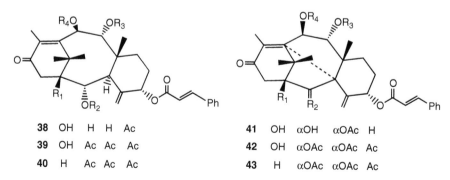

38	OH	H	H	Ac
39	OH	Ac	Ac	Ac
40	H	Ac	Ac	Ac

41	OH	αOH	αOAc	H
42	OH	αOAc	αOAc	Ac
43	H	αOAc	αOAc	Ac

Figure 8.11 Photocyclization.

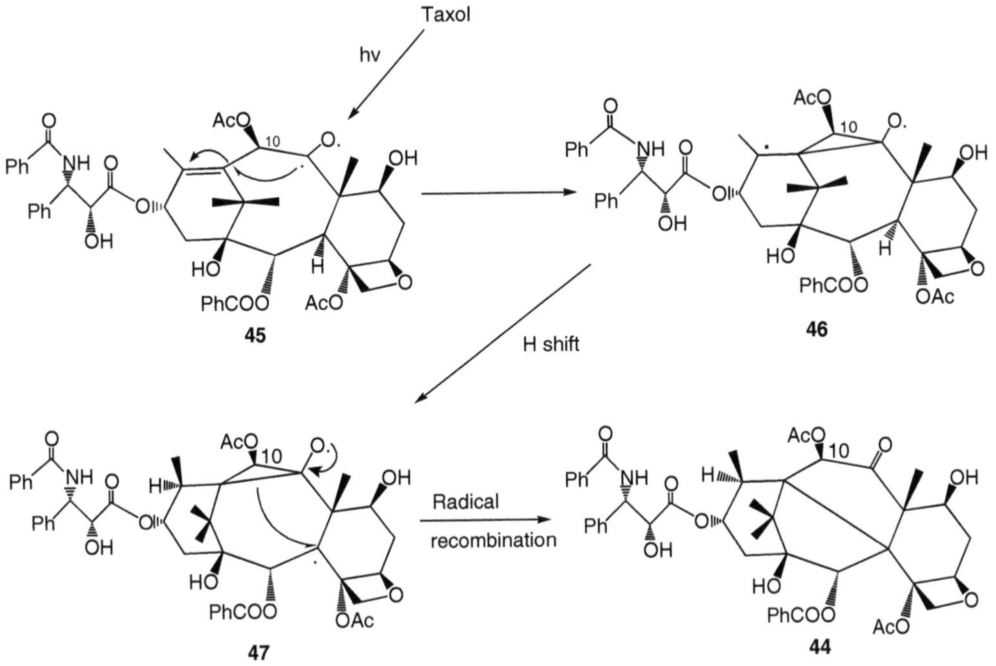

Figure 8.12 The photochemistry of taxol.

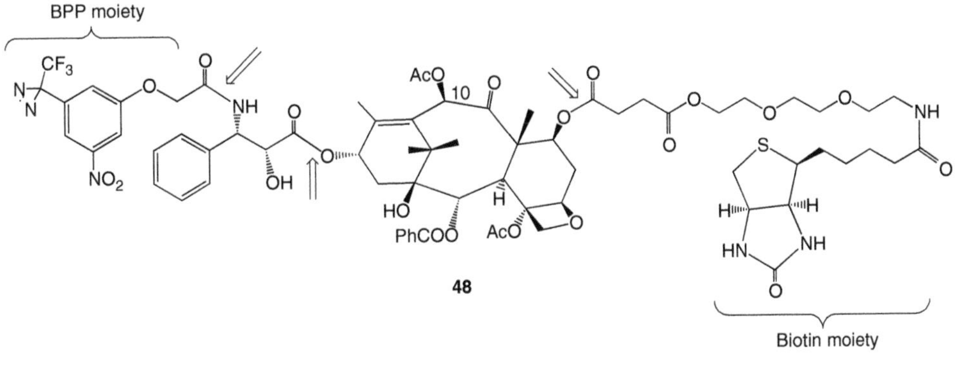

Figure 8.13 Taxoid photoaffinity labeling probe.

structures such as taxol can be constructed employing noninvasive REC strategy and novel solid phase synthesis techniques. The first 400-membered taxoid library was synthesized successfully in a discrete format and in quantities of multimilligrams/member (Xiao *et al.*, 1997) (Figure 8.14).

Two diastereomeric 2″,3″-dibromo-cephalomannines and their two corresponding 7-epimers were obtained by treatment of extracts of *T. yunnanensis* with bromine under mild conditions.

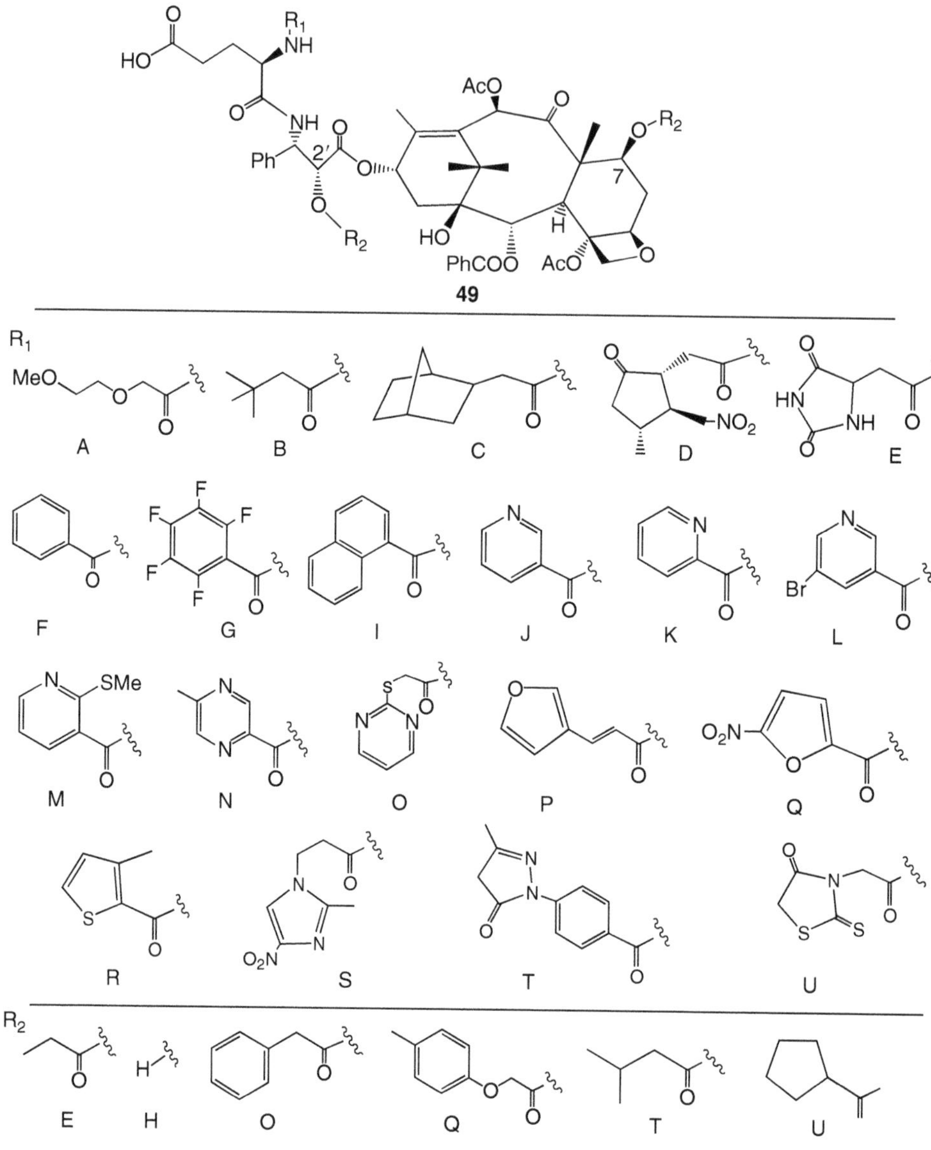

49

(Building blocks A, C, D, F, G, I, J, K, L, M, N, P, R, and S for R$_2$ are the same as for R$_1$)

Figure 8.14 Taxoid library.

Treatment of the same extract with chlorine solution yielded four diastereomeric 2″,3″-dichlorocephalomannines, with no 7-epimers. The diastereomeric mixtures were separated into the individual components by preparative HPLC on C18 reversed-phase silica gel. A more efficient analytical separation was obtained on a pentafluorophenyl-bonded phase. The reaction products were isolated and fully identified spectroscopically. Slight differences were observed in the NMR spectra of the 7-epimers when compared to their 7β-OH analogues. On the basis of

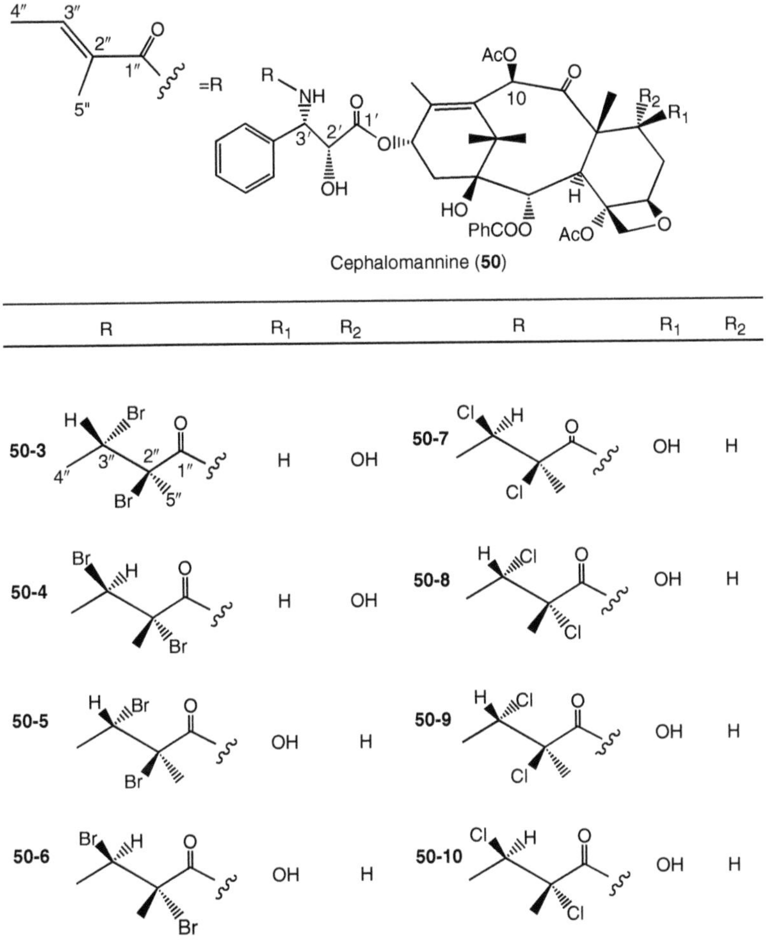

Figure 8.15 Epimerization.

a comparison of physicochemical data, the bromo compounds were identified as (2″R, 3″S)-dibromo-7-*epi*-cephalomannine (50–3), (2″S, 3″R)-dibromo-7-epi-cephalomannine (50–4), (2″R, 3″S)-dibromocephalomannine(50–5), and (2″S, 3″R)-dibromocephalomannine (50–6), and the chloro compounds as (2″R, 3″R)-dichlorocephalomannine (50–7), (2″S, 3″S)-dichlorocephalomannine (50–8), (2″R, 3″S)-dichlorocephalomannine (50–9), and (2″S, 3″R)-dichlorocephalomannine (50–10). Cytotoxic activity was tested against the NCI 60 human tumor cell line panel in comparison with taxol (1), and encouraging results were obtained in some cases (Pandey *et al.*, 1998) (Figure 8.15).

A stereoselective synthesis of (4S, 5R)-2,4-diphenyloxazoline-5-carboxylic acid, a precursor of the taxol C-13 side-chain, was achieved using palladium-catalyzed oxazoline formation reaction from commercially available amino acid (Lee *et al.*, 1998b) (Figure 8.16).

A conformationally constrained analogue of the paclitaxel side-chain was synthesized from ethyl 1H-indene-2-carboxylate (56). Synthesis of the asymmetric dihydroxylation of the trisubstituted double bond of 56 took place with satisfactory enantioselectivity. After conventional

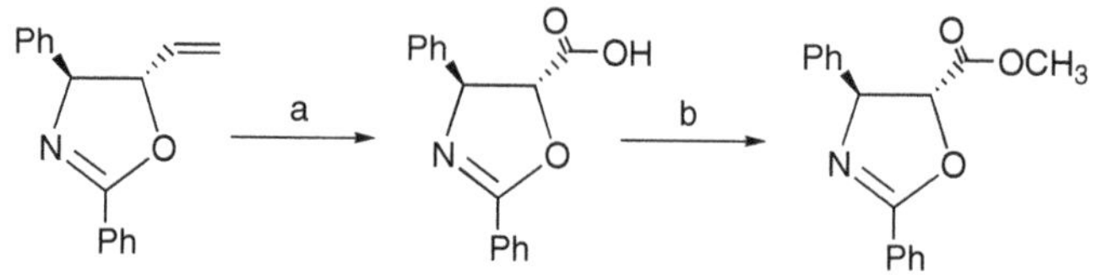

Figure 8.16 (a) RuCl$_3$, NaIO$_4$, NaHCO$_3$, H$_2$O/CCl$_4$/CH$_3$CN, 2 days, 78%; (b) CH$_2$N$_2$, (C$_2$H$_5$)$_2$O, 100%.

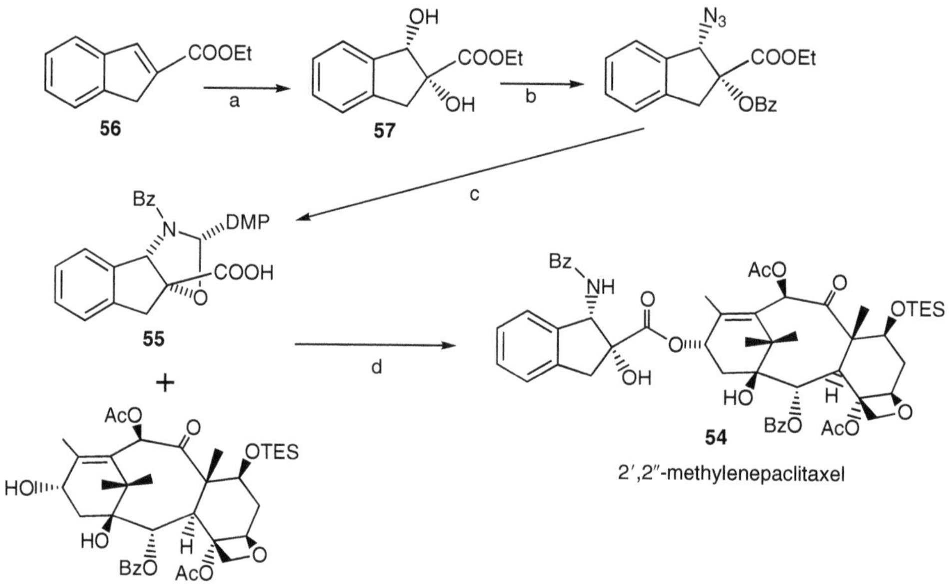

Figure 8.17

functional group manipulation, the protected oxazolidine **55** was obtained in overall 10% yield from **56**. In spite of the increased hindrance, this carboxylic acid was smoothly coupled to 7-triethylsilylbaccatin III (**58**). After deprotection, 2″, 2″-methylenepaclitaxel (**54**) was obtained in a 35% yield from 7-triethylsilylbaccatin III (Barboni *et al.*, 1998) (Figure 8.17).

Reduction of methyl 3-chloro-2-oxo-3-phenylpropanoate with various reducing agents gave *syn*- and *anti*-3-chloro-2-hydroxy-3-phenylpropanoates **59**, which underwent an efficient lipase-catalyzed resolution. All four diastereomers were subsequently converted to *N*-benzoyl-(2*R*, 3*S*)-3phenylisoserine methyl ester, C-13 side-chain analogues of paclitaxel (Hamamoto *et al.*, 2000) (Figure 8.18).

Reduction of ethyl 2-chloro-3-phenyl-3-oxopropionate with borohydride predominately effects the *syn*-chlorohydrin. Resolution of this ester with the lipase MAP-10 gives (2*S*, 3*R*)-2-chloro-3-hydroxypropionic acid, which after esterification with MeOH/HCl is converted to the *cis*-epoxide with potassium carbonate and DMF. Aminolysis of the epoxide using aqueous ammonia results in ring opening and amide formation. The amide is converted to an isobutyl ester upon treatment with isobutyl alcohol and HCl(g) at 100°C. Neutralization of the

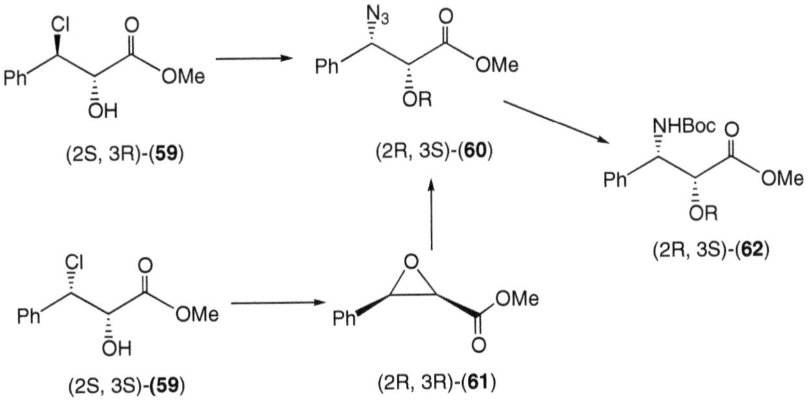

Figure 8.18 Reagents and conditions: (a) NaN₃, DMF, 60°C, 42 h, 82%; (b) NaOMe, MeOH, 0°C, 8 h, 27%; (c) NaN₃, NH₄Cl, acetone/water, reflux, 24 h, 75%; (d) (Boc)₂O, Pd/C, EtOAc, rt, 48 h, 70%.

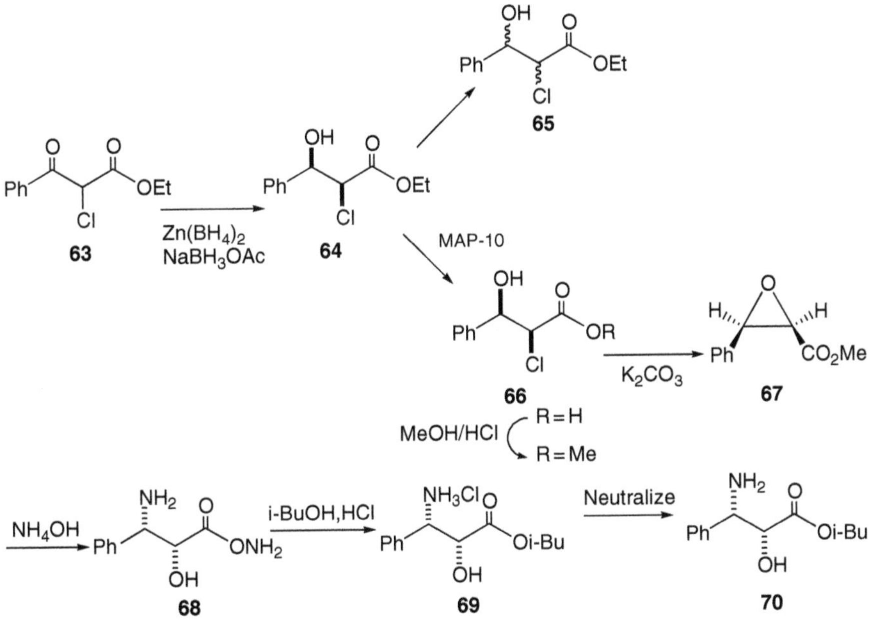

Figure 8.19

ammonium salt then affords the taxol side-chain as the free amine (Wuts *et al.*, 2000) (Figure 8.19).

Addition of dimethyl phosphite to racemic *N*-Boc-phenylglicinal led to a 75 : 25 mixture of *syn*- and *anti*-dimethyl 2-[(*tert*-butoxycarbonyl)amino]-1hydoxy-2-phenylethylphosphonates (72). The *syn*-diastereoisomer was obtained in a 50% yield after a single crystallization. Resolution of the *syn*-isomer was achieved via the (*S*)-*O*-methylmandelate esters. Racemization-free

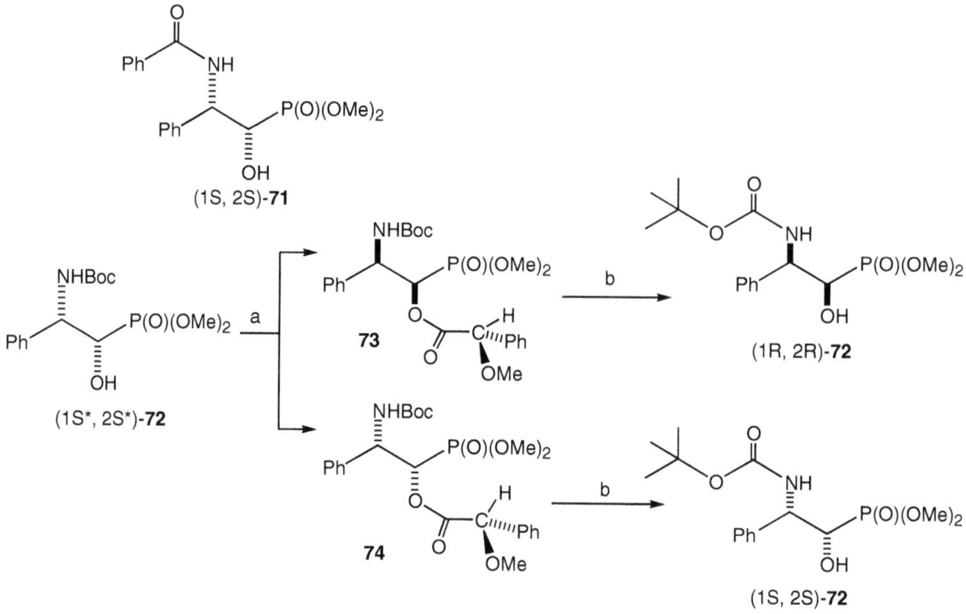

Figure 8.20 Reagent and conditions: (a) (S)-Ph(MeO)CHCOOH (75), DCC, DMAP, (b) aqueous NH₃, THF.

ammonolysis gave both enantiomers in high enantiomeric excess. Benzoates of both *N*-Boc *syn*-enantiomers were transformed into dimethyl (1*R*, 2*R*)- and (1*S*, 2*S*)-2-(benzoylamino)-1-hydroxy-2-phenylethylphosphonates in good yields (Wroblewski and Pitrowska, 2000).

The resolution of (1*S**, 2*S**)-72 was achieved via esters with (*S*)-*O*-methylmandelic acid (75). Racemic 72 was quantitatively esterified with 75 by the route of (a) and pure esters 73 (less polar) and 74 (more polar) were obtained. Treatment of 73 with aqueous ammonia (b) followed by separation of *O*-methylmandelamide by column chromatography gave (1*R*, 2*R*)-72 in 94% yield. In a similar manner, (1*S*, 2*S*)-72 was obtained from 74 in 92% yield (Figure 8.20).

Cycloaddition to taxane skeleton

Many cycloadditions of taxoids have been reported. The treatment of keto aldehyde 76 with low-valent Ti results in a stereoselective intramolecular pinacol coupling that produces the taxane intermediate 77 (Swindell *et al.*, 1993) (Figure 8.21).

A palladium(0) catalyzed intramolecular Heck olefination was used to establish the C10−C11 connection in a highly functionalized taxane analog. In the key step, reaction of 78 with palladium(II) acetate, afforded an 80% yield of cyclized 79. But interestingly 80, the corresponding C1−C2 diastereomer of 78, failed to undergo any discernable intramolecular Heck reaction (Masters *et al.*, 1993) (Figure 8.22).

The intramolecular Ni(II)/Cr(II)-mediated coupling reaction of activated olefins with aldehydes is studied in order to identify ideal arrangement of functionality for construction of the taxane ring system. Under the standard conditions 81a and 81b, afforded 82a and 82b respectively, each as a single diastereomer (Kress *et al.*, 1993a,b) (Figure 8.23).

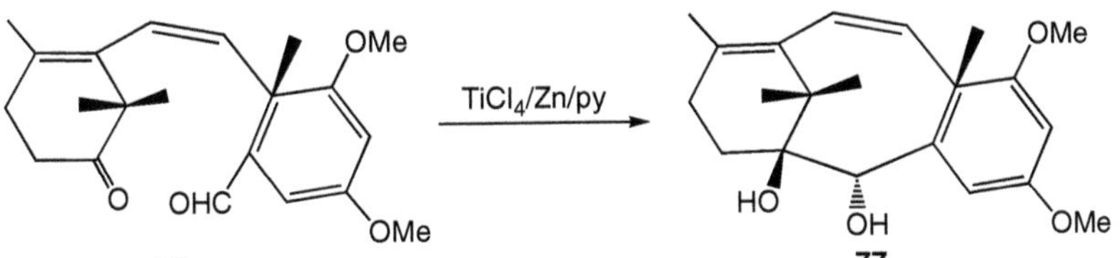

76 → 77   TiCl₄/Zn/py

Figure 8.21

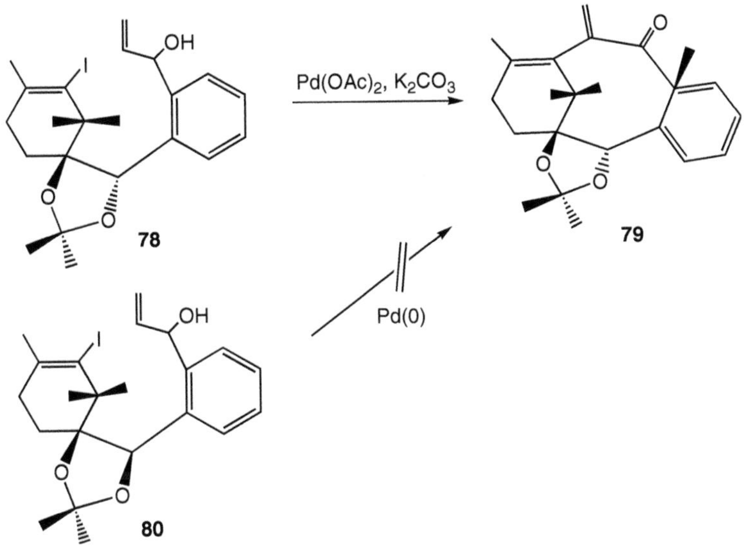

Figure 8.22 An intramolecular Heck olefination reaction.

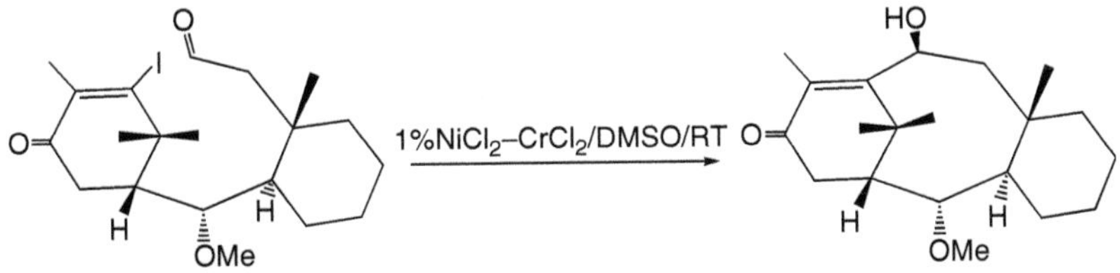

81a: R = H
81b: R = Me

82a: R = H
82b: R = Me

Figure 8.23 Intramolecular Ni(II)/Cr(II)-mediated coupling reaction.

A direct, intramolecular Diels–Alder route to functionalized tricyclic taxanes is described. This will facilitate the preparation of potentially useful analogues and ultimately the total synthesis of taxol. A cyclohexenone is converted to the trimethylcyclohexene, then the diene and acetylenic side-chains attached by sequential nucleophilic additions. Removal of a trimethylsilyl protecting group and Dess-Martin oxidation afforded the triene 83. Microwave assisted thermal

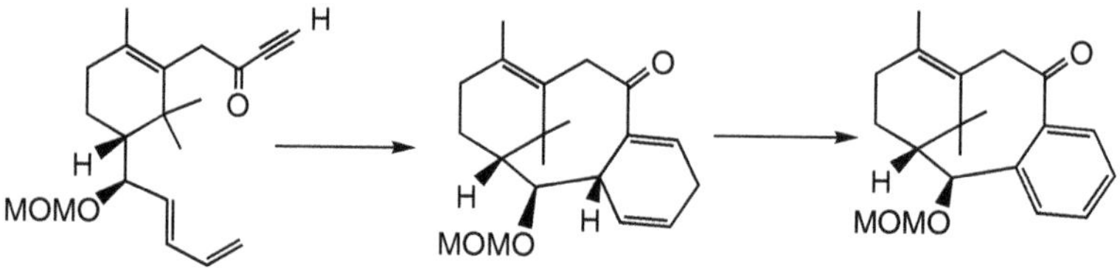

Figure 8.24 An intramolecular Diels–Alder approach to tricyclic taxoid skeleton.

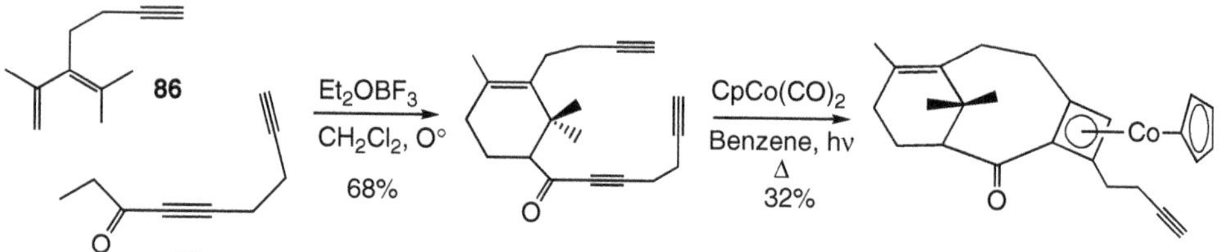

Figure 8.25

cyclization generated the tricyclic ketone **84** stereoselectively (Hoz *et al.*, 2000). Its structure was conclusively established by conversion to the aromatic system **85** upon treatment with DDQ (Lu and Fallis, 1993) (Figure 8.24).

In various intramolecular [4 + 2] cycloaddition strategies directed toward the eventual total synthesis of taxol, the initial model studies employed a substituted cyclohexene to hold the dienophile and the diene in proximity for the cycloaddition to yield the requisite tricyclo [9,3,1,0^{3,8}] pentadecene skeleton. Routes to ring A building blocks and the appropriate dienes were developed and elaborated into more highly substituted Diels–Alder precursors. Unfortunately, these failed to cyclize and a new strategy was selected in which a tartrate "tether control group" both directed the initial cycloaddition to a substituted decalin and imposed the required asymmetry. Ring cleavage followed by selective functional group manipulation then afforded a new diene–dienophile combination from which the desired tricyclic nucleus could be constructed by Lewis acid catalyzed cycloaddition (Fallis, 1997).

The AB taxane ring system has constructed through a [4 + 2] reaction and Co(I)-[2 + 2] cyclization. For the first time, CpCo(CO)$_2$ was used to catalyze ring closure of a strained polyunsaturated triynic compound into an eight-membered ring. The intermolecular Diels–Alder cycloaddition of **86** with **87** proceeded with BF$_3$Et$_2$O catalysis at 0°C in CH$_2$Cl$_2$ to give cyclohexene **88** in 68% yield. The addition of a stoichiometric amount of CpCo(CO)$_2$(ζ-cyclopentadienyldicarbonyl cobalt) and irradiation either in refluxing benzene or toluene furnished the red crystalline compound **89** as a single diastereomer in 32% yield (Phansavath *et al.*, 1998) (Figure 8.25).

Treating Hagmann's ester derivative **90** with Me$_2$CuLi, together with *in situ* trapping by TMSCl, furnished silyl enol ether **91**. Mukaiyama-type cyclization of the latter compound gave the tricyclic derivative **92**, which is structurally related to the taxane core (Cave *et al.*, 1998) (Figure 8.26).

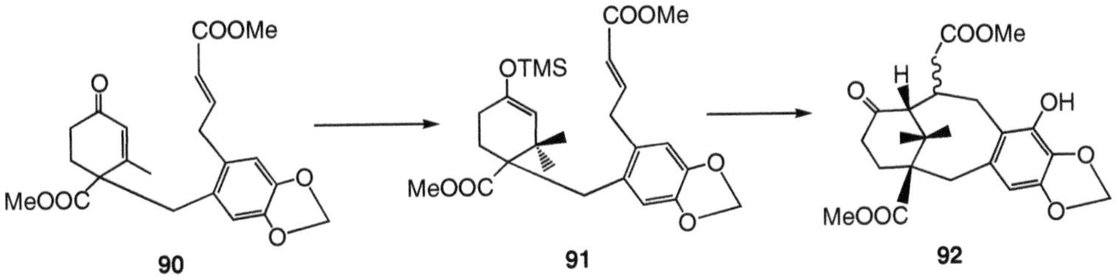

Figure 8.26

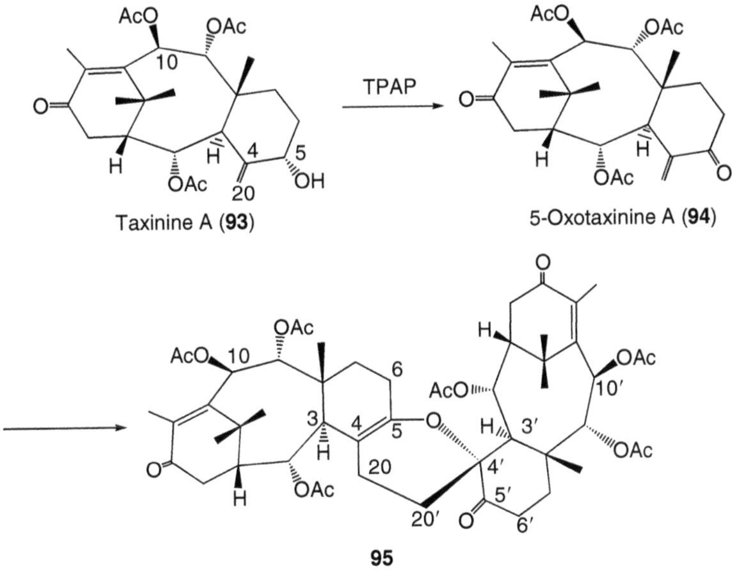

Taxinine A (**93**)

5-Oxotaxinine A (**94**)

95

Figure 8.27 A dimer through Diels–Alder cycloaddition.

Cycloaddition and formation of dimer

Oxidation of taxinine A (**93**) with tetrapropylammonium perruthenate afforded 5-oxotaxinine A (**94**). Subsequently a new dimeric compound (**95**) was obtained through regio- and stereospecific Diels–Alder cycloaddition. The relative stereostructure of **95** was established from spectral data and X-ray analysis (Hosoyama *et al.*, 1998) (Figure 8.27).

Boron trifluoride-catalyzed reaction of β-4(20)-epoxy-5-triethylsilyltaxinine A (**96**) gave both 3,5-diene (**98**) and 3,8-cyclopropane (**99**) derivatives, while the similar reaction of the corresponding α-4(20)-epoxide (**97**) afforded the ring contracted derivative (**100**) and its hemiacetal dimer (**101**). Plausible mechanisms were proposed, including 1,2-hydride shift (**98** and **99** from **96**) or pinacol-type rearrangement (**100** and **101** from **97**) (Hosoyama *et al.*, 1999) (Figure 8.28).

5-Hydroxytaxinine B (**102**) was readily prepared following known methodology (Bathini *et al.*, 1994) from taxinine B. Epoxidation of **102** using *m*-chloroperbenzoic acid (*m*CPBA) gave a sole product which was determined to be α-4(20)-epoxy-5-hydroxytaxinine **103**. Dehydration of

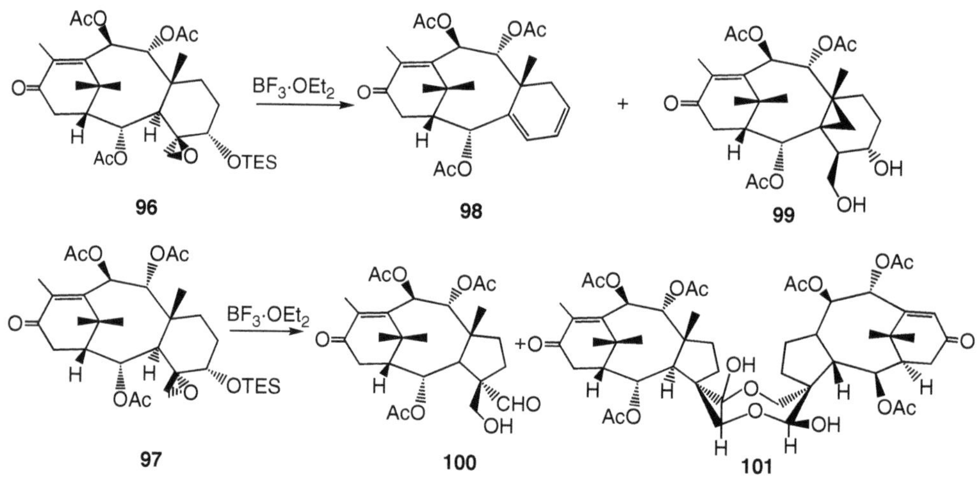

Figure 8.28

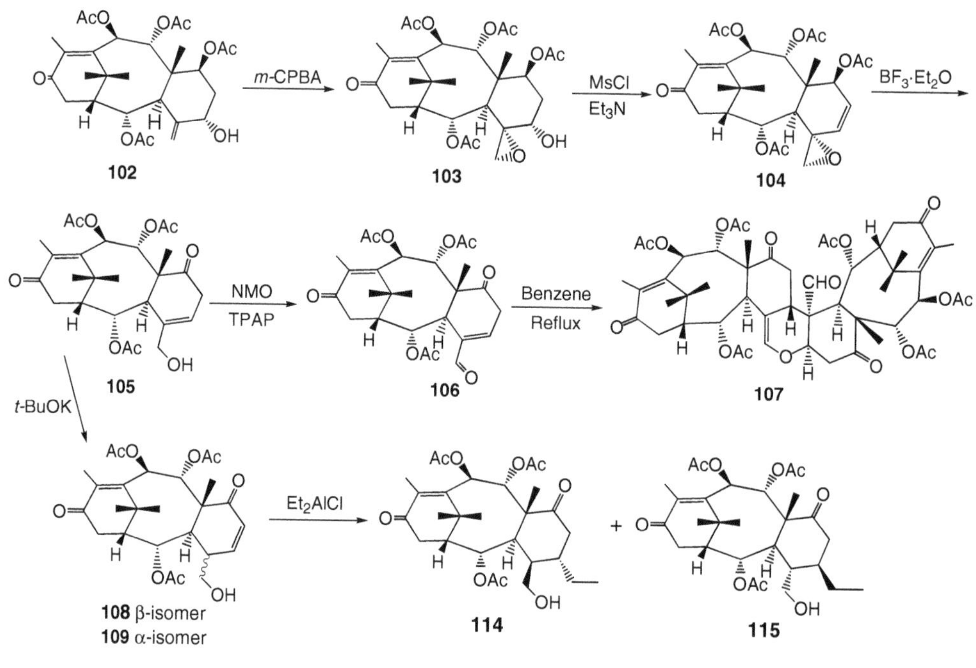

Figure 8.29

103 using methanesulfonyl chloride led to α-4(20)-epoxy-5,6-*ene*-taxinine B **104**, which subsequently was treated with boron trifluoride diethyl etherate in dichloroethane at −78°C for 2 h, resulting in the rearranged alcohol **105** in 62% yield. Oxidation of **105** with tetrapropylammonium perruthenate together with 4-methylmorpholine N-oxide (TPAP/NMO) easily yielded α,β-unsaturated aldehyde **106**. Regio- and stereospecific Diels–Alder cycloaddition in benzene for 8 h under reflux. readily afforded a new dimeric compound **107** as a sole product in

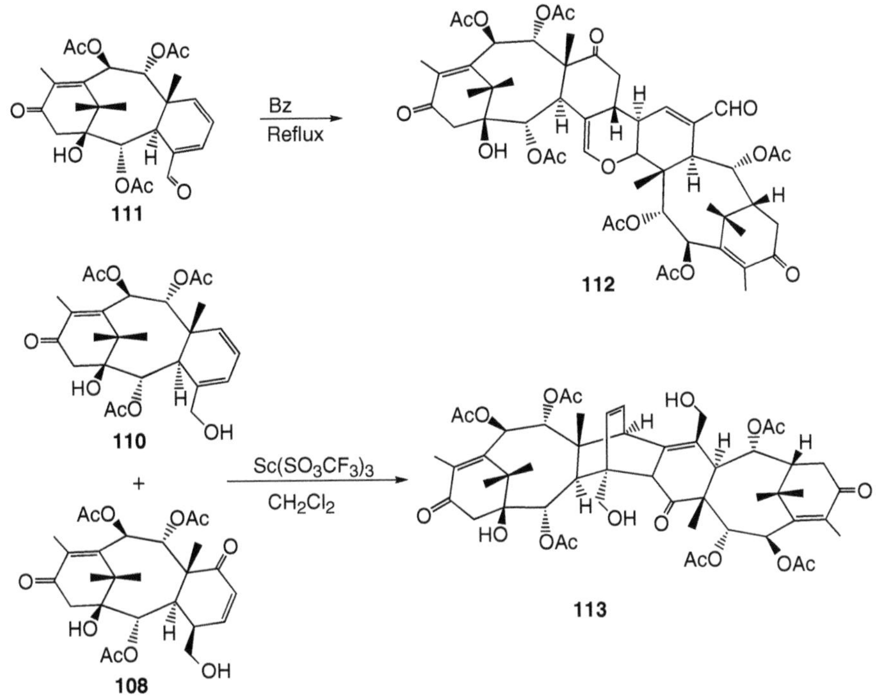

Figure 8.30

a 90% yield. Only 50% conversion of 106 to 107 was observed when 106 was allowed to stand at room temperature for two days (Cheng *et al.*, 2000a) (Figure 8.29).

Treatment of 105 with potassium *t*-butoxide gave a mixture of β-isomer 108 and α-isomer 109 in a 4:1 ratio. A mixture of 108 and 109 underwent stereoselective Michael addition instead of Diels–Alder reaction in the presence of diethylaluminum chloride leading to 4β-hydroxymethylene-5α-ethyl-7-oxotaxinine B 114 and 4α-hydroxymethylene-5β-ethyl-7-oxotaxinine B 115.

Similar hetero Diels-Alder reaction of 4:5,6:7-diene aldehyde 111 gave two new dimeric taxoids 112. Lewis acid-catalyzed Diels–Alder cycloaddition between 108 and 110 took place in the presence of Sc(Otf)$_3$ for 12 h, to give *exo*-cycloadduct 113 (Cheng *et al.* 2000a) (Figure 8.30).

Interconversion

Interconversions between structurally related taxoids has also been reported.

7,10-Dideoxy taxol 119 was prepared from baccatin III in 4 steps via the Barton deoxygenation reaction. The key reaction was the tributyltin hydride-mediated direct reduction of the C-10 acetate. Treatment of xanthate 116 with five equivalents of Bu$_3$SnH in benzene at 80°C yielded the corresponding 7-deoxy derivative (117). However, under slightly different conditions (10 equiv. Bu$_3$SnH, toluene, 100°C, 12 h), 7,10-dideoxy baccatin (119) was produced in a high yield. Furthermore, treatment of 7-deoxy derivative (117) with six equivalent of Bu$_3$SnH (100°C in toluene) afforded the 7,10-dideoxy baccatin derivative (118) in a 90% yield (Chen *et al.*, 1993) (Figure 8.31). Similarly, 10-deoxy-7-*epi* taxol (121) was obtained in an excellent yield from 7-*epi* taxol (120) in one step. These result showed that, after reduction at C-7,

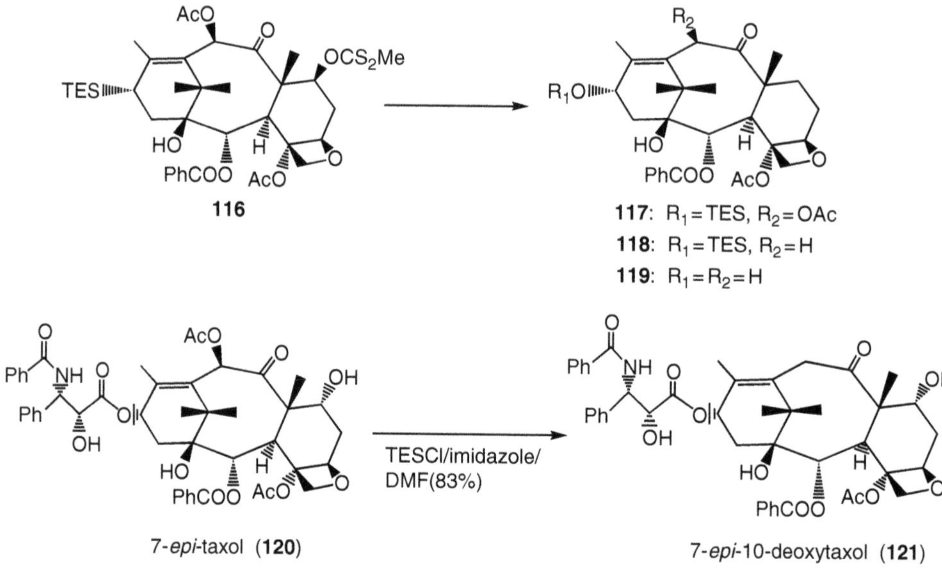

Figure 8.31 Synthesis of 7,10-dideoxytaxol and 7-*epi*-10-deoxytaxol.

deoxygenation at C-10 can be readily effected under typical Barton conditions (Khoo and Lee, 1968) (Figure 8.31).

Two C-2' hydroxymethyl analogues of docetaxel (taxotere) were synthesized from 10-desacetyl baccatin III and enantiopure (3S)-3-(N-*tert*-butoxycarbonylamino)-2-methylene-3-phenylpropanolic acid. The latter compound was prepared using a new method, which is discussed in the report, as is the biological evaluation of the two new analogues (Genisson *et al.*, 1996).

In the presence of DCC and DMAP (warm toluene solution, for 72 h), the acid underwent smooth coupling with the 7,10-*bis*(trichloroethoxycarbonyl) derivative of 10-DAB (122) to provide ester 123, in an 83% yield. On treatment with a catalytic amount of OsO_4 and excess trimethylamine N-oxide, the double bond of this acrylate underwent vicinal dihydroxylation to give a separable 70:30 mixture of diols 124 and 125 in a 82% combined yield. Deprotection of the C-7 and C-10 hydroxyl groups in 124 and 125 under the usual conditions proceeded uneventfully to yield the targeted diastereomeric docetaxel analogues 126 and 127, respectively, each in c.70% yield (Figure 8.32).

An efficient and regioselective method for the N-debenzoylation of paclitaxel has been developed and applied to the conversion of paclitaxel to 10-acetyldocetaxel and to docetaxel (Jagtap and Kingston, 1999). Treatment of paclitaxel with benzylchloroformate and dimethylaminopyridine (DMAP) in dichloromethane, followed by the addition of chlorotriethylsilane and imidazole, gave the 2',7-O-protected derivative 128 in a 95% yield in a one-pot reaction. Treatment of 128 with di-*tert*-butyl dicarbonate and DMAP in acetonitrile provided the N-protected compounds 129 and 130. Compound 129 was found to be very labile in acid, and was converted to 130 on standing in the presence of slightly acidic $CHCl_3$ for 2–3 h; the overall yield for the conversion of 128 to 130 was 86%. Debenzoylation and deacetylation of 130 was achieved using magnesium methoxide in methanol to give 7-O-triethylsilyl-10-acetyldocetaxel (131); Selective N-debenzoylation was achieved to give 131 in 72–75% isolated yield based on unrecovered starting material. The reaction requires careful monitoring to avoid the formation of 7-O-triethylsilylbaccatin III (132) (Jagtap and Kingston, 1999) (Figure 8.33).

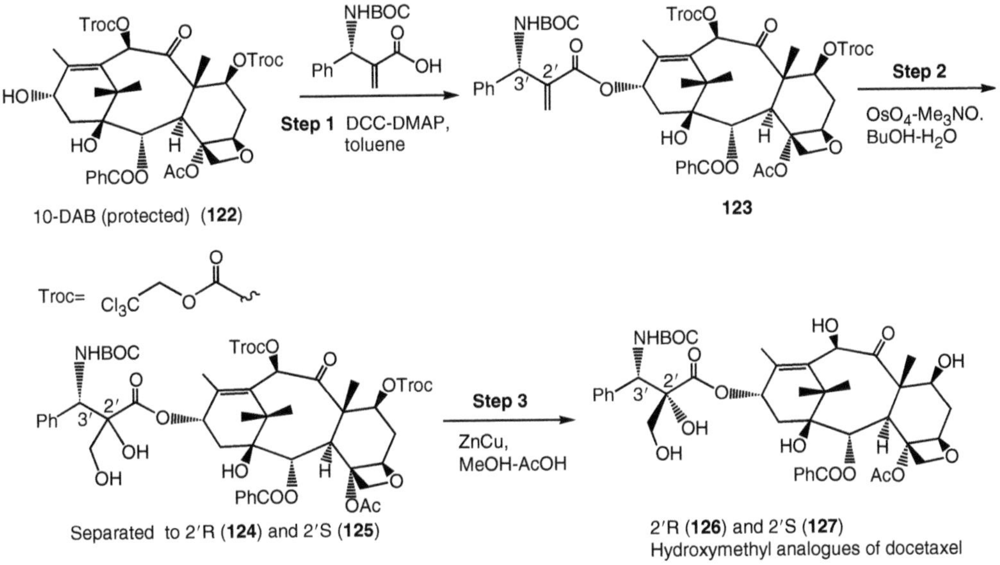

Figure 8.32

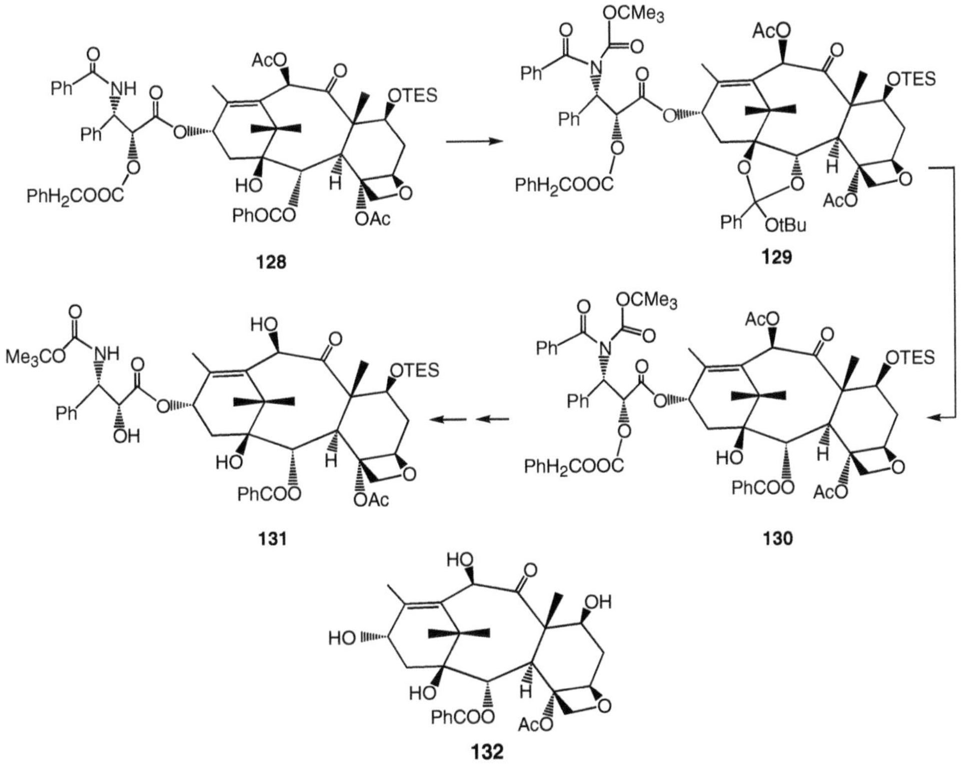

Figure 8.33 Taxol to docetaxel.

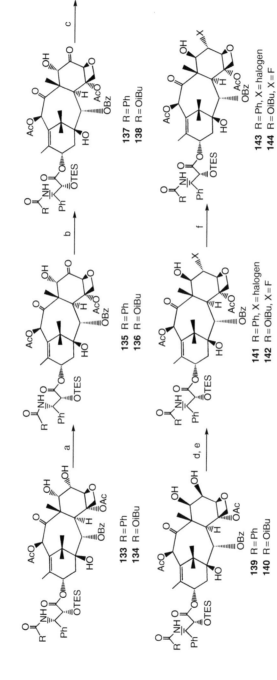

133 R=Ph
134 R=OiBu

135 R=Ph
136 R=OiBu

137 R=Ph
138 R=OiBu

139 R=Ph
140 R=OiBu

141 R=Ph, X=halogen
142 R=OiBu, X=F

143 R=Ph, X=halogen
144 R=OiBu, X=F

Figure 8.34 Reagents and condition: (a) 4-BzO-TEMPO (2%), KBr, Chlorox®; (b) silica gel, CH₂Cl₂; (c) Na(OAc)₃BH, HOAc, CH₃CN (75% overall yield for a–c); (d) SOCl₂, Et₃N; NaIO₄, RuCl₃, CCl₄/CH₃CN/H₂O (77%); (e) Bu₄NX (F, Cl, Br), THF (76%); (f) 1 N HCl, acetonitrile (85%).

The stereospecific syntheses of the metabolically blocked 6-α-F, Cl, and Br taxols, and 6-α-F-10-acetyldocetaxel has been described together with *in vitro* and *in vivo* activity (Wittman *et al.*, 2001a) (Figure 8.34).

Semisynthesis

10-Deacetoxytaxol (146) and 10-deoxytaxotere (147) were prepared from 10-deacetylbaccatin III (145) by attaching the C-13 side-chain and deoxygenation. 10-Deacetoxytaxol and taxol are comparable to taxol in their cytotoxicity against P-388 cells, but 10-deoxytaxotere is significantly more cytotoxic than taxol (Chaudhary and Kingston, 1993) (Figure 8.35).

Taxotere (docetaxel) was synthesized through an efficient coupling of 7,10-diTroc-10-deacetylbaccatin III (149) with enantiomerically pure (3R, 4S)-1-*t*BOC-3-EEO-4-phenylazetidin-2-one, which was obtained via a chiral ester enolate – imine cyclocondensation (Ojima *et al.*, 1993) (Figure 8.36).

A novel deacylation reaction was used to prepare 9-dihydrotaxol (151). This analogue exhibits increased activity in the microtuble assay (Klein, 1993).

Arseniyadis *et al.* (1998a) described an aldol-annelation–fragmentation strategy culminating in the stereoselective synthesis of the entire 20-carbon A-seco taxane framework. This concise

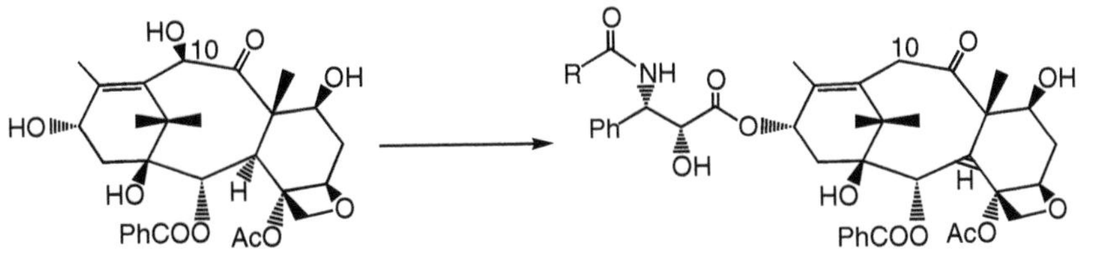

10-Deacetylbaccatin III (145)

146 10-Deacetoxytaxol R = Ph
147 10-Deoxytaxotere R = (CH₃)₃CO

Figure 8.35 Synthesis of 10-deacetoxytaxol and 10-deoxytaxotere.

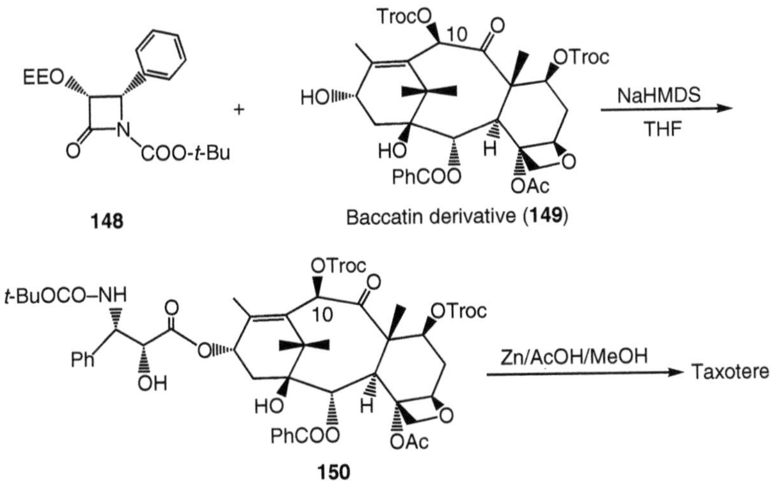

148

Baccatin derivative (149)

150

Figure 8.36 A highly efficient route to taxotere by the β-lactam synthon method.

route to an optically pure A-*seco* taxoid framework took only 12 linear steps starting from the two achiral aldol partners. In addition, they reported a four-step A + C route to an optically pure taxoid ABC framework also containing suitable functionalities for further elaboration (Arseniyadis *et al.*, 1998b,c). Docetaxel was synthesized by a novel route using esterification of (2*R*, 3*R*)-glycidic acid with 7,10-bis-O-(2,2,2-trichloroethoxycarbonyl)-10-deacetylbaccatin (Yamaguchi *et al.*, 1998). Starting from R-carvone, enantiospecific synthesis of functionalized chiral C-ring taxane derivatives was described (Srikrishna *et al.*, 1998).

A new method for the semisynthesis of paclitaxel from baccatin III via a dioxo-oxathiazolidine intermediate was reported. Compound 153 was coupled 7-(triethylsilyl)baccatin III (152) in the presence of dicyclohexylcarbodiimide (DCC) and 4-(dimethylamino)pyridine (DMAP). The product was not the expected dioxothiazolidine derivative, but rather the previously prepared oxazoline derivative 154, which then was converted to taxol (1) (Baloglu and Kingston, 1999) (Figure 8.37).

The synthesis and biological activity of new 7-deoxypaclitaxel analogues 160 and 161 in which c-AMP is present at C-9 the hydroxy group at C-2′ of the side-chain and C-10 in the B-ring are substituted by cAMP and benzoyloxy group. 9-cAMP taxoid 157 was prepared in a 84% yield from 155 by a coupling with (−)-adenosine 3′,5′-cyclic phosphate chloride 156. Compound 157 was then treated with TESCl followed by selective reduction using $NaBH_4$ to afford the desired 13α-hydroxy-9-cAMP taxoid 159. Finally, the 9-cAMP-7-deoxypaclitaxel 160 was accomplished in a 74% yield in overall through esterification of 159 with 161 followed by acidic hydrolysis (Cheng *et al.*, 2000b) (Figure 8.38).

4(20),11(12)-Taxadiene 162 is specifically converted to 4(5),11(12)-taxadiene 166 via a Claisen-like rearrangement and Barton radical desulfurization (cleavage of C—S bond) in high yield. Treatment of 162 with hydroxylamine at 80°C afforded 5-hydroxy derivative 163 in high yield. Treatment of 163 with carbon disulfide and sodium hydride in THF gave the 20-dithiocarbonate derivative 165 in excellent yield. The proposed mechanism involves an intermediate 164, which undergoes Claisen-like rearrangement to give the desired product 165. Dithiocarbonate 165 was reduced with tributyltin hydride upon heating at 150°C in *p*-cymene or xylene to produce 4(5),11(12)-taxadiene 166 in good yield. This reaction most likely proceeds

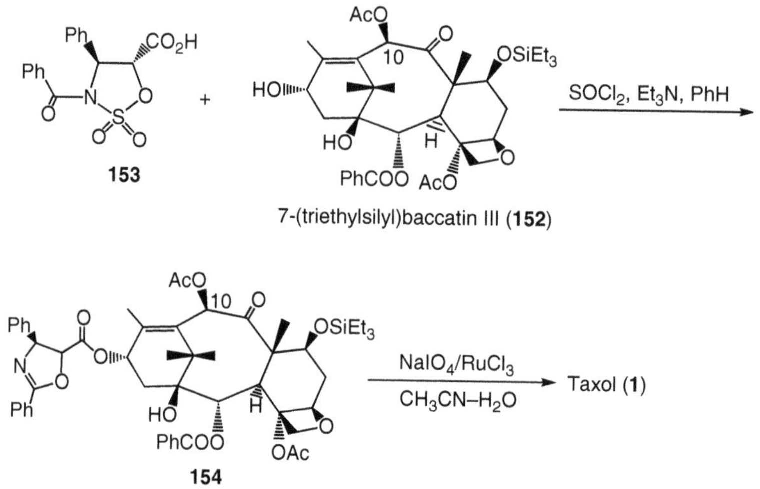

Figure 8.37 Semisynthesis of paclitaxel from baccatin III.

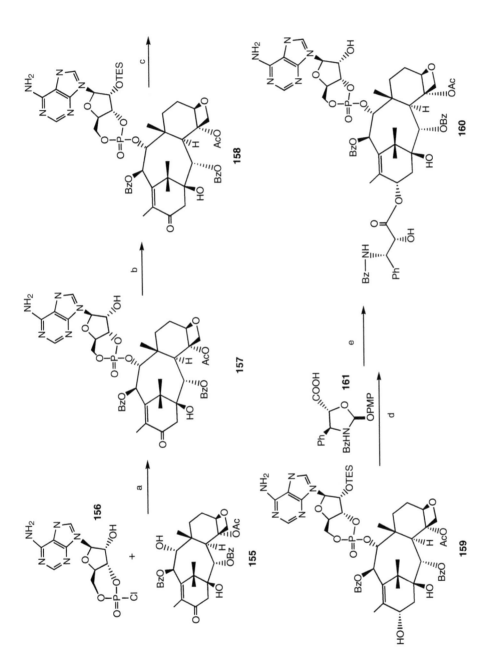

Figure 8.38 Reagents and conditions: (a) DMAP, DMF, rt, 84%; (b) TESCl, imidazole, DMAP, rt, 89%; (c) NaBH₄, CH₃OH, rt, 77%; (d) DCC, DMAP, toluene, 75°C; (e) TFA, H₂O, 0° to rt, 74% from 159.

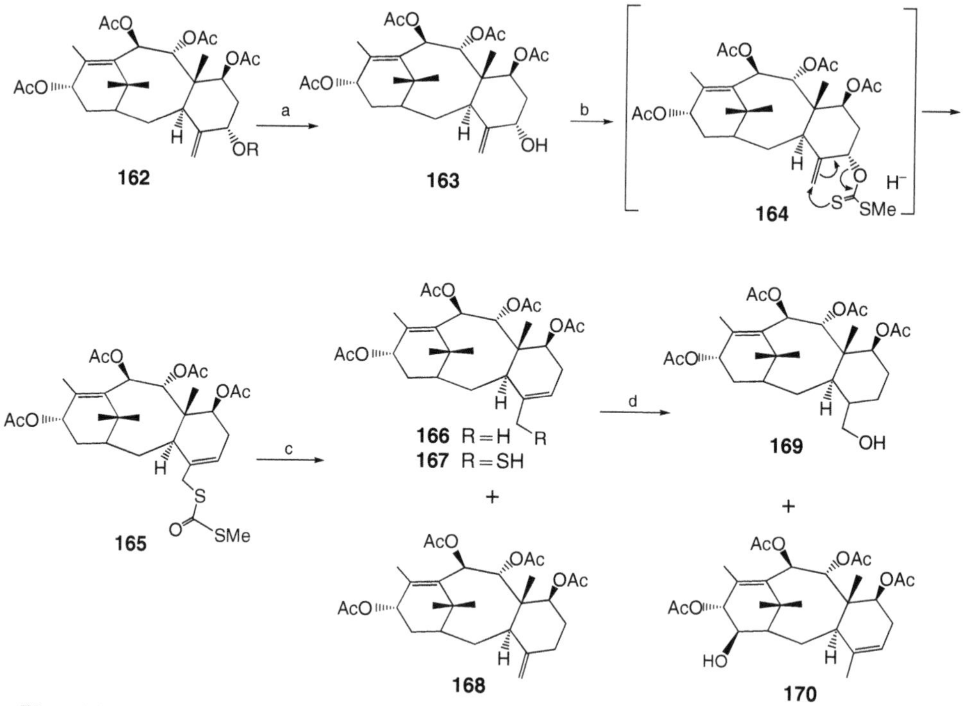

Figure 8.39

by tributyltin mediated radical fission of the allylic C—S bond, rather than the usual C—O bond cleavage. The rearranged product, 4(20),11(12)-taxadiene **168** is also produced in low yield. The minor product **167** corresponds with thioester bond cleavage, further supporting the Claisen-like rearrangement mechanism. Taxoid **166** and its deoxygenated derivatives are potential biosynthetic precursors of paclitaxel and other taxoids. 20-hydroxylated metabolite **169** is the major product from microbially mediated hydroxylation of **166**.

Microbial enzymatic systems may be useful tools to mimic biosynthetic steps of biosynthesis, including extensive oxidation of the taxane skeleton. In the case of compound **166**, hydroxylation was expected to proceed either at the 5-position followed by an allylic shift of the double bond from 4(5) to 4(20) to produce compound **163**, or at the 20-position to produce compound **169**. Substandard incubation conditions and biotransformaton of **166** by *Absidia coerula*, a filamentous fungus, gave one major product **169** in moderate yield. Recently, this 20-hydroxylated taxoid (**169**) has been isolated from *T. mairei*. (Shi *et al.*, 1999). In addition, a minor metabolite, 14β-hydroxylated derivative **170** was also produced (Hu *et al.*, 2000) (Figure 8.39).

Conjugation with other active components

C-4 acylation of **171** was effected using monobenzyl glutamate under DCC/DMAP condition, affording the new compound **172** in a 86% yield. Compound **172** was treated with OsO$_4$ and N-methylmorpholine-*N*-oxide (NMO) in THF/H$_2$O for 24 h, affording a more polar product **173** in a 81% yield. Compound **173** was hydrogenated to afford the acid **174** in a 87% yield and reaction of **174** in THF at −78°C with 10 equivalents of phenyl lithium proceeded regioselectively to give the desired benzoate **175** in good yield. Compound **175** was protected as its

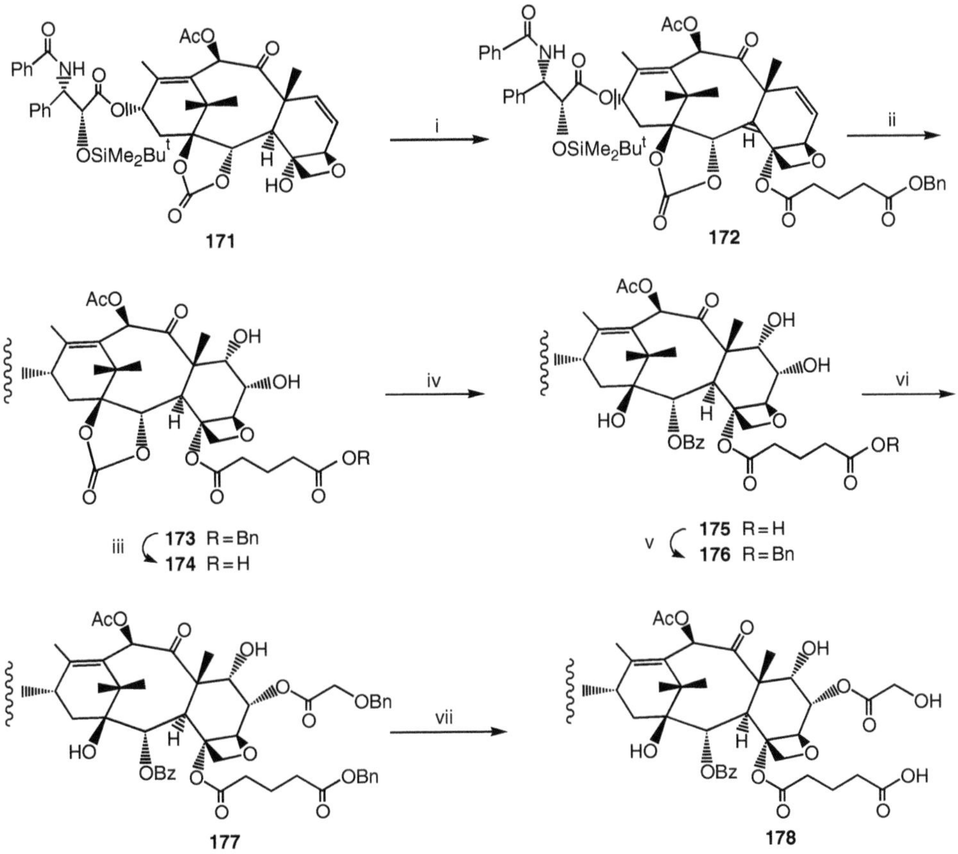

Figure 8.40 (i) PhCH$_2$OOC(CH$_2$)$_3$COOH, DCC, DMAP, toluene, 86%; (ii) OsO$_4$, NMO, THF/H$_2$O, 81%; (iii) H$_2$, Pd/C, EtOAc, 87%; (iv) PhLi, THF, $-78°$C, 78%; (v) PhCH$_2$OH, DCC, DMAP, toluene, 94%; (vi) BnOCH$_2$COCl, DMAP, CH$_2$Cl$_2$, 85%; (vii) H$_2$, Pd/C, EtOAc, 71%.

benzyl ester to give **176**. Reaction of **176** with 2-benzyloxyacetyl chloride proceeded smoothly and cleanly in a 85% yield to give the C-6α ester **177**. Hydrogenation of compound **177** unmasked both of the benzyl protecting groups and afforded the key intermediate hydroxy acid **178** in a 71% yield (Figure 8.40).

The novel taxol analog **181**, which contains a bridge between the C-4 and C-6 positions, was synthesized and found to be less active than taxol in *in vitro* tublin assembly and cytotoxicity assays. Slow addition of a solution of the hydroxy acid **179** and triethylamine in acetonitrile to a solution of acetonitrile containing 2-chloro-1-methylpyridinium iodide refluxing at 80°C over 8 h afforded two less polar products, along with starting material and other minor products that were not isolated. One of the major products isolated in a 10% yield was identified as the desired lactone **180**, and the other major product isolated in a yield of 12% was tentatively identified as the dilactone **182** (Yuan and Kingston, 1999) (Figure 8.41).

The synthesis and antitumor activity of three paclitaxel-chlorambucil hybrids were reported. Hybrid **183** showed significant *in vivo* efficacy (Wittman *et al.*, 2001) (Figure 8.42).

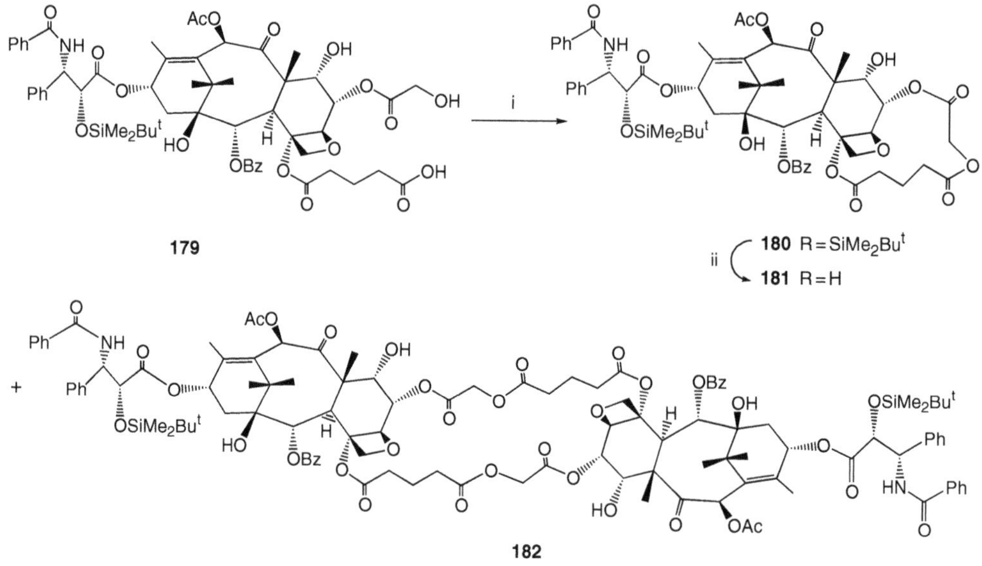

Figure 8.41 (i) 2-Chloro-1-methylpyridinium iodide, Et$_3$N, CH$_3$CN, 80°C, 14 (10%), 16 (12%); (ii) HF/pyridine, 64%.

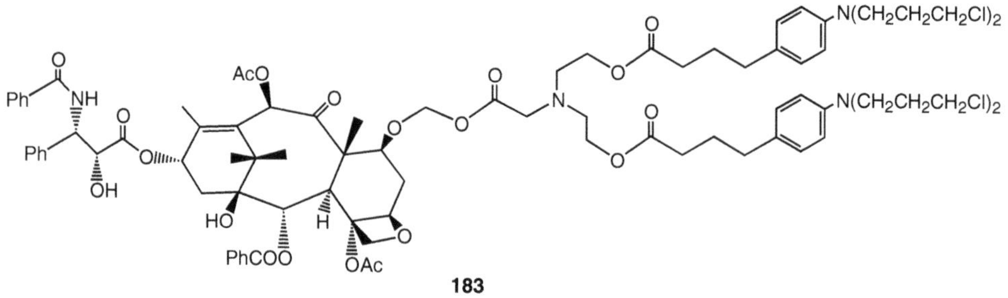

Figure 8.42

Bifunctional, heterodimeric compounds were synthesized to test their ability to create polyvalent arrays between DNA and microtubles in cells. Each dimer was examined for the capacity to bind to microtubles and cytotoxicity against MES-SA and MES-SA/Dx/5 cell lines.

The procedure of Nicolau *et al.* (Guy *et al.*, 1996), was used to prepare the amino-functionalized taxol, resulting in 7-β-alanyltaxol **184**. Daunorubicin (**185**) was functionalized with the linking groups by reacting the aminosugar with one end of a dicarboxylic acid or its equivalent to afford daunorubicin acids **186a–f**. Dimer synthesis was achieved in the final step by coupling **184** and **186a–f**. The target molecules **187a–f** were obtained in a 40–70% yield by chromatographic purification. The chromatographic mobilities, polarities, and solubilities of the heterodimeric molecules were more similar to those of taxol than those of daunorubicin (Kar *et al.*, 2000) (Figure 8.43).

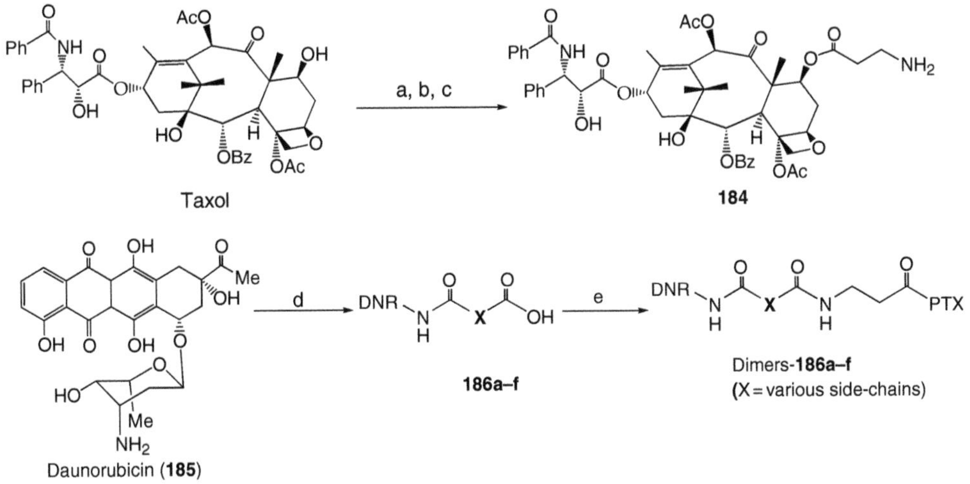

Figure 8.43 (a) Benzylchloroformate, pyr, CH$_2$Cl$_2$, 65%; (b) N-Cbz-β-alanine, DCC, DMAP, CH$_2$Cl$_2$, 95%; (c) H$_2$, Pd/C, MeOH, 95%; (d) Condition 1: succinic, glutamic or phthalic anhydride (5 equiv.), 2,6-lutidine (2 equiv.), MeOH, rt, overnight, yield: 56–90%. Condition 2: terephthalic acid or trans-β-hydromuconic acid (2 equiv.), PyBrop (1.1 equiv.), DIPEA (2 equiv.), DMF, rt 30 min, yields; 35–50%. Condition 3: adipic acid monoethyl ester (2 equiv.), PyBroP (1.1 equiv.), DIPEA (2 equiv.), DMF, rt, 30 min, then LiOH (3 equiv.), THF H$_2$O (3 : 1 v/v), 40%; (e) PTX (184) (0.9 equiv.), PyBroP (1.1 equiv.), DIPEA (2 equiv.), THF, rt, 30 min, yields; 40–70%.

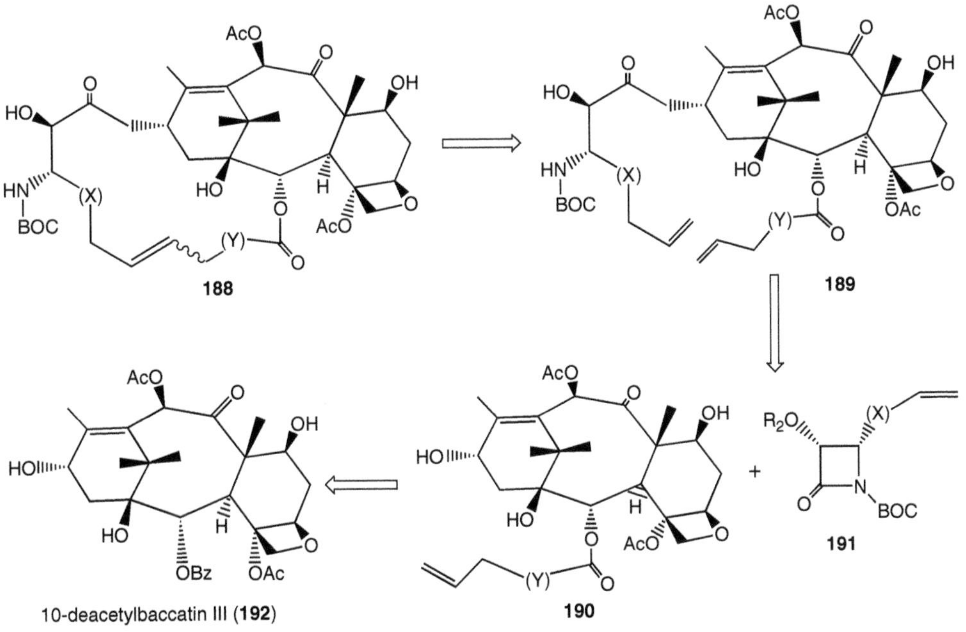

Figure 8.44

A series of novel macrocyclic taxoids **188** and **188′** bearing 16-, 17-, and 18-membered rings connecting the substituents at the C-2 and C-3′ positions were designed and synthesized. The synthesis of these macrocycles **188** and **188′** were accomplished using the highly efficient ruthenium-catalyzed ring-closing metathesis (RCM) of taxoid-ω,ω′-dienes **189** in the key step. Taxoid–taxoid-ω,ω′-dienes **189** were obtained through the ring-opening coupling of 4-alkenyl-β-lactams **191** with 2-alkenoylbaccatins **190** in good to high yield. Although various novel pentacyclic maclocycles **188** and **188′** were successfully synthesized, in some cases, the desired RCM did not proceed. The scope and limitations, of RCM in its application to highly functionalized complex substrates were discussed. All macrocyclic taxoids **188** and **188′** were found to be cytotoxic, with some exhibiting submicromolar IC_{50} values against a human breast cancer cell line (Ojima *et al.*, 2000) (Figure 8.44).

References

Appendino, G., Fenoglio, I., and Vander Velde, D.G. (1997) Novel base-catalyzed rearrangement of the taxane skeleton. *J. Nat. Prod.*, **60**, 464–466.

Appendino, G., Lusso, P., Garibold, P., Bombardelli, E., and Gabetta, B. (1992) A 3,11-cyclotaxane from *Taxus baccata. Phytochemistry*, **31**(12), 4259–4262.

Arseniyadis, S., Ferreira, M.R., Moral, J.Q.del, Yashunsky, D.V., and Potier, P. (1998a) Studies towards the total synthesis of taxoids , the aldol-annelation–fragmentation strategy. *Tetrahed. Lett.*, **39**, 571–574.

Arseniyadis, S., Ferreira, M.R., Moral, J.Q.del, Hernando, J.I.M., and Potier, P. (1998b) Stereoselective synthesis of an A-seco-taxoid subunit using the aldol-annelation–fragmentation strategy. *Tetrahed. Lett.*, **39**, 7291–7294.

Arseniyadis, S., Hernando, J.I.M., Moral, J.Q.del, Ferreira, M.R., Birlirakis, N., and Potier, P. (1998c) An expedient synthesis of a conveniently functionalized optically homogeneous taxoid ABC ring system. *Tetrahed. Lett.*, **39**, 9011–9014.

Baloglu, E. and Kingston, D.G.I. (1999) A new semisynthesis of paclitaxel from baccatin III. *J. Nat. Prod.*, **62**, 1068–1071.

Barboni, L., Lambertucci, C., Appendino, G., and Bombardelli, F. (1998) Synthesis of a conformationally analogue of paclitaxel. *Tetrahed. Lett.*, **39**, 7177–7180.

Bathini, Y., Micetich, R.G., and Daneshtalab, M. (1994) Synthetic studies towards taxol analogs: chemoselective cleavage of C-5 cinnamoyl group in taxane group of diterpenoids with hydroxylamine. *Synth. Commun.*, **24**, 1513–1517.

Boge, T.C., Hepperle, M., Vander Verde, D.G., Gunn, C.W., Grunewald, G.L., and George, G.I. (1999) The oxetane conformational lock of paclitaxel: structural analysis of D-secopaclitaxel. *Bioorg. & Med. Chem. Lett.*, **9**, 3041–3046.

Cave, C., Valancogne, I., Casas, R., and d'Angelo, J. (1998) Mukaiyama-type eight-membered ring closure: access to a tricyclic system related to taxanes. *Tetrahed. Lett.*, **39**, 3133–3136.

Chaudhary, A.G. and Kingston, D.G.I. (1993) Synthesis of 10-deacetoxytaxol and 10-deoxytaxotere. *Tetrahed. Lett.*, **34**(31), 4921–4924.

Chen, S.-H., Combs, C.M., Hill, S.E., Farina, V., and Doyle, T.W. (1992) The photochemistry of taxol: synthesis of a novel pentacyclic taxol isomer. *Tetrahed. Lett.*, **33**, 7679–7680.

Chen, S.-H., Wei, J.-M., Vyas, D.M., Doyle, T.W., and Farina, V. (1993) A facile synthesis of 7,10-dideoxy taxol and 7-*epi*-10-deoxy taxol. *Tetrahed. Lett.*, **34**(43), 6845–6848.

Cheng, Q., Oritani, T., Horiguchi, T., Yamada, T., and Hassner, A. (2000a) Novel dimeric taxoids via highly regio- and stereospecific Diels-Alder cycloadditions of taxinine B and taxicine I derivatives. *Tetrahedron*, **56**, 1181–1192.

Cheng, Q., Oritani, T., and Horiguchi, T. (2000b) The synthesis and biological activity of 9- and 2′-cAMP 7-deoxypaclitaxel analogues from 5-cinnamoyltriacetyltaxicin-I. *Tetrahedron*, **56**, 1667–1679.

Chiang, H.C., Woods, M.C., Nakadaira, Y., and Nakanishi, K. (1967) Structures of four new taxinine congeners, and a photochemical transannular reaction. *J. Chem. Soc., Chem. Commun.*, **23**, 1201.

Cravallee, C., Didier, E., and Pecquet, P. (1998) Methyleniminium salts as acylating agent – One step synthesis of baccatin III from 10-deacetylbaccatin III with high selectivity. *Tetrahed. Lett.,* 39, 4263–4266.

Damen, E.W.P., Braamer, L., and Scheeren, H.W. (1998) Lanthanide trifluoromethanesulfonate catalysed selective acylation of 10-deacetylbaccatin III. *Tetrahed. Lett.,* 39, 6081–6082.

Fallis, A.G. (1997) From tartrate to taxoids: a double, intramolecular Diels-Alder strategy. *Pure & Appl. Chem.,* 69(3), 495–500.

Fang, K., Hashimoto, M., Jokusch, S., Turro, N.J., and Nakanishi, K. (1998) A bifunctional photoaffinity probe for ligand/receptor interaction studies. *J. Am. Chem. Soc.,* 120, 8543–8544.

Genisson, Y., Massardier, C., Gautier-Luneau, I., and Greene, A.E. (1996) Effective enantioselective approach to α-aminoalkylacrylic acid derivatives via a synthetic equivalent of an asymmetric Baylis–Hillman reaction: application to the synthesis of two C-2′ hydroxymethyl analogues of docetaxel. *J. Chem. Soc., Perkim Trans. 1,* 2869–2872.

Guy, R.K., Scott, Z.A., Sloboda, R.D., and Nicolau, K.C. (1996) Fluorescent taxoids. *Chem. Biol.,* 3, 1021.

Hamamoto, H., Mamedov, V.A., Kitamoto, M., Hayashi, N., and Tsuboi, S. (2000) Chemoenzymatic synthesis of the C-13 side-chain of paclitaxel (taxol) and docetaxel (taxotere). *Tetrahed. Asymm.,* 11, 4485–4497.

Hernando, J.I.M., Ferreira, M.del R.R., Lena, J.I.C., Birlirakis, N., and Arseniyadis, S. (2000a) Studies towards the diterpene ABC-ring system: practical access to highly functionalized enantiomerically pure analogues of major group representatives. *Tetrahed. Assymm.,* 11, 951–973.

Hernando, J.I.M., Ferreira, M.del R.R., Lena, J.I.C., Birlirakis, N., and Arseniyadis, S. (2000b) A rapid entry into major groups of taxoids. *Tetrahed. Lett.,* 41, 863–866.

Hosoyama, H., Shigemori, H., and Kobayashi, J. (1999) Unusual boron trifluoride-catalyzed reactions of taxinine derivatives with α- and β-4(20)-epoxides, *Tetrahed. Lett.,* 40, 2149–2152.

Hosoyama, H., Shigemori, H., In, Y., Ishida, T., and Kobayashi, J. (1998) Occurrence of a new dimeric compound of 5-oxotaxinine A through Diels-Alder cycloaddition. *Tetrahed. Lett.,* 39, 2159–2162.

Hoz, A. de la, Diaz-Ortis, A., Moreno, A., and Langa, F. (2000) Cycloadditions under microwave irradiation conditions: methods and applications. *Eur. J. Org. Chem.,* 3659–3673.

Hu, S., Sun, D., and Scott, I. (2000) An efficient synthesis of 4(5),11(12)-taxadiene derivatives and microbial mediated 20-hydroxylation of taxoids. *Tetrahed. Lett.,* 41, 1703–1705.

Jagtap P.G. and Kingston, D.G.I. (1999) A facile N-debenzoylation of paclitaxel: conversion of paclitaxel to docetaxel. *Tetrahed. Lett.,* 40, 189–192.

Kar, A.K., Braun, P.D., and Wandless, T.J. (2000) Synthesis and evaluation of daunorubicin–paclitaxel dimers. *Bioorg. & Med. Chem.,* 10(3), 261–264.

Kingston, D.G. (2000) Recent advances in the chemistry of taxol. *J. Nat. Prod.,* 63, 726–734.

Kingston, D.G., Molinero, A.A., and Rimoldi, J.M. (1993) "The taxane diterpenoids", in *Progress in the Chemistry of Organic Natural Products,* 61, 1–206.

Klein, L.L. (1993) Synthesis of 9-dihydrotaxol: a novel bioactive taxane. *Tetrahed. Lett.,* 34(13), 2047–2050.

Khoo, L.E. and Lee, H.H. (1968) Hydrostannolysis reactions. Part I. Reduction of esters. *Tetrahed. Lett.,* 9, 4351–4354.

Kress, M.H., Ruel, R., Miller, W.H., and Kishi, Y. (1993a) Synthetic studies toward the taxane class of natural products. *Tetrahed. Lett.,* 34(38), 5999–6002.

Kress, M.H., Ruel, R., Miller, W.H., and Kishi, Y. (1993b) Investigation of the intramolecular Ni(II)/Cr(II)-mediated coupling reaction: application to the taxane ring system. *Tetrahed. Lett.,* 34(38), 6003–6006.

Lee, D., Kim, K.-C., and Kim, M.-J. (1998a) Selective acylation of 10-deacetylbaccatin III. *Tetrahed. Lett.,* 39, 9039–9042.

Lee, K.-Y., Kim, Y.-H., Park M.-S., and Ham, W.-H. (1998b) A highly stereocontrolled asymmetric synthesis of the taxol C-13 side chain; (4S, 5R)-2,4-diphenyloxazoline 5-carboxylic acid. *Tetrahed. Lett.,* 39, 8129–8132.

Lin, S., Fang, K., Hashimoto, M., Nakanishi, K., and Ojima, I. (2000) Design and synthesis of a novel photoaffinity taxoid as a potential probe for the study of paclitaxel-microtubles. *Tetrahed. Lett.,* 41, 4287–4290.

Lu Y.F. and Fallis, A.G. (1993) An intramolecular Diels-Alder approach to tricyclic taxoid skeleton. *Tetrahed. Lett.,* 34(21), 3367–3370.

Masters, J.J., Jung, D.K., Bomman, W.G., and Danishefsky, S.J. (1993) A concise synthesis of a highly functionalized C-aryltaxol analog by an intramolecular Heck olefination reaction. *Tetrahed. Lett.*, 34(45), 7253–7256.

Ojima, I., Sun, C.M., Zucco, M., Park, Y.H., Duclos, O., and Kuduk, S. (1993) A highly efficient route to taxotere by the β-lactam synthon method. *Tetrahed. Lett.*, 34(26), 4149–4152.

Ojima, I., Lin, S., Inoue, T., Miller M.L., Borella, C.P., Geng, X., and Walsh, J.J. (2000) Macrocycle formation by ring-closing methathesis. Application to the synthesis of novel macrocyclic taxoids. *J. Am. Chem. Soc.*, 122, 5343–5353.

Paquette, L.A., Zhao, M., and Friedrich, D. (1992) An unprecedented silyl triflate-promoted hydride shift within a taxane derivative. *Tetrahed. Lett.*, 33(48), 7311–7314.

Pandey, R., Yankov, L.K., Poulev, A., Nair, R., and Caccamese, S. (1998) Synthesis and separation of potential antitumor active dihalocephalomannine diastereomers from extract of *Taxus yunnanensis*. *J. Nat. Prod.*, 61, 57–63.

Phansavath, P., Aubert, C., and Malacria, M. (1998) The [4+2], [2+2] strategy for the construction of the AB taxane ring system. *Tetrahed. Lett.*, 39, 1516–1564.

Raber, D.J. and Guida, W.C. (1974) Preparation of ethylene acetals (1,3-dioxolanes) from 2-methoxyethyl carboxylates via 1,3-dioxolanium ions. *Synthesis*, 808–809.

Rao, K.V. and Johnson Jr., J.H. (1998) Selective nitration of paclitaxel and related taxanes. *Tetrahed. Lett.* 39, 4611–4614.

Samaranayake, G., Magri, N.F., Jitrangsri, C., and Kingston, D.G.I. (1991) Modified taxols. 5. Reaction of taxol with electrophillic reagents and preparation of a rearranged taxol derivative with tublin assembly activity. *J. Org. Chem.*, 56, 5114–5119.

Shi, Q.-W., Oritani, T., and Cheng, Q. (1999) Three new bicyclic taxane diterpenoids from the needles of Japanese yew, *Taxus cuspidata* Sieb. et Zucc. *Nat. Prod. Lett.*, 13, 305–312.

Srikrishna, A., Kumar, P.P., and Reddy, T.J. (1998) Carvone based approaches to chiral functionalised C-ring derivatives of taxanes. *Tetrahed. Lett.*, 39(32), 5815–5818.

Swindell, C.S., Chandler, M.C., Heerding, J.M., Klimko, P.G., Rahman, L.T., Raman, J.V., and Venkataraman, H. (1993) An AC → ABC approach to taxol involving B-ring closure at C-1-C-2. *Tetrahed. Lett.*, 34(44), 7005–7008.

Van Rosendaal, E.L.M., Veldhuis, H., van Veldhuizen, B., van Beek, T.A., de Groot, A., Beusker, P.H., and Scheeren, H.W. (2000) Rearrangement of O-cinnamoyltaxicin I to a novel C-13 spiro-taxane. *J. Nat. Prod.*, 63, 179–181.

Wittman, M.D., Altstadt, T.J., Fairchild, C., Hansel, S., Johnston, K. Kadow, J.F., Long, B.H., Rose, W.C., Vyas, D.M., Wu, M-J., and Zoeckler, M.E. (2001a) Synthesis of metabollically blocked paclitaxel analogues. *Bioorg. & Med. Chem. Lett.*, 11, 809–810.

Wittman, M.D., Kadow, J.F., Vyas, D.M., Lee, F.L., Rose, W.C., Long, B.H., Fairchild, C., and Johnson, K. (2001b) Synthesis and antitumor activity of novel paclitaxel-chlorambucil hybrids. *Bioorg. & Med. Chem. Lett.,* 11, 811–814.

Wolff, S., Schreiber, W.L., Smith III, A.B., and Agosta, C. (1972) *J. Am. Chem. Soc.*, 94, 7797.

Wroblewski, A.E. and Pitrowska, D.G. (2000) Enantiomeric phosphonate analogs of the docetaxel C-13 side chain. *Tetrahed. Asymm.*, 11, 2615–2624.

Wuts, P.G.M., Gu, R.L., and Northuis, J.M. (2000) Synthesis of (2R, 3S)-isobutyl phenylisoserinate, the taxol side chain, from ethyl benzoylacetate. *Tetrahed. Asymm.*, 11, 2117–2123.

Xiao, X.-Y., Parandoosh, Z., and Nova, M.P. (1997) Design and synthesis of a taxoid library using radiofrequency enclosed combinatorial chemistry. *J. Org. Chem.*, 62(17), 6029–6033.

Yamaguchi, T., Harada, N., Ozaki, K., and Hashiyama, T. (1998) Synthesis of taxoids 3. A novel and efficient method for preparation of taxoids by employing *cis*-glycidic acid. *Tetrahed. Lett.,* 39, 5575–5578.

Yuan, H. and Kingston, D.G.I. (1998) Synthesis of 6α-hydroxypaclitaxel, the major human metabolite of paclitaxel. *Tetrahed. Lett.,* 39, 4967–4970.

Yuan, H. and Kingston, D.G.I. (1999) Synthesis and biological activity of a novel C4–C6 bridged paclitaxel analog. *Tetrahedron*, 55, 9709–9716.

9 Total synthesis of taxoids

Zhiyan Xiao, Hideji Itokawa, and Kuo-Hsiung Lee

Introduction

Taxoids are a family of diterpene alkaloids sharing a tricyclic taxane core (Figure 9.1A). The most well recognized member of the taxoid family is taxol (Figure 9.1B), a minor cytotoxic component isolated from yew bark (Wani *et al.*, 1971). Taxol was reported as an antimitotic agent exhibiting its biological activity by promoting the irreversible assembly of tubulin into microtubules (Schiff *et al.*, 1979). It was approved by the FDA for the treatment of ovarian and breast cancers in the early 1990s.

The unique molecular mechanism and remarkable clinical efficacy of taxol have stimulated intensive efforts to procure novel taxol analogs in the hope of improving pharmacological and pharmacokinetic profiles. Besides the isolation of over 350 taxoids from various natural sources (Baloglu and Kingston, 1999), partial and total syntheses have also expanded the research scope.

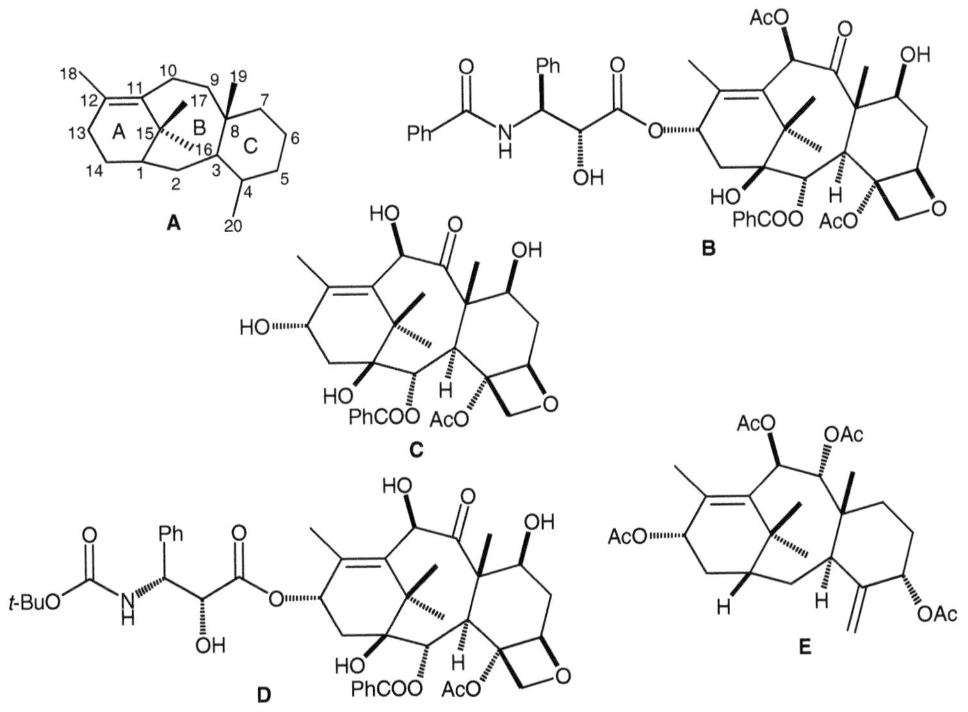

Figure 9.1 The taxoid family.

Semisynthesis has served as a practical solution to taxol's supply problem. 10-Deacetyl baccatin III (Figure 9.1C) is efficiently converted to taxol (Ojima *et al.*, 1991) through esterification of the C_{13} hydroxyl with an N-benzoyl-β-lactam. This approach has provided an optimistic outlook for taxol's escalating market. The development of Taxotere (Figure 9.1D), a novel taxane diterpenoid used clinically as an alternative antimitotic agent (Bissery *et al.*, 1995), has also been attributed to the semisynthetic approach.

In contrast to partial synthesis, the total synthesis of taxoids seems to be more academically than practically attractive. Because of the lengthy steps and unfavorable overall yield, total synthesis is unlikely to be an efficient solution for taxol supply. However, the complexity of the taxoid molecules (the unique tricyclic core, the numerous chiral centers and the high oxygenation level) presents a formidable challenge, as well as a precious opportunity for organic chemists to demonstrate their synthetic skill and ingenuity. In addition, studies in this field do offer potential access to novel taxoids with better pharmaceutical values. Motivated by both scientific and utilitarian objectives, over 50 research groups have confronted the demanding task of taxoid total synthesis. Decades of global endeavor were finally rewarded when the first total synthesis of taxusin (Figure 9.1E), a relatively simple taxoid, was achieved in 1988 (Holton *et al.*, 1988). This breakthrough was soon followed by the realization of the synthesis of taxol. In 1994, Holton's and Nicolaou's groups simultaneously reported the total syntheses of the daunting taxol molecule (Holton *et al.*, 1994a,b; Nicolaou *et al.*, 1994b, 1995a,b,c,d). Thereafter, several other groups successfully accomplished their syntheses of taxusin and taxol (For taxusin: Hara *et al.*, 1996, 1999; Paquette and Zhao, 1998; Paquette *et al.*, 1998. For taxol: Masters *et al.*, 1995; Danishefsky *et al.*, 1996; Wender *et al.*, 1997a,b; Mukaiyama *et al.*, 1999; Kusama *et al.*, 2000; Kuwajima and Kusama, 2000a,b). These achievements not only marked tremendous successes in the organic synthesis of complex, natural molecules, but also offered accessible synthetic approaches toward the diverse family of taxoids.

In context of the synthetic efforts toward taxoids, this chapter will provide a summary of the impressive body of studies on taxane core construction, a description of currently available syntheses toward taxoids, and a brief conclusion suggesting the perspective in this field. For a historical overview or elaborate discussion on related facets in this area, the readers are referred to reviews by Wender *et al.* (1995), Boa *et al.* (1994), Nicolauo *et al.* (1994a), Kingston *et al.* (1993) and Swindell (1991).

Considerations in synthetic design

A quick examination of the taxoid molecules suggests that the abundant stereochemical details, high oxygenation level and arduous tricyclic skeleton construction would be the most serious synthetic challenges.

The absolute stereochemistry is handled through one of the several alternative methods: resolution of a racemic mixture, introduction of enantiocontrol elements and use of enantiomerically pure starting materials. The first solution is often used initially to examine the viability of a synthetic strategy; however, the other two solutions are elegantly applied in later stages to settle stereochemical details and improve synthetic effectiveness.

The diverse peripheral oxygenation implies vast range of functional reactivities, which makes the order of functionalization a key issue for synthetic success. It is strategically desirable to delay the introduction of more reactive functionalities, so as to minimize the necessity of protection/deprotection and the possibility of byproduct formation.

The construction of the tricyclic taxane core requires intensive synthetic efforts. Synthetic convergency is strategically beneficial for the synthesis of such complicated molecules.

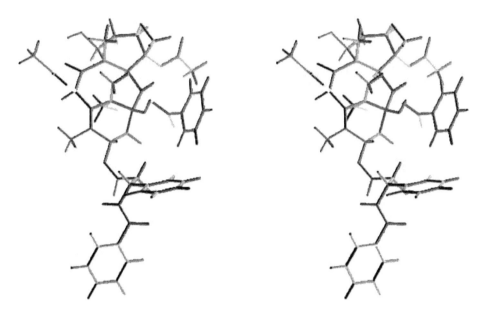

Figure 9.2 Crossed-eye stereo view of taxol. (Generated by Sybyl program on a Silicon Graphics Octane workstation using the standard Tripos forcefield.)

The advantages of convergent approaches over linear ones are higher throughput and more flexibility, because individual sequences involved in a convergent plan can be developed simultaneously and independently. However, most early synthetic efforts applied linear rather than convergent strategies. This observation can be readily understood by recognizing the demands imposed by synthesis of the unique, eight-membered B-ring from highly functionalized A- and C-rings in the final steps of convergent approaches. In contrast, linear approaches usually involve annulation of six-membered rings in later stages, which should be much more resourceful than eight-membered ring construction.

A close inspection of the three-dimensional (3-D) representation of taxoids provides important information regarding the structural and reactive features of taxoids (Figure 9.2).

The "inverted-cup" 3-D shape brings the dense flanking functionalities into extreme proximity. The resultant group interactions and reactivity gradient become a critical issue in the synthetic design. The relative reactivities of the C_1, C_2, C_7 and C_{13} hydroxyl groups must be carefully considered in the peripheral functionalization. The mutual repulsion between taxoid's C_{17} and C_{19} methyl groups would make late introduction of them a formidable task. The different steric environments around the $C_{11,12}$ double bond and C_4 exomethylene are responsible for regioselective dihydroxylation during taxol synthesis. Additional examples on how group-interactions would influence reactivity can be readily gleaned from the literature.

Approaches toward the construction of taxane skeleton

The construction of the taxane core is pivotal for the total synthesis of taxoids. Some representative retrosynthetic dissections of the tricyclic skeleton are illustrated in Figure 9.3.

These retrosynthetic approaches can be classified into linear approaches and convergent approaches. In the linear approaches, the three rings of the taxane skeleton are built sequentially

248 *Zhiyan Xiao, Hideji Itokawa, and Kuo-Hsiung Lee*

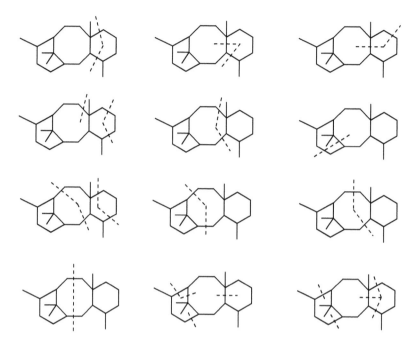

Figure 9.3 Representative retrosynthetic dissections of the taxane core.

(either A → ABC or C → ABC); while, in the convergent approaches, the B-ring is established in the final stages from a conjoined A- and C-ring precursor. Because the synthetic efforts toward taxane core have been reviewed comprehensively elsewhere (Swindell, 1991; Kingston *et al.*, 1993; Boa *et al.*, 1994; Wender *et al.*, 1995), only a brief overview of successful approaches will be provided herein.

Linear approaches

Linear approaches can be subgrouped into left to right (A → ABC) and right to left (C → ABC) approaches, according to the synthetic sequences.

Left to right approaches

PATTENDEN'S APPROACH (SCHEME 9.1)

Pattenden's group (Hitchcock and Pattenden, 1992) adopted a tandem radical macrocyclization–transannulation strategy to generate the taxane core. A functionalized A-ring precursor (6) was obtained from a Diels–Alder reaction between a diene and an acrolein and was further modified to an iodide intermediate (7). Upon treatment with tributyltin hydride, the iodide produced two radical conformational isomers 8a and 8b, which readily underwent 12-endo trig cyclization to offer the corresponding macrocyclic radical products (9a, 9b). The tricyclic products (10, 11) were obtained in 25% total yield through 6-exo trig transannulation. The formation of 10, the stereoisomer with desired C_3 stereochemistry, was more favorable than

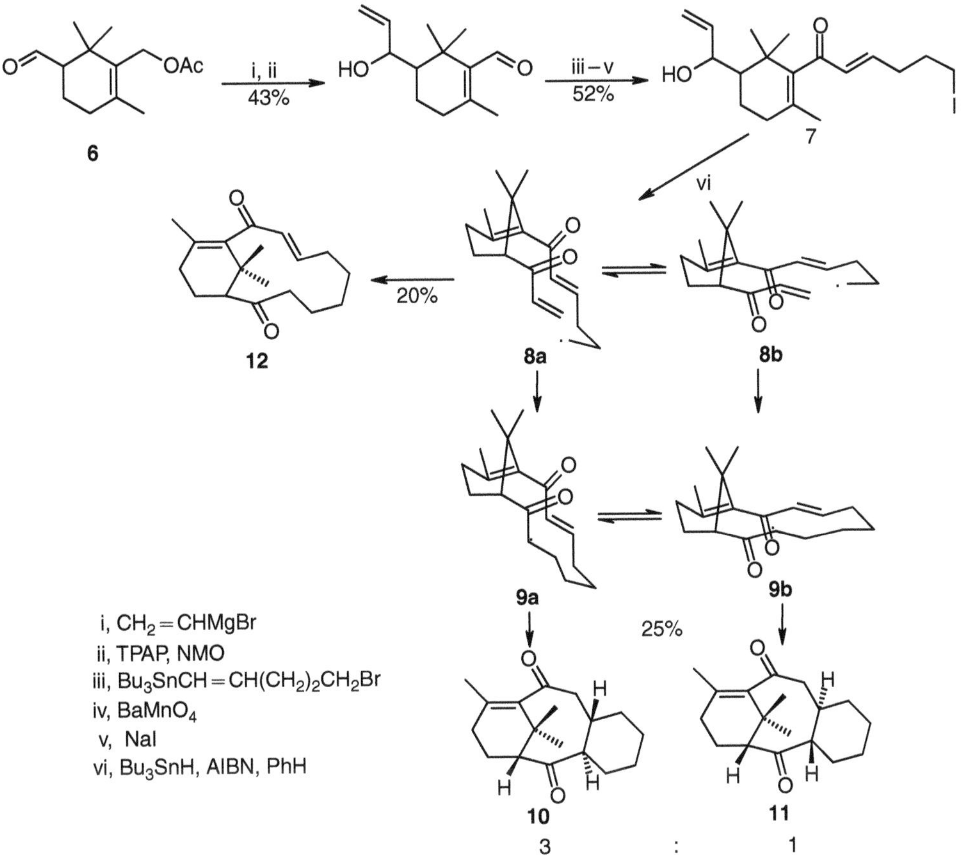
OAc
i, ii
43%
HO
iii–v
52%
HO
I
6
7
vi
20%
12
8a
8b
9a
9b
25%
10
11
3 : 1
i, $CH_2=CHMgBr$
ii, TPAP, NMO
iii, $Bu_3SnCH=CH(CH_2)_2CH_2Br$
iv, $BaMnO_4$
v, NaI
vi, Bu_3SnH, AIBN, PhH

Scheme 9.1 Pattenden's approach.

that of 11. The bicyclic product 12, which was produced by untimely quenching of the radical intermediates **9a** and **9b**, was the major by-product.

A merit of this radical approach is that its mild reaction conditions should make the intermediates more tolerant to early introduction of flanking functionalities. However, a major challenge of this pathway is the subsequent introduction of the C_8 methyl in any real taxoid. The unfavorable overall yield also limits the application of this approach in taxoid synthesis.

SAKAN'S APPROACH (SCHEME 9.2)

This approach (Sakan and Craven, 1983; Sakan *et al.*, 1991) represents one of the early attempts to complete the taxane core through intramolecular Diels–Alder cyclization. The starting material 13 (prepared from commercially available materials in 65% yield) was converted to trienes 14 and 15 in two steps and was subsequently transformed to the bicyclic intermediate 16 via an intermolecular Diels–Alder reaction. Repeated functional manipulation, which involved sequential aldehyde generation and Wittig reaction, furnished 16 with proper diene and dienophile moieties to provide the Diels–Alder substrate 18. Two stereoisomers could be obtained in a condition-dependent manner. Upon heating to 160°C, the desired taxane-like

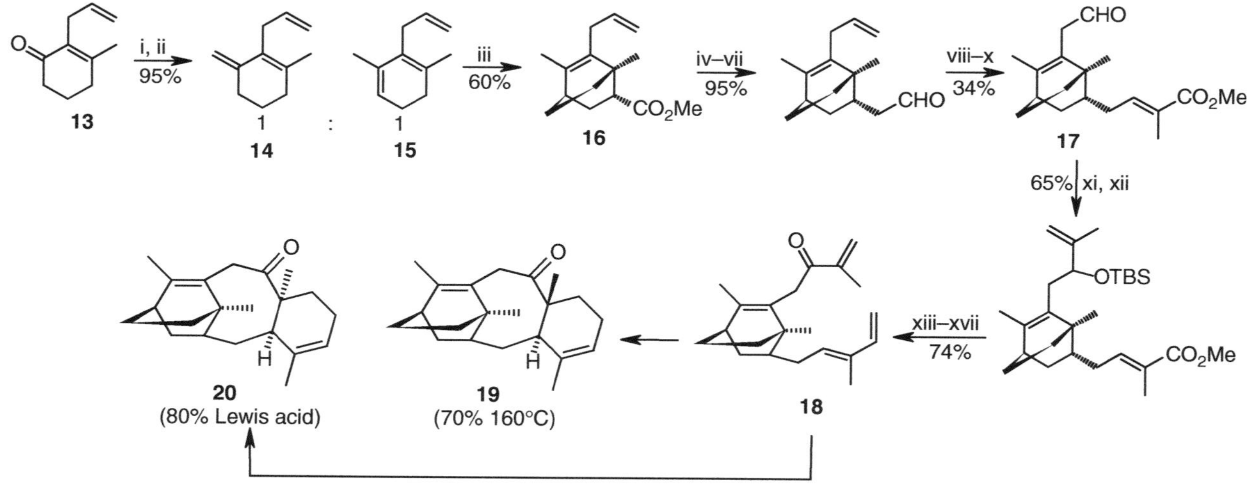

Scheme 9.2 Sakan's approach.

i, MeMgBr   ii, MgSO₄   iii, methyl acrylate, AlCl₃   iv, LiAlH₄   v, MsCl, Et₃N   vi, NaI, NaCN
vii, DIBAL   viii, Ph₃P=C(Me)CO₂Me   ix, OsO₄   x, NaIO₄   xi, CH₂=C(Me)MgBr
xii, t-BuMe₂Cl, DMAP   xiii, DIBAL   xiv, PDC   xv, CH₂=PPh₃
xvii, TFAA, DMSO, Hunig's Base (DIEA)

system 19 was yielded via an *exo* transition state. However, *endo* cycloaddition was promoted by treatment with Lewis acid to produce the other isomer 20.

This approach generated a ring system containing all 20 taxane carbons in 22 steps with a 3.6% overall yield (calculated from the commercially available materials). The A-ring bridge was introduced intentionally to facilitate the intramolecular Diels–Alder reaction. However, a C–C bond cleavage was required to release the C_{17} methyl group. Increased bridge lability would help to solve this synthetic dilemma. Another impressive feature of this approach was that the A- and C-ring alkenes and B-ring ketone were suitably situated for further functional elaboration.

FALLIS' APPROACH (SCHEME 9.3)

Fallis' approach (Lu and Fallis, 1993) completed the taxane core through an intramolecular Diels–Alder reaction. The A-ring precursor 21 was prepared from a cyclohexenone in nine steps and was converted to 22 by sequential coupling of a diene unit to the aldehyde moiety, MOM-protection of the resultant alcohol, reduction of nitrile moiety to aldehyde, and addition of TMS-acetylene to the carbonyl. After desilylation and Dess-Martin oxidation, the key intermediate 23 was produced and was stimulated to a thermal Diels–Alder reaction by using microwave irritation. The desired product 24 was obtained in 35–40% yield via an *endo* intramolecular cyclization.

Although the low yield of oxidation and cyclization steps undermined the efficiency of the approach, this novel and flexible method smartly circumvented the overwhelming problem of B-ring construction.

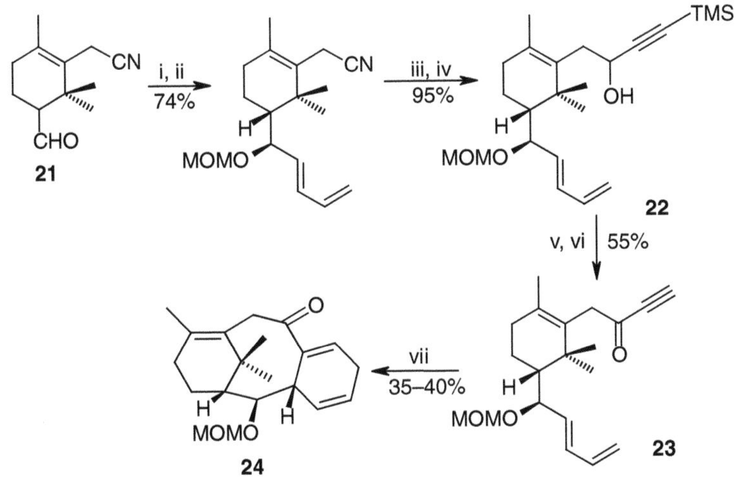

i, (E)–Bu₃SnCH=CH–CH=CH₂, n-BuLi, THF, –78°C
ii, MOM–Cl, DIPEA, DCM
iii, DIBAL–H
iv, HC≡CTMS, n-BuLi, –78°C
v, KOH, MeOH, DCM
vi, Dess-Martin Periodinane (DMP)
vii, 0.05 M in PhMe, microwave, 1% hydroquinone

Scheme 9.3 Fallis' approach.

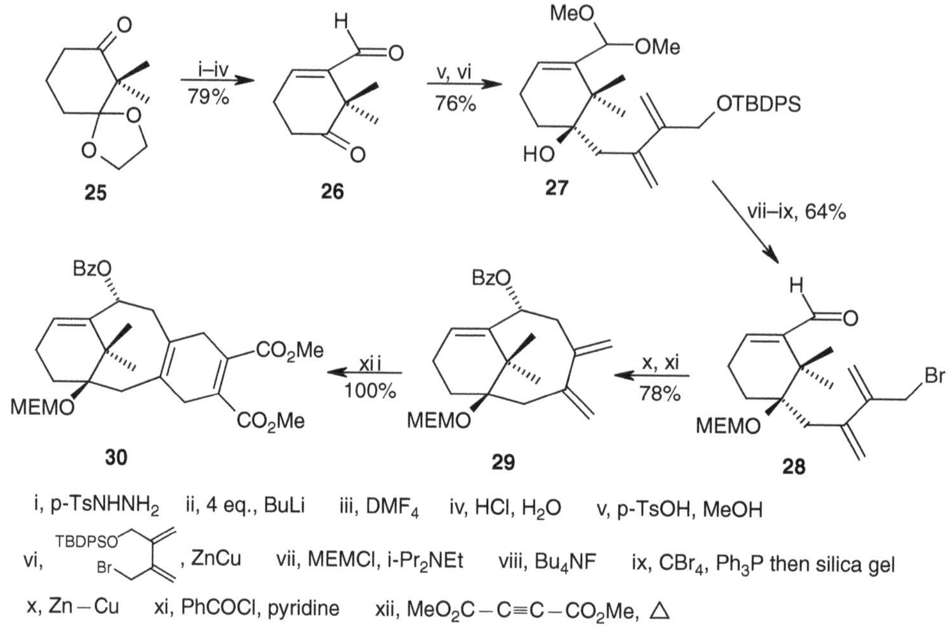
i–iv
79%
v, vi
76%
MeO
OMe
OTBDPS
HO
25
26
27
vii–ix, 64%
BzO
CO2Me
CO2Me
MEMO
xii
100%
x, xi
78%
Br
30
29
28
i, p-TsNHNH2 ii, 4 eq., BuLi iii, DMF4 iv, HCl, H2O v, p-TsOH, MeOH
vi, TBDPSO Br , ZnCu vii, MEMCl, i-Pr2NEt viii, Bu4NF ix, CBr4, Ph3P then silica gel
x, Zn–Cu xi, PhCOCl, pyridine xii, MeO2C–C≡C–CO2Me, Δ

Scheme 9.4 Wang's approach.

WANG'S APPROACH (SCHEME 9.4)

Starting from a mono-protected cyclohexadione 25, Wang's group (Wang *et al.*, 1993) established the tricyclic skeleton via sequential intramolecular acyclic closure and intermolecular Diels–Alder reactions. A Shapiro reaction (Shapiro, 1976) was followed by acidic hydrolysis to convert 25 to ketoaldehyde 26. After acetal protection of the aldehyde, the keto carbonyl was coupled to a diene moiety to stereoselectively produce tertiary alcohol 27. This alcohol was transformed to a bromide intermediate 28 via sequential operation of hydroxyl protection, aldehyde deprotection and bromo substitution. Upon treatment with zinc–copper reagent, 28 underwent intramolecular closure to provide AB-ring fragment 29, which was exposed to a disubstituted ethyne to generate the desired tricyclic system 30 through intermolecular thermal Diels–Alder reaction.

This approach provided a highly oxygenated tricyclic ring system by using acyclic closure and Diels–Alder cyclization to establish the $C_9{-}C_{10}$ bond and C-ring respectively. Starting from 25, this concise and efficient approach produced 30 in 12 steps with an overall 30% yield. However, its omission of the C_{19} methyl and C_{18} allylic methyl groups severely hindered its further development.

BLECHERT'S APPROACH (SCHEME 9.5)

Blechert reported (Neh *et al.*, 1984, 1989; Blechert *et al.*, 1992) a photochemical [2 + 2] cycloaddition–fragmentation strategy to generate the taxane ring system. The starting material 31 was converted to dione 32 in seven steps through an intramolecular nucleophilic cyclization. An allyloxycarbonyl moiety was added to 32 to produce intermediate 33, a prelude to [2 + 2] cycloaddition. Upon exposure to cyclohexene, 33 underwent intermolecular [2 + 2] cycloaddition and basic elimination to give the pentacyclic intermediate 34. Acidic hydrolysis of 34

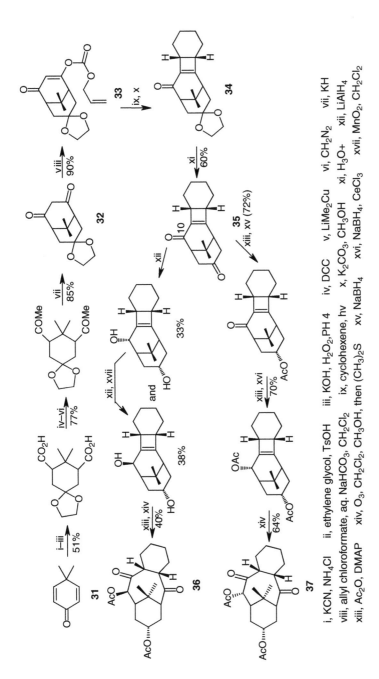

i, KCN, NH$_4$Cl   ii, ethylene glycol, TsOH   iii, KOH, H$_2$O$_2$, PH 4   iv, DCC   v, LiMe$_2$Cu   vi, CH$_2$N$_2$ vii, KH
viii, allyl chloroformate, aq. NaHCO$_3$, CH$_2$Cl$_2$ ix, cyclohexene, hv   x, K$_2$CO$_3$, CH$_3$OH   xi, H$_3$O+   xii, LiAlH$_4$
xiii, Ac$_2$O, DMAP   xiv, O$_3$, CH$_2$Cl$_2$, CH$_3$OH, then (CH$_3$)$_2$S   xv, NaBH$_4$ xvi, NaBH$_4$, CeCl$_3$ xvii, MnO$_2$, CH$_2$Cl$_2$

Scheme 9.5 Blechert's approach.

formed diketone 27. Alternate sequences produced opposite stereochemistry at C_{10}. Exhaustive reduction of diketone 27, followed by acetylation and ozonolysis provided a 10β-oxygenated product 36. However, regioselective reduction of the C_{13} carbonyl, together with subsequent acetylation and fragmentation, gave the 10α congener 37.

This approach was highlighted by its elegant B-ring construction and high-level of peripheral oxygenation. Epimerization of C_3 and introduction of the C_{18}, C_{19} methyls are the main objectives to be met in future development.

INOUYE'S APPROACH (SCHEME 9.6)

Inouye and co-workers (Kojima *et al.*, 1985) also took advantage of sequential intramolecular [2 + 2] photocycloaddition and fragmentation to circumvent the problem of B-ring construction. Similarly to Blechert's approach, a [2,1,2] bicyclic diketone 38 was prepared and coupled to C-ring precursor 39 to provide substrate 40. Photocycloaddition then afforded the tetracyclic product 41. Oxidation of the cyclic ether to lactone 42 was followed by alkaline treatment to generate simultaneous fragmentation and oxidative hydrolysis, and afford the tricyclic structure 43.

This approach produced the C_{19} angular methyl group with correct stereochemistry and provided two carbonyl and one carboxyl group to facilitate flanking functionalization. However, crucial issues in future efforts would be the introduction of C_{16}, C_{17} and C_{18} methyl groups and the epimerization of C_3.

HOLTON'S APPROACH (SCHEME 9.7)

Holton's approach (Holton, 1984; Holton and Kennedy, 1984) started from the commercially available (−)-β-patchoulene oxide 44. Skeletal rearrangement of 44 via a 1,2-*gem*-dimethyl shift was followed by epoxidation and fragmentation to afford the key intermediate 45. MOM protection and enolization of 45 provided the enol ether 46, which was subjected to sequential alkylation and intramolecular aldol reaction to accomplish C-ring annelation. Due to its instability, the resultant tricyclic product 47 was reduced to the more stable alcohol form 48.

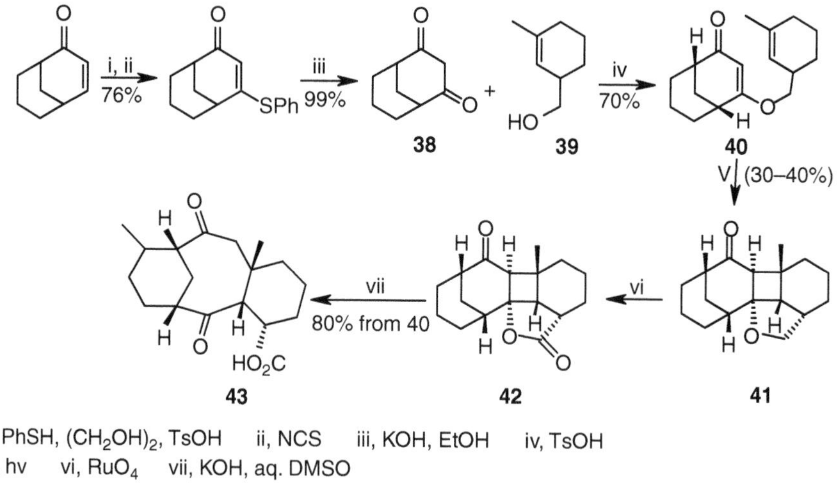

i, PhSH, (CH₂OH)₂, TsOH   ii, NCS   iii, KOH, EtOH   iv, TsOH
v, hv   vi, RuO₄   vii, KOH, aq. DMSO

Scheme 9.6 Inouye's approach.

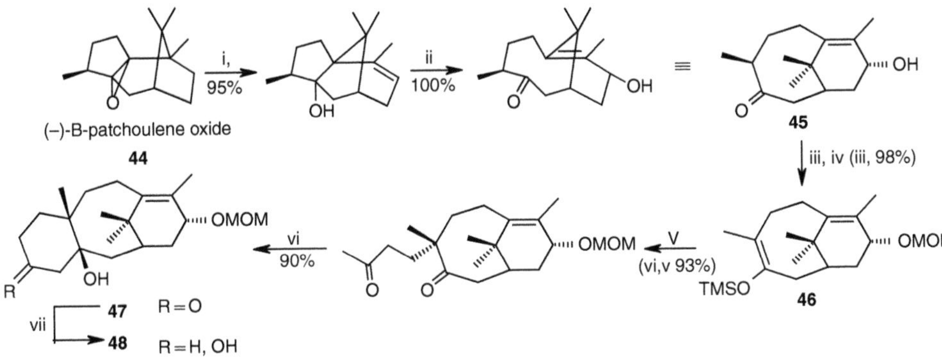

i, BF$_3$.OEt$_2$ ii, (i-PrO)$_4$Ti, t-BuOOH, Me$_2$S/Δ   iii, MOMCl, Hunig's Base (DIEA)   iv, (i-Pr)$_2$MgBr, TMSCl, Et$_3$N
v, MeLi, CH$_2$=C(TMS)COCH$_3$ vi, (i-Pr)$_2$MgBr or (c-hex)$_2$MgBr   vii, Red-Al

Scheme 9.7 Holton's approach.

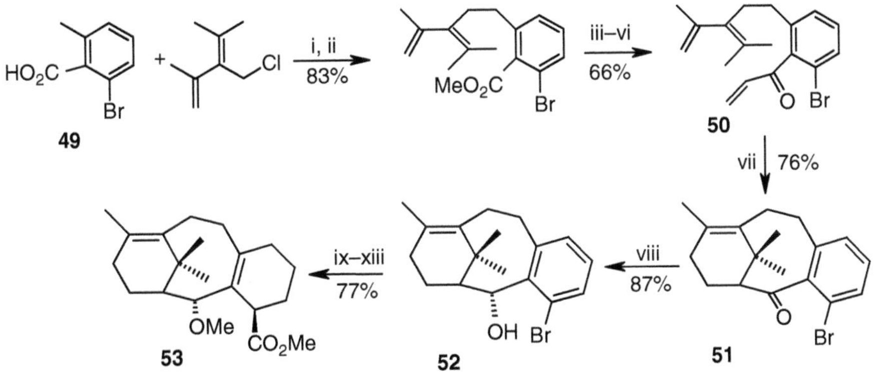

i, LDA   ii, CH$_2$N$_2$ iii, DIBAL   iv, PCC   v, CH$_2$=CHMgBr   vi, BaMnO$_4$ vii, Et$_2$AlCl
viii, DIBAL   ix, NaH, CH$_3$I   x, t-BuLi, CO$_2$ xi, Li/NH$_3$, EtOH   xii, CH$_2$N$_2$ xiii, H$_2$, PtO$_2$, EtOH

Scheme 9.8 Shea's approach.

This approach represented a concise and efficient synthesis toward the tricyclic model. Starting from a natural terpenoid, this synthesis smartly bypassed the challenge of B-ring construction. However, due to the opposite stereochemistry of the commercially available (−)-β-patchoulene oxide 44, the established tricyclic ring system was enantiomeric to the natural taxane core.

Right to left approaches

SHEA'S APPROACH (SCHEME 9.8)

A Lewis-acid catalyzed intramolecular Diels–Alder reaction served as a key step in this approach (Jackson *et al.*, 1992a,b). Starting from an aromatic bromide 49, this approach suitably positioned the diene and enone moieties to allow proper [4 + 2] intramolecular cycloaddition.

The selectivity between *endo* and *exo* cyclization was determined by the reaction conditions. Atropisomerism studies revealed that, in the presence of Lewis acid (e.g. AlCl₃), the *endo* isomer would be strongly preferred over the *exo* isomer (Shea and Davis, 1983). Accordingly, upon treatment with Lewis acid, substrate **50** underwent *endo* cyclization to afford the single isomer **51**, which was subjected to stereoselective reduction to give **52** as a major product. The alcohol **52** was further transformed to non-aromatic ester **53** by carboxylation at C-4 and reduction of the aromatic C-ring.

This approach was one of the early attempts to apply intramolecular Diels–Alder cycloaddition to furnish the tricyclic taxane core. It demonstrated the impressive efficiency of the Diels–Alder reaction in simultaneous contruction of the [3, 1, 5] bicyclic fragment.

JENKINS' APPROACH (SCHEME 9.9)

Jenkins and co-workers (Brown *et al.*, 1984; Brown and Jenkins, 1986; Bonnert and Jenkins, 1987; Bonnert and Jenkins, 1989) also utilized a Diels–Alder approach to generate the taxane core. Reductive silylation of **54** was followed by ozonolysis and esterification to yield ester **55**. Functionalization of the two side-chains provided cyclization substrate **58**, which is quite similar to substrate **50** in Shea's approach. The two compounds vary only in the C-ring precursor: Shea's route used an aromatic C-ring precursor, while the Jenkins' route applied an alicyclic C-ring. Like in Shea's approach, Lewis acid was used to promote the intramolecular cycloaddition and a single diastereoisomer **59** was obtained via an *endo* transition state.

This attractive approach included all methyl groups and allowed for early introduction of C-ring fuctional groups. A later extension of this approach (Bonnert and Jenkins, 1987; Bonnert *et al.*, 1991) used a chiral pool derived C-ring to produce a highly functionalized C-ring precursor **60**. Coupling of diene and dienophile components to **60** and subsequent intramolecular Diels–Alder reaction may generate more elaborate ring systems.

YADAV'S APPROACH (SCHEME 9.10)

Yadav's approach (Yadav and Ravishankar, 1991; Yadav, 1993) to the taxane core also applied an intramolecular Diels–Alder reaction. However, this approach was characterized by its ring

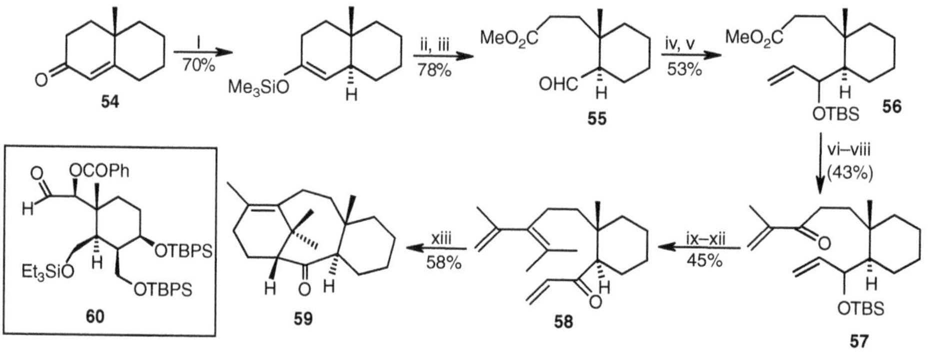

i, Li, NH₃, then Me₃SiCl, Et₃N   ii, O₃, MeOH, CH₂Cl₂, Sudan red   iii, CH₂N₂   iv, CH₂=CHMgBr
v, TBSOTf, 2,6-lutidine   vi, DIBAL–H, toluene, –78°C   vii, CH₂=C(Me)MgBr   viii, Collins' reagent
ix, Me₂C(SePh)Li   x, SOCl₂, Et₃N   xi, HF, aq. MeCN   xii, Collins' reagent   xiii, BF₃.OEt₂

Scheme 9.9 Jenkins' approach.

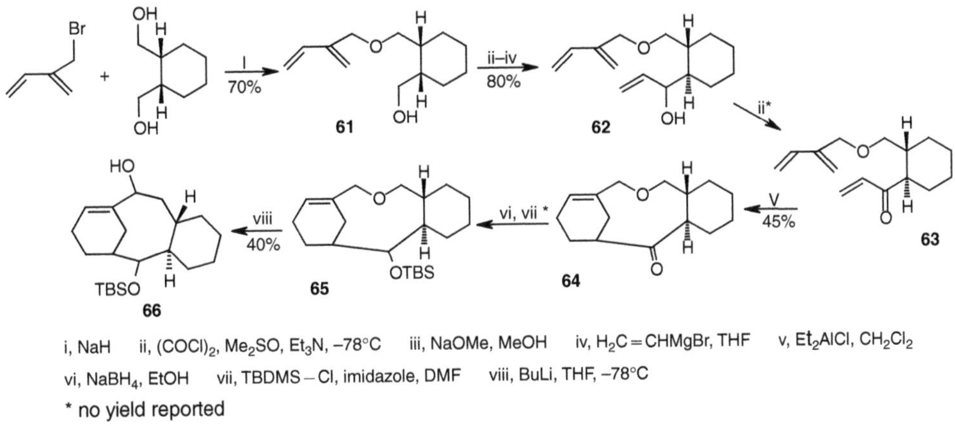
Br
OH
H
OH
H
i
70%
61
ii–iv
80%
62
ii*
63
v
45%
64
vi, vii *
65
OTBS
viii
40%
66
HO
TBSO
i, NaH ii, (COCl)2, Me2SO, Et3N, –78°C iii, NaOMe, MeOH iv, H2C=CHMgBr, THF v, Et2AlCl, CH2Cl2
vi, NaBH4, EtOH vii, TBDMS–Cl, imidazole, DMF viii, BuLi, THF, –78°C
* no yield reported

Scheme 9.10 Yadav's approach.

contraction step (65–66) via a [2,3] Wittig rearrangement. After construction of diene and dienophile moieties onto a C-ring precursor, the Diels–Alder substrate 63 readily underwent intramolecular cycloaddition upon treatment with Lewis acid. The resulting ether 64 was subjected to sequential reduction, TBS protection and ring contraction to afford the tricyclic taxane core 66.

However, the introduction of the $C_{16,17}$ *gem*-dimethyl group and C_8 angular methyl would seriously challenge further synthetic efforts.

SWINDELL'S APPROACH (SCHEME 9.11)

Swindell has prepared (Swindell and DeSolms, 1984; Swindell *et al.*, 1985, 1987; Swindell and Patel, 1990) a highly functionalized taxane skeleton (73) through sequential photocycloaddition, fragmentation and A-ring annulation. A commercially available cyanohydrin was coupled to a dimedone and the resulting product 67 was subsequently transformed to carbamate 68. An N-bridged photoproduct 69 was afforded via [2 + 2] intramolecular cycloaddition. This product was then subjected to Rubottom oxygenation (Rubottom *et al.*, 1974), stereoselective reduction, and mesylation to give the fragmentation substrate 70, which was converted to the BC-fragment 71 in three steps. A-ring annulation was achieved via a series of functional manipulations to give epoxide 72 in a 26% yield, which underwent an amazing closure reaction to provide the tricyclic taxane core 73.

The final product 73 bears all the peripheral methyl groups and ample flanking oxygenation, which makes it a promising precursor for further elaboration.

WINKLER'S APPROACH (SCHEME 9.12)

This approach established the eight-membered B-ring via fragmentation of a [2.0.4] bicyclic octane (Winkler *et al.*, 1992). The starting material 74 was first converted to enone 75 and was subsequently subjected to a series of functional manipulations (conjugate addition, alkylation, transesterification, protection and ozonolysis) to afford aldehyde 76 in an 18% yield. Addition of

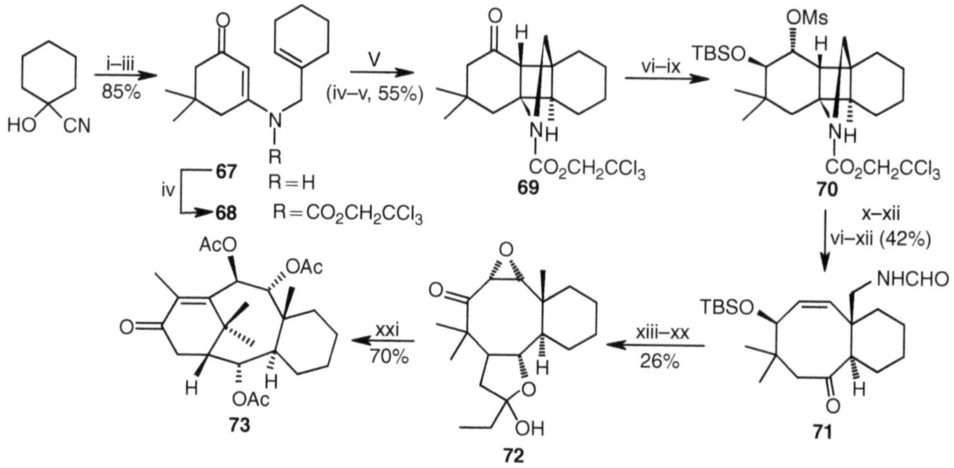

i, LiAlH$_4$ ii, SOCl$_2$ iii, dimedone, C$_6$H$_6$, reflux   iv, Cl$_3$CCH$_2$OCOCl, NaOH, Adogen$_{464}$, H$_2$O, CH$_2$Cl$_2$
v, hv, C$_6$H$_6$ vi, TBSOTf, Et$_3$N   vii, m-CPBA   viii, K-Selectride   ix, MsCl, Et$_3$N   x, Zn, THF
xi, aq. HOAc   xii, AcOCHO, py.   xiii, H$_3$CCH=C(OMe)CH$_2$I, LDA   xiv, MsCl, py.   xv, Na, NH$_3$, NH$_4$Cl
xvi, TBAF   xvii, aq. HOAc x   viii, aq. HCl   xix, MnO$_2$ xx, t-BuOOH, Triton-B   xxi, DBU,LiCl, Ac$_2$O,THF

Scheme 9.11 Swindell's approach.

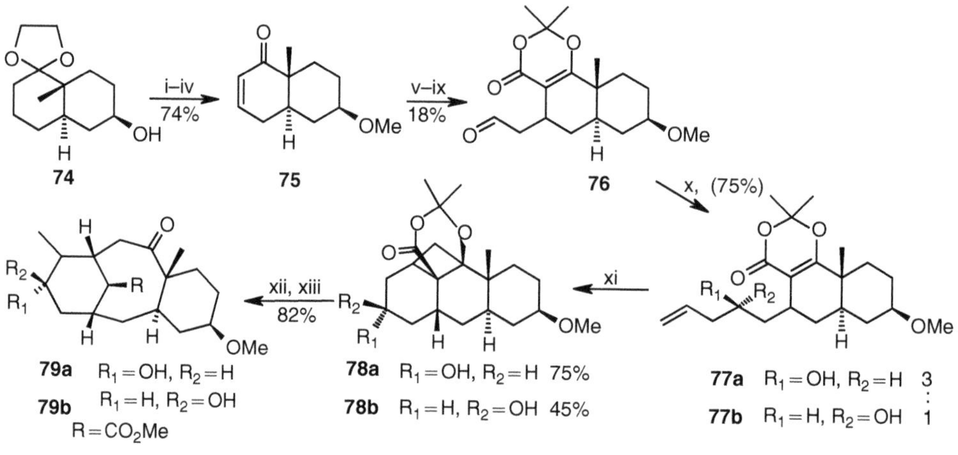

i, KH, MeI   ii, aq. HCl, THF   iii, Br$_2$, HOAc   iv, CaCO$_3$, DMA   v, CH$_2$=CHCH$_2$TMS, TiCl$_4$, CH$_2$Cl$_2$
vi, LDA, MeOCOCN   vii, p-MeOC$_6$H$_4$CH$_2$OH, DMAP   viii, TFAA, TFA, acetone
ix, O$_3$, Ph$_3$P, CH$_2$Cl$_2$ x, CH$_2$=CHCH$_2$, TMS, TiCl$_4$ xi, hv   xii, KOH   xiii, CH$_2$N$_2$

Scheme 9.12 Winkler's approach.

allyltrimethyl silane to the aldehyde carbonyl produced two stereoisomers **77a** and **77b**, which
served as a substrate for internal [2 + 2] photocycloaddition to provide the bicyclic octane (**78a**,
78b). After fragmentation and deprotection, highly functionalized taxane ring systems, **79a** and
79b, were generated.

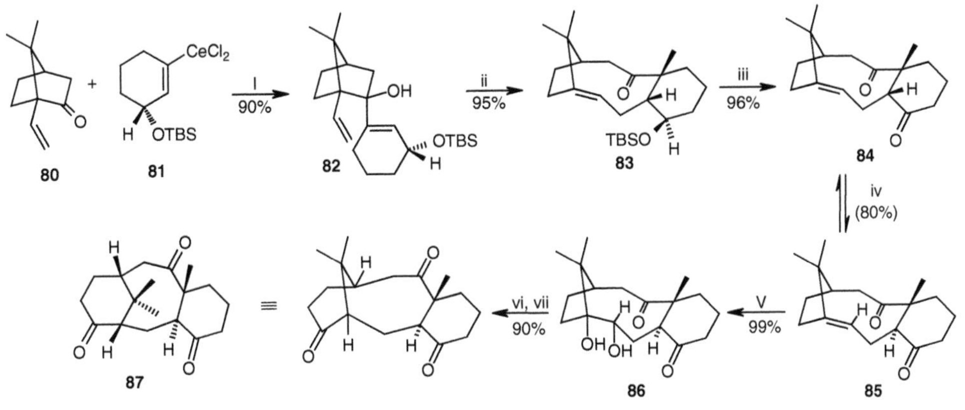

i, THF, –78°C   ii, KH, 18-crown-6   iii, Bu$_4$NF, then PDC   iv, NaOMe, MeOH, separate and recycle
v, OsO$_4$, NaHSO$_3$, pyridine, water   vi, CH$_3$SO$_2$Cl, pyridine   vii, Et$_2$AlCl

Scheme 9.13 Paquette's approach.

This approach afforded a fully saturated taxane ring system with dense peripheral functionalization. Using this route, simpler taxoid analogs could be readily prepared. Their biological evaluation was crucial to identify minimal structural requirements for taxol's remarkable activity.

PAQUETTE'S APPROACH (SCHEME 9.13)

Paquette developed (Elmore *et al.*, 1991; Paquette *et al.*, 1992a,b,c) an efficient approach to construct the taxane core via oxy-Cope rearrangement. The rearrangement substrate **82** was obtained from coupling between readily available chiral synthons **80** and **81**. Sigmatropic rearrangement of **82** through an *endo* transition state resulted in formation of the tricyclic product **83**, which contains a five-membered A-ring and a nine-membered B-ring. The silyl ether of **83** was subjected to deprotection and oxidation to generate the C$_4$ carbonyl of **84**. Epimerization of the C$_3$ chiral center (**85**) and osmylation of the B-ring olefin yielded diol **86**. The secondary hydroxyl was selective mesylated, then exposure to Lewis acid promoted a 1,2 shift of the *gem*-dimethyl bridge. Thus, the desired AB-ring system **87** was produced via a simultaneous ring expansion and contraction.

Its efficiency highlighted this approach. Starting from the chiral synthons, the taxane core was completed in seven steps with an overall yield of 60%. In addition, all peripheral methyl groups and angular hydrogen atoms were positioned in correct stereochemistry.

Convergent approaches

Although they are characterized for higher efficiency and more flexibility, the application of convergent strategies was retarded due to the extreme difficulty to construct an eight-membered B-ring from highly functionalized A- and C-ring precursors. With the rapid development of synthetic methods inside and outside the taxoid field, the tricyclic taxane core has finally been accomplished through convergent approaches.

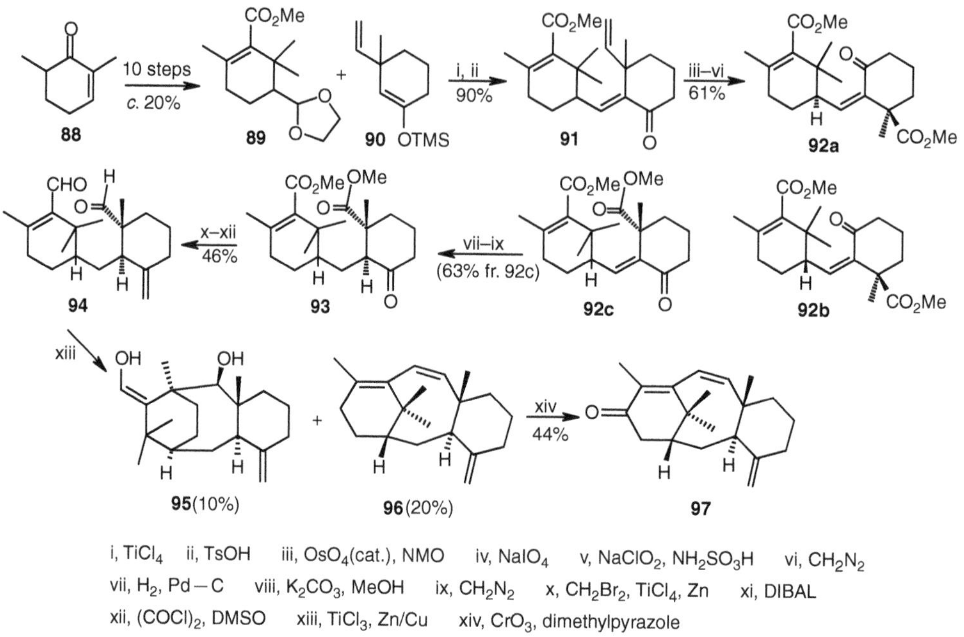
10 steps
c. 20%
88
89
90 OTMS
i, ii
90%
91
iii–vi
61%
92a
x–xii
46%
94
93
vii–ix
(63% fr. 92c)
92c
92b
xiii
95(10%)
96(20%)
xiv
44%
97
i, TiCl4 ii, TsOH iii, OsO4(cat.), NMO iv, NaIO4 v, NaClO2, NH2SO3H vi, CH2N2
vii, H2, Pd–C viii, K2CO3, MeOH ix, CH2N2 x, CH2Br2, TiCl4, Zn xi, DIBAL
xii, (COCl)2, DMSO xiii, TiCl3, Zn/Cu xiv, CrO3, dimethylpyrazole

Scheme 9.14 Kende's approach.

Kende's approach (Scheme 9.14)

This approach (Kende *et al.*, 1986) started from coupling between an A-ring precursor 89 (prepared from 2,6-dimethylcyclohexenone 88 in ten steps) and a C-ring precursor 90. The resultant product 91 underwent regioselective oxidative cleavage to afford a diketone intermediate (not shown), which was oxidized and esterified to yield three inseparable isomers **92a**, **b** and **c**. Hydrogenation of the isomeric mixture followed by epimerization of the C_3 center and reesterification of the hydrolized carboxylates provided diester 93. After methylenation, the ester was reduced to give aldehyde 94, the substrate for the ring closure reaction. Under McMurry's conditions (McMurry and Kees, 1977; McMurry, 1983), 94 underwent cyclization to give 95 and 96 in 10% and 20% yields, respectively. Kende amplified the significance of this approach by utilizing an allylic oxidation of 96 to introduce a C_{13} ketone, which was suitable for subsequent functional manipulation to furnish the C_{13} side-chain.

Kende's approach represented a milestone in taxane synthesis, because it was the first report to furnish the entire 20 carbon taxane core and also the first successful attempt through convergent strategy.

Funk's approach (Scheme 9.15)

This approach (Funk *et al.*, 1988) constructed the taxane core through a Claisen rearrangement of macrocyclic ketene-acetals, which ingeniously bypassed the difficult B-ring closure. The dianion derivative 99 was prepared from hydrazone 98 and coupled to aldehyde 100 to form the dihydroxyl ester 101. Selective protection of the alcohol and deprotection of the carboxyl acid provided 102. This intermediate underwent intramolecular cyclization according to Mukaiyama

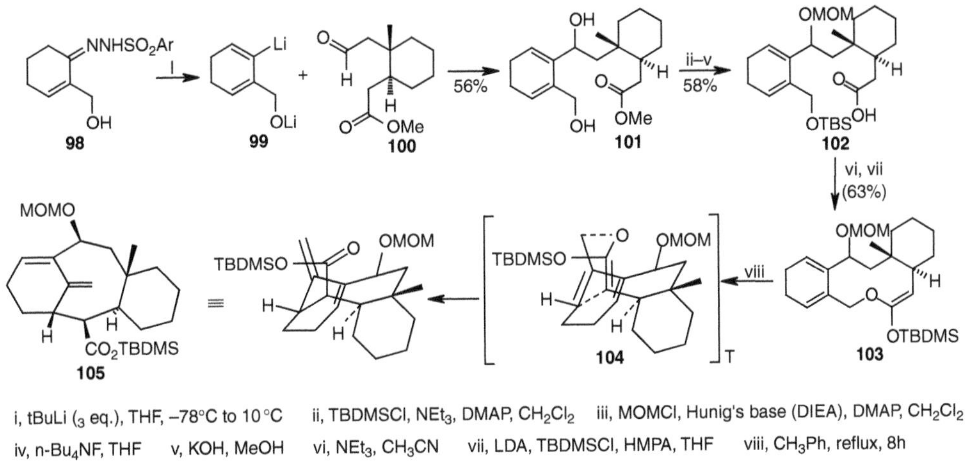
NNHSO2Ar
i
OH
98
Li
OLi
99
+
H
H
OMe
100
56%
OH
OH
OMe
101
ii–v
58%
OMOM
OTBS
OH
102
vi, vii
(63%)
MOMO
H
H
CO2TBDMS
105
TBDMSO
OMOM
H
TBDMSO
O
OMOM
H
H
104
T
viii
OMOM
H
OTBDMS
103
i, tBuLi (3 eq.), THF, –78°C to 10°C ii, TBDMSCl, NEt3, DMAP, CH2Cl2 iii, MOMCl, Hunig's base (DIEA), DMAP, CH2Cl2
iv, n-Bu4NF, THF v, KOH, MeOH vi, NEt3, CH3CN vii, LDA, TBDMSCl, HMPA, THF viii, CH3Ph, reflux, 8h

Scheme 9.15 Funk's approach.

protocol (Mukaiyama *et al.*, 1976), followed by enolization and silylation to afford the corresponding macrolide **103**, the substrate for the key-ring contraction step. The stereoselective rearrangement of ketene-acetal **103** formed the tricyclic ring system **105** via chair-like transition state **104**.

This approach was characterized by its novelty, efficiency and conciseness. Starting from the commercially available starting materials **98** and **100**, it produced the taxane core **105** in eight steps with 17% overall yield. Because of the high level of convergency and flexibility, using more functionalized A- and C-ring precursors should reasonably avoid the complication arising from peripheral functionalization and provide easy access to diverse taxoid analogs. However, ring contraction step might not tolerate steric congestion and dense oxygenation around the A- and C-rings.

Kuwajima's approach (Scheme 9.16)

Kuwajima completed (Horiguchi *et al.*, 1989) the synthesis of the taxane core through a Lewis acid catalyzed intramolecular cyclization. The starting material **106** served as an aromatic C-ring precursor, which was transformed to the corresponding anion and coupled with the allylic bromide **107** to provide the allylic acetate **108**. The A-ring fragment was built by deprotection and oxidation of the secondary hydroxyl group and subsequent Michael addition–condensation to give the β-diketone **109**. Enolization and Peterson olefination then provided compounds **111a**, **b** and **c**. Upon exposure to Lewis acid, these compounds underwent intramolecular cyclization to form the C_9-C_{10} bond and afford tricyclic products **112 a**, **b** and **c**, respectively. Alternatively, treatment with $TMSCH_2Li$ and mild acid followed by exposure to Lewis acid also converted keto-acetal **110** to a tricyclic product via an intramolecular Mukaiyama-type cyclization (not shown).

Kuwajima established the taxane ring system (**112a**, **b** and **c**) in seven steps with an overall yield between 25% and 30%. However, future challenges included the hydrogenation of the C-ring or, alternatively, the use of a non-aromatic C-ring precursor, in addition to the introduction

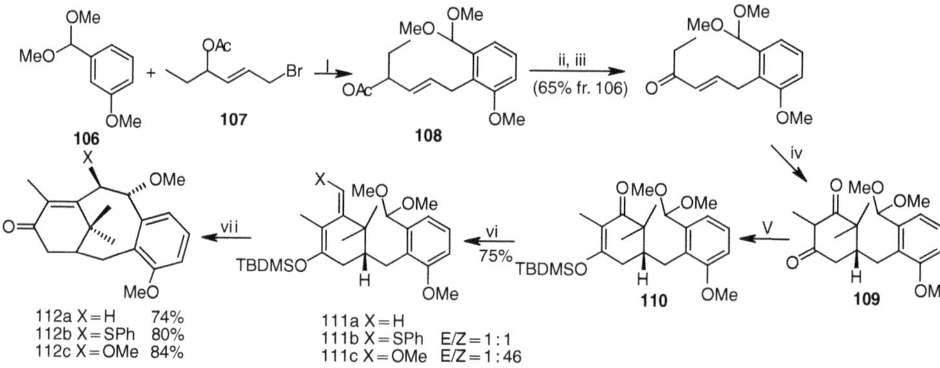

i, t-BuLi, then Cu—Br—Me₂S   ii, K₂CO₃, MeOH   iii, PCC, NaOAc   iv, [structure] OLi / OEt, t-BuOK   v, TBDMSCl, Et₃N
vi, Peterson olefination (H₃SiCH₂CH₂O-)   vii, TiCl₄

Scheme 9.16 Kuwajima's approach.

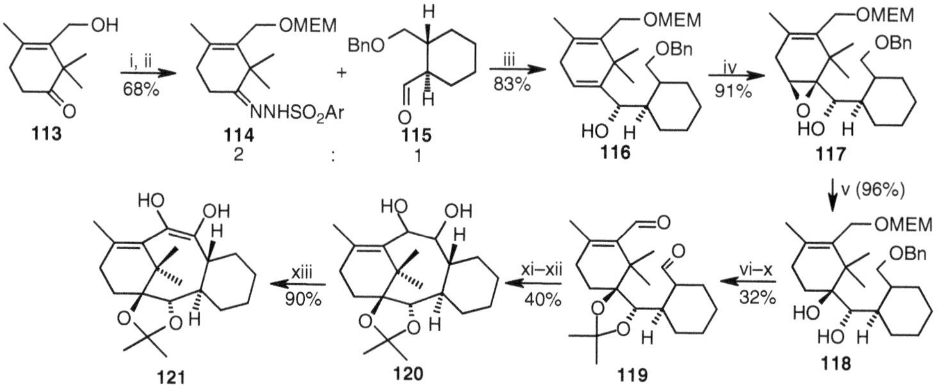

i, i-Pr₂EtN / MEMCl, CH₂Cl₂   ii, ArSO₂NHNH₂, Ar = [structure with i-Pr groups], MeOH   iii, n-BuLi, THF, −78 → 0°C

iv, TBHP, VO(acac)₂, PhH   v, LiAlH₄, Et₂O   vi, 2,2-DMP, CSA, CH₂Cl₂   vii, H₂,20% Pd(OH)₂-C, EtOAc
viii, Ac₂O, 4-DMAP, CH₂CCl₂   ix, TiCl₄, CH₂Cl₂, −78°C → 20°C   x, K₂CO₃, MeOH
xi, 5 mol% TPAP, NMO, mol sieves, CH₂Cl₂   xii, TiCl₃-(DME), Zn-Cu, DME   xiii, MnO₂, CH₂Cl₂

Scheme 9.17 Nicolaou's approach.

of the C_{19} methyl group. As we will discuss later in this chapter (sections on "Kuwajima's route toward taxusin" and "Kuwajima's approach toward taxol"), Kuwajima's group eventually created strategies to overcome these problems and achieve the total synthese of taxusin and taxol.

Nicolaou's approach (Scheme 9.17)

Nicolaou's group exploited (Nicolaou *et al.*, 1993) a series of stereoselective syntheses of A- and C-ring fragments and aimed their synthetic efforts at coupling these fragments through C_9-C_{10}

bond formation. Due to space limitation, the approaches related to the synthesis of A- and C-ring fragments are not included. Readers are referred to Wender's (Wender *et al.*, 1995) review and citations therein.

Scheme 9.17 illustrates one of the early attempts to furnish the tricyclic structure. A-ring precursor 114 was prepared from ketone 113 and was coupled to C-ring precursor 115 to provide the cyclohexadiene 116. The secondary hydroxyl mediated a Sharpless epoxidation (Sharpless and Verhoeven, 1979) to give epoxide 117 regioselectively. Reduction to 118 created the correct stereochemistry at C_1 and C_2. The vicinal diol 118 was transformed to an acetonide and was further oxidized to produce the dialdehye intermediate 119. Under modified McMurry conditions (McMurry, 1989), the substrate 119 underwent intramolecular cyclization through a pinacol type coupling (Li, 1993). The resultant product 120 was exposed to MnO_2 oxidation to afford the enediol 121.

This approach was based on a strategy similar to that of Kende's. However, it was highlighted by its versatility and efficiency. The defect of the omitted C_{19} methyl group was corrected in their later efforts as will be discussed in the section on "Nicolaou's approach toward taxol."

Kishi's approach (Scheme 9.18)

The key step of Kishi's approach (Kress *et al.*, 1993) was an acyclic closure reaction to generate the $C_{10}-C_{11}$ bond. The A-ring fragment 122 and C-ring fragment 123 were coupled together through the formation of the C_1-C_2 bond to provide intermediate 124. Seperate protocols were used to convert compound 124 to either diol 125a or alcohol 125b. Ketal 125b was subsequently subjected to methylation, hydrolysis and alkylation to produce intermediate 126. Sequential iodization and oxidation of 126 afforded iodide 127, which was transformed to the cyclization substrate 128 through deprotection and oxidation of the primary hydroxyl group. Upon treatment with Lewis acid, iodoenone 128 underwent a ring closure reaction to provide the tricyclic structure 129.

This approach explored the possibility to construct the B-ring from an acyclic closure at $C_{10}-C_{11}$ and successfully achieved the synthesis of a taxane core 129 with proper regio- and stereochemistry of all peripheral methyl groups.

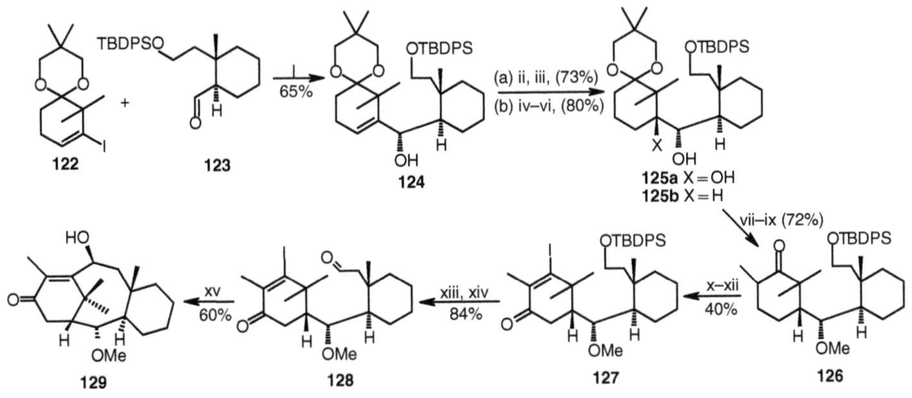

i, t-BuLi, THF   ii, m-CPBA, CH_2Cl_2 iii, LAH, Et_2O iv, TBAF, THF   v, H_2, Rh on Al_2O_3 vi, TBDPSCl, imid., DMF

vii, MeI, NaH, DMF   viii, pTSA, acetone   ix, LDA, THF, MeI   x, KHMDS, THF,PhNTf$_2$ xi, (nBu$_3$Sn)(nBu)CuCNLi$_2$, THF, then I$_2$

xii,CrO$_3$-3-5-dimethylpyrazole, CH_2Cl_2 xiii, TBAF, THF   xiv, Swern oxidation, (COCl)$_2$, Me$_2$SO; Et$_3$N   xv, 1%NiCl$_2$/CrCl$_2$, DMSO

Scheme 9.18 Kishi's approach.

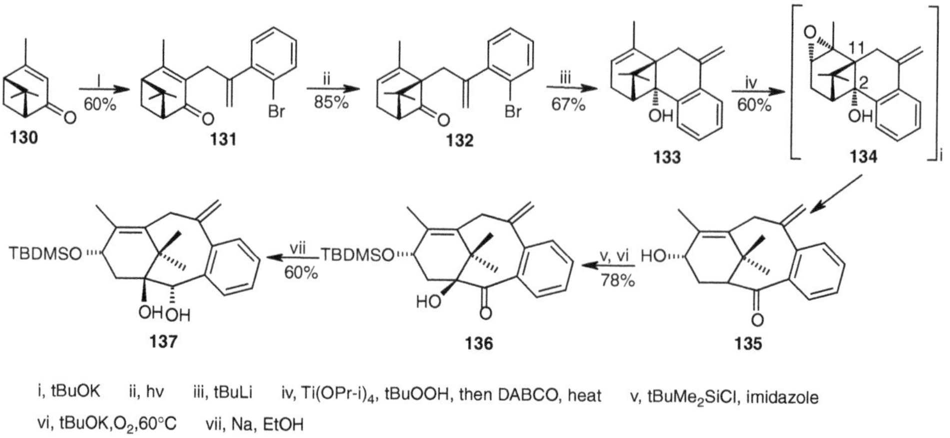
i
60%
ii
85%
iii
67%
iv
60%
Br
Br
OH
OH
O
11
2
130
131
132
133
134
i
TBDMSO
vii
60%
TBDMSO
v, vi
78%
HO
HO
O
O
OH OH
137
136
135
i, tBuOK ii, hv iii, tBuLi iv, Ti(OPr-i)4, tBuOOH, then DABCO, heat v, tBuMe2SiCl, imidazole
vi, tBuOK,O2,60°C vii, Na, EtOH

Scheme 9.19 Wender's approach.

Wender's approach (Scheme 9.19)

Wender developed (Wender and Mucciaro, 1992) an elegant approach to circumvent the difficult issue of B-ring cyclization through fragmentation of a bicyclo [4.2.0] octane system. The commercially available (+)-α-pinene (not shown) could be easily oxidized to the starting material, (+)-verbenone 130. Subsequent alkylation produced enone 131, and incorporated an aromatic C-ring precursor. Upon irradiation, 131 underwent a 1,3-alkyl shift to give cyclobutanone 132, which was subjected to sequential intramolecular cyclization and epoxidation to provide the fragmentation substrate 134. The C_2-C_{11} bond of the cyclobutane was readily cleaved to afford the tricyclic product 135. The C1β and C2α hydroxyl groups (137) were added through keto-directed oxidation and stereoselective reduction.

By using enantiomeric (+)-α-pinene as a starting material, this approach efficiently procured taxanes in enantiomerically enriched form. However, hydrogenation of the aromatic C-ring and elaboration of peripheral functionalization would be major issues beween the tricyclic model 137 and any natural taxane. To overcome this problem, Wender later developed a linear-like approach to construct the C-ring in the last steps (section on "Wender's route toward taxol").

Swindell's approach (Scheme 9.20)

After accomplishing the BC→ABC synthetic approach to taxanes (section on "Swindell's approach"), Swindell's group encountered serious challenges in adapting their protocol to incorporate the more complicated C-ring precursors required by natural taxanes. In their search for alternative strategies, they disclosed an AC→ABC route (Swindell *et al.*, 1993), in which the formation of the C_1-C_2 bond was the critical step.

A-ring precursor 140 was prepared from the corresponding monoketal 138 and was coupled to aromatic C-ring precursor 141 to produce intermediate 142. Sequential *trans* reduction, dehydration, and re-protection of 142 yielded a *trans* diene 143, which underwent photochemical isomerization to give 144. Deprotection of 144 was followed by oxidation to provide keto aldehyde 146, the substrate for pinacol closure (Li, 1993). Upon treatment with low-valent titanium, the C_1-C_2 bond was established to generate the tricylic product 147 as a single diastereomer.

The major concern in the application of this protocol was whether it could accommodate a non-aromatic and more functionalized C-ring precursor.

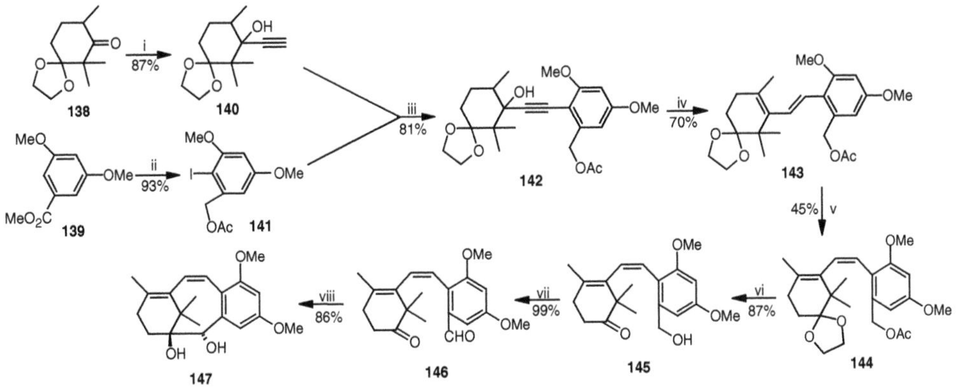

i, HC≡CLi-(CH₂NH₂)₂   ii, (a) LiAlH₄, (b) Ac₂O/py, (c) I₂/H₂O₂   iii, (a) Pd(OAc)₂, PPh₃/Et₃N
iv, (a) Red/Al, (b) Ac₂O/py, (c) POCl₃, (d) (CH₂OH)₂, CH(OMe)₃, TsOHv, hv   vi, (a) aq. TsOH, (b) aq. KOH
vii, MnO₂   viii, TiCl₄/Zn/py

Scheme 9.20 Swindell's approach.

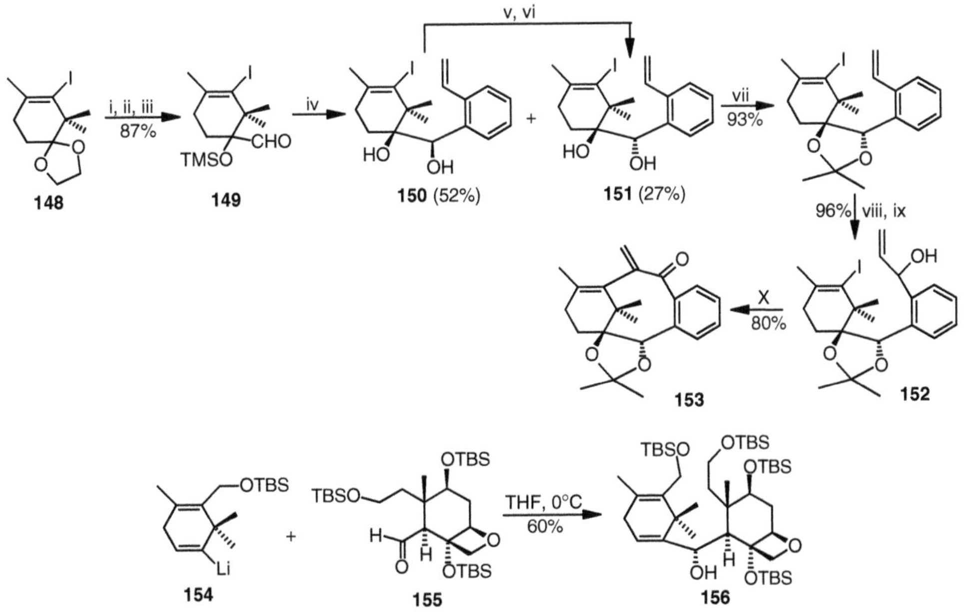

i,1:1 HCl : THF, Δ   ii, TMSCN, KCN (cat.), 18-crown-6, CH₂Cl₂, 0°C   iii, DIBAL-H, hexane, −78°C
iv, (a) 2-lithiostyrene, THF, −78°C; (b) TBAF, THF, rt   v, (COCl)₂, DMSO, Et₃N   vi, NaBH₄, EtOH, 0°C
vii, DMP, (R)-(−)-CSA (cat.), 70°C   viii, O₃, CH₂Cl₂, −78°C, PPh3   ix, CH₂=CHMgBr, THF, −78°C
x, Pd(OAc)₂, K₂CO₃, 80°C, 46 h

Scheme 9.21 Danishefsky's approach.

Danishefsky's approach (Scheme 9.21)

Masters *et al.* (1993) generated a tricyclic model using a palladium-catalyzed intramolecular Heck olefination (de Meijere and Meyer, 1994) in the pivotal step to close ring at $C_{10}-C_{11}$. Initially, iodoketal 148 was converted to a TMS protected aldehyde alcohol 149, which was coupled with an aromatic C-ring precursor to produce a mixture of isomeric alcohols, 150 and 151. Only diastereoisomer 151 possessed the desired stereochemistry. Protection of the vicinal diol in 151 was followed by ozonolysis and nucleophilic addition to provide the vinyl carbinol 152. Treatment with palladium (II) acetate resulted in regiospecific Heck reaction and concomitant oxidation to afford the tricyclic structure 153.

Danishefsky's group applied this method to the synthesis of taxol using appropriately substituted A- and C-rings. An initial attempt is illustrated in Scheme 9.21. Under the conditions shown, A-ring precursor 154 was successfully coupled to C-ring precursor 155 to furnish the C_1-C_2 connection in a stereospecific manner. A modification of this protocol together with an intramolecular Heck reaction eventually led to successful taxol synthesis (section on "Danishefsky's route toward taxol").

Routes for total synthesis toward taxusin

The first synthesis of taxusin (Figure 9.1E), a relatively simple taxoid, was achieved by Holton's group (Holton *et al.*, 1988) and regarded as a breakthrough in the longlasting efforts toward taxoid synthesis. Besides the construction of tricyclic taxane core, the introduction of $C_{11,12}$ and C_4-*exo* olefinic bonds, oxygenation of peripheral carbons, and adjustment of stereochemistry at C_5, C_9, C_{10} and C_{13} are required to establish the target structure. This section will describe the three available approaches toward taxusin.

Holton's route toward taxusin

Holton's approach (Scheme 9.22; Holton *et al.*, 1988) started from the commercially available $(-)$-β-patchoulene oxide 44, and established the tricyclic skeleton in $(-)$-taxusin 172 through fragmentation of a [3.3.0] bicyclic system and A-ring annulation. Upon treatment with *tert*-butyllithium, 44 was converted to allylic alcohol 157, which underwent epoxidation to give the epoxy alcohol 158. Subsequent skeletal rearrangement of 158 was followed by selective oxidation of the secondary alcohol to give ketone 160, which was transformed to hydroxy enone 161 in three steps. Carbons 6 and 7 were added through sequential radical cyclization, hydrolysis and oxidation to afford an intermediate keto lactone. Stereoselective oxidation generated a C-9α hydroxyl group. The Red-Al reduction of this hydroxy ketone 163 gave tetraol 164 and was followed by selective protection of the primary alcohol to yield triol 165. Oxidative fragmentation of the [3.3.0] bicyclic system gave the taxane AB-ring fragment 166. C_4 and C_5 were introduced by nucleophilic addition of α-methoxyvinyllithium to the protected triol 167. The directing effect of MEM protecting group stereoselectively (β-OH : α-OH > 30 : 1) afforded hydroxy ketone 168 via an *in situ* hydrolysis. SmI_2-induced reduction of 168 removed the α-hydroxyl and generated ketone 169. The latter compound underwent C-ring annulation through sequential deprotection, tosylation and cyclization to yield the tricyclic product 170. Silylation of enolated 170 was followed by stereospecific $C_{5\alpha}$ hydroxylation, C_4 enol deprotection, and C_5 alcohol acetylation to produce tetraacetoxy ketone 171. Olefination of 171 gave the final product $(-)$-taxusin, the enantioisomer of natural $(+)$-taxusin. This synthesis was highlighted by its high efficiency: it proceeded in >20% yield from $(-)$-β-patchoulene oxide 44.

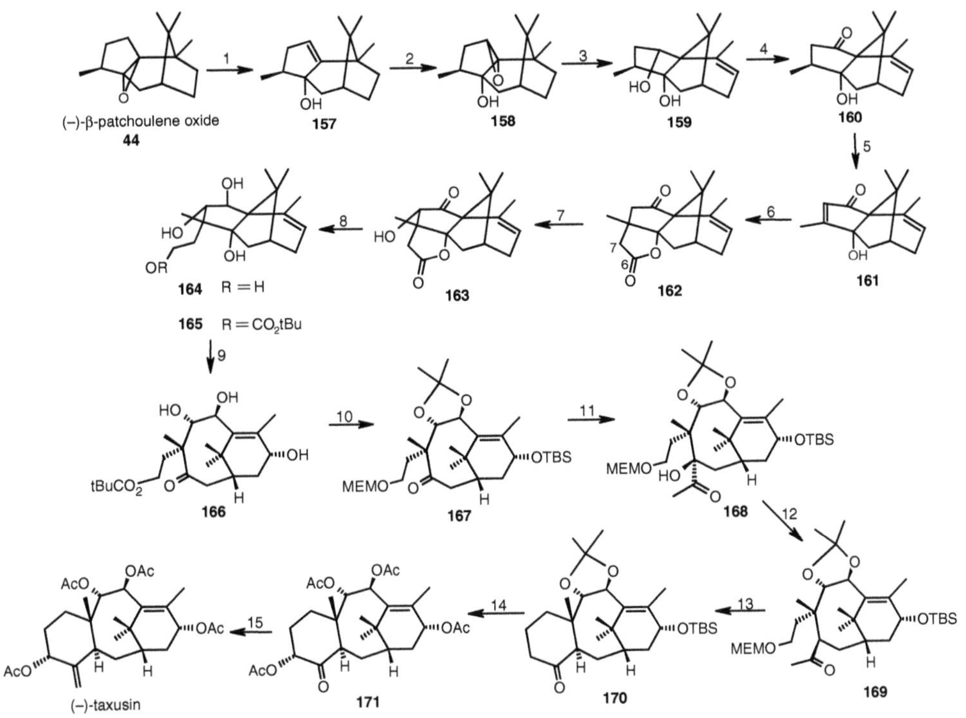

Scheme 9.22 Holton's approach toward taxusin.

1, t-BuLi, hexane reflux, 5 h
2, t-BuOOH, Ti(OiPr)$_4$, CH$_2$Cl$_2$, 2 h
3, BF$_3$.Et$_2$O, CF$_3$SO$_3$H, CH$_2$Cl$_2$, −80°C, 22 h
4, PDC, DME
5, (a) LDA/TMSCl; (b) PhSeCl; c)H$_2$O$_2$, MeOH, K$_2$CO$_3$
6, (a) CH$_2$BrCHBrOCH$_3$, PhN(Me)$_2$, CH$_2$Cl$_2$, 25°C, 2–3 days;
   (b) Bu$_3$SnH, AIBN, PhH, reflux 3 h
   (c) hydrolysis
   (d) Jones' oxidation (CrO$_3$, aq. Acetone)
7, (a) LDA/TMSCl; (b) CH$_3$CO$_3$H, CH$_2$Cl$_2$, 25°C, 2 h
8, (a) Red-Al, 1 : 9 THF : PhH, reflux, 12 h; (b) t-BuCOCl, pyridine
9, (a) anhydrous CH$_3$CO$_3$H, CH$_2$Cl$_2$, reflux, 45 min;
   (b) Ti(OiPr)$_4$, CH$_2$Cl$_2$, reflux, 45 min

10, (a) CH$_3$C(OCH$_3$)$_2$CH$_3$, p-TsOH, TBSOTf, pyridine
    (b) K$_2$CO$_3$, CH$_3$OH
    (c) MEMCl, (iPr)$_2$NEt, CH$_2$Cl$_2$
11, (a) a-methoxyvinyllithium, hexane, −13°C, 12 h
    (b) 2 : 1 : 1 THF : HOAc : H$_2$O
12, SmI$_2$, THF, 0°C, 2 h
13, (a) FeCl$_3$, Ac$_2$O, −45°C, 4 h
    (b) NaOCH$_3$, CH$_3$OH, 25°C, 1 h
    (c) (Ts)$_2$O, pyridine
    (d) NaOt-Bu, THF, 25°C
14, (a) LDA/TMSCl; (b) mCPBA, CH$_2$Cl$_2$;
    (c) Bu$_4$NF, THF, 0.25 N HCl/THF; (d) Ac$_2$O,pyridine
15, Ph$_3$PCH$_2$, 1 : 1 toluene : hexane, 25°C, 3 h

This route achieved the total synthesis of (−)-taxusin from the enantiomeric (−)-β-patchoulene oxide. Considering the reported protocol to convert (+)-camphor to (−)-β-patchoulene oxide (Buechi *et al.*, 1964) and the abundance of (−)-camphor, Holton's method could provide natural (+)-taxusin and (−)-taxol from (−)-camphor.

Kuwajima's route toward taxusin

As illustrated in the section on "Kuwajima's approach," Kuwajima prepared a tricyclic model through an intramolecular vinylogous aldol reaction, which established the eight-membered B-ring at C$_9$, C$_{10}$ (Horiguchi *et al.*, 1989; Furukawa *et al.*, 1992; Seto *et al.*, 1993; Morihira *et al.*, 1993; Nakamura *et al.*, 1994; Seto *et al.*, 1994). By extending this method to non-aromatic, allyl ester-type C-ring derivatives and adding the C$_{19}$ methyl group through a cyclopropane cleavage, Kuwajima's group achieved the total synthesis of a racemic mixture of (±)-taxusin in 1996 (Hara *et al.*, 1996). This route (Scheme 9.23) began from the non-aromatic C-ring precursor 172,

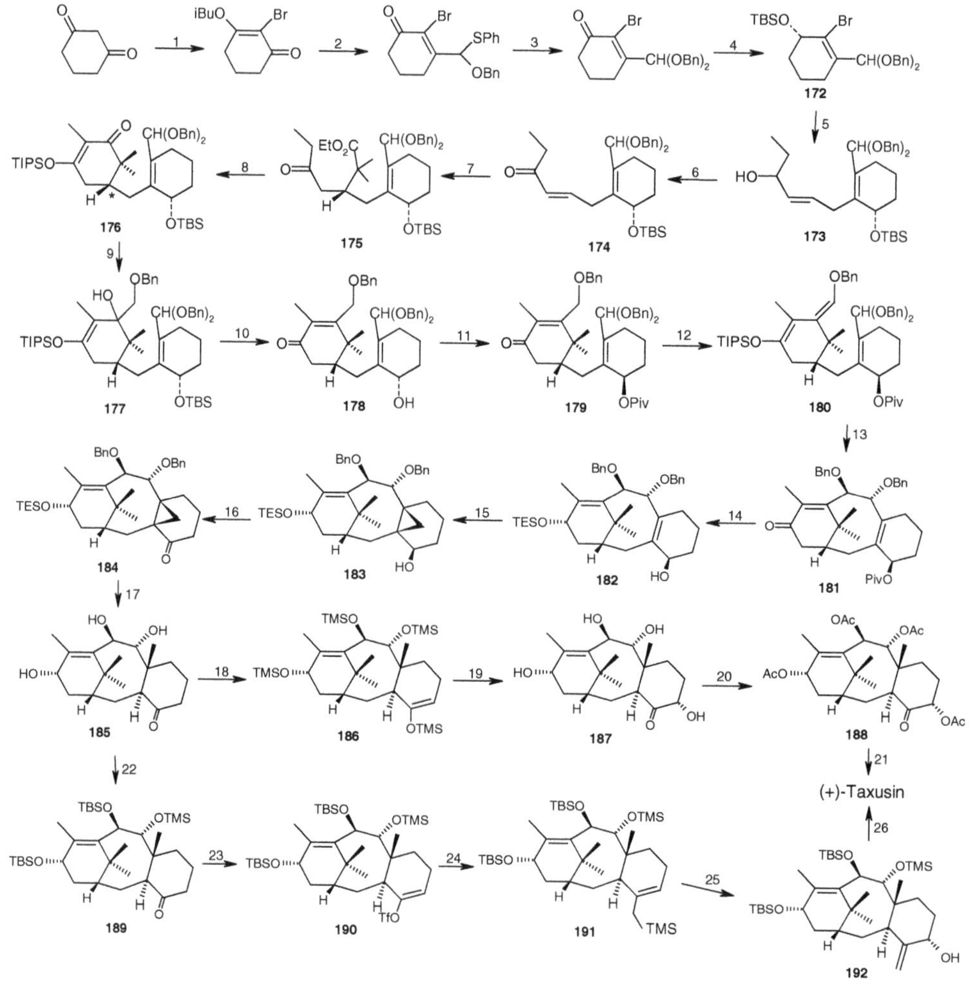

Scheme 9.23 Kuwajima's approach toward taxusin. 1, (a) i-BuOH, TsOH, benzene, reflux, 7 h, 97%; (b) NBS, (ClCH₂)₂, 0°C, 3 h, 93%; 2, PhSCH(OBn)Li, THF−TMEDA, −78°C, 1 h acidic workup, 93%; 3, CuCl₂−CuO, BNOH, CH₃CN, 40°C, 4 h, 76%; 4, (a) BH₃·SMe₂, [structure] THF-toluene, −23 to 0°C, 2 h; (b) DMF, rt, overnight, (a–b, 88%); 5, (a) t-BuLi, Et₂O, −78°C, 1.5 h; (b) CuCN, −45°C, 1 h; (c) 3,4-epoxy-1-hexene, −23°C, 2 h; 6, PDC, MS 4 Å, CH₂Cl₂, rt, 1.5 h, (5–6, 66%); 7, LDA/Me₂CHCO₂Et, THF, −78 to 5°C, 7 h, 100%; 8, (a) t-BuOK, THF, 0°C, 1 h; (b) TIPSCl, Et₃N, 0°C, 4 h, 60%; 9, BnOCH₂Li, THF, −78°C, 2.5 h, 86%; 10, (a) K10, MS 4 Å, BnOH, CH₂Cl₂, −45°C, 1 h, 79%; (b) TBAF, THF, rt, overnight, 83%; 11, DEAD, Ph₃P, PivOH, THF, rt, 1 week, 67%; 12, (a) t-BuOK, THF, 0°C, 1 h; (b) TIPSCl, −78°C, 1 h, 100%; 13, MeAlOTf, CH₂Cl₂, −45°C, 62%; 14, (a) Li(t-BuO)₃AlH, THF, rt, overnight; (b) TESOTf, 2,6-lutidine, CH₂Cl₂, −23°C, 1 h; (c) DIBAL, CH₂Cl₂, −78°C, 1 h, (a–c, 87%); 15, CH₂I₂, ZnEt₂, Et₂O, rt, 6 h, 100%; 16, PDC, MS 4 Å, CH₂Cl₂, rt, 1.5 h, 85%; 17, (a) Li, t-BuOH, NH₃ (l), THF, −78°C, 1 h; (b) MeOH, rt, 1 h, 91%; 18, TMSCl, LDA, THF, −78°C, 10 min, then 0°C, 30 min; 19, m-CPBA, KHCO₃, CH₂Cl₂, 0°C, 10 min; 20, Ac₂O, DMAP, Et₃N, CH₂Cl₂, rt, 1.5 h, (18–20, 80%); 21, Ph₃P=CH₂, benzene, hexane, 0°C, 1.5 h, 53%; 22, (a) TBSOTf, 2,6-lutidine, CH₂Cl₂, 0°C, 1 h; (b) TMSOTf, −78°C, 1 h, 95%; 23, KHMDS, PhNTf₂, −45°C, 1 h; 24, TMSCH₂MgCl, Pd(PPh₃)₄, THF, rt, 1 h, (23–24, 70%); 25, m-CPBA, MeOH, rt, 30 min, 95%; 26, (a) TBAF, THF-HMPA, rt, 2 h, 93%; (b) Ac₂O/DMAP, CH₂Cl₂−Et₃N, rt, 98%.

which was coupled with 3,4-epoxy-1-hexene to give 173 through an S_N2' reaction. Oxidation and asymmetric conjugate addition were followed by a Dieckmann-type cyclization (Banerjee, 1974) and *in situ* silylation to give enone 176, which established both A- and C-ring fragments. 1,2-Addition of (benzyloxy)-methyllithium to the enone carbonyl in 176 was followed by simultaneous dehydration and desilylation to yield product 178. Subsequent epimerization at C_4 through a Mitsunobu reaction (Mitsunobu, 1981) produced 179, which was subjected to sequential enolization and silylation to afford the cyclization substrate 180. Upon treatment with Me_2AlOTf, 180 underwent ring closure to give the desired eight-membered ring 181. Reduction of the C_{13} ketone, silylation and reductive cleavage of the pivalate group produced allyl alcohol 182. The C_{19} methyl group was introduced by a hydroxyl-group-directed cyclopropanation according to the Dauben protocol (Dauben and Deviny, 1966). The resultant methylene derivative 183 was oxidized with PDC to provide cyclopropyl ketone 184. Under Birch conditions (Birch, 1996), reductive cleavage of the cyclopropane ring and liberation of the three peripheral hydroxyl groups were achieved concomitantly to yield triol 185, which contained correct stereochemistry at the C_3 and C_8 sites. Upon exposure to excess LDA and TMSCl, sequential enolization and silylation produced enol silyl ether 186. Subsequent oxidation, hydrolysis and acetylation afforded 188, and methylenation of the C_4 carbonyl furnished (+)-taxusin. The overall yield of taxusin from the commercially available 1,3-cyclohexanedione was 2%.

Later, enantioselective total synthesis of (+)-taxusin was accomplished by incorporating an asymmetric center at C_1 (Hara, 1999). Starting from an optically active 2-bromo-3-siloxycyclohexenecarbacetal 172, the AC-ring fragment 176 was prepared containing a chiral center corresponding to the C_1 site. In addition, an alternative sequence was developed to improve the efficiency of the conversion of 185 to (+)-taxusin (Scheme 9.23). Regiospecific silylation of the C_{10} and C_{13} hydroxyls with $TBSOT_f$/lutidine was followed by *in situ* protection of the C_9 alcohol as a silyl ether to transform 185 to 189. The enol triflate form of ketone 189 underwent a cross-coupling reaction with $TMSCH_2MgCl$ to efficiently yield allylsilane 191. Selective $C_5\alpha$-hydroxylation (192) was followed by desilylation and acetylation to generate the target (+)-taxusin.

Paquette's route toward taxusin

Paquette exploited (Paquette, 1992a,b,c) the oxy-Cope rearrangement and subsequent 1,2-migration of the *gem*-dimethyl-substituted carbon to generate the taxane core (Scheme 9.13 and the section on "Paquette's approach"). In their recent reports (Paquette and Zhao, 1998; Paquette *et al.*, 1998), Paquette's group accomplished total synthesis of taxusin through the formation of a *trans*-$\Delta^{9,10}$-tricyclic olefinic intermediate and chemoselective oxidation/reduction within the confines of the A-ring (Scheme 9.24). The trione intermediate 87 underwent selective enolization to give the enol ether ketone 193. The regioselectivity was attributed to the steric hindrance around the C_9 carbonyl. Oxidation of 193 with dimethyl dioxirane afforded hemiacetal 194 as the major product (96%), which could be converted to the minor product 195 through alkaline treatment. Sequential MOM-protection and Wittig olefination produced 197. The methylene moiety was introduced at this early stage because the C_9 ketone was screened from nucleophilic attack by the bulky phosphorane reagent and much less reactive to olefination. Reduction and dehydration of 197 produced *trans*-$\Delta^{9,10}$-tricyclic olefinic intermediate 199, which was subsequently oxidized to *trans* vicinal diol 200. After protection of the diol as acetonide 201 and desilylation of the enol ether to form ketone 202, the following A-ring functionalizations were required to complete the (+)-taxusin structure: introduction of $\Delta^{11,12}$ double bond, installation of C_{12} methyl group, oxygenation at C_{13} and deoxygenation at C_{14}. After a trial-and-error procedure, Paquette finally achieved the total synthesis of taxusin through chemoselective oxidation and reduction in the A-ring. α-Oxygenation of the enolate derived from 202 produced α-diketone 203 (in an enol ketone form). To adjust the oxidation level of A-ring, the C_{14} carbonyl 203 was reduced regioselectively and the resulting C_{14} hydroxyl group benzoylated to provide 204. The benzoyl moiety was introduced to guide the regioselective deprotonation at C_{12}. Sequential phenylseleneylation (205), elimination (206) and debenzoylation of 204 provided enone 207. The C_{12} allylic methyl group could be introduced by adding a methyl anion to a preformed carbonyl (210 → 211). This goal was achieved via a dihydroxylation pathway. Dihydroxylation of the bridgehead enone in 207 was followed by acetal protection, stereoselective reduction, SEM protection and PMP removal to give the diol 209. In the reductive step, the dioxolane ring facilitated α-hydride attack to predominantly produce the desired C_{13}-β derivative. Selective oxidation of the secondary hydroxyl group with TPAP reagent transformed 209 to 210. Finally, 1,2-addition of methyllithium to 210 introduced the C_{12} methyl group and provided diol 211. C_{13} oxidation, reductive cleavage and β-elimination generated the $C_{11,12}$ double bond (213). The C_{13} carbonyl was stereoselectively reduced to an α-hydroxyl (214) by reacting with DIBAL-H. Deprotection and acetylation then afforded the final product (+)-taxusin. The overall yield from 202 to (+)-taxusin was 1.3%.

Routes for total synthesis toward taxol

Because of its structural complexity, unique cytotoxic mechanism and remarkable clinical efficacy, taxol (Figure 9.1B) has attracted intensive research efforts on both chemical and biological aspects. The major challenges for the total synthesis of taxol include construction of the highly functionalized tricarbocycle and stereocontrol of the numerous asymmetric centers. This imposing goal was achieved independently by Holton and Nicolaou's groups in 1994 (Holton, 1994a,b; Nicolaou, 1994b, 1995a–d), and marked a milestone in this research area. Since then, four other groups (Masters, 1995, Danishefsky, 1999; Wender 1997a,b; Shiina, 1998a,b; Mukaiyama, 1999; Kusama, 2000, Kuwajima and Kusama, 2000a,b) have accomplished total syntheses of this daunting molecule. These six synthetic approaches will be briefly described herein.

Holton's route toward taxol

Holton's strategy for the total synthesis of taxol (Holton *et al.*, 1994a,b) is illustrated retrosynthetically in Scheme 9.25. The starting diol 159 was also an intermediate in the synthesis of taxusin and was readily available from camphor in either enantiomeric form. It was prepared from β-patchouline oxide 44 via a skeletal rearrangement as described previously and in Scheme 9.22. The crucial intermediate 215 was generated through successive hydroxyl-directed epoxidation, epoxy alcohol fragmentation (Holton, 1984) and TBS protection of 159. During the subsequent operations of C-ring annulation and peripheral functionalization, multiple conformational control elements were applied to control stereoselectivity at the asymmetric centers and eventually establish the tricyclic skeleton 216. This highly functionalized tricarbocycle was furnished with a D-ring and finally converted to the C-7 protected baccatin III (217), which was readily transformed to taxol.

The complete synthesis is shown in Scheme 9.26. In the later stage, C-ring annulation would require C-8α deprotonation. However, if a C-10β oxygenated subsituent was present in combination with a C-8β methyl group and a C-3 ketone, the conformational equilibrium would shift strongly toward a chair–boat conformation, which would not undergo C-8α deprotonation.

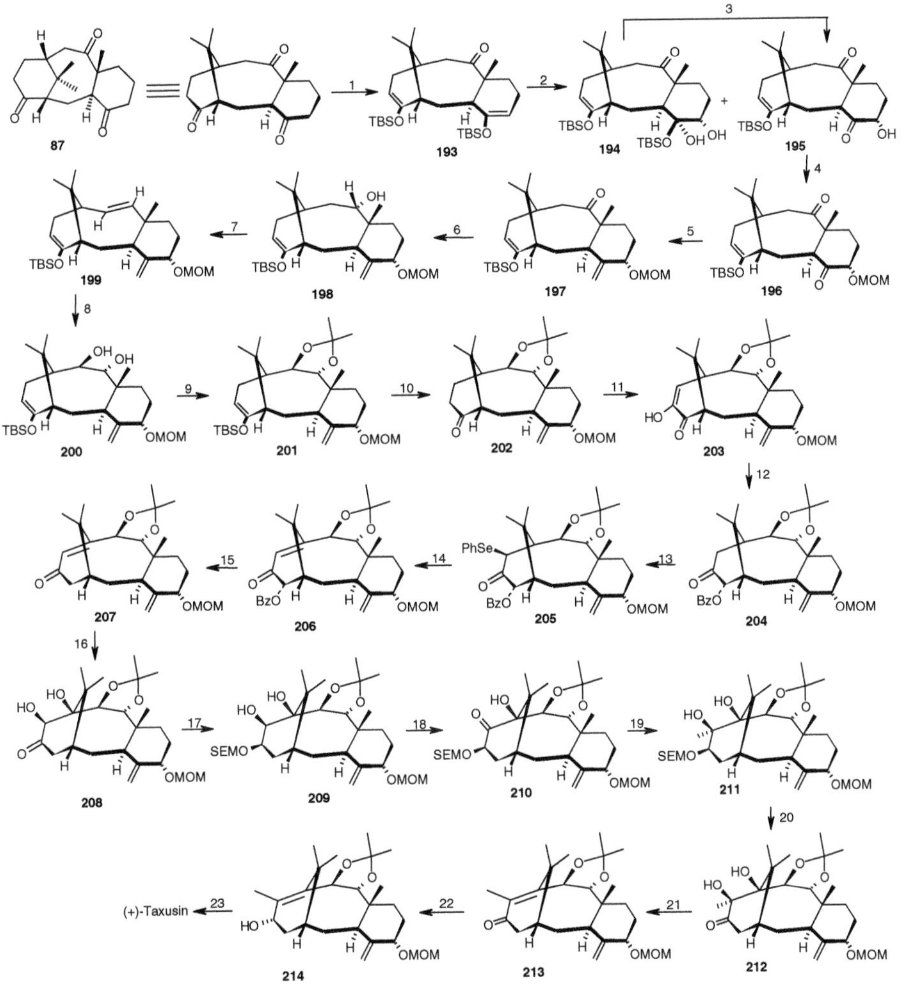

Scheme 9.24 Paquette's approach toward taxusin. 1, TBSOTf, Et₃N, CH₂Cl₂, −78 to 0°C (95%)°C
2, ⤨ acetone, CH₂Cl₂, −78 to 10°C (93%) 3, K₂CO₃, CH₃OH (92%) 4, MOMCl, (iPr)₂NEt, CH₂Cl₂
(97%) 5, PhPCH3⁺I⁻, n-BuLi, THF, 0–25°C (97%) 6, (i-Bu)₂AlH, C₆H₆, 6°C (100%) 7,
[PhC(CF₃)₂O]₂SPh₂, C₆H₆, 25°C (98%) 8, OsO₄, py., NaHSO₃, H₂O (91%) 9, (CH₃)₂C(OCH₃)₂, TsOH,
DMF (97%) 10, (n-Bu)₄N⁺F⁻, THF, H₂O, −78 to 0°C (96%) 11, KN(SiMe₃)₂, 18-crown-6, THF, 0°C;
O₂ (80%) 12, LiAlH₄, ether (65%); C₆H₅COCl, py., DMAP, 25°C (85%) 13, LiN(i-Pr)₂, THF, PhSeBr,
−78 to 25°C (92%) 14, 30% H₂O₂, HOAc, 0°C (92%) 15, SmI₂, THF, CH₃OH, −78°C (82%) 16, OsO₄,
py., NaHSO₃ (66%) 17, (a) p-CH₃OC₆H₄CH(OCH₃)₂, CSA, DMF, 50°C (100%) (b) LiAlH₄, ether,
SEMCl, (i-Pr)₂NEt (81%) (c) DDQ, PPTS, CH₃OH, H₂O (91%) 18, TPAP, NMO, CH₂Cl₂ (92%)
19, CH₃Li, CeCl₃, THF, −78°C (83%) 20, (a) TBAF, THF (86%) (b) TPAP, NMO, CH₂Cl₂ (92%)
21, (a) SmI₂, THF, MeOH (72%) (b) SOCl₂, py., DMAP (64%) 22, (i-Bu)₂AlH, CH₂Cl₂ (67%) 23,
(a) LiBF₄, H₂O, MeCN, 45°C, 2h (51%) (b) Ac₂O, DMAP, Et₃N, CH₂Cl₂ (85%).

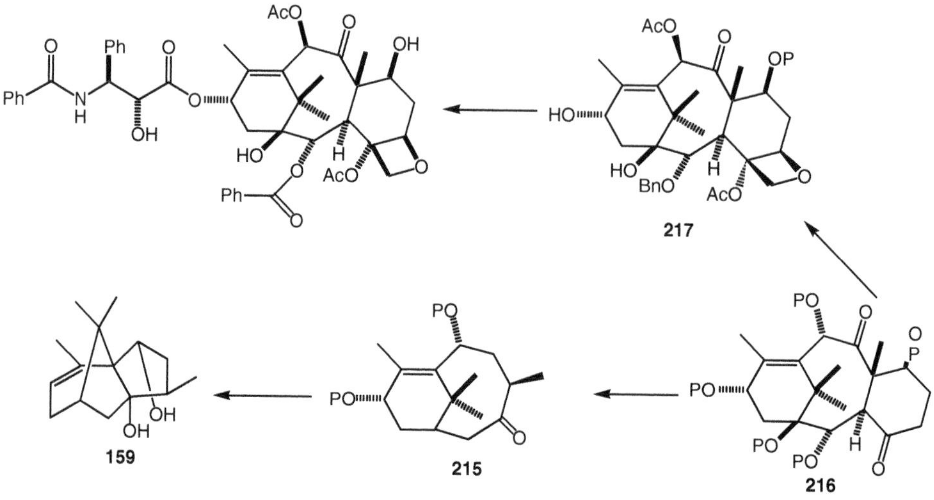

Scheme 9.25 Retrosynthetic analysis of taxol by Holton.

Therefore, a C-10α silyloxy substituent was introduced (218) as a conformational control element. Subsequent epoxy alcohol fragmentation (Holton and Kennedy, 1984) and TBS protection produced bicyclic ketone 219. The magnesium enolate of 219 was subjected to aldol condensation and carbonate protection to give ketone 220. α-Hydroxylation of 220 was followed by stereoselective reduction of the C_3 carbonyl to afford a triol (not shown), which underwent facile intramolecular cyclization to produce the cyclic carbonate 222. The next goals were to establish the functionalization and stereochemistry of C_1, C_2 and C_3.

Necessary steps were the introduction of a bulky, epimerizable C-3α substituent to shift the equilibrium to an enolizable boat–chair conformation and of a C_2 ketone to generate the C-1β and C-2α oxygenation through α-hydroxylation and stereoselective reduction. Swern oxidation (Tidwell, 1990) of 222 provided the C2-ketone derivative 223, which was in a chair–chair conformation. Upon treatment with LTMP, 223 underwent skeletal rearrangement to give hydroxy lactone 224. Conversion of 224 to the stable enol 225. Exposure to silica gel afforded a mixture of *cis-* and *trans-*fused lactones 226a and b (*cis* : *trans* = 6 : 1). The *trans-*lactone 226b could be converted to 226a, the desired stereoisomer, through a recycling procedure (Scheme 9.26, inset). Thus, the *cis-*fused lactone 226a in the desired boat–chair conformation was obtained in 91% yield from 224. Deprotonation and oxidation at C_1 in 226a provided the *cis-*fused α-hydroxyl ketone 227a in an 88% yield, along with 8% of its *trans-*fused isomer 227b. Notably, the deprotonation of 226a occurred almost exclusively at the C_1 proton, although the C_3 proton should be more acidic. Red-Al reduction of the C_2 ketone in 227a was followed by phosgene treatment to provide the carbonate 228. The cyclization substrate 229 was obtained via sequential ozonolysis, oxidation and esterification of 228. Dieckmann cyclization (Banerjee, 1974) of 229 followed by treatment with LDA afforded enol ester 230. After MOP-protection of the C_7 hydroxyl, 230 underwent decarbomethoxylation to give 231. The MOP protection was then replaced by BOM protection (232) to prevent C_7 epimerization and ensure that the C_7-β oxygenation survived the remaining chemistry. TMS enolation of 232 was followed by exposure to *m*-CPBA oxidation to afford ketone 233. Addition of methylmagnesium bromide to the C_4 carbonyl in 233 produced tertiary alcohol 234, which was subjected to dehydration and deprotection to provide allylic

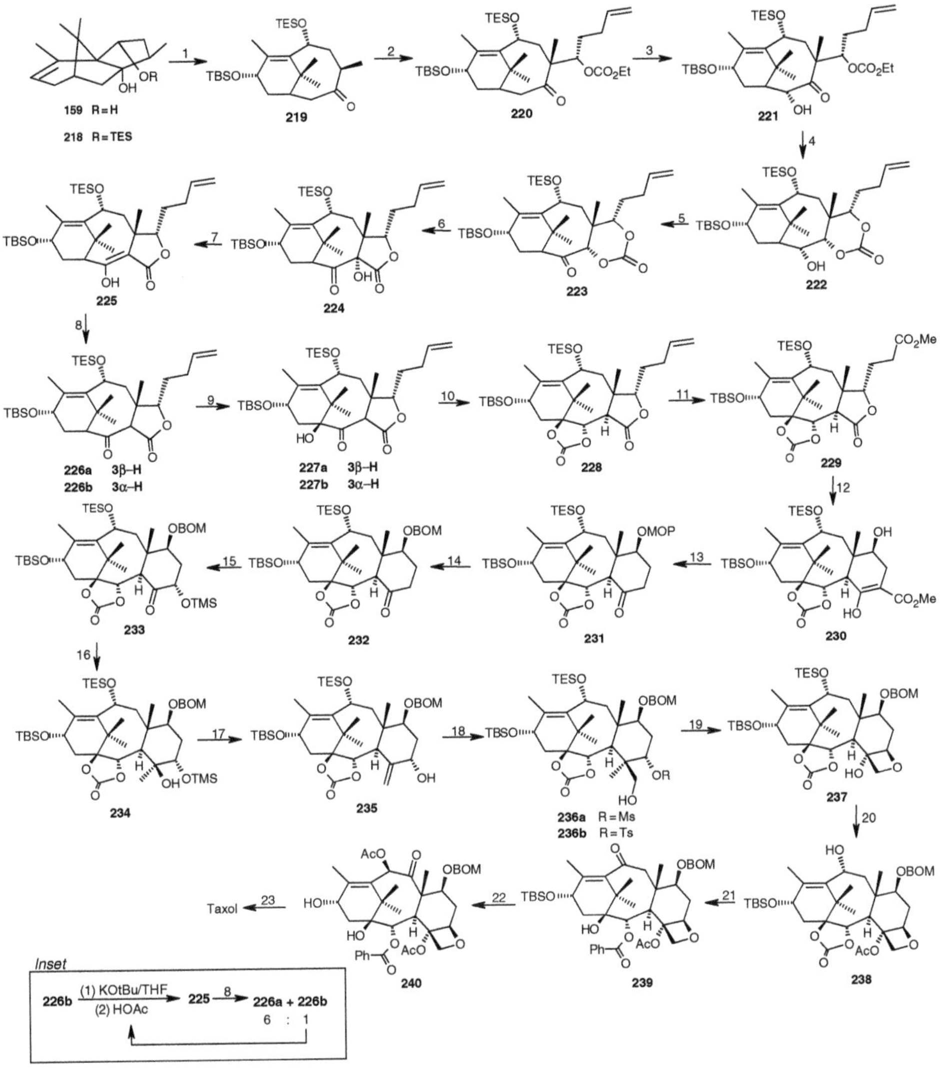

159 R=H
218 R=TES

219

220

221

222

223

224

225

226a 3β–H
226b 3α–H

227a 3β–H
227b 3α–H

228

229

230

231

232

233

234

235

236a R = Ms
236b R = Ts

237

238

239

240

Taxol

Inset

226b $\xrightarrow[(2)\ HOAc]{(1)\ KOtBu/THF}$ **225** $\xrightarrow{8}$ **226a + 226b**
6 : 1

Scheme 9.26 Holton's approach toward taxol. 1, (a) (i-PrO)$_4$Ti, t-BuOOH, Me$_2$S/Δ; (b) TBSOT$_f$, 2,6-lutidine, CH$_2$Cl$_2$, 0°C, 1 h (a–b 93%); 2, (a) HN(iPr)$_2$, THF, MeMgBr, 25°C, 3 h; then **219**, 1.5 h; (b) 4-pentenal, THF, −23°C ,1.5 h; (c) COCl$_2$, py., CH$_2$Cl$_2$, −10°C, 0.5 h; then, ethanol 0.5 h (a–c 75%); 3, LDA, THF, −35°C, 0.5 h; then, −78°C 1.0 eq. (+)-camphorsulfonyl oxaziridine (85%); 4, (a) 20 eq. Red-Al, toluene, −78°C, 6 h → 25°C, 6 h; (b) COCl$_2$, py., CH$_2$Cl$_2$, −78°C to 25°C, 1 h (a–b 97%); 5, Swern oxidation (95%); 6, 1.05 eq. LTMP −25~−10°C (90%); 7, SmI$_2$; 8, Silica gel (7–8 91%); 9, (a) 4 eq. LTMP −10°C; (b) 5 eq. (±)-camphorsulfonyl oxaziridine, −40°C (a–b 88%); 10, (a) Red-Al, THF, −78°C, 1.5 h, basic condition (88%); (b) 10 eq. Phosgene, py., CH$_2$Cl$_2$, −23°C, 0.5 h, (100%); 11, (a) ozonolysis; (b) KMnO$_4$, KH$_2$PO$_4$; (c) CH$_2$N$_2$; (a–c 93%); 12, (a) Dieckmann cyclization (t-BuOK, THF, 0°C, 1 h); (b) LDA, THF, −78°C, 0.5 h; then, HOAc, THF; (a–b 93%); 13, (a) p-TsOH, 2-methoxypropene (100%); (b) PhSK, DMF, 86°C, 3 h, (92%); 14, (a) acidic hydrolysis; (b) EtN(iPr)$_2$, CH$_2$Cl$_2$, (Bu)$_4$NI, reflux, 32 h (a–b 92%); 15, (a) LDA, THF, TMSCl, −78°C; (b) m-CPBA, hexane, 25°C, 5 h (a–b 86%); 16, 10 eq. MeMgBr, −45°C, 15 h; 17, (a) Burgess' reagent; (b) acidic hydrolysis (16–17 65%); 18, for **236a**: (a) MsCl, py. (100%); (b) OsO$_4$, NaHSO$_3$, py. (60–65%); for **236b**: (a) OsO$_4$, NaHSO$_3$ (ether/py., 0°C, 12 h, 80%); (b) TMSCl, Et$_3$N, −78°C; (c) LDA, TsCl, −35°C, 3 h; then, HOAc, 0°C, 14 h (b–c 85%) 19, DBU, toluene, 105°C, 2 h, 80–85%; 20, (a) Ac$_2$O, py., DMAP, 24 h, 25°C (70–75%); (b) HF pyridine complex, MeCN, 0°C, 11 h (100%); 21, (a) 2.1eq. PhLi, THF, −78°C, 10 min; (b) TPAP oxidation (NMO, molecular sieves, CH$_2$Cl$_2$, 25°C, 15 min) (a–b 85%); 22, (a) 4eq. KOtBu, THF, −78 to 0°C, 0.5 h, (b) (PhSeO)$_2$O, 8 eq. THF, 0°C, 40 min (c) 4 eq. KOtBu, THF, −78°C, 10 min (d) Ac$_2$O, py., DMAP, 20 h, 25°C (100%), (e) TASF, THF, 25°C, 1 h (94%); 23, (a) LHMDS, THF, (b) , THF, 0°C, 1 h (241), THF, 0°C, 1 h, (c) HF, pyridine, MeCN, 0°C, 1 h, (d) H$_2$, Pd/C, EtOH, reflux, 1 h (a–d 93%).

alcohol 235. Compound 236a was obtained through the mesylation and osmylation of 235. Alternatively, 236b was obtained via several functional manipulations, including osmylation, selective protection of the C_{20} hydroxyl, tosylation of the C-5α hydroxyl and deprotection of the C_{20} hydroxyl. Both 236a and 236b could undergo intramolecular cyclization to generate the oxetanol D-ring (237). Acetylation of the C_4 hydroxyl and removal of the C_{10} TES protection afforded 238. Hydrolytic addition of phenyllithium to the carbonate ring furnished the C_2 benzoate and was followed by oxidation of the C_{10} hydroxyl to provide ketone 239. Oxidative modification of 239 proceeded nicely to elaborate regio- and stereochemistry at C_9 and C_{10}. Upon treatment with benzeneseleninc anhydride, the enolate of 239 underwent C_9 oxidation and rearrangement to yield an α-hydroxy ketone (not shown) with correct stereochemistry. Subsequent acetylation of the C_{10} hydroxyl and removal of C_{13} TBS protection provided 7-BOM baccatin III (240). The lithium alkoxide of 240 was treated with TES-protected β-lactam 241 to generate the C_{13} side-chain. The resulting product was subjected to desilylation and hydrogenolysis to furnish the final product taxol. The overall yield of taxol from diol 159 was *c*.4–5%.

Holton's route took advantage of rearrangement of tricyclic precursor 159 to build the AB-ring system and elegantly circumvented the problematic issue of B-ring construction. The introduction of conformational control elements was a versatile and efficient strategy to adjust stereochemical details at multiple chiral centers. By using alternative starting materials, enantiomeric (−)-borneol or (−)-patchino, either natural (−)-taxol or its enantiomer (+)-taxol could be synthesized, respectively, through the established protocol.

Nicolaou's approach toward taxol

Concurrent to Holton's achieving a linear synthesis of taxol, Nicolaou's group developed a convergent stategy toward taxol. The target structure was retrosynthetically dissected into four parts (Scheme 9.27). Appropriately substituted A- and C-ring precursors were prepared from Diels–Alder reactions and were coupled together through a Shapiro reaction (Chamberlin and Bloom, 1990) and McMurry coupling (McMurry, 1989). These reactions were followed by installation of the oxetane D-ring and peripheral functionalization of the B- and C-rings. At the last stage, oxygenation and esterification at C_{13} furnished the side-chain and completed the taxol structure (Nicolaou 1994b, 1995a–d).

The preparation of the A-(242) and C-(243) ring fragments is illustrated in Scheme 9.28. The synthesis of 242 proceeded from diene 244 and dienophile 1-chloroacrylonitrile. The major concern was the regioselectivity of the Diels–Alder reaction, because steric hindrance and electronic induction would promote opposite regio-preferences. However, in practice, electronic induction overrode steric hindrance, and the desired regioisomer 245 was produced predominantly. The latent carbonyl group in 245 was released through the Shiner protocol (Shiner *et al.*, 1983) and the primary alcohol was concomitantly deprotected to give the hydroxy ketone 246. Subsequent TBS-protection and the addition of (triisopropylsulfonyl)-hydrazine afforded intermediate 242, which was a suitable substrate for the Shapiro reaction (Chamberlin and Bloom, 1990). The multiple stereocenters and high level of oxygenation in the C-ring complicated the synthesis of its precursor with correct regio- and stereochemical details. The pioneering work by Narasaka *et al.* (1991) suggested that 3-hydroxy-2-pyrone (248) could be used as the diene in the Diels–Alder reaction to obtain the desired regiochemical outcome. Accordingly, dienophile 249 was coupled with diene 248 in the presence of phenylboronic acid to yield the desired regioisomer 250 via formation of a boro-complex. Compound 250 underwent skeletal rearrangement through intramolecular acyl transfer to provide bicyclic product 251. Relief of strain likely facilitated rearrangement the [2.2.2] cycloaddition product 250 to the [3.4.0] bicyclic system 251.

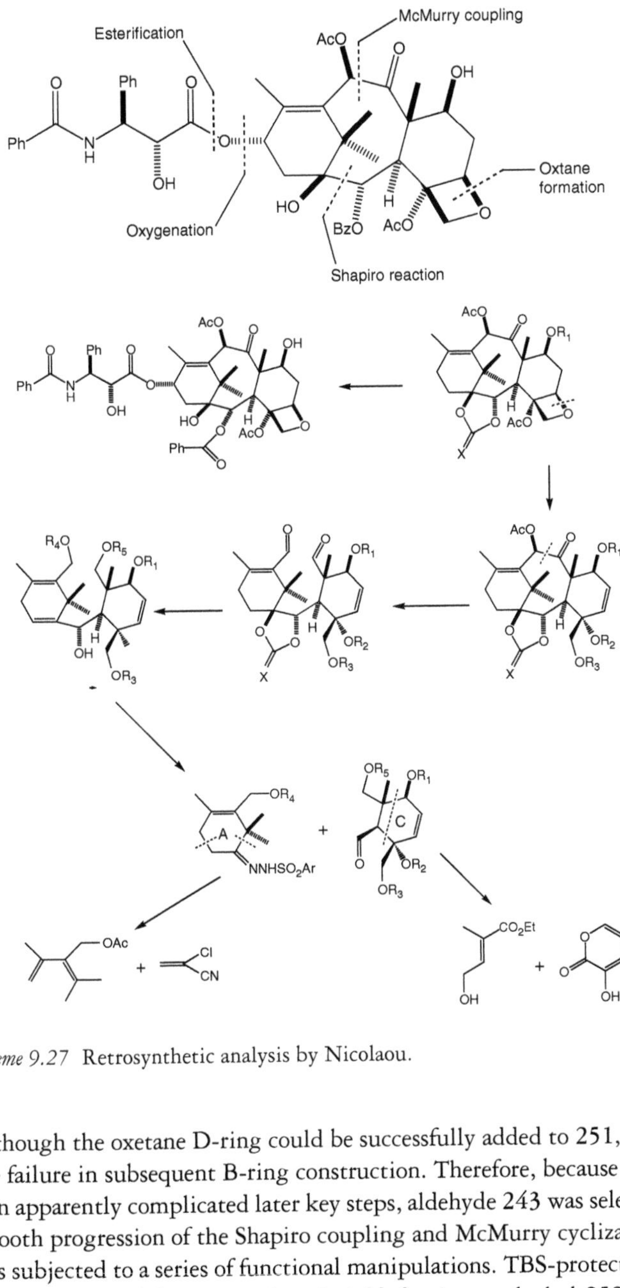

Scheme 9.27 Retrosynthetic analysis by Nicolaou.

Although the oxetane D-ring could be successfully added to 251, these efforts were thwarted by the failure in subsequent B-ring construction. Therefore, because the abundant C-ring oxygenation apparently complicated later key steps, aldehyde 243 was selected as the target to ensure the smooth progression of the Shapiro coupling and McMurry cyclization. To achieve this goal, 251 was subjected to a series of functional manipulations. TBS-protection of the hydroxyls and selective reduction of the ester carbonyl yielded primary alcohol 252. Selective deprotection of the secondary alcohol produced a diol (not shown), then the primary and secondary alcohols were differentially protected to give 253. Reductive cleavage of the lactone ring, along with desilylation of the tertiary alcohol provided triol 254. Finally, acetonide protection of the vicinal diol and oxidation of primary alcohol yielded aldehyde 243.

Shapiro coupling (Shapiro, 1976) of hydrazone 242 and aldehyde 243 produced allylic alcohol 256 as a single diastereoisomer. The stereoselectivity probably resulted from the formation of

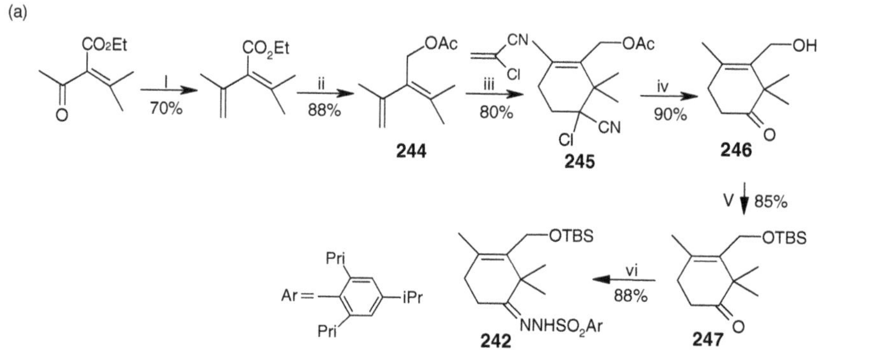

(a)

i, MeMgBr, Et$_2$O, p-TsOH, benzene;   ii, i-Bu$_2$AlH, CH$_2$Cl$_2$; Ac$_2$O, Et$_3$N, DMAP, CH$_2$Cl$_2$;   iii, 130°C;
iv, KOH, t-BuOH, 70°C;   v, TBSCl, imidazole, CH$_2$Cl$_2$;   vi, (2,4,6-triisopropylbenzenesulfonyl)hydrazine, THF

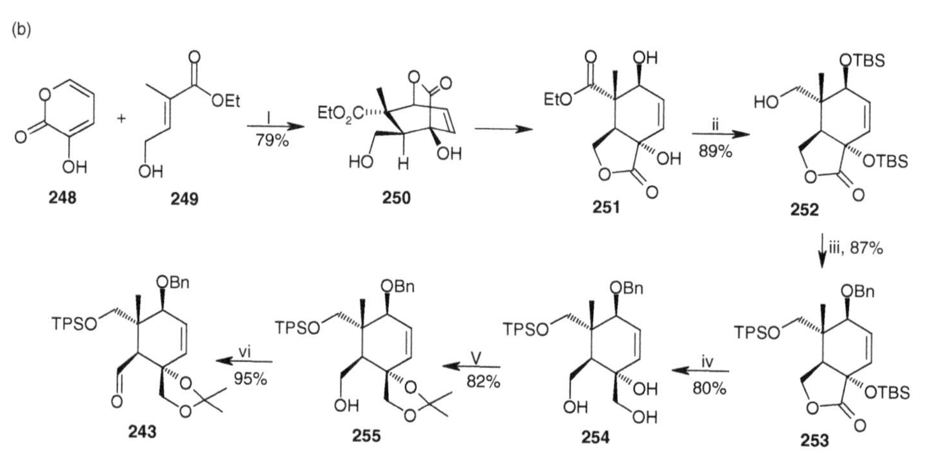

(b)

i, PhB(OH)$_2$, PhH, reflux; then, 2,2-dimethyl-1,3-propanediol;   ii, (a) t-BuMe$_2$SiOTf, 2,6-lutidine; (b) LiAlH$_4$, Et$_2$O;
iii, (a) CSA, MeOH, (b) KH, Et$_2$O, BnBr, n-Bu$_4$NI (cat.):   iv, LiAlH$_4$, Et$_2$O;   v, 2,2-dimethoxypropane, CSA;   vi, TPAP, NMO

Scheme 9.28 (a) Synthesis of hydrazone and (b) Synthesis of aldehyde.

chelated intermediate **243c**. In this fixed conformation, nucleophilic attack occurred at the *re* face at the aldehyde, because the *si* face was blocked by the C$_8$ methyl group. Regioselective epoxidation of the less crowded C$_1$=C$_{14}$ bond gave **257**. Subsequent reductive opening of the epoxide ring provided *trans*-diol **258**, which was subjected to cyclic carbonate protection, desilylation and oxidation to afford dialdehyde **259**, the substrate for McMurry coupling (McMurry, 1989). The intramolecular cyclization proceeded readily to give tricyclic intermediate **260**, which contained both C-9β and C-10β hydroxyl groups.

Further elaboration of these C$_9$ and C$_{10}$ oxygen substituents required differentiation of the two hydroxyl groups. Fortunately, such distinction was readily achieved due to the much higher reactivity of the allylic C$_{10}$ hydroxyl. Regioselective acetylation of **260** was followed by TPAP–NMO oxidation to afford ketoacetate **261**, with correct regio- and stereochemistry at C$_9$ and C$_{10}$. In the initial attempts to saturate the C-ring and install D-ring, both hydroboration of the C$_5$=C$_6$ bond and cyclization of D-ring proved to be quite substrate-dependent. After a thorough investigation of possible candidates, C-5α hydroxy compounds were chosen as promising intermediates. Hydroboration of **261** was followed by treatment with basic hydrogen peroxide to produce a mixture of C-5α alcohol and its C$_6$ regio-isomer. Acidic removal of the acetonide group and chromatographic seperation gave triol **262** as the major product. The oxetane ring was constructed by following routes developed by Potier's (Ettouati *et al.*, 1991) and Danishefsky's (Megee *et al.*, 1992) groups, which involved chemoselective manipulations of the C-4α, C-5α and C-20 hydroxyl groups. After selective acetylation of the primary alcohol in **262**, the benzyl-protection at C$_7$ was replaced by TES-protection (**264**), which could survive the later stages of the synthesis (the C$_{20}$ hydroxyl was concomitantly deacetylated). The oxetane ring was completed by sequential monosilylation of the primary OH, triflation of the secondary OH and mild acid treatment to yield tetracyclic **265**, which was acetylated to provide the important intermediate **266**.

The conversion of enantiomerically pure **266** to the target taxol involved the elaboration of peripheral functionalities at C$_1$, C$_2$, C$_7$ and C$_{13}$. The carbonate ring was attacked by phenyl-lithium to give the C$_1$ and C$_2$ functionalization. Sequential PCC oxidation and stereoselective reduction at C$_{13}$ produced the C-13α oxygenated product 7-TES-baccatin III. The C$_{13}$ side-chain was attached by using the Ojima–Holton method (Ref. Scheme 9.26). Subsequent liberation of the C-7 hydroxyl with HF-pyridine completed the synthesis of (−)-taxol. The overall yield from **243** to (−)-taxol was *c.*0.15%.

Although it appears less efficient than Holton's route in view of the overall yield, Nicolaou's route is likely more adaptable for the synthesis of taxol derivatives. The key operations in this synthesis were the coupling between A- and C-ring precursors **242** and **243** to assemble the ABC-ring skeleton, which contained sufficient functionalization to allow entry to the final target.

Danishefsky's route toward taxol

Shortly after the first total syntheses of taxol, Danishefsky reported (Masters *et al.*, 1995; Danishefsky *et al.*, 1996) another convergent approach toward this awesome molecule by using an intramolecular Heck reaction (de Meijere and Meyer, 1994) as the key step. The retrosynthetic analysis (Scheme 9.30) was based on a suitable confluence of dione **267** and Wieland-Miescher ketone **268** to eventually form baccatin III, which could then be transformed to taxol via well-known protocols. The starting materials **267** and **268** could be converted to appropriately substituted A- and C- (or CD) ring precursors, respectively, furnishing 19 of the 20 carbons in the taxane core and locating the peripheral functionalities in correct regio- and stereochemistry. The two fragments would be coupled via sequential formation of C$_1$−C$_2$ and C$_{10}$−C$_{11}$ bonds and generate the tricyclic taxane core. Subsequent adjustment of flanking oxygenation would provide protected baccatin III and eventually taxol itself. This stategy is distinguished from Nicolaou's and other routes by early incorporation of the oxetane D-ring.

As suggested in their initial exploration (Masters *et al.*, 1993; Scheme 9.21), the C$_1$−C$_2$ connection proceeded smoothly with complicated A- and CD-ring precursors. This observation encouraged the attempts to construct the B-ring from highly functionalized precursors to improve convergency. The preparations of A-ring fragment **269** and CD-ring fragment **270** are illustrated in Scheme 9.31. The dione **267** was converted to its monohydrazone **271**, which upon treatment with iodine in a Barton reaction (Barton *et al.*, 1988), gave rise to monoene iodide **272**. Iododienone **273** was produced by the iodine-induced dehydrogenation of **272**. 1,2-Addition of TMSCN afforded cyanohydrin **274** and reaction with *t*-BuLi gave lithiated **269**, suitable for coupling with the CD-ring precursor (**270**). Wieland-Miescher ketone **268** was converted to the latter precursor through an arduous procedure. It was first stereoselectively reduced to hydroxy enone **275** and subsequently transformed to ketal **276**. After TBS protection of the

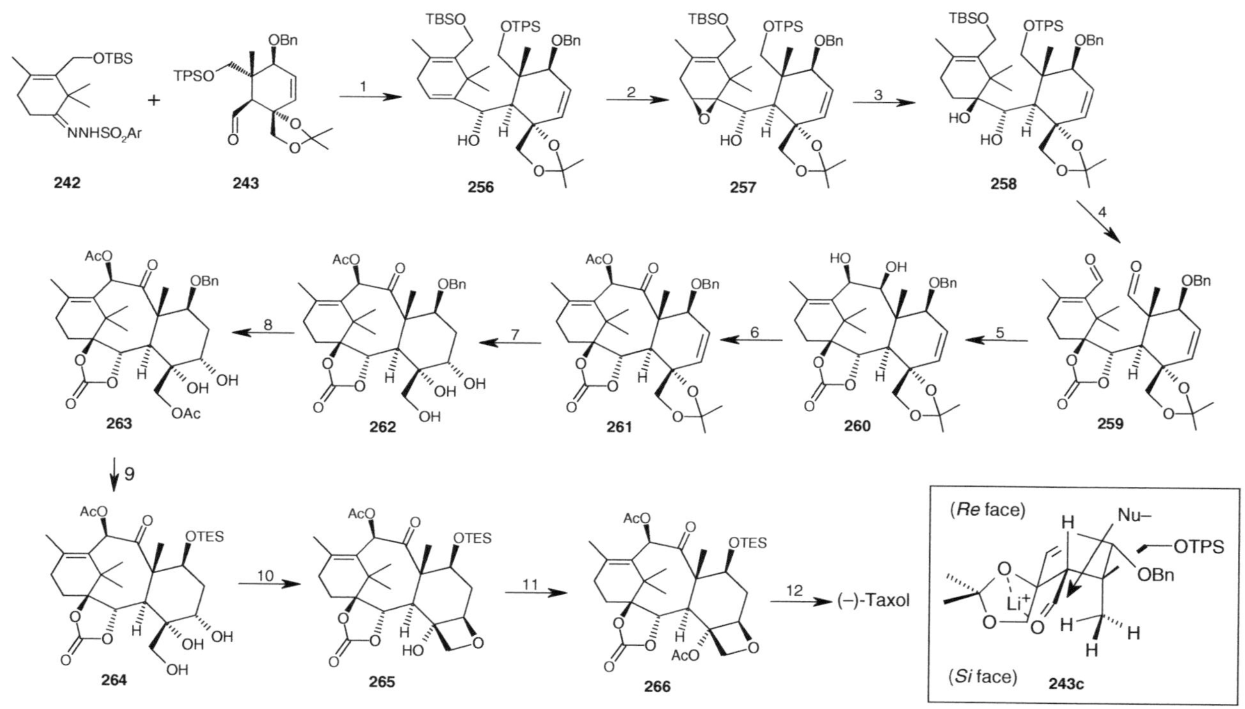

Scheme 9.29 Nicolaou's route toward taxol. 1, 242, n-BuLi, THF, −78°C→25°C→0°C; 243 (THF), 0.5 h, 82%; 2, VO(acac)$_2$, t-BuOOH, molecular sieves, benzene, 25°C, 12 h, 87%; 3, LiAlH$_4$, Et$_2$O, 25°C, 7 h, 76%; 4, (a) KH, HMPA/Et$_2$O, COCl$_2$, 25°C, 2 h, 48%; (b) TBAF, THF, 25°C, 7 h, 80%; (c) TPAP, NMO, MeCN : CH$_2$Cl$_2$ 2 : 1, 25°C, 2H, 82%; 5, (TiCl$_3$)$_2$—(DME); Zn–Cu, DME, 70°C, 1 h, 23%; 6, (a) Ac$_2$O, 4-DMAP, CH$_2$Cl$_2$, 25°C, 2 h, 95%; (b) TPAP, NMO, MeCN, 25°C, 2 h, 93%; 7, (a) BH$_3$—THF, THF, 0°C, 2 h; then, H$_2$O$_2$, aq NaHSO$_3$, 0.5 h, 55%; (b) conc. HCl, MeOH, H$_2$O, 25°C, 5 h, 80%; 8, Ac$_2$O, 4-DMAP, CH$_2$Cl$_2$, 25°C, 0.5 h, 95%; 9, (a) H$_2$, 10% Pd(OH)(C), EtOAc, 25°C, 0.5 h, 97%; (b) Et$_3$SiCl, py., 25°C, 12 h, 85%; (c) K$_2$CO$_3$, MeOH, 0°C, 15 min, 95%; 10, (a) Me$_3$SiCl, py., CH$_2$Cl$_2$, 0°C, 15 min, 96%; (b) Tf$_2$O, i-Pr$_2$NEt, CH$_2$Cl$_2$, 25°C, 0.5 h, 70%; (c) CSA, MeOH, 25°C, 10 min; then, silica gel, CH$_2$Cl$_2$, 25°C, 4 h, 60%; 11, Ac$_2$O, 4-DMAP, CH$_2$Cl$_2$, 25°C, 4 h, 94%; 12, (a) PhLi, THF, −78°C, 10 min, 80%; (b) PCC, NaOAc, celite, benzene, reflux, 1 h, 75%; (c) NaBH$_4$, MeOH, 25°C, 5 h, 83%; (d) NaN(SiMe$_3$)$_2$, β-lactam 241, THF, 0°C, 87%; (e) HF·pyridine, THF, 25°C, 1.5 h, 80%.

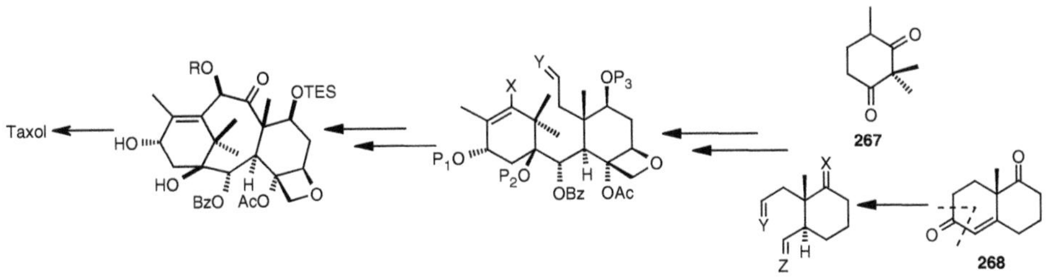

Scheme 9.30 Retrosynthetic analysis by Danishefsky.

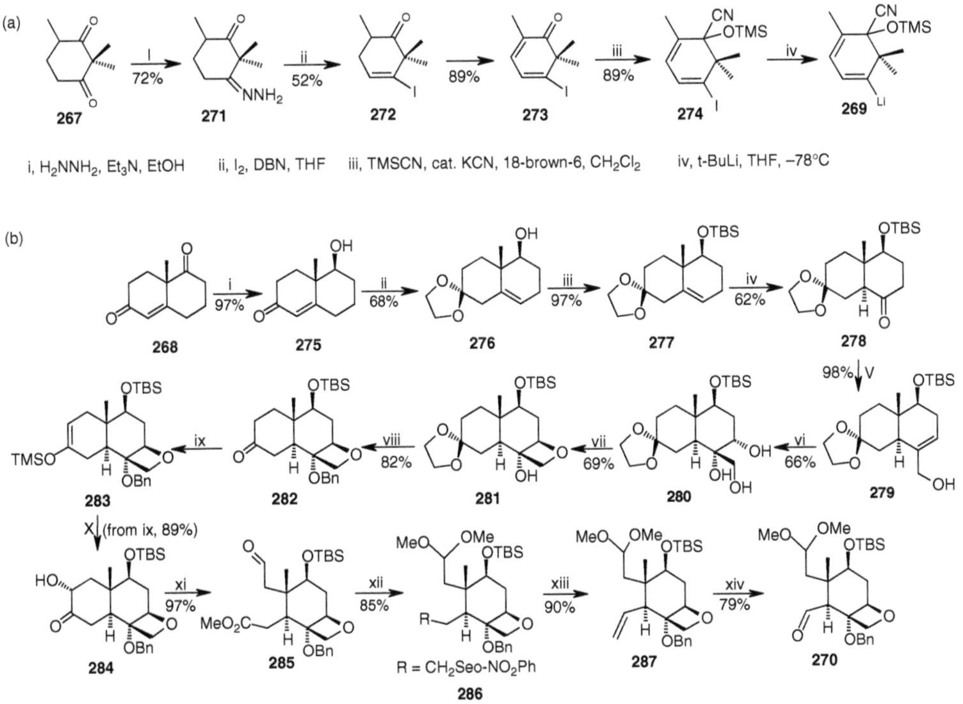

Scheme 9.31 (a) Synthesis of A ring precursor **269** and (b) Synthesis of C ring precursor **270**: i, NaBH₄, EtOH, 0°C; ii, (a) Ac₂O, DMAP, py., CH₂Cl₂, 0°C; (b) (CH₂OH)₂, PhH, naphthalenesulfonic acid, reflux; (c) NaOMe, MeOH, THF; iii, TBSOTf, 2,6-Lutidine, CH₂Cl₂, 0°C; iv, (a) BH₃.THF, THF, 0°C → rt; (b) H₂O₂, NaOH, H₂O; (c) PDC, CH₂Cl₂, 0°C → r.t.; (d) NaOMe, MeOH, THF; v, (a) Me₃S⁺I⁻, KHMDS, THF, 0°C; (b) Al(OiPr)₃, PhMe, reflux; vi, OsO₄, NMO, acetone, H₂O; vii, (a) TMSCl, py., CH₂Cl₂, −78°C → r.t.; (b) Tf₂O, −78°C → r.t.; (c) (CH₂OH)₂, 40°C; viii, (a) BnBr, NaH, TBAI, THF, 0°C → r.t.; (b) TsOH, acetone, H₂O; ix, (a) TMSOTf, Et₃N, CH₂Cl₂, −78°C; x, (a)3,3-dimethyl-dioxirane, CH₂Cl₂, 0°C; (b) CSA, acetone r.t.; xi, Pb(OAc)₄, MeOH, PhH, 0°C; xii, (a) MeOH, CPTS, 70°C; (b) LiAlH₄, THF, 0°C; (c) o-NO₂C₆H₄SeCN, PBu₃, THF, r.t.; xiii, 30% H₂O₂, THF, r.t.; xiv, O₃, CH₂Cl₂, −78°C; then, PPh₃.

secondary alcohol, hydroboranation of the double bond was followed by PDC oxidation to provide the *trans*-fused ketone **278**. To elaborate the oxetane D-ring, the allylic alcohol **279** was set as the next interim goal. By following Corey's sulfonium ylide methodology (Corey and Chaykorsky, 1965), ketone **278** was converted to a spiroepoxide (not shown) and Lewis acid-induced epoxide cleavage was followed by dehydration to yield the desired allylic alcohol **279**. Osmium tetroxide oxidation of **279** afforded triol **280** as the major product. Although complete α-face stereospecificity was expected because of higher hindrance in β-face, 15% of the β-alcohol (not shown) was obtained, along with its desired diastereoisomer **280**. The primary, secondary and tertiary hydroxyl groups could be readily differentiated to allow regioselective functional manipulation. The primary alcohol was selectively protected by silylation, then the secondary alcohol was activated by triflation. The diprotected form of **280** was treated with ethylene glycol at reflux to give the desired oxetane **281**. A hydroxy ketone byproduct (not shown) was obtained concomitantly via a pinacol-like rearrangement. The tertiary alcohol in **281** was protected as the benzyl ether and the ketal deprotected to provide ketone **282** without cleaving the oxetane moiety or the silyl protecting group. Enolization and silylation of **283** led to the silyl enol ether **283**, which underwent regio- and stereospecific hydroxylation to provide α-hydroxy ketone **284**. Ring fragmentation of **284** gave the methyl ester **285**. The aldehyde was protected with dimethyl acetal, the ester reduced to the carbinol, and the hydroxyl group substituted with O-nitrophenyl selenide to afford compound **286**. Upon treatment with hydrogen peroxide, **286** was converted to alkene **287**. Ozonolysis of **287** gave aldehyde **270**, which would serve as the CD-ring fragment.

With compounds **269** and **270** in hand, the C_1-C_2 connection was readily established through 1,2 addition of the aldehyde carbonyl and subsequent liberation of the latent ketone in A-ring fragment to provide intermediate **288**. Epoxide **289**, produced from directed epoxidation of **288**, underwent hydrogenation to give the *trans*-diol **290**. The diol was protected as the cyclic carbonate **291** and the double bond was reduced to afford ketone **292**. Despite its dense array of functional groups, **292** was converted successfully to vinyl triflate **293**. Hydrolysis and Wittig olefination provided the cyclization substrate **294**, which was subjected to an intramolecular Heck reaction (de Meijere and Meyer, 1994) to yield the tricyclic intermediate **295**. The TBS group was replaced by the more labile protecting group TES to facilitate deprotection in the presence of mutiple functional groups. Regioselective epoxidation of **295**'s tetrasubstituted double bond protected it against cleavage in later stages and was followed by benzoate hydrogenolysis and acetylation to produce **296**. The cyclic carbonate was opened with phenyllithium and exocyclic methylene cleaved via sequential reation with osmium tetroxide and lead tetraacetate to form **297**, which then contained the C_1 hydroxyl and C_2 benzoate. The A-ring double bond was restored upon treatment with samarium iodide in the presence of acetic anhydride. By following the powerful chemistry of Holton's group (Scheme 9.26), **298** was obtained with proper oxygenation at C_9 and C_{10}. The C-13α hydroxyl was introduced via allylic oxidation and stereoselective reduction to generate intermediate **299**, which was also an intermediate in Nicolaou's approach. Removal of the C_7-TES protection gave rise to baccatin III, which could be converted readily to taxol through previously described protocols (Ojima *et al.*, 1991). The overall yield from **270** to baccatin III was *c*.0.3% (Scheme 9.32).

Besides the early introduction of the oxetane ring previously mentioned, this approach was also characterized by its catalytically induced enantiotopic control, which made it independent of racemic resolution and procural of chiral starting materials. Similar to Nicolaou's strategy, the key to success was the realization of coupling highly functionalized A- and CD-ring precursors to construct the B-ring and permit further elaboration of peripheral functionalization. The original $C_{10}-C_{11}$ ring closure strategy was accomplished through the crucial intramolecular Heck reaction step (from **294** to **295**).

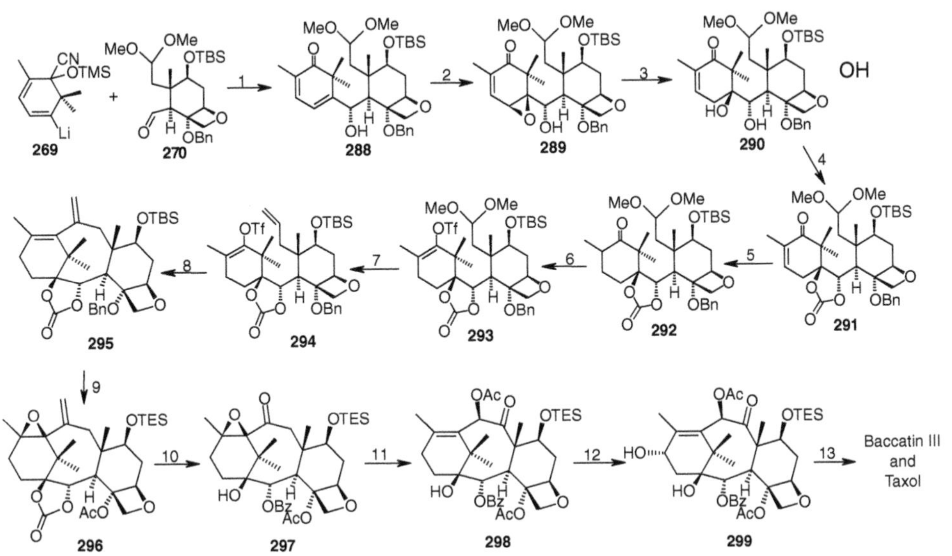

Scheme 9.32 Danishefsky's route toward taxol. 1, (a) 269, THF, −78°C; then 270, 93%; (b) TBAF, THF, −78°C, 80%; 2, (b) *m*-CPBA, CH$_2$Cl$_2$, r.t., 80%; 3, H$_2$, Pd/C, −5°C, EtOH, 65%; 4, CDI, NaH, DMF, 81%; 5, L-Selectride, THF, −78°C, 93%; 6, PhNTf$_2$, KHMDS, THF, −78°C, 98%; 7, (a) PPTS, acetone, H$_2$O, 96%; (b) Ph$_3$P=CH$_2$, THF, −78°C → 0°C, 77%; 8, Pd(PPh$_3$)$_4$, K$_2$CO$_3$, MeCN, molecular sieves, 90°C, 49%; 9, (a) TBAF, THF, r.t., 92%; (b) TESOTf, Et$_3$N, CH$_2$Cl$_2$, −78°C, 92%; (c) *m*-CPBA, NaHCO$_3$, CH$_2$Cl$_2$, r.t., 45%; (d) H$_2$, Pd/C, EtOH, r.t., 82%; (e) Ac$_2$O, DMAP, py., r.t., 66%; 10, (a) PhLi, THF, −78°C, 93%; (b) OsO$_4$, py., 105°C; Pb(OAc)$_4$, PhH, MeOH, 0°C, 61%; 11, (a) SmI$_2$, Ac$_2$O, THF, −78°C, 92%; (b) KOtBu, (PhSeO)$_2$O, THF, −78°C; KOtBu, THF, −78°C, 81%; (c) Ac$_2$O, DMAP, py., 76%; 12, PCC, NaOAc, PhH, reflux, 64%; NaBH$_4$, MeOH, 79%; 13, HF.pyr, THF, 85% of baccatin II.

Wender's route toward taxol

Wender's group developed a concise, stereocontrolled synthesis of taxol via a pinene pathway (Wender *et al.*, 1997a,b). Pinene is a monoterpene readily available from pine trees and the industrial solvent turpentine. Wender's synthesis of taxol started from verbenone (300), the air oxidation product of pinene. This compound supplied 10 of the 20 needed carbons and the desired chirality of the taxane core. As illustrated in Scheme 9.33, by forming C$_{10}$−C$_{11}$ and C$_2$−C$_3$ linkages, an additional six-membered ring was coupled to 300, along with an intramolecular 1,3-alkyl transfer from C$_{13}$ to C$_{11}$. Subsequent fragmentation of the *gem*-dimethyl containing butane ring generated the eight-membered B-ring. After C$_3$ was elaborated with an appropriately substituted aldehyde side-chain, C-ring annulation was achieved through an aldol reaction. D-ring formation and peripheral functionalization followed to complete the target structure, taxol.

Implementation of this strategy is elaborated in Scheme 9.34. The starting material 300 was coupled with prenyl bromide to give the C$_{11}$-alkylated product (not shown), which was converted to aldehyde 301 by selective ozonolysis of the more electon-rich double bond. The carbon−carbon connectivity of the taxane A-ring was established through a photo-induced 1,3-alkyl shift to give 312. A two-carbon connector between the C$_2$ and C$_9$ carbonyls was introduced

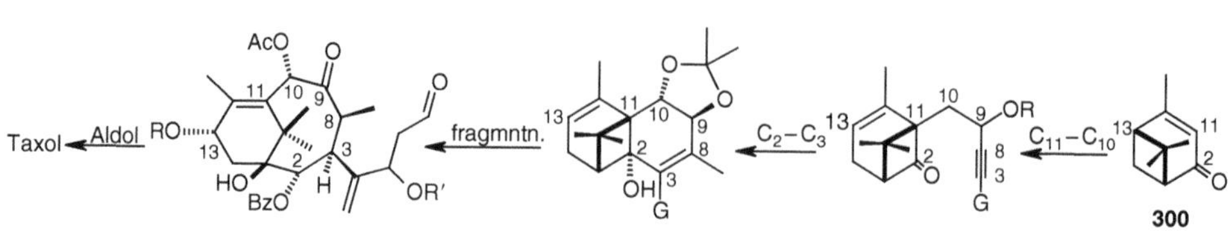

Scheme 9.33 Retrosynthetic analysis by Wender.

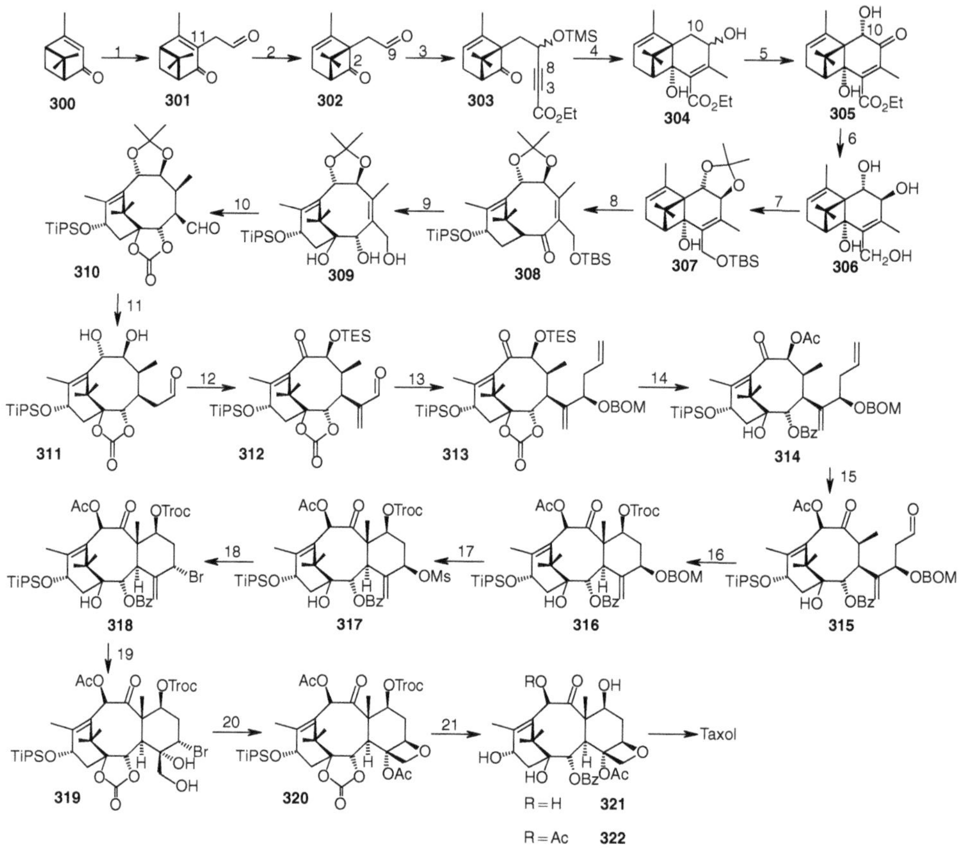

Scheme 9.34 Wender's route toward taxol. 1, (a) KOtBu, 1-bromo-3-methyl-2-butene, DME, $-78°C \rightarrow$ r.t., 79%; (b) O_3, CH_2Cl_2, MeOH, 85%; 2, *hv*, MeOH, 85%; 3, LDA, ethyl propiolate, THF, $-78°C$; TMSCl, 89%; 4, Me_2CuLi, Et_2O, $-78°C \rightarrow$ r.t.; AcOH, H_2O, 97%; 5, (a) $RuCl_2(PPh_3)_3$, NMO, acetone, 97%; (b) KHMDS, Davis' oxaziridine (2-(phenylsulfonyl)-3-phenyloxaziridine), THF, $-78°C \rightarrow -20°C$, 97%; 6, $LiAlH_4$, Et_2O, 74%; 7, TBSCl, imid; PPTS, 2-methoxypropene, r.t., 91%; 8, (a) *m*-CPBA, Na_2CO_3, CH_2Cl_2; (b) DABCO (cat.), CH_2Cl_2, Δ; TIPSOTf, 2,6-lutidine, $-78°C$, (a–b, 85%); 9, KOt–Bu, O_2, $P(OEt)_3$, THF, $-40°C$; NH_4Cl, MeOH, r.t.; $NaBH_4$ 91%; 10, (a) H_2, Crabtree's catalyst (**[Ir(cod)pyr(PCy$_3$)]PF$_6$**), CH_2Cl_2, r.t., TMS–Cl, py., $-78°C$; triphosgene, 0°C, 98%; (b) PCC, 4 Å molecular sieves, CH_2Cl_2, 100%; 11, (a) $Ph_3PCHOMe$, THF, $-78°C$, 91%; (b) 1 N HCl (aq.), NaI, dioxane, 94%; 12, (a) TESCl, pt., CH_2Cl_2, $-30°C$, 92%; (b) Dess-Martin periodinane, CH_2Cl_2; Et_3N, Eschenmoser's salt, 97%; 13, (a) allyl-MgBr, $ZnCl_2$, THF, $-78°C$, 89%; (b) BOMCl, (i-Pr)$_2$NEt, 55°C; 14, (a) NH_4F, MeOH, r.t., (13b–14a, 93%); (b) PhLi, THF, $-78°C$; Ac_2O, DMAP, py., 79%; (b) O_3, CH_2Cl_2, $-78°C$; $(POEt)_3$, 86%; 15, (a) [structure], CH_2Cl_2, r.t., 1 h, 80%; 16, DMAP, CH_2Cl_2; TrocCl, 62%; 17, (a) NaI, HCl(aq.), acetine, 97%; (b) MsCl, py., DMAP, CH_2Cl_2, 83%; 18, LiBr, acetone, 79%; 19, (a) OsO_4, py., THF, $NaHSO_3$; imid, $CHCl_3$, 76%; (b) triphosgene, py., CH_2Cl_2, 92%; (c) KCN, EtOH, 0°C, 76%; 20, (a) (i-Pr)$_2$NEt, toluene, 110°C, 95%; (b) Ac_2O, DMAP, 89%; 21, TASF, THF, 0°C; PhLi, $-78°C$, 46% 10-deacetylbaccatin III, 33% baccatin III.

by selective 1,2-addition of lithium ethyl propiolate to the C_9 carbonyl. The resulting alkoxide (not shown) was trapped *in situ* with TMSCl to produce 303. The C_8 methyl group was added by conjugate addition of Me_2CuLi to 303. After the generation of a C_3 carbanion, intramolecular C_2-C_3 coupling was proceeded readily to establish the taxane B-ring in the bicyclo[4.2.0]-octene 304. After sequential oxidation and deprotonation of 304, addition of Davis' oxaziridine introduced the C_{10} oxygen from the less hindered enolate face, providing α-hydroxyketone 305 as the major product. Stereoselective reduction of 305 with excessive LAH gave tetraol 306, which was converted to 307 by masking the *trans*-vicinal C_9, C_{10} diol as the acetonide and selectively protecting the primary alcohol as the TBS ether. Fragmentation to the bicyclic octane and addition of the A-ring oxygen substitution were achieved by chemoselective epoxidation of the trisubstituted alkene from the less encumbered α-face and DABCO-induced fragmentation of the resultant hydroxy-epoxide, followed by *in situ* protection of the C_{13} alcohol to afford the AB fragment 308. The presence of the acetonide was crucial to control subsequent C_1 oxidation and stereogenesis at C_2, C_3 and C_8. The acetonide ring forced 308 to assume a conformation in which the C_2 carbonyl and C_1 hydrogen were aligned for enolization as required to introduce the C_1 hydroxyl. This conformational influence was also expected to cause selectivity in reduction of the C_2 ketone and hydrogenation of the C_3-C_8 alkene. Accordingly, α-hydroxylation of ketone 308 was followed by *in situ* removal of the TBS group and stereoselective reduction of the C_2 ketone to give triol 309. Treatment with PCC gave aldehyde 310, which served as a versatile taxane precursor, containing the complete carbon framework and oxygenation pattern of taxol's A- and B-rings.

The conversion of 310 to the ABC-tricyclic taxane core proceeded through self-assembly of an AB-bicyclic ketoaldehyde precursor (e.g. 315) under mild conditions. To meet this goal, aldehyde 310 was subjected to homologation with $Ph_3PC(H)OMe$ and hydrolysis of the enol ether and acetonide groups to provide aldehyde 311. Selective TES protection of the C_9 hydroxyl, Dess–Martin periodinane oxidation, and treatment with Eschenmoser's salt (to add C_{20}) yielded enone 312. Two more carbons were added to complete the tricyclic taxane core by sequential operations of 1,2 addition of allylmagnesium bromide to the enone carbonyl and BOM protection of the enol hydroxyl to afford ether 313. The additional olefin could later be cleaved to provide a labile aldehyde handle. After removing the C_9 silyl group, the carbonate ring was opened with PhLi to provide the C_2 benzoate, and subsequent *in situ* acetylation produced acetate 314. The timing of the carbonate hydrolysis was crucial for C_8 deprotonation, since the B-ring could assume a conformation suitable for C-8α deprotonation only after C_1-C_2 cyclic carbonate was cleaved to hydroxybenzoate. Transposition of the acetoxyketone was achieved by exposure to guanidinium base. Regioselective ozonolysis of the monosubstituted double bond yielded aldehyde 315, which was converted to tricyclic intermediate 316 through DMAP-induced aldol cyclization and subsequent Troc-protection of the C_7 hydroxyl. Initial attempts to introduce the oxetane ring by displacing a C_{20} leaving group with the β-oriented C_5 hydroxyl were unsuccessful (Holton *et al.*, 1995). A complementary closure strategy (nucleophilic C_{20} hydroxyl, C_5 leaving group) successfully furnished 316 with the oxetane D-ring. To meet this end, the C_5 BOM ether was transformed to the labile mesylate 317, then bromination afforded an α-bromide at C_5 (318). Stereoselective oxidation of the terminal double bond provided the C_4, C_{20} *trans*-diol. Direct closure of the diol resulted in benzoyl migration from C_2 to C_{20} rather than the desired formation of oxetane D-ring. Therefore, to ensure smooth D-ring closure, the C_1-C_2 diol was sequestered as the cyclic carbonate to afford 319. Upon treatment with Hünig's base, D-ring cyclization was achieved in correct regio- and stereochemistry and subsequent acetylation of the C_4 hydroxyl yielded the tetracyclic product 320. Finally, the C_7 and C_{13} protecting groups were removed and the carbonate ring was cleaved to provide a mixture of baccatin III (321a) and 10-deacetyl baccatin III (321b), which could be converted to taxol through known sequences (Ojima *et al.*, 1992; Holton, 1994b; Holton *et al.*, 1995). The overall yield from verbenone 300 to baccatin III (321a) or 10-deacetyl baccatin III (321b) was *c*.0.6%.

This strategy represented a concise access to taxol and its key analog in a stereocontrolled manner. Verbenone (300), the synthetic starting point-provided exact A-ring connectivity and key B-ring fragments. The B- and C-rings were established by fragmentation and aldol reactions, respectively, and the final product, taxol, could be procured efficiently in the correct enantiomeric form.

Mukaiyama's approach toward taxol

In contrast to the convergent approaches, in which B-ring closure is the last synthetic stage, Mukaiyama reported a taxol synthesis (Shiina, 1998a,b; Mukaiyama *et al.*, 1999) toward taxol, that began with construction of the B-ring. As described in the retrosynthetic analysis (Scheme 9.35), the B-ring precursor was prepared from optically active polyoxy-unit 323 (Mukaiyama *et al.*, 1995). Then, A- and C-ring annulations established the tricyclic core. The synthetic efforts could be logically divided into four stages: (1) construction of functionalized B-ring; (2) annulation of A- and C-ring systems; (3) introduction of oxetane D-ring and adjustment of the peripheral functionalization; and (4) attachment of C_{13} side-chain.

An asymmetric synthesis of taxol's fully functionalized B-ring system could be achieved through an intramolecular Reformatsky-type reaction (Tabuchi *et al.*, 1986) from the optically active precursor 323. The latter compound was obtained through successive stereoselective aldol reactions (Scheme 9.36). The commercially available starting material 324 was converted to benzyl protected aldehyde 325 to provide two versatile handles for further functional manipulations. After acetal protection of the aldehyde in 325, the benzyl group was removed and the resulting exposed hydroxyl group oxidized. The resulting prochiral aldehyde 326 was subjected to an asymmetric aldol reaction with ketene silyl acetal 327 to yield the desired optically active ester 328 with good selectivity (*anti/syn* = 79/21). Sequential reduction of the ester to a primary alcohol and differential protection of the primary and secondary alcohols with TBS and PMB, respectively, were followed by acetal deprotection to liberate the latent aldehyde (329). Another diastereoselective aldol reaction with 327 afforded 2,3,5-*anti, anti* diastereomer 330 as the major product (selectivity of 330 over the other three possible diastereomers was 81:19:0:0). After TBS protection of the newly produced secondary alcohol (331), DIBAL reduction of the ester

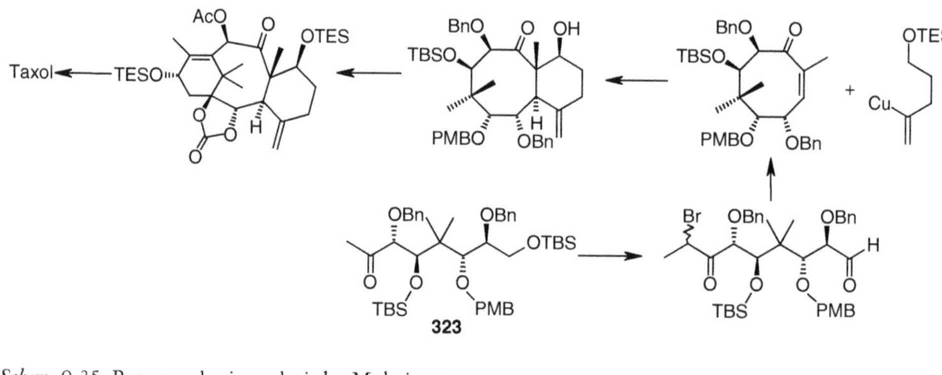

Scheme 9.35 Retrosynthetic analysis by Mukaiyama.

was followed by Swern oxidation to give an aldehyde (not shown), which was subjected to subsequent alkylation and Swern oxidation to produce methyl ketone 323.

Although an α-bromoketo aldehyde could be generated directly from 323 and the functionalized B-ring completed via intramolecular ring closure (Shiina *et al.*, 1995), the thus obtained B-ring precursor would lack the C_8 methyl group. The omission would present serious challenges in later synthetic efforts toward taxol. Bearing this in mind, Mukaiyama determined to introduce an additional carbon to 323 before the ring closure. To meet this end, 323 was subjected to successive bromination and methylation to provide an α-bromo ketone (not shown). Regioselective removal of TBS protection and Swern oxidation then gave α-bromoketo aldehyde 332. Upon treatment with SmI_2, 332 facile underwent intramolecular aldol cyclization to provide a mixture of β-hydroxycyclooctanones with good stereoselectivity to the α-methyl, β-hydroxyl diastereomer (83 : 17 : 0 : 0). Acetylation of the mixture yielded the α-methyl, β-acetoxy isomer (333) as the major product, which upon exposure to DBU readily underwent 1,2 elimination to afford the fully functionalized B-ring in the enone form (334). Michael addition (Lawler and Simpkins, 1988) of the cuprate reagent 335 to B-ring precursor 334 produced the desired ketoaldehyde with C_3, C_8 *cis* configuration (not shown). The diastereoselectivity likely resulted from the α-face selectivity of the enolate anion hydrolysis. Subsequent deprotection and oxidation of the Michael adduct gave rise to aldehyde 336, which upon treatment with base, underwent aldol reaction to accomplish C-ring annulation (337a, 337b) (Shiina *et al.*, 1997). The stereoselectivity was in favor of the desired diastereomer 337b.

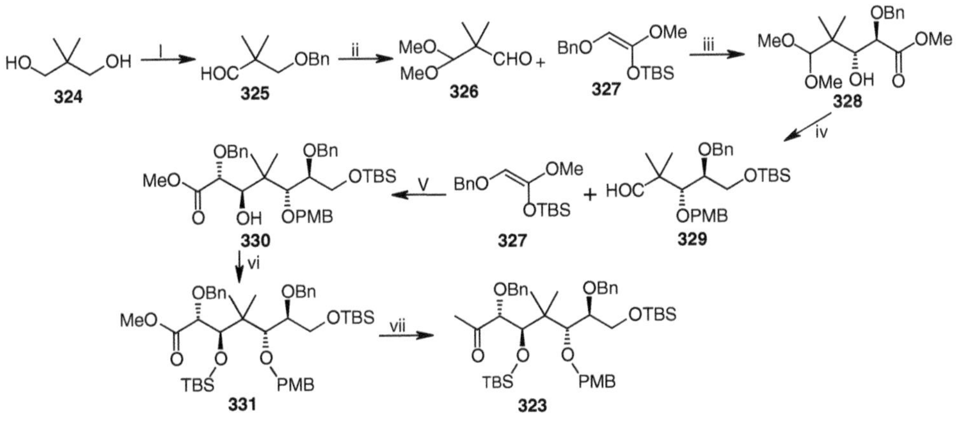

i, (a) PhCH(OMe)₂, CSA, CH₂Cl₂, rt., 97%; (b) LiAlH₄, AlCl₃, CH₂Cl₂ : Et₂O = 1 : 1, reflux, 95%;
(c) (COCl)₂, DMSO, Et₃N, CH₂Cl₂, −78°C to rt., 98%;

ii, (a) HC(OMe)₃, TsOH, MeOH, rt., 95%; a (b) H₂, Pd/C, EtOH, rt., 99%;
(c) (COCl)₂, DMSO, Et₃N, CH₂Cl₂, −78°C to rt., 86%;

iii, [structure], Sn(OTf)₂, Bu₂Sn(OAc)₂, CH₂Cl₂, −23°C, 45%;

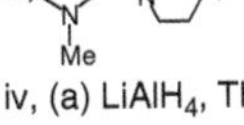

iv, (a) LiAlH₄, THF, 0°C; (b) TBSCl, imidazole, CH₂Cl₂, 0°C; (a–b, 62%); (c)PMBOC(CCl₃)=NH, EtOH, Et₂O, rt.;
(d) AcOH, H₂O, THF, rt. (c–d, 80%);

v, MgBr₂.OEt₂, toluene, −15°C, 87%; vi, TBSOTf, 2,6-lutidine, CH₂Cl₂, 0°C, 95%;

vii, (a) Dibal, toluene, −78°C, 96%; (b) (COCl)₂, DMSO, Et₃N, CH₂Cl₂, −78°C to rt., 98%;

(c) MeMgBr, THF, −78°C, 98%; (d) (COCl)₂, DMSO, Et₃N, CH₂Cl₂, −78°C to 0°C, 89%;

Scheme 9.36 Synthesis of intermediate (323).

The flexible bicyclic intermediate **337b** was expected to exist as a mixture of conformational isomers. Diastereoselective reduction of **337b** with AlH_3 provided a 1,3-diol, which was protected as an acetal to enhance the rigidity of the BC-ring system and "lock" the conformation in that (**338**) required for A-ring annulation. PMB group deprotection and subsequent PDC oxidation afforded **339**. Stereoselective alkylation of the C_1 carbonyl with homoallyllithium reagent in benzene was followed by TBS deprotection to give β-alcohol **340**. The diastereoselectivity of the alkylation was found to be solvent-dependent (Shiina *et al.*, 1998a). The *cis*-diol **340** was treated with dialkylsilyl compounds to yield silylene compounds **341a–c**. Alkylation of **341a–c** with methyllithium provided compounds **342a–c** with desirable regioselectivity. Oxidation of the liberated secondary hydroxyl group at C_{11} was followed by C_{12} oxygenation under forced Wacker oxidation conditions (Tsuji, 1984) to produce the diketones **343a–c**. Upon treatment with low-valent titanium reagent, **343a–c** underwent intramolecular pinacol coupling to give the ABC ring systems **344a–c**. Deprotection of **344a–c** afforded pentaol **345**, which was subjected to selective protection to provide the C_{10} acetoxy, C_1-C_2 carbonate **346**. Five steps were needed to convert **346** to **347**. Removal of the acetonide protection was followed by regioselective TES protection of the C_7 hydroxyl. Regioselective oxidation of the resultant triol then furnished C_9 ketone. Finally, the vicinal diol was converted to a thionocarbonate intermediate, which upon desulfurization, afforded intermediate **347**.

Regioselective oxygenation of **347** was followed by K-Selectride mediated reduction to give the desired C13α-alcohol stereoselectively. TES protection of the C13α-hydroxyl gave the tetracyclic compound **348**, which possessed all peripheral functionalities necessary for the synthesis of baccatin III and taxol, except for the oxetane D-ring. The introduction of the D-ring could be achieved by sequential allylic bromination, osmylation and DBU-induced cyclization. The allylic bromination could give two possible allylic bromides (**349** and **350**) and could be obtained through the allylic bromination. However, under thermodynamic conditions, the equilibrium was favorably shifted to the desired bromide **350**. Selective α-dihydroxylation of **350** provided the dihydroxyl bromide **351**, which underwent intramolecular cyclization to furnish the oxetane ring. Acetylation of the resulting tertiary alcohol provided the acetate **352**. Sequential benzoylation at the C_2 position and desilylation of the C_7 and C_{13} hydroxyl groups afforded baccatin III. Although many synthetic procedures were available to convert baccatin III to taxol, Mukaiyama developed novel sequences for the synthesis and attachment of the side-chain moiety (Shiina *et al.*, 1998b). Due to the space limitation, these procedures will not be discussed in depth herein. The readers are referred to the literature cited above. In summary, Mukaiyama synthesized the side-chain by using enantioselective aldol reaction from two achiral starting materials. The side-chain was attached to the taxoid nucleus by dehydration condensation between a protected N-benzoylphenylisoserine and 7-TES baccatin III. The overall yield from the enantiomeric starting material **323** to the final product taxol was *c.*0.1–0.2% (Scheme 9.37).

This route was distinct for its B → BC → ABC → ABCD sequences. A fully functionalized B-ring was constructed through intramolecular aldol cyclization at the initial stage. Such a strategy would be readily applicable to the synthesis of taxol derivatives from chiral linear precursors of the A- and C-rings.

Kuwajima's approach toward taxol

Kuwajima's group developed (Kusama *et al.*, 2000; Kuwajima and Kusama, 2000a,b) an enantioselective approach to taxol through a convergent strategy. Retrosynthetically (Scheme 9.38), the synthesis started from a nucleophilic addition of A-ring precursor **353** to C-ring precursor **354**

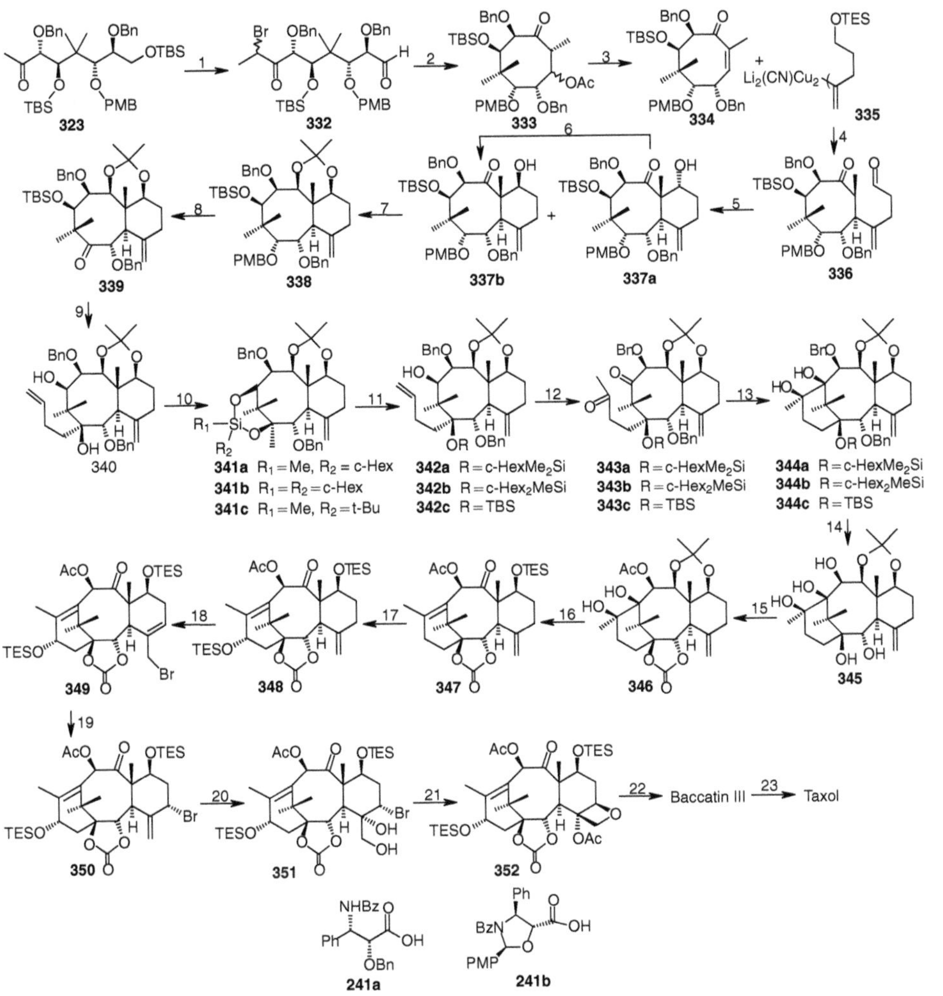

Scheme 9.37 Mukaiyama's approach toward taxol. 1, (a) LHMDS, TMSCl, THF, −78°C to 0°C; (b) NBS, THF, 0°C, (a–b, 100%); (c) LHMDS, MeI, HMPA, THF, −78°C, 100%; (d) 1 N HCl, THF, rt., 84%; (e) (COCl)$_2$, DMSO, Et$_3$N, CH$_2$Cl$_2$, −78°C to rt., 98%; 2, (a) SmI$_2$, THF, −78°C, 70%; (b) Ac$_2$O, DMAP, pyridine, rt., 85%; 3, DBU, benzene, 60°C, 91%; 4, (a) Et$_2$O, −23°C, 92%; (b) 0.5 N HCl, THF, 0°C, 100%; (c) TPAP, NMO, MS 4 Å, CH$_2$Cl$_2$, 0°C, 92%; 5, NaOMe, MeOH, THF, 0°C (98%, 337a : 337b = 8 : 92); 6, NaOMe, THF, 0°C, 90%; 7, (a) AlH$_3$, toluene, −78°C, 94%; (b) Me$_2$C(OMe)$_2$, CSA, CH$_2$Cl$_2$, rt., 100%; 8, (a) DDQ, H$_2$O, CH$_2$Cl$_2$, rt., 97%; (b) PDC, CH$_2$Cl$_2$, rt., 90%; 9, (a) homoallyl-I, s-BuLi, c-hexane, benzene, −23°C to 0°C, 96%; (b) TBAF, THF, 50°C, 100%; 10, (a) c-HexMeSiCl$_2$, imidazole, DMF, rt., 99% of 341a; (b) c-Hex$_2$Si(OTf)$_2$, pyridine, 0°C, 100% of 341b; (c) t-BuMeSi(OTf)$_2$, pyridine, 0°C, 100% of 341c; 11, MeLi, HMPA, THF, −45°C (95% of 342a, 96% of 342b, 96% of 342c); 12, (a) TPAP, NMO, MS4 Å, CH$_2$Cl$_2$, CH$_3$CN, rt (80%, 91%, 85% from 342a,b,c resp.); (b) PdCl$_2$, H$_2$O, DMF, rt., (98% of 343a, 92% of 343b, 91% of 343c); 13, TiCl$_2$, LiAlH$_4$, THF, 40°C (71–43% of 344a); 35°C (63–51% of 344b); 35°C(52% of 344c); 14, (a) Na, liq. NH$_3$, −78°C to 45°C; (b) TBAF, rt., (100% from 344a, 83% from 344b, 61% from 344c); 15, (a) (CCl$_3$O)$_2$CO, py., CH$_2$Cl$_2$, −45°C, 100%; (b) Ac$_2$O, DMAP, benzene, 35°C, 84%; 16, (a) 3N HCl, THF, 60°C; (b) TESCl, pyridine, rt., (a–b, 83%); (c) TPAP, NMO, MS 4 Å, CH$_2$Cl$_2$, rt., 76%; (d) TCDI, DMAP, toluene, 100°C, 86%; (e) P(OMe)$_3$, 110°C, 63%; 17, (a) PCC, NaOAc, celite, benzene, 95°C, 78%; (b) K-Selectride, THF, −23°C, 87%; (c) TESOTf, pyridine, −23°C, 98%; 18, CuBr, PhCO$_3$t-Bu, CH$_3$CN, −23°C (62% of 349, 15% of 350); 19, CuBr, CH$_3$CN, 55°C, (25% of 349, 64% of 350); 20, OsO$_4$, py., THF, rt., 92%; 21, (a) DBU, py., toluene, 50°C, 77%; (b) Ac$_2$O, py., DMAP, rt., 91%; 22, (a) PhLi, THF, −78°C, 94%; (b) HF.py., THF, rt., 96%; 23, (a) TESCl, py., rt., 92%; (b) 241a, DPTC, toluene, 73°C, four times, 100%; Pd(OH)$_2$/C, H$_2$, EtOH, rt., 76%; HF. pyridine, THF, rt., 100% or 5% HCl, EtOH, rt., 100% Alternatively, (b) 241b, DPTC, toluene, 73°C, 95%; TFA, H$_2$O, 0°C, 100%.

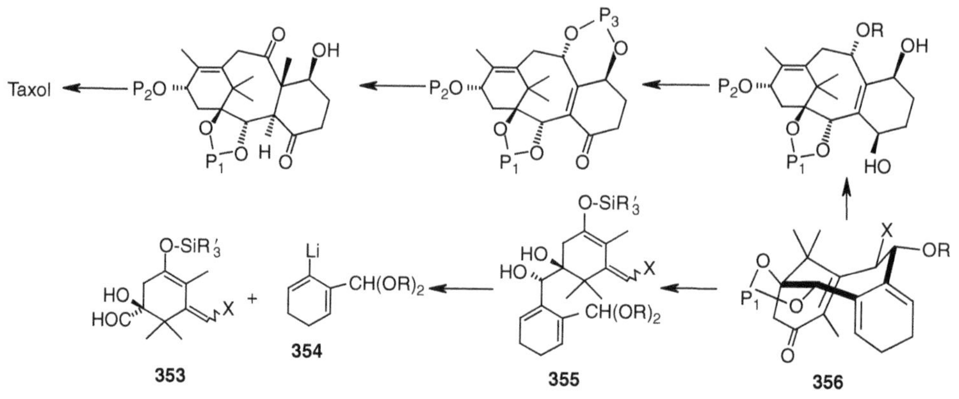

Scheme 9.38 Retrosynthetic analysis by Kuwajima.

to afford the diol AC fragment 355. The aldol-like C—C bond formation between C_9 and C_{10} led to the diastereoselective formation of the *endo* tricarbocyclic intermediate 356, which possessed correct stereochemistry at the C_1 and C_2 sites. A quick examination of the structural features of *endo* intermediate 356 demonstrated that C_{19}-methyl and C_4, C_7 oxygenation could be introduced from the convex β-face. Appropriate functional group elaboration on the B- and C-rings could complete the taxol structure.

The synthesis of A-ring fragment 353 and C-ring fragment 354 are illustrated in Scheme 9.39. The synthesis of 353 started from addition of lithiated propargyl ether to propionaldehyde, which was followed by Lindlar reduction and Swern oxidation to give enone 357. Conjugate addition of isobutylric ester enolate to 357 produced keto ester 358. Upon exposure to base, 358 underwent Claisen-like cyclization and was further converted to pivaloate 359. After removal of the THP protection, Swern oxidation provided aldehyde 360. Subsequent enolization and silylation of aldehyde 360 yielded enol silyl ether 361. Due to the steric effect of the geminal methyl groups, 361 was obtained exclusively as the E-isomer, which served as a suitable substrate for Sharpless's asymmetric dihydroxylation (Kolb *et al.*, 1994). Upon treatment with the chiral reagent DHQ–PHN, 361 was converted to α-hydroxy aldehyde 362. After protecting the aldehyde moiety as an aminal, the pivaloyl group was replaced by a TIPS group in a one-pot operation. Elution of the reaction mixture through a silica gel column removed the aminal moiety to give aldehyde 363. Following a similar procedure, Peterson olefination (Ager, 1990) of 363 afforded dienol silyl ether 353 as a mixture of geometric isomers. This compound served as A-ring precursor in the synthesis.

The synthesis of cyclohexadiene 354 started from 2-bromocyclohexenone 364. The TES protected enolate form of 364 was oxidized to 365. Nucleophilic addition to the carbonyl occurred from the opposite side of the TES protected hydroxyl group to give the *cis*-diol 366 as the major product. After conversion of the dimethyl acetal to a dibenzyl acetal, the vicinal diol was protected as a thionocarbonate (367) and transformed to 368 by following Corey's protocol (Corey and Hopkins, 1982). Upon treatment with *tert*-butyl lithium, vinyllithium drivative 354, the C-ring precursor, was produced.

A chelation-controlled coupling between 353 and 354 afforded the *trans*-diol (not shown) as a single isomer. The ^{1}H NMR spectrum of this diol showed broad signals indicative of a mixture

(a)

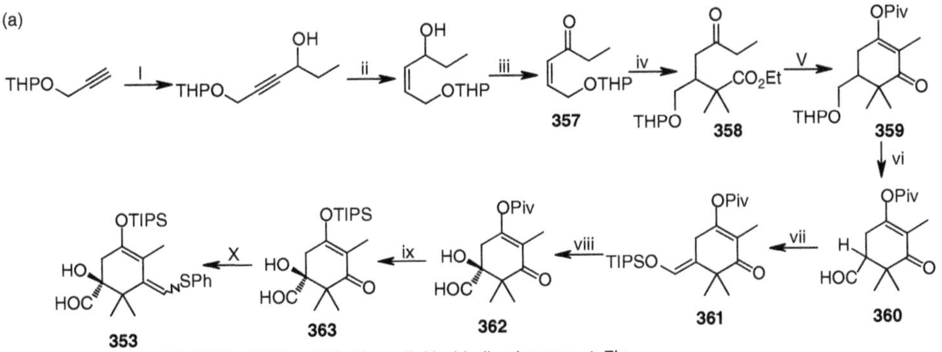

353   **363**   **362**   **361**   **360**

i BuLi, C₂H₅CHO, THF, –78°C to 0°C, 3h;   ii, H₂, Lindlar, hexane, rt, 7h;

iii, (COCl)₂, DMSO, Et₃N, CH₂Cl₂, –78°C to rt, overnight;,

iv, $\overset{OLi}{\underset{OEt}{\diagup}}$, THF, –78°C, 0.5h; 0°C, 0.5 h;   v, t-BuOK, Et₂O, 0°C, 3h; PivCl, CH₂Cl₂, rt, overnight;

vi, (a) p-TsOH, MeOH, rt, 3h, 52% for i–vi a; (b) (COCl)₂, DMSO, Et₃N, CH₂Cl₂, –78°C to 0°C, overnight, 92%;

vii,TIPSOTf, DBU, DMA, CH₂Cl₂, –78°C to rt, 4h, 71%;

iii, (a) K₂OsO₂(OH)₄, K₂Fe(CN)₆, K₂CO₃, DHQ-PHN, t-BuOH, 0°C, 11h, 98%;   (b) benzene, reflux.0.5h, 98%;

ix, (a) (MeNHCH₂)₂, benzene, reflux, 0.5h; (b) KOMe, TIPSCl,THF, –23°C, 6h; (c) SiO₂, 62%;

x, (a)(MeNHCH₂)₂, benzene, reflux, 0.5h; (b) PhSCH(Li)TMS, THF, 0°C, 3h; (c) SiO₂, 64%;

(b)

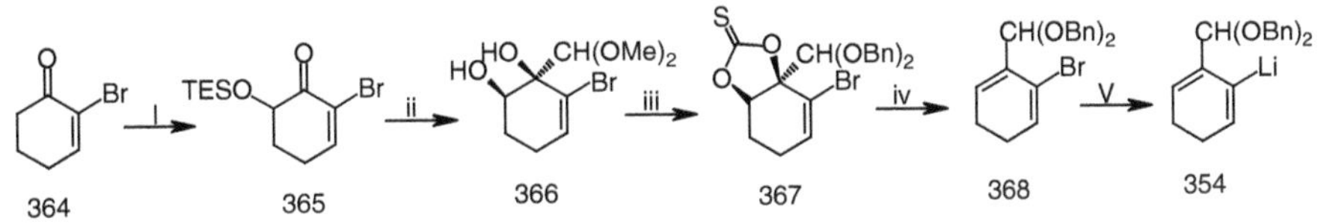

364   **365**   **366**   **367**   **368**   **354**

i, (a) LDA, TESCl, THF, –78°C, 0.5h; (b) m-CPBA, KHCO₃, CH₂Cl₂, 0°C, 0.5h, 96%;

ii, (a) PhSCH(Li)OMe, THF,-78°C, 2h; (b) CuO, CuCl₂, MeOH, THF-MeOH, reflux, 4h, 54%;

iii, (a)p-TsOH, BnOH, benzene, reflux, 6h; (b) Cl₂C=S, DMAP, CH₂Cl₂, 0°C to rt, 0.5h, 64%;

iv, MeN–P(Ph)–NMe, t-Bu–[OH aryl]–t-Bu, THF,rt, 2days, 82%;

Scheme 9.39 (a) Synthesis of intermediate **353** and (b) synthesis of intermediate **354**.

of atropisomers. The C_1, C_2 diol was protected as boronic ester **369**, the desirable (P)-isomer, with the acetal moiety and the dienol ether terminus situated closely to facilitate the B-ring closure. A Lewis acid mediated B-ring cyclization followed by removal of boronate provided the tricyclic product **370** stereoselectively. After protecting the C1β, C2α-diol with di-*tert*-butylsilyene, the C_{13}-keto was reduced then silylated to give **371**. Upon exposure to singlet oxygen, the diene moiety was selectively oxygenated from the β-face of the C-ring. The resulting product was treated with Bu₃SnH and AIBN to simultaneously cleave the peroxide bond and remove the phenylthio group. Removal of the benzyl group and protection of the C_7, C_9 diol afforded **372** as a mixture of diastereomers (α : β = c.1 : 4). The β-isomer of **372** underwent methylenation and Dess–Martin oxidation to give cyclopropyl ketone **373**, while, the α-isomer could not undergo methylene transfer. The conversion of **373** to diol **374** involved successive protection and deprotection of the C_7,C_9 and $C_{1,2}$ diols. The $C_{7,9}$ benzylidene protecting group was replaced by carbonate to allow acidic hydrolysis of the $C_{1,2}$ silyl ether. After protection of the vicinal $C_{1,2}$ diol with benzylidene, basic hydrolysis of the carbonate liberated the C_7 and C_9 hydroxyl groups to give diol 374. In the presence of SmI_2, the cyclopropane ring cleavage proceeded smoothly to yield a stable enol, which, after the removal of the TBS group, underwent base-induced C3α protonation to provide the desired C3α protonated ketone 376. The enol/keto isomerization produced an almost 1:1 equilibrium mixture of 375 and 376. Ketone 376 was enriched by

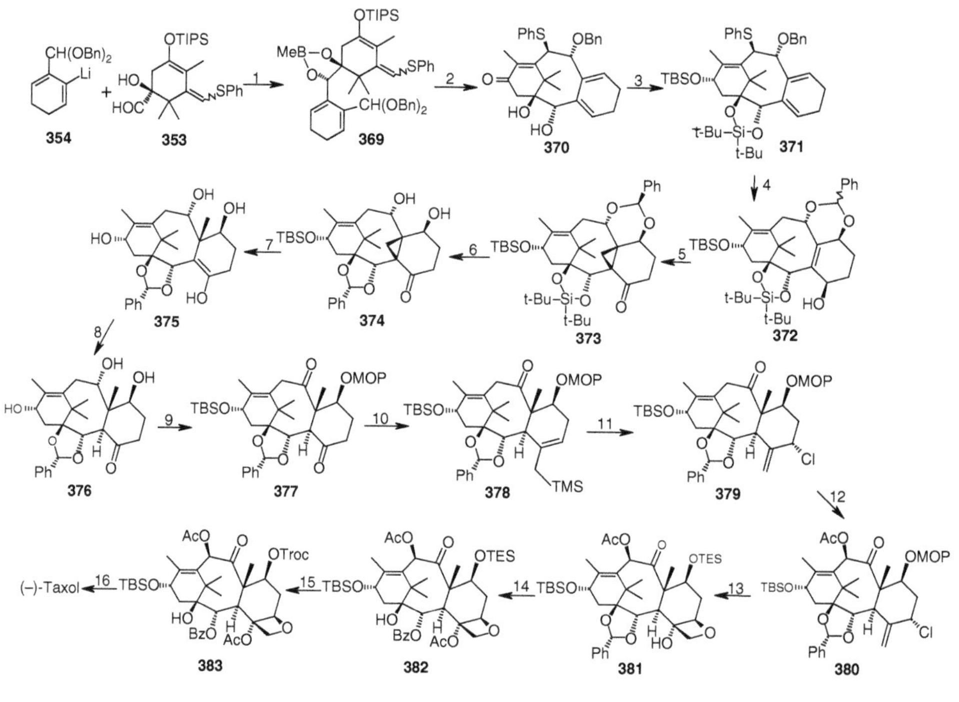

Scheme 9.40 Kuwajima's approach toward taxol. 1, (a) 353, *t*-BuMgCl, THF, −78°C, then, 301, −78°C, 68%; (b) (MeBO)₃, py., benzene, 77%; 2, (a) TiCl₂(OiPr)₂, CH₂Cl₂, −78°C to 0°C; (b) pinacol, DMAP, benzene, 59%; 3, (a) BuLi, t-Bu₂Si(H)Cl, THF, −78°C to 0°C; (b) DIBAL, CH₂Cl₂, −78°C; (c) TBSOTf, 2,6-lutidine, CH₂Cl₂, −23°C, 62%; 4, (a) O₂, TPP, hv, CH₂Cl₂; (b) Bu₃SnH, AIBN, benzene, reflux, (a–b, 80%); (c) Pd/C, H₂, EtOH; (d) PhCH(OMe)₂, CSA, CH₂Cl₂, −23°C, (c–d, 79%); 5, (a) Et₂Zn, ClCH₂I, toluene, 0°C, 66%; (b) Dess-Martin, CH₂Cl₂, 77%; 6, (a) Pd(OH)₂, H₂, EtOH, 84%; (b) Triphosgene, py., CH₂Cl₂, −45°C; (c) TBAF, AcOH, THF, (b–c, 95%); (d) PhCH(OMe)₂, PPTS, benzene, reflux, 86%; (e) K₂CO₃, MeOH–THF, 100%; 7, (a) SmI₂, THF-MeOH–HMPA, 100%; (b) TBAF, BHT, THF; 8, NaOMe, BHT, degassed MeOH, (7b–8, 45%); 9, (a) PhB(OH)₂, CH₂Cl₂; (b) TBSOTf, 2,6-lutidine, CH₂Cl₂, −45°C; (c) H₂O₂, NaHCO₃, AcOEt, (a–c, 70%); (d) Dess-Martin, CH₂Cl₂, 92%; (e) 2-Methoxypropene, PPTS, CH₂Cl₂, 97%; 10, (a) KHMDS, PhNTf₂, −78°C, 89%; (b) Pd(PPh₃)₄, TMSCH₂MgCl, Et₂O, 91%; 11, (a) NCS, MeOH, 88%; (b) 2-methoxypropene, PPTS, CH₂Cl₂, 89%; (c) LDA, MoOPH, THF, −23°C, 80%; (b) Ac₂O, DMAP, CH₂Cl₂, 92%; (c) DBN, toluene, reflux, 68%; 13, (a) OsO4, py., Et₂O, 86%; (b) DBU, toluene, reflux, 86%; (c) PPTS, MeOH; (d) TESCl, imidazole, DMF, (c–d, 97%); 14, (a) Pd(OH)₂, H₂, EtOH, 97%; (b) triphosgene, py., CH₂Cl₂, −78–0°C, 94%; (c) Ac₂O, DMAP, CH₂Cl₂, 66%; (d) PhLi, THF, −78°C, 83%; 15, (a) HF-pyr, THF, 88%; (b) trocCl, pyr, CH₂Cl₂, 94%; 16, (a) TASF, THF, 88%; (b) LHMDS, 241, THF, −78°C to 0°C, 77%; (c) Zn, AcOH–H₂O, 84%.

repeating the isomerization procedure. The C_7,C_9 diol was temporarily protected as a boronate to allow silylation of C_{13} with TBSOT$_f$. Boron protection was then removed and subsequent selective Dess–Martin oxidation of the C_9-OH and MOP protection of C_7-OH provided intermediate 377. A C_4 exomethylene moiety and a C_5 functionality were needed to install the oxetane D-ring. Accordingly, regioselective $\Delta^{4,5}$-enolation was followed by a cross-coupling reaction with TMSCH$_2$MgCl to give the corresponding allylsilane 378. Chlorination of 378 with NCS and TMS removal gave rise to the C5α-chloride 379. At this stage, attempts to generate the C_4,C_{20} diol by OsO$_4$ oxidation exclusively formed the unexpected $C_{11,12}$ diol. Thus, C_{10} functionalization was performed before oxetane D-ring construction to create a crowded-environment around the $\Delta^{11,12}$-double bond and decrease its reactivity. LDA-mediated $\Delta^{9,10}$-enolation was followed by stereoselective oxidation to afford the C10α-alcohol exclusively. The stereochemistry at C_{10} was reversed by sequential acetylation of the C_{10}-hydroxyl and treatment with DBN to yield the thermodynamically favorable C10β-acetate 380 as the major product. Regioselective dihydroxylation of the $\Delta^{4,20}$-double bond followed by DBU-induced cyclization provided a tetracyclic intermediate. Its MOP protecting group was replaced by a TES group to give 381. Acetylation of the tertiary alcohol was hindered due to the influence of the phenyl group. Thus, the $C_{1,2}$ diol protecting group was converted from benzylidene to carbonate to allow smooth acetylation of the C_4 α-OH. Subsequent hydrolysis of the carbonate ring with PhLi furnished the C_2 benzoyl group and provided intermediate 382. The C_7-TES protecting group was replaced by a Troc group (383), and the C_{13} TBS group removed to liberate the hydroxyl group and allow attachment of side-chain moiety 241 using-known protocols. Subsequent removal of protecting groups produced the final product (−)-taxol in an overall yield of 0.06–0.07% from 354 (Scheme 9.40).

This strategy was highlighted by enantioselective total synthesis, which incorporated the initial introduction of an asymmetric C_1 center to allow control of all aymmetric sites at later stages. Other features included: the facile and efficient construction of an ABC *endo*-tricarbocycle through Lewis acid mediated ring closure and the introduction of the C_{19}-methyl via cleavage of a cyclopropyl ketone.

Conclusion

This chapter has provided a retrospective view of the longstanding, worldwide endeavors toward total synthesis of taxoids. We end with the following quotation from Nicolaou (1994a): "The achievement is an academic one at the moment, but it could eventually pave the way to a more practical synthesis. We feel the main advance lies in our ability to construct designed taxols that have better biological activity than the natural one."

References

Ager, D.J. (1990) The Peterson olefination reaction. *Org. React.*, 38, 1–223.

Baloglu, E. and Kingston, D.G.I. (1999) The taxane diterpenoids. *J. Nat. Prod.*, 62, 1448–72.

Banerjee, D.K. (1974) Dieckmann cyclization and utilization of the products in the synthesis of steroids. *Proc. Indian Acad. Sci., Sect., A* 79, 282–309.

Barton, D.H.R., Bashiardes, G. and Fourrey, J.-L. (1988) Studies on the oxidation of hydrazones with iodine and with phenylselenenyl bromide in the presence of strong organic bases; an improved procedure for the synthesis of vinyl iodides and phenyl vinyl selenides. *Tetrahedron*, 44, 147–62.

Birch, A.J. (1996) The Birch reduction in organic synthesis. *Pure Appl. Chem.*, 68, 553–6.

Bissery, M.C., Nohynek, G., Sanderink, G.T. and Lavelle, F. (1995) Docetaxel (Taxotere): a review of preclinical and clinical experience. Part I: preclinical experience. *Anti-Cancer Drugs*, 6, 339–55.

Blechert, S., Muller, R. and Beitzel, M. (1992) Stereoselective synthesis of taxol derivatives. *Tetrahedron*, 48, 6953–64.

Boa, A.N., Jenkins, P.R. and Lawrence, N.J. (1994) Recent progress in the synthesis of taxanes. *Contemp. Org. Synth.*, 1, 47–75.

Bonnert, R.V. and Jenkins, P.R. (1987) A new synthesis of substituted dienes and its application to an alkylated taxane model system. *J. Chem. Soc., Chem. Commun.*, 20, 1540–1.

Bonnert, R.V. and Jenkins, P.R. (1989) A synthesis of an alkylated taxane model system. *J. Chem. Soc., Perkin Trans. 1*, 3, 413.

Bonnert, R.V., Howarth, J., Jenkins, P.R. and Lawrence, N.J. (1991) Robinson annulation on a carbohydrate derivative. *J. Chem. Soc., Perkin Trans. 1*, 5, 1225–9.

Brown, P.A., Jenkins, P.R., Fawcett, J. and Russell, D.R. (1984) A stereocontrolled route to the tricyclo [9.3.1.03,8]pentadecane ring system of taxane. *J. Chem. Soc. Chem. Commun.*, 4, 253.

Brown, P.A. and Jenkins, P.R. (1986) Synthesis of the taxane ring system using an intramolecular Diels–Alder reaction of a 2-substituted diene. *J. Chem. Soc. Perkin Trans. 1*, 7, 1303–9.

Buechi, G., MacLeod, W.D., Jr. and Padilla, J. (1964) Terpenes. XIX. Synthesis of patchouli alcohol. *J. Am. Chem. Soc.*, 86, 4438.

Chamberlin, A.R. and Bloom, S.H. (1990) Lithioalkenes from arenesulfonylhydrazones. *Org. React.*, 39, 1–83.

Corey, E.J. and Chaykovsky, M. (1965) Methylsulfinyl carbanion. Formation and application to organic synthesis. *J. Am. Chem. Soc.*, 87, 1353.

Corey, E.J. and Hopkins, P.B. (1982) A mild procedure for the conversion of 1,2-diols to olefins. *Tetrahedron Lett.*, 23, 1979–82.

Danishefsky, S.J., Masters, J.J., Young, W.B., Link, J.T., Snyder, L.B., Magee, T.V., Jung, D.K. *et al.* (1996) Total synthesis of baccatin III and taxol. *J. Am. Chem. Soc.*, 118, 2843–59.

Dauben, W.G. and Deviny, E.J. (1966) Reductive opening of conjugated cyclopropyl ketones with lithium in liquid ammonia. *J. Org. Chem.*, 31, 3794–8.

Elmore, S.W., Combrink, K.D. and Paquette, L.A. (1991) A convenient means for controlling the oxidation level of bridgehead carbon C-1 in functionalized tricyclo[9.3.1.03,8] pentadecanes. *Tetra. Lett.*, 32, 6679–82.

Ettouati, L., Ahond, A., Poupat, C. and Potier, P. (1991) First semisynthesis of a taxane-type compound with an oxetane group in position 4(20),5. *Tetrahedron*, 47, 9823–38.

Funk, R.L., Daily, W.J. and Parvez, M. (1988) Convergent approach to the taxane class of compounds. *J. Org. Chem.*, 53, 4141–3.

Furukawa, T., Morihira, K., Horiguchi, Y. and Kuwajima, I. (1992) Synthetic studies on taxane carbon frame-work. A highly efficient eight-membered ring cyclization with complete stereocontrol. *Tetrahedron*, 48, 6975–84.

Hara, R., Furukawa, T., Horiguchi, Y. and Kuwajima, I. (1996) Total synthesis of (±)-taxusin. *J. Am. Chem. Soc.*, 118, 9186–7.

Hara, R., Furukawa, T., Kashima, H., Kusama, H., Horiguchi, Y. and Kuwajima, I. (1999) Enantioselective total synthesis of (+)-taxusin. *J. Am. Chem. Soc.*, 121, 3072–82.

Hitchcock, S.A. and Pattenden, G. (1992) A tandem radical macrocyclization. Radical transannulation strategy to the taxane ring system. *Tetra. Lett.*, 33, 4843–6.

Holton, R.A. (1984) Synthesis of the taxane ring system. *J. Am. Chem. Soc.*, 106, 5731–2.

Holton, R.A. and Kennedy, R.M. (1984) Stereochemical requirements for fragmentation of homoallylic epoxy alcohols. *Tetra. Lett.*, 25, 4455–8.

Holton, R.A., Juo, R.R., Kim, H.B., Williams, A.D., Harusawa, S., Lowenthal R.E. and Yogai, S. (1988) A synthesis of taxusin. *J. Am. Chem. Soc.* 110, 6558–60.

Holton, R.A., Somoza, C., Kim, H.B., Liang, F., Biediger, R.J., Boatman, P.D., Shindo, M. *et al.* (1994a) First total synthesis of taxol. 1. Functionalization of the B Ring. *J. Am. Chem. Soc.*, 116, 1597–8.

Holton, R.A., Kim, H.B., Somoza, C., Liang, F., Biediger, R.J., Boatman, P.D., Shindo, M. *et al.* (1994b) First total synthesis of taxol. 2. Completion of the C and D rings. *J. Am. Chem. Soc.*, 116, 1599–600.

Holton, R.A., Somoza, C., Kim, H.-B., Liang, F., Biediger, R.J., Boatman, P.D., Shindo, M., Smith, C.C., Kim, S. *et al.* (1995) In *Taxane Anticancer Agents: Basic Science and Current Status*. Georg, G.I., Chen, T.T., Ojima, I., Vyas, D.M. (eds), ACS Symposium Series 583; American Chemical Society, Washington, DC, 288–301.

Horiguchi, Y., Furukawa, T. and Kuwajima, I. (1989) A highly efficient eight-membered-ring cyclization for construction of the taxane carbon framework. *J. Am. Chem. Soc.*, 111, 8277–9.

Jackson, R.W., Higby, R.G., Gilman, J.W. and Shea, K.J. (1992a) The chemistry of C-aromatic taxane derivatives atropisomer control of reaction stereochemistry. *Tetrahedron*, 48, 7013–32.

Jackson, R.W., Higby, R.G. and Shea, K.J. (1992b) Stereoselective elaboration of the tricyclo [9.3.1.03,8]pentadecane ring system. Atropisomeric control of stereochemistry. *Tetra. Lett.*, 32, 4695–8.

Kende, A.S., Johnson, S., Sanfilippo, P., Hodges, J.C. and Jungheim, L.N. (1986) Synthesis of a taxane triene. *J. Am. Chem. Soc.*, 108, 3513–5.

Kingston, D.G.I., Molinero, A.A. and Rimoldi, J.M. (1993) *Progress in the Chemistry of Organic Natural Products*. Springer-Verlag, New York, Vol. 61.

Kojima, T., Inouye, Y. and Kakisawa, H. (1985) Synthesis of a (±)-3β-trinortaxane derivative. *Chem. Lett.*, 3, 323–6.

Kolb, H.C., VanNieuwenhze, M.S. and Sharpless, K.B. (1994) Catalytic asymmetric dihydroxylation. *Chem. Rev.*, 94, 2483–547.

Kress, M.H., Réjean, R., Miller, W.H. and Kishi, Y. (1993) Synthetic studies toward the taxane class of natural products. *Tetra. Lett.*, 34, 5999–6002.

Kusama, H., Hara, R., Kawahara, S., Nishimori, T., Kashima, H., Nakamura, N., Morihira, K. and Kuwajima, I. (2000) Enantioselective total synthesis of (−)-taxol. *J. Am. Chem. Soc.*, 122, 3811–20.

Kuwajima, I. and Kusama, H. (2000a) Enantioselective total synthesis of taxusin and taxol. *J. Syn. Org. Chem. Jpn.*, 58, 172–82.

Kuwajima, I. and Kusama, H. (2000b) Synthesis studies on taxoids. Enantioselective total synthesis of (+)-taxusin and (−)-taxol. *Synlett.*, 10, 1385–401.

Lawler, D.M. and Simpkins, N.S. (1988) A concise synthesis of (□)-chokol A. *Tetra. Lett.*, 29, 1207–8.

Li, C.J. (1993) Organic reaction in aqueous media – with a focus on carbon –carbon bond formation. *Chem. Rev.*, 93, 2023–35.

Lu, Y.-F. and Fallis, A.G. (1993) An intramolecular Diels–Alder approach to tricyclic taxoid skeletons. *Tetra. Lett.*, 34, 3367–70.

Masters, J.J., Jung, D.K., Bornmann, W.G. and Danishefsky, S.J. (1993) A concise synthesis of a highly functionalized C-aryl taxol analog by an intramolecular Heck olefination reaction. *Tetra. Lett.*, 34, 7253–6.

Masters, J.J., Link, J.T., Snyder, L.B., Young, W.B. and Danishefsky, S.J. (1995) A total synthesis of taxol. *Angew. Chem. Int. Ed. Engl.*, 34, 1723–6.

McMurry, J.E. and Kees, K.L. (1977) Synthesis of cycloalkenes by intramolecular titanium-induced dicarbonyl coupling. *J. Org. Chem.*, 42, 2655–6.

McMurry, J.E. (1983) Titanium-induced dicarbonyl-coupling reactions. *Accta Chem. Res.*, 16, 405–11.

McMurry, J.E. (1989) Carbonyl-coupling reactions using low-valent titanium. *Chem. Rev.*, 89, 1513–24.

de Meijere, A. and Meyer, F.E. (1994) Clothes make the people: the Heck reaction in new clothing. *Angew. Chem. Int. Ed. Engl.*, 33, 2379–411.

Mitsunobu, O. (1981) The use of diethyl azodicarboxylate and triphenylphosphine in synthesis and transformation of natural products. *Synthesis*, 1, 1–28.

Morihira, K., Seto, M., Furukawa, T., Horiguchi, Y. and Kuwajima, I. (1993) The seven-membered ring intermediate to control the stereochemistry on the eight-membered taxane B ring cyclization. *Tetra. Lett.*, 34, 345–8.

Mukaiyama, T., Usui, M. and Saigo, K. (1976) The facile synthesis of lactones. *Chem. Lett.*, 1, 49–50.

Mukaiyama, T., Shiina, I., Sakata, K., Emura, T., Seto, K. and Saitoh, M. (1995) A novel approach to the synthesis of taxol. A synthesis of optically active 3,7-dibenzyloxy-4,8-di-tert-butyl-dimethylsiloxy-5, 5-dimethyl-6-p-methoxy-benzyloxy-2-octanone by way of stereo-selective aldol reactions. *Chem. Lett.*, 3, 179–80.

Mukaiyama, T., Shiina, I., Iwadare, H., Saitoh, M., Nishimura, T., Ohkawa, N., Sakoh, H. *et al.* (1999) Asymmetric total synthesis of taxol. *Chem. Eur. J.*, 5, 121–61.

Nakamura, T., Waizumi, N., Tsuruta, K., Horiguchi, Y. and Kuwajima, I. (1994) Synthesis of the taxol derivatives: control of atropisomerism. *Synlett*, 8, 584–6.

Narasaka, K., Shimada, S., Osoka, K. and Iwasawa, N. (1991) Phenylboronic acid as a template in the Diels–Alder reaction. *Synthesis*, 12, 1171–2.

Neh, H., Blechert, S., Schnick, W. and Jansen, M. (1984) A new route to taxane structure. *Angew. Chem. Int. Ed. Engl.*, 23, 905.

Neh, H., Kuhling, A. and Blechert, S. (1989) Stereoselective synthesis of taxane derivatives. *Helv. Chim. Acta.*, 72, 101–9.

Nicolaou, K.C., Yang, Z., Sorensen, E.J. and Nakada, M. (1993) Synthesis of ABC taxoid ring systems via a convergent strategy. *J. Chem. Soc. Chem. Commun.*, 12, 1024–6.

Nicolauo, K.C., Dai, W.M. and Guy, R.K. (1994a) Chemistry and biology of taxol. *Angew. Chem., Int. Ed. Engl.*, 33, 15–44.

Nicolaou, K.C., Yang, Z., Liu, J.J., Ueno, H., Nantermet, P.G., Guy, R.K., Claiborne, C.F. *et al.* (1994b) Total synthesis of taxol. *Nature*, 367, 630–4.

Nicolaou, K.C., Nantermet, P.G., Ueno, H., Guy, R.K., Couladouros, E.A. and Sorensen, E.J. (1995a) Total synthesis of taxol. 1. Retrosynthesis, degradation and reconstitution. *J. Am. Chem. Soc.*, 117, 624–33.

Nicolaou, K.C., Liu, J.J., Yang, Z., Ueno, H., Sorensen, E.J., Claiborne, C.F., Guy, R.K. *et al.* (1995b) Total synthesis of taxol. 2. Construction of A and C ring intermediates and initial attempts to construct the ABC ring system. *J. Am. Chem. Soc.*, 117, 634–44.

Nicolaou, K.C., Yang, Z., Liu, J.J., Nantermet, P.G., Claiborne, C.F., Renaud, J., Guy, R.K. *et al.* (1995c) Total synthesis of taxol. 3. Formation of taxol's ABC ring skeleton. *J. Am. Chem. Soc.*, 117, 645–52.

Nicolaou, K.C., Ueno, H., Liu, J.J., Nantermet, P.G., Yang, Z., Renaud, J. *et al.* (1995d) Total synthesis of taxol. 4. The final stages and completion of the synthesis. *J. Am. Chem. Soc.*, 117, 653–59.

Ojima, I., Habus, I., Zhao, M., Georg, G.I. and Jayasinghe, L.R. (1991) Efficient and practical asymmetric synthesis of the taxol C-13 side chain, N-benzoyl-(2R,3S)-3-phenylisoserine, and its analogs via chiral 3-hydroxy-4-aryl-β-lactams through chiral ester enolate-imine cyclocondensation. *J. Org. Chem.*, 56, 1681–3.

Ojima, I., Habus, I., Zhao, M., Zucco, M., Park, Y.H., Son, C.M. and Brigaud, T. (1992) New and efficient approaches to the semisynthesis of taxol and its C-13 side chain analogs by means of β-lactam synthon method. *Tetrahedron*, 48, 6985–7012.

Paquette, L.A., Elmore, S.W., Combrink, K.D., Hikey, E.R. and Rogers, R.D. (1992a) An enantioselective approach to the taxanes: direct access to functionalized cis-tricyclo[9.3.1.03,8]pentadecanes via α-hydroxy ketone and Wagner–Meerwein rearrangements. *Helv. Chim. Acta.*, 75, 1755–71.

Paquette, L.A., Combrink, K.D., Elmore, S.W. and Zhao, M. (1992b) Setting the bridgehead oxidation level in trans-tricyclo[9.3.1.03,8]pentadecanes as a prelude to the dual synthesis of taxol and taxusin. *Helv. Chim. Acta.*, 75, 1772–91.

Paquette, L.A., Zhao, M. and Friedrich (1992c) An unprecedented silyl triflate-promoted hydride shift within a taxane derivative. *Tetra. Lett.*, 33, 7311–14.

Paquette, L.A. and Zhao, M. (1998) Enantioselective total synthesis of natural (+)-Taxusin. 1. Retrosynthesis, advancement to diastereomeric *trans*-$\Delta^{9,10}$-tricyclic olefinic intermediates, and the stereoocontrol attainable because of intrinsic rotational barriers therein. *J. Am. Chem. Soc.*, 120, 5203–12.

Paquette, L.A., Wang, H.L., Su, Z., Zhao, M. (1998) Enantioselective total synthesis of natural (+)-Taxusin. 1. Functionalization of the A ring and arrival at the target. *J. Am. Chem. Soc.*, 120, 5213–25.

Rubottom, G.M., Vazquez, M.A. and Pelegrina, D.R. (1974) Peracid oxidation of trimethylsilyl enol ethers. Facile α-hydroxylation procedure. *Tetrahedron Lett.*, 49/50, 4319–22.

Sakan, K. and Craven, B.M. (1983) Synthetic studies on the taxane diterpenes. Utility of the intramolecular Diels–Alder reaction for a single-step, stereocontrolled synthesis of a taxane model system. *J. Am. Chem. Soc.*, 105, 3732–4.

Sakan, K., Smith, D.A., Babirad, S.A., Fronczek, F.R. and Houk, K.N. (1991) Stereoselectivities of intramolecular Diels–Alder reactions. Formation of the taxane skeleton. *J. Org. Chem.*, **56**, 2311–7.

Schiff, P.B., Fant, J. and Horwitz, S.B. (1979) Promotion of microtubule assembly *in vitro* by taxol. *Nature*, **277**, 665–7.

Seto, M., Morihira, K., Katagiri, S., Furukawa, T., Horiguchi, Y. and Kuwajima, I. (1993) Synthesis of C-aromatic taxinine derivatives. *Chem. Lett.*, **1**, 133–6.

Seto, M., Morihira, K., Horiguchi, Y. and Kuwajima, I. (1994) An efficient approach toward taxane analogs: atrop- and diastereoselective eight-membered B ring cyclizations for synthesis of aromatic C-ring taxinine derivatives. *J. Org. Chem.*, **59**, 3165–74.

Shapiro, R.H. (1976) Alkenes from tosylhydrazones. *Org. React.*, **23**, 405–507.

Sharpless, K.B. and Verhoeven, T.R. (1979) Metal-catalyzed, highly selective oxygenations of olefins and acetylenes with tert-butyl hydroperoxide. Practical considerations and mechanisms. *Aldrichemica Acta*, **12**, 63–74.

Shea, K.J. and Davis, P.D. (1983) The tricyclo[9.3.1.0³,⁸]pentadecane ring system. A short synthesis of the C-aromatic taxane skeleton. *Angew. Chem. Int. Ed. Engl.*, **22**, 419.

Shiina, I., Uoto, K., Mori, N., Kosugi, T. and Mukaiyama, T. (1995) An asymmetric synthesis of fully functionalized B ring system of taxol. *Chem. Lett.*, **3**, 181–2.

Shiina, I., Iwadare, H., Sakoh, H., Tani, Y.-I., Hasegawa, M., Saitoh, K. and Mukaiyama, T. (1997) Stereocontrolled synthesis of the BC ring system of taxol. *Chem. Lett.*, **11**, 1139–40.

Shiina, I., Iwadare, H., Sakoh, H., Hasegawa, M., Tani, Y. and Mukaiyama, T. (1998a) A new method for the synthesis of baccatin III. *Chem. Lett.*, **1**, 1–2.

Shiina, I., Saitoh, M., Frechard-Ortuno, I. and Mukaiyama, T. (1998b) Total asymmetric synthesis of taxol by dehydration condensation between 7-TES baccatin III and protected N-benzoyl phenylisoserines prepared by enantioselective aldol reaction. *Chem. Lett.*, **1**, 3–4.

Shiner, C.S., Fisher, A.M. and Yacobi, F. (1983) Intermediacy of α-chloro amides in the basic hydrolysis of α-chloro nitriles to ketones. *Tetra. Lett.*, **24**, 5687–90.

Swindell, C.S. and DeSolms, S.J. (1984) Synthesis of the taxane diterpenes: construction of a BC ring intermediate for taxane synthesis. *Tetra. Lett.*, **25**, 3801–4.

Swindell, C.S., Isaacs, T.F. and Kanes, K.J. (1985) Bicyclo[5.3.1]undec-1(10)-ene bridgehead olefin stability and the taxane bridgehead olefin. *Tetra. Lett.*, **26**, 289–92.

Swindell, C.S., Patel, B.P., deSolms, S.J. and Springer, J.P. (1987) A route for the construction of the taxane BC substructure. *J. Org. Chem.*, **52**, 2346–55.

Swindell, C.S. and Patel, B.P. (1990) Stereoselective construction of the taxinine AB system through a novel tandem aldol-Payne rearrangement annulation. *J. Org. Chem.*, **55**, 3–5.

Swindell, C.S. (1991) Taxane diterpene synthesis strategies. *Org. Prep. Proced. Int.*, **23**, 465–543.

Swindell, C.S., Chander, M.C., Heerding, J.M., Klimko, P.G., Rahman, L.T., Rahman, J.V. and Venkataraman, H. (1993) An AC → ABC approach to taxol involving B-ring closure at C-1/C-2. *Tetra. Lett.*, **34**, 7005–8.

Tabuchi, T., Kawamura, K., Inanaga, J. and Yamaguchi, M. (1986) Preparation of medium- and large-ring lactones. SmI_2-induced cyclization of ω-(α-bromoacyloxy) aldehydes. *Tetra. Lett.*, **27**, 3889–90.

Tidwell, T.T. (1990) Oxidation of alcohols to carbonyl compounds via alkoxysulfonium ylides: the Moffat, Swern, and related oxidations. *Org. React. (N.Y.)*, **39**, 297–572.

Tsuji, J. (1984) Synthetic applications of the palladium-catalyzed oxidation of olefins to ketones. *Synthesis*, **5**, 369–84.

Wang, Z., Warder, S.E., Perrier, H., Grimm, E.L., Bernstein, M.A. and Ball, R.G. (1993) A straightforward approach to the synthesis of the tricyclic core of taxol. *J. Org. Chem.*, **58**, 2931–2.

Wani, M.C., Taylor, H.L., Wall, M.E., Coggon, P. and McPhail, A.T. (1971) Plant antitumor agents. VI. The isolation and structure of taxol, a novel antileukemic and antitumor agent from Taxus brevifolia. *J. Am. Chem. Soc.*, **93**, 2325–7.

Wender, P.A. and Mucciaro, T.P. (1992) A new and practical approach to the synthesis of taxol and taxol analogs: the pinene path. *J. Am. Chem. Soc.*, **114**, 5878–9.

Wender, P.A., Natchus, M.G. and Shuker, A.J. (1995) Toward the total synthesis of taxol and its analogues. In, *Taxol® Science and Applications*. Suffness, M. (ed.), CRC Press, New York, 123–87.

Wender, P.A., Badham, N.F., Conway, S.P., Floreancig, P.E., Glass, T.E., Gränicher, C., Houze, J.B. *et al.* (1997a) The pine path to taxanes. 5. Stereocontrolled synthesis of a versatile taxane precursor. *J. Am. Chem. Soc.*, 119, 2755–56.

Wender, P.A., Badham, N.F., Conway, S.P., Floreancig, P.E., Glass, T.E., Houze, J.B. *et al.* (1997b) The pine path to taxanes. 6. A concise stereocontrolled synthesis of taxol. *J. Am. Chem. Soc.*, 119, 2757–8.

Winkler, J.D., Subrahmanyam, D. and Hsung, R.P. (1992) Studies directed towards the synthesis of taxol: preparation of C-13 oxygenated taxane congeners. *Tetrahedron*, 48, 7049–56.

Yadav, J.S. and Ravishankar, R. (1991) A novel approach towards the synthesis of functionalized taxane skeleton employing Wittig rearrangement. *Tetra. Lett.*, 32, 2629–32.

Yadav, J.S. (1993) Synthesis of antitumor agents. *Pure App. Chem.*, 65, 1349–56.

Abbreviations

Adogen$_{464}$	tri-C8-10-alkylmethylammonium chlorides
AIBN	2,2′-azobis(isobutyronitrile)
BHT	butylated hydroxytoluene
BOMCl	benzyloxymethyl chloride
Burgess' reagent	methyl N-(trimethylammonium-sulfonyl)carbamate
CDI	N,N′-carbonyldiimidazole
Collins reagent	trioxobis(pyridine)-chromium
m-CPBA	*m*-chloroperbenzoic acid
CSA	10-camphorsulfonic acid
CPTS	2,4,6-collidine *p*-toluenesulfonate
DABCO	1,4-diazabicyclo[2.2.2]octane
DBN	1,5-diazabicyclo[4.3.0]non-5-ene
DBU	1,8-diazabicyclo[5.4.0]undec-7-ene
DCC	N,N′-dicyclohexylcarbodiimide
DCM	4-(Dicyanomethylene)-2-methyl-6-(*p*-dimethylaminostyryl)-4H-pyran
DDQ	2,3-dichloro-5,6-dicyano-1,4-benzoquinone
DEAD	diethyl azodicarboxylate
Dess–Martin periodinane	1,1,1-tris(acetyloxy)-1,1-dihydro-1,2-Benziodoxol-3(1H)-one
DHQ-PHN	dihydroquinine 9-O-phenanthryl ether
DIBAL(-H)	diisobutylaluminum hydride
DIEA	DIPEA
Dimedone	5,5-dimethyl-1,3-cyclohexanedione
DIPEA	diisopropylethylamine
DMA	dimethylacetamide
DMAP	4-(dimethylamino)pyridine
DME	dimethoxyethane
DMF	dimethylformamide
DMP	dimethylolpropyleneurea
DMSO	dimethyl sulfoxide
DPTC	O,O′-di(2-pyridyl)-thiocarbonate
Eschenmoser's salt	N-methyl-N-methylene-methanaminium iodide
HMPA	hexamethylphosphoric triamide
KHMDS	potassium hexamethyldisilazide
K-Selectride	potassium tris(*s*-butyl)hydroborate
LAH	lithium aluminum hydride
LDA	lithium diisopropylamide
LHMDS	lithium hexamethyldisilazide
Lindlar	Pd–Pb/CaCO$_3$
L-selectride	lithium tri-*s*-butylborohydride
LTMP	lithium 2,2,6,6-tetramethylpiperidide
MEM-Cl	β-methoxyethoxymethyl chloride

MOM-Cl	methoxymethyl chloride
MsCl	methanesulfonyl chloride
NBS	N-bromosuccinimide
NCS	N-chlorosuccinimide
NMO	N-methylmorpholine N-oxide
PCC	pyridinium chlorochromate
PDC	pyridinium dichromate
Piv	pivaloyl
PivCl	pivaloyl chloride
PMB	*p*-methoxybenzyl
PPTS	pyridinium *p*-toluenesulfonate
Red-Al	sodium bis(2-methoxyethoxy)hydride
SEMCl	β-trimethylsilylethoxymethyl chloride
SmI$_2$	samarium iodide
TASF	tris-(diethylamino)sulfonium difluorotrimethylsilicate
TBAF	tetra-*n*-butylammonium fluoride
TBAI	tetra-*n*-butylammonium iodide
TBDMS-Cl	TBS-Cl, *t*-butyldimethylsilyl chloride
TBDPS	*t*-butyldiphenylsilyl
TBDPS-Cl	*t*-butyldiphenylsilyl chloride
TBHP	*t*-butyl hydroperoxide
TBSCl	*t*-butyldimethylsilyl chloride
TBSOT$_f$	*t*-butyldimethylsilyl trifluoromethanesulfonate
TCDI	1,1'-thiocarbonyldiimidazole
TESCl	triethylsilyl chloride
TESOT$_f$	triethylsilyl trifluoromethanesulfonate
THF	tetrahydrofuran
Tf	trifluoromethanesulfonyl
TFA	trifluoroacetic acid
TFAA	trifluoroacetic anhydride
THP	tetrahydropyran
TIPSCl	1,3-dichloro-1,1,3,3-tetraisopropyldisiloxane
TIPSOT$_f$	1,1,3,3-tetraisopropyldisiloxane trifluoromethanesulfonate
TMEDA	tetramethylethylenediamine {1,2-bis(dimethylamino)ethane]
TMS	trimethylsilyl
TMSCl	trimethylsilyl chloride
TMSOT$_f$	trimethylsilyl trifluoromethanesulfonate
TPAP	tetra-*n*-propyl ammonium perruthenate
TPP	tetraphenylporphyrin
Triton-B	N,N,N-trimethyl-benzenemethanaminium hydroxide
TrocCl	2,2,2-trichloroethyl chloroformate
Ts	*p*-toluenesulfonyl
p-TSA	*p*-toluenesulfonamide
VO(acac)$_2$	vanadyl acetylacetonate

10 Structure–activity relationships of taxoids

Xihong Wang, Hideji Itokawa, and Kuo-Hsiung Lee

Introduction

Natural products and related derivatives, including their semi-synthetically modified or totally synthetic analogs, play an important role in the development of novel therapeutic agents. It was reported that approximately one-third of the top-selling drugs in the world are natural products or their derivatives (Strohl, 2000). Many anticancer drugs currently in clinical use, such as Vinca alkaloids, camptothecins, derivatives of podophyllotoxin and others, are derived from natural products (Georg *et al.*, 1995d). Taxol® or paclitaxel (1) and its semi-synthetic analog taxotere® or docetaxel (2) (Figure 10.1) are other examples of natural products used as a source of anti-cancer drugs. Taxol® was first isolated from the bark of *Taxus brevifolia* in 1971 by Wall and Wani. It was approved by the FDA for the treatment of the advanced ovarian cancer and breast cancer in December 1992 and April 1994, respectively. Taxotere® was approved by the FDA in May 1996 for the treatment of breast cancer (Ojima *et al.*, 1997a). Due to its unique mechanism of action and its potent antitumor activity, taxol has stimulated research activities in pharmacology, chemistry and medicine over the last two decades (Cheng *et al.*, 2000b).

Taxol is an antimitotic agent. However, unlike other plant-derived antimicrotubule agents such as colchicines, podophyllotoxin and the Vinca alkaloids that inhibit microtubule assembly, taxol possesses a unique mechanism of action: it promotes tubulin assembly and inhibits micro-tubule disassembly (Manfredi and Horwitz, 1984). In addition, taxol can also induce the expression of tumor necrosis factor-α and interleukin-1 in macrophages, stimulate MAP-2 kinase activity and increase tyrosine phosphorylation. All these effects might influence taxol's antitumor activity. In addition to their role in mitosis, microtubules possess multiple functions in the regulation of cellular processes; thus taxol could conceivably be used in therapeutic areas other than cancer chemotherapy (Georg *et al.*, 1995d).

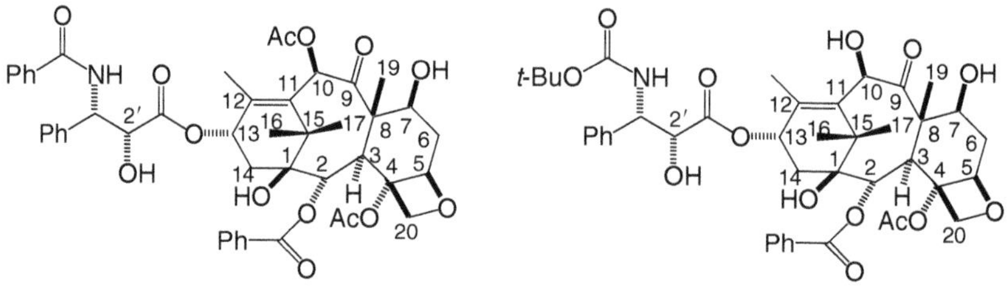

Figure 10.1 Structures of taxol (1) and taxotere (2).

In this chapter, we will review the structure–activity relationships (SARs) of the natural and modified taxoids and the design and the activity of water-soluble taxol prodrugs, which show promise for increasing both the water solubility and selectivity of taxol. We will also review the current status of the synthesis and biological evaluation of taxol photoaffinity labels, which can be used to determine the taxol binding site on tubulin and p-glycoprotein. Lastly, we will discuss the conformational properties of taxol and its analogs in polar and non-polar solvents.

Discussion of SARs of taxol derivatives is complicated because different assay systems have been used by the various investigators. The results depend on the exact assay conditions. Three major classes of assays are briefly described below (Kingston, 1991).

(1) *In vitro* mammalian microtubule assays. Three microtubule assays are frequently used to evaluate taxol analogs: initial slope of tubulin polymerization (Horwitz method), extent of microtubule assembly (Himes method), and microtubule disassembly (Potier method). The results are usually presented as relative data to minimize differences caused by the different methodologies.

(2) *In vitro* cytotoxicity against cancer cell lines. *In vitro* cytotoxicity of taxol and taxol analogs has been determined with the following cell lines: KB cell (human carcinoma of the nasopharynx), J774-2 cell (a mouse macrophage-like cell line), P388 (a mouse lymphocytic leukemia cell line), A121 (ovarian cancer), A549 (non-small cell lung cancer), PC-6 (human small cell lung cancer), PC-6/VCR29-9 (PC-6's variant, vincristine-resistant cell line), PC-12 (human non-small cell lung cancer), CA46 (human Burkitt lymphoma), B16 (melanoma), MCF-7 (human breast cancer), L1210 (Leukemia), HCT116 (human colon carcinoma), HT-29 (human colon adenocarsinoma), LCC6-MDR and MCF7-R (drug-resistant human breast cancer), and others (Kingston, 1991; Rowinsky and Donehower, 1991).

(3) *In vivo* growth inhibition of mouse and human tumor xenografts. The activities of taxol and its analogs have been tested in P388 leukemia, B16 melanoma and other mouse-based assays, including human tumors as xenografts in athymic mice (Kingston, 1991).

Naturally occurring taxoids with antitumor activity

Until 1992, approximately 100 taxoids had been isolated and characterized. But in the last 10 years, over 250 additional taxoids have been isolated and characterized from various Taxus species such as *T. baccata, T. wallichiana, T. cuspidata, T. canadensis, T. chinesensis, T. Yunnanensis* and others (Kingston, 1998a, 2000; Baloglu and Kingston, 1999).

In this section, we will review the SARs of the naturally occurring taxoids. We define taxanes as compounds possessing the A-, B-, C- ring system of taxol, taxoids are taxane-like derivatives, differing in the structure of the diterpene skeleton.

Cephalomannine and N-alkanoyl-N-debenzoyltaxol derivatives

When screening plants for antitumor activity, the polar extracts of the leaves, stem and root of *T. wallichiana* showed activity against KB and P388 cell lines (Mclaughlin *et al.*, 1981; Miller *et al.*, 1981). Cephalomannine (3) (Figure 10.2) was obtained from the cytotoxic fraction. It is a taxane closely related to taxol, but with a *N*-tigloyl group instead of the *N*-benzoyl group in taxol. Compared with taxol, cephalomannine (3) showed similar but slightly reduced cytotoxicity (Mclaughlin *et al.*, 1981) and microtubule disassembly properties (Georg *et al.*, 1995d)

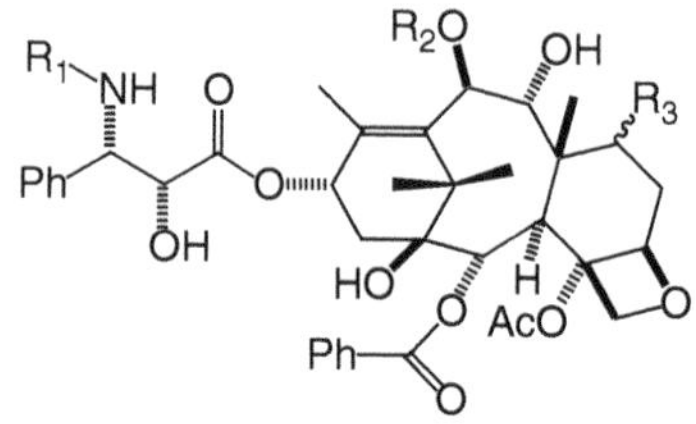

Compound	R_1	R_2	R_3
1. Taxol	Bz	Ac	β-OH
3. Cephalomannine	⋀CO	Ac	β-OH
4. Taxol C	n-C_5H_{11}-CO	Ac	β-OH
5. 10-Deacetyltaxol C	n-C_5H_{11}-CO	H	β-OH
7. Taxol D	n-C_3H_7-CO	Ac	β-OH
8. N-debenzoyl-*N*-cinnamoyl paclitaxel	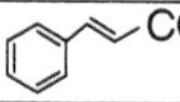 ⌬⋀CO	Ac	β-OH
9. N-debenzoyl-*N*-α-methylbutyryl paclitaxel	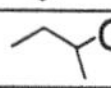 ⋀CO	Ac	β-OH
10. 10-Deacetyltaxol	Bz	H	β-OH
11. 10-Deacetylcephalomannine	⋀CO	H	β-OH
12. 7-*epi*-taxol	Bz	Ac	α-OH
13. 10-Deacetyl-7-*epi*-taxol	Bz	H	α-OH
14. 10-Deacetyl-7-*epi*-cephalomannine	⋀CO	H	α-OH
16. 7-Xylosyltaxol	Bz	Ac	β-xylosylO
17. 7-Xylosylcephalomannine	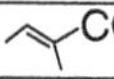 ⋀CO	Ac	β-xylosylO
18. 10-Deacetyl-7-xylosyltaxol	Bz	H	β-xylosylO
19. 10-Deacetyl-7β-xylosyltaxol D	n-C_3H_7-CO	H	β-xylosylO

Figure 10.2 Structures of naturally occurring taxanes.

(Table 10.1). This result was the early indication that the N-benzoyl group of taxol may not be essential for cytotoxicity and could be structurally modified without loss of activity. Taxol C (4), 10-deacetyltaxol C (5) (Figure 10.2), N-methyltaxol C (6) (Figure 10.4) and taxol D (7) (Figure 10.2) were first isolated from cell cultures of *T. baccata* (Ma *et al.*, 1994) and the roots of *T. Yunnanensis* (Zhang *et al.*, 1994). All four compounds showed comparable activity to taxol in the tubulin assembly assay, in particular, taxol C (4) was as active as taxol in this assay and possessed potent and selective cytotoxicity in several human cancer cell lines (Table 10.1) (Ma *et al.*, 1994; Georg *et al.*, 1995d). Compound **8** and **9** (Figure 10.2) were isolated from the roots of *Taxus x media*. Compound **8** showed cytotoxity and tubulin binding comparable to those of taxol. This result indicates that the vinylation of the N-benzoyl group is tolerated. Compound **9** showed decreased cytotoxicity and tubulin binding compared to taxol D (7) and taxol (1); thus it was speculated that α-branching on the N-acyl group is detrimental for activity (Gabetta *et al.*, 1999). These bioactive natural N-alkanoyl taxoids and the semisynthetic taxol analog taxotere (2) verify that the N-benzoyl group can be replaced by other N-alkenoyl groups (Gabetta *et al.*, 1999).

10-Deacetyl taxol analogs

10-Deacetyltaxol (10) and 10-deacetylcephalomannine (11) (Figure 10.2) were first isolated from the polar extracts of *T. wallichiana*, and both compounds showed activity against KB,

Table 10.1 Biological activity of compounds 1, 3–24

Compound	KB[a]	P388[b]	J774.2[c]	MCF-7[d]	Tubulin activity
1 Taxol	0.001	+	S		1[e]
3 Cephalomannine	0.004	+	M/S		1.5[e]
4 Taxol C					1.1[f]
5 10-Deacetyltaxol C					2.6[f]
6 N-methyltaxol C					1.7[f]
7 Taxol D					2.0[f]
8 N-debenzoyl-N-cinnamoyl paclitaxel				1.3	1.2[f]
9 N-debenzoyl-N-α-methylbutyryl paclitaxel				>10	6.7[f]
10 10-Deacetyltaxol	0.003	+	M		1.3[e]
11 10-Deacetylcephalomannine	0.03	+	M		5[e]
12 7-*epi*-taxol	3×10^{-5}[g]				3.0[e]
13 10-Deacetyl-7-*epi*-taxol	0.03				
14 10-Deacetyl-7-*epi*-cephalomannine	0.05				
15 10-Deacetoxy-7-*epi*-10-oxo-taxol	0.02[g]				
16 7-Xylosyltaxol					0.4[e]
17 7-Xylosylcephalomannine					0.5[e]
18 10-Deacetyl-7-xylosyltaxol					0.6[e]
19 10-Deacetyl-7β-xylosyltaxol D					c.1.0
20 2-Debenzoyl-2-tigloyl paclitaxel				12	1.1[f]
21 Baccatin VI					140[e]
22 Baccatin III	2.0				52[e]
23 10-Deacetylbaccatin III	1.0				46[e]
24 19-Hydroxybaccatin III	2.2	−	0		

Notes
a ED$_{50}$ in μg/ml for growth inhibition of KB cells.
b Activity against P388 leukemia *in vivo*; + = increse in life span >25%, − = negative.
c Growth inhibition of macrophage-like J774.2 cells relative to taxol. S: strong, M: medium, 0: none.
d Growth inhibition of MCF-7 cell line relative to Taxol ED$_{50}$/ED$_{50(ptaxol)}$.
e Microtubule disassembly assay ID$_{50}$/ID$_{50(Taxol)}$.
f Microtubule assembly assay ED$_{50}$/ED$_{50(taxol)}$.
g Taxol ED$_{50}$ = 1×10^{-5} μg/ml.

J774.2 (Parness *et al.*, 1982), P388 leukemia (McLaughlin *et al.*, 1981) cell lines, and in the tubulin disassembly assay (Georg *et al.*, 1995d) (Table 10.1). 7-Xylosyltaxol (16) and 10-deacetyl-7-xylosyltaxol (18) (Figure 10.2) showed similar activity in the tubulin disassembly assay (Lataste *et al.*, 1984) (Table 10.1). These results indicate that the acetylation of the C-10 hydroxyl group is not essential for the antitumor activity. Taxotere (2), which also lacks the C-10 acetyl group and is more active than taxol, further supports this conclusion.

7-epi-taxol derivatives

7-*epi*-taxol (12), 10-deacetyl-7-*epi*-taxol (13), and 10-deacetyl-7-*epi*-cephalomannine (14) (Figure 10.2) are respectively the C-7 epimers of taxol (1), 10-deacetyltaxol (10), 10-deacetyl-cephalomannine (11), and also showed potent growth inhibition of KB cells (Table 10.1). Thus, the stereochemistry of C-7 does not have a significant influence on the biological activity of taxanes (Georg *et al.*, 1995d).

10-Deacetoxy-7-*epi*-10-oxo-taxol (15) (Figure 10.4) was isolated from *T. brevifolia* (Huang *et al.*, 1986). It showed decreased cytotoxicity against KB cells (Table 10.1), perhaps due to the 10-keto rather than hydroxy group (Huang *et al.*, 1986).

7-Xylosyl taxol analogs

7-Xylosyltaxol (16), 7-xylosylcephalomannine (17), 10-deacetyl-7-xylosyltaxol (18), 10-deacetyl-7β-xylosyltaxol D (19) (Figure 10.2) were isolated from *T. baccata* and showed potent activity in the microtubule disassembly assay (Table 10.1) (Lataste *et al.*, 1984; Zee-Cheng and Cheng, 1986; Guo *et al.*, 1995). This result suggested that the C-7 hydroxyl group of taxol could be replaced by bulky and/or hydrophilic substituents without loss of activity. 10-Deacetyl-7β-xylosyltaxol C, 7β-xylosyltaxol C, 10-deacetyl-7β-xylosylcephalomannine, 10-deacetyl-10-(3-hydroxybutyrate)taxol and 10-deacetyl-10-(3-hydroxybutyrate)cephalomannine were also isolated from the bark of *T. baccata*, but no biological data were reported (Sénilh *et al.*, 1984; Guo *et al.*, 1995).

2-Debenzoyl taxol derivatives

2-Debenzoyl-2-tigloyl paclitaxel (20) (Figure 10.4), the first natural analog of taxol with a modified ester group at C-2, retained tubulin-binding activity but displayed decreased cytotoxity compared to taxol (1) and 2-debenzoyl-2-senecioyldocetaxel analog (201, $ED_{50(201)}/ED_{50(taxol)} = 0.65$) (Table 10.1). This result suggests that α-branching on the C-2 acyl moiety is also detrimental to activity (Gabetta *et al.*, 1999).

The importance of the C-13 side chain for biological activity

Naturally occurring taxanes without a C-13 *N*-benzyol-3'-phenylisoserine side-chain usually possess either a C-13 acetyl group, as in baccatin VI (21) (Figure 10.4) (Lataste *et al.*, 1984), or a free C-13 hydroxyl group, as in baccatin III (22) (Miller *et al.*, 1981; Lataste *et al.*, 1984), 10-deacetylbaccatin III (23) (Kingston *et al.*, 1982; Lataste *et al.*, 1984), 19-hydroxybaccatin III (24), 14β-hydroxy-10-deacetylbaccatin III (25) (Figure 10.3) (Appendino *et al.*, 1992) and taxuspine E (26) (Figure 10.4) (Kobayashi *et al.*, 1995) or C-13 keto as in taxuspine F (27), taxuspine G (28) (Figure 10.5) (Kobayashi *et al.*, 1995).

Baccatin III (22) was first isolated from the roots of *T. baccata* in 1965 (Preuss and Orth, 1965). It can also be obtained from taxol by base-catalyzed methanolysis (Georg *et al.*, 1995d). Baccatin VI (21), Baccatin III (22) and its related derivatives (23–25) showed significantly reduced activity against KB cells and in the microtubule disassembly assay (Lataste *et al.*, 1984;

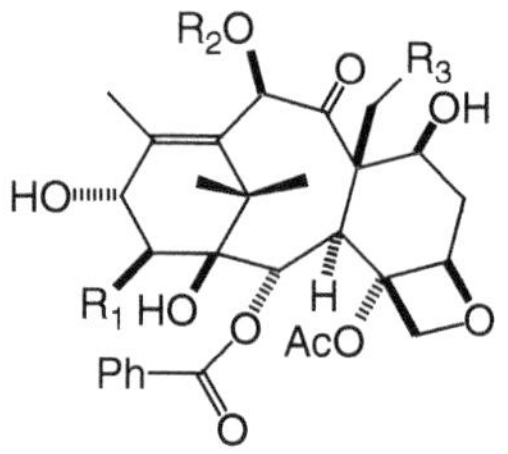

Compound	R₁	R₂	R₃
22. baccatin III	H	Ac	H
23. 10-deacetylbaccatin III	H	H	H
24. 19-hydroxybaccatin III	H	Ac	OH
25. 14β-hydroxy-10-deacetylbaccatin III	OH	H	H

Figure 10.3 Structures of baccatin III derivatives.

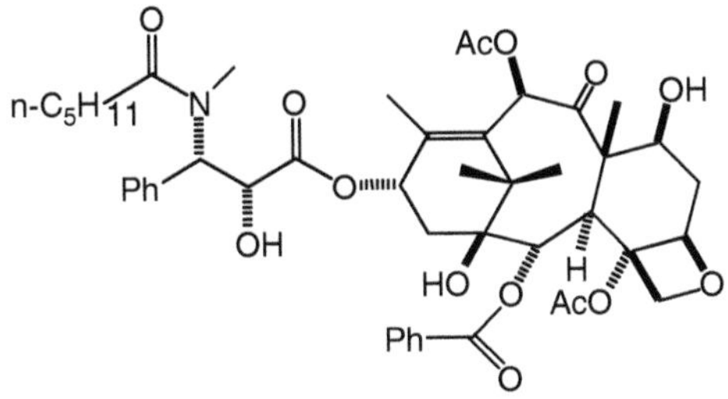

N-methyltaxol C (**6**)

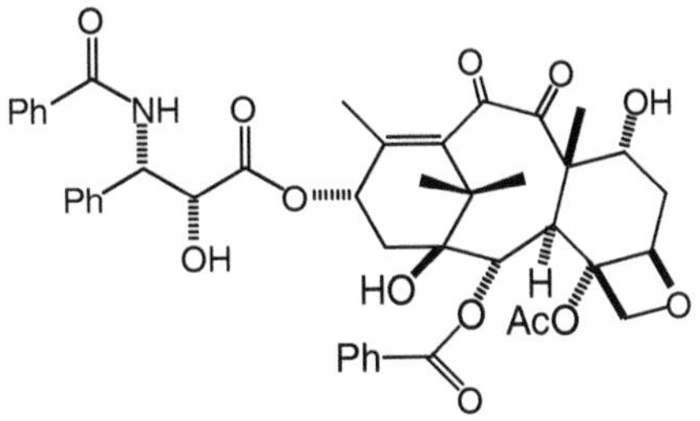

10-Deacetoxy-7-*epi*-10-oxo-taxol (**15**)

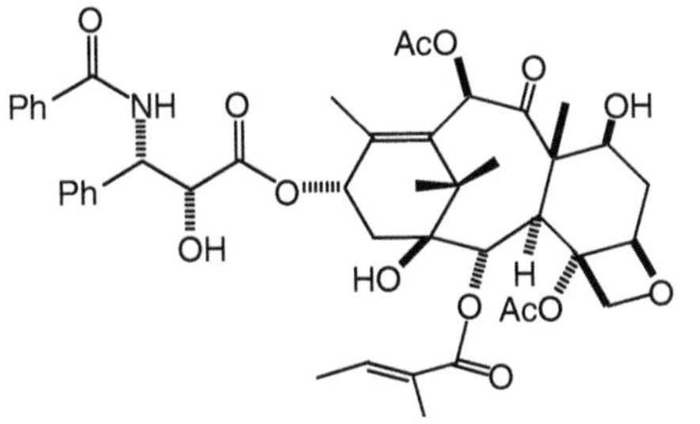

2-Debenzoyl-2-tigloylpaclitaxel (**20**)

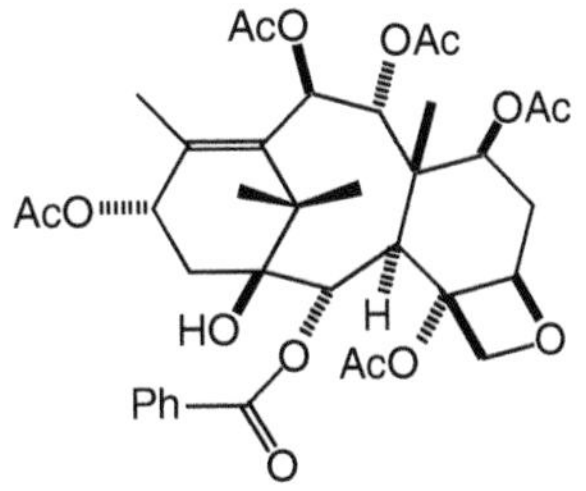

Baccatin VI (**21**)

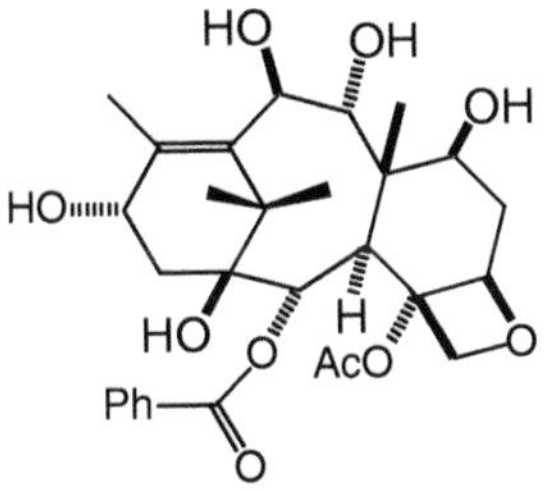

Taxuspine E (**26**)

Figure 10.4 Structures of compound 6, 15, 20, 21 and 26.

Miller *et al.*, 1981) (Table 10.1). These data indicate that the C-13 side-chain is very important for the antitumor activity of taxol. Taxuspine E (**26**) was isolated from the Japanese yew *T. cuspidata* (Kobayashi *et al.*, 1995). Interestingly, although baccatin III analogs are usually reported to be *ca.*1700-fold less cytotoxic against KB cells than taxol (Kingston, 1991), taxuspine E (**26**) shows potent cytotoxicity against KB cells and L1210 cells (Table 10.2). The main difference between taxuspine E and other baccatin III derivatives is the functional group at C-9: hydroxyl in taxuspine E, ketone in baccatin III. This indicates 9-OH group is important for cytotoxicity (Kobayashi *et al.*, 1995).

Compounds 27–47 (Figure 10.5) are unique taxoids with open oxetane ring or contracted A-ring. Compounds 27, 28, 32–36, 40–47 were isolated from the Japanese yew *T. cuspidata* and showed no cytotoxicity against L1210 and KB cells (Table 10.2) (Kobayashi *et al.*, 1995, 1997; Morita *et al.*, 1998). Compounds 29–31 were isolated from *T. baccata* and 30 was much less active than taxol (Guo *et al.*, 1995) (Table 10.2). Compounds 37–39 were isolated from *Taxus*

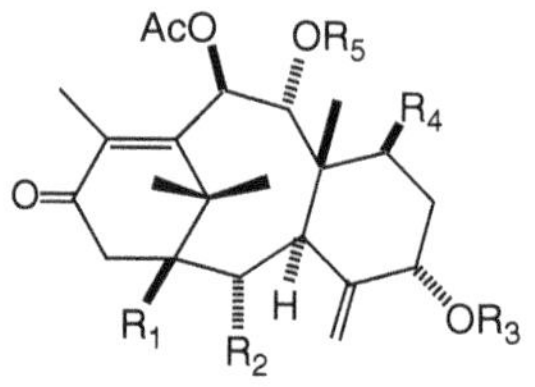

27 $R_1 = H$, $R_2 = OAc$, $R_3 = H$, $R_4 = OAc$, $R_5 = Ac$
28 $R_1 = H$, $R_2 = OH$, $R_3 = H$, $R_4 = H$, $R_5 = Ac$
29 $R_1 = OH$, $R_2 = OH$, $R_3 = COCH_2CH(NMe_2)Ph$, $R_4 = H$, $R_5 = H$

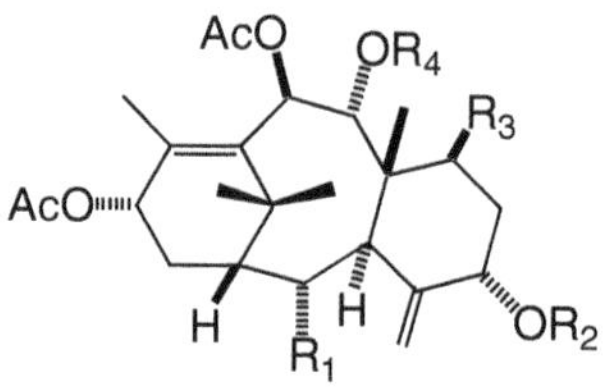

30 $R_1 = H$, $R_2 = COCH_2CH(NMe_2)Ph$, $R_3 = OAc$, $R_4 = Ac$
31 $R_1 = OH$, $R_2 = COCH_2CH(NMe_2)Ph$, $R_3 = R_4 = H$
32 $R_1 = H$, $R_2 = COCH_2CH(NMe_2)Ph$, $R_3 = H$, $R_4 = Ac$
33 $R_1 = OAc$, $R_2 = Ac$, $R_3 = H$, $R_4 = OAc$
34 $R_1 = OAc$, $R_2 = H$, R3 $= OAc$, $R_4 = OAc$

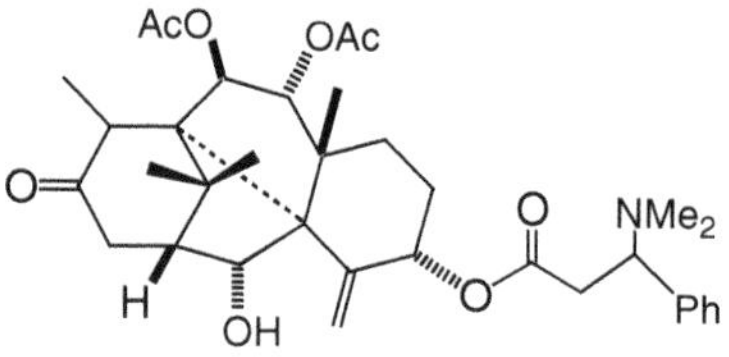

35

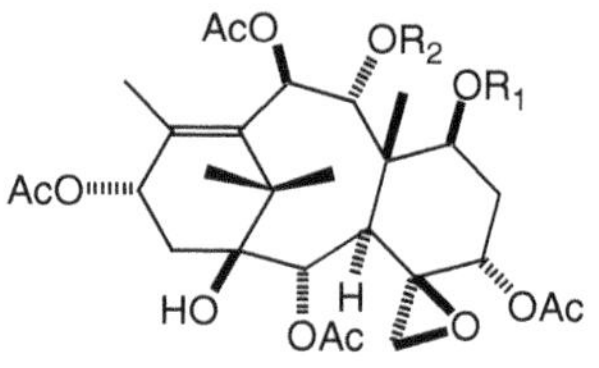

36 $R_1 = R_2 = H$
39 $R_1 = R_2 = Ac$

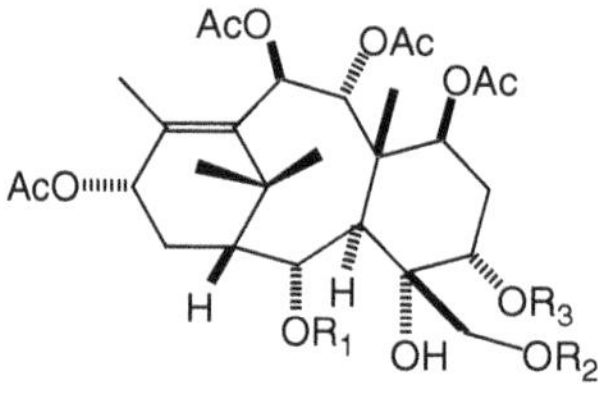

37 $R_1 = H$, $R_2 = COPh$, $R_3 = Ac$
38 $R_1 = COPh$, $R_2 = H$, $R_3 = Ac$

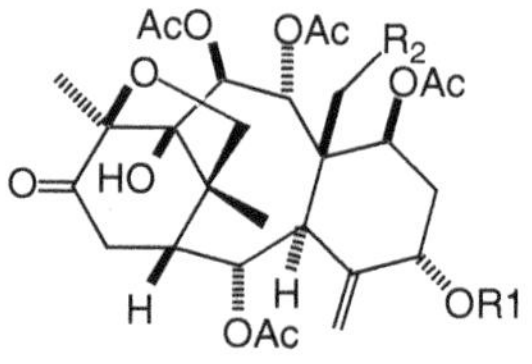

40 $R_1 = COCH = CHPh$, $R_2 = H$
41 $R_1 = COCH = CHPh$, $R_2 = OCOPh$
42 $R_1 = H$, $R_2 = H$
43 $R_1 = H$, $R_2 = OCOPh$

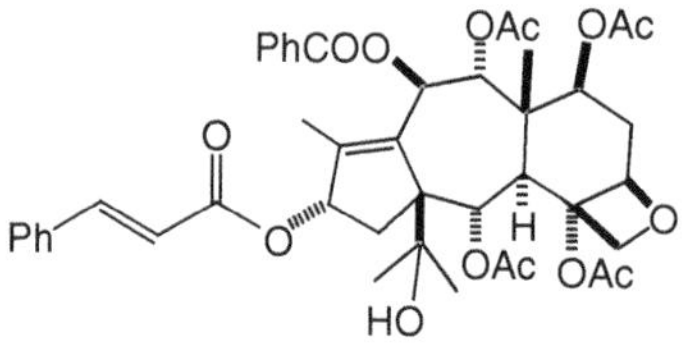

44

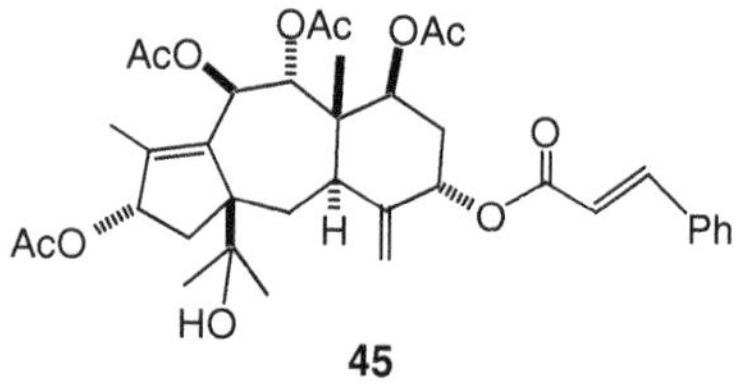

45

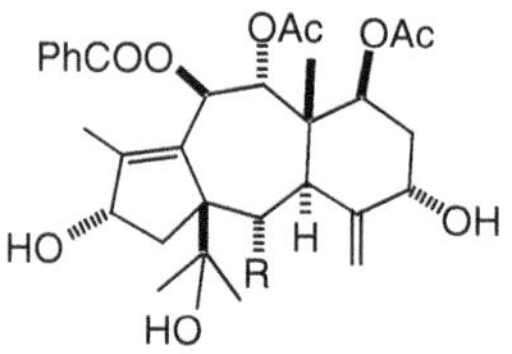

46. $R = H$
47. $R = OAc$

Figure 10.5 Structures of compounds 27–47.

Table 10.2 Cytotoxicity of compounds 26–47 (IC$_{50}$ μg/ml)

Compound	L1210	KB	P388	A549	HT-29
1 Taxol	0.33	0.0088			
26 Taxuspine E	0.27	0.08			
27 Taxuspine F	>10	>10			
28 Taxuspine G	>10	>10			
32 7,2′-Didesacetoxyaustrospipicatine	>10	>10			
33 12α-acetoxytaxusin	>10	>10			
34 Decinnamoyltaxinine J	>10	>10			
35 Taxuspine H	>10	1.6			
36 1β-Hydroxybaccatin I	>10	>10			
37 Taxumairol A		0.4	0.05	1.7	1.0
38		1.0	2.4	4.8	1.4
39 Taxumairol B		1.4	>50	>50	>50
40 Taxagifine	1.3	0.86			
41 Taxacin	>10	>10			
42 Decinnamoyltaxagifine	>10	>10			
43 Taxinine M	>10	9.4			
44 Taxchinin B	3.8	>10			
45 Taxuspine J	>10				
46 Brevifoliol	>10	0.4			
47 2-Acetooxybrevifoliol	>10	>10			

mairei, and compounds **37** and **38** were less active than taxol against several cancer cell lines while **39** showed no activity (Shen *et al.*, 1996). All these compounds lack not only the C-13 side-chain but also the oxetane ring. The cytotoxicity results indicate that an intact taxane ring system and C-13 side-chain are important to the biological activity. In addition, compounds **44–47** have a contracted A-ring and **45–47** also lack the oxetane D-ring. These results indicate that the contraction of ring A can reduce cytotoxicity, perhaps due to instability of the contracted A-ring in the tumor cell culture media (Kingston, 1991).

Summary

The SARs obtained from naturally occurring taxoids are summarized in Figure 10.6. The C-13 side-chain is very important to the biological activity (Wani *et al.*, 1971). However, the *N*-benzoyl group in this side-chain can be replaced by tigloyl or other alkanoyl groups and vinyl or extended by a group without significantly reducing the activity (Mclanghlin *et al.*, 1981; Miller *et al.*, 1981; Ma *et al.*, 1994; Zhang *et al.*, 1994; Georg *et al.*, 1995d; Gabetta *et al.*, 1999). Compounds with epimerization of the C-7 hydroxy group or with a xylosyl group at C-7 show comparable or better activity than taxol (Lataste *et al.*, 1984; Zee-Cheng and Cheng, 1986; Georg *et al.*, 1995d; Guo *et al.*, 1995). Compounds with or without an acetyl group on the C-10 hydroxyl group have similar activity (Parness *et al.*, 1982; Lataste *et al.*, 1984). All these results indicate that modification can be made at C-7, C-10 and C-13 *N*-benzoyl without loss of activity. Compounds with an opened oxetane ring or contracted A-ring have no activity; thus, intact oxetane and A-rings are essential to the activity (Kobayashi *et al.*, 1995; Guo *et al.*, 1995; Shen *et al.*, 1996).

SAR of taxoids with modifications at the diterpene moiety

Modifications of the diterpene moiety and subsequent analysis of the SARs of the taxol derivatives can give us more information on the taxol pharmacophore(s). Based on this information, novel analogs can be designed that are more effective and have fewer side-effects.

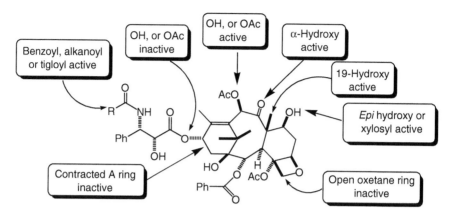

Figure 10.6 Summary of SARs from naturally occurring taxoids.

Modifications at the 7-hydroxyl group

The SAR studies on naturally occurring taxoids demonstrated that the 7-hydroxyl group could be modified or epimerized without significant loss of biological activity. Based on this fact and because the 7th position of taxol can be modified easily and selectively, most of the initial structural modifications were made at the 7-hydroxyl group (Georg *et al.*, 1995d). Modifications at C-7 include the formation of an ether or ester at the C-7 hydroxyl group, reduction to 7-deoxy analogs, introduction of a fluorine atom or azido group at C-7, oxidation to a 7-oxotaxol, and 7,8,19 cyclization.

The most common modifications at the C-7 position are to form an ether or ester at the 7-OH group, such as 7-α-trimethylsilyether (**48**), 7-α-dimethylphenylsilyether (**49**), 7-β-triethylsilyether (**50**) (Chen *et al.*, 1994a), 7-acetyltaxol (**56**) (Lataste *et al.*, 1984; Grover *et al.*, 1995), 7-benzoyltaxol (**57**) (Kingston *et al.*, 1990), 7-mesylate (**58**), 7-chloromethylester (**59**), 7-ethylcarbonate (**60**), 7-carbamate (**61–63**) (Chen *et al.*, 1994a), 7-glutaryl derivatives (**64** and **68**) (Guéritte-Voegelein *et al.*, 1991) and C-7 amino acid esters (**65–67** and **69**) (Guéritte-Voegelein *et al.*, 1991, Mathew *et al.*, 1992) (Tables 10.3 and 10.4).

Most of the 7-ether analogs showed no significant activity in either the microtubule or cytotoxicity assays, and the two C-7-*epi*-silyether analogs **48** and **49** were completely inactive. It was noteworthy that compound **51**, BMS-184476, showed similar potency to paclitaxel in cytotoxicity and tubulin assays (Table 10.3) and is currently in Phase I clinical trials. Compounds **52–55** are 7-sulfur-substituted paclitaxel analogs (Table 10.3). Compound **52** showed less activity than paclitaxel in both tubulin polymerization and cytotoxicity assay. In contrast, **53** showed better activity than paclitaxel in both assays, while its epimer, **54**, was significantly less cytotoxic than **53**. Compound **55** (7-SCH$_2$OCH$_3$) was four-fold more cytotoxic than its transposed isomer (compound **51**, 7-OCH$_2$SCH$_3$) (Rose *et al.*, 2000; Mastalerz *et al.*, 2001). However, most of the 7-ester derivatives showed excellent activity in both microtubule assays (Rose *et al.*, 2000; Mastalerz *et al.*, 2001). Compared to taxol, three C-7 carbamates (**61–63**) showed significantly reduced cytotoxicity but retained activity in the tubulin polymerization assay (Chen *et al.*, 1994a) (Table 10.3).

The synthesis and biological activity of 7-dehydroxyl taxoids have also been investigated by several groups. Compounds are included 7-dehydroxylpaclitaxel (**70**), 6,7-olefinpaclitaxel (**71**), 7-dehydroxyldocetaxel (**72**), 7-deoxy-3'-N-bocpaclitaxel (**74**) (Chaudhary *et al.*, 1993b; Chen *et al.*, 1993c, 1994a; Poujol *et al.*, 1997a,b). These compounds proved to be less active in the tubulin assay but equally cytotoxic to taxol in the HCT116 cell line (Table 10.3). Pulicani

Table 10.3 C-7 Hydroxyl modified taxol and taxotere derivatives

Compound	R_1	R_2	R_3	Tubulin poly. init. rate ratio	IC_{50} (nM) HCT 116
1. Paclitaxel				1.00	4.0
2. Docetaxel				2.00	3.0
48. 7α-Trimethylsilyether paclitaxel	Ph	Ac	α-OTMS	0	>60
49. 7α-Dimethylphenylsilyether paclitaxel	Ph	Ac	α-OSiMe₂Ph	0	>78
50. 7β-Trimethylsilyether paclitaxel	Ph	Ac	β-OTES	0.53	>78
51. BMS-184476	Ph	Ac	β-OCH₂SCH₃	1.1	2.1
52.	Ph	Ac	β-SH	1.8	13.5
53.	Ph	Ac	β-SMe	0.89	0.2
54.	Ph	Ac	α-SMe	16	13.1
55.	Ph	Ac	β-SCH₂OCH₃	1.9	0.5
56. 7-Acetyltaxol	Ph	Ac	OAc	2.0ᵃ	
57. 7-Benzoyltaxol	Ph	Ac	OBz	1.7ᵃ	
58. 7-Mesylate paclitaxel	Ph	Ac	OMs	1.10	
59. 7-Chloromethylester paclitaxel	Ph	Ac	OCOCH₂Cl	0.81	7.0
60. 7-Ethylcarbonate paclitaxel	Ph	Ac	OCOOEt	0.72	6.0
61. 7-Butylcarbamate paclitaxel	Ph	Ac	OCONHBu	0.96	46
62.	Ph	Ac	OCONH(CH₂)₃COOH	0.71	>3900
63.	Ph	Ac	OCONH(CH₂)₂NMe₂	0.40	>100
64.	Ph	Ac	Glutarate	1ᵃ	
65.	OBuᵗ	H	Phenylalanate	1ᵃ	
66.	Ph	Ac	N,N-diglycinate	Active	
67.	Ph	Ac	Alanate	Active	
70. 7-Deoxypaclitaxel	Ph	Ac	H	0.23	4.0
71. 6,7-Dehydropaclitaxel	Ph	Ac	6,7-olefin	1.31	5.0
72. 7-Deoxydocetaxel	OBuᵗ	H	H	0.8ᵃ	0.95ᵇ
74. 7-Deoxy-3′-N-Boc paclitaxel	OBuᵗ	Ac	H	0.80	1.6
79.				1.3ᵃ	8.25ᵇ
80. 7-*epi*-fluoro-3′-N-Boc-paclitaxel	OBuᵗ	Ac	α-F	0.80	4.7
81. 7-*epi*-fluoro-paclitaxel	ph	Ac	α-F	1.45	11
82. 7-Deoxy-7α-azido paclitaxel	Ph	Ac	α-N₃		1.0ᶜ
83. 7,19-Cyclopropane paclitaxel	Ph	Ac		0.56	8.0
84. 10-Acetyl-7,19-cyclopropane docetaxel	OBuᵗ	Ac			0.5ᵈ

Notes

a Microtubule disassembly assay ($ED_{50}/ED_{50(taxol)}$).

b $IC_{50}/IC_{50(taxotere)}$ against P388 leukemia cells.

c $IC_{50}/IC_{50(taxol)}$ against CA46 human Burkitt lymphoma cells.

d $IC_{50}/IC_{50(taxol)}$ against HT29.

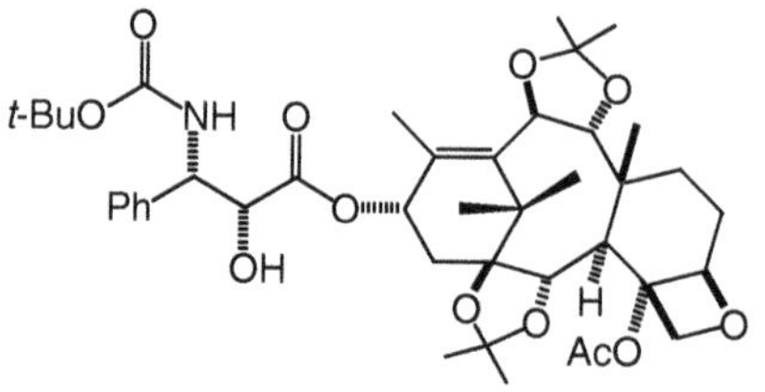

Figure 10.7 Structures of 7-deoxydocetaxol derivatives.

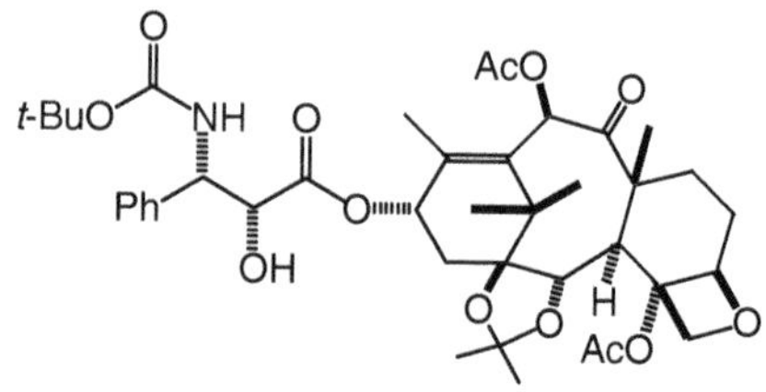

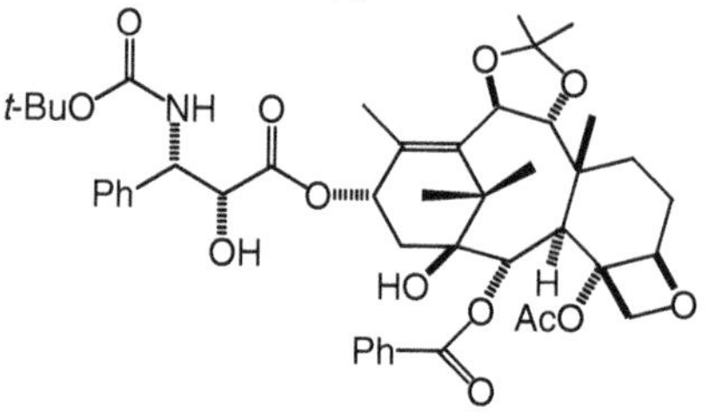

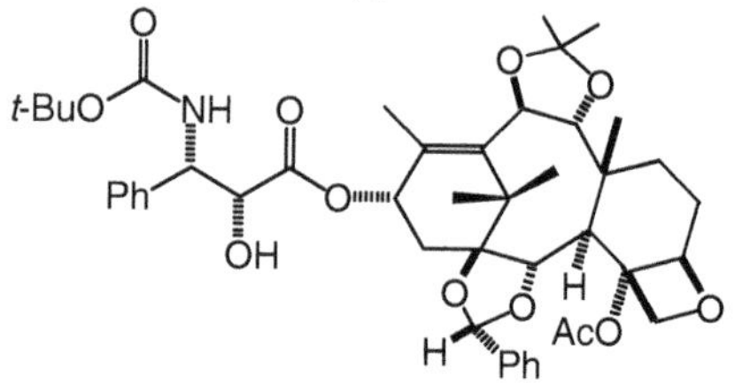

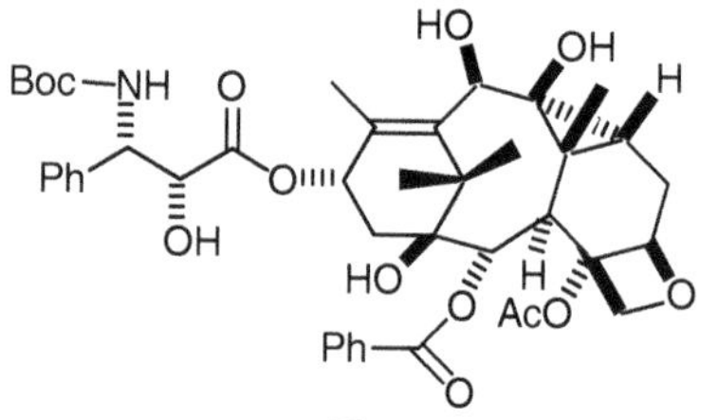

et al. (1994a) and Poujol *et al.* (1997a,b) further synthesized a series of 7-deoxydocetaxel derivatives (Figure 10.7). Compounds 75–78 showed no cytotoxicity and weak microtubule disassembly inhibitory activity. Compound 79 is less cytotoxic against P388 but showed excellent inhibitory activity in the microtubule disassembly assay (Pulicani *et al.*, 1994a).

Another modification at the C-7 position is to introduce a fluorine atom or azido group in place of the OH group. 7-*epi*-fluoro-3′-N-Boc-taxol (80) and 7-dehydroxy-7α-fluorotaxol (81) showed similar activity to taxol in tubulin polymerization assay and showed potent cytotoxicity. 7-Deoxy-7α-azidopaclitaxel (82) was not effective at promoting tubulin assembly but did stabilize polymerized tubulin as well as did paclitaxel but has the same activity as paclitaxel in the cytotoxicity assay against human Burkitt lymphoma CA46 cells (Table 10.3) (Liang *et al.*, 1995a).

7,8,19-cyclopropyl analogs (83,84) (Chen *et al.*, 1993c, 1994a) are more or comparably active than paclitaxel (Figure 10.8 and Table 10.3). Compound 83 was as potent as taxol against HCT116 ($IC_{50}/IC_{50(taxol)} = 2$) tumor cell line, and compound 84 showed excellent cytotoxicity against HT29 ($IC_{50}/IC_{50(taxol)} = 0.5$), B16F10 ($IC_{50}/IC_{50(taxol)} = 0.12$), and P388 ($IC_{50}/IC_{50(taxol)} = 0.11$) cells (Klein *et al.*, 1994).

Kingston's group (1990) reported the synthesis of 7-oxotaxol (85) (Figure 10.8), which showed greatly reduced activity ($ED_{50} = 0.5$ µg/ml to KB cell line) compared with taxol. The decreased cytotoxicity is thought to be due to the instability of 7-oxotaxol under the assay conditions.

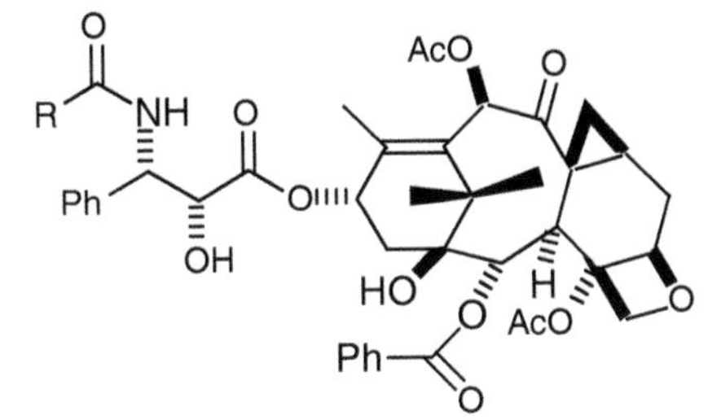

Figure 10.8 Structures of compounds 83–85.

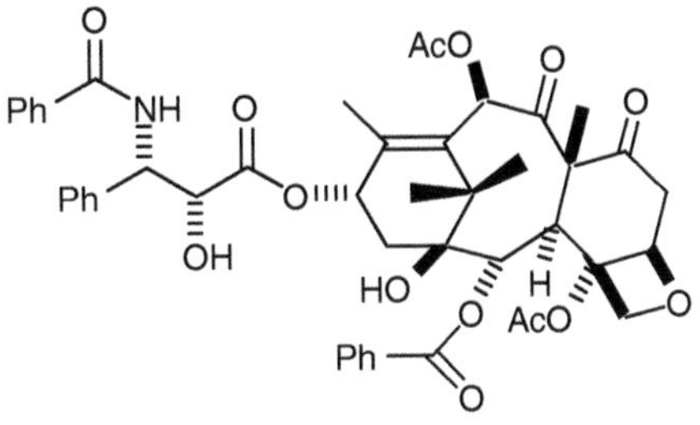

Compounds **60**, **70**, **71**, **74**, **80**, **81**, and **83** were also evaluated *in vivo* against the murine M109 lung carcinoma and all the analogs were active (T/C ⩾ 125%) (Chen *et al.*, 1994a).

In summary, 7-ester, 7-deoxy, 6,7-olefin, 7-fluro and 7-azido analogs may improve cytotoxicity against cancer cell lines. Thus it was concluded that replacing the C-7 hydroxyl group with a small substituent does not significantly affect the activity. In addition prodrugs and photoaffinity labels can be developed by modification at C-7.

Modifications at C-10

Previous SAR studies have revealed that modifications at C-10 can be made without loss of biological activity. Modifications at C-10 include 10-deoxy analogs, simple carbamates, acetylated (Kant *et al.*, 1994) and alkylated derivatives (Nakayama *et al.*, 1998) of taxoids.

Taxotere (**2**), 10-deacetyltaxol (**10**), 10-acetyltaxotere (**88**), 7,10-diglutaryl (**68**) and 7,10-diglycyl (**69**) derivatives show similar activity in the microtubule disassembly assay (Table 10.4) (Guéritte-Voegelein *et al.*, 1991; Kant *et al.*, 1994). This fact indicates that taxol's C-10 acetoxy group may not play an important role in microtubule binding. This conclusion was confirmed through the biological evaluation of various synthetic analogs (Guéritte-Voegelein *et al.*, 1991; Kant *et al.*, 1994). All compounds (**90–98**) (Table 10.4) showed comparable or better activity than taxol in both assays, with C-10 carbamate (**92**) being the most potent analog. The synthesis and biological evaluation of 10-deacetoxytaxol (**89**) was reported by several groups (Chaudhary *et al.*, 1993a; Chen *et al.*, 1993b,c; Georg *et al.*, 1994c; Holton *et al.*, 1994). This compound possessed only slightly reduced activity against HCT 116 ($ED_{50}/ED_{50(taxol)} = 1.75$) (Chen *et al.*, 1993b) and equal activity against B16 melanoma cells ($ED_{50}/ED_{50(taxol)} = 1$) (Georg *et al.*, 1995d) compared to taxol. 7,10-Dideoxy-docetaxel (**73**) and 10-deacetyloxy-7-dehydroxytaxol (**86**) showed similar activity in tubulin assay but was slightly less cytotoxic (Chen *et al.*, 1993c). 10-Benzoyloxy-7-deoxypaclitaxel (**87**) showed similar cytotoxicity against three human tumor cell lines (MCF, SK-OV3 and WIDR) compared to taxol (Cheng *et al.*, 2000a) (Table 10.4).

Because of taxol and taxotere's extremely low water solubility, many research groups have synthesized water-soluble taxoids such as esterase- or phosphatase-cleavable prodrugs at C-2′ and C-7. However these prodrugs can exhibit unstable efficacy because of variation in the enzymatic activity among patients. Soga's group synthesized a series of water-soluble nonprodrug docetaxel analogs (Nakayama *et al.*, 1998; Uoto *et al.*, 1998; Iimura *et al.*, 2000, 2001). Nakayama *et al.* (1998)

Table 10.4 Structure and cytotoxicity of C-10 and C-10, C-7 *bis*-modified taxol and taxotere analogs

Compound	R_1	R_2	R_3	Tubulin activity $ED_{50}/ED_{50(taxol)}$	HCT116 $ED_{50}/ED_{50(taxol)}$
2 Taxotere	OBu^t	OH	OH	0.5[a]	
10 10-deacetyltaxol	Ph	OH	OH	1.3[a]	
68	OBu^t	Glutarate	Glutarate	2[a]	
69	OBu^t	Glycinate	Glycinate	1.2[a]	
73 7,10-dideoxy-docetaxel	OBu^t	H	H	1.0[a]	8.0[c]
86 10-deacetoxy-7-dehydrotaxol	Ph	H	H		7.5
87	Ph	OBz	H		0.93[d]
88 10-acetyltaxotere	OBu^t	OAc	OH	0.7[a]	1.0
89 10-deacetoxytaxol	Ph	H	OH	0.5[b]	1.75
90	Ph	OCOBu	OH	1.5	1.7
91	Ph	OCOcyclopro	OH	1.1	1.1
92	Ph	OCON(Me)$_2$	OH	1.0	0.5
93	Ph	OMe	OH	1.0	6
94	OBu^t	OMe	OH	0.3	0.6
95	Ph	OCOOMe	OH	1.1	1.5
96	OBu^t	OCOOMe	OH	0.8	0.7
97	Ph	COPh	OH	19	1.1
98	OBu^t	COPh	OH	2.1	1.0

Notes
a Microtubule disassembly assay.
b Microtubule assembly assay.
c $IC_{50}/IC_{50(taxotere)}$ against P388 leukemia cells.
d $ED_{50}/ED_{50(taxol)}$ against MCF-7.

used a radical coupling reaction to synthesize analogs with a polar substituent or at the end of a C-10 alkyl moiety (Table 10.5). The analogs with polar substituents, such as the amino (103), acetamido, (104), carboxy (106) groups, were significantly less active than docetaxel, while the cyano (100) and hydroxy (102) analogs were moderately less active. The ethoxycarbonyl (99), methyloxo (101) and methoxycarbonyl (105) analogs showed fairly good activity, and methoxycarbonyl (105) was particularly more active than docetaxel in the three cell lines tested. These results indicate that the activity is dependent on the substituent's polarity rather than on its steric demands (Nakayama *et al.*, 1998). Compound 107 (morpholine) showed significantly improved activity than compound 103 (NH$_2$) but less activity than docetaxel (Table 10.5). Most of the *sec*-aminoethyl analogs (108–115) showed similar activity to

Table 10.5 Structure and cytotoxicity of 10-alkylated docetaxel analogs

Compound	n	R_1	R_2	P388 (GI_{50} (ng/ml))	PC-6 (GI_{50} (ng/ml))	PC-12 (GI_{50} (ng/ml))	PC-6/VCR29-9 (GI_{50} (ng/ml))
Paclitaxel				25.3	4.64	172	
Docetaxel				6.7	1.1	53.4	39.6-230
99	2	COOEt	OH	4.31	1.35	12.0	
100	2	CN	OH	18.4	0.681	76.9	
101	2	COCH$_3$	OH	7.12	0.266	32.4	
102	3	OH	OH	15.5	0.777	67.0	
103	3	NH$_2$	OH	1520	236	6820	
104	3	NHAc	OH	71.8	10.5	634	
105	3	COOMe	OH	3.30	0.771	24.2	
106	3	COOH	OH	1870	934	>10000	
107	3	—N(morpholine)	OH	20.6	5.5	432	
108	2	—N(morpholine)	OH	3.1	0.6	13.5	76.3
109	2	—N(thiomorpholine, S)	OH	3.1	4.8	32.9	
110	2	—N(dimethylamino)	OH	17.3	1.2	299	
111	2	—N(piperidine)	OH	3.8	1.4	193.5	
112	2	—N NMe (N-methylpiperazine)	OH	5.8	2.0	293	
113	2	—N(pyrrolidine)	OH	11.2	1.8	414	
114	2	—N(azepane)	OH	3.6	2.7	91.2	916
115	2	—N(dimethylmorpholine)	OH	1.1	1.3	5.6	79.6
116	1	—N(dimethylmorpholine)	OH	1.1	1.2	3.3	62.0
117	2	—N(morpholine)	OMe		0.27	0.76	9.01
118	2	—N(morpholine)	H		0.76	1.47	20.4
119	2	—N(morpholine)	6,7-olefin		0.50	0.82	18.0
120	2	—N(morpholine)	α-F		0.38	2.55	16.2
121	2	—N(morpholine)	methano		0.87	3.56	38.6
122	2	—N(morpholine)	OCH$_2$F		0.25	1.50	40.4

docetaxel, and **108** and **115** were more potent than docetaxel. Among **107**, **108**, and **116**, the morphilinomethyl analog (**116**) is the most active compound. This result indicated that the chain length of the alkyl spacer greatly influences the activity and shortening of the chain length markedly improves the cytotoxicity (Iimura *et al.*, 2000). Although 10-deoxy-10-*C*-morphilinoethyl docetaxel analog **108** showed excellent cytotoxicity, the cytotoxicity against drug-resistant cell lines was not sufficiently potent. Thus, it was further modified at the 7-position to give compounds **117–122** (Table 10.5) (Iimura *et al.*, 2001). Compared with **108**, all the new compounds showed improved cytotoxicity against two human lung cancer and resistant cell lines.

Table 10.6 Structure and cytotoxicity of 10-*O*-*sec*-aminoethyl docetaxel analogs

Compound	R_1	R_2	R_3	Tubulin $IC_{50}/IC_{50(1)}$	P388 GI_{50} (ng/ml)	PC-6 GI_{50} (ng/ml)	PC-12 GI_{50} (ng/ml)
Paclitaxel				1	25.3	4.64	172
Docetaxel				0.44	6.74	1.13	53.4
123	—N⟨⟩O	Ph	Me	0.65	24.1	4.79	215
124	—N⟨⟩S	Ph	Me		2.90	2.54	65.4
125	—N⟨⟩	Ph	Me	0.42	74.6	6.93	>1000
126	—N⟨⟩	Ph	Me	0.44	171	5.38	>1000
127	—N⟨⟩S	Ph	Et		4.21	2.25	4.06
128	—N⟨⟩O	2-furyl	Et	0.55	2.21	2.08	5.89
129	—N⟨⟩OCH$_3$SO$_3$H	2-furyl	Et		3.00	2.86	6.27
130	—N⟨⟩S	2-furyl	Et		1.00	0.84	5.58
131	—N⟨⟩SCH$_3$SO$_3$H	2-furyl	Et		1.25	1.45	6.01
132	—N⟨⟩SCH$_3$SO$_3$H	2-furyl	Pr	0.60	1.82	2.62	9.37

This result is consistent with the previous study showing that 7-methoxy, 7-deoxy, 6,7-olefin modification can improve activity. Among this series, the 7-methoxy analog (117) showed the most potent activity. Compound 117 was tested *in vivo* by both iv and po administration and was highly active against B16 melanoma BL-6 (data not shown) (Iimura *et al.*, 2001). The same group also studied the activity of 10-*O-sec*-aminoethyl docetaxel analogs (Uoto *et al.*, 1998) (Table 10.6). Among these analogs, the morphilino (123) and thiomorphilino (124) analogs were as or more potent than docetaxel, while the piperidino (125) and pyrrolidinyl (126) analogs showed less cytotoxicity but equivalent microtubule disassembly inhibitory activity as docetaxel. This result indicates that the *sec*-amino group at C-10 can significantly influence the cytotoxicity. The increased activity of 127 compared with 124 showed the influence on modification at C-4. Compounds 128, 130, and 132 showed greatly increased cytotoxicity against all cell lines tested. Furthermore the methanesulfonic acid salts (129, 131) showed fairly good activity and water solubility.

Two additional C-10 modified derivatives, the cyclic carbonates 133 and 134 (Figure 10.9) had good microtubule assembly properties ($ED_{50}/ED_{50(taxol)}$ of 2.4 and 0.5, respectively), but displayed greatly reduced cytotoxicity against B16 melanoma cells (Datta *et al.*, 1994a). This reduced cytotoxicity may be due to a difference in uptake and/or metabolism compared to taxol (Georg *et al.* 1995d).

Modifications at C-2

Modifications at C-2 mainly include removing or replacing the 2-benzoyl group by aroyl, heteroaroyl, and other groups, C-2 epimerization, and preparing benzoylamido, carbonate, and $C_{1,2}$ cyclic carbonate analog.

2-Debenzoyloxytaxol (135) (Figure 10.9) is the first analog modified at C-2 and was shown to be inactive in both microtubule binding and *in vitro* cytotoxicity assays [HCT116 cells ($ED_{50}/ED_{50\ (taxol)} = 120$)]. This result indicated that the 2-benzoyloxy group is essential for microtubule binding and cytotoxocity (Chen *et al.*, 1993a), and triggered more investigations on C-2 modified taxol analogs.

Many groups systematically investigated the influence of aromatic substitution on the bioactivity of taxol (136–187, Table 10.7) and taxotere analogs (193–198, Table 10.8) (Chaudhary *et al.*, 1994; Chen *et al.*, 1994b; Georg *et al.*, 1994a, 1995a; Grover *et al.*, 1995; Pulicani *et al.*, 1994b; Kingston *et al.*, 1998b; He *et al.*, 2000). The introduction of a para substituent in the 2-benzoate group (154–163, 196–198) greatly reduced microtubule polymerization activity and cytotoxicity. Moreover, introduction of para-substituents into 2,5- or 3,5-active disubstituted analogs (165, 168) gave 166, 169 and 175 and reduced the interaction with tubulin. The negligible activity of para-substituted derivatives was independent of the electronic nature of

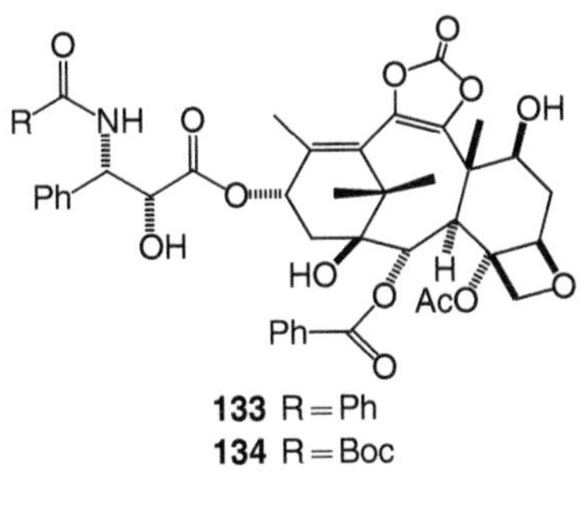

133 R = Ph
134 R = Boc

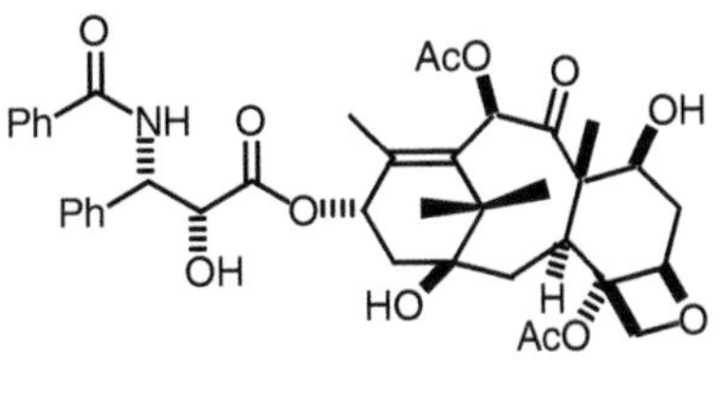

135

Figure 10.9 Structures of two cyclic carbonates 133, 134 and 135.

Table 10.7 Structure and activity of C-2 modified taxol analogs

Compound	R	Microtuble assembly $(ED_{50}/ED_{50(taxol)})$	CA46 $(ED_{50}/ED_{50(taxol)})$	A121 $(ED_{50}/ED_{50(taxol)})$	HCT116 $(ED_{50}/ED_{50(taxol)})$	B16 $(ED_{50}/ED_{50(taxol)})$
136	2-N$_3$Ph	0.70[b]		0.77	0.36	
137	2-OMePh	1.5[b]	0.67	1.4	1.9	
138	2-ClPh	Inactive				
139	2-MePh	1.1				0.84
140	3-N$_3$Ph	0.37[b]	0.17	0.18	1.4	
141	3-HOPh	1.0				
142	3-OMePh	0.50[b]	0.27	0.095	0.35	
143	3-OEtPh	1.1[b]	1.0	1.7	1.6	
144	3-OiPrPh	>140[b]	1.3	15	17	
145	3-FPh	0.70[b]		0.40	0.45	
146	3-ClPh	1.1 (0.30[b])	0.67	0.05	1.7	0.57
147	3-BrPh	0.80[b]		0.32	0.32	
148	3-IPh	0.79[b]	2.3	1.6	0.85	
149	3-CNPh	1.0[b]		0.32	1.1	
150	3-NO$_2$Ph	1.3 (0.69[b])	1.0	1.3	0.88	0.73
151	3-SMePh		1.0	3.3		
152	3-CF$_3$Ph	0.64 (1.30[b])	0.67	1.4	3.0	2.7
153	3-MePh	1.7				1.6
154	4-N$_3$Ph	Inactive				
155	4-OMePh	6.0			>44	22
156	4-OHPh	19[b]	3.0	31	52	
157	4-FPh	3.2				>36
158	4-ClPh	0.79	1.3	7.6		3.1
159	4-CNPh	Inactive				
160	4-NO$_2$Ph	>8.8			>42	
161	4-MePh	1.6				5.1
162	4-N(Me)$_2$Ph			2200		Inactive
163	4-NO$_2$PhNH	16[b]			>42	
164	2,3-FPh	0.8[b]		14	0.55	
165	2,5-FPh	0.34[b]	0.67	0.67	0.44	
166	2,4,5-FPh	1.5[b]	1.0	3.1	1.9	
167	2,3-OMePh	>270[b]	>3.3	>48	59	
168	2,5-OMePh	0.5[b]	0.67	0.12	0.14	
169	3,4,5-OMePh	32[b]	>3.3	>48	110	
170	3,5-OMePh	2.6[b]	1.0	3.3	5.4	
171	3,4-FPh	2.2[b]	0.33	3.4	2.7	
172	3,5-N$_3$Ph	0.67[b]	1.0	1.3	1.6	
173	3,4-ClPh	0.41 (0.76[b])		1.0	0.77	2.7
174	3,5-FPh	0.60[b]		0.12	0.36	
175	3,4,5-FPh	1.8[b]	1.3	3.2	3.4	
176	3,5-ClPh	0.98 (0.44[b])	0.67	0.38	0.51	1.8
177	3,5-NO$_2$Ph	4.2[b]	2.3	20	3.1	
178	2-thienyl	1.1 (4.1[b])	0.67	4.8	4.5	3.8
179	3-thienyl	1.4 (1.9[b])	0.33	4.3	4.5	2.1
180	2-furyl	2.4 (26[b])	2.0	13	52	14

Table 10.7 (Continued)

Compound	R	Microtuble assembly ($ED_{50}/ED_{50(taxol)}$)	CA46 ($ED_{50}/ED_{50(taxol)}$)	A121 ($ED_{50}/ED_{50(taxol)}$)	HCT116 ($ED_{50}/ED_{50(taxol)}$)	B16 ($ED_{50}/ED_{50(taxol)}$)
181	3-furyl	2.1 (7.8[b])	1.0	12	18	8.2
182	2-pyridyl	2.8				8.7
183	3-pyridyl	4.0				>36
184	4-pyridyl	7.3				>36
185	2-N-methylpyrrole	2.6				2.6
186	α-naphthalene			1.4×10^4		
187	β-naphthalene			1.4×10^4		3.5×10^4
188	Cyclohexyl	130 (1.7)	56[a]		>45	
189	CH_3					
190	$CH_3(CH_2)_3$					
191	$PhOCH_2$	4.7				
192	$PhSCH_2CH_2$		Poor activity[a]	4.1×10^3		

Notes
a $IC_{50}/IC_{50(taxol)}$ against P388.
b Tubulin polymerization ratio of analog relative to taxol.

the substituent. Thus, it seems likely that a significant steric effect reduces the binding of para-substituted C-2 benzenoid analogs to the paclitaxel site on tubulin. However, the introduction of small substituents such as azido (140), methoxy (142) and chloro (146 and 194) groups at the meta position provided derivates significantly more active than taxol. Significant steric trends are evident in the alkoxy series (methoxy > ethoxy > propoxy) but not in the halide series (Br > Cl > F > I). 2-Debenzoyl-2-(*m*-azidobenzoyl) paclitaxel (140) was more potent than paclitaxel in enhancing tubulin polymerization and inhibiting cell growth in murine P388 leukemia, human HL-60 leukemia, and human Burkitt lymphoma CA46 lines. 2-Debenzoyl-2-(3-hydroxybenzoyl)taxol (141) was isolated and identified as one of the major metabolites in rat bile (Monsarrat *et al.*, 1990). It was as active as taxol in the microtubule disassembly assay but less active against L1210 cells ($ED_{50}/ED_{50(taxol)} = 41$). Only a small number of ortho-substituted benzoyl analogs (136–139) were prepared. They were less active than the meta-substituted analogs. It can be concluded activity is dependent on the electronic nature and the steric effect (position) of the substituents (Chordia and Kingston, 1996; Kingston *et al.*, 1998b). Compounds 186 (Nicolaou *et al.*, 1995) and 187 (Nicolaou *et al.*, 1994c), which bear naphthalene groups, showed no activity against several cancer cell lines (Table 10.7).

Analogs with 2-heteroaroyl groups were also investigated (Nicolaou *et al.*, 1994c, 1995; Georg *et al.*, 1995b; Kingston *et al.*, 1998b). The thiophene analogs (178, 179) and furanyl analogs (180, 181) showed comparable cytotoxicity to that of taxol, while the others (182–185) showed decreased activity (Table 10.7). Compounds 178 and 179 also showed equivalent inhibition of *in vivo* tumor growth compared to taxol.

Several non-aromatic C-2 esters, such as acetyl (189), valeryl (190) (Kingston *et al.*, 1998b), cyclohexyl (188, 199 and 200) (Chen *et al.*, 1994b; Ojima *et al.*, 1994a, 1997b; Boge *et al.*, 1994), phenoxymethyl (191) (Grover *et al.*, 1995) and (phenylthio)ethyl (192) (Nicolaou *et al.*, 1995), were significantly less active than paclitaxel, while the 3,3-dimethylnacrylic ester (201) (Ojima *et al.*, 1997b) showed better cytotoxicity to paclitaxel against several cell lines (Tables 10.7 and 10.8).

The C-2 carbamate analog (163, Table 10.7) and $C_{1,2}$ carbonate analog (202, Figure 10.10) (Chen *et al.*, 1994b) were inactive in both tubulin polymerization and HCT116 cytotoxicity assays.

Table 10.8 Structure and activity of C-2 modified taxotere analogs

Compound	R_1	R_2	Microtubule disassembly assay $IC_{50}/IC_{50(taxol)}$	P388 $IC_{50}/IC_{50(taxol)}$
193	3-F—Ph	H	1.2	0.95
194	3-Cl—Ph	H	0.75	5.1
195	3-CF$_3$—Ph	H	0.7	9.3
196	4-F—Ph	H	0.8	12.8
197	4-MeO—Ph	H	5	100
198	3-CF$_3$—Ph	H	50	660
199	Cyclohexyl	H	0.85	11[b]
200	Cyclohexyl	Ac	0.67[a]	1.1[c]
201	(CH$_3$)$_2$C=CH—	Ac		0.65[d]

Notes
a Microtubule assembly assay.
b $ED_{50}/ED_{50(taxotere)}$.
c $ED_{50}/ED_{50(taxol)}$ against B16 melanoma cells.
d $IC_{50}/IC_{50(taxol)}$ against MCF-7.

2-*epi*-paclitaxel (**203**) (Figure 10.10) was prepared through a 2-debenzoylpaclitaxel intermediate and showed no activity in tubulin-assembly and HCT116 cytotoxicity assays (Chordia and
Kingston, 1996). This result provides further evidence that the nature and stereochemistry of
the 2-aroyl group is very important for the bioactivity of paclitaxel.

1-Benzoyl–2-[(*S*-methyldithio)carbonyl]paclitaxel (**204**) and 1-benzoyl–2-des(benzoyloxy)
paclitaxel (**205**) (Figure 10.10) (Chaudhary *et al.*, 1995) were less active than paclitaxel in the
P-388 leukemia cytotoxicity assay ($ED_{50}/ED_{50(taxol)}$ = 9.3 for **204**, $ED_{50}/ED_{50(taxol)}$ = 56.7
for **205**).

Recently (Fang *et al.*, 2001), the 2α-benzoylamido analog of docetaxel (**206**) (Figure 10.10)
was synthesized and retained cytotoxicity against three tumor cell lines, A-549, KB and A278
($IC_{50}/IC_{50(paclitaxel)}$ = 20.5, 16.8, 0.8, respectively).

Modifications at C-9

Because of the low reactivity of the C-9 carbonyl group, modifications at C-9 have been explored
extensively in recent years (Kingston, 1991). As discussed before, two cyclic carbonates (**133** and
134) (Figure 10.9) had good microtubule-assembly properties but no significant cytotoxicity.
Klein *et al.* (1993) synthesized 9(*R*)-dihydrotaxol (**207**) from 13-acetyl-9-dihydrobaccatin III,
which was isolated from *Taxus candidensis*. Compound **207** exhibited potent tubulin binding

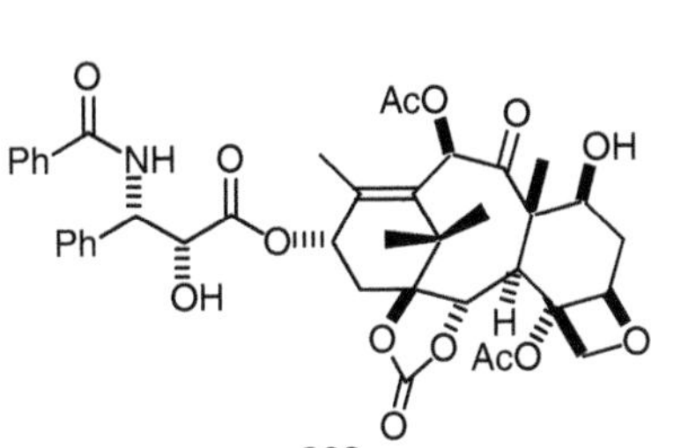

Figure 10.10 Structures of compounds 202–206.

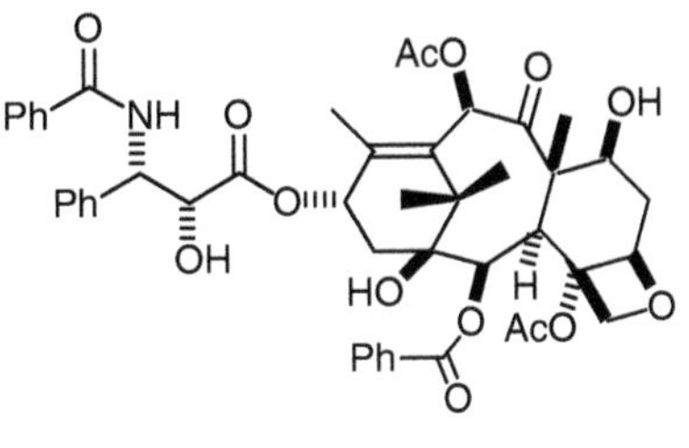

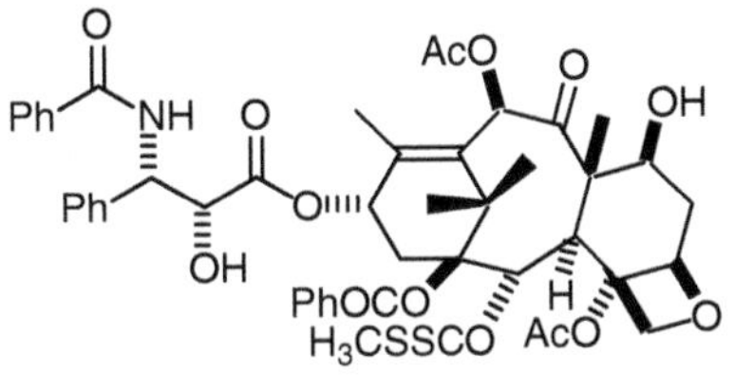

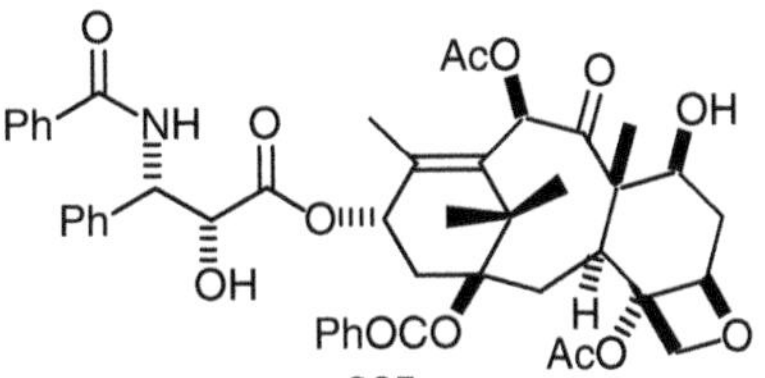

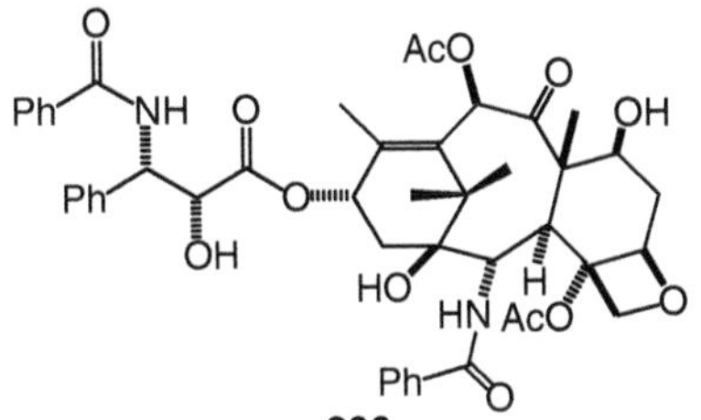

activity (Klein *et al.*, 1993, 1995). A wide variety of 9(*R*)-dihydrotaxane analogs were subsequently prepared (Figure 10.11, Table 10.9) (Klein *et al.*, 1995). Methylation of the C-7 (211, 212) or of the C-9 (213, 214) hydroxyl group of 9(*R*)-dihydrotaxane increased the potency, and they were the most potent compounds in both assays. This result confirms that decreased polarity at C-7, either by excision (such as 7-deoxytaxanes) or alkylation, improves the activity. The size of a C-7 alkoxy substituent is not limited to methoxy because others such as the allyl analog (215) are equally active to taxol. Whereas, adding polar groups into the ether side-chain as in 216, 218 and 219 led to greatly decreased activity, particularly cytotoxicity (Klein *et al.*, 1995). Diacetate compound 210 and the cyclic derivatives 220–222 bearing both C-7 and C-9 substituents showed slightly reduced activity (Table 10.9).

The 9(*R*)-dihydrotaxanes showed great promise in that they have greater stability and water solubility while exhibiting similar *in vitro* activity to their C-9 carbonyl analogs. Polar functionality at C-7 or C-9 generally decreased the biological activity, while alkylation or acylation of these groups increased activity.

Compounds 223–226 (Figure 10.11, Table 10.9) are derived from paclitaxel, taxotere, cephalomannine and 3′-dephenyl-3′-alkyl taxoid, respectively. They showed substantial retention of activity compared to the corresponding starting material. In this way, a second class of active analogs of paclitaxel was obtained, which can be further modified in the side-chain (Appendino, 1997).

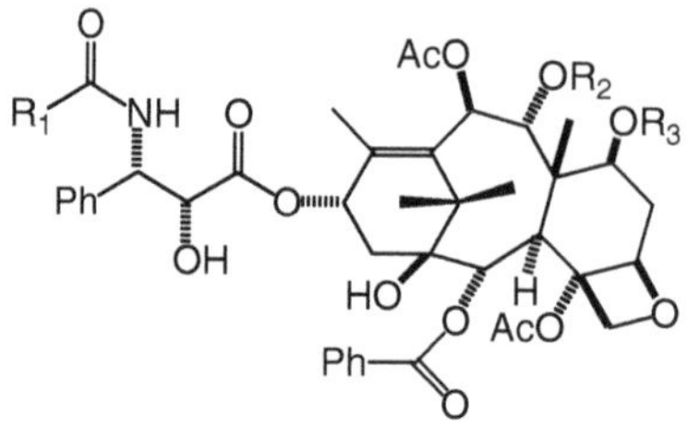

207 R$_1$ = Ph, R$_2$ = H, R$_3$ = H
208 R$_1$ = Ph, R$_2$ = Ac, R$_3$ = H
209 R$_1$ = t-BuO, R$_2$ = H, R$_3$ = H
210 R$_1$ = Ph, R$_2$ = Ac, R$_3$ = Ac
211 R$_1$ = Ph, R$_2$ = H, R$_3$ = Me
212 R$_1$ = t-BuO, R$_2$ = H, R$_3$ = Me
213 R$_1$ = Ph, R$_2$ = Me, R$_3$ = H
214 R$_1$ = t-BuO, R$_2$ = Me, R$_3$=H

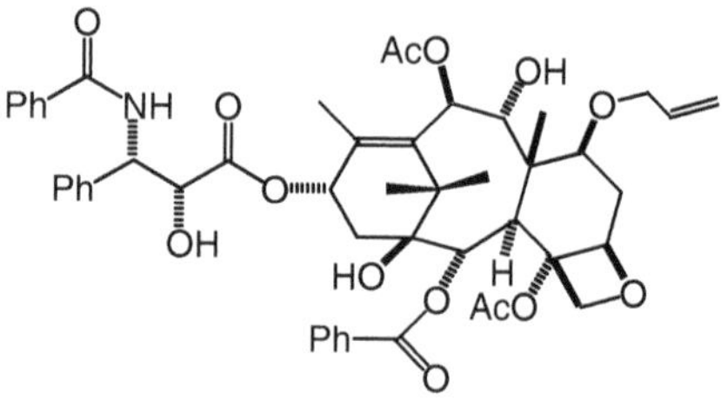

215

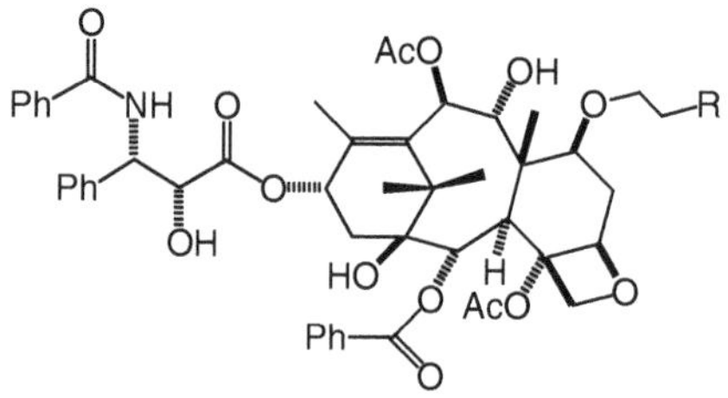

216. R = OH
217. R = OAc
218. R = NEt$_2$

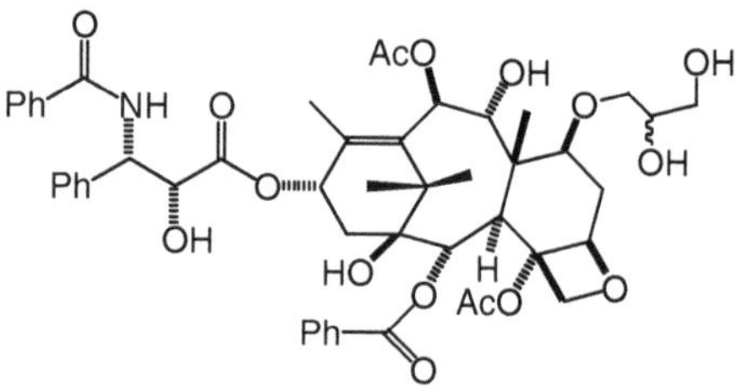

219

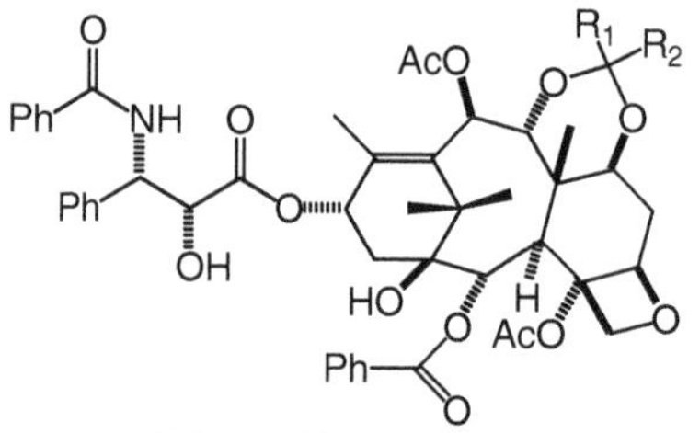

220 R$_1$ = R$_2$ = CH$_3$
221 R$_1$ = R$_2$ = S
222 R$_1$ = R$_2$ = CH$_2$CH$_3$

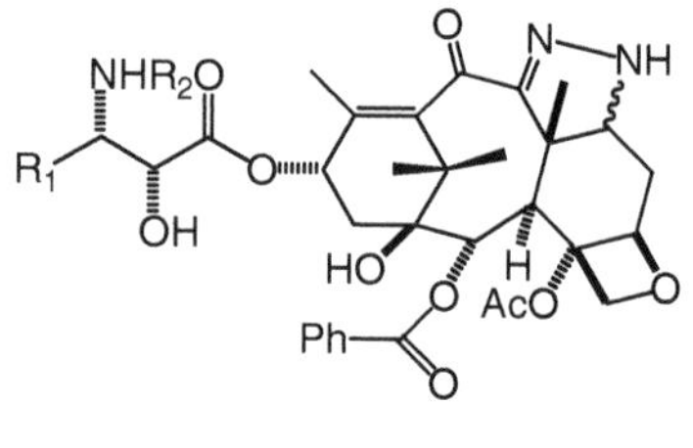

223 R$_1$ = Ph, R2 = Bz
224 R$_1$ = Ph, R2 = Boc
225 R$_1$ = Ph, R2 = Tigl
226 R$_1$ = iBu, R2 = Boc

Figure 10.11 Structures of compounds 207–226.

Modifications at C-19

10-Deacetyl-19-hydroxytaxotere (227) (Figure 10.12) was derived from naturally occurring 10-deacetyl-19-hydroxybaccatin III (Margraff *et al.*, 1994) and was very active in both the microtubule disassembly assay (ID$_{50}$/ID$_{50(taxol)}$ = 0.4) and *in vitro* cytotoxicity assay against P388 leukemia cells (ED$_{50}$ = 0.07 μg/ml). In addition, as mentioned previously C-7, 8, 19 cyclopropyl

Table 10.9 Activity of analogs of 9(R)-dihydrotaxanes

Compound	Tubulin assembly ($ED_{50}/ED_{50(taxol)}$)	A549	HT-29	$B_{16}F_{10}$	P388	MDA-MB321 ($ED_{50}/ED_{50(taxol)}$)
Taxol	1.00	2.5	1.8	3.4	8.8	
Taxotere	0.7	0.18	0.21	0.6	1.5	
207	0.75	16	6.4	25	49	
246	>10	>1000	>1000	>1000	>1000	
208	0.83	11	1.9	39	140	
209	1.31	0.26	0.65	0.4	2.8	
210	0.91	15	3.3	13		
211	0.92	1.2	1.4	1.5	3.9	
212	1.74	4.7	3.1	4.8	7.8	
213	0.35	0.27	0.15	0.2	0.6	
214	0.51	0.1	0.5	0.9	1.3	
215	1.4	1.0	1.2	2.7	5.3	
216	1.77	39	10	119	170	
217	2.47	120	31	43	79	
218	2.61	400	108	110	310	
219	2.02	626	310	>1000	>1000	
220	0.76	34	29	34	42	
221	2.78	19	11	20	35	
222	—	23	16	15	22	
223						3.17
224						0.58
225						1.08
226						2.83

analogs of taxol and taxotere (**83** and **84**) (Figure 10.8) have been reported to be as active as taxol in various cell lines, including HCT116 (Table 10.3). Thus, structural modifications at C-19 may be tolerated without significant loss of bioactivity.

Modifications at C-6

Before 1994, no synthetic modifications at the 6-position of taxol had been reported. Harris *et al.* (1994) reported that the major metabolite produced in humans contained a hydroxyl group at the diterpene 6α-position. This metabolite (**228**) (Figure 10.12) showed significantly less activity than taxol in MOLT4 and U937 cell lines (Harris *et al.*, 1994). Thus, 6-hydroxylation of paclitaxel represents a route for the detoxication and elimination of paclitaxel in man. Blockade of this metabolic pathway should improve the therapeutic efficacy and reduce clearance. Wittman *et al.* (2001a) synthesized several metabolically blocked paclitaxel and docetaxel analogs by introducing 6α-halogen (**229–232**) (Figure 10.12, Table 10.10). These compounds showed equivalent or better activity to taxol in tubulin polymerization and *in vitro* (HCT116) and *in vivo* (M109) cytotoxicity assays. This result indicates that the halogenation of the 6-α position does not affect the *in vitro* or *in vivo* efficacy of paclitaxel analogs. The authors also evaluated the analogs and paclitaxel in a human liver S9 fraction, which is capable of hydroxylating paclitaxel, and the results showed that 6-α-halogenation did block the major metabolic pathway to some degree (Wittman *et al.*, 2001a).

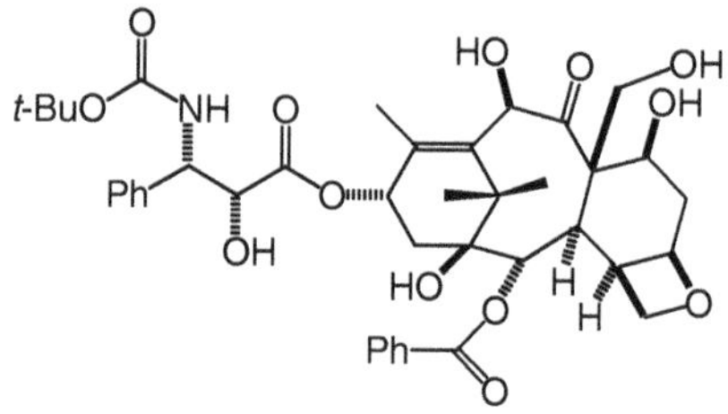

227

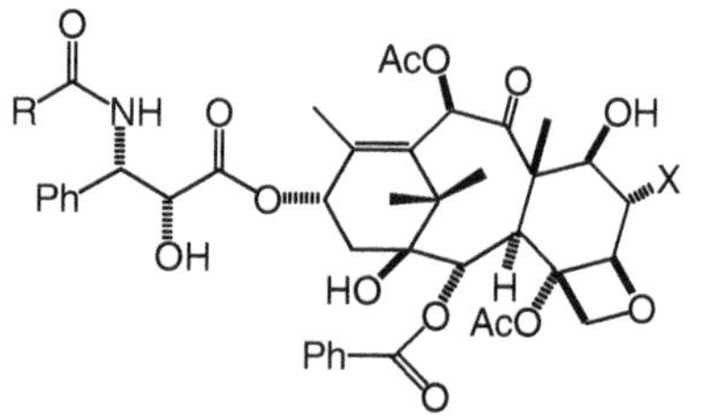

228 R = Ph, X = OH
229 R = Ph, X = F
230 R = Ph, X = Cl
231 R = Ph, X = Br
232 R = O-*t*Bu, X = F

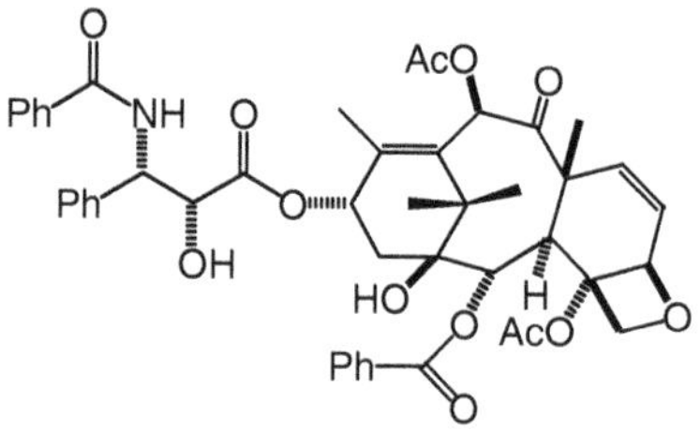

71

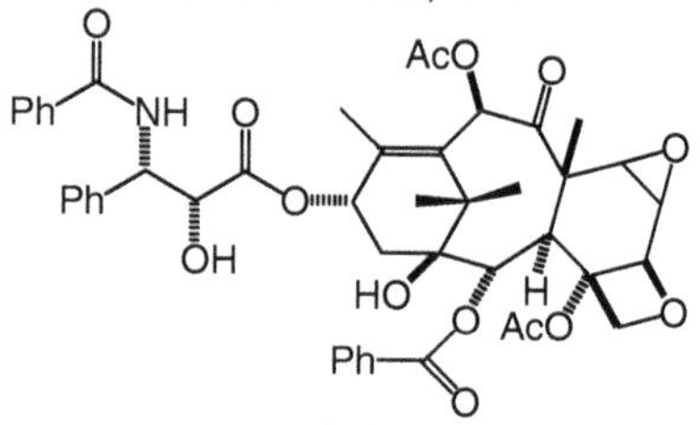

241

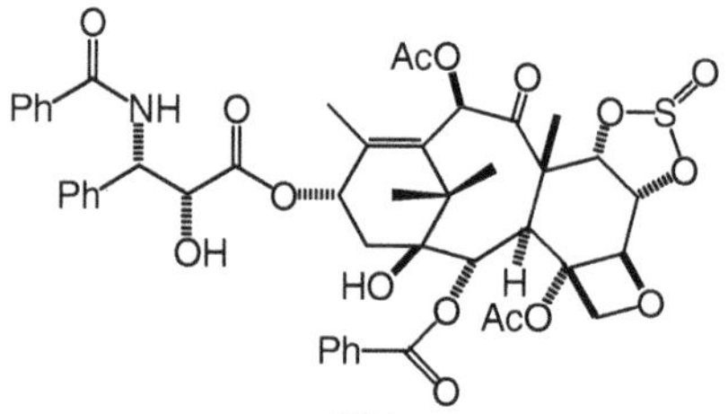

242

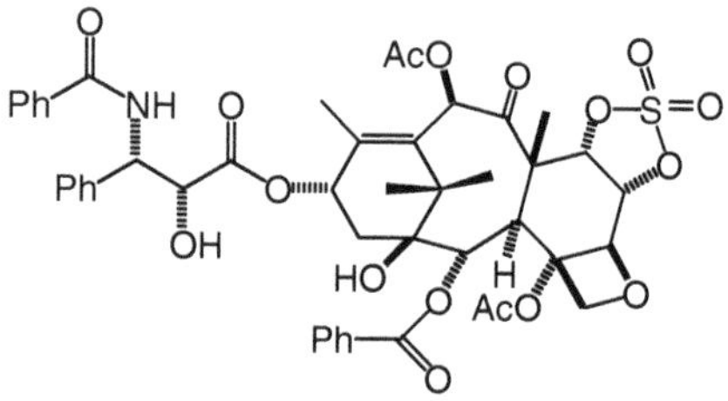

243

Figure 10.12 Structures of compound 227–232, 241–243.

Recently, 7-*epi*-taxol has been modified at the 6-α-position. Compound 238 was 40-fold less potent than paclitaxel in the tubulin polymerization assay. Compounds 71, 233–237 and 241 were not as effective as paclitaxel in promoting tubulin assembly but, except for 233 and 236, stabilized polymerized tubulin as well as or better than paclitaxel. All the analogs except 235–237 showed comparative cytotoxicity to paclitaxel (Table 10.10) (Liang *et al.*, 1995a). Therefore, the presence of bulky substituents on the α-face of the molecule is detrimental to receptor binding.

Compared with paclitaxel, the β-amino analog (239) was significantly less active than paclitaxel, but the β-azido analog (240) showed 2–3 times more cytotoxicity (Table 10.10). The cyclic sulfite analog (242) was about 10-fold less active than paclitaxel in the HCT116 cell line, but the cyclic sulfate (243) was inactive (Figure 10.12) (Yuan *et al.*, 2000).

Modifications at C-4

C-4 modified analogs include 4-deacetylpaclitaxel, 4-deacetoxypaclitaxel, and 4-ether, ester, carbonate and carbamate derivatives.

Table 10.10 Structures and biological activity of compounds modified at 6-position

Compound	R	Tubulin assembly	Tubulin disassembly (maximum rate)	CA46 $IC_{50}/IC_{50(taxol)}$	HCT116 $IC_{50}/IC_{50(taxol)}$	A2780 $IC_{50}/IC_{50(taxol)}$
Paclitaxel		1.0	70	1	1	1
229		1.0			0.66	
230		0.8			0.57	
231		1.1			1.23	
232		0.6			0.09	
233	α-OH	2.1	140	3		
234	α-OCOCH$_3$	0	70	2		
235	α-OCOC$_6$H$_5$	0	47	5		
236	α-OCO-*o*-CH$_3$C$_6$H$_4$	0	120	20		
237	α-OCO-cyclo-C$_3$H$_5$	0	45	7		
238	α-OTs	40				
239	β-NH$_2$				38	13.5
240	β-N$_3$				0.35	0.62
71		2.0	60	3		
241		1.7	73	2		
242					9.0	3.29
243					>51	>31

4-Deacetyltaxol (244) (Datta *et al.*, 1994b; Neidigh *et al.*, 1994) and 10-acetyl-4-deacetyltaxotere (245) (Datta *et al.*, 1994b) were inactive in the microtubule assembly assay (Figure 10.13, Table 10.11). Compound 244 also showed no cytotoxicity against several cell lines (Neidigh *et al.*, 1994). These results illustrate the importance of the 4-acetyl moiety for microtubule binding. Comparison of compound 246 with 207 (Figures 10.11 and 10.13, Table 10.9) can further confirm this conclusion (Klein *et al.*, 1995). Chordia *et al.* (1994) synthesized two 4-deacetoxypaclitaxel derivatives (247 and 248), which were significantly less active than paclitaxel in tubulin assembly and cytotoxicity assays. This result again indicates that an ester substituent at C-4 is essential to the biological activity. 4-Deacetyl-4-methyl ether paclitaxel analogs (249 and 250) were slightly less cytotoxic than paclitaxel (Chen *et al.*, 1996a); thus, removal of the carbonyl from the C-4 substituent is detrimental to the biological activity, suggesting that the C-4 acyl moiety is an important element in receptor binding (Chen *et al.*, 1996a). A series of paclitaxel analogs with 4-ester, carbonate, and carbamate groups instead of the 4-acetate have been investigated (Chen *et al.*, 1994c, 1995a,b; Georg *et al.*, 1994a). C-4 benzoate bearing derivatives (251, 252) were less active than paclitaxel both in tubulin polymerization and cytotoxicity assays. However, all of the aliphatic esters showed good to excellent activities in both assays. Among the six straight aliphatic esters (taxol, 255, 257, 258, 262, 264),

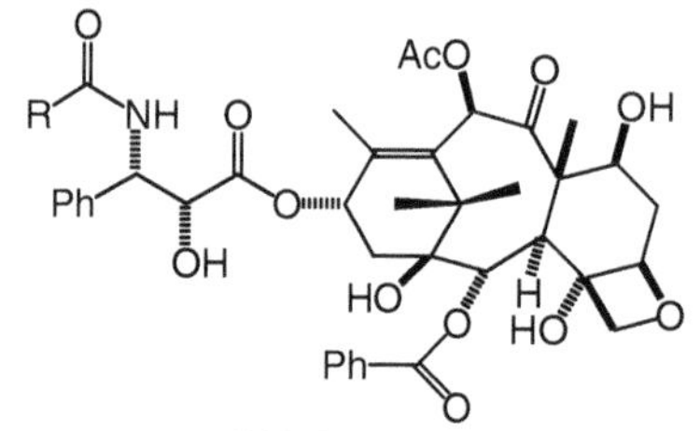

244 R = Ph
245 R = *t*-BuO

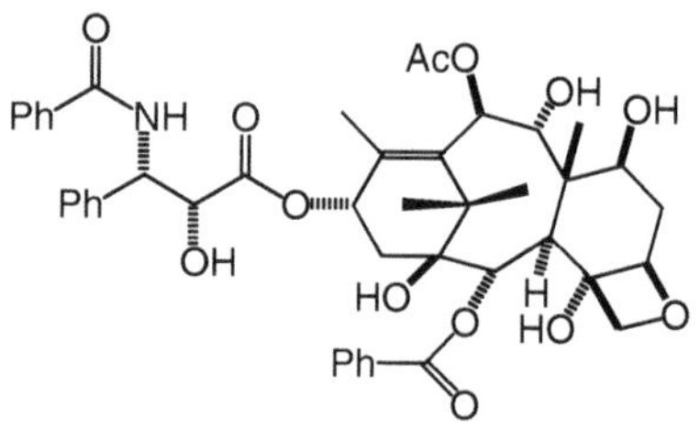

246

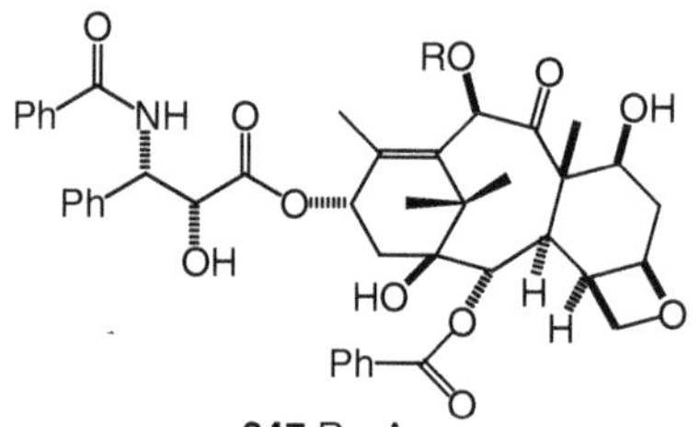

247 R = Ac
248 R = H

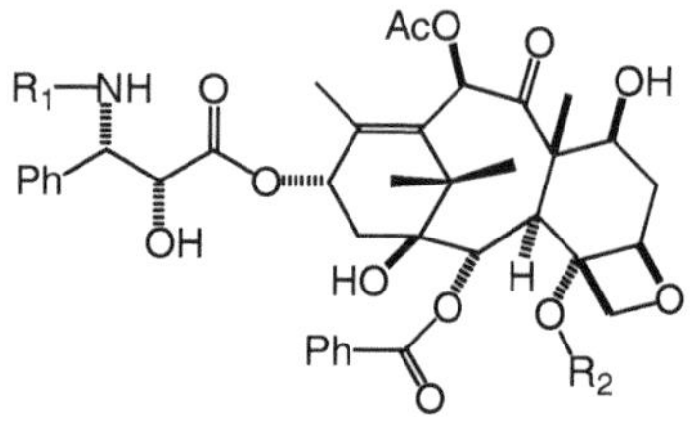

249 R_1 = Bz, R_2 = Me
250 R_1 = Boc, R_2 = Me

Figure 10.13 Structures of compounds 244–250.

the 4-butyrate ester (257) is the most potent analog. Therefore, the four-carbon chain is probably the optimal size for effective receptor binding. The 4-cyclopropyl ester (261) is even more potent than 257, its equivalent straight chain analog. Three C-4 carbonate analogs (266–268) were as or more potent than paclitaxel in both assays. Among the C-4 carbamate analogs (269–271), the C-4-imidozole (270) was inactive, the C-4-aziridine (271) carrying analog showed slightly weaker activity in both the tubulin and cytotoxicity assays, while the C-4 cyclobutylcarbamate (269) showed comparable cytotoxicity to paclitaxel and is more potent than paclitaxel in tubulin assay (Table 10.11) (Chen *et al.*, 1995b).

Interestingly, replacing the 4-cyclopropyl of 261 with an isosteric aziridine ring (271) resulted in a nine-fold loss of potency in the tubulin polymerization assay, confirming that the C-4 substituent is involved in the intimate binding with tubulin. The 4-cyclopropyl group (261) is more sterically demanding than the aziridine (271), and may fill the binding pocket better than the relatively flat aziridine ring (Table 10.11) (Chen *et al.*, 1995a).

Synthesis and bioactivity evaluation of a series of 2,4-diacyl analogs of paclitaxel (272–289) (Table 10.12) (Chordia *et al.*, 1994, 2001; Samaranayake *et al.*, 1991) also confirmed the results from C-2 and C-4 modifications. Several analogs had improved tubulin assembly ability compared with paclitaxel, particularly the methoxycarbonyl (274–278) and cyclopropylcarbonyl (279–284) analogs.

Modifications at the oxetane ring

Oxetane ring modifications include disrupting the D-ring to generate *seco*-analogs or a 5(20)-deoxydocetaxel analog or replacing the oxygen by nitrogen, sulfur and other heteroatoms.

Compounds 290 (D-secopaclitaxel) and 291 (Figure 10.14) have an opened D-ring but also lack a 4-acetyl group; they were inactive in both cytotoxicity and tubulin assembly assays

Table 10.11 Structures and biological activity of compounds modified at C-4

Compound	R	Tubulin poly. ratio	Tubulin assembly $(ED_{50}/ED_{50(taxol)})$	HCT-116 IC_{50} (nM)	B16 $(ED_{50}/ED_{50(taxol)})$
Taxol	CH_3	1.0	1.0	2.4	1
244		200	>10	>1000	>294
245			>32		
251	C_6H_5	>600		410	
252	C_6H_5F-p	61		790	
253	CH_2F	1.2		7.0	
254	CCl_3	0.68		4.0	
255	CH_2CH_3	1.5		2.0	
256	$CH=CH_2$	0.42		6.0	
257	$CH_2CH_2CH_3$	0.61		1.1	
258	$CH(CH_3)_2$	0.46	2.6	5.0	4.5
259	$C(Me)=CH_2$	0.71		4.5	
260	$CH=CH(Me)$trans	0.70		2.3	
261	C_3H_5-c	0.24		1.0	
262	$(CH_2)_3CH_3$	1.04		2.0	
263	C_4H_7-c	0.44		1.5	
264	$(CH_2)_4CH_3$	3.4		6.0	
265	C_5H_9-c	1.54		2.0	
266	OMe	0.41		2.0	
267	OEt	0.64		1.0	
268	Opr-n	0.76		2.6	
269	NHC_4H_7-c	0.76		9.2	
270	Imidozole	263		803	
271	Aziridine	2.8		15.6	

(Chordia *et al.*, 1994; Samaranayake *et al.*, 1991). However, their lack of bioactivity may be due to the absence of the 4-acetyl group rather than just the oxetane ring. Compounds 292–295 (Figure 10.14) are D-secopaclitaxel analogs with acetate groups at C-4 and C-20 and various oxidation states at C-7 and of the C5–C6 bond. All were inactive in both cytotoxicity and tubulin assembly assays. The minireceptor model proposed by Barboni *et al.* (2001a) predicts that bulky groups at C-5 and C-20 are detrimental to the activity of paclitaxel analogs. Compounds 296–297 were also inactive against several cell lines. The low activity was postulated to be due to the lack of an inflexible C-ring (Beusker *et al.*, 2001). These results indicate that the oxetane ring is important for biological activity.

However, 5(20)deoxydocetaxel (298) (Figure 10.14, Table 10.13) is as active as paclitaxel in inhibiting the microtubule disassembly and contains a 3-membered cyclopropyl D-ring rather than an oxetane ring. The authors concluded that the oxetane is not essential for the interaction

Table 10.12 Structures and bioactivity of 2,4-diacyl analogs of paclitaxel

Compound	R_1	R_2	Tubulin data	HCT116(nM)
Taxol	Bz	Ac	6.9	2.0
272	Bz	β-alanyl		8.12
273	Bz	glutaryl		>108
274	3-methoxybenzoyl	methoxycarbonyl	6.2	6.7
275	3-methylbenzoyl	methoxycarbonyl	5.5	2.1
276	3-chlorobenzoyl	methoxycarbonyl	1.7	2.2
277	3-azidobenzoyl	methoxycarbonyl	NT	1.3
278	3,3-dimethylacryloyl	methoxycarbonyl	NT	4.0
279	3-methoxybenzoyl	cyclopropylcarbonyl	9.0	2
280	3-azidobenzoyl	cyclopropylcarbonyl	0.55	1.7
281	3-chlorobenzoyl	cyclopropylcarbonyl	0.41	1.0
282	2,4-difluorobenzoyl	cyclopropylcarbonyl	1.3	2.1
283	2,5-dimethoxybenzoyl	cyclopropylcarbonyl	3.4	1.8
284	2,4-dichlorobenzoyl	cyclopropylcarbonyl	5.4	2.5
285	3-azidobenzoyl	S-methylxanthyl	250	16.8
286	3-chlorobenzoyl	S-methylxanthyl	21	5.7
287	3-methoxybenzoyl	S-methylxanthyl	53	4.7
288	3-chlorobenzoyl	3-chlorobenzoyl	420	2430
289	3-methoxybenzoyl	t-butoxycarbonyl	NT	>106

with microtubules. Instead, the cyclopropyl ring might act as the oxetane ring to rigidify the C-ring and point the C-4 acetoxy group in the appropriate direction for binding. A comparison of the inhibitory activity of docetaxel and 298 (Table 10.13) indicates that the presence of oxygen is not absolutely required but it could participate in the stabilization of the taxoid–tubulin complex (Dubois *et al.*, 2000).

In order to clarify the role of the oxetane oxygen atom, several groups synthesized paclitaxel analogs with nitrogen, sulfur and other heteroatoms (Marder-Karsenti *et al.*, 1997; Gunatilaka *et al.*, 1999; Dubois *et al.*, 2000). The compounds with nitrogen (299–303) were inactive in the *in vitro* cytotoxicity assay, and while 300 showed some tubulin polymerization activity, it was 16 times less active than docetaxel (Table 10.13). These results suggested a specific interaction of the heteroatom with the protein. The presence of the nitrogen atom introduced the problem of basicity; it would be protonated at neutral pH and would presumably interact with tubulin in a very different way than a neutral oxygen atom (Gunatilaka *et al.*, 1999). However, the sulfur and selenium derivatives (304, 305) were also found to be less active than paclitaxel in biological assays (Table 10.13), raising the possibility that the activity loss may not be due to basicity of the heteroatom and the oxygen atom in the oxetane ring could play an important role in the

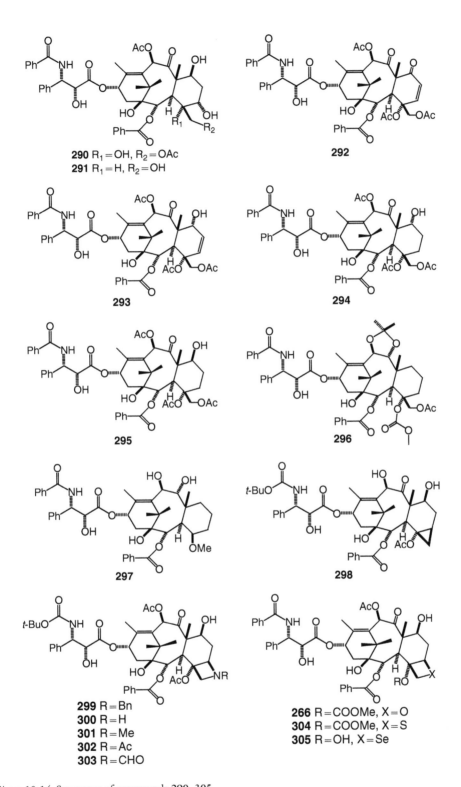

290 R₁=OH, R₂=OAc
291 R₁=H, R₂=OH

292

293

294

295

296

297

298

299 R=Bn
300 R=H
301 R=Me
302 R=Ac
303 R=CHO

266 R=COOMe, X=O
304 R=COOMe, X=S
305 R=OH, X=Se

Figure 10.14 Structures of compounds 290–305.

Table 10.13 Biological activity of compounds with modified oxetane ring

Compound	Tubulin polymerization $(ID_{50}/ID_{50(taxol)})$	Microtubule disassembly	CA46	PC3	MCF-7	1A9	HCT116	KB
Paclitaxel	1	1	1	1	1	1	1	1
Docetaxel	0.62	0.5	0.6	0.23	0.25	0.5		0.53
298		1.2						
299		Inactive						
300		8						Inactive
266	0.5		0.4	0.25	0.1	0.5	0.83	Inactive
304	>5.95		>200	>625				
305							Inactive	

mechanism by which paclitaxel exhibits its anticancer activity the tubulin binding region may be either very sensitive to steric effects or the oxetane ring may act as a hydrogen bond acceptor.

Modifications at A-ring and B-ring

A-ring modifications can induce a significant loss of cytotoxicity without effect on the activity in microtubule binding. The A-ring modified taxol derivatives 306 and 307 (Figure 10.15) were found to polymerize tubulin at a faster rate than taxol; however, they were *ca.*10 times less active than taxol against HCT116 cell lines (Chen *et al.*, 1993b). Nor-secotaxol (308) and nor-secotaxotere (309) (Figure 10.15) and their derivatives with a C-13 N-amide linkage were studied by Ojima *et al.* (1994b, 1998). Compounds 308 and 309 were 20–40 times less cytotoxic than taxol against several cancer cell lines, but had comparable activity in the MCF7-R cell line. These results also indicate the importance of the A-ring for the optimal cytotoxicity of taxoids and the reduced activity of the nor-seco taxoids maybe due to an enhanced conformational flexibility of the phenylisoserine moiety caused by the loss of the constrained taxane A-ring.

Treatment of taxol and taxotere led to the formation of A- and B-ring contracted rearrangement products (310–317) (Figure 10.15). Compound 310 and 311 were synthesized from brevifoliol (46) and 2-acetooxybrevifoliol (47), respectively (Georg *et al.*, 1993, 1995d). They did not show microtubule binding or cytotoxicity. The lack of activity may be due to the lack of the 4-acetyl and the oxetane ring, which are essential to the activity. Compound 312 and 313 possess the brevofoliol ring system (but with intact oxetane moiety) and were as active as taxol in the tubulin depolymerization assay (Samaranayake *et al.*, 1991; Wahl *et al.*, 1992). Compound 312 showed no cytotxicity in KB cell culture ($ED_{50}/ED_{50(taxol)} = 2 \times 10^5$), indicating that cell absorption or metabolism may play a role in the absence of cytotoxicity (Samaranayake *et al.*, 1991). Compounds 314–317 showed no cytotoxicity against HCT116, while only 316–317, which contain a lactonized B-ring and contracted C-ring, respectively, were active in the tubulin assembly assay (Table 10.14). Thus many significant changes in the ring system do not greatly affect tubulin binding activity, but may affect cytotoxicity (Yuan *et al.*, 1999a). In addition, contracted B-ring compounds 318 and 319 were prepared from a rearrangement reaction on the 9-dihydrotaxane skeleton. Compound 318 was 4- to 40-fold less cytotoxic than taxol against various cell lines, while 319 showed nearly equal potency to taxol (Klein *et al.*, 1994). Finally, a photochemically induced radical cascade resulted in the formation of a covalent link between C-3 and C-11 giving the pentacyclic derivative 320. This compound was 100-fold less active than taxol in the microtubule assembly assay (Chen *et al.*, 1992).

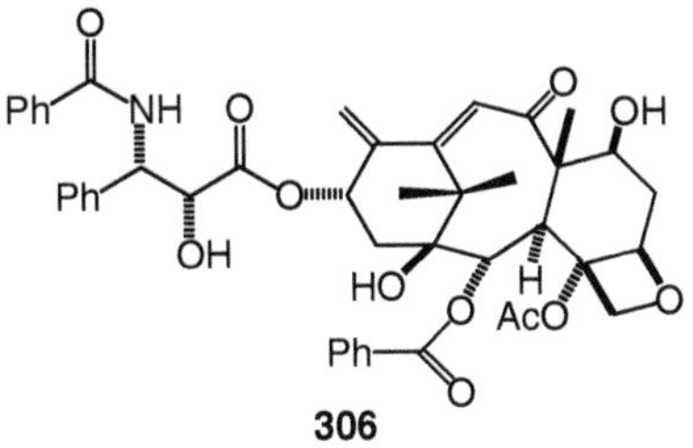

306

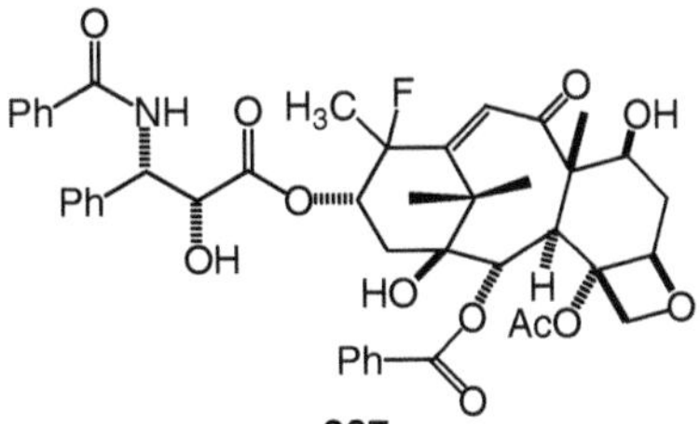

307

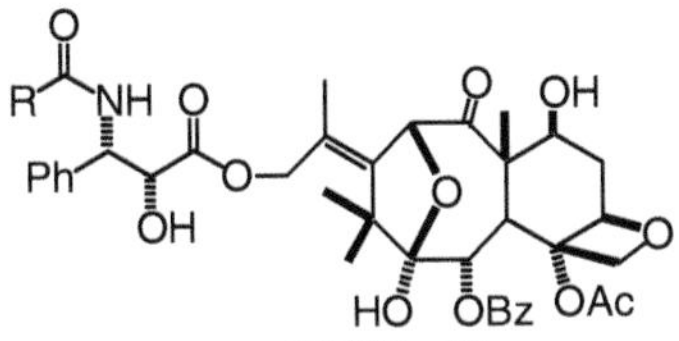

308 R = Ph
309 R = t-BuO

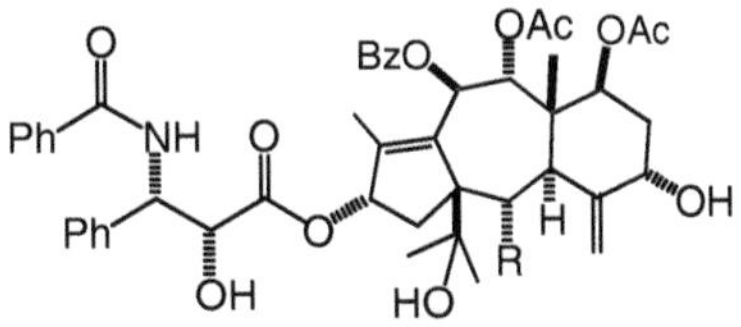

310 R = H
311 R = OAc

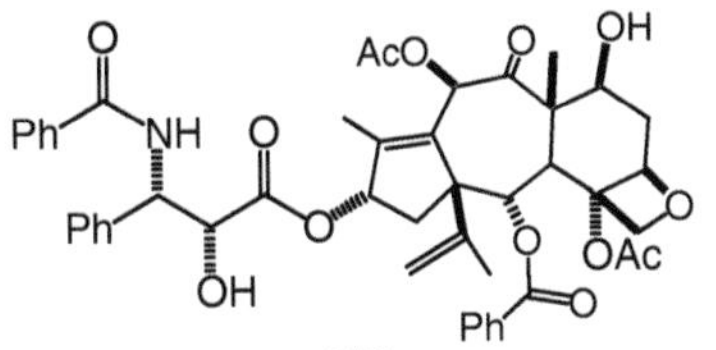

312

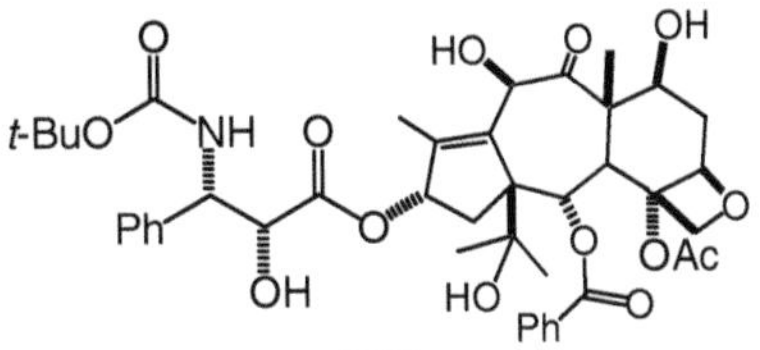

313

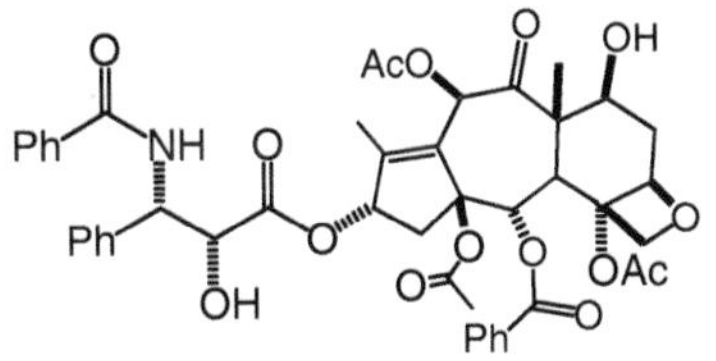

314

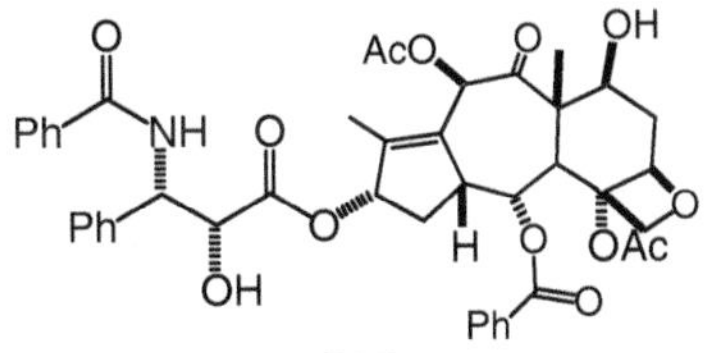

315

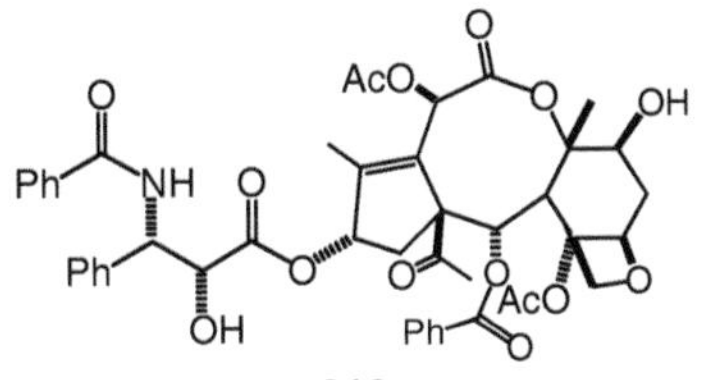

316

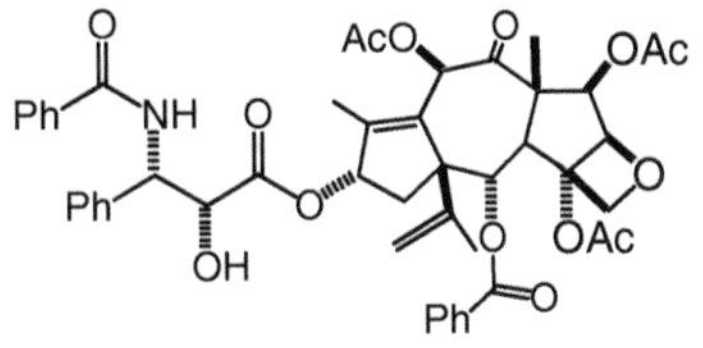

317

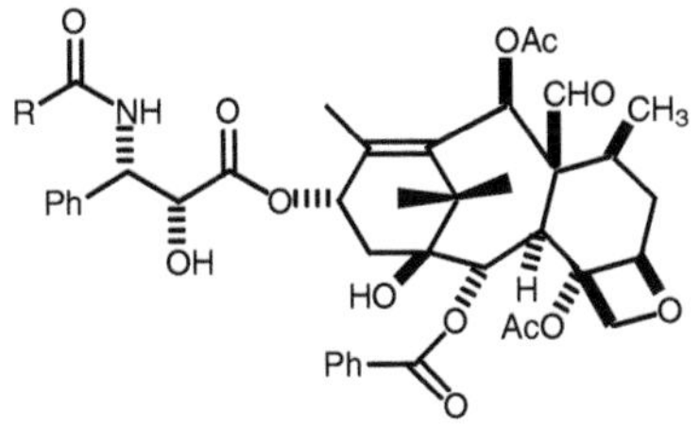

318 R = Ph
319 R = t-BuO

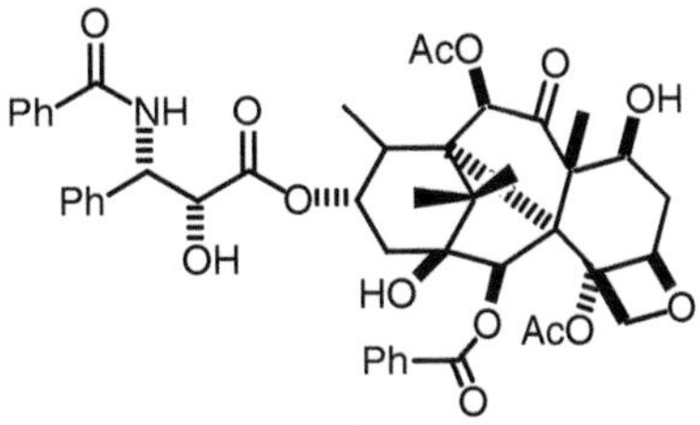

320

Figure 10.15 Structures of compounds 306–320.

Table 10.14 Biological activity of compounds modified at A-B-ring

Compound	*Tubulin assembly*	*A121*	*A549*	*HT-29*	*MCF-7*	*MCF7-R*	*HCT116*
Paclitaxel	1	1	1	1	0.0017	1	1
308		18.6	36.9	37.2	46.5	1.57	
309		20.8	46.9	47.5	59.4	1.2	
314	>172						>78
315	>172						>83.3
316	0.91						>78
317	1.59						81.1

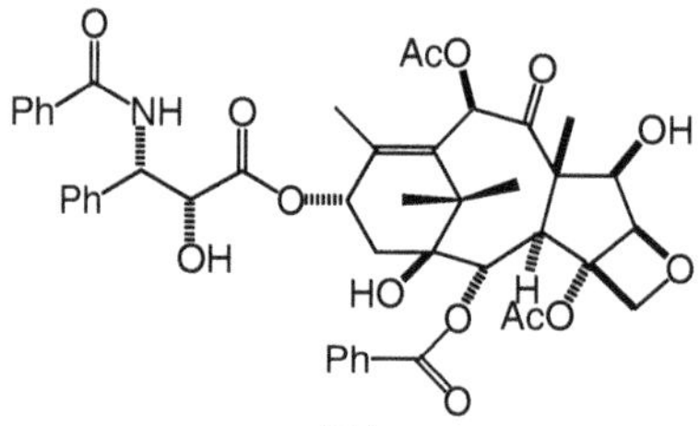

321

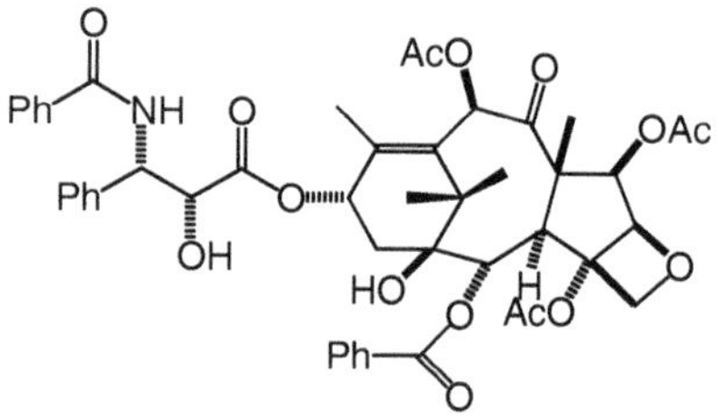

322

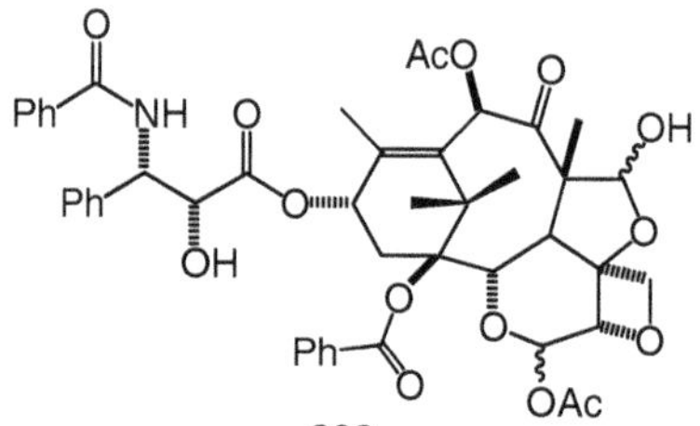

323

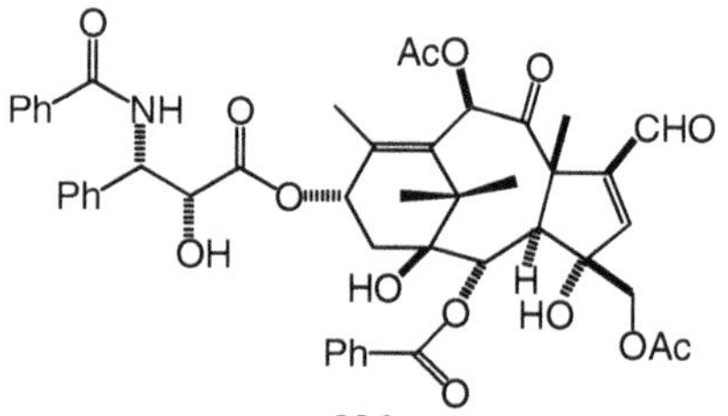

324

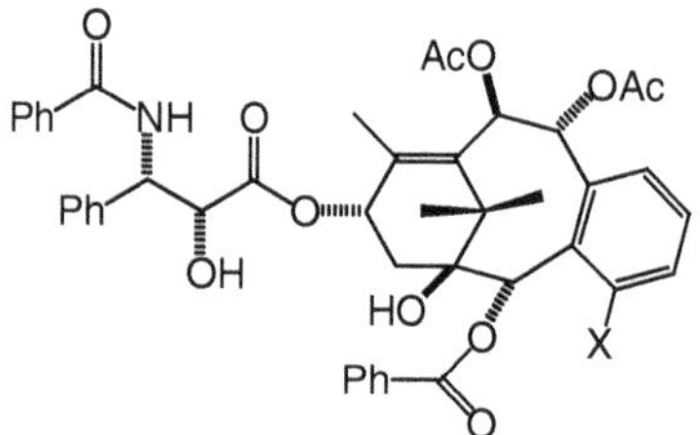

325 X = H
326 X = OBn
327 X = OMe

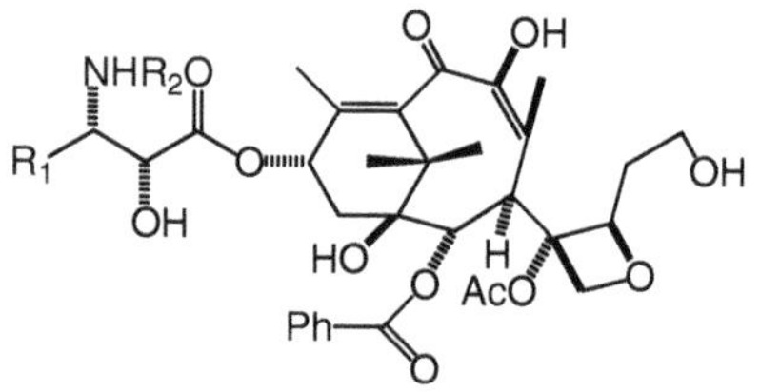

328 R_1 = Ph, R_2 = Bz
329 R_1 = Ph, R_2 = Boc
330 R_1 = *i*-But, R_2 = Boc
331 R_1 = *i*-But, R_2 = *n*-hexanoyl

Figure 10.16 Structures of compounds with modification at C-ring.

Modifications at C-ring

Ring C modifications include contraction with or without changes on the D-ring, replacement (together with ring D) by a phenyl group and ring opening (C-seco paclitaxel).

The contracted C-ring analogs (321–324) (Figure 10.16, Table 10.15) were significantly less active in the tubulin assembly and *in vitro* cytotoxicity assays (Chen *et al.*, 1995c;

Table 10.15 Activity of compounds with modification at C-ring

Compound	Tubulin assembly	HCT116 ($IC_{50}/IC_{50(taxol)}$)	MDA-MB321 ($IC_{50}/IC_{50(taxol)}$)	MCF-7 ADRr ($IC_{50}/IC_{50(taxol)}$)
Paclitaxel	1.0	1	1	1
Docetaxel			0.32	0.30
321	8.3	2.17		
322	47	6.38		
323	>230	134		
324	>230	139		
328			112	>3.84
329			10	>1.92
330			2	2.31
331			13.2	0.38

Liang *et al.*, 1995b, 1997). These results indicate that changes in the size and conformation of rings C and D make significant effects on the activity of paclitaxel.

Aryl C-ring taxoids (325–327) (Figure 10.16) showed lower cytotoxicity than taxol against several cell lines, for example, compound 325, $IC_{50}/IC_{50(taxol)}$ = 450 against Ovcar-3, $IC_{50}/IC_{50(taxol)}$ = 25 against HT-29, $IC_{50}/IC_{50(taxol)}$ = 46 against UCLA-P3 (Nicolaou *et al.*, 1997).

Several C-seco paclitaxel derivatives (328–331) (Figure 10.16) were investigated. Compound 328 showed greatly reduced activity, indicating the importance of the C-ring. However, modifications of the 3′-amide and at the 4′-position (329–331) could increase activity. Compound 331 was even more potent than paclitaxel in the adriamycin resistant breast tumor cell line (Table 10.15) (Appendino *et al.*, 1997).

Modifications at C-1

The function of the C-1 hydroxyl group has received little investigation. However, recently, Kingston *et al.* (1999) synthesized a series of 1-deoxypaclitaxel analogs (332–340) (Figure 10.17) from the naturally occurring taxoid baccatin III. Except for 336 and 339, which have large substituents on the "northern hemisphere," most of the compounds displayed reasonable tubulin–assembly activity and cytotoxicity (Table 10.16). Compounds 332 and 210 are identical except for the absence of the C-1 hydroxyl group in the former, which resulted in a two-fold loss in tubulin–assembly activity and a greater loss in cytotoxicity (Kingston *et al.*, 1999; Yuan *et al.*, 1999b).

Modifications at C-14

The conversion of the naturally occurring taxane 14β-hydroxy-10-deacetylbaccatin III (25) the active taxol and taxotere analogs 341a and 341b (Figure 10.17) demonstrated that the 14β-hydroxyl group can increase bioactivity (Georg *et al.*, 1995d). Compound 342a and 342b are structurally similar to 341a and 341b with the exception of the cyclic carbonate moiety at C-1 and C-14 (Figure 10.17). Compound 342a was less active than taxol, while taxotere analog 342b was as active as taxol (Ojima *et al.*, 1994d).

Summary

The SAR of the diterpene moiety of taxol obtained through the semisynthetic analogs is summarized in Figure 10.18. The 4-acetyl group and the 2-benzoyloxy moiety of taxol are essential

Table 10.16 Bioactivity of 1-deoxypaclitaxel analogs

Compound	Tubulin assembly assay $(IC_{50}/ IC_{50(taxol)})$	Cytotoxicity against HCT116 $(IC_{50}/ IC_{50(taxol)})$
Paclitaxel	1	1
210	0.9	0.9–1.8[a]
332	2.0	11.2
333	2.4	2.9
334	5.8	2.7
335	2.0	2.3
336	>200	6.6
337	1.6	1.2
338	1.8	2.0
339	>200	>70
340	2.2	2.8

Note

a $(IC_{50}/ IC_{50(taxol)})$ against HT-29.

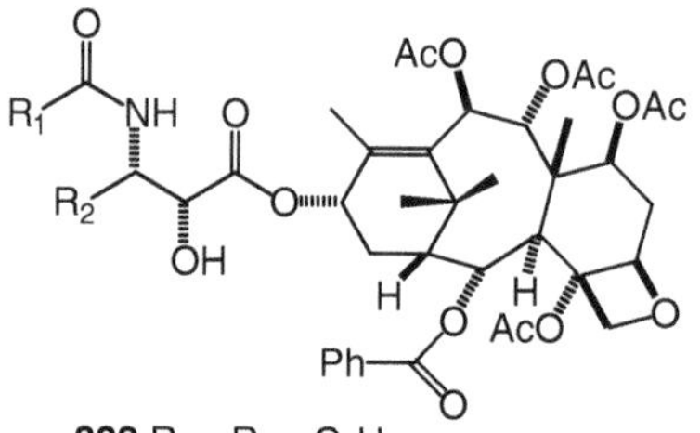

332 $R_1 = R_2 = C_6H_5$
333 $R_1 = t$-BuO, $R_2 = C_6H_5$
334 $R_1 = t$-BuO, $R_2 = CH=CMe_2$

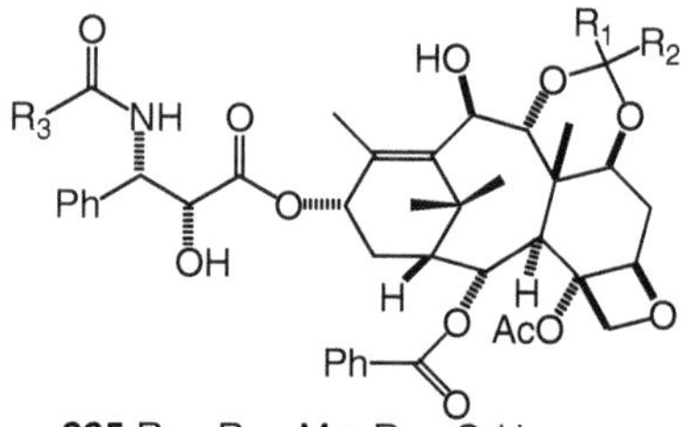

335 $R_1 = R_2 = Me$, $R_3 = C_6H_5$
336 $R_1 = C_6H_5$, $R_2 = H$, $R_3 = t$-BuO
337 $R_1 = R_2 = Me$, $R_3 = t$-BuO

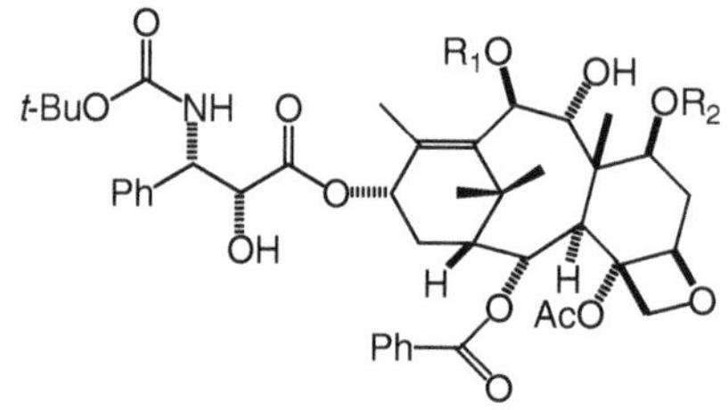

338 $R_1 = R_2 = H$
339 $R_1 = (CH_2)_4OMe$, $R_2 = Me$
340 $R_1 = H$, $R_2 = Me$

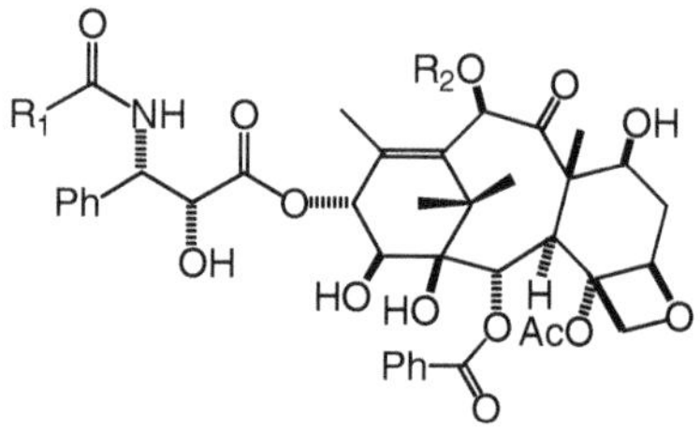

341a $R_1 = Ph$, $R_2 = Ac$
341b $R_1 = t$-BuO, $R_2 = H$

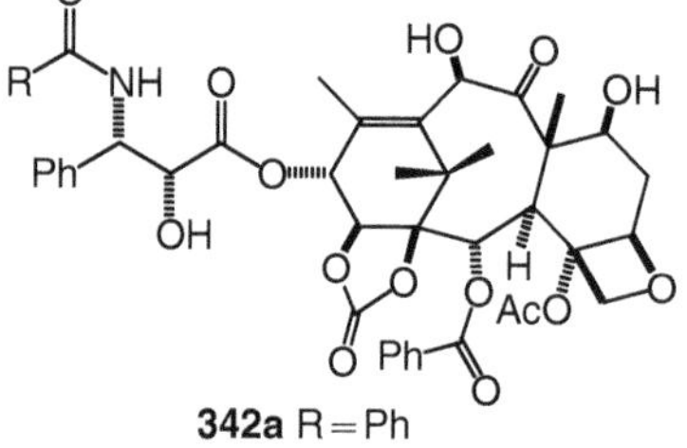

342a $R = Ph$
342b $R = t$-BuO

Figure 10.17 Structures of 1-deoxy- and 14β-hydroxypaclitaxel analogs.

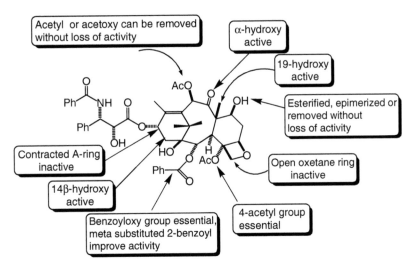

Figure 10.18 Summary of SAR for the diterpene moiety of taxol from semisynthetic analogs.

for cytotoxicity (Chen *et al.*, 1993a, 1996a; Chordia *et al.*, 1994; Neidigh *et al.*, 1994; Datta *et al.*, 1994b; Klein *et al.*, 1995). Introduction of meta substituents at the 2-benzoate group provided highly active analogs, whereas para substituents greatly reduced activity (Chordia and Kingston, 1996; Kingston *et al.*, 1998b). Modifications of the A- and B-ring did not significantly affect microtubule binding, but cytotoxicity usually greatly decreased (Ojima *et al.*, 1994b; Ojima and Lin, 1998). The oxetane ring is essential to the activity. 10-Acetyl or acetoxy group can be removed without significant loss of activity. C-10 hydroxyl group can be esterified by other acids and provide active analogs (Kant *et al.*, 1994; Chaudhary *et al.*, 1993a; Chen *et al.*, 1993b,c). Esterification, epimerization or deletion of C-7 hydroxyl group did not affect activity (Chen *et al.*, 1994a; Liang *et al.*, 1995a).

Over all, the southern perimeter part of taxol, including carbons 1-5 and 14 with oxygen functions at C-1, C-2 and C-4, and the oxetane ring, are crucial to taxol's activity, While the northern perimeter part of taxol, including carbons 6–12 with oxygen functions at C-7, C-9 and C-10, is tolerated to the structure modifications (Kingston, 1994; Nicolaou *et al.*, 1994a).

Modifications at the C-13 side-chain of taxol

Previous SAR studies indicate that the C-13 (2′R, 3′S)-*N*-benzoyl-3′-phenylisoserine side-chain is essential for taxol cytotoxicity. Because of its biological importance and the efficient development of synthetic routes, this side-chain has been more extensively investigated than any other part of taxol.

Simplified C-13 side-chain analogs

Investigations of the minimal structural requirement for the C-13 side-chain have not generated any simplified derivatives that are as active as taxol (Georg *et al.*, 1995d). Taxol derivatives containing an intact diterpene moiety but carrying simplified C-13 side-chains derived from acetic acid (343) (Lataste *et al.*, 1984), 3-phenylpropionic acid (344) (Guéritte-Voegelein *et al.*, 1991), crotonic acid (345) (Guéritte-Voegelein *et al.*, 1991), (R)- or (S)-lactic acid (347 and 348) (Swindell *et al.*, 1991), (R)- or (S)-phenyllactic acid (349–350) (Guéritte-Voegelein *et al.*, 1991, Swindell *et al.*, 1991), and N-benzoyl (R)- or (S)-isoserine (351, 352) (Swindell *et al.*, 1991) were found to possess significantly reduced activity in either or both the microtubule binding or cytotoxicity assays (Figure 10.19). A comparison of the microtubule disassembly activity of

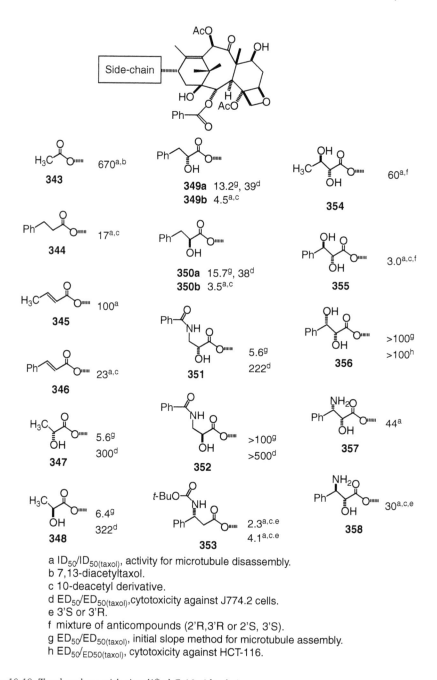

a ID$_{50}$/ID$_{50(taxol)}$, activity for microtubule disassembly.
b 7,13-diacetyltaxol.
c 10-deacetyl derivative.
d ED$_{50}$/ED$_{50(taxol)}$, cytotoxicity against J774.2 cells.
e 3'S or 3'R.
f mixture of anticompounds (2'R,3'R or 2'S, 3'S).
g ED$_{50}$/ED$_{50(taxol)}$, initial slope method for microtubule assembly.
h ED$_{50}$/ED50(taxol), cytotoxicity against HCT-116.

Figure 10.19 Taxol analogs with simplified C-13 side-chains.

3-phenylpropionic acid analog 344 (ID$_{50}$/ID$_{50(taxol)}$ = 17) with that of phenyllactic acid derivatives 349a and 350b (ID$_{50}$/ID$_{50(taxol)}$ = 4.5 and 3.5, respectively) demonstrates that the 2′-hydroxyl group is necessary for good bioactivity (Georg *et al.*, 1995d). Interestingly, deletion of the 2′-hydroxyl group of taxotere yielded an analog (353) that was only two to four times less active than taxol (Guéritte-Voegelein *et al.*, 1991). Replacing the 3′-phenyl group of the cinnamic acid moiety in 346 by a methyl group in the crotonic acid derivative 345 and replacing the 3′-phenyl group of the 2′, 3′-dihydroxy-3′-phenylpropionate derivative 355 with a methyl group in 354 considerably reduced the biological activity (Guéritte-Voegelein *et al.*, 1991). Deletion of the phenyl group altogether as in taxol analogs with C-13 N-benzoyl-isoserine side-chains (351, 352) produced a very significant drop in cytotoxicity (Swindell *et al.*, 1991). Thus, the 3′-phenyl group or a closely related substituent is likely to be important for strong microtubule binding and cytotoxicity. Also of note is the taxol derivative carrying a 2,3-dihydroxy-3-phenylpropionate side-chain (355 and 356); compound 355 possessed considerable activity in the microtubule assay (ID$_{50}$/ID$_{50(taxol)}$ = 3) but 356 was devoid of any activity in both the tubulin polymerization assay and cytotoxicity (Guéritte-Voegelein *et al.*, 1991; Chen *et al.*, 1997). The *syn* and *anti* 3′-amino-2′-hydroxy-3′-phenylpropionates (357) and (358) were considerably less active in the microtubule assay (ID$_{50}$/ID$_{50(taxol)}$ = 44 and 30, respectively) (Guéritte-Voegelein *et al.*, 1991).

Stereochemistry of C-13 side-chain analogs

The configuration at C-2′ and C-3′ has a moderate influence on the activity of the taxol analogs (Guéritte-Voegelein *et al.*, 1991). Taxol (1), taxotere (2), 10-deacetyltaxol (10), and 10-acetyl-taxotere (88), which all have (2′R, 3′S) side-chain configuration (taxol stereochemistry), were more active than their corresponding (2′S, 3′R) isomers 359 and 360 in the microtubule disassembly assay. The (2′R, 3′R) and (2′S, 3′S) isomers 361–363 were only slightly less potent than taxol (Figure 10.20). This indicates that the taxol stereochemistry (2′R, 3′S) is favorable to the bioactivity (Georg *et al.*, 1995d). Several side-chain conformationally restricted alkylidene taxoids (364–368) were synthesized by insertion of a carbon linker between the 2′-carbon and the ortho-position of 3′-phenyl ring. Compounds 364 showed comparable activity to taxol in tubulin binding and cytotoxicity assays. Epimerization of 364 at C2′, C3′ afforded inactive compound 368 in both assays. This result again indicates the configuration at C2′, C3′ is important for the activity (Table 10.17) (Barboni *et al.*, 2001b).

C-13 Side-chain homologues

A one-carbon homologation of the C-13 side-chain was detrimental to biological activity. The homologues (369, 370) (Figure 10.20) of 10-deacetyltaxol and taxotere, respectively, showed greatly reduced abilities to induce microtubule formation (ED$_{50}$/ED$_{50(taxol)}$>27) (Jayasinghe *et al.*, 1994; Georg *et al.*, 1995d). By comparing the activities of compounds 364 and 366 with 365 and 367, respectively, it can be concluded that the activity is dependent on the length of the tether (Table 10.17) (Barboni *et al.*, 2001b).

Taxol side-chain analogs with 2′, 3′ transposed substituents

Several α-amino-β-hydroxy acid derivatives were synthesized (Guéritte-Voegelein *et al.*, 1991). This substituent arrangement constitutes a formal interchange of 2′ and 3′ substituents of the taxol side-chain. All of the derivatives showed reduced activity in the microtubule disassembly assay (Figure 10.21). Derivatives possessing the natural taxol (2′R, 3′S) stereochemistry (371, 373, 374) were about 10 times less active than taxol and those with (2′S, 3′R) stereochemistry

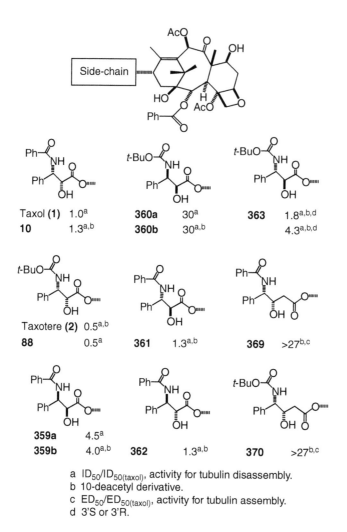

a $ID_{50}/ID_{50(taxol)}$, activity for tubulin disassembly.
b 10-deacetyl derivative.
c $ED_{50}/ED_{50(taxol)}$, activity for tubulin assembly.
d 3'S or 3'R.

Figure 10.20 Stereochemistry of C-13 side-chain taxol analogs and C-13 side-chain homologues.

(372 and 375) were more than 100 times less active (Guéritte-Voegelein *et al.*, 1991; Georg *et al.*, 1995d). These results also indicate the importance of the stereochemistry in the C-13 side-chain.

Modifications at the 3'-phenyl group

Replacement of the 3'-phenyl group by substituted aryl or heteroaryl moieties

Because previous studies on simplified side-chain analogs had underscored the importance of a 3'-phenyl group rather than a 3'-methyl group for good activity, many groups synthesized taxol and docetaxel derivatives with variously substituted 3'-phenyl groups (Georg and Cheruvallath, 1992a,b; Georg *et al.*, 1992c, 1995d, 1996; Boge *et al.*, 1994; Bourzat *et al.*, 1995; Ojima *et al.*, 1996b,c, 1998, 2000; Appendino *et al.*, 1997). With the exception of the *p*-fluoro derivative

Table 10.17 Structure and biological activity of compounds 364–368

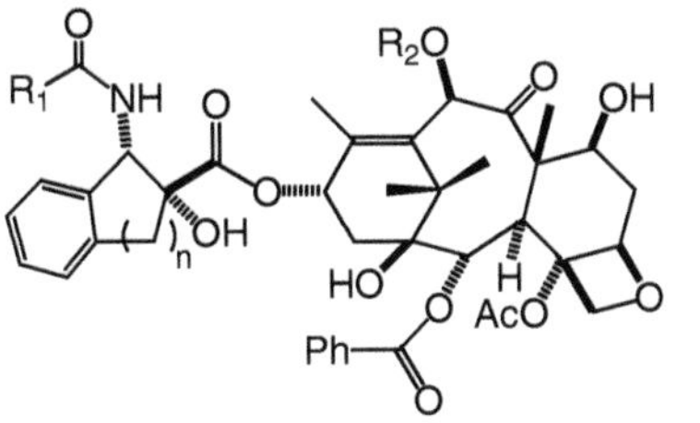

364–367 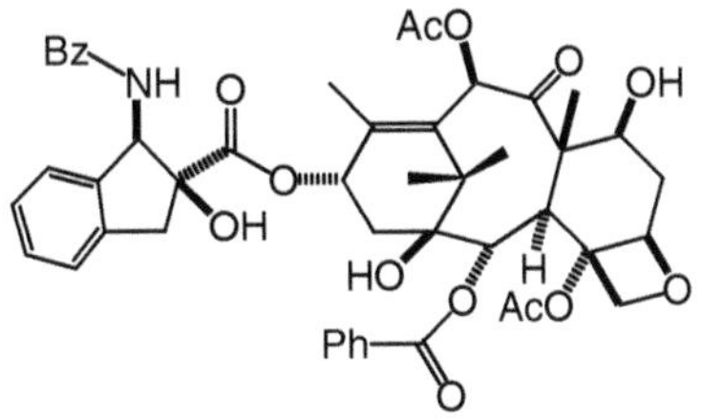   **368**

Compound	R_1	R_2	n	Tubulin binding ($ED_{50}/ED_{50(taxol)}$)	MCF7 ($ED_{50}/ED_{50(taxol)}$)
364	Bz	Ac	1	4	9
365	Bz	Ac	2	42	333
366	Boc	H	1	7	2
367	Boc	H	2	30	239
368				Inactive	1604

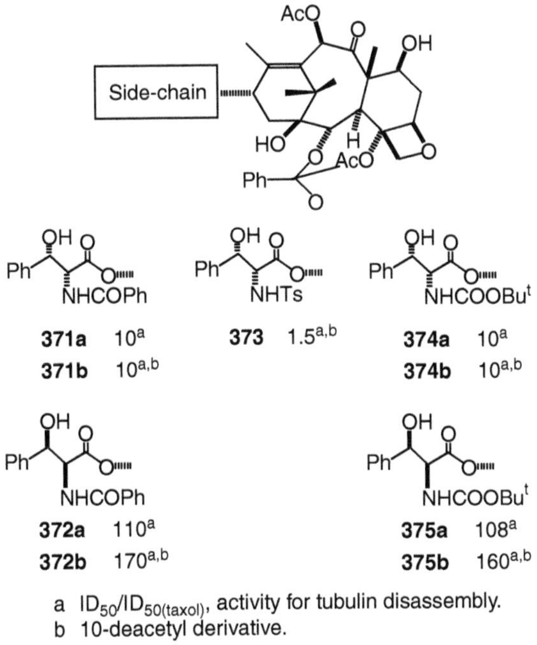

371a 10[a]	**373** 1.5[a,b]	**374a** 10[a]
371b 10[a,b]		**374b** 10[a,b]

372a 110[a]	**375a** 108[a]
372b 170[a,b]	**375b** 160[a,b]

a $ID_{50}/ID_{50(taxol)}$, activity for tubulin disassembly.
b 10-deacetyl derivative.

Figure 10.21 Taxol analogs with 2', 3' transposed substituents.

(376), most of the taxol derivatives (377–382) showed slightly decreased activity in the tubulin assembly assay and were less cytotoxic against several cancer cell lines (Table 10.18) (Georg and Cheruvallath, 1992a,b; Georg *et al.*, 1992c; Ojima *et al.*, 1996b,c). In, contrast to the taxol analogs, most of the docetaxel derivatives (389–398) showed equivalent or greater potency in

Table 10.18 Structures and biological activity of C-3′ modification of taxol

Compound	R	Tubulin assembly	B16	A121	A549	HT-29	MCF7	MCF7R
Paclitaxel	Ph	1	1	1	1	1	1	1
Docetaxel				0.32	0.19	0.28	0.33	0.59
376	F—⟨⟩—	1.1	1.2	1	1.17	4.03	3	>3.33
377	Cl—⟨⟩—	1.9	2.2					
378	Cl (meta)	4.4	6.7					
379	H₃C—⟨⟩—	2.4	3.0					
380	HO—⟨⟩—	0.8ᵃ				10ᶜ		
381	Me₂N (meta)	4.6	5.8					
382	naphthyl	7.1	6.7					
383	2-pyridyl	0.7	0.8					
384	3-pyridyl	0.5	27					
385	4-pyridyl	0.4	1.3					
386	2-furyl	0.9	0.3	1.0ᵇ		0.2ᵈ		
387	3-furyl	0.9	3.3					
388	2-thienyl			1.5ᵇ		0.2ᵈ		
406	(CH₃)₂CH=CH			0.23	0.13	0.30	0.32	0.38
407	*t*-butyl	1.8	7.5ᵉ					

Notes
a $ID_{50}/ID_{50(taxol)}$, microtubule disassembly assay.
b $ED_{50}/ED_{50(taxol)}$, cytotoxicity against HCT116.
c $ED_{50}/ED_{50(taxol)}$, cytotoxicity against L1210 leukemia cells.
d $ED_{50}/ED_{50(taxol)}$, cytotoxicity against HCTVM46 human colon carcinoma cells.
e $ED_{50}/ED_{50(taxol)}$, cytotoxicity against B16.

Table 10.19 Structure and biological activity of C-3′ modification of taxotere

Compound	R_1	R_2	Tubulin assembly $(ED_{50}/ED_{50(taxol)})$	P388 (µg/ml)
Paclitaxel			1	
Docetaxel	Ph	H	0.64	0.04
389	F-⟨⟩-	H		0.28[a] (0.14[b])
390	F-⟨⟩-	Ac		0.24[a] (0.13[b])
391	H₃C-⟨⟩-	H	1.3	0.04
392	⟨⟩-⟨⟩-	H	2.0	0.65
393	H₂C=HC-⟨⟩-	H	0.7	0.06
394	H₃COC-⟨⟩-	H	0.95	0.19
395	⟨⟩-CO-⟨⟩-	H	20	>1
396	HC≡C-⟨⟩-	H	0.6	0.085
397	CHO-⟨⟩-	H	0.75	0.13
398	HOH₂C-⟨⟩-	H	0.4	0.06
399	2-furyl	Ac	0.20	0.41[c]

Notes
a $IC_{50}/IC_{50(taxol)}$, cytotoxicity against MCF7.
b $IC_{50}/IC_{50(taxol)}$, cytotoxicity against A549.
c $ED_{50}/ED_{50(taxol)}$, cytotoxicity against B16 melanoma.

both assays; compound **395** was the sole exception (Table 10.19) (Bourzat *et al.*, 1995; Georg *et al.*, 1995; Ojima *et al.*, 1996b,c). In particular, 3′-Dephenyl-3′-(4-fluorophenyl)docetaxel (**389**) and 3′-dephenyl-3′-(4-fluorophenyl)-10-acetyldocetaxel (**390**) were much more active than paclitaxel. Molecular modeling and an NMR conformational study indicates that the fluorine atom contributes significantly to the formation of a "hydrophobic cluster" (Ojima *et al.*, 1996b,c).

Table 10.20 Structure and activity of 4-aziridine analogs

Compound	R_1	R_2	Tubulin polymerization $(ED_{50}/ED_{50(taxol)})$	HCT-116 $(IC_{50}/IC_{50(taxol)})$
1 Paclitaxel			1.0	1
271 4-aziridine	Ph	Bz	2.8	6.5
400 3'-furyl, 4-aziridine	furyl	Bz	0.7	2.8
401 3'-furyl,3'-*N*-Boc, 4-aziridine	furyl	Boc	0.8	0.8

All the 3'-heteroaromatic taxol analogs (383–388) possessed similar or better activity than taxol in both tubulin polymerization and cytotoxicity assays. 3'-Dephenyl-3'-(2-furyl)paclitaxel (386) was the most potent analog in the cytotoxicity assays although the 4-pyridyl derivative (385) was most potent in the microtubule assay (Table 10.18) (Georg *et al.*, 1996). 3'-Dephenyl-3'-(2-furyl)-10-acetyldocetaxel (399) was more potent than taxol in both the microtubule and cytotoxicity assays (Table 10.19) (Ali *et al.*, 1995).

Interestingly, in a study of C-4 aziridine analogs (see Table 10.11), 4-aziridine (271) showed decreased activity in both a tubulin polymerization assay and an *in vitro* cytotoxicity assay (HCT-116). Its 3'-furyl, 4-aziridine (400) and 3'-furyl, 3'-*N*-Boc, 4-aziridine (401) derivatives showed similar activity compared to paclitaxel (Table 10.20). Thus, both C-3'-furyl and C-3'-*N*-Boc can enhance the *in vitro* potencies of paclitaxel analogs (Georg *et al.*, 1994b; Chen *et al.*, 1995a).

Replacement of the 3'-phenyl group by aliphatic moieties

Replacing the 3'-phenyl with a cyclohexyl group in the side chain of taxol (402) and taxotere (405) led to strong activity in the microtubule assay. Similar substitution of the phenyl of the C-3' and C-2 benzoyl group also gave an extremely active compound (403). Compared with taxol, the C-3' taxotere analog (405) remained more active while the C-3', C-2 dicyclohexyl derivative (404) was two-fold less potent (Table 10.21) (Boge *et al.*, 1994; Chen *et al.*, 1995a; Georg *et al.*, 1995d). Compound 406 was the only reported 3'-alkenyl paclitaxel analog. It showed excellent cytotoxicity against the cancer cell lines tested while compound 407 showed decreased activity (Table 10.18) (Ojima *et al.*, 1994c).

Taxotere analogs (408–409) bearing a difluoromethyl group at C-3' had better activity than paclitaxel and docetaxel against LCC6-WT and its drug resistant cancer cell line LCC6-MDR (Table 10.22) (Ojima *et al.*, 2000). Compound 410 bearing trifluoromethyl at C-3' exhibited excellent cytotoxicity against all cell lines tested, while 411 and 412 with trifluoropropyl at C-3' were less active (Ojima *et al.*, 1996c). Compound 413 was synthesized from 10-deacetyl

Table 10.21 Structure and activity of compounds 402–405

Compound	R_1	R_2	R_3	$ED_{50}/ED_{50(taxol)}$
402	Ph	Ac	Ph	0.2^a
403	C_6H_{11}	Ac	C_6H_{11}	0.4^a
404	OBu^t	H	C_6H_{11}	2.0^b
405	OBu^t	H	Ph	0.7^b

Notes
a Microtubule assembly assay.
b Microtubule disassembly assay.

baccatin III and showed better cytotoxicity against several human tumor cell lines, in particular, it had *ca.*20 times stronger activity against human colon cancer cell lines (WiDr and Colon302) (Yamaguchi *et al.*, 1999). Compounds 414 and 415 (Ali *et al.*, 1995) were very active in both tubulin and cytotoxicity assays. Compounds 416 contained cyclopropyl rings in the C-3′ substituents and exhibited extremely high activity against a drug-resistant human breast cancer cell line LCC6-MDR (Table 10.22) (Ojima *et al.*, 1998).

Numerous 10-modified 3′-alkyl and 3′-alkenyl-3′-dephenyldocetaxel analogs were synthesized by means of the β-lactam synthon method (417–445) (Ojima *et al.*, 1994c, 1996a). All compounds showed excellent cytotoxicity against all cell lines tested (Table 10.22). It is noteworthy that most of the compounds showed much better activity than paclitaxel and docetaxel against MCF7 drug-resistant cancer cells (MCF7-R). This fact opens a new avenue for the development of a second generation of taxoids that are more potent than paclitaxel and docetaxel against drug-resistant cancer cells.

From the results, it was also concluded that the C-10 modifications were tolerated for cytotoxicity against normal cancer cell lines while the MCF7-R activity was very sensitive to the structure of the C-10 modifier. The activity against MCF7-R is thought to be related to the ability of the taxoids to bind p-glycoprotein; thus the C-10 position is crucial for p-glycoprotein to recognize and bind taxoid antitumor agents. However, the activity of C-10-methoxy analogs 425 and 444 decreased by two orders of magnitude relative to taxol, implying the importance of a carbonyl functionality at C-10 (Ojima *et al.*, 1994c, 1996a).

Compounds 417, 419 and 421 were tested for *in vivo* antitumor activity against B16 melanoma in B6D2F$_1$ mice. Compounds 417 and 419 were as active as docetaxel, while 421 showed greater *in vivo* antitumor activity than docetaxel (Ojima *et al.*, 1994c, 1996a). The 3′-alkenyl group was quite favorable to activity and numerous analogs (419–439) were made. Synthesis and biological evaluation of 3′-alkenyl 1,14-carbonate docetaxel analogs (446–469, Table 10.23) also indicated that 3′-alkenyl is favorable to the biological activity (Ojima *et al.*, 1996a, 1997a).

Table 10.22 Structure and biological activity of C-3′ modification of taxotere with saturated substituents ($IC_{50}/IC_{50(taxol)}$)

Compound	R_1	R_2	Tubulin assembly	LCC6	A121	A549	HT-29	MCF7	MCF7R
408	CF_2H	H	0.31						
409	CF_2H	Ac	0.16						
410	CF_3	Ac			0.06	0.07	0.11	0.15	0.06
411	$CF_3CH_2CH_2$	H			12.4	10	12.2	21.2	>3.33
412	$CF_3CH_2CH_2$	Ac			1.59	3.89	1.97	5.47	0.73
413	cyclopropyl	Ac			0.04[e]	0.06[f]			
414	i-But	Ac		0.52[b]					0.03
415	t-BuO	Ac	0.38	0.40[c]					
416		Ac		0.008[d]					
417	$(CH_3)_2CH-CH_2$	H	0.78[a]						
418	$(CH_3)_2CH-CH_2$	Ac			0.60	0.27	0.89	2.35	0.12
419	$(CH_3)_2C=CH$	H	0.64[a]						
420	$(CH_3)_2C=CH$	Ac			0.07	0.07	0.17	0.32	0.04
421	$(CH_3)_2C=CH$	CH_3CH_2-CO			0.02	0.08	0.09	0.10	0.007
422	$(CH_3)_2C=CH$	Cyclopropane$-CO$			0.04	0.12	0.1	0.12	0.007
423	$(CH_3)_2C=CH$	$(CH_3)_2N-CO$			0.05	0.16	0.14	0.07	0.01
424	$(CH_3)_2C=CH$	CH_3O-CO			0.04	0.09	0.08	0.08	0.01
425	$(CH_3)_2C=CH$	CH_3			0.11	0.25	0.25	0.22	0.41
426	$(CH_3)_2C=CH$	$CH_3(CH_2)_3-CO$			0.09	0.17	0.17	0.29	0.1
427	$(CH_3)_2C=CH$	$CH_3(CH_2)_4-CO$			0.08	0.16	0.25	0.23	0.06
428	$(CH_3)_2C=CH$	$(CH_3)_2CHCH_2-CO$			0.09	0.11	0.17	0.23	0.03
429	$(CH_3)_2C=CH$	$(CH_3)_3CCH_2-CO$			0.05	0.14	0.14	0.23	0.03
430	$(CH_3)_2C=CH$	Cyclohexane$-CO$			0.08	0.21	0.29	0.27	0.07
431	$(CH_3)_2C=CH$	(E)$-(CH_3)_2C=CH-CO$			0.07	0.47	0.17	0.15	0.01
432	$(CH_3)_2C=CH$	$(CH_3CH_2)_2N-CO$			0.03	0.05	0.11	0.12	0.03
433	$(CH_3)_2C=CH$	Morpholine-4$-CO$			0.01	0.06	0.05	0.05	0.04
434	$(CH_3)_2C=CH$	CH_3NH-CO			0.19	0.19	0.33	0.23	0.07
435	$(CH_3)_2C=CH$	CH_3CH_2NH-CO			0.05	0.04	0.11	0.18	0.54
436	$(CH_3)_2C=CH$	$CH_3(CH_2)_2NH-CO$			0.05	0.06	0.15	0.21	0.29
437	$(CH_3)_2C=CH$	$(CH_3)_2CHNH-CO$			0.05	0.1	0.14	0.27	0.22
438	$(CH_3)_2C=CH$	$CH_2=CHCH_2NH-CO$			0.07	0.1	0.15	0.26	0.25
439	$(CH_3)_2C=CH$	Cyclohexyl$-NH-CO$			0.04	0.05	0.13	0.19	0.16
440	$(CH_3)_2CH-CH_2$	CH_3CH_2-CO			0.06	0.15	0.15	0.20	0.009
441	$(CH_3)_2CH-CH_2$	Cyclopropane$-CO$			0.08	0.30	0.22	0.3	0.01
442	$(CH_3)_2CH-CH_2$	$(CH_3)_2N-CO$			0.06	0.14	0.17	0.21	0.02
443	$(CH_3)_2CH-CH_2$	CH_3O-CO			0.03	0.1	0.12	0.16	0.02
444	$(CH_3)_2CH-CH_2$	CH_3			0.11	0.16	0.17	0.19	0.71
445	$(CH_3)_2CH-CH_2$	Cyclohexyl$-NH-CO$			0.79	1.03	1.39	0.94	0.11

Notes

a $ID_{50}/ID_{50(taxol)}$, microtubule disassembly assay.

b $IC_{50}/IC_{50(taxol)}$, against MDA-MB321 cells.

c $ED_{50}/ED_{50(taxol)}$, against B16.

d $IC_{50}/IC_{50(taxol)}$, against LCC6-MDR cells.

e $IC_{50}/IC_{50(taxotere)}$, against Colon320.

f $IC_{50}/IC_{50(taxotere)}$, against WiDr.

Table 10.23 Structure and biological activity of 1,14-carbonate taxotere analogs

Compound	R₁	R₂	LCC6	A121	A549	HT-29	MCF-7	MCF7-R
Paclitaxel			1	1	1	1	1	1
Docetaxel			0.32	0.20	0.28	0.38	0.58	0.78
446	CF₂H	H	1.03					
447	CF₂H	Ac	0.61					
448	CF₂H	Me₂NCO	0.90					
449	CF₂H	EtCO	1.03					
450	CF₂H	Me₂CCH₂CO	1.03					
451	CF₂H	c-PrCO	1.80					
452	2-furyl	H		0.06	0.14	0.19	0.29	0.45
453	2-furyl	Ac		0.06	0.14	0.19	0.29	0.16
454	Me₂C=CH	H		0.28	0.06	0.19	0.29	0.24
455	Me₂C=CH	Ac		0.24	0.39	0.75	1.94	0.12
456	Me₂C=CH	EtCO		0.11	0.14	0.18	0.12	0.09
457	Me₂C=CH	c-PrCO		0.08	0.14	0.31	0.23	0.09
458	Me₂C=CH	Me₂NCO		0.11	0.17	0.37	0.23	0.11
459	Me₂C=CH	(E)-MeCH=CHCO		0.13	0.14	0.47	0.23	0.05
460	Me₂C=CH	CH₃OCO		0.09	0.14	0.22	0.18	0.13
461	(E)-MeCH=CH	H		0.43	0.33	0.59	0.94	2.55
462	(E)-MeCH=CH	Ac		0.39	0.11	0.94	0.94	0.24
463	Me₃CCH₂	H		17.4	14.2	16.2	28.2	>3.34
464	Me₂CHCH₂	H		0.41	0.39	1.31	1.76	0.63
465	Me₂CHCH₂	Ac		0.20	0.19	0.47	0.65	0.12
466	Me₂CHCH₂	c-PrCO		0.18	0.33	1.03	0.41	0.07
467	Me₂CHCH₂	Me₂NCO		0.10	0.17	0.41	0.35	0.07
468	CH₃CH₂CH₂	H		0.34	1.22	1.19	2.47	1.29
469	CH₃CH₂CH₂	Ac		0.38	1.30	1.5	2.70	0.67

From a previous study, 9(R)-dihydrotaxol showed promising antitumor activity; accordingly, Li *et al.* (1994) synthesized a series of C-3′-modified anlogs of 9(R)-dihydrotaxol (470–483) (Table 10.24). Among these derivatives, compound 482 with a 3′-isobutyl group was the most active. These results also confirmed that the 3′-phenyl group was not essential for activity and could be replaced by heteroaromatic rings and alkyl and alkenyl groups.

Modification at the 3′-nitrogen

Modifications at the N-benzoyl group of the C-13 taxol side-chain, including the introduction of polar, ionizable groups, are generally well tolerated.

Table 10.24 Structure and biological activity of 9(*R*)-dihydrotaxoid analogs

Compound	R_1	R_2	Tubulin ($ED_{50}/ED_{50(taxol)}$)	A549 IC_{50} (ng/ml)	HT-29 IC_{50} (ng/ml)	B16F10 IC_{50} (ng/ml)	P388 IC_{50} (ng/ml)
207 9(*R*)-dihydrotaxol	Ph	Ph	0.86	19	80	25	53
10-deacetyl-9(*R*)-dihydrotaxol	Ph	Ph	0.87	3	0.16	0.4	2.5
470	(epoxide structure)	Boc	7.91	>100	>100	86	>100
471	(diol structure)	Boc	2.71	>100	>100	>100	>100
472	MeOCH$_2$	Bz	3.14	>100	79	>100	>100
473	PhOCH$_2$	Bz	5.81	92	39	84	>100
474	Benzyl	Boc	>17.1	>100	>100	>100	>100
475	4-thiazolyl	Boc	1.36	6.3	0.92	0.3	1.4
476	Methyl	Boc	1.08	15	8.3	19	43
477	Vinyl	Boc	0.92	1.1	1.8	4.4	16
478	Ethyl	Boc	0.61	0.83	1.4	3.2	11
479	Butyl	Boc	2.15	9.5	8.8	9.7	18
480	(3'*R*)-butyl	Boc	9.74	>100	67	>100	>100
481	Pentyl	Boc	2.00	12	16	14	16
482	Isobutyl	Boc	0.95	0.12	0.38	0.9	0.36
483	Cyclohexyl	Boc	0.57	6.5	5	5.7	14

Aliphatic N-acyl analogs

Taxotere (2) is a notable aliphatic *N*-acyl (*t*-butoxy) analog of taxol. It was more active than taxol in the microtubule disassembly assay (Guéritte-Voegelein *et al.*, 1991), and also more active as an inhibitor of cell replication. It was more than two-fold as active against J774.2 and P388 cells and about five-fold more active in taxol-resistant cells as compared to taxol (Ringel *et al.*, 1991). In addition, both *N*-(*t*-butoxycarbonyl)(10-acetyltaxotere) (88) and *N*-(*n*-butoxycarbonyl) (488) taxol derivatives had slightly better activity than taxol in microtubule assembly and cytotoxicity assays (Table 10.25). Although derivatives with lipophilic substituents (484, 485, 88, 488, 491) were the most active compounds, the glutaryl derivative (494) with an acidic substituent at the nitrogen was also remarkably potent (Guéritte-Voegelein *et al.*, 1991; Georg *et al.*, 1995d; Roh *et al.*, 1999).

Previous study has demonstrated that replacement of 3'-phenyl by 2-furyl or *t*-butyl can enhance *in vitro* activity. Several groups synthesized 3'-*N*-phenyl derivatives of 3'-dephenyl-3'-(2-furyl)paclitaxel (399, 495–496) and 3'-dephenyl-3'-(*t*-butyl)paclitaxel (407, 415, 497–506)

Table 10.25 Structure and biological activity of 3'N-phenyl replaced by aliphatic groups

Compound	R	Tubulin assembly	A549	SK-OV3	MCF-7	B16
Paclitaxel	Ph	1	1	1	1	1
484	cyclopentyl		0.4	1.67		
485	cyclopentenyl		<0.05	0.13		
486	cyclohexyl		4.44	5.68		
487	cyclohexenyl		0.56	0.83		
88. 10-acetyltaxotere	t-BuO	0.55				0.23
488	n-BuO	0.8				0.6
489	$(CH_3)_2CHCH_2$	1.18				2.65
490	$(CH_3)_3CCH_2$	0.74				3.13
491	$(CH_3)_2C=CH_2$		0.08		0.35	
492	t-Bu	2.61				22.4
493	$PhOCH_2$	1.86				
494[a]	$(CH_2)_3COOH$	1.0[b]				

Notes
a 10-deacetyl derivative.
b Tubulin disassembly.

(Table 10.26) (Ali *et al.*, 1995; Georg *et al.*, 1996). Most derivatives showed similar or better activity than paclitaxel in both microtubule and cytotoxicity B16 melanoma assays. Compounds 399 and 415 [N-(t-butyoxy)carbonyl], 500 [N-(t-amyloxy)carbonyl], 503 (N-3'-phenylurea) and 506 (N-3'-t-butylurea) showed much better activity in both assays.

Aromatic N-acyl analogs

Except for heteroaromatic analogs 529–532, none of the aromatic N-acyl analogs was more active than taxol in microtubule assays (Table 10.27). The 4-chlorobenzoyl (510), 3-chlorobenzoyl (511), 4-toluoyl (507), and 3-dimethylaminobenzoyl (518) analogs were relatively potent in the microtubule assembly assay and against B16 melanoma cell growth (Georg *et al.*, 1992a,b, 1994d). The 3,4-dichlorobenzoyl (524), 2-toluoyl (508), 4-fluorobenzoyl (512), and 4-nitrobenzoyl (517) derivatives were slightly less effective than taxol in the microtubule assembly assays and considerably less cytotoxic than taxol against B16 melanoma cells (Georg *et al.*, 1994d).

Table 10.26 Structure and activity of compounds 495–506

495–496   **497–506**

Compound	R	Tubulin assembly $ED_{50}/ED_{50(taxol)}$	B16 melanoma $ED_{50}/ED_{50(taxol)}$
Paclitaxel		1	1
Docetaxel		0.45	0.41
399	*t*-BuO	0.20	0.41
495	$CH_3(CH_2)_4$	1.6	0.50
496	$CH_3(CH_2)_8$	2.1	2.6
407		1.8	7.5
415		0.38	0.40
497	isobutoxy	0.54	1.5
498	*n*-butoxy	0.86	2.1
499	*n*-hexyloxy	2.3	12
500	*tert*-amyloxy	0.45	0.58
501	*tert*-butyl	1.3	32
502	*tert*-butyl-CH_2	0.48	3.6
503	*tert*-butyl-NH	0.60	0.90
504	*tert*-butyl-NHCS	3.6	>41
505	PhO	1.4	>32
506	PhNH	0.32	0.70

However compounds 507, 510, 512, 514–516 and 523 were as cytotoxic as taxol against J774.1 cells. Even though four heteroaromatic analogs [3-pyridyl (529), 4-pyridyl (530), 2-furyl (531), 3-furyl (532)] exhibited strong microtubule assembly properties, only 531 and 532 were active in the B16 melanoma assay (Table 10.27) (Ojima *et al.*, 1997).

Isosteric replacement of oxygen by sulfur

Recently, Xue *et al.* (2000) reported the synthesis of sulfur containing paclitaxel side-chain analogs (534–544) (Table 10.28). These compounds were tested first in a tubulin polymerization assay and in an *in vitro* cytotoxicity assay against human colon cancer cell lines HCT116. Those with *in vitro* activity were then tested in an *in vivo* screen against murine M109 lung carcinoma (Xue *et al.*, 2000). All of the 3'-N-thiocarbamate analogs (534–539) were 2–10 times more potent than paclitaxel in the tubulin polymerization assay and, except for 539, were much more cytotoxic than paclitaxel. However, 536–538 were less active *in vivo* than paclitaxel. Of the 3'-N-thiourea analogs, only the two 3'-furyl analogs 543 and 544 were as active as paclitaxel in the tubulin polymerization assay. All N-thiourea analogs were less

Table 10.27 Structure and biological activity of 3'N-phenyl replaced by aromatic groups

Compound	R	Tubulin assembly $(ED_{50}/ED_{50(taxol)})$	J774.1 $(IC_{50}/IC_{50(taxol)})$	B16
Paclitaxel	phenyl	1	1	
507	H_3C–C6H4–	1.6	0.77	1.4
508	2-CH_3-C6H4–	1.9		7.5
509	F_3C–C6H4–	6.0		17.7
510	Cl–C6H4–	2.4	1.36	1.5
511	3-Cl-C6H4–	2.0		1.8
512	F–C6H4–	1.2	0.95	4.3
513	H_3CO–C6H4–		2.68	
514	C_2H_5–C6H4–		0.68	
515	C_3H_7–C6H4–		0.81	
516	$(H_3C)_3C$–C6H4–		1.09	
517	O_2N–C6H4–	2.0		2.1
518	3-Me_2N-C6H4–	1.4		1.6
519	2-SO_3H-C6H4–	5.5[a,b]		
520	H_3C–C6H4–SO_2–	5.0[a,b]		
521	3,4-(H_3CO)$_2$-C6H3–		2.82	
522	3-OCH_3,4-H_3CO-C6H3–		3.45	
523	2,4-F$_2$-C6H3–		1.12	

Table 10.27 (Continued)

Compound	R	Tubulin assembly $(ED_{50}/ED_{50(taxol)})$	J774.1 $(IC_{50}/IC_{50(taxol)})$	B16
524		2.1		11
525			3.36	
526			29.54	
527		4.6		14.8
528	2-pyridyl	3.5		4.6
529	3-pyridyl	0.6		26
530	4-pyridyl	2.0		>36
531	2-furyl	0.8		1.0
532	3-furyl	0.4		0.5

Notes
a 10-deacetyl derivative.
b Tubulin disassembly.

cytotoxic than paclitaxel. Comparison of the structure and activity of urea (533) and thiourea (542) derivatives indicated that the carbonyl moiety attached to the 3'-N is very important to the bioactivity.

The 3'-carbamate taxotere analogs (545 and 546) were slightly less active than paclitaxel in both tubulin and cytotoxicity assays but compound 545 showed equivalent *in vivo* activity as taxol (Table 10.29) (Chen *et al.*, 1997).

Modifications at 2'-hydroxyl group

The 2'-hydroxyl group is important for optimal biological activity. The 2'-methoxy (547) (Kant *et al.*, 1993), 2'-*t*-butyldimethylsilyl ether (548) (Magri and Kingston, 1988) and 2'-deoxy (549) taxol analogs showed no *in vitro* cytotoxicity (Table 10.30). Acetylation of the 2'-hydroxyl group (2'-acetyltaxol) (550) (Table 10.30) leads to greatly decreased activity in the microtubule disassembly assay, but similar cytotoxicity as paclitaxel. This fact suggested that endogenous esterases readily cleave 2'-hydroxyl esters and release taxol *in vivo* (Kant *et al.*, 1993; Magri and Kingston, 1988; Lataste *et al.*, 1984). Therefore, the 2'-hydroxyl group may be a suitable site for the development of paclitaxel prodrugs. This topic will be discussed later in more detail. Replacing the 2'-hydroxyl group by fluorine (551) also reduced cytotoxicity significantly (Table 10.30) (Kant *et al.*, 1993).

Replacement of the C-13 ester linkage by amide linkage

None of the C-13 amidopaclitaxel analogs (552–555) (Table 10.31) showed any activity in the tubulin polymerization or the *in vitro* cytotoxicity assays (Chen *et al.*, 1996b). Three amide

Table 10.28 Structure and activity of sulfur containing taxoids

Compound	R1	R2	X	Tubulin assembly $(EC_{50}/EC_{50(taxol)})$	HCT116 $(IC_{50}/IC_{50(taxol)})$	In vivo ip M109 T/C (mg/kg/inj)
Paclitaxel	Ph	Ph	O	1.0	1.0	145–218 (60)
10-acetyltaxotere	Ph	t-BuO	O	0.66	0.60	177 (32)
533	Ph	t-BuNH	O	0.72	1.0	n.d.
534	Ph	n-BuS	O	0.6	0.88	n.d.
535	Ph	t-BuS	O	0.3	0.30	n.d.
536	2-Furyl	n-BuS	O	0.2	0.17	139 (40)
537	2-Furyl	t-BuS	O	0.1	0.30	134 (1.25)
538	2-Furyl	i-BuS	O	0.2	0.28	145 (50)
539	2-Furyl	i-PrS	O	0.2	14.7	n.d.
540	Ph	BnNH	S	12	>40	n.d.
541	Ph	n-BuNH	S	18	21.8	n.d.
542	Ph	t-BuNH	S	19	~50	n.d.
543	2-Furyl	n-BuNH	S	1.0	4.6	n.d.
544	2-Furyl	t-BuNH	S	1.0	4.9	n.d.

Table 10.29 Structure and activity of compounds 545–546

Compound	R	Tubulin polym. ratio	HCT-116	In vivo T/C(%) M-109 (mg/kg/inj)
Paclitaxel		1.0	1.0	166–228(60)
545	Ac	1.5	3.3	166(50)
546	H	7.8	5.1	

Table 10.30 Structures and activity of 2′-modified taxol analogs

Compound	R	Tubulin disassembly $(ID_{50}/ID_{50(taxol)})$	HCT-116 $(ED_{50}/ED_{50(taxol)})$
Taxol	OH		1
547	OMe		216
548	OSiMe$_2$Bu^t		3×10^{4a}
549	H		74
550	OAc	30	2^a
551	F		118

Note
a $ED_{50}/ED_{50(taxol)}$, cytotoxicity against KB cell.

Table 10.31 Structures and activity of C-13 amide analogs

Compound	R_1	R_2	X	Tubulin Polym.	HCT-116
Paclitaxel	Ph	Ac	O	1.0	1.0
244 4-deacetylpaclitaxel	Ph	H	O	200	>24.2
266	Ph	COOMe	O	0.41	0.83
552	Ph	H	NH	60	>40
553	Ph	COOMe	NH	>200	22.5
554	2′-Furyl	COOMe	NH	>200	34.2
555	Ph	Ac	NH	>200	>41.7

linked A-norpaclitaxel analogs (556–558) were significantly less active than paclitaxel in the P388 cytotoxicity assay and 558 was inactive in tubulin assembly assay. Compound 556 and 558 showed less cytotoxicity than their corresponding A-norpaclitaxel (559 and 560) (Table 10.32) (Chordia and Kingston, 1997).

Table 10.32 Structures and cytotoxicity of A-norpaclitaxel

Compound	Ar	X	Tubulin assembly ($ED_{50}/ED_{50(taxol)}$)	P-388 ($ED_{50}/ED_{50(taxol)}$)
556	C_6H_5	NH	NT	135
557	m-ClC_6H_4	NH	NT	19
558	m-$CH_3OC_6H_4$	NH	Inactive	5
559	C_6H_5	O	Less active	25
560	m-$CH_3OC_6H_4$	O	Less active	1.5

Note
NT: no test

Table 10.33 Structures and cytotoxicity of nor-seco taxoids (IC_{50} μM)

Compound	R_1	R_2	R_3	A121	A549	HT-29	MCF-7	MCF7-R
Paclitaxel				0.0063	0.0036	0.0036	0.0017	0.300
308				0.117	0.133	0.134	0.079	0.471
309				0.131	0.169	0.171	0.101	0.360
561	Ph	Ph	H	>1.0	>1.0	>1.0	>1.0	>1.0
562	Boc	Ph	H	>1.0	>1.0	>1.0	>1.0	>1.0
563	Boc	$Me_2C=CH$	H	>1.0	>1.0	>1.0	>1.0	>1.0
564	Boc	Me_2CHCH_2	H	>1.0	>1.0	>1.0	>1.0	>1.0
565	Ph	Ph	CH_3	0.535	0.708	0.292	0.200	>1.0

Nor-secotaxol (308) and Nor-secotaxotere (309) were 20–40 times weaker than taxol against several cancer cell lines. The Nor-seco taxiods with a N-H amide linkage (561–565) were virtually inactive; thus replacement of C-13 ester linkage with an N-H amide linkage is deleterious to cytotoxicity. But interestingly, methylation of the amide (565) recovers the cytotoxicity substantially (Table 10.33) (Ojima *et al.*, 1994b, 1998).

Summary

The C-13 side-chain and the natural (2'R, 3'S) stereochemistry are essential to the biological activity of taxoids (Lataste *et al.*, 1984; Guéritte-Voegelein *et al.*, 1991; Swindell *et al.*, 1991; Barboni *et al.*, 2001b). Aromatic substitution in the 3'-phenyl group or the *N*-benzoyl group led to derivatives with slightly reduced activity (Georg and Cheruvallath, 1992a,b; Ojima *et al.*, 1996b,c; Georg *et al.*, 1992a,b,c, 1994d). Replacing either of the two phenyl groups in the side-chain by heteroaromatic groups such as furoyls provided highly active analogs (Ali *et al.*, 1995; Chen *et al.*, 1995a; Georg *et al.*, 1994b). The most flexible site for side-chain modifications is the *N*-benzoyl group, which can be replaced by lipophilic and hydrophilic groups without any dramatic change in bioactivity. The 2'-hydroxyl group is essential for microtubule binding and cytotoxicity (Magri and Kingston, 1988; Kant *et al.*, 1993; Lataste *et al.*, 1984). The C-13 ester linkage is important for microtubule binding and cytotoxicity. C-13-amidopaclitaxel analogs dramatically reduce the activity (Chordia and Kingston, 1997; Ojima *et al.*, 1994b, 1998).

Bifunctional paclitaxel hybrid molecules

Biological macromolecules possess multiple ligand binding sites. Whiteside and others have shown that covalently tethering many ligands together to form a polyvalent array can greatly improve the binding of low affinity, small extracellular molecules to biological macromolecules (Kar *et al.*, 2000). In general, hybrid molecules are not as effective as combination therapy, because it is not possible to optimize the individual drug concentration and schedule. But there may be potential advantages if both drugs can be deactivated as a result of the linking function and then released together (Wittman *et al.*, 2001b). Recently, several groups investigated the biological activity of 7-deoxy-9-dihydropaclitaxel-AZT (567–570) (Figure 10.22) (Cheng *et al.*, 2000b), paclitaxel-chlorambucil (571–574) (Figure 10.22) (Wittman *et al.*, 2001b), paclitaxel-daunorubicin (575–580) (Figure 10.22) (Kar *et al.*, 2000) hybrid molecules. The compounds were evaluated for *in vitro* cytotoxicity against several human tumor cell lines and activity in the microtubule binding assay (Table 10.34). Most of the compounds were less active than taxol in both assays. The decreased activity was likely due to decreased target affinity because of the increased bulk of the dimers. Compounds 571–574 showed *in vivo* antitumor activity in the paclitaxel-sensitive M109 model, in which compound 573 was the most active, suggesting that the hybrid retains both paclitaxel- and chloroambucil-like activity (Wittman *et al.*, 2001b).

Taxol prodrugs

Although taxol is an interesting and potent anticancer drug, some disadvantages still limit its clinical use. As an injectable chemotherapeutic agent, the major problems of taxol are its poor solubility in water (0.25 µg/ml) and lack of selectivity (Georg *et al.*, 1995d). Prodrug strategy can overcome these impediments.

Prodrug strategies are temporary chemical modification of a compound to alter its aqueous solubility and biotransformation, while the inherent pharmacological properties of the parent drug remain intact (Greenwald *et al.*, 1996). *In vivo*, the prodrug can release the active drug by either an enzymatic mechanism or simple hydrolysis at physiological pH. Enzymatic conversion to the parent drug is the most common approach (Georg *et al.*, 1995d).

Several requirements are crucially important for the development of a prodrug. For an injectable prodrug, it is critical that the prodrug is converted rapidly to the parent compound after injection, before other secondary metabolic processes can take place. Another important aspect in prodrug design involves chemical stability. This aspect is particularly important for a highly insoluble drug such as taxol (Georg *et al.*, 1995d).

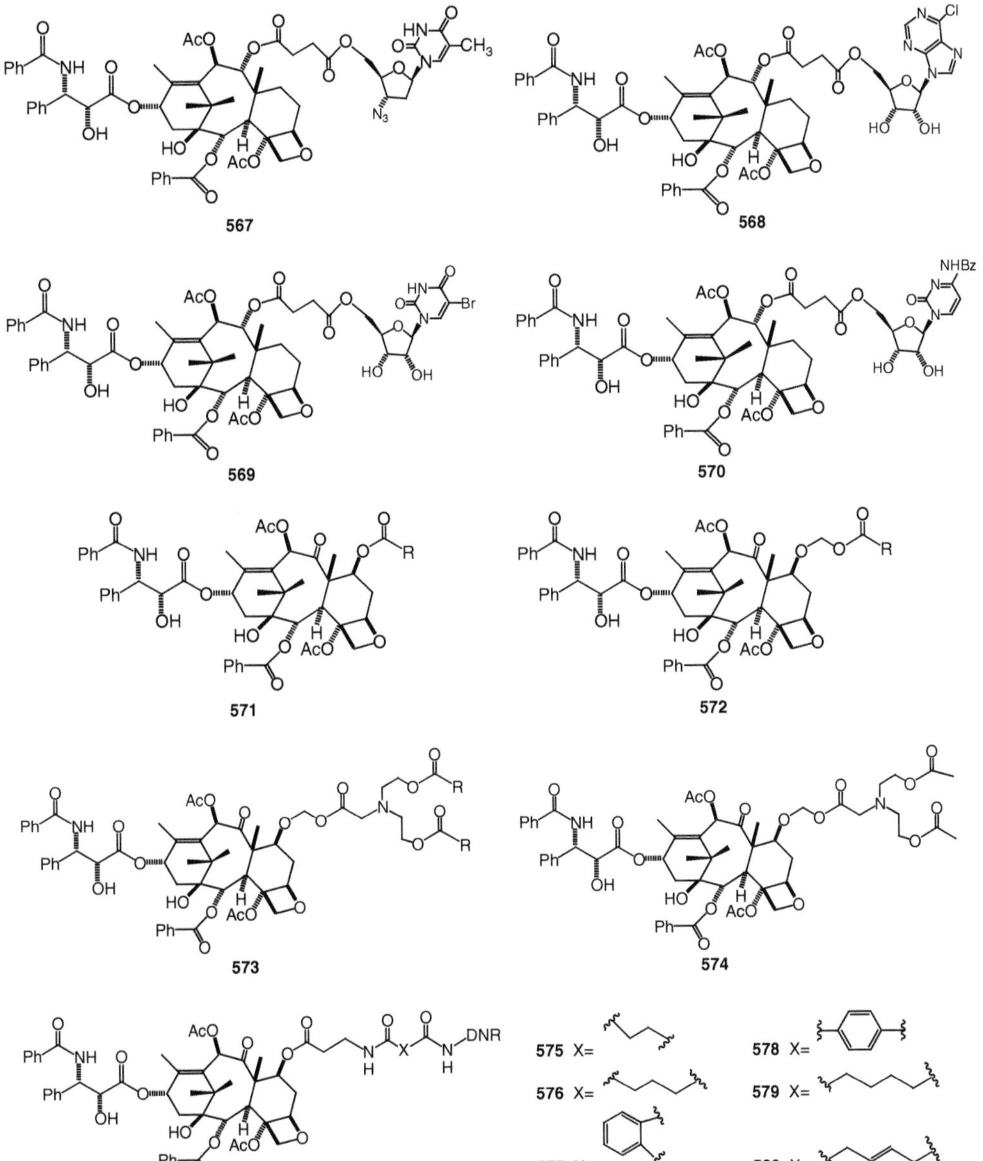

Figure 10.22 Structures of the hybrid molecules (567–580). (For 571–574: R = CH$_2$CH$_2$CH$_2$C$_6$H$_4$–
4–N(CH$_2$CH$_2$Cl)$_2$.

In order to synthesize a taxol prodrug, possible sites for modifications were investigated. The two most accessible sites for chemical modification are the 7-hydroxyl and the 2′-hydroxyl. As discussed earlier in this chapter, several naturally occurring, as well as semisynthetic taxol analogs with modified 7-hydroxyl groups, were highly cytotoxic. Taxol analogs esterified at the 2′-hydroxyl group lack *in vivo* tubulin binding but retain cytotoxicity, presumably due to cleavage of the ester group in the latter assay. Thus, ester formation at the 2′-hydroxyl, the 7-hydroxyl, and possibly at the 10-hydroxyl group holds promise for prodrug design.

Table 10.34 Activity of the hybrid molecules (567–580)

Compound	Cytotoxicity against tumor cells ($ED_{50}/ED_{50(taxol)}$)					Tubulin
	HCT-166	MESSA	MCF-7	BEL-7402	Eca-109	
Taxol	1	1	1.00	1.00	1.00	1
567			3.13	7.32	3.54	>50 [a]
568			3.68	0.24	0.18	>37 [a]
569			2.64	0.13	0.16	>37 [a]
570			1.63	1.45	1.03	2.1 [a]
571	80					53 [b]
572	2.3					62 [b]
573	12					70 [b]
574	2.0					6.5 [b]
575		1166				
576		1583				
577		165				
578		1666				
579		3833				
580		466				

Notes
a $ED_{50}/ED_{50(taxol)}$, microtubule assembly assay.
b Initial rate of polymerization.

Water-soluble prodrugs of taxol contain ionizable substituents such as amines, carboxylic acids, sulfonic acids, amino acids and phosphates (Georg *et al.*, 1995d).

Water-soluble prodrug of taxol

Succinate and glutarate derivatives

Early in 1988, Kingston's group synthesized 2'-succinyltaxol (581a), which showed reduced cytotoxicity against KB cells compared to taxol ($ED_{50}/ED_{50(taxol)} = 10^3$) (Magri and Kingston, 1988). Deutsch *et al.* (1989) prepared several esters of taxol containing acidic moieties (581a and 582a).

The sodium salts 581b and 582b had improved antitumor activity in the *in vivo* B16 melanoma system compared to the free acids (Magri and Kingston, 1988; Deutsch *et al.*, 1989). N-Methylglucammoniaium salt 581d and especially triethanolammonium salt 581c were more potent than the parent acid and sodium salt. Glutaric acid sodium salt 582b and its tri-ethanolammonium salt 582c were more active antitumor agents than the corresponding succinic acid derivatives. Salts 581c and 581d showed greatly improved aqueous solubility of up to *ca.*1% concentration (Georg *et al.*, 1995d) (Figure 10.23).

Ammonium salt 583b was the most potent derivative (Deutsch *et al.*, 1989). At a dose of 10 mg/kg of 583b, a T/C of 352 was observed in the *in vivo* B16 melanoma system. It showed excellent antitumor activity in the MX-1 breast xenograft assay. At doses of 20 and 40 mg/kg, all live animals were tumor free. In the context of the their synthetic work, Deutsch *et al.* also observed the formation of oxazolone 584 (Figure 10.23), which had also been isolated by Magri and Kingston (1988). Compound 584 showed low cytotoxicity against KB cells ($ED_{50} = 0.2 \mu g/ml$) and was inactive in the *in vivo* B16 assay. Thus, this cyclic carbamate was not labile enough to be useful as a prodrug (Georg *et al.*, 1995d).

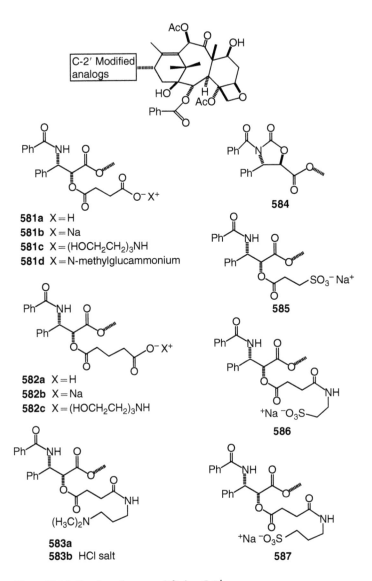

Figure 10.23 Taxol prodrugs modified at C-2'.

2'-Succinyl paclitaxel and salts thereof (581) had a slightly improved *in vivo* activity, but they were unstable in slightly basic aqueous solution. In order to solve this problem, Damen *et al.* (2000) synthesized more stable malic acid derivatives of paclitaxel (588–591) (Table 10.35) in which an intramolecular hydrogen bond could be formed between the additional hydroxyl group and either carbonyl. These prodrugs (588–591) (Table 10.35) showed improved water solubility (20–60 times) and were stable in PBS buffer at 37°C for 48 h. 2'-Malyl paclitaxel (588) and its sodium salt (589) can generate paclitaxel in plasma and these can be used as prodrugs, while 7-malyl paclitaxel (591) and 2',7-dimalyl paclitaxel (590) do not. Compounds 588 and 589 showed similar *in vitro* cytotoxicity to paclitaxel while 590 and 591 were significantly less

Table 10.35 Structures and properties of malic acid paclitaxel derivatives

Compound	R_1	R_2	Water solubility (mg/ml)	$T_{1/2}$ (h) PH7.4	$T_{1/2}$ (h) plasma
Paclitaxel	H	H	0.01		
588	(structure)	H	0.2	>24	20
589	(structure)	H	0.6	No pacl.[a]	4
590	(structure)	(structure)	0.5	No pacl.[a]	No pacl.[a]
591	H	(structure)	0.3	No pacl.[a]	No pacl.[a]

Note
a No paclitaxel detected (HPLC).

Table 10.36 Cytotoxicity of malic acid paclitaxel derivatives (IC_{50} ng/ml)

Compound	MCF-7	EVSA-T	WIDR	IGROV	M19	A498
Paclitaxel	<3	<3	<3	10	<3	<3
588	<3	<3	<3	<3	3	93
589	<3	<3	<3	233	6	13
590	69	59	167	49	311	436
591	390	300	589	241	1344	1485

cytotoxic (Table 10.36). Prodrug **589** was tested for *in vivo* activity against a murine P388 tumor model and was much more active than paclitaxel.

Sulfonic acid derivatives

Zhao and Kingston (1991) prepared three sulfonic acid salts (**585–587**) (Figure 10.23). The partition coefficients (octanol/water) for the three derivatives were 210, 198 and 118 times greater, respectively, than taxol. The sulfonate derivatives display reduced activity in comparison to taxol in the *in vitro* cytotoxicity against four different cell lines and *in vivo* against P388 (Georg *et al.*, 1995d).

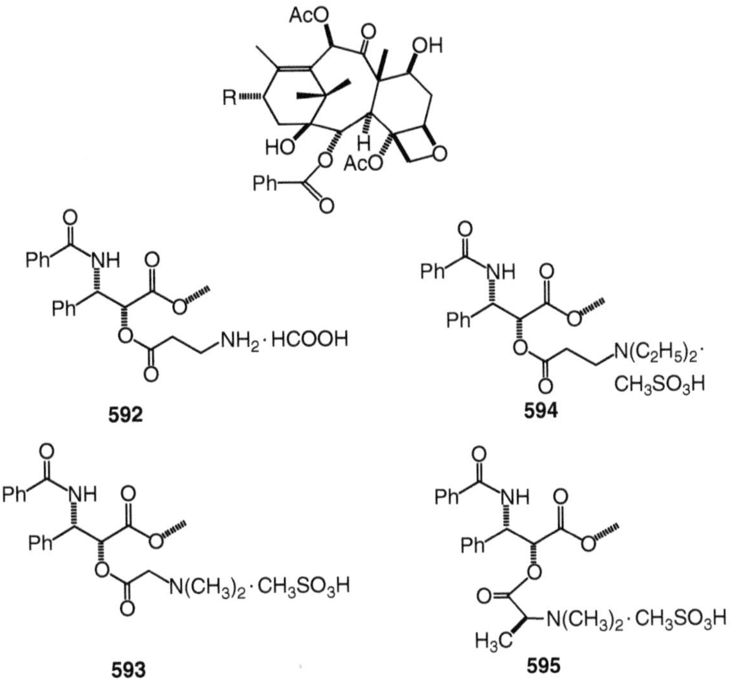

Figure 10.24 2'-Amino acid derivatives as taxol prodrugs.

Amino acid derivatives

Several 2'-amino acid derivatives (Figure 10.24) were prepared as potential water-soluble pro-drugs. Magri and Kingston (1988) reported the synthesis of the formic acid salt of 2'-(β-alanyl)taxol 592 (Figure 10.24). This derivative was not suitable as a prodrug, because it was weakly cytotoxic and reverted back to taxol rapidly in aqueous solution. Zhao *et al.* (1991) synthesized 2'-glycyltaxol (593) and 2'-(γ-aminobutyryl)taxol (594) and found them to be even more unstable than 2'-(β-alanyl)taxol(592). Mathew *et al.* (1992) also reported the synthesis of C-2' amino acid derivatives of taxol (593–595) as potential prodrugs. Various ammonium salts of these compounds were prepared, in which the methanesulfonic acid salts showed the greatest water solubility. The methanesulfonic salt of 2'-(N,N-dimethylglycyl)taxol 593 had reasonable stability with a half-life of 96 h at pH 3.5 and a half-life of 6 h at pH 7.4. Its water solubility was >2 mg/ml. The enhanced stability of 593 in comparison to 2'-glycyltaxol can be attributed to the introduction of the two electron donating methyl groups in the N,N-dimethylglycyl moiety, thus reducing the strong electron withdrawing effect of the ammonium group. Methane sulfonic acid salt 2'-[3-(N,N-diethylamino)propionyl]taxol 594 was more stable than 593 at pH 3.5 with a half-life of 438 h but had a half-life of only 0.25 h at pH 7.4. The solubility of 594 was >10 mg/ml. 2'-(L-alanyl)taxol (595) was prepared but possessed poor chemical stability (Mathew *et al.*, 1992). In human plasma, 593 and 594 were substrates for enzymatic cleavage of the C-2' ester and had half-lives of 3 h and 4 min, respectively (Georg *et al.*, 1995d).

The amino acid derivatives 593 and 594 demonstrated potent cytotoxicity against B16 melanoma cells; 594 was even more toxic than taxol. In the MX-1 mammary tumor screen, 594

Table 10.37 Structure and *in vivo* antitumor activity of compound 413, 596–600

Compound	R	Solubility in saline (mg/ml)	B16 Optimal dose (mg/kg/day)	ILS(%)	Body weight change on day 10 (%)
Docetaxel			20	149	−13.0
413			6.3	187	−7.5
596	H₂NOC (NH₂) COO	5	6.3	165	−3.2
597	H₂NOC (NH₂) COO	50	6.3	127	11.1
598	MeHNOC (NH₂) COO	10	6.3	185	−0.5
599	HOOC (NH₂) COO	1	25	212	3.3
600	HOOC (NH₂) COO	10	12.5	213	−2.9

was again more potent than **593** and showed complete tumor remission at a dose of 4.5 mg/kg. It also had reasonable solid-state stability upon storage. From these results, it appears that **594** has favorable chemical and enzymatic stability properties as a prodrug (Mathew *et al.*, 1992; Georg *et al.*, 1995d).

Yamaguchi *et al.* (1999) designed 2′-substituted prodrugs that contain terminal amino acid derivatives on a glycolate spacer (Table 10.37). The parent compound **413** (3′-desphenyl-3′-cyclopropyl analog of taxotere) showed *ca*.20 times stronger activity against human colon cancer cell lines than docetaxel and also was more potent *in vivo* (Table 10.37). Prodrugs **596–600** showed antitumor activity superior to **413** and Docetaxel. From the body weight change, **596**, **599** and **600** appeared to be less toxic than **413** (Table 10.37). The prodrugs **596–600** had greatly improved solubility and stability in saline. The glycolate spacer was found to be an effective moiety to make the prodrugs with amino acids at 2′-OH stable (Yamaguchi *et al.*, 1999).

Phosphate prodrugs

A water-soluble phosphate prodrug has been successfully used with another poorly soluble anti-cancer drug, etoposide (Doyle and Vyas, 1990). The phosphate prodrug can be converted into the active drug in animals via the action of plasma, serum or tissue phosphatases (Mamber *et al.*, 1995). Paclitaxel C-2′, C-7, and C-2′ and C-7-phosphate analogs (601–603) (Figure 10.25) were synthesized by Vyas *et al.* (1993). None of them showed activity in the microtubule assembly assay. Paclitaxel C-2′, and C-7-phosphate analog (601, 602) were not activated by alkaline phosphatase. *In vivo* evaluation of 601–603 against the murine M109 solid tumor model, only 601 showed anti-tumor activity (T/C = 140), however it was lower than taxol (T/C = 270). It was concluded that 601–603 are not suitable prodrugs of taxol (Vyas *et al.*, 1993). Because the phosphate groups are located too close to the taxane nucleus, steric hindrance to the enzyme approach might result in the lack of enzymatic susceptibility of these prodrugs. Accordingly, the phosphate prodrugs 604 and 605 were designed (Ueda *et al.*, 1993). After cleavage of the more distant phosphate group, it was expected that "trimethyl lock" acceleration would result in the release of taxol (Figure 10.26). Derivatives 604 and 605 were found to be highly soluble (>10 mg/ml) and stable in water. Their prodrug moiety was readily cleaved by several alkaline phosphatase preparations, and 604 and 605 were active in the microtubule assembly assay after incubation with phosphatase. But when incubated with rat, dog and human plasma, 604 and 605 did not release taxol. It was concluded that these taxol prodrugs bind tightly to plasma proteins, thereby making them inaccessible for phosphatase. Thus, these derivatives are not suitable as taxol prodrugs. *In vivo* evaluation of 604 and 605 in the M109 tumor model demonstrated that derivative 604 was marginally active (T/C = 144 for 100 mg/kg per injection) in comparison to taxol (T/C = 275 for 30 mg/kg per injection). Compound 605, however, had better activity (T/C = 153 for 70 mg/kg per injection). The taxol standard in the later assay showed a T/C = 144 for 40 mg/kg per injection (Ueda *et al.*, 1993; Georg *et al.*, 1995d).

Paclitaxel-2′-ethylcarbonate (606) (Figure 10.25) showed remarkable *in vivo* antitumor activity, but was as poorly water soluble as paclitaxel (Ueda *et al.*, 1994). In order to improve its water solubility, Ueda *et al.* (1995) introduced a phosphate moiety into the 7-position (Figure 10.25). Analogs 607–609 were much more soluble in water (2.5–5 mg/ml H$_2$O) than paclitaxel. Compounds 607 and 608 were stable at pH 7.4 and could generate paclitaxel-2′-ethylcarbonate rapidly (T$_{1/2}$ < 5 min) when exposed to alkaline phosphatase; No data were available on 609.

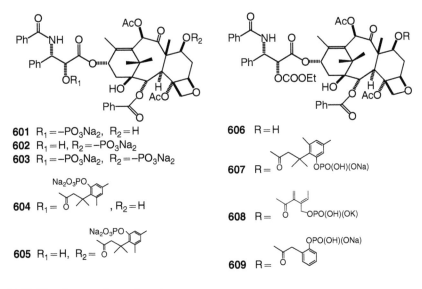

601 R$_1$ = –PO$_3$Na$_2$, R$_2$ = H
602 R$_1$ = H, R$_2$ = –PO$_3$Na$_2$
603 R$_1$ = –PO$_3$Na$_2$, R$_2$ = –PO$_3$Na$_2$

604 R$_1$ = [structure], R$_2$ = H

605 R$_1$ = H, R$_2$ = [structure]

606 R = H

607 R = [structure] OPO(OH)(ONa)

608 R = [structure] OPO(OH)(OK)

609 R = [structure] OPO(OH)(ONa)

Figure 10.25 Phosphate prodrugs of taxol.

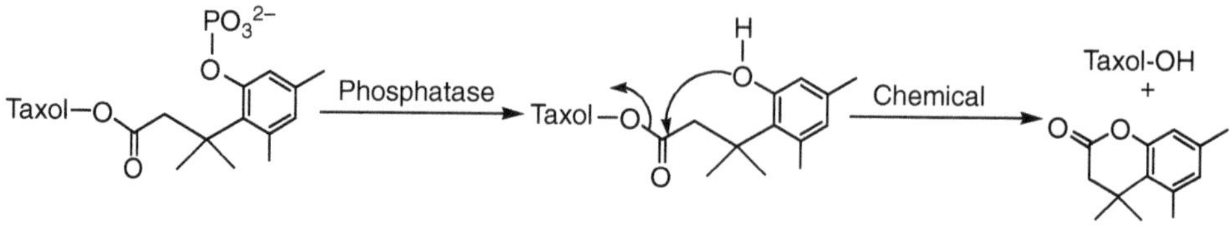

Figure 10.26 Mechanism of release taxol from phosphate prodrugs.

Table 10.38 Structures and cytotoxicity of MPA prodrugs

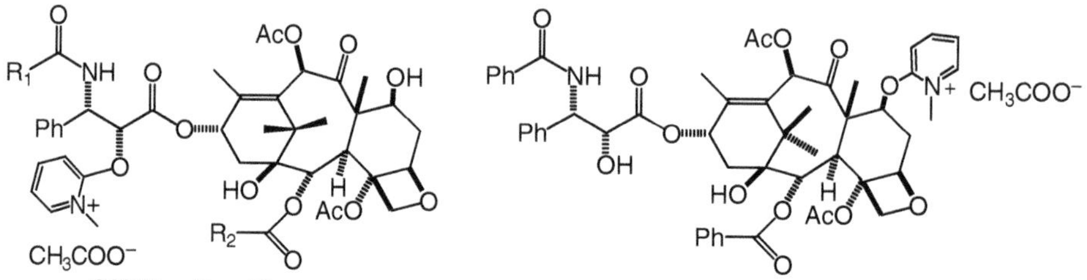

610 R$_1$ = R$_2$ = Ph
611 R$_1$ = ph, R$_2$ = 3-thiophene
612 R$_1$ = t-BuO, R$_2$ = Ph

613

Compound	L1210	SK-MEL28	HL-60	PC-3	Ovcar-3
Paclitaxel	1	1	1	1	1
610	9.8	632	0.14	9.2	1.05
611	0.82	5.69	1.11	8.6	0.35
612	0.01	49	0.50		11.4
613	27	5050	3.25	100	11.8

In the M109 tumor model, the phosphonoxyphenyl-propionate ester 607 and phospho-noxymethylbenzoate derivate 608 proved to be useful prodrugs of paclitaxel-2'-ethylcarbonate and showed comparable *in vivo* antitumor activity to paclitaxel. Compound 609 was inactive. It was speculated that paclitaxel or paclitaxel-2'-ethylcarbonate was not generated under the assay conditions. Thus, 609 was too stable to be an effective prodrug (Ueda *et al.*, 1995).

Methylpyridinium actate (MPA) prodrugs

Taxol-2'-MPA (610), 3-thieophene taxol-2'-MPA (611), taxotere-2'-MPA (612), and taxol-7-MPA (613) (Table 10.38) were investigated (Nicolaou *et al.*, 1994b, 1995). They exhibited promising *in vitro* cytotoxicities equivalent to taxol (Table 10.38) together with high aqueous solubility and stability. Compound 610 was also tested in the *in vivo* assay but display no *in vivo* activity (Nicolaou *et al.*, 1995).

Ester and carbonate as potential prodrugs

A series of potential prodrugs (614–622) (Figure 10.27) was prepared by Nicolaou *et al.* (1993). Derivatives 614–618 had improved water solubility and most of the prodrugs had significant

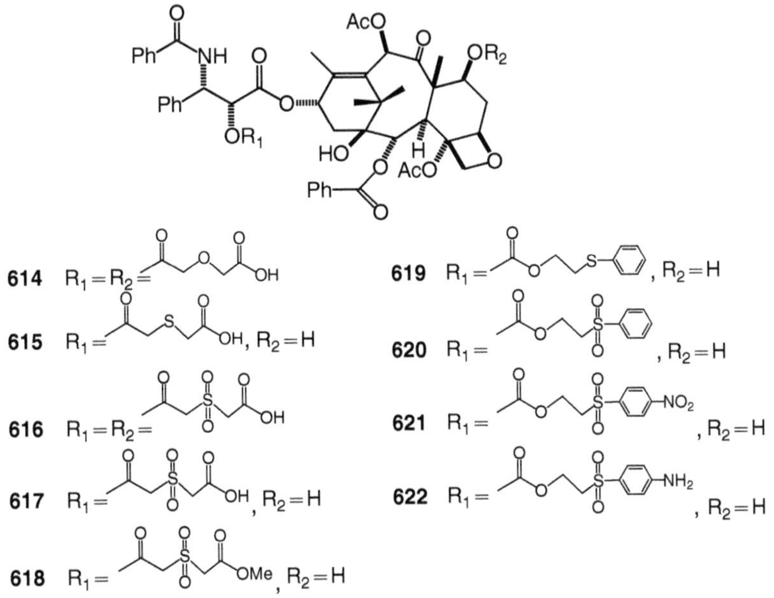

Figure 10.27 Esters and carbonates of taxol prodrugs.

stability in water with half-lives of >500 min at pH $= 7.5$. Prodrug **618** is the most promising derivative. Its solubility and stability in aqueous media are good and it can release taxol rapidly in human blood plasm (half-life $= 100$ min). Several of these prodrugs had comparable cytotoxic activity and selectivity to taxol in various cancer cell lines. Based on these results and the fact that taxol could be isolated from the cytotoxicity studies, it was concluded that the cytotoxicity of these analogs was a result of the release of taxol and not due to the intrinsic cytotoxicity of these derivatives (Nicolaou *et al.*, 1993; Georg *et al.*, 1995d).

2′-Methoxyacetate(MAC)taxol (R$=$COCH$_2$OCH$_3$, Table 10.39) (Greenwald *et al.*, 1996) was synthesized and showed good activity, but was unstable. The same authors (1996) introduced PEG into the 2′-MAC taxol to give derivatives (**623–625**) with good solubility and increased stability. All of the 2′-PEG esters showed similar cytotoxicity to that of paclitaxel in which **625** was demonstrated equivalent *in vivo* activity in a P388 model to that of paclitaxel (Table 10.39) (Greenwald *et al.*, 1996). Pendri *et al.* (1998) designed a PEG tripartite prodrug (**626**) containing a glycine spacer that resulted in enhanced circulatory retention. *In vitro* cytotoxicities of PEG-paclitaxel and PEG-Gly-paclitaxel were similar and only slightly lower than that of paclitaxel. Both prodrugs stored *in vivo* activity against P388/0 murine leukemia, with PEG-Gly-paclitaxel having better activity (Paclitaxel showed an ILS of 33% at 75 mg/kg and PEG-Gly-paclitaxel showed ILS of 44% at the same dose).

The dipeptide Z-Phe-Lys-PABC-DOX was studied as a prodrug of the anticancer agent Doxorubicin. PABC, a self-immolative spacer that aids drug release, is stable in human plasma, and can be cleaved by the ubiquitous lysosomal cystein protease cathepsin B. Recently, Dubowchik *et al.* (1998) designed Boc-Phe-Lys-PABC-7-paclitaxel (**627**) and Z-Phe-Lys-PABC-2′-paclitaxel (**628**) (Figure 10.28). These two compounds are stable in human plasma, however they are also relatively stable to cathepsin B (T$_{1/2}$ for **627** is 40 min, for **628** is 9.0 h) and in rat liver lysosomes (T$_{1/2}$ for **627** is 66 min, for **628** is 19 min). The stability may occur because

Table 10.39 Structures and properties of PEG prodrugs

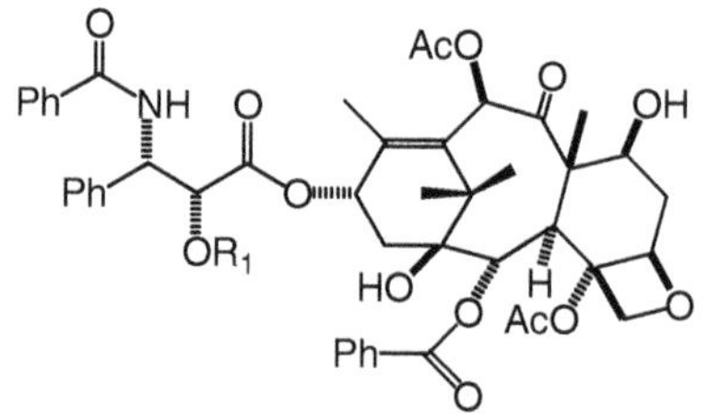

Compound	R	MW	P388 ($IC_{50}/IC_{50(taxol)}$)	$T_{1/2}$ in PBS (h)	$T_{1/2}$ in Rat plasma (h)
Paclitaxel			1		
623	$-O-CO-CH_2-O-PEG_{MW}$	5000	2.5	5.5	0.5
624	$-O-CO-CH_2-O-PEG_{MW}$	20000	2.8	5.5	0.5
625	$-O-CO-CH_2-O-PEG_{MW}$	40000	1.6	5.5	0.4
626	$-O-CO-CH_2-O-NHCO-CH_2OPEG$		2.3	7.0	0.4

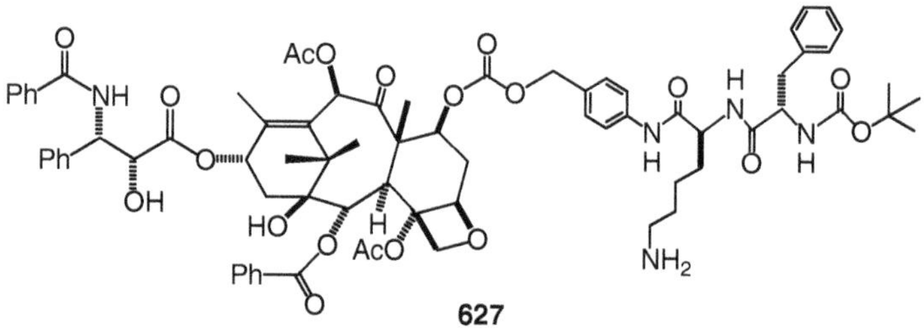

627

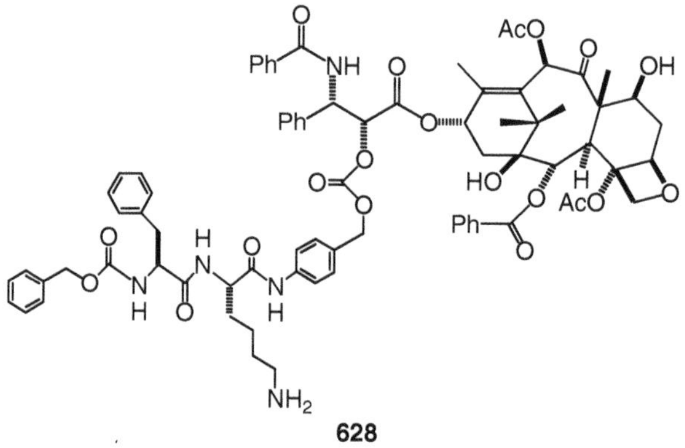

628

Figure 10.28 Structures of dipeptide prodrugs 627–628.

paclitaxel is a large and bulky molecule, and therefore the prodrug is less readily processed by cathepsin B.

Selective water soluble prodrugs of paclitaxel

The previous prodrugs all increased the water solubility of taxol. However, the lack of selectivity of chemotherapeutic anticancer agents is also a serious drawback in conventional cancer therapy (de Groot *et al.*, 2000). Tumor-associated monoclonal antibodies-enzyme conjugates, tumor-associated enzymes and receptor-mediated bioconjugates can be used to increase selectivity of water soluble prodrugs (Safavy *et al.*, 1999).

β-Glucuronyl carbamate prodrugs are potential candidates for antibody-directed enzyme prodrug therapy (ADEPT) (Leenders *et al.*, 1995). In ADEPT, enzymes attached to mAbs selectively bind to the tumor site. The enzymes convert a prodrug into a cytotoxic drug at the site of the tumor. ADEPT has been successfully used with many cytotoxic agents such as methotrexate, etoposide, or doxorubicine. Human β-glucuronidase is a lysosomal enzyme with no activity in blood at neutral pH. de Bont *et al.* (1997) designed taxol prodrugs with β-glucuronides coupled to a spacer moiety via a carbamate linkage (**629–631**) (Figure 10.29). β-Glucuronidase catalyzed hydrolysis and taxol release was facilitated by further reaction of the spacer with the 2'-OH ester on paclitaxel (Figure 10.30).

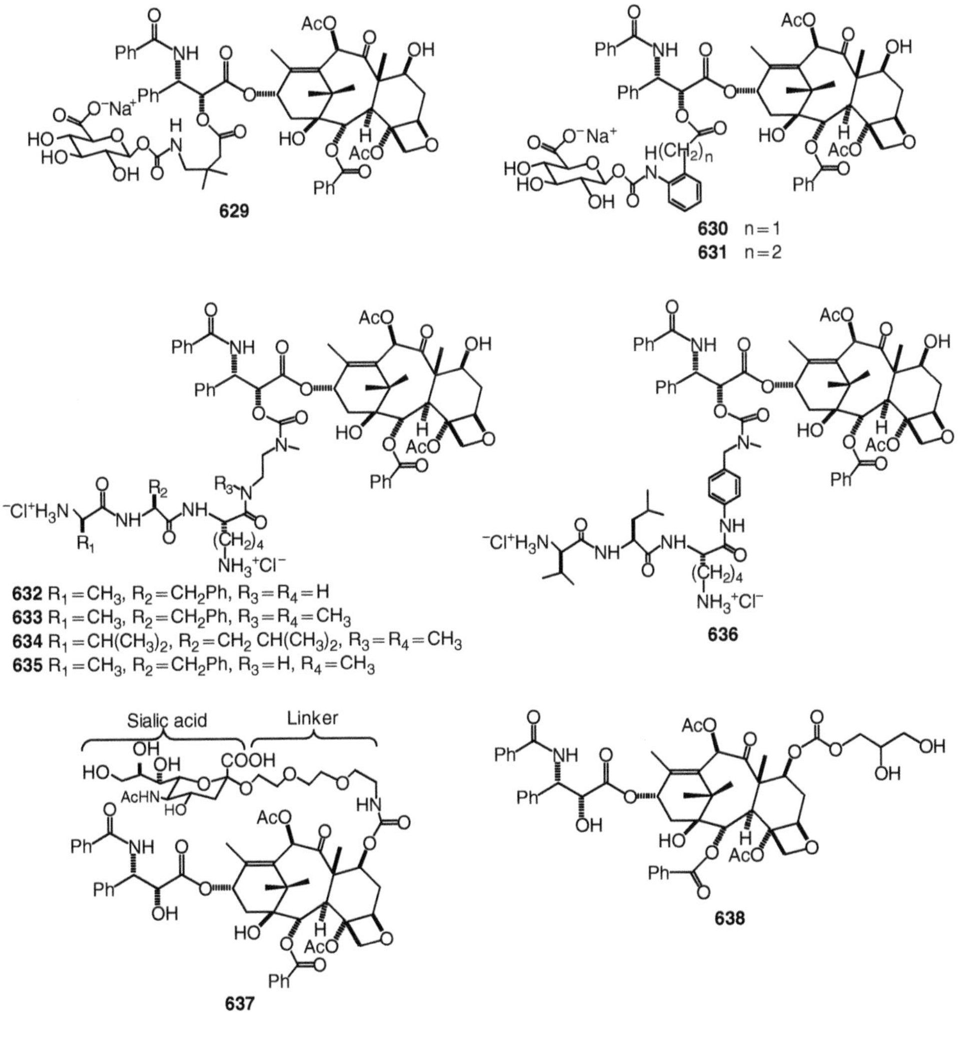

Figure 10.29 Structures of compounds 629–638.

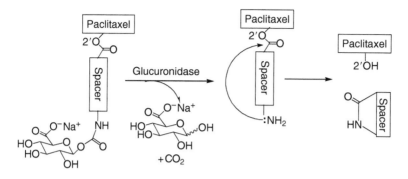

Figure 10.30 Mechanism of the activation of prodrugs by β-glucuronidase.

Table 10.40 Cytotoxicity of prodrugs in the absence and
presence of β-glucuronidase

Compound	$IC_{50}/IC_{50(taxol)}$	$IC_{50}/IC_{50(taxol)}$ *with* β-*gucuronidase*
Paclitaxel	1	1
629	95	3
630	1.1	1.1
631	135	1.8

Compounds 629–631 showed more water solubility (>10 mM) than paclitaxel (<0.005 μM). While 629 and 631 are stable at PH 6.8 at 37°C, 630 hydrolyzed to paclitaxel under the same condition with a half-life of 3.5 h. β-Glucuronidase catalyzed half-lives of prodrug 629 and 631 is 2 h and 45 min, respectively.

The cytotoxicities of 629–631 were evaluated against OVCAR-3 cells in the absence and in the presence of β-glucuronidase (Table 10.40). The prodrugs 629 and 631 were 100 times less cytotoxic than paclitaxel in the absence of β-glucuronidase, but when activated by β-glucuronidase, were nearly as active as paclitaxel.

Using tumor-associated enzymes to convert the prodrug into active parent drug in a site specific manner can increase selectivity of the anticancer drugs for tumor cells. Protease plasmin plays important roles in tumor invasion and metastasis. Previous studies demonstrated D-Ala-Phe-Lys and D-Ala-Leu-Lys were suitable plasmin substrates. de Groot *et al.* (2000) prepared taxol produgs by coupling taxol to the spacer end of tripeptide spacer moieties (632–636) (Figure 10.29). Prodrug 632 was hydrolyzed completely in 24 h. The prodrugs 633–636 were highly stable in a tris buffer solution (pH 7.3). In human plasma, prodrugs 633 and 634 remained stable for at least three days. The enzyme hydrolysis half-life for prodrug 635 was 3.5 min, and for prodrug 636 was 42 min. The half-life of spacer cyclization after peptide cleavage for prodrug 635 was 23 h, and for prodrug 636 was instantaneous. All prodrugs showed strongly reduced cytotoxicity in seven human cancer cell lines compared to paclitaxel (Table 10.41). Prodrugs 635 and 636 are the most nontoxic paclitaxel prodrugs now reported. No information is available on the *in vivo* activity of these prodrugs.

Table 10.41 Cytotoxicity of prodrugs in various tumor cell lines (IC$_{50}$ ng/ml)

Compound	MCF-7	EVSA-T	WIDR	IGROV	M19	A498	H226
Paclitaxel	<3	<3	<3	10	<3	<3	<3
632	16×10^3	11×10^3	13×10^3	19×10^3	45×10^3	40×10^3	41×10^3
633	6.6×10^3	7.2×10^3	10×10^3	11×10^3	17×10^3	38×10^3	21×10^3
634	6.9×10^3	5.2×10^3	11×10^3	26×10^3	24×10^3	45×10^3	25×10^3
635	$>63 \times 10^3$	$>63 \times 10^3$	47×10^3	45×10^3	47×10^3	$>63 \times 10^3$	48×10^3
636	9.2×10^3	9.3×10^3	11×10^3	27×10^3	20×10^3	$>60 \times 10^3$	16×10^3

Neuraminidase is another tumor-associated enzyme, with greatly enhanced activity at the cancer cell surface. Takahashi *et al.* (1998) designed and synthesized a "taxol-sialic acid" hybrid using triethylene glycol (TEG) as linker to a potential neuraminidase substrate (637) (Figure 10.29). Its water solubility was 28 mg/ml, ED$_{50}$ in the tubulin assembly assay was 3.86 μM (taxol:0.86 μM), and cytotoxicity *in vitro* against human lung carcinoma A427 and A427/VCR were 1.0×10^{-7} M and 1.0×10^{-6} M. Thus, this sialic acid derivative to modify the 7-position of taxol is water soluble and maintains binding affinity and cytotoxicity. No data is available about the stability of the drug in the presence of neurominidase.

Hyaluronic acid (HA) acts as a signaling molecule in cell motility, inflammation and cancer metastasis. HA receptors are overexpressed in transformed human breast epithelial cells and other cancers. Coupling antitumor agents to HA not only can increase water solubility but also target the bioconjugate to tumor cells and tumor metastases by receptor-mediated uptake. Luo and Trestwich (1999) synthesized HA-taxol bioconjugates by linking the taxol 2′-OH to adipic dihydrazine-modified HA (ADH) via a succinate ester. Uptake experiments demonstrated that HA binds specifically to tumor cell surfaces and is rapidly taken up via a HA receptor-mediated pathway. HA–taxol conjugates were cytotoxic against several cancer cell lines, but were not cytotoxic against nontransformed NIH 3-T-3 cells. The selective cytotoxicity of the conjugates may be due to receptor-mediated binding and uptake (Luo and Trestwich, 1999).

Because the tumor microenviroment is acidic, Niethammer *et al.* (2001) designed and synthesized prodrug 638 to give a slow and pH-dependent release of the parent drug (Figure 10.29). Prodrug 638 is 50-fold more water soluble than paclitaxel and it is stable at ambient temperature in an iv infusion formulation.

Summary

Several taxol prodrugs were prepared and evaluated. Modifications were made mainly at C-2′ and C-7. Most C-7 modified produgs showed excellent chemical stability, but were not suitable as prodrugs. Many selective prodrugs have been investigated recently, and some showed very promising results. In the future the design of more modified spacer and the discovery of more tumor-associated enzymes can be used to increase the rates of enzyme hydrolysis and improve the chemotherapy selectivity.

Approaches toward identification of the taxol binding site: synthesis and biological activity of taxol photoaffinity labels

Identification of binding site on tubulin

Taxol and its analogs exhibit unique properties of promoting tubulin assembly and preventing microtubules from cold-induced depolymerization. These actions are thought to be the basis for their cytotoxicity (Loeb *et al.*, 1997). But the precise binding site on tubulin at the molecular level has not been fully elucidated. Such information will be useful for the development of specific taxol analogs that would precisely fit the binding site and thus show significant biological activity.

Photoaffinity labeling is an excellent and frequently used tool to study the interaction of ligands with polypeptides such as microtubules or enzymes (Bayley, 1983). Incubation of a taxol photoaffinity analog with microtubules followed by photoactivation is expected to cause the formation of covalent bond between taxol and the peptide binding domain on microtubules. The photoactivated group can crosslink to the amino acids that are in close proximity to the taxol binding domain. In order to identify the binding site, the covalently linked taxol–microtubule complex is digested enzymatically to provide several small peptides. Once the oligopeptide containing the label is identified, sequencing can provide information on the amino acids to which the taxol photoaffinity analog is bound (Georg *et al.*, 1995d). Photoaffinity analogs contain groups that, upon photoactivation, generate highly reactive species, which are trapped in or near the binding site (Bayley, 1983). Diazo, diazirine and azides exemplify groups that are used to generate photoaffinity labels.

The ideal photoaffinity label would resemble the original molecule, be easy to synthesize, bind to the receptor, possess good photolysis properties, and be stable under the experimental condition (Georg *et al.*, 1995d).

Attachment of photoreactive moieties at the 7-hydroxyl group

For the synthesis of a taxol photoaffinity label, it is important to choose an appropriate label and site of attachment to taxol (Rimoldi *et al.*, 1993). The most reactive functional groups of taxol are the best sites to attach the photoreactive moieties, and the most easily accessible functional groups of taxol are the hydroxyl groups at C-2′ and C-7. From the previous SAR studies, taxol analogs with modification at C-2′ do not bind to microtubules; however, taxol analogs with modification at C-7 retain most of the activity. Therefore, most of the taxol photoaffinity labels are taxol analogs modified at the C-7 hydroxyl group (Georg *et al.*, 1995d).

Romoldi *et al.* (1993) reported the synthesis and biological activity of taxol analogs containing carbene-precursors at the C-7 position (Compounds 639–642) (Figure 10.31). Carbenes can be generated from diazo compounds and diazirines and can react with nucleophilic groups and undergo insertion chemistry with hydrocarbons and aromatic compounds (Young and Platz, 1989). Compared with taxol, 639–642 exhibited only slightly reduced activity in the microtubule disassembly assay (Table 10.42), but were different with regard to their ability to induce microtubule formation. Incubation of 639–642 with MAP free tubulin in the presence of glutamate (polymerization inducer) at 37°C initiated the formation of stable microtubule polymers. However, at reduced temperature (15°C) or low ionic strength, these derivatives did not induce microtubule formation, although taxol can promote the formation of microtubules under these conditions (Georg *et al.*, 1995d).

Several taxol photoaffinity analogs with azidobenzoyl groups at C-7 were investigated (Georg *et al.*, 1992d, 1994b, 1997; Carboni *et al.*, 1993). Compounds 643 and 644 showed greatly diminished activity in the microtubule assembly assay (Georg *et al.*, 1992d, 1994b, 1997). Because the previous SAR studies indicated that most esters at the 7-hydroxyl group have activity similar to taxol, the steric requirement of the 4-azido group may prevent effective binding to microtubules (Georg *et al.*, 1997). Compounds 646 and 647 with the azido moiety at the meta position also possessed extremely poor activity in the microtubule assembly assay. Among the aminoaryl analogs (649, 651–653), 3-(dimethylamino)benzoyl analog 649 showed greatly reduced microtubule assembly activity, but its activity was sufficient for the fluorescent studies. The bioisostere analog of compound 648 was inactive. Compounds 651–653 were equivalent with their parent compound 53 and they were useful for the fluorescent studies. Interestingly, 645 induced the assembly of stable microtubules *in vitro* and showed comparable cytotoxicity to taxol against J774.2 cell lines. Competitive binding assays with [^{3}H]taxol demonstrated that 645 has the same binding domain on microtubules but with a lower binding affinity than taxol. It is proposed that the activity is due to the conformational modifying *ortho*-nitro substituent (Carboni *et al.*, 1993; Georg *et al.*, 1997).

7-*O*-Iodoacetate 650 showed similar activity to taxol in both assays. Its synthetic intermediate 7-*O*-chloroacetylpaclitaxel 59 was as active as taxol and is promising as a probe to map the paclitaxel binding site (Georg *et al.*, 1997).

Compound 654 showed only slightly reduced cytotoxicity against several tumor cell lines compared to that of paclitaxel ($ED_{50}/ED_{50(paclitaxel)}$ = 1.4 for A121, 3.6 for A549, 1.9 for HT29, 1.4 for MCF-71.5 for MCF7-R). Compound 654 does not promote the polymerization of tubulin to microtubules, but it does inhibit the disassembly process of preformed microtubules.

Compounds 655 and 656 were less active than paclitaxel but their cytotoxicity was sufficient in intracellular fluorescence mapping (Rao *et al.*, 1998).

Attachment of photoreactive moieties at C-13 side-chain

From the previous SAR studies of taxol it is well known that the *N*-benzoyl group can be substituted with many various substituents without loss of activity. Therefore it is also a suitable site to attach a photoaffinity label.

N-(4-Azidobenzoyl)-*N*-debenzoyltaxol (659) (Figure 10.32) showed reduced activity in microtubule assembly and cytotoxicity (J774.2 cells) assays compared to taxol (Table 10.42) (Dasgupta *et al.*, 1994; Swindle *et al.*, 1994). However, competitive binding studies with taxol demonstrated that 659 does bind to the taxol binding site (Dasgupta *et al.*, 1994; Swindell *et al.*, 1994). *N*-(4-Azido-2,3,5,6-tetrafluorobenzoyl)-*N*-debenzoyltaxol (527) (Figure 10.32) was synthesized by Georg *et al.* (1994d) and showed reduced activity in microtubule assembly and cytotoxicity (B16 cells) assays compared to taxol (Table 10.42). *N*-(3-Azidobenzoyl)-*N*-debenzoyltaxol (662) possessed excellent activity in microtubule assembly and cytotoxicity assays, while *N*-(3-azido-5-nitrobenzoyl)-*N*-debenzoyltaxol (663) showed decreased activity (Georg *et al.*, 1995c). Thus, azido groups at the meta position were better than those at the para position. These results are consistent with the data with C-7 photoaffinity labels.

TaxAPU (660) was less potent on tubulin assembly than taxol, but competition binding studies demonstrated it retains sufficient affinity for the taxol binding site to be used as a photoaffinity probe (Combeau *et al.*, 1994).

Compound 661 showed reduced activity in microtubule assembly and cytotoxicity (J774.2 cells) assays compared to 659 (Table 10.42) (Swindle *et al.*, 1994). This result indicates that the large benzoyl substituent is detrimental for the activity. However, 3′-*N*-BzDC-3′-*N*-debenzoylpaclitaxel (664), which has an ethyl spacer, showed the same properties as paclitaxel in the tubulin/microtubule system. Competitive binding studies with taxol demonstrated that 661 (Swindell *et al.*, 1994) and 664 (Ojima *et al.*, 1995) do bind to the taxol binding site.

Attachment of photoreactive moieties at C-2 and C-10 position

Compounds 140 and 665 are analogs of paclitaxel and docetaxel, respectively, with 3-azidobenzoyl moiety at the C-2 position (Figure 10.33). They showed excellent activity in both the

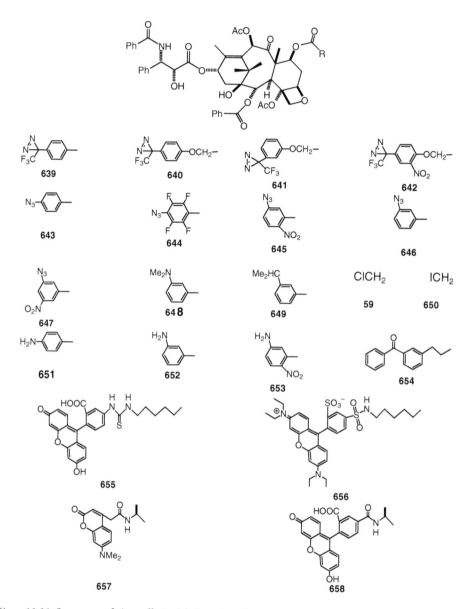

Figure 10.31 Structures of photoaffinity labels at C-7 of taxol and baccatin III analogs.

microtubule assembly and cytotoxicity (B16 cells) assays. Compound 140 was also at least 10-fold more active than taxol against several cancer cell lines, and is the most active photoaffinity labels reported (Chaudhary *et al.*, 1994; Georg *et al.*, 1995c). 10-BzDC-paclitaxel (666) (Figure 10.33) displayed several-fold weaker cytotoxicity against several human cancer cell lines than taxol and 7-BzDC-paclitaxel (654) (Figure 10.31) (Ojima *et al.*, 1999).

Table 10.42 Biological activity of taxol photoaffinity analogs

Compound	Microtubule assembly	Microtubule disassembly	Initial slope for microtubule assembly	B16	J774.2
Taxol	1	1	1	1	1
639		3.4			
640		1.9			
641		2.8			
642		3.1			
643	21			30	
644	>22				
645			24.2		3.2
646	4.1			9.7	
647	>24				
648	>24				
649	22				
650	1.6			0.92	
651	3.3			3.7	
652	1.7			17	
653	1.5			7.1	
654	4.4			16	
659	2.0			4.6	3.1
527	4.6			15	
661					16.6
662	0.81			2.6	
663	3.0			17.4	
664	0.40			0.65	
140	1.2			0.37	

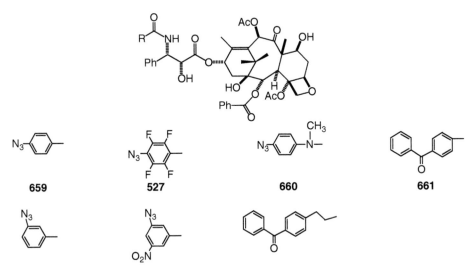

Figure 10.32 Structures of photoaffinity analogs at C-13 side-chain.

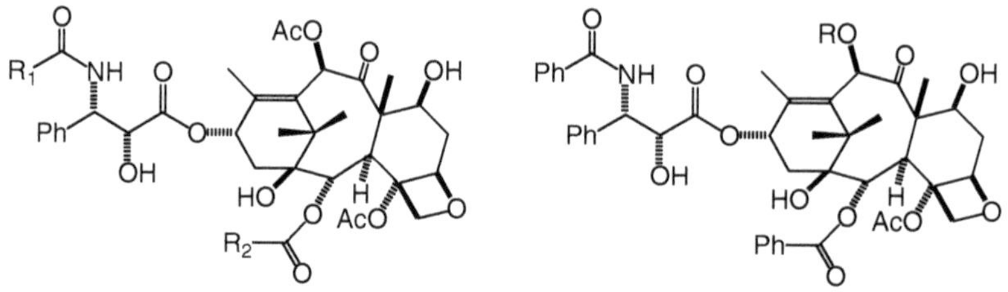

140 R₁ = Ph, R₂ = 3-azidophenyl

665 R₁ = t-BuO, R₂ = 3-azidophenyl

666 R =

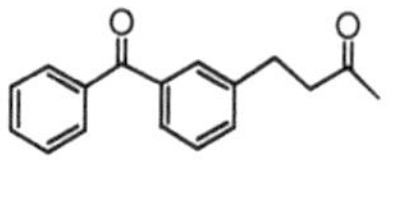

Figure 10.33 Structures of photoaffinity analogs at C-2 and C-10.

Studies on photoaffinity labeling of tubulin

Based on the photoaffinity labeling of microtubules with [^{3}H]-taxol, Rao *et al.* (1992) proposed that the taxol binding site is on the β-subunit of the tubulin dimer, as no binding to the α-subunit was detected. However, the low extent of photoincorporation limited the use of [^{3}H]-taxol as a probe for defining the taxol binding site on β-tubulin (Rao *et al.*, 1994). Several groups have investigated suitable photoaffinity labels. Rao *et al.* (1994 and 1995) reported that [^{3}H]3′-(p-azidobenzoyl)taxol ([^{3}H]-659) covalently bound to the 31 amino acid N-terminal of β-tubulin. Taxol competes with [^{3}H]-659 binding, suggesting that they bind at the same or an overlapping site. But Georg and coworkers (1995) reported that this compound labeled both the α-subunit and β-subunit, even though 80% labeling was found in the β-subunit (Dasgupta *et al.*, 1994). It was concluded that the paclitaxel binding site may overlap the tubulin subunits. This result was confirmed by results with [^{3}H]-660 ([^{3}H]-TaxAPU) (Combeau *et al.*, 1994; Leob *et al.*, 1997) and [^{3}H]-662 (Ojima *et al.*, 1995, 1999). [^{3}H]-660 photolabeled amino acids 281–304 of α-tubulin (Loeb *et al.*, 1997), and with [^{3}H]-662, *ca.*70% of the label was on β-tubulin. [^{3}H]2-(*m*-Azidobenzoyl)taxol ([^{3}H]-140) specifically photolabeled amino acids 217–231 of β-tubulin. This photolabeling was competitive with taxol and unlabeled 140, suggesting a common binding domain (Rao *et al.*, 1995).

Although [^{14}C]-labeled paclitaxel derivatives might be more useful because of the high sensitivity and long half-life of the isotope, [^{14}C]paclitaxel derivatives are difficult to synthesize. Recently, 7-([carbonyl-^{14}C]-acetyl)paclitaxel (**667**) was synthesized (Figure 10.34) (Rao *et al.*, 1998), and its specific activity was sufficient for drug accumulation studies.

Identification of binding site on p-glycoprotein

Drug resistance is a serious disadvantage of antitumor drugs. When mammalian cells are treated with paclitaxel or other antitumor drugs, they often display the multidrug resistance (MDR) phenotype, which is characterized by overproduction of p-glycoprotein. p-Glycoprotein acts as a drug-efflux pump so that the intracellular concentration of the drugs is lower than the cytotoxic level (Ojima *et al.*, 1995). Study of the binding site of paclitaxel on p-glycoprotein is important for the development of new taxol analogs with fewer side-effects and less resistance.

[^{3}H]-664 and [^{3}H]-666 successfully photolabeled p-glycoprotein. A competitive binding study demonstrated that p-glycoprotein has a specific binding site for paclitaxel (Ojima *et al.*, 1995, 1999). An extensive study by the same group showed that the site of photoincorporation is in the C-terminal half of murine p-glycoprotein (Wu *et al.*, 1998).

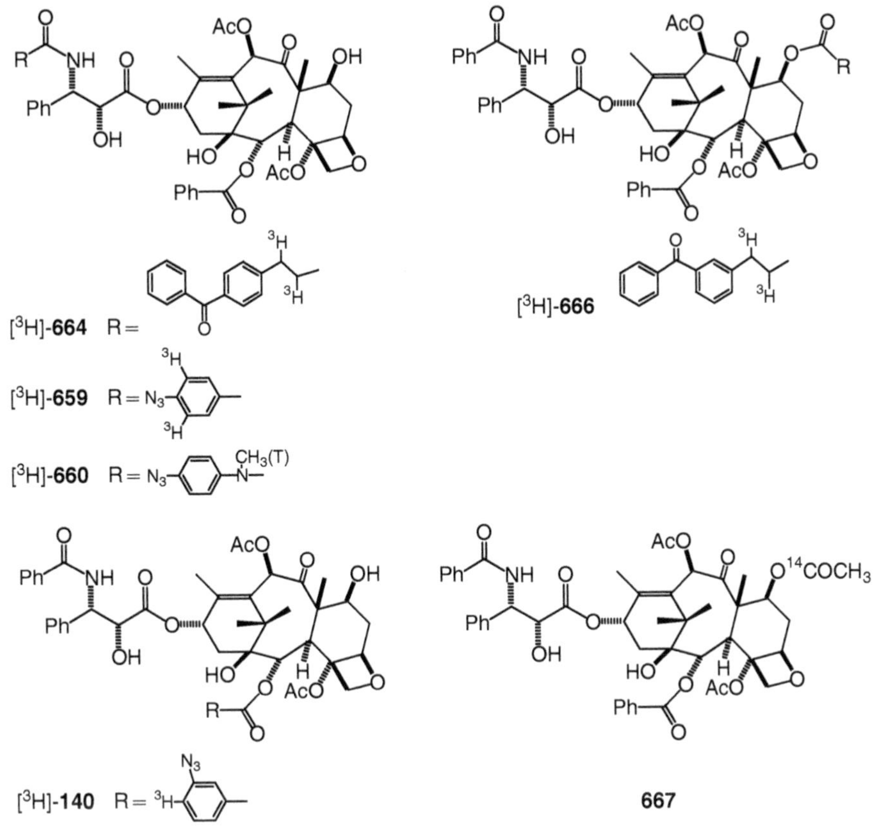

Figure 10.34 Structures of radiolabeled analogs of paclitaxel.

Conformational analysis of taxol and its analogs

The development of a three-dimensional taxol pharmacophore model is very important for rational drug design. This model would allow more precise modification and synthesis of optimal taxol analogs. Previous SAR studies on natural and semisynthetic taxol analogs have demonstrated which parts of the molecule are important for activity. The "northern hemisphere" portion of the molecule (C-7 to C-10) can be altered without much effect on the biological activity. The C-13 side-chain and the "southern hemisphere" portion of the molecule are very crucial to the activity. This result suggested that the pharmacophore is located within these groups. An analysis of taxol's conformation can give us more detailed information (Georg *et al.*, 1995d).

The conformation of taxol has been investigated by nuclear magnetic resonance spectroscopy (NMR), X-ray diffractometry and comparative molecular field analysis (CoMFA). Each method has its advantages and limitations. X-ray diffractometry determines the conformation of a molecule as it exists in the crystalline state, but this conformation may not precisely describe the biological situation. Conformational information obtained by NMR spectroscopy is important because taxol interacts with tubulin in solution. However, taxol's extremely low water solubility limits the application of NMR. Theoretical consideration of conformations using computational methods (molecular modeling) are valuable in the analysis of taxol. In combination with NMR and X-ray diffractometry, molecular modeling is a powerful method to determine taxol's conformation (Georg *et al.*, 1995d).

The X-ray structures of taxol and its analogs have been reported. Docetaxel is the first example of X-ray analysis of a taxane diterpenoid containing both an oxetane ring and the side-chain (Guéritte-Voegelein *et al.*, 1990). In the solid state, the six-membered A-ring assumes a distorted boat conformation flattened by the endocyclic olefin and the flagpole interaction between H-13 and the C-16 methyl group. The six-membered C-ring adopts an envelope-like conformation distorted by the planar oxetane D-ring. The central eight-membered B-ring approximates a boat–chair conformation. The overall conformation of the diterpene ring system can be described as "cup-" or "cage-like" (Guéritte-Voegelein *et al.*, 1990; Georg *et al.*, 1995d; Boge *et al.*, 1999a) (Figure 10.35). The conformation of the C-13 side-chain was also determined from the X-ray structure of taxotere. The ester adopts the sterically preferred *s*-Z conformation (C-1′–O-13), resulting in the side-chain extending under the concave face of the diterpene ring system (Figure 10.35). The intramolecular hydrogen bonding between 1′ ester carbonyl, the 2′-hydroxyl and the 3′ NH within the side-chain is regarded to structure the side-chain in a specific orientation relative to the core (Gao *et al.*, 1995).

The crystal structure of taxol was reported in 1995 (Mastropaolo *et al.*, 1995). The research demonstrated that the monoclinic crystals contained two independent paclitaxel molecules (molecules A and B). The taxane ring conformation is very similar in both paclitaxel molecules and is similar to that of docetaxel. The conformations of the side-chain of the two molecules were different from each other and from that of the side-chain of docetaxel. The differences in the torsion angle about the C1′–C2′ bond is the most dramatic differences between the molecules and this leads to the positions and orientations of the C3′ phenyl groups are markedly different in the two molecules. Similarly, the differences in the torsion angle about the C3′–N and C5′–C1 bonds affect the positions and orientations of the benzamide group. The orientation of the C2 benzoate and the acetate groups at C4 and C10 are very similar in the two molecules: C2 benzoate and C10 acetate groups extend away from the ring structure in opposite directions approximately perpendicular to the C13 side-chain, the C4 acetate points away from the C2 benzoate group. Molecule B showed the same hydrophobic clustering of the C2 benzoyl, C4 acetyl, and the C3′ phenyl groups as the results of the NMR polar solvent studies. But the

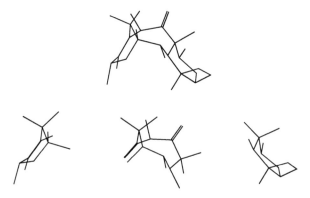

Figure 10.35 Diterpene moiety and ring fragments of taxol.

conformation of molecule A is different and hitherto unobserved. It is noteworthy that no intramolecular hydrogen bonding between the 2′ hydroxyl and the C1′ carbonyl oxygen or between the O2′ hydroxyl and the amide nitrogen was observed, which was found in some of NMR studies of taxol and in an analysis of crystal structure of docetaxel.

The comparison of the conformation of 2-debenzoyl, 2-acetoxy paclitaxel, an inactive paclitaxel analog, and docetaxel, an active paclitaxel analog, showed that the conformation of the tetracyclic ring systems is essentially identical and the conformation of the C-13 side-chain and the orientation of the side-chain relative to the core are similar (Gao *et al.*, 1995). The only slight differences are the conformations of substituents at C2, C4 and C10. The pattern of hydrogen bonding in 2-debenzoyl, 2-acetoxy paclitaxel is also similar with that of docetaxel. The only difference is a hydrogen bond between N3′ and the isopropyl alcohol molecule in 2-debenzoyl, 2-acetoxy paclitaxel, but the same atom in docetaxel does not form any intermolecular hydrogen bonding. This might explain the relative deviation on the conformation of 2-debenzoyl, 2-acetoxy paclitaxel from docetaxel at the end of the side-chain.

Gao and Golik (1995) compared the conformation of 2′-carbamate taxol with taxotere and revealed the conformation of the tetracyclic ring system is identical to that of docetaxel. The slight differences in the conformation of the substitutes at C2 and C4 were observed. But the C-13 side-chain possessed a different conformation from that of taxotere. The changes of torsion angles about O13-C1′, C1′-C2′ and C3′-N4′ moves the O5′ of 2′-carbamate taxol away from the position of O5′ in taxotere and the end of the side-chain points away from the benzoyl group at C2 and the acetyl group at C4. These changes may explain the loss of activity of 2′-carbomate taxol and it is speculated that O5′ is important in receptor binding.

Gao and Chen (1996a) and Gao and Parker (1996b) studied the conformation of 10-deacetyl-7-epitaxol and 7-mesylpaclitaxel. In the crystal study of paclitaxel and 10-deacetyl-7-epitaxol, the authors found that in solid state, these bioactive paclitaxel analogs can adopt a "hydrophobic collapse" conformation, which is usually held only in an aqueous medium through the process of hydrophobic collapse. More importantly, the authors demonstrated that the "hydrophobic collapse" conformation could exit in a non-aqueous environment. N3′-benzoyl or 3′-phenyl group forms hydrophobic interactions with the 2-benzoyl and 4-acetyl groups (Gao and Chen, 1996a). Another conformation is "apolar" conformation, which occurs in non-polar organic solvents. Interestingly the two conformations were observed in crystal structure of paclitaxel, docetaxel and 10-deacetyl-7-epitaxol. It is noteworthy that in all the X-ray diffraction studied bioactive paclitaxel analogs, they showed essentially identical conformation of the tetracyclic ring systems. The only difference is in the C13 side-chain. The torsion angles for the side-chain of conformer A and B of paclitaxel, conformer A and B of 10-deacetyl-7-epitaxol, docetaxel and 7-mesylpaclitaxel are listed in Table 10.43.

The conformations of taxol and its analogs were also investigated by NMR spectroscopy and molecular modeling studies by several groups. Conformational studies of taxol and taxotere in non-polar solvents such as methylene chloride, or chloroform showed that these compounds have similar conformations as in the crystal form (Falzone *et al.*, 1992). The coupling constant of the diterpene ring system in non-polar solvents are consistent with the torsion angles obtained from the crystal study (Georg *et al.*, 1995d). The flexible C-13 side-chain also has the same conformation as in the solid state, folding under the diterpene ring system. The coupling constant $J_{H2'-H3'}$ is about 2.7 Hz with a gauche $<H-C2'-C3'-H>$ torsion angle. No NOE between the side-chain and the taxane ring-system have been observed which suggests that the side-chain is rather extended (Williams *et al.*, 1993, 1994). VCD studies identified one additional minor conformation which contributes significantly to the conformation equilibrium of taxol in

Table 10.43 Selected torsion angle (°) for the side-chain of conformer A and B of paclitaxel and 10-deacetyl-7-epitaxol, docetaxel, 7-mesylpaclitaxel

| | Paclitaxel | | 10-deacetyl-7-epitaxol | | Docetaxel | 7-mesylpaclitaxel |
	A	B	A	B		
C13-O13-C1'-O1'	2	4	11.8	4.3	−6.6	12.3
C13-O13-C1'-C2'	180	−177	−166.6	−176.9	168.0	−166.4
C13-C1'-C2'-O2'	−84	−138	−134.0	−122.4	−176.7	−159.5
C13-C1'-C2'-C3'	159	103	108.3	118.0	60.2	77.0
C1'-C2'-C3'-C31'	−64	−58	−66.4	−53.6	−179.4	−174.4
C1'-C2'-C3'-N3'	176	179	168.8	−176.7	56.4	61.3
H2'-C2'-C3'-H3'	−174	−179	173.2	−172.2	57.3	64.2
O1'-C1'-C2'-O2'	93	41	47.5	56.5	−2.2	21.8
O1'-C1'-C2'-C3'	−24	−77	−70.0	−63.0	−125.3	−101.6
C2'-C3'-N3'-N4'	−118	−155	−143.0	−109.4	−141.3	−98.6
C2'-C3'-C31'-C32'	−166	102	106.8	131.2	83.6	118.0
O2'-C2'-C3'-N3'	60	61	51.7	63.7	−64.7	−60.2
O2'-C2'-C3'-C31'	180	−175	176.4	−173.3	59.5	64.2
C3'-N3'-C4'-O4'	1	−1	−3.0	−6.3	12.8	1.4
C3'-N3'-C4'-C41'[a]	−178	−178	177.0	172.3	−172.4	179.7
C31'-C3'-N3'-C4'	120	83	92.4	125.5	97.3	134.6
C32'-C31'-C3'-N3'	−73	−137	−130.5	−107.4	−154.6	−117.6
H3'-C3'-N3'-H'(N3')	158	123	153.6	−173.0	159.4	−163.0

Note
a In docetaxel, the corresponding torsion angle is C3'-N3'-C4'-O5'.

chloroform. Its coupling constant $J_{H2'-H3'}$ is about 4.1 Hz (Williams *et al.*, 1993). Recently, the NMR analysis of molecular flexibility in solution (NAMFIS) methodology was used to study the conformation of taxol in chloroform and eight taxol conformations. The most populated form (35%) is the "non-polar" form. The fifth most populated conformer (4%) is the so-called "polar" form usually observed in aqueous solution (Snyder *et al.*, 2000).

But in hydrophilic media, such as water-d_6-DMSO, the side-chain H—C2'—C3'—H anti-conformation was found to be dominant by ^{1}H NMR and hydrophobic clustering of the 2-benzoyl, 3'-phenyl, and 4-acetyl groups which formed the "hydrophobic collapse." The nonpolar side-chain amide groups (benzoyl for taxol, *t*-BOC for taxotere) do not seem to be involved. In this conformation the intramolecular hydrogen bonds were lost (Dubois *et al.*, 1993; Vander Velde *et al.*, 1993; Willams *et al.*, 1993, 1994; Paloma *et al.*, 1994; Snyder *et al.*, 2000). It was speculated that the bound ligand conformation recognized by the microtubular taxol binding site might be implicated by the taxol and taxotere solution conformation (Swindell and Krauss, 1991; Guéritte-Voegelein *et al.*, 1991, Guérnard *et al.*, 1993; Williams *et al.*, 1994). In the solid state structure of the side-chain methylester, several prominent intramolecular hydrogen bonds and van der Waals contacts were observed and speculated that such interactions could be crucial for the binding of taxol analogs to the microtubular taxol binding site (Peterson *et al.*, 1991). The conformation of the side-chain methyl ester was also studied by NMR spectroscopy and molecular modeling in hydrophobic (CDCl$_3$) and hydrophilic (water, d$_6$-DMSO) solvents (Williams *et al.*, 1993; Milanesio *et al.*, 1999). Substantial solvent-induced conformational changes are observed for the side-chain from the guache torsion angle in chloroform to the anti torsion angle in aqueous solvent as recorded by the coupling constant $J_{H2'-H3'}$. This similar conformational change was also found in taxol.

Recently, Ojima *et al.* (1997c) proposed a new interpretation of the conformational changes in paclitaxel in polar solvents by using the fluorine probe approach and indicated a new conformer with an almost eclipsed arrangement of the substituents around the C2′–C3′ bond. This conformation might be the molecular structure first recognized by tubulin binding site (Ojima *et al.*, 1997c). This conformation can also be found in the solution structure of paclitaxel-7-MPA in D_2O (Paloma *et al.*, 1994).

Recently, the conformations of C-2′ modified paclitaxel analogs, such as 2′-deoxypaclitaxel, 2′-methoxypaclitaxel and 2′-acetate paclitaxel were studied by NMR and molecular modeling (Williams *et al.*, 1996; Moyna *et al.*, 1997). The studies showed that the 2′-modified paclitaxel took on solution conformations similar to those of paclitaxel in similar media. The conformational study of 2′-OH acetylation of 7-O-[N-(4′-Fluoresceincarbonyl)-L-ananyl]taxol also obtained similar result (Jiménez-Barbero *et al.*, 1998). These results indicate that neither removal nor methylation or acetylation of paclitaxel's 2′-OH significantly affects the conformation. It was speculated that 2′-OH must act through the interaction with microtubule binding site either by hydrogen bonding to polar amino acid residues or interaction with water bound to the protein (Williams *et al.*, 1996; Moyna *et al.*, 1997; Jiménez-Barbero *et al.*, 1998). It was also suggested that the taxane ring system is recognized first by the receptor, followed by additional interactions of the binding site with the side-chain. So the C-13 side-chain is considered to enhance the drug-receptor interaction (Dubois *et al.*, 1993).

A NMR and molecular modeling study of taxol and the side-chain deletion analogs 349b, 350b, 351 and 352 (Figure 10.36) in chloroform and DMSO-water indicates a correlation between biological activity and the chloroform conformation of taxol (Williams *et al.*, 1994). Biologically active analogs 349b, 350b and 351 possess similar chloroform conformations of taxol, while the inactive analog 352 does not. No similar correlation can be detected for those analogs in DMSO-water. Based on these results it was hypothesized that the chloroform conformation of taxol is important for recognition by the hydrophobic microtubule binding site (Williams *et al.*, 1994; Georg *et al.*, 1995d).

Analysis of the NMR data of paclitaxel and D-secopaclitaxel indicated that opening the D-ring might induce A-ring conformation change from a distorted boat conformation in paclitaxel to a pseudo envelope in D-secopaclitaxel and slight B-ring conformational change. The overall conformational changes bring the C13 side-chain closer to the diterpene ring system. In the NOESY experiment of D-secopaclitaxel in DMSO/water, no aromatic inter-ring NOE cross peaks were observed. From these results, it was concluded that the oxetane could serve as a conformational locker for the diterpene moiety and the side-chain (Boge *et al.*, 1999a).

In order to testify the bound conformation of taxol, Boge *et al.* (1999b) synthesized conformational constricted paclitaxel analogs by connecting the 3′-phenyl and 2-benzoate moiety with two-atom tethers to mimic the "hydrophobic collapse" conformation. None of the analogs

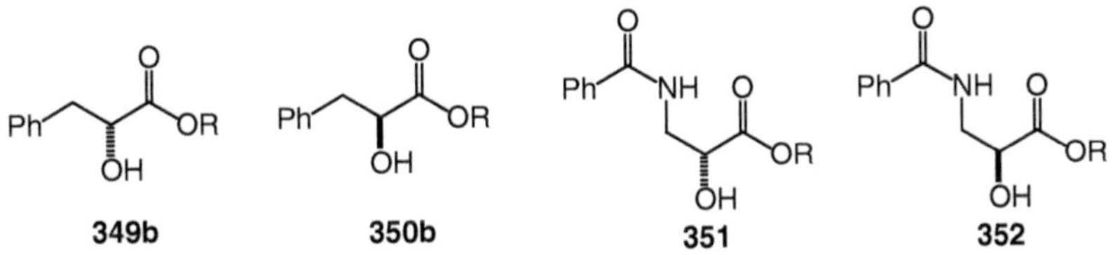

Figure 10.36 C-13 side-chain simplified taxol analogs investigated for their solution conformation (R=baccatin III).

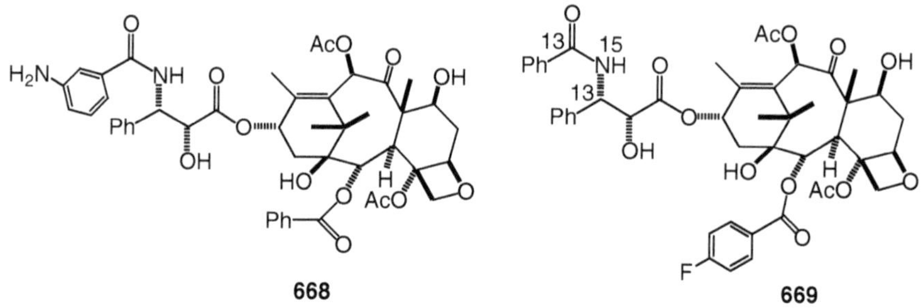

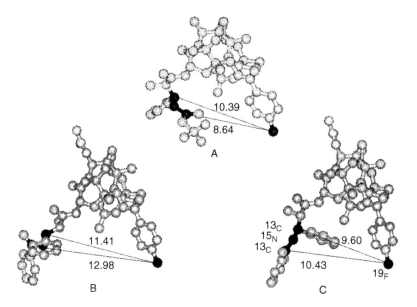

668   **669**

Figure 10.37 Structures of compounds **668–669**.

Figure 10.38 Three possible conformations of microtubule-bound paclitaxel. Reprinted from Li *et al.*,
Conformation of microtubule-bound paclitaxel determined by fluorescene spectroscopy and
REDOR NMR. *Biochemistry*, **39**, 281 (2000). (Copyright 2000 with permission from
American Chemical Society.)

showed activity in a tubulin assembly assay (Boge *et al.*, 1999a). This conformation does not fit
the electron crystallographic density at the binding cavity of β-tubulin (Barboni *et al.*, 2001).
These results provide evidence that the bound state of paclitaxel is unlike the "hydrophobic
collapse." And it was speculated that the bound state is in fact highly extended, with only the
remaining taxane intramolecular hydrophobic contact involving the 3'-phenyl and 4-acetyl
groups (Boge *et al.*, 1999a).

Li *et al.* (2000a,b) studied the conformation of microtubule-bound paclitaxel using the
fluorescence spectroscopy and REDOR NMR. The fluorescent derivatives **671** and **672**
(Figure 10.37) were used to probe paclitaxel–microtubule interactions. The results indicated

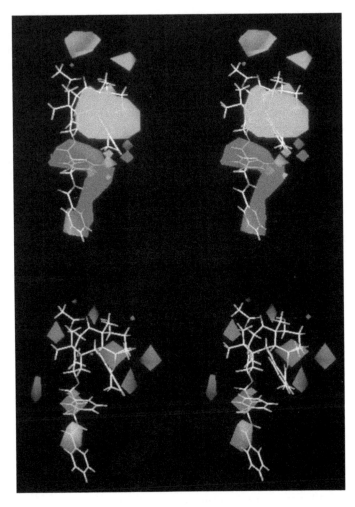

Figure 10.39 Stereoview of the CoMFA contour plot from PLS analysis. Above: steric, below: electrostatic. Reprinted from Morita *et al.*, 3D QSAR analysis of Taxoids from *Taxus cuspidata* var. *nana* by Comparative molecular field approach. *Bioorg. Med. Chem. Lett.*, 7(18), 2387 (1997). (Copyright 1995 with permission from Elsevier Science Ltd.) (See Color Plate I.)

that microtubule contains two types of binding sites of paclitaxel, and the binding affinity depends on the nucleotide content of tubulin (Li *et al.*, 2000b). Three possible conformations for microtubule-bound paclitaxel are illustrated in Figure 10.38 (Li *et al.*, 2000a). Structure A is docetaxel using the X-ray coordinates and it illustrates the conformation of paclitaxel observed in aprotic solvents. Structure C closely resembles one of the conformations of paclitaxel observed in protic solvents. The combined results of fluorescence and REDOR measurement identify structure C as the best representation of the conformation of microtubule-bound paclitaxel.

A series of naturally occurring taxoids from *Taxus cuspidata* var. *nana* were investigated 3D QSAR by CoMFA analysis (Morita *et al.*, 1997). The major steric and elctrostatic features of the QSAR are represented in Figure 10.39 in the form of three-dimensional contour maps. In the steric CoMFA map (above), as indicated previously, the side-chain at C-13 would enhance the growth inhibition. The side-chain at C-5 had a small contribution to the action. Interestingly, the result showed that the side-chain at C-2 of a smaller bulk would enhance the activity. The result also indicated that smaller bulks at C-7 and C-9 would enhance the activity. In the electrostatic CoMFA map (below), the regions where a positive electrostatic interaction would improve the growth inhibition are located around the N-acylphenylisoserine ester at C-13, while those where negative electrostatic interaction would occur are around each of the ester side-chains and oxetane rings (Morita *et al.*, 1997). A series of paclitaxel analogs were also studied using CoMFA by Zhu *et al.* (1997) and got similar result as above.

Overall, the major models for paclitaxel conformation are determined. One is "hydrophobic clapse" in DMSO-d_6/water in which the 3'-phenyl, 2-benzoyl and 4-acetyl moieties are in close contact. The second model is called "apolar" conformation in CDCl3 in which the 2-benzoyl group is in proximity to the tert-butyl group (docetaxel) or the phenyl group of the N-benzoyl moiety (paclitaxel) (Boge *et al.*, 1999a). Until recently, structural information on the microtubular taxol binding site and the taxol–microtubule complex is not available. Which one of the taxol conformation in hydrophobic or hydrophilic media, or neither, is recognized by microtubules remains debatable (Williams *et al.*, 1994).

References

Ali, S.M., Hoemann, M.Z., Aube, J., Mitscher, L.A. and Georg, G.I. (1995) Novel cytotoxic 3'-(tert-butyl)3'-dephenyl analogs of paclitaxel and docetaxel. *J. Med. Chem.*, 38, 3821–3828.

Appendino, G. (1997) Biologically active diterpenoids. Synthesis of analogs of paclitaxel and resinifera toxin. *Gazzetta Chimica Italiana*, 127, 461–469.

Appendino, G., Danieli, B., Jakupovic, J., Belloro, E., Scambia, G. and Bombadelli, E. (1997) Synthesis and evaluation of C-seco paclitaxel analogs. *Tetrahedron Letters*, 38(24), 4273–4276.

Appendino, G., Gariboldi, P., Gabetta, B., Pace, R., Bombadelli, E. and Viterbo, D., (1992) 14β-hydroxy-10-deacetylbaccatin III, a new taxane from Himalayan yew (*Taxus willichiana Zucc.*). *J. Chem. Soc. Perkin Trans.*, 1, 2925.

Baloglu, E. and Kingston, D.G.I. (1999) The taxane diterpenoids. *J. Nat. Prod.*, 62, 1448–1472.

Barboni, L., Datta, A., Dutta, D., Georg, G.I. and Vander Velde D.G., Himes, R.H., Wang, M. and Snyder, J.P. (2001a) Novel D-seco paclitaxel analogs: synthesis, biological evaluation, and model testing. *J. Org. Chem.*, 66, 3321–3329.

Barboni, L., Lambertucci, C., Appendino, G., Velde D.G.V., Himes, R.H. and Bombardelli, E. *et al.* (2001b) Synthesis and NMR driven conformational analysis of taxol analogs conformationally constrained on the C13 side chain. *J. Med. Chem.*, 44, 1576–1587.

Bayley, H., *Photogenerated Reagents* in *Biochemistry and Molecular Biology*. Elsevier, New York, 1983.

Beusker, P.H., Veldhuis, H., van den Bossche, B.A.C. and Scheeren, H.W. (2001) Semisynthesis of flexible 5,7-dideoxypaclitaxel derivative from taxine B. *Eur. J. Org. Chem.*, 9, 1761–1768.

Boge, T.C., Himes, R.H., Vander Velde, D.G. and Georg, G.I. (1994) The effect of the aromatic rings of taxol on biological activity and solution conformation: synthesis and evaluation of saturated taxol and taxotere analogs. *J. Med. Chem.*, 37, 3337–3343.

Boge, T.C., Wu, Z.J., Himes, R.H., Vander Velde, D.G. and Georg, G.I. (1999a) Conformationally restricted paclitaxel analogs: macrocyclic mimics of the "hydrophobic collapse" conformation. *Bioorg. Med. Chem. Lett.*, 9, 3047–3052.

Boge, T.C., Hepperle, M., Vander Velde, D.G., Gunn, C.W., Grunewald, G.L. and Georg, G.I. (1999b) The oxetane conformational lock of paclitaxel: structural analysis of D-secopaclitaxel. *Bioorg. Med. Chem. Lett.*, 9, 3041–3046.

Bourzat, J.-D., Lavelle, F. and Commercon, A. (1995) Synthesis and biological activity of *para*-substituted 3′-phenyl docetaxel analogs. *Bioorg. Med. Chem. Lett.*, 5, 809–814.

Carboni, J.M., Farina, V., Rao, S., Hauch, S.B. and Ringel, I. (1993) Synthesis of a photoaffinity of analog of taxol as an approach to identify the taxol binding site on microtubules. *J. Med. Chem.*, 36, 513–515.

Chaudhary, A.G. and Kingston, D.G.I (1993) Synthesis of 10-deacetoxytaxol and 10-deoxytaxotere. *Tetrahedron Letters*, 34, 4921–4924.

Chaudhary, A.G., Rimoldi, J.M., Kingston, D.G.I. (1993) Modified taxols. 10. Preparation of 7-deoxytaxol, a highly bioactive taxol derivative, and interconversion of taxol and 7-*epi*-taxol. *J. Org. Chem.*, 58, 3798–3799.

Chaudhary, A.G., Gharpure, M.M., Rimoldi, J.M., Chordia, M.D., Gunatilaka, A.A.L. and Kingston, D.G.I. (1994) Unexpectedly facile hydrolysis of the 2-benzoate group of taxol and syntheses of analogs with increased activities. *J. Am. Chem. Soc.*, 116, 4097–4098.

Chaudhary, A.G., Chordia, M.D., Kingston, D.G.I. (1995) A novel benzoyl group migration: synthesis and biological evaluation of 1-benzoyl-2-des(benzoyloxy) paclitaxel. *J. Org. Chem.*, 60, 3260–3262.

Chen, S.-H. (1996) First syntheses of C-4 methyl ether analogs and the unexpected reactivity of 4-deacetyl-4-methyl ether baccatin III. *Tetrahedron Letters*, 37, 3935–3938.

Chen, S.-H., Combs, C.M., Hill, S.E., Farina, V. and Doyle, T.W. (1992) The photochemistry of taxol: synthesis of a novel pentacyclic taxol isomer. *Tetrahedron Letters*, 33, 7679.

Chen, S.-H., Wei, J.-M., Farina, V. (1993a) Taxol structure–activity relationships: synthesis and biological evaluation of 2-deoxytaxol. *Tetrahedron Letters*, 34, 3205–3206.

Chen, S.-H., Fairchild, C., Mamber, S.W., Fairna, V. (1993b) Taxol structure–activity relationships: synthesis and biological evaluation of 10-deoxytaxol. *J. Org. Chem*, 58, 2927–2928.

Chen, S.-H., Huang, S., Kant, J., Fairchild, C., Wei, J., Farina, V. (1993c) Synthesis of 7-deoxy- and 7,10-dideoxytaxol via radical intermediates. *J. Org. Chem.*, 58, 5028–5029.

Chen, S.-H., Kant, J., Mamber, S.W., Roth, G.P., Wei, J.-M., Marshall, D. *et al.* (1994a) Taxol structure–activity relationships: synthesis and biological evaluation of taxol analogs modified at C-7. *Bioorg. Med. Chem. Lett.*, 4, 2223–2228.

Chen, S.-H., Farina, V., Wei, J.-M., Long, B., Fairchild, C., Mamber, S.W. *et al.* (1994b) Structure–activity relationships of taxol: synthesis and biological evaluation of C2 taxol analogs. *Bioorg. Med. Chem. Lett.*, 4, 479–482.

Chen, S.-H., Kadow, J.F. and Farina, V. (1994c) First syntheses of novel paclitaxel (taxol) analogs modified at the C4-position. *J. Org. Chem.*, 59, 6156–6158.

Chen, S.-H., Fairchild, C. and Kingston, D.G.I. (1995a) Synthesis and biological evaluation of novel C-4 aziridine-bearing paclitaxel (taxol) analogs. *J. Med. Chem.* 38, 2263–2267.

Chen, S.-H., Wei, J.-M., Long, B., Fairchild, C., Carboni, J. and Mamber, S.W. *et al.* (1995b) Novel C-4 paclitaxel (taxol®) analogs: potent antitumor agents. *Bioorg. Med. Chem. Lett.*, 5, 2741–2746.

Chen, S.-H., Huang, S. and Roth, G.P. (1995c) An interesting C-ring contraction in paclitaxel (taxol). *Tetrahedron Letters*, 36, 8933–8936.

Chen, S., Farina, V., Vyas, D.M. and Doyle, T.W. (1996) Synthesis and biological evaluation of C-13 amide-linked paclitaxel (taxol) analogs. *J. Org. Chem.*, 61, 2065–2070.

Chen, S.H., Xue, M., Huang S., Long, B.H., Fairchild, C.A. and Rose W. C. *et al.* (1997) Structure–activity relationships study at the 3′-N position of paclitaxel – Part 1: Synthesis and biological evaluation of the 3′-(*t*)-butylaminocarbonyloxy bearing paclitaxel analogs. *Bioorg. Med. Chem. Lett.*, 7, 3057–3062.

Cheng, Q., Oritani, T. and Horiguchi, T. (2000a) The synthesis and biological activity of 9- and 2′-cAMP 7-deoxypaclitaxel analogs from 5-cinnamoyltriacetyltaxicin-I. *Tetrahedron*, 56, 1667–1679.

Cheng, Q., Oritani, T., Horiguchi, T., Yamada, T. and Mong, Y. (2000b) Synthesis and biological evaluation of novel 9-functional heterocyclic coupled 7-deoxy-9-dihydropaclitaxel analog. *Bioorg. Med. Chem.
Lett.*, **10**, 517–521.

Chordia, M.D. and Kingston, D.G.I. (1996) Synthesis and biological evaluation of 2-epi-paclitaxel. *J. Org.
Chem.*, **61**, 799–801.

Chordia, M.D. and Kingston, D.G.I. (1997) Synthesis and biological evaluation of amide-linked
A-norpaclitaxel. *Tetrahedron*, **53**, 5699–5710.

Chordia, M.D., Chaudhary, A.G., Kingston, D.G.I., Jiang Y.Q., and Hamel, E. (1994) Synthesis and biological evaluation of 4-deacetoxypaclitaxel. *Tetrahedron Letters*, **35**, 6843–6846.

Chordia, M.D., Yuan, H., Jagtap, P.G., Kadow, J.F., Long, B.H., Fairchild, C.R. and Johnston, K.A.
et al. (2001) Synthesis and bioactivity of 2,4-diacyl analogs of paclitaxel. *Bioorg. Med. Chem.*, **9**,
171–178.

Combeau, C., Commercon, A., Mioskowski, C., Rousseau, B., Aubert, F. and Goeldner, M. (1994)
Predominant labeling of β- over α-tubulin from porcine brain by a photoactivatable taxoid derivative.
Biochemistry, **33**, 6676–6683.

Damen, E.W.P., Wiegerinck, P.H.G., Braamer, L., Sperling, D., de Vos D. and Scheeren, H.W. (2000)
Paclitaxel esters of malic acid as prodrugs with improved water solubility. *Bioorg. Med. Chem.*, **8**,
427–432.

Dasgupta, D., Park, H., Harriman, G.C.B., Georg, G.I. and Himes, R.H. (1994) Synthesis of a photoaffinity taxol analog and its use in labeling tubulin. *J. Med. Chem.*, **37**, 2976–2980.

Datta, A., Aube, J., Georg, G.I., Mitscher, L.A. and Jayasinghe, L.R. (1994a) The first synthesis of a C-9
carbonyl modified baccatin III derivative and its conversion to novel taxol and taxotere analogs. *Bioorg.
Med. Chem. Lett.*, **4**, 31.

Datta, A., Jayasinghe, L.R. and Georg, G.I. (1994b) 4-Deacetyltaxol and 10-acetyl-4-deacetyltaxotere:
synthesis and biological evaluation. *J. Med. Chem.*, **37**, 4258–4260.

de Bont, D.B.A., Leenders, R.G.G., Haisma, H.J., Meulen-Muileman, I. and Scheeren, H.W. (1997)
Synthesis and biological activity of β-glucuronyl carbamate-based prodrugs of paclitaxel as potential
candidates for ADEPT. *Bioorg. Med. Chem.*, **5**, 405–414.

de Groot, F.M.H., van Berkom, L.W.A. and Scheeren, H.W. (2000) Synthesis and biological evaluation of
2′-carbamate-linked and 2′-carbanate-linked prodrugs of paclitaxel: selective activation by the tumor-
associated protease plasmin. *J. Med. Chem.*, **43**, 3093–3102.

Deutsch, H.M., Glinski, J.A., Hernandez, M., Haugwitz, R.D., Narayanan, V.L. and Suffness, M. (1989)
Synthesis of congeners and prodrugs. 3. Water–soluble prodrugs of taxol with potent antitumor activity.
J. Med. Chem. Lett., **32**, 788–792.

Doyle, T.W. and Vyas D.M. (1990) Second generation analogs of etoposide and mitomycin C. *Cancer
Treatment Review*, **17**, 127–131.

Dubois, J., Guénard, D., Guéritte-Voegelein, F., Guedira, N., Potier, P. and Gillet, B. *et al.* (1993)
Conformation of Taxotere® and analogs determined by NMR spectroscopy and molecular modeling
studies. *Tetrahedron*, **49**, 6533–6544.

Dubois, J., Thoret, S., Guéritte, F. and Guénard, D. (2000) Synthesis of 5(20)deoxydocetaxel, a new active
docetaxel analog. *Tetrahedron Letters*, **41**, 3331–3334.

Dubowchik, G.M., Mosure, Kathleen, Knipe,J.O. and Firestone, R.A. (1998) Cathepsin B-sensitive diterpene prodrugs. 2. Models of anticancer drugs paclitaxel (Taxol®), mitomysin C and doxorubicin. *Bioorg.
Med. Chem. Lett.*, **8**, 3347–3352.

Falzone, C.J., Benesi, A.J. and Lecomte, J.T.J. (1992) Characterization of taxol in methylene chloride by
NMR spectroscopy. *Tetrahedron Letters*, **33**, 1169–1172.

Fang W.-S., Fang, Q.-C. and Liang, X.-T. (2001) Synthesis of the 2α-benzoylamido analog of docetaxel.
Tetrahedron Letters, **42**, 1331–1333.

Gabetta, B., Fuzzati, N., Orsini, P., Peterlongo, F., Appendino, G., Vander Velde, D.G. (1999) Paclitaxel
analogs from taxus *x* media cv. Hicksii. *J. Nat. Prod.*, **62**, 219–223.

Gao, Q. and Chen, S.-H. (1996a) An unexpected side chain conformation of paclitaxel (taxol) crystal
structure of 7-mesylpaclitaxel. *Tetrahedron Letters*, **37**, 3425–3428.

Gao, Q. and Golik, J. (1995) 2′-carbamate taxol. *Acta Cryst.*, C51, 295–298.

Gao, Q. and Parker, W.L. (1996b) The "hydrophobic collapse" conformation of paclitaxel (taxol) has been observed in a non-aqueous environment: crystal structure of 10-deacetyl-7-epitaxol. *Tetrahedron*, 52, 2291–2300.

Gao, Q., Wei, J.-W. and Chen, S.-H. (1995) Crystal structure of 2-debenzoyl, 2-acetoxy paclitaxel (Taxol): conformation of the paclitaxel side chain. *Pharmaceut. Res.*, 12, 337–341.

Georg, G.I. and Cheruvallath Z.S. (1992a) Synthesis and biologically active taxol analogs with modified phenylisoserine side chains. *J. Med. Chem.*, 35, 4230–4237.

Georg, G.I. and Cheruvallath Z.S. (1992b) Semisynthesis and biological activity of taxol analogs: baccatin III 13-(N-benzoyl)-(2′R,3′S)-3′-(p-tolyl)isoserinate), baccatin III 13-(N-(p-toluoyl)-(2′R,3′S)-3′-phenylisoserinate), baccatin III 13-(N-benzoyl)-(2′R,3′S)-3′-(p-trifluoromethylphenyl)isoserinate) and baccatin III 13-(N-(p-trifluoromethylbenzoyl)-(2′R,3′S)-3′-phenylisoserinate). *Bioorg. Med. Chem. Lett.*, 2, 1751–1754.

Georg, G.I., Cheruvallath Z.S., Himes, R.H. and Mejillano, M.R. (1992c) Novel biologically active taxol analogs: baccatin III 13-(N-(p-chlorobenzoyl)-(2′R,3′S)-3′-phenylisoserinate) and baccatin III 13-(N-benzoyl)-(2′R,3′S)-3′-(p-Chlorophenyl)isoserinate). *Bioorg. Med. Chem. Lett.*, 2, 295–298.

Georg, G.I., Harriman, G.C.B., Himes, R.H. and Mejillano, M.R. (1992d) Taxol photoaffinity labels: 7-(p-azidobenzoyl)taxol synthesis and biological evaluation. *Bioorg. Med. Chemi. Lett.*, 2, 735–738.

Georg, G.I., Cheruvallath, Z.S., Vander Velde, D., Ye, Q.-M., Mitscher, L.A. and Himes, R.H. (1993) Semisynthesis and biological evaluation of brevofoliol 13-[N-benzoyl-(2′R,3′S)-3′-phenylisoserinate]. *Bioorg. Med. Chem. Lett.*, 3, 1349–1350.

Georg, G.I., Ali, S.M., Boge, T.C., Datta, A. and Falborg, L. (1994a) Selective C-2 and C-4 deacylation and acylation of taxol: the first synthesis of a C-4 substituted taxol analog. *Tetrahedron Letters*, 35, 8931–8934.

Georg, G.I., Harriman, G.C.B., Park, H. and Himes, R.H. (1994b) Taxol photoaffinity labels. 2. Synthesis and biological evaluation of N-(4-azidobenzoyl)-N-debenzoyltaxol, N-(4-azido-2,3,4,5-tetrafluorobenzoyl)-N-debenzoyltaxol, and 7-(4-azido-2,3,4,5-tetrafluorobenzoyl)taxol. *Bioorg. Med. Chem. Lett.*, 4, 487–490.

Georg, G.I. and Cheruvallath, Z.S. (1994) Samarium diiodide–mediated deoxygenation of taxol: a one step synthesis of 10-deacetoxytaxol. *J. Org. Chem.*, 59, 4015.

Georg, G.I., Boge, T.C., Cheruvallath, Z.S., Harriman, G.C.B., Hepperle, M. and Park, H. *et al.* (1994c) Scotten–Baumann acylation of N-debenzoyltaxol: an efficient route to N-acyl taxol analogs and their biological evaluation. *Bioorg. Med. Chem. Lett.*, 4, 335–338.

Georg, G.I., Ali, S.M., Boge, T.C., Datta, A., Falborg, L. and Park, H. (1995a) Synthesis of biologically active 2-benzoyl paclitaxel analogs. *Bioorg. Med. Chem. Lett.*, 5, 259–264.

Georg, G.I., Harriman, G.C.B., Ali, S.M., Datta, A. and Hepperle, M. (1995b) Synthesis of 2-O-heteroaroyl taxanes: evaluation of microtuble assembly promotion and cytotoxicity. *Bioorg. Med. Chem. Lett.*, 5, 115–118.

Georg, G.I., Boge, T.C., Park, H. and Himes, R.H. (1995c) Paclitaxel and docetaxel photoaffinity labels. *Bioorg. Med. Chem. Lett.*, 5, 615–620.

Georg, G.I., Boge, T.C., Cheruvallath, Z.S., Clowers, J.S., Harriman, G.C.B., Hepperle, M. and Park, H. (1995d) The medicinal chemistry of taxol, Chapter 13 in *Taxol® Science and Applications*, edited by M. Suffness, pp. 317–375. Boca Raton, New York, London, Tokyo.

Georg, G.I., Harriman, G.C.B., Hepperle, M., Clowers, J.S. and Velde D.G.V. (1996) Synthesis, conformational analysis, and biological evaluation of heteroaromatic taxanes. *J. Org. Chem.*, 61, 2664–2676.

Georg, G.I., Liu, Y., Boge, T.C. and Himes, R.H. (1997) 7-O-acylpaclitaxel analogs: potential probes to map the paclitaxel binding site. *Bioorg. Med. Chem. Lett.*, 7, 1829–1832.

Greenwald, R.B., Gilbert, C.W., Pendri, A., Conover, C.D., Xia, J. and Martinez, A. (1996) Drug delivery system: water soluble taxol 2′-poly(ethylene glycol) ester prodrugs-design and *in vivo* effectiveness. *J. Med. Chem.*, 39, 424–431.

Grover, S., Rimoldi, J.M., Molinero, A.A., Chaudhary, A.G., Kingston, D.G.I. and Hamel, E. (1995) Differential effects of paclitaxel (taxol) analogs modified at position C-2, C-7, and C-3' on tubulin polymerization and polymer stabilization: identification of a hyperactive paclitaxel derivative. *Biochemistry*, 34, 3927–3934.

Guéritte-Voegelein, F., Guénard, D., Mangatal, L., Potier, P., Guilhem, J. and Cesario, M. *et al* (1990) Structure of a synthetic taxol precursor: *N-tert*-butoxycarbonyl-10-deacetyl-*N*-debenzoyltaxol. *Acta Cryst.*, C46, 781–784.

Guéritte-Voegelein, F., Guénard, D., Mangatal, L., Lavelle F., Le Goff, M. and Potier, P. (1991) Relationship between the structure of taxol analogs and their antimitotic activity. *J. Med. Chem.*, 34, 992–998.

Gunatilaka, A.A.L., Ramdayal, F.D., Sarragiotto, M.H., Kingston, D.G.I., Sackett, D.L. and Hamel, E. (1999) Synthesis and biological evaluation of novel paclitaxel (taxol) D-ring modified analogs. *J. Org. Chem.*, 64, 2694–2703.

Guo, Y., Diallo, B., Jaziri, M., Vanhaelen-fastre, R. and Vanhaelen, M. (1995) Two new taxoids from the stem bark of *Taxus baccata*. *J. Nat. Prod.* 58, 1906–1912.

Harris, J.W., Katki, A., Anderson, L.W., Chmurny, G.N., Paukstelis, J.V. and Collis, J.M. (1994) Isolation, structural determination, and biological activity of 6α-hydroxytaxol, the principle human metabolite of taxol. *J. Med. Chem.*, 37, 706.

He, L., Jagtap, P.G., Kingston, D.G.I., Shen, H., Orr, G.A. and Horwitz, S.B. (2000) A common pharmacophore for taxol and the epothilones based on the biological activity of a taxane molecule lacking a C-13 side chain. *Biochemistry*, 39, 3972–3978.

Holton, R.A., Somoza, C. and Chai, K.B. (1994) A simple synthesis of 10-deacetoxytaxol derivatives. *Tetrahedron Letters*, 35, 1665.

Huang, C.H.O., Kingston, D.G.I., Magri, N.F. and Samaranayake, G. (1986) New taxanes from *Taxus brevifolia*. *J. Nat. Prod.*, 49, 665.

Iimura, S., Ohsuki, S., Chiba, J., Uoto, K., Iwahana, M., Terasawa, H. and Soga, T. (2000) Synthesis and antitumor activity of non-prodrug water soluble taxoid:10-C-aminoalkylated docetaxel analogs. *Heterocycles*, 53, 2719–2737.

Iimura, S., Uoto, K., Ohsuki, S., Chiba, J., Yoshino, T., Iwahana, M., Jimbo, T., Terasawa, H. and Soga, T. (2001) Orally active docetaxel analog: synthesis of 10-deoxy-10-C-morpholinoethyl docetaxel analogs. *Bioorg. Med. Chem. Lett.*, 11, 407–410.

Jayasinghe, L., Datta, A., Ali, S.M., Zygmunt, J., Vander Velde, D.G. and Georg, G.I. (1994) Structure–activity studies of antitumor taxanes: synthesis of novel C-13 side chain homologated taxol and taxotere analogs. *J. Med. Chem.*, 37, 2981–2984.

Jiménez-barbero, J., Souto, A.A., Abal, M., Barasoain, I., Evangelio, J.A. and Acuña, A.U. *et al.* (1998) Effect of 2'-OH acetylation on the bioactivity and conformation of 7-O-[N-(4'-fluoresceincarbonyl)-L-alanyl]taxol. A NMR-fluorescence microscopy study. *Bioorg. Med. Chem.*, 6, 1875–1863.

Kant, J., Huang, S., Wong, H., Fairchild, C., Vyas, D. and Farina, V. (1993) Studies towards structure–activity relationships of Taxol® analogs with C-2' modified phenylisoserine side chains. *Bioorg. Med. Chem. Lett.*, 3, 2471–2474.

Kant, J., Okeeffe, W.S., Chen, S., Farina, V., Fairchild, C. and Johnston, K. (1994) A chemoselective approach to functionalize the C-10 position of 10-deacetylbaccatin III. Synthesis and biological properties of novel C-10 taxol analogs. *Tetrahedron Letters*, 35, 5543–5546.

Kar, A.K., Braun, P.D. and Wandless, T.J. (2000) Synthesis and evaluation of daunorubicin-paclitaxel dimers. *Bioorg. Med. Chem. Lett.*, 10, 261–264.

Kingston, D.G.I. (1991) The chemistry of taxol. *Pharmac. Ther.*, 52, 1–34.

Kingston, D.G.I. (1994) Taxol: the chemistry and structure–activity relationships of a novel anticancer agent. *Trends Biotech.*, 12, 222–227.

Kingston, D.G.I. (1998) Studies on the chemistry of taxol. *Pure Appl. Chem.*, 70, 331–334.

Kingston, D.G.I. (2000) Recent advances in the chemistry of taxol. *J. Nat. Prod.*, 63, 726–734.

Kingston, D.G.I, Hawkins, D.R. and Ovington, L. (1982) New taxanes from *Taxus brevifolia*. *J. Nat. Prod.*, 45, 466.

Kingston, D.G.I., Samaranayake, G. and Ivey, C.A. (1990), The chemistry of taxol, a clinically useful anti-cancer agent. *J. Nat. Prod.*, 53, 1.

Kingston, D.G.I., Chaudhary, A. G., Chordia, M.D., Gharpure, M., Gunatilaka, A.A.L., Higgs, P.I. *et al.* (1998) Synthesis and biological evaluation of 2-acyl analogs of paclitaxel (taxol). *J. Med. Chem.*, 41, 3715–3726.

Kingston, D.G.I., Chordia, M.D., Jagtap, P.G., Liang, J., Shen, Y. and Long, B.H. *et al.* (1999) Synthesis and biological evaluation of 1-deoxypaclitaxel. *J. Org. Chem.*, 64, 1814–1822.

Klein, L.L. (1993) Synthesis of 9-dihydrotaxol: a novel bioactive taxane. *Tetrahedron Letters*, 34, 2047–2050.

Klein, L.L., Maring, C.J., Li, L., Yeung, C.M., Thomas, S.A. and Grampovnik, D.J. *et al.* (1994) Synthesis of ring B-rearranged taxane analogs. *J. Org. Chem.*, 59, 2370–2373.

Klein, L.L., Li, L., Maring, C.J., Yeung, C.M., Thomas, S.A. and Grampovnik, D.J. *et al.* (1995) Antitumor activity of 9(*R*)-dihydrotaxane analogs. *J. Med. Chem.*, 38, 1482–1492.

Kobayashi, J., Inubushi, A., Hosoyama, H., Yoshida, N., Sasaki, T. and Shigemori, H. (1995) Taxuspines E–H and J, new taxoids from the Japanese yew *Taxus cuspidata*. *Tetrahedron*, 51, 5971–5978.

Kobayashi, J., Hosoyama, H., Wang, X.-X., Shigemori, H., Koiso, Y. and Iwasaki, S. *et al.* (1997) Effects of Taxoids from *Taxus cuspidata* on microtubule depolymerization and vincristine accumulation in MDR cells. *Bioorg. Med. Chem. Lett.*, 7, 393–398.

Lataste, H., Sénilh, V., Wright, M., Guénard, D., Potier, P. (1984) Relationships between the structures of taxol and baccatine III derivatives and their *in vitro* action on the disassembly of mammalian brain and physarum amoebal microtubules. *Proc. Natl. Acad. Sci. USA*, 81, 4090–4094.

Leenders, R.G.G., Gerrits, K.A.A., Ruijitenbeek, R., Scheeren, H.W., Haisman, H.J. and Boven, E. (1995) β-Glucuronyl carbamate-based pro-moieties designed for prodrugs in ADEPT. *Tetrahedron Letters*, 36, 1701–1704.

Li, L.P., Thomas, S.A., Klein, L.L., Yeung, C.M., Maring, C.L. and Grampovnik, D.J. *et al.* (1994) Synthesis and biological evaluation of C-3′-modified analogs of 9(*R*)-dihydrotaxol. *J. Med. Chem.*, 37, 2655–2663.

Li, Y., Poliks, B., Cegelski, L., Poliks, M., Gryczynski, Z. and Piszczek. G. (2000a) Conformation of microtubule-bound paclitaxel determined by fluorescent spectroscopy and REDOR NMR. *Biochemistry*, 39, 281–291.

Li, Y., Edsall, R., Jagtap, P.G., Kingston, D.G.I. and Bane, S. (2000b) Equilibrium studies of a fluorescent paclitaxel derivative binding to microtubule. *Biochemistry*, 39, 616–623.

Liang, X., Kingston, D.G.I., Lin, C.M. and Hamel, E. (1995a) Synthesis and biological evaluation of paclitaxel analogs modified in ring C. *Tetrahedron Letters*, 36, 2901–2904.

Liang, X., Kingston, D.G.I., Long, B.H., Fairchild, C.A. and Johnston, K.A. (1995b) Synthesis, structure elucidation and biological evaluation of C-norpaclitaxel. *Tetrahedron Letters*, 36, 7795–7798.

Liang, X., Kingston, D.G.I., Long, B.H., Fairchild, C.A. and Johnston, K.A. (1997) Paclitaxel analogs modified in ring C: synthesis and biological evaluation. *Tetrahedron*, 53, 3441–3456.

Loeb, C., Combeau, C., Ehret-Sabatier, L., Breton-Gilet, A., Faucher, D., Rousseau, B. and Commercon, A. *et al.* (1997) [^{3}H](Azidophenyl)ureido taxoid photolabels peptide amino acids 281–304 of α-tubulin. *Biochemistry*, 36, 3820–3825.

Luo, Y. and Prestwich, G.D. (1999) Synthesis and selective cytotoxicity of a hyaluronic acid–antitumor bioconjugate. *Bioconjugate Chem.*, 10, 755–763.

Ma, W., Park, G.L., Gomez, G.A., Nieder, M.H., Adams, T.L. and Aynsley, J.S. *et al.* (1994) New bioactive taxoids from cell cultures of *Taxus baccata*. *J. Nat. Prod.*, 57, 116–122.

Magri, N.F. and Kingston, D.G.I. (1988) Modified taxols, IV. Synthesis and biological activity of taxols modified in the side chain. *J. Nat. Prod.*, 51, 298–306.

Mamber, S.W., Mikkilineni, A.B., Pack, E.J., Rosser, M.P., Wong, H. and Ueda, Y. (1995) Tubulin polymerization by paclitaxel (taxol) phosphate prodrugs after metabolic activation with alkaline phosphatase. *J. Pharmacol. Exp. Ther.*, 274, 877–883.

Manfredi, J.J. and Horwitz, S.B. (1984) Taxol: an antimitotic agent with a new mechanism of action. *Pharmaco. Ther.*, 25, 83.

Margraff, R., Bezard, D., Bourzat, J.D. and Commercon, A., (1994) Synthesis of 19-hydroxy docetaxel from a novel baccatin. *Bioorg. Med. Chem. Lett.*, 4, 233.

Mastalerz, H., Zhang, G., Kadow, J., Fairchild, C., Long, B. and Vyas, D.M. (2001) Synthesis of 7-β-sulfur analogs of paclitaxel utilizing a novel epimerization of the 7α-thiol group. *Org. Lett.*, 3(11), 1613–1615.

Mastropaolo, D., Camerman A., Luo, Y., Brayer, G.D. and Camerman. N. (1995) Crystal and molecular structure of paclitaxel (taxol). *Proc. Natl. Acad. Sci. USA*, 92, 6920–6924.

Mathew, A.M., Mejillano, M.R., Nath, J.P., Himes, R.H. and Stella, V.J. (1992) Synthesis and evaluation of some water-soluble prodrugs and derivatives of taxol with antitumor activity. *J. Med. Chem.*, 35, 145–151.

Marder-Karsenti, R., Dubois, J., Bricard, L., Guénard, D. and Guéritte-Voegelein, F. (1997) Synthesis and biological evaluation of D-ring-modified taxanes: 5(20)-azadocetaxel analogs. *J. Org. Chem.*, 62, 6631–6637.

Mclaughlin, J.L., Miller, R.W., Powell, R.G., Smith, C.R. Jr. (1981) 19-Hydroxybaccatin III, 10-deacetyl-cephalomannine, and 10-deacetyltaxol: new antitumor taxanes from *Taxus wallichiana. J. Nat. Prod.*, 44, 312–319.

Milanesio, M., Ugliengo, P., Viterbo, D. and Appendino, G. (1999) *Ab initio* conformation study of the phenylisoserine side chain of paclitaxel. *J. Med. Chem.*, 42, 291–299.

Miller, R.W., Powell, R.G., Smith, C.R. Jr. (1981) Antileukemic alkaloids from *Taxus wallichiana* zucc. *J. Org. Chem.*, 46, 1469–1474.

Monsarrat, B., Mariel, E., Cros, S., Gares, M., Guénard, D., Gueritte,. F., *et al.* (1990) Taxol metabolism isolation and identification of three major metabolites of taxol in rat bile. *Drug Metabol. Disposition*, 18, 895–901.

Morita, H., Gonda, A., Wei, L., Takeya, K. and Itokawa, H. (1997) 3D QSAR analysis of taxoids from *Taxus cuspidata var nana* by comparative molecular field approach. *Bioorg. Med. Chem. Lett.*, 7, 2387–2392.

Morita, H., Gonda, A., Wei, L., Yamamura, Y., Wakabayashi, H., Takeya, K. *et al.* (1998) Taxoids from *Taxus cuspidata var nana. Phytochemistry*, 48, 857–862.

Moyna, G., Williams, H.J., Scott, A.I., Ringel, I., Gorodetsky, R. and Swindell, C.S. (1997) Conformational studies of paclitaxel analogs modified at the C-2′ position in hydrophobic and hydrophilic solvent systems. *J. Med. Chem.*, 40, 3305–3311.

Nakayama, K., Terasawa, H., Mitsui, I., Ohsuki, S., Uoto, K., Iimura, S. *et al.* (1998) Synthesis and cytotoxic activity of novel 10-alkylated docetaxel analogs. *Bioorg. Med. Chem. Lett.*, 8, 427–432.

Neidigh, K.A., Gharpure, M.M., Rimoldi, J.M. and Kingston, D.G.I. (1994) Synthesis and biological evaluation of 4-deacetylpaclitaxel. *Tetrahedron Letters*, 35, 6839–6842.

Nicolaou, K.C., Riemer, C., Kerr, M.A., Rideout, D. and Wrasidlo, W. (1993) Design, synthesis and biological activity of protaxols. *Nature*, 364, 464–466.

Nicolaou, K.C., Dai, W.M. and Guy, R.K. (1994a) Chemistry and biology of taxol. *Angew. Chem. Int. Ed. Engl.*, 33, 15–44.

Nicolaou, K.C., Guy, R.K., Pitsinos, E.N. and Wrasidlo, W. (1994b) A water soluble prodrug of taxol with self-assembling properties. *Angew. Chem. Int. Ed. Engl.*, 33, 1583–1585.

Nicolaou, K.C., Couladouros, E.A., Nantermet, P.G., Renaud, J., Guy, R.K. and Wrasidlo W. (1994c) Synthesis of C-2 taxol analogs. *Angew. Chem. Int. Ed. Engl.*, 33, 1581–1583.

Nicolaou, K.C., Renaud, J., Nantermet, Couladouros, E.A., Guy, R.K., Wrasidlo, W. (1995) Chemical synthesis and biological evaluation of C-2 taxoids. *J. Am. Chem. Soc.*, 117, 2409–2420.

Nicolaou, K.C., Claiborne, C.F., Paulvannan, K., Postema, M.H.D. and Guy, R.K. (1997) The chemical synthesis of C-ring aryl taxoids. *Chem. Eur. J.*, 3, 399–409.

Niethammer A., Gaedicke, G., Lode, H.N. and Wrasidlo, W. (2001) Synthesis and preclinical characterization of a paclitaxel prodrug with improved antitumor activity and water solubility. *Bioconjugate Chem.*, 12(3), 414–420.

Ojima, I. and Lin, S. (1998) Efficient asymmetric syntheses of β-lactams bearing a cyclopropane or an epoxide moiety and their application to the syntheses of novel isoserines and taxoids. *J. Org. Chem.*, 63, 224–225.

Ojima, I., Duclos, O., Zucco, M., Bissery, M., Vrignaud, P., Riou, J.F. *et al.* (1994a) Synthesis and structure–activity relationships of new antitumor taxoids, effects of cyclohexyl substitution at the C-3′ and/or C-2 of taxotere (docetaxel). *J. Med. Chem.*, 37, 2602–2608.

Ojima, I., Fenoglio, I., Park, Y.H., Sun, C.-M., Appendino, G. and Pera, P. *et al.* (1994b) Synthesis and structure–activity relationships of novel nor-seco analogs of taxol and taxotere. *J. Org. Chem.*, 59, 515–517.

Ojima, I., Duclos, O., Kuduk, S.D., Sun, C.-M., Slater, J.C., Lavelle, F., Veith, J.M. and Bernacki, R.J. (1994c) Synthesis and biological activity of 3′-alkyl- and 3′-alkenyl-3′-dephenyldocetaxels. *Bioorg. Med. Chem. Lett.*, 4, 2631–2634.

Ojima, I., Park, Y.H., Sun, C.-M., Fenogli, I., Appendino, G., Pera, P. and Bernacki, R.J. (1994d) Structure–activity relationships of new taxoids derived from 14β-hydroxy-10-deacetylbaccatin III. *J. Med. Chem.*, 37, 1408.

Ojima, I., Duclos, O., Dorman, G., Simomnot, B., Prestwich, G.D. and Rao, S. (1995) A new paclitaxel photoaffinity analog with a 3-(4-benzolphenyl)propanoyl probe for characterization of drug-binding site on tubulin and p-glycoprotein. *J. Med. Chem.*, 38, 3891–3894.

Ojima, I., Slater, J.C., Michaud, E., Kuduk, S.D., Bounaud, P.-Y. and Vrignaud, P. *et al.* (1996a) Synthesis and structure–activity relationships of the second-generation antitumor taxoids: exceptional activity against drug-resistance cancer cells. *J. Med. Chem.*, 39, 3889–3896.

Ojima, I., Kuduk, S.D., Slater, J.C., Gimi, R.H. and Sun, C.-M. (1996b) Synthesis of novel fluorine-containing taxoids by means of β-lactam synthon method. *Tetrahedron*, 52, 209–224.

Ojima, I., Kuduk, S.D., Slater, J.C., Gimi, R.H., Sun, C.-M. and Chakravarty, S. *et al.* (1996c) Syntheses, biological activity, and conformational analysis of fluorine-containing taxoids. In *Biomedical Frontiers of Fluorine Chemistry*, edited by Ojima, I., McCarthy, J.R., Welch, J.H., PP228-2–43.

Ojima, I., Slater, J.C., Kuduk, S.D., Takeuchi, C.S., Gimi, R.H. and Sun, C.-M. *et al.* (1997a) Syntheses and structure–activity relationships of taxoids derived from 14β-hydroxy-10-deacetylbaccatin III. *J. Med. Chem.*, 40, 267–278.

Ojima, I., Kuduk, S.D., Pera, P., Veith, J.M., Bernacki, R.J. (1997b) Synthesis and structure–activity relationships of nonaromatic taxoids: effects of alkyl and alkenyl ester groups on cytotoxicity. *J. Med. Chem.*, 40, 279–285.

Ojima, I., Kuduk, S.D., Chakravarty, S., Ourevitch, M. and Bégué, J.-P. (1997c) A novel approach to the study of solution structures and dynamic behavior of paclitaxel and docetaxel using fluorine-containing analogs as probes. *J. Am. Chem. Soc.*, 119, 5519–5527.

Ojima, I., Bounaud, P.-Y. and Ahern, D.G. (1999) New photoaffinity analogs of paclitaxel. *Bioorg. Med. Chem. Lett.*, 9, 1189–1194.

Ojima, I., Lin, S., Slater, J.C., Wang, T., Pera, P. and Bernacki, R.J. *et al.*, (2000) Syntheses and biological activity of C-3′-difluoromethyl-taxoids. *Bioorg. Med. Chem.*, 8, 1619–1628.

Paloma, L.G., Guy, R.K., Wrasidlo, W. and Nicolaou, K.C. (1994) Conformation of a water-soluble derivative of taxol in water by 2D-NMR spectroscopy. *Chem. Biol.*, 1, 107–112.

Parness, J., Kingston, D.G.I., Powell, R.G., Harracksingh, C. and Horwitz, S.B. (1982) Structure–activity study of cytotoxicity and microtubule assembly *in vitro* by taxol and related taxanes. *Biochem. Biophys. Res. Commun.*, 105, 1082–1083.

Pendri, A., Conover, C.D. and Greenwald, R.B. (1998) Antitumor activity of paclitaxel-2′-glycine conjugate to poly(ethylene glycol): a water soluble prodrug. *Anti-Cancer Drug Design*, 13, 387–395.

Peterson, J.R., Do, H.D. and Rogers, R.D. (1991) X-ray structure and crystal lattice interaction of the taxol side-chain methyl ester. *Pharmaceut. Res.*, 8, 908–912.

Poujol, H., Ahond, A., Mourabit, A.A., Chiaroni, A., Poupat, C., Riche, C. *et al.* (1997a) Taxoides: nouveaux analogues du 7-deshydroxydocetaxel prepares a partir des alcaloides de iif. *Tetrahedron*, 53, 5169–5184.

Poujol, H., Mourabit, A.A., Ahond, A., Poupat, C. and Potier, P. (1997b) Taxoides: 7-deshydroxy-10-acetyldocetaxel et nouveaux analogues prepares a partir des alcaloides de l'if. *Tetrahedron*, 53, 12575–12594.

Preuss, F.R. and Orth, H. (1965) Uber Inhaltsstoffe der Eibe (*Taxus baccata* L.). *Pharmazie*, 20, 698.

Pulicani, J., Bouchard, H., Bourzat, J. and Commercon, A. (1994a) Preparation of 7-modified docetaxel analogs using electrochemistry. *Tetrahedron Letters*, 35, 9709–9712.

Pulicani, J. Bezard, D., Bourzat, J. Bouchard, H., Zucco, M., Deprez, D. *et al.* (1994b) Direct access to 2-debenzoyl taxoids by electrochemistry, synthesis of 2-modified docetaxel analogs. *Tetrahedron Letters*, 35, 9717–9720.

Rao, C.S., Chu, J.-J., Liu, R.-S. and Lai, Y.-K. (1998) Synthesis and evaluation of [C]-labelled and fluorescent-tagged paclitaxel derivatives as new biological probes. *Bioorg. Med. Chem.*, 6, 2193–2204.

Rao, S., Horwitz, S.B. and Ringel, I. (1992) Direct photoaffinity labeling of tubulin with taxol. *J. National Cancer Institute*, 84, 785–788.

Rao, S., Krauss, N.E., Heerding, J.M., Swindell, C.S., Ringel, I., Orr, G.A. and Horwitz, S.B. (1994) 3'-(p-azidobenzamido) taxol photolabels the N-terminal 31 amino acids of β-tubulin. *J. Biol. Chem.*, 269, 3132–3134.

Rao, S., Orr, G.A., Chaudhary, A.G., Kingston, D.G.I. and Horwitz, S.B. (1995) Characterization of the taxol binding site on the microtubule: 2-(m-azidobenzoyl)taxol photolabels a peptide (amino acids 217–231) of β-tubulin. *J. Biol. Chem.*, 270, 20235–20238.

Rose, W.C., Lee, F.Y., Fairchild, C.R. and Kadow, J.F. (2000) Preclinical antitumor activity of a new paclitaxel analog, BMS-184476. *Clin. Cancer Res.*, 6 (Suppl.), 4577s.

Rimoldi, J.M., Kingston, D.G.I., Chaudhary, A.G., Samaranayake, G., Grover, S. and Hamel, E. (1993) Modified taxols, 9. Synthesis and biological evaluation of 7-substituted photoaffinity analogs of taxol. *J. Nat. Prod.*, 53, 1313–1330.

Roh, E.J., Song, C.E., Kim, D., Pae, H.-O., Chung, H.-T. and Lee, K.S. *et al.*, (1999) Synthesis and biology of 3'-N-acyl-N-debenzoylpaclitaxel analogs. *Bioorg. Med. Chem.*, 7, 2115–2119.

Rowinsky, E.K. and Donehower, R.C. (1991) The clinical pharmacology and use of antimicrotubule agents in cancer chemotherapeutics. *Pharmacol. Ther.*, 52, 35

Safavy, A., Raisch, K.P., Khazaeli, M.B., Buchsbaum, D.J. and Bonner, J.A. (1999) Paclitaxel derivatives for targeted therapy of cancer: toward the development of smart taxanes. *J. Med. Chem.*, 42, 4919–4924.

Samaranayake, G., Magri, N.F., Jitrangsri, C. and Kingston, D.G.I. (1991) Modified taxols. 5. Reaction of taxol with electrophilic reagents and preparation of a rearranged taxol derivative with tubulin assembly activity. *J. Org. Chem.*, 56, 5114–5119.

Sénilh, V., Blechert, S., Colin, M., Guénard, D., Picot, F., Potier, P. *et al.* (1984) Mise en evidence de nouveaux Analogs du Taxol extraits de *Taxus baccata*. *J. Nat. Prod.*, 47, 131–137.

Shen, Y.-C., Tai, H.-R. and Chen, C.-Y. (1996) New taxane diterpenoids from the roots of *Taxus mairei*. *J. Nat. Prod.*, 59, 173–176.

Snyder, J.P., Nevins, N., Cicero, D.O. and Jansen, J. (2000) The conformations of taxol in chloroform. *J. Am. Chem. Soc.*, 122, 724–725.

Strohl, W.R. (2000) The role of natural products in a modern drug discovery program. *Drug Discovery Today*, 5, 39–41.

Swindell, C.S. and Krauss, N.E. (1991) Biological active taxol analogs with deleted A-ring side chain substitutents and variable C-2' configurations. *J. Med. Chem.*, 34, 1176–1184.

Swindell, C.S., Heerding, J.M., Krauss, N.E., Horwitz, S.B., Rao, S. and Ringel, I. (1994) Characterization of two taxol photoaffinity analogs bearing azide and benzophenone-related photoreactive substituents in the A-ring side chain. *J. Med. Chem.*, 37, 1446–1449.

Takahashi, T., Tsukamoto, H. and Yamada, H. (1998) Design and synthesis of a water-soluble taxol analog: taxol-sialyl conjugate. *Bioorg. Med. Chem. Lett.*, 8, 113–116.

Ueda, Y., Mikkilineni, A.B., Knipe, J.O., Rose, W.C., Casazza, A.M. and Vyas, D.M. (1993) Novel water-soluble phosphate prodrugs of Taxol® possessing *in vivo* antitumor activity. *Bioorg. Med. Chem. Lett.*, 3, 1761–1766.

Ueda, Y., Wong, H., Matiskella, J.D., Mikkilineni, A.B., Farina, V., Fairchild, C. *et al.*, (1994) Antitumor evaluation of 2'-oxycarbonylpaclitaxels (paclitaxel-2'-carbonates). *Bioorg. Med. Chem. Lett.*, 4, 1861–1864.

Ueda, Y., Matiskella, J.D., Mikkilineni, A.B., Farina, V., Knipe, J.O., Rose, W.C. *et al.* (1995) Novel, water-soluble phosphate derivatives of 2′-ethoxy carbonylpaclitaxel as potential prodrugs of paclitaxel: synthesis and antitumor evaluation. *Bioorg. Med. Chem. Lett.*, 5, 247–252.

Uoto, K., Takenoshita, H., Yoshino, T., Hirota, Y., Ando, S., Mitsuo, I., Terasawa, H. and Soga, T. (1998) Synthesis and evaluation of water-soluble non-prodrug analogs of docetaxel bearing *sec*-aminoethyl group at the C-10 position. *Chem. Pharm. Bull.*, 46(5), 770–776

Vander Velde, D.G., Georg, G.I., Grunewald, G.L., Gunn, C.W. and Mitscher, L.A. (1993) "Hydrophobic collapse" of taxol and taxotere solution conformation in mixture of water and organic solvent. *J. Am. Chem. Soc.*, 115, 11650–11651.

Vyas, D.M., Wong, H., Crosswell, A.R., Casazza, A.M., Knipe, J.O., Mamber, S.W. and Doyle, T.W. (1993) Synthesis and antitumor evaluation of water soluble taxol phosphates. *Bioorg. Med. Chem. Lett.*, 3, 1357.

Wahl, A., Guéritte-Voegelein, F., Guénard, D., Le Goff, M.-T. and Potier, P. (1992) Rearrangement reactions of taxanes: structural modifications of 10-deacetylbaccatin III. *Tetrahedron*, 48(34), 6965–6974.

Wani, M.C., Taylor, H.L., Wall, M.E., Coggon, P. and McPhail, A.T. (1971) Plant antitumor agents. VI. The isolation and structure of taxol, a novel antileukemia and antitumor agent from *Taxus brevifoloa*. *J. Am. Chem. Soc.*, 93, 2325–2327.

Williams, H.J., Scott, A.I., Dieden, R.A., Swindell, C.S., Chirlian, L.E., Francl, M.M. *et al.*, (1993) NMR and molecular modeling study of the conformations of taxol and of its side chain methylester in aqueous and non-aqueous solution. *Tetrahedron*, 49, 6545–6560.

Williams, H.J., Scott, A.I., Dieden, R.A., Swindell, C.S., Chirlian, L.E., Francl, M.M. *et al.*, (1994) NMR and molecular modeling study of active and inactive taxol analogs in aqueous and non-aqueous solution. *Can. J. Chem.*, 72, 252–260.

Williams, H.J., Moyna, G., Scott, A.I. (1996) NMR and molecular modeling study of the conformation of taxol 2′-acetate in chloroform and dimethyl sulfoxide solutions. *J. Med. Chem.*, 39, 1555–1559.

Wittman, M.D., Altstadt, T.J., Fairchild, C., Hansel, S., Johnston, K., Kadow, J.F., Long, B.H., Rose, W.C., Vyas, D.M., Wu, M.-J. and Zoeckler, M.E. (2001a) Synthesis of metabolically blocked paclitaxel analogs. *Bioorg. Med. Chem. Lett.*, 11, 809–810.

Wittman, M.D., Kadow, J.F., Vyas, D.M., Lee, F.L., Rose, W.C., Long, B.H., Fairchild, C. and Johnston, K. (2001b) Synthesis and antitumor activity of novel paclitaxel–chlorambucil hybrids. *Bioorg. Med. Chem. Lett.*, 11, 811–814.

Wu, Q., Bounaud, P.-Y., Kuduk, S.D., Yang, C.-P.H., Ojima, I. and Horwitz, S.B. *et al.*, (1998) Identification of the domains of photoincorporation of the 3′- and 7-benzophenone analogs of taxol in the carboxyl-terminal half of murine mdr1b P-glycoprotein. *Biochemistry*, 37, 11272–11279.

Xue, M., Long, B.H., Fairchild, C.A., Johnston, K., Rose, W.C. and Kadow, J.F. *et al.*, (2000) Structure–activity relationships study at the 3′-N position of paclitaxel-part 2: Synthesis and biological evaluation of the 3′-N-thiourea- and 3′-N-thiocarbamate-bearing paclitaxel analogs. *Bioorg. Med. Chem. Lett.*, 10, 1327–1331.

Yamaguchi, T., Harada, N., Ozaki, K., Arakawa, H., Oda, K. and Nakanishi, N. (1999) Synthesis of taxoids 5. Synthesis and evaluation of novel water-soluble prodrugs of a 3′-desphenyl-3′-cyclopropyl analog of docetaxel. *Bioorg. Med. Chem. Lett.*, 9, 1639–1644.

Young, M.J.T. and Platz, M.S. (1989) Polyfluorinated azides as photoaffinity labeling reagents; the room temperature CH insertion reaction of singlet pentafluorophenyl nitrenes. *Tetrahedron Letters*, 30, 2199.

Yuan, H. and Kingston, D.G.I. (1999) Synthesis and biological activity of a novel C4–C6 bridged paclitaxel analog. *Tetrahedron*, 55, 9707–9716.

Yuan, H. and Kingston, D.G.I., Long, B.H., Fairchild, C.R. and Johnston, K.A. (1999) Synthesis and biological evaluation of C-1 and ring modified A-norpaclitaxel. *Tetrahedron*, 55, 9089–9100.

Yuan, H., Fairchild, C.R., Liang, X. and Kingston, D.G.I. (2000) Synthesis and biological activity of C-6 and C-7 modified paclitaxels. *Tetrahedron*, 56, 6407–6414.

Zee-Cheng, R.K.Y. and Cheng, C.C. (1986) Taxol. *Drugs Future*, 11, 45.

Zhang, H., Takeda, Y., Matsumoto, T., Minami, Y., Yoshida, K., Xiang, W., Mu, Q. and Sun, H. (1994) Taxol related diterpenes from the roots of *taxus yunnanensis*. *Heterocycles*, **38**, 975

Zhao, Z., Kingston, D.G.I. (1991) Modified taxols, 6' preparation of water-soluble prodrugs of taxol. *J. Nat. Prod.*, **54**, 1607–1611.

Zhu, Q., Guo, Z., Huang, N., Wang, M. and Chu, F. (1997) Comparative molecular field analysis of a series of paclitaxel analogs. *J. Med. Chem.*, **40**, 4319–4328.

11 Preclinical and clinical studies of the taxanes

David T. Brown

Introduction

This chapter concerns the lead compound paclitaxel (trade name: Taxol) derived from the bark of the pacific yew, *Taxus brevifolia* and its close relative, docetaxel (trade name: Taxotere). Both drugs share a broadly common mode of action that is unique in cancer chemotherapy; yet there are subtle differences in their formulation and pharmacokinetics, and even nuances in their action at sub-cellular level, that have earned them both an important place in cancer medicine. Such differences may also help to explain their differing side-effect profiles, potency and potential for beneficial and harmful interactions with other drugs.

Preclinical development of the taxanes

As Figure 11.1 illustrates, paclitaxel was the first taxane to be developed. Even by modern standards, its development took a long time. The reader is referred to the detailed and fascinating account of the 30-year passage of the drug from tree to clinic by Suffness and Wall (1995). First collections of the bark were made in 1962 as part of a plant screening programme run jointly between the US National Cancer Institute (NCI) and the US Department of Agriculture (USDA). Cytotoxicity was demonstrated in crude extracts of the bark *in vitro* in 1964. It was not until 1967 that an isolation procedure and unique molecular structure were announced and paclitaxel finally met its public (Wall and Wani, 1967). A more detailed report followed in 1971 (Wani *et al.*, 1971) where significant activity against L1210, P388 and P1534 leukaemia cells, Walker 256 carcinosarcoma and 180 and Lewis lung sarcomas in mice were reported. Early development work with paclitaxel was hampered by low yields from bark samples (typically 0.01–0.02% by weight) and bedevilled by its poor aqueous solubility. In addition, early *in vitro* cancer models produced less than impressive results. In the early 1970s, new cancer models that better reflected *in vivo* solid tumour conditions were studied at the NCI and paclitaxel demonstrated sufficient antitumour activity in one of these – the B16 melanoma mouse model – to meet NCI criteria for further development in 1977. Paclitaxel still had its problems; notably low potency, low water solubility in conjunction with the relatively high doses that needed to be delivered, the lack of suitable co-solvents and cost. Antitumour activity appeared to be limited to intraperitoneal injection against localised tumours in mice.

In 1979, the development of paclitaxel received a boost from the announcement that the drug appeared to have a unique mode of action (Schiff *et al.*, 1979). It prevented cell division at the G2-M phase in cell growth, inhibiting mitosis and subsequent tumour growth and thus had the same sort of effect as the existing naturally derived anti-tumour drugs vincristine and vinblastine; but it did this by stabilising microtubules, discouraging their depolymerisation to tubulin, rather than binding to soluble tubulin and preventing the reformation of microtubules, as is the

1962　Collection of bark from *Taxus brevifolia* as part of the NCI/USDA screening programme.
1964　Cytotoxicity of crude extract demonstrated against KB cells *in vitro*.
1966　First isolation of pure paclitaxel and confirmation of cytotoxicity in KB cells.
1967　Isolation of paclitaxel reported at 0.02% concentration.
1971　Chemical structure of paclitaxel published.
1974　Activity of paclitaxel versus B16 mouse melanoma reported.
1975　Confirmation of earlier activity in B16 melanoma model; pacitaxel meets NCI therapeutic agent development criteria.
1977　Preclinical development of paclitaxel commenced.
1978　Confirmed activity of paclitaxel against MX-1 breast xenograft.
1979　Mode of action published.
1980　Route and dose ranging studies for paclitaxel completed. Ethanol – Cremophor-EL formulation chosen for subsequent development. Toxicological studies commenced.
1982　Toxicological studies with paclitaxel completed.
1983　Investigational New Drug Application (INDA) filed for paclitaxel, prior to entering Phase I clinical trials.
1984　FDA approval for initiation of paclitaxel clinical trials granted. Phase I trails with paclitaxel begin.
1985　Semisynthesis of *docetaxel* from 10-deacetylbaccatin III, extracted from the needles of the European yew, *Taxus baccata*.
1985　Phase II trials start with paclitaxel.
1987　*Docetaxel* selected for clinical development.
1989　First study of paclitaxel in refractory ovarian cancer reported (30% response).
1990　Phase I trials with *docetaxel* begin.
1991　US government grants CRADA to Bristol Myers to develop paclitaxel commercially. Trials announce good response (57%) in metastatic breast cancer.
1992　FDA grants marketing approval for paclitaxel in refractory metastatic, ovarian cancer.
1992　Phase II trials with *docetaxel* begin.
1994　FDA grants marketing approval for paclitaxel in metastatic breast cancer, refractory or insensitive to anthracyclines.
1994　Full synthesis of paclitaxel published.
1996　FDA grants marketing approval for *docetaxel* for anthracycline resistant breast cancer.
1998　FDA grants marketing approval for *docetaxel* in metastatic breast cancer, second line.

Figure 11.1 Development timeline for the taxanes. (See text for further details.)

case with the vinca alkaloids. The recognition that paclitaxel had both a unique chemical structure and a novel mode of action rekindled interest in the compound as an antitumour agent; but it was not until the introduction of yet another new cancer model into the barrage of tests that paclitaxel had to undergo that the drug began to justify the expectations of those who championed its cause at the NCI.

In 1976, human tumour xenografts (human tumours grafted internally in mice incapable of rejecting them because of the lack of a functioning immune system) had been introduced. Paclitaxel was tested in three important xenograft systems: LX-1 (lung), MX-1 (breast) and CX-1 (colon). The drug caused regression of tumour growth in LX-1 and 80–90% tumour growth inhibition in MX-1 and CX-1. It did this after subcutaneous injection against tumours implanted in the subrenal capsule, thus demonstrating favourable distribution properties from a distal injection site (Suffness and Wall, 1995). From late 1978, the place of paclitaxel as an antitumour agent was confirmed and further development could proceed with confidence.

A prime concern was the development of a pharmaceutically stable formulation. As mentioned earlier, paclitaxel is sparingly soluble in water; this, coupled with the relatively high doses required due to low potency, presented a problem. Various formulations were investigated, including liposomes and emulsions (Adams *et al.*, 1993) together with largely unsuccessful structural modifications to give more water-soluble prodrugs (Georg *et al.*, 1995; Straubinger, 1995). Two lead formulations emerged, one involving an ethanol – Cremophor-EL (surfactant) cocktail and the other using polyethylene glycol 400. The former presentation was preferred because, while the solubility of paclitaxel was lower, antitumor activity in the implanted B16 melanoma model showed greater efficacy coupled with lower toxicity.

As Suffness and Wall (1995) have pointed out, the choice had important repercussions for paclitaxel. First, Cremophor-EL was (and is) associated with serious hypersensitivity reactions which dogged early clinical trials; these almost caused the withdrawal of the drug but eventually led to more cautious, longer administration times that ultimately enhanced efficacy. Second, far from being an inert excipient, Cremophor-EL appeared to increase the susceptibility of multiple drug-resistant cancer cells to paclitaxel in cancer cell cultures (Chervinski *et al.*, 1993) and adequate levels were achieved to do likewise in patients (Webster *et al.*, 1993); furthermore, Cremophor-EL may have some cytotoxic activity of its own (Fjallskog *et al.*, 1993).

Thus, by 1980, researchers were in possession of sufficient supplies of a stable formulation of a drug with promising *in vitro* antitumour activity demonstrated in an appropriate model that met the NCI criteria for further development. In late 1980, paclitaxel was embarked on a programme of toxicological studies. Of prime interest were the observations from experiments with the Cremophor-EL vehicle in beagles, showing that relatively large volume, single doses produced serious side-effects, including hypotension, respiratory distress, lethargy and death associated with high peak serum levels of Cremophor-EL; but that these could be minimised by using smaller, repeated doses (National Cancer Institute, 1983). This provided an important practical pointer to the fact that if such effects were to be minimised in patients, the paclitaxel formulation should be administered in a similar way, in smaller, divided doses or as an infusion over a longer time. The toxicity profile of paclitaxel itself was unsurprising. The drug produced myelosupression and lymphoid depletion in both Sprague-Dawley rats and beagle dogs. Emesis, diarrhoea and gastrointestinal ulceration were also observed in dogs. Damage to male reproductive organs was also seen in CD2F1 mice, rats and beagles. From the results of five-day dosage regimens, all toxicities appeared to be cumulative.

Further preclinical trials showed that paclitaxel has good activity against human breast, endometrial, ovarian, lung, tongue and pancreatic tumours transplanted into animal models (Riondel *et al.*, 1986; Sternberg *et al.*, 1987; Rose, 1995) with only minor effects on cardiovascular, nervous and renal tissues. The drug was also found to have a significant inhibitory effect in a human tumour stem cell model (Von Hoff, 1991), including breast, lung, ovary, prostate and testis, at concentrations achievable *in vivo*. In a similar *in vitro* model, paclitaxel compared well to standard cytotoxic agents in 36 varied human tumour specimens at concentrations with acceptably mild toxicity to granulocyte-macrophage colony forming cells (Fan *et al.*, 1987).

In an extensive review of preclinical antitumour activity of paclitaxel and docetaxel, Rose (1995) concluded that they both have broadly similar activities and that in the *in vitro* models used to predict *in vivo* activity (tubulin binding, tubulin polymerisation, inhibition of tubulin disassembly and cytotoxicity) both drugs worked well. Docetaxol was approximately twice as active in most *in vitro* models, notably in inducing tubulin polymerisation and in the human tumour stem cell assay and was more potent *in vivo*.

The US Food and Drug Administration (FDA) approval for initiating trials of paclitaxel in man was obtained in the Spring of 1984. Full details of these Phase 1 trials are covered elsewhere

(Rowinsky *et al.*, 1990, 1992a; Markman, 1991; Slichenmeyer and Von Hoff, 1991; Arbuck *et al.*, 1993a; Arbuck and Blaylock, 1995). Even at this early stage, it was appreciated that while toxicological studies had shown single injections would be preferable to avoid cumulative toxicity of the drug, this would mean injecting high single doses of the Cremophor EL vehicle which had its own dose-related toxicity profile. Thus, a variety of dosages were studied, including 1-, 3-, 6-, and 24-h infusions every 21 days and 1- to 6-h infusions daily for five days repeated every 21 days. Most side-effects seen were of the Type A variety, being predictable, dose-related and reversible: myelosupression, peripheral neuropathy and mucositis. Hypersensitivity reactions were also noted; these occurred on all dosage regimens and caused termination of some Phase 1 studies. Such reactions had a Type B profile, in that they were unpredictable, not related to dose and relatively serious; anaphylaxis and two deaths were recorded (Weiss *et al.*, 1990). Investigators felt, with some degree of certainty and not a little trepidation (Suffness and Wall, 1995), that these reactions were caused by the Cremophor EL that was so essential for a stable formulation. However, antitumour activity against melanoma had been observed in some Phase 1 trials (Wiernik *et al.*, 1987a) and this encouraged investigators to persevere, with a selection of pre-medications (dexamethasone, cimetidine and diphenhydramine) designed to block the effects of histamine and other inflammatory mediators released as a result of hypersensitivity. In addition, the infusion times were extended to 24 h in order to reduce peak plasma concentrations of both paclitaxel and Cremophor-EL (Wiernik *et al.*, 1987; Weiss *et al.*, 1990). This strategy proved very successful in reducing the incidence and severity of hypersensitivity reactions. Suffness and Wall (1995) suggested that this strategy was critical to the subsequent development of paclitaxel and surmised that the longer infusion time adopted (24 h) enhanced clinical efficacy by extending tumour exposure time. There is now considerable support for this hypothesis, in that both the cytotoxicity and *in vivo* antitumour activity of taxanes are in general, greater on prolonged exposure or more leisurely dosage regimens (Rose, 1995). For example, at relatively low doses, paclitaxel infusions of 24-h appear to have some antitumour benefit over 3-h infusions when used to treat refractory ovarian cancer (Bokkel Huinink *et al.*, 1993).

The range of cancers studied in early clinical trials was quite limited. However, important and impressive activity against refractory ovarian cancer was observed between 1988 and 1989 (McGuire *et al.*, 1989). This encouraged the NCI to seek a much larger supply of paclitaxel and prompted the pharmaceutical company Bristol Myers to sign a Cooperative Research and Development Agreement (CRADA) with the NCI to supply and develop the drug (then widely known as Taxol) with a view to marketing in 1991. This collaboration led ultimately to the FDA marketing approval for paclitaxel in refractory ovarian cancer in 1992.

At about the same time, Phase 2 clinical trials revealed that paclitaxel had impressive activity in primary metastatic breast cancer, with single agent response rates of up to 56% (Holmes *et al.*, 1991). Subsequently, the FDA marketing approval for the treatment of this condition, refractory or insensitive to anthracyclines, was granted in 1994.

In vitro, cytotoxicity and *in vivo* antitumour screening identified the bark of the Pacific yew as the best natural source of paclitaxel and it remained a major source of the drug until 1994 (Suffness and Wall, 1995). However, the costs of production, poor yields and the potential and actual environmental damage caused by removing the bark and thus killing large numbers of yew trees, spurred the search for a semi-synthetic method of production from a sustainable natural source. For detailed accounts of this process, see elsewhere in this book; Witherup *et al.* (1990); and Croom (1995). Summarising, a sustainable, reproducible, semi-synthetic route has been established through isolation of the intermediary compound, 10-desacetylbaccatin III from needles of two related Taxus species, *T. baccata* and *T. wallachiana* (Bissery *et al.*, 1995; Holton *et al.*, 1995).

In the mid-1980s, French workers announced a number of semi-synthetic taxane derivatives of 10-desacetylbaccatin III with antitumour activity, extracted from *T. baccata*. One of these, docetaxel, was selected for clinical development in 1987 as a competitor to paclitaxel, because of its increased solubility, potency (docetaxel has a higher microtubule binding site affinity than paclitaxel and is more potent at encouraging microtubule polymerisation, Diaz and Andreu, 1993) and because it could be produced by semi-synthesis from a renewable source (Arbuck and Blaylock, 1995). Preclinical *in vivo* results indicated incomplete cross-resistance between docetaxel and paclitaxel, suggesting subtle differences between the cytotoxic activity of the two drugs that could be exploited therapeutically (Hill *et al.*, 1992; Untch *et al.*, 1994). Preclinical evaluation of docetaxel (Bissery *et al.*, 1995; Cortes and Pazdur, 1995; Lavelle *et al.*, 1995) revealed that the drug was cytotoxic at clinically relevant concentrations against fresh human tumour biopsy specimens from breast, lung, ovary, colorectal and melanoma tissue. Docetaxel also had significant *in vivo* antitumour activity against B16 melanoma, C38 colon adenocarcinoma, C51, PO3, MA16/C, MA44, Lewis lung and GOS murine solid tumours, and caused complete regression of advanced stage tumours. Docetaxel was also more potent than paclitaxel as a cytotoxic agent *in vitro* and in tumour xenografts (Ringel and Horwitz, 1991; Rose, 1995). Impressive activity was also demonstrated in nude mice against human xenografts of colon, melanoma, lung, breast and ovary tumours. In combination, synergism was observed with established anticancer drugs 5-fluorouracil, cyclophosphamide, etoposide, vinorelbine and methotrexate. Dose limiting toxicities in animals were haematologic and gastrointestinal in mice and dogs. For the B16 tumour, docetaxel was shown to be twice as active as paclitaxel.

Phase I dose ranging studies of docetaxel commenced in 1990, showed dose limiting toxicity of neutropenia, mucositosis and skin reactions and a maximum tolerated dose range in adults of 80–115 mg/m^2 (van Oosterom and Schrijvers, 1995; von Hoff, 1997). Other less severe reactions included hypersensitivity reactions, neurotoxicity, skin rashes, alopecia and asthenia. Responses were observed in breast, lung, bladder and ovary carcinoma. The subsequent dosage recommendations for Phase II trials in Europe and America were for 100 mg/m^2 as a 1-h infusion every three weeks; this schedule balanced likely efficacy and acceptable tolerability against allowing full recovery from neutropenia between courses. Preliminary Phase II studies, started in 1992, confirmed activity against breast, non-small cell and small cell lung, ovarian, head and neck, gastric cancers and melanoma and soft tissue sarcomas. Neutropenia was again the main dose-limiting side-effect.

In a similar developmental pattern to paclitaxel that seems to have lagged behind by approximately two years, docetaxel received FDA marketing approval for anthracycline resistant breast cancer in 1996 and second-line in metatstatic breast cancer in 1998.

Mode of action and resistance

Mode of action

The mode of action of paclitaxel and docetaxel is described in detail elsewhere (Rowinsky, 1997). Taxanes inhibit the proliferation of cancer cells by blocking mitosis between metaphase and anaphase. They do this by binding to and stabilising microtubules, the hollow tubes present in all nucleated cells, thus encouraging their polymerisation (Manfredi and Horwitz, 1994; Rao *et al.*, 1994). The actual binding site is the N-terminal of the 31 amino acids of the beta-tubulin subunit of the tubulin polymer. Docetaxel is approximately twice as potent as paclitaxel in this respect (Ringel and Horwitz, 1991; Diaz and Andreu, 1993). Other anti-tumour agents that exert their action by microtubule binding, such as the vinca alkaloids vincristine and vinblastine, prevent microtubule assembly; thus the action of the taxanes is unique among established antitumour drugs. Taxane action also results in the formation of an incomplete metaphase chromosomal plate and abnormal array of the spindle microtubules. After such disruption, the cells die although the reasons for this are still unclear; it has been suggested that apoptosis – programmed cell death – occurs (Jordan *et al.*, 1996). Both taxanes have been shown to enhance the cytotoxic effects of ionising radiation *in vitro* at clinically achievable concentrations (Schiff *et al.*, 1995) and in a variety of tumours *in vivo* (Milas *et al.*, 1999). This is unsurprising given that the taxanes appear to act at the most radiosensitive phases of the cell cycle. A third mechanism that may be important in the activity of the taxanes is their ability to inhibit angiogenesis (Belotti *et al.*, 1996; Klauber *et al.*, 1997).

Resistance to taxanes

Based on the above mechanism, two main explanations of taxane resistance have been proposed (Parekh and Simpkins, 1997; Rowinsky, 1997):

1   Some tumours contain tubulin subtypes that polymerise relatively slowly to form microtubules, thus diminishing the effects of the taxanes.
2   Amplification of cellular efflux pumps, conferring multiple drug resistance by drug exclusion. This effect may be overcome with the taxane formulations, due to the high levels of surfactant solubilising agents (Cremophor EL – polyoxyethylated castor oil in the case of docetaxel and Tween 80 in the case of docetaxel) used in the formulation (Webster *et al.*, 1993; Millward *et al.*, 1996).

Other mechanisms, as yet unproven, may include: expression of abnormal tubulin isotypes, decreased microtubule bundle formation, and decreased expression of anti-apoptotic proteins with which the taxanes would normally react.

Real and potential differences between the taxanes

In a conceptually important study, docetaxel was shown to have activity in patients with breast cancer who had failed on paclitaxel (Valero *et al.*, 1998). Variation in both *in vitro* and *in vivo* potency between the taxanes has fuelled speculation that their modes of action may be, at least in part, different. Earhart (1999) suggests that the difference might be multifactorial, but rooted in the greater effect of docetaxel on tubulin polymerization and stabilisation, related to its higher tubulin binding affinity. In addition, cancer cell uptake is faster and efflux is slower with docetaxel, leading to longer intracellular residence; this is supported by the observation that the activity of docetaxel is relatively independent of regimen whereas that of paclitaxel is dependent. Docetaxel is about 100 times more potent than paclitaxel in inducing certain mechanisms that encourages apoptosis in cancer cells and thus hasten cell death (Earhart, 1999).

Pharmacokinetic profiles

Adults

The pharmacokinetics of the taxanes in animals is reviewed in detail elsewhere (Sparreboom *et al.*, 1998). The pharmacokinetic behaviour of paclitaxel and docetaxel in man is summarised in Table 11.1.

Table 11.1 Important human pharmacokinetic characteristics of the taxanes

Characteristic	Paclitaxel	Docetaxel	Comment
Relationship to dose	Non-linear	Linear	Saturation at high therapeutic doses
Compartments	2–3	2–3	
Plasma protein binding	>95%	>95%	Plus extensive extravascular tissue binding; especially tubulin
Specific, steady state volume of distribution	Approx. 87 L/m^2	Approx. 74 L/m^2	
Cmax in plasma	5 μM (175 mg/m^2 over 3 h) 10 μM (250 mg/m^2 over 3 h) 0.5 μM (175 mg/m^2 over 24 h)	3–4.7 μM (100 mg/m^2 over 1 h)	
Tissue distribution	Extensive, except CNS and testes	Extensive, except CNS	
Clearance – all routes, (composite range)	Approx. 394 ml/min/m^2 (100–993)	Approx. 350 ml/min/m^2 (267–597)	Large variation will lead to differences in related parameters
Renal clearance	<10%	<10%	Minor importance
Hepatic clearance	70–80% in faeces	70–80% in faeces	Via hydroxylation and inactivation. CYP3A cyctochrome P450 enzyme mainly responsible
Distribution half-life	0.3 hours	0.5 hours	
Elimination half-life			Prolonged terminal half-life probably related to binding affinity to tubulin
2 compartments	5–7 h	12 h	
3 compartments	20 h	13 h	

Source: Cumulated data from: Rowinsky, 1995; van Oosterom and Schrijvers, 1995; Straubinger, 1995; Arbuck and Blaylock, 1995; Rowinsky and Donehower, 1996.

Although there is considerable inter-patient variation, in general, both have large volumes of distribution and are taken up rapidly by most tissues, with the exception of traditional tumour sanctuary sites such as the testes and central nervous system. The maximum plasma concentration during short infusion regimens and the steady state concentrations achieved during 24-h infusions indicate that palcitaxel exceeds, for a significant time, the minimum effective concentrations required to exert cytotoxic or at least cytostatic effects shown *in vitro* (0.1–10 µM; Rowinsky and Donehower, 1996). Both drugs are extensively bound to plasma proteins, including albumin, alpha-1-acid glycoprotein and lipoproteins. Taxanes are cleared extensively by hepatic metabolism followed by biliary excretion; renal clearance accounts for less than 10% of an administered dose. The main metabolite of paclitaxel in man is 6-alpha-hydroxy paclitaxel; this was not found in animal studies (Sonnichsen and Relling, 1994). Tests in human tumour cell lines showed that this metabolite is some 30 times less cytotoxic than paclitaxel (Harris *et al.*, 1994). Studies with docetaxel have shown a different array of metabolites produced by oxidation rather than hydroxylation (Eisenhauer and Vermorken, 1998), but in both cases, the metabolites are inactive or much less cytotoxic than the parent taxanes. The extensive hepatic metabolism of the taxanes has implications for dosing in patients with hepatic disease and raises the possibility of interactions with drugs that either induce, or compete for, hepatic metabolic enzymes (see below). One important difference is that metabolism of paclitaxel is saturable at high therapeutic doses and care should be exercised, particularly when giving short infusions or high doses. Both drugs have high specific volumes of distribution that probably reflect binding to tubulin and other proteins. Paclitaxel has been shown to accumulate in ascites fluid in one patient to a concentration that was 40% higher than plasma (Wiernik *et al.*, 1987a). There is poor penetration into the CSF; the highest concentrations are seen in the lung, liver, kidney and spleen (Straubinger, 1995). Extensive inter-patient variability is demonstrated by the wide range of clearance values for paclitaxel that may be found in the literature (100–993 ml/min/m^2; with a mean of 496 ml/min/m^2). As expected, such large variations lead to appreciable differences in other related parameters such as steady state drug level after infusion and area under the time/plasma concentration curve. Such variation may have important implications for tissue exposure time and thus cytotoxic success or the development of side-effects.

Children

Early clinical studies indicated that the doses of paclitaxel that could be administered to children safely were significantly greater than those for adults (Sonnichsen and Relling, 1994); but it appears that increased drug clearance is not the basis for increased paclitaxel tolerability (Sonnichsen *et al.*, 1994). Median systemic clearances, apparent steady state volumes of distribution, and elimination half-lives are similar to adult values; although the same wide inter-patient variability is observed. There is also the same saturation of metabolism at high doses.

Drug interactions

Paclitaxel is extensively metabolised in the liver, mainly by both the CYP3A4 and CYP2C8 cytochrome subtypes; hydroxylation of paclitaxel involves mainly CYP2C8 and no drug interactions of clinical importance are anticipated in this system. CYP3A4 plays a minor role in paclitaxel metabolism but a major role in the metabolism of docetaxel. The potential for clinically relevant drug interactions between the taxanes and other drugs that are substrate for, or inhibit/induce this enzyme should be appreciated. While it is known that at least one drug (ketoconazole) metabolised by CYPA4 does not inhibit the elimination of paclitaxel, there have been no formal *in vivo* studies of drug interactions with docetaxel. *In vitro* studies have shown that concomitant administration of compounds that induce, inhibit or are metabolised by CYP3A4 may modify the metabolism of docetaxel. Such drugs, for example, cyclosporin, terfenadine, ketokonazole, erythromycin and troleandomycin should be used with caution when co-administered with docetaxel (Taxotere UK Summary of Product Characteristics, August 2000); caution is also advised with paclitaxel (Taxol UK Summary of Characteristics, December 2000).

Cisplatin is another drug that may interact with the taxanes in this way. Both drugs may be given in combination for the treatment of refractory ovarian cancer; but the sequence of administration has been shown to affect the degree of resultant neutropenia. This was much less when cisplatin was administered after paclitaxel rather than before (Rowinsky *et al.*, 1991a) and was attributed to a 25% reduction in paclitaxel hepatic clearance. The paclitaxel-then-cisplatin sequence also showed greater antitumour activity *in vitro* (Rowinsky *et al.*, 1993a) and was therefore selected for subsequent clinical studies (Rowinsky and Donehower, 1995). Similar sequence dependency between docetaxel and cisplatin has not been observed.

Rowinsky (1997) also lists sequence-dependent interactions of a similar nature with paclitaxel and doxorubicin, or paclitaxel and cyclophosphamide, but in these cases, toxicity was greater when paclitaxel preceded the other drug in a 24-h infusion. Reference is also made to sequence effects observed *in vitro* between paclitaxel and melphalan, etoposide, 5-fluorouracil, estramustine and gallium nitrate. Pharmacokinetic studies have shown an approximate 30% decrease in doxorubicin clearance when paclitaxel is administered first in doublet therapy (Holmes *et al.*, 1994; Gianni *et al.*, 1997).

There are no clinically significant interactions between either taxane and other drugs with the potential to displace highly bound drugs from their protein binding sites.

Clinical applications

The taxanes have been hailed as the most important cytotoxic agents to reach the clinic in the last 20 years. The fact that they are derived from natural sources, albeit now by semi-synthesis, makes them extremely interesting to anybody with more than a passing interest in phytopharmaceuticals. These agents have proven their worth in a range of cancers, notably metastatic breast cancer and ovarian cancer; but their story is far from complete, as new information on their efficacy against a range of other cancers, new dose regimens and combinations with other drugs is expanded by publication almost on a weekly basis. The major trials and conclusions drawn from them appear below under the various indications.

Key criteria for evaluating the efficacy of cytotoxic agents are the overall response rate (RR – the proportion of patients with greater than 50% reduction in tumour size after therapy), progression-free survival and quality of life, including toxicity and other side-effects. Response rate can be further subdivided into complete response (CR) defined as a total disappearance of detectable malignant disease for a set period of time, usually four weeks; and a partial response (PR) defined as at least 50% reduction in tumour size for at least four weeks without an increase in size of any area of known malignant disease or the appearance of new lesions. Phase I studies mainly provide information on toxicity, particularly with respect to dose ranging; while Phase II studies are most useful for providing response and survival rates. Phase III studies provide additional information on response, quality of life and side-effects. Phase III trials are usually randomised and controlled, thus providing the best evidence on which to base clinical guidelines.

Metastatic breast cancer

Paclitaxel – first line therapy

Early phase II studies involving 166 evaluable patients, demonstrated impressive activity for paclitaxel in this disease. Using the 24-h infusion technique and doses of 200–250 mg/m^2, response rates of 57% and 62% were achieved (Holmes *et al.*, 1991; Reichman *et al.*, 1993). Where doses of 175–250 mg/m^2 were given over three hours, responses were less impressive (32–43%) (Mamounas *et al.*, 1995; Gianni *et al.*, 1995a; Seidman *et al.*, 1995a), suggesting that cytotoxic exposure time was an important determinant of success. Swain *et al.* (1995) achieved a response of just 32% when the dose was reduced to 135 mg/m^2.

To date, there have been four randomised, controlled Phase III trials using paclitaxel first-line in advanced breast cancer. With one exception (Bishop *et al.*, 1997, 1999), full details have still to be published. Details of the other three may be gleaned from disparate published abstracts (Piccart-Gebhart *et al.*, 1996; Awada *et al.*, 1997; Paridaens *et al.*, 1997, 2000; Sledge *et al.*, 1997; Pluzzanska and Jassem, 1999). The available results from these trials have been reviewed in detail by Lister-Sharp *et al.* (2000), who viewed pooling of the results to be inappropriate because of differences in study design (e.g. inclusion criteria, controls, paclitaxel dosage regimen, adjuvant chemotherapy, crossover criteria). They concluded that there is little evidence that single agent paclitaxel is superior to control in terms of response, progression free survival or overall survival in the first-line treatment of metastatic breast cancer. However, their analysis makes a valuable contribution.

In total, there were 1,425 patients who had had no previous therapy for advanced disease. The most common paclitaxel dosage regimen when used as a single agent, was 200 mg/m^2 as a 3-h infusion for seven or eight, three-week cycles. Where the drug was given with doxorubicin, the dose was 220 mg/m^2. None of the trials demonstrated a superior overall response rate for paclitaxel alone compared to the controls, most of which contained an anthracycline – the standard first-line therapy for metastatic breast cancer. Indeed in one trial (Paridaens *et al.*, 2000) there was a statistically significant difference in favour of the control (25% on paclitaxel vs 41% on doxorubicin). Furthermore, doxorubicin produced a significantly longer median progression-free survival compared to paclitaxel (7.5 vs 3.9 months). When crossed over for second-line therapy, response rates of 30% and 16% were achieved for doxorubicin and paclitaxel respectively. The median survival times for the two treatments were not significantly different (18.3 vs 15.6 months). Doxorubicin produced greater toxicity, notably haematologic, gastrointestinal and cardiotoxicity, but better symptom control, particularly pain, as detailed in quality of life questionnaires; although overall, there was no significant difference in quality of life.

The ways of reporting adverse events also varied between trials, but their spectrum and severity is comparable with data obtained from trials where paclitaxel was used as second line. However, combination of paclitaxel with doxorubicin resulted in a greater median length of survival (8.3 months vs 6.2 months) compared with a fluorouracil/anthracycline control (Pluzzanska and Jassem, 1999) and a longer median time to treatment failure (eight months versus six months) compared to doxorubicin on its own (Sledge *et al.*, 1997). Even in the most fully reported trial (Bishop *et al.*, 1999) where control was a cocktail of cyclophosphamide, methotrexate, 5-fluorouracil and prednisolone, single agent paclitaxel was no more than equivalent to the control in terms of overall response rate, median progression free survival and median length of survival.

Docetaxel – first line

Docetaxel (75–100 mg/m^2) showed promise in early trials as first-line therapy against metastatic disease with response rates of 52–68% (Dieras *et al.*, 1994; Chevallier *et al.*, 1995; Ravdin and

Valero, 1995; van Oosterom and Schrijvers, 1995; Hudis *et al.*, 1996; Trudeau *et al.*, 1996; Fumoleau *et al.*, 1996a). Response rates of 50% or more were seen consistently with docetaxel administered as a 100 mg/m^2 bolus every three-weeks. Fluid retention accompanied success however; this was sufficiently serious to cause cessation of therapy in up to 30% of patients. Lower doses (75 mg/m^2) produced a lower response rate without appreciably changing the incidence or severity of the fluid retention (Dieras *et al.*, 1994).

Crown (1997) reported the results of a Phase III trial of 326 first-line patients randomised to receive either docetaxel or doxorubicin. The response rate was higher with docetaxel (48% vs 33%) but the time to progression and overall survival were similar. Interim results are available for a randomised trial comparing docetaxel vs methotrexate plus 5-fluorouracil in 199 evaluable patients (Sjostrom *et al.*, 1998). Docetaxel produced a higher response rate (42% vs 19%) and a significantly longer time to progression (six- *versus* three-months).

Details of one additional trial are available in abstract only (Nabholtz *et al.*, 1999b). Docetaxel (75 mg/m^2) plus doxorubicin (50 mg/m^2) was compared with cyclophosphamide (600 mg/m^2) plus doxorubicin (60 mg/m^2), both given as three-week cycles. There were 213 and 210 evaluable patients in the two respective groups. None of the participants had undergone previous chemotherapy for advanced disease. The paclitaxel/doxorubicin combination produced significantly greater overall response (60% vs 47%) but progression free or overall survival were not reported.

Palcitaxel – second line or in combination

There is evidence to support the use of both taxanes in advanced breast cancer after failure of conventional therapies, particularly anthracycline-resistant disease and where patients are intolerant to the anthracyclines. This remains the standard indication for both drugs in most countries where the products are licensed. More evidence is available for docetaxel than paclitaxel.

Early trial work with paclitaxel had been conducted in women with little or no previous chemotherapy for their disease. Soon however, the activity of paclitaxel in often heavily pre-treated patients showed good activity, although the response was inversely correlated to the extent of this (Seidman *et al.*, 1993, 1995b; Seidman, 1995).

Detailed reviews of Phase II studies with paclitaxel are available (Seidman, 1994; Miller and Sledge, 1999). Summarising, 13 Phase II trials with paclitaxel in previously treated metastatic disease, particularly cases of anthracycline resistance, have been conducted (Holmes *et al.*, 1991; Munzone *et al.*, 1993; Vermoken *et al.*, 1993; Wilson *et al.*, 1994; Abrams *et al.*, 1995; Riseberg *et al.*, 1995; Gianni *et al.*, 1995a; Seidman *et al.*, 1995a,b; Fountzilas *et al.*, 1996; Geyer *et al.*, 1996; Nabholtz *et al.*, 1996; Seidman *et al.*, 1996). These trials involved 1,057 evaluable patients and gave response rates ranging from 0% (at 140 mg/m^2 given over 96 h) to 55% (at 250 mg/m^2 over 24 h); success was generally lower than in first-line use.

In a Phase II, randomised, controlled, but open-label study, Dieras *et al.* (1995) compared paclitaxel 175 mg/m^2 given as a 3-h infusion in a three-week cycle to mitomycin (12 mg/m^2) given as a slow bolus injection in a six-week cycle. All 81 patients had received prior treatment with one (metastatic) or two (adjuvant and metatstatic disease) chemotherapy regimens. Patients had a range of metastatic sites, but the distribution of these was similar in both arms of the trial. Seventy-two patients were available for evaluation of efficacy. While no patient showed a complete response, patients in the paclitaxel arm had a significantly longer duration of disease control (3.5 months vs 1.6 months; $P = 0.026$). Patients were allowed to cross over to the other arm of the trial on disease progression and more than half the subjects originally in the mitomycin arm did so. In these patients, the disease control on paclitaxel proved superior. The median length of survival in the paclitaxel arm was also longer (12 months vs 8.4 months). Just two patients responded to mitomycin. Sixty per cent of patients receiving paclitaxel suffered neu-tropenia compared with 3% of mitomicin patients. Anaemia and leucopenia occurred in 27% and 21% of patients in the paclitaxel arm respectively. Nausea and vomiting were rare with paclitaxel (3% of patients) and peripheral neuropathy was equally scarce (4%); arthralgia/ myal-gia were observed in 11% of patients taking paclitaxel. On its own, this open label trial provides rather weak evidence of efficacy for paclitaxel.

Paclitaxel has been investigated in combination with a range of other cytotoxic agents in this disease, mainly in patients previously treated with other cytotoxic agents.

PACLITAXEL WITH ANTHRACYCLINES

Miller and Sledge (1999) reviewed 21 Phase II trials where paclitaxel was used in combination with the anthracylines, doxorubicin and epirubicin or the anthracenedione, mitoxantrone. Response rates in 895 evaluable patients with advanced or metastatic disease, given a range of doses and combinations was never lower than 46% and achieved 70% or greater in 11 trials: seven with doxorubicin and four with etoposide (See Table 11.2).

In a safety review of the combination, involving eight single treatment trials and two randomised studies, Gianni *et al.* (1998) found the incidence of clinical cardiotoxicity caused by the combination to be similar to that seen with doxorubicin alone (approximately 4%). They concluded that the combination could be delivered safely as long as the cumulative dose of doxorubicin was limited to 360 mg/m^2.

Despite good response rates, the combination appears to have little effect on overall survival. This was demonstrated in a large trial (Neuberg *et al.*, 1997) involving 739 first-line patients who were randomised in three-arms to receive either, doxorubicin (60 mg/m^2), or paclitaxel (175 mg/m^2 over 24 h, or a combination of doxorubicin (50 mg/m^2) with paclitaxel (150 mg/m^2 over 24 h). Response rates in the three groups were 34%, 33% and 46%, respectively, and time to progression was 6.2, 5.9 and 8.0 months. Overall survival times were not significantly different (20.1, 22.2 and 22.4 months, respectively). Quality of life scores were also not improved with the combination. Kaufman (1999) discusses the preliminary results of an ongoing trial compar-ing the combination of a losoxantrone, a structural analogue of mitoxantrone, with paclitaxel. While clinical efficacy was not discussed, safety proved to be broadly similar to that of other paclitaxel/anthracycline combinations.

PACLITAXEL WITH PLATINUM COMPOUNDS

Preliminary trials indicated that paclitaxel should be given first when combined with cisplatin, because cisplatin decreases the clearance of paclitaxel by some 25% with a consequent increased risk of myelosuppression (Citardi *et al.*, 1990; Rowinsky *et al.*, 1991a). When the combination was given every three weeks, a response rate of 52% was achieved (Wasserheit *et al.*, 1996); sim-ilar response rates were achieved with carboplatin/paclitaxel combinations (Fountzilas *et al.*, 1998; Perez *et al.*, 1998). A magnified response (85%) was achieved with bi-weekly administra-tion (Gelmon *et al.*, 1996) but this could not be replicated in two subsequent investigations where response rates were 60% (McCaskill-Stevens *et al.*, 1998) and 21% (Sparano *et al.*, 1996).

PACLITAXEL WITH 5-FLUOROURACIL

The addition of paclitaxel to conventional, three-day regimens of 5-fluorouracil and leucovorin has been reported to increase response rates to greater than 60% in three, small Phase II studies

Table 11.2 Phase II trials of paclitaxel/anthracycline combinations in metastatic or locally advanced breast cancer

Reference	P dosage regimen (mg/m^2) Infusion time (h)	Evaluable patients	CR	PR	Overall response (%)
Gianni *et al.*, 1995	P, 200, 3 h with doxorubicin	32	13	17	94
Fisherman *et al.*, 1996	P, 180, 72 h with doxorubicin	39	3	25	72
Gehl *et al.*, 1996	P, 155–200, 3 h with doxorubicin	29	7	17	83
Cazap *et al.*, 1996	P, 200, 3 h with doxorubicin	27	1	12	48
Schwartzmann *et al.*, 1996	P, 250, 3 h with doxorubicin	25	7	13	80
Frassineti *et al.*, 1997	P, 130–250, 3 h with doxorubicin	32	10	15	78
Martin *et al.*, 1997	P, 175–225, 3 h with doxorubicin	57	10	30	70
Dombernowsky *et al.*, 1997	P, 200, 3 h with doxorubicin	32	3	19	69
Moliterni *et al.*, 1997	P, 200, 3 h with doxorubicin	73	25	63	88
Sledge *et al.*, 1997	P, 150, 24 h with doxorubicin	230	Nr	Nr	46
Baldini *et al.*, 1996	P, 100–150, 3 h with etoposide	32	4	20	75
Luck *et al.*, 1997	P, 175, 3 h with etoposide	25	3	14	68
Conte *et al.*, 1996	P, 135–225, 3 h with etoposide	22	4	14	83
Capri *et al.*, 1996	P, 100–150, 3 h with etoposide	27	5	20	93
Ries *et al.*, 1997	P, 135–180, 3 h every 14 days with etoposide	14	4	8	85
Carmichael *et al.*, 1997	P, 200, 3 h with etoposide	30	2	12	47
Alabiso *et al.*, 1998	P, 175–200, 3 h with etoposide	39	8	16	62
Kohler *et al.*, 1997	P, 80, 1 h, every week for 6 weeks with etoposide	35	7	11	51
DiConstanzo *et al.*, 1996	P, 175, 3 h, with mitoxantrone	23	4	11	65
Panagos, 1997	P, 200, 3 h with mitoxantrone	27	1	14	56
Greco and Hainsworth, 1997	P, 135, 1 h with mitoxantrone, 5-fluorouracil and leukovorin	45	2	21	51

Notes
P – paclitaxel; CR – complete response; PR – partial response; Nr – not reported; All paclitaxel doses given in 21-day cycles unless stated. All subjects had metastatic disease except in Moliterni *et al.* (1997).

(Paul and Garrett, 1995; Nicholson *et al.*, 1996; Johnson *et al.*, 1997). Other combinations with notable response rates include paclitaxel, 5-fluorouracil, leucovorin and mitoxantrone (RR = 51%; Hainsworth *et al.*, 1996); and paclitaxel, 5-fluorouracil and cisplatin (RR = 82%; Klassen *et al.*, 1997).

PACLITAXEL WITH THE VINCA ALKALOIDS

A combination of paclitaxel with the vinca alkaloids that prevent polymerization of tubulin seems logical and a paclitaxel vinorelbine combination met with some success (RR = 40–57%) in five, small Phase II, second-line trials (Tortoriello *et al.*, 1996; Ellis *et al.*, 1998; Gardin *et al.*, 1998; Martin *et al.*, 1998; Willey *et al.*, 1998).

OTHER COMBINATIONS

Other combinations that have been investigated with some success are paclitaxel with cyclophosphamide as first line (RR = 50%; Sessa *et al.*, 1994) with more modest results in second-line therapy (RR = 28%; Kennedy *et al.*, 1996). A combination of paclitaxel with ifosfamide produced a response of 48% when used as second line (Murad *et al.*, 1997).

Docetaxel – second line or in combination

In Phase II trials of docetaxel second line, doses of $100 \, \text{mg/m}^2$ have produced RRs of 50–58% in patients who had progressed after anthracyclines (ten Bokkel *et al.*, 1994; Ravdin *et al.*, 1995; Valero *et al.*, 1995); with more modest activity (RR = 44%) at a lower dose of $60 \, \text{mg/m}^2$ (Taguchi *et al.*, 1994). Much lower RRs (19%) have been reported in more heavily pre-treated patients (Trandafir *et al.*, 1996). A dose response relationship was observed when the drug was used at either $100 \, \text{mg/m}^2$ (RR = 53%) or $75 \, \text{mg/m}^2$ (RR = 33%) (Leonard *et al.*, 1996).

The most common toxicity was neutropenia. A bothersome and novel reaction noted in these trials was fluid retention (moderate in 27% and serious in 8%, with some patient withdrawals), with a median onset of four–five cycles; this was not cardiac or renal in origin but hypersensitivity-based and could be managed with steroid and antihistamine pre-medication and, in some cases, with diuretics.

Recently completed trials show that docetaxel is superior to doxorubicin as single agent therapy for breast cancer patients in whom prior treatment with alkylating agents has failed as measured by RR and time to response (Chan *et al.*, 1999). Lower than first-line RRs were reported in 392, second-line patients randomised to receive either docetaxel or a combination of vinblastine with mitoxantrone (Nabholtz *et al.*, 1997a); although the results for docetaxel were more favourable (30% vs 11.6%).

In a further investigation, docetaxel was also shown to be superior to a combination of vinblastine and mitomycin C when used second line, judged by RR, time to progression and survival (Nabholtz *et al.*, 1999a). Combination with doxorubicin appears to be more effective than a doxorubicin/cyclophosphamide doublet and is not accompanied by increased cardiotoxicity (Nabholtz *et al.*, 1999a). Potentially of great importance, docetaxel was shown to have activity in patients with breast cancer who had failed on paclitaxel (Valero *et al.*, 1998).

Four randomised, controlled, phase III trials, involving 1,092 patients showed significantly longer progression-free survival compared to controls. The median progression-free survival was approximately six months and the gain over controls ranged from 5 to 16 weeks. The four trials used different comparator drugs.

Data from these trials may be viewed in a variety of publications and will be referred to as Trial 1: (Chan 1997a,b; Chan *et al.*, 1997, 1999) Trial 2: (Nabholtz *et al.*, 1997b,c; Nabholtz *et al.*, 1999a). Trial 3: (Sjostrom *et al.*, 1998) and Trial 4: (Bonneterre *et al.*, 1997), which represents a preliminary analysis of an ongoing study. Trial details, therapies and overall RRs are presented in Table 11.3.

Table 11.3 Summary of randomised Phase III trials of docetaxel as second-line therapy for advanced breast cancer

Trial	Key reference	Disease studied	Previous therapy	Treatment (mg/m^2) Arm 1	Control (mg/m^2) Arm 2	Evaluable patients Arm 1 Arm 2	Treatment response (%)[a] Arm 1 Arm 2
Trial 1	Chan et al., 1999	Metastatic breast cancer	Alkylating agents permitted but anthracycline, anthracene and taxanes excluded	D, 100, 1 h,[b] up to 7, 21-day cycles Dexamethasone and antiemetic pre-med. No GCSF	Doxorubicin, 75, up to 7, 21-day cycles. Antiemetics. No GCSF.	161 165	77(48) 55(33)
Trial 2	Nabholtz et al., 1999	Metastatic breast cancer	Anthracyclines permitted but mitomycin vinca alkaloids and taxanes excluded	D, 100, 1 h, up to 10, 21-day cycles Dexamethasone and antiemetic pre-med No GCSF	Mitomycin C, 12, 42-day cycle with vinblastine, 6, 21-day cycle; up to 10 cycles Antiemetics No GCSF	203 189	61(30) 22(12)
Trial 3	Sjostrom et al., 1998	Primary breast cancer	Anthracyclines permitted but taxanes excluded	D, 100, 1 h, at least 6, 21-day cycles. Dexamethasone pre-med No antiemetics or GCSF	Methotrexate, 200, with 5-fluorouracil, 600, at least 6 cycles No antiemetics or GCSF	143 139	61(43) 29(21)
Trial 4	Bonneterre et al., 1997	Metastatic breast cancer	Anthracycline permitted	D, 100 1 h, 21-day cycles	5-fluorouracil, 750 with navelbine, 25	46 45	25(54) 20(44)

Notes
GCSF – granulocyte colony stimulating factor; D – docetaxel.
a overall response (complete plus partial responses combined).
b length of infusion in hours. See text for further details.

All patients had undergone previous chemotherapy and in Trials 2, 3 and 4 had received previous anthracycline therapy. In Trial 3, patients receiving anthracycline therapy were excluded, but must have received alkylating agents to be eligible. All trials except Trial 4 excluded patients who had received taxane therapy previously. Although all four trials employed different conventional treatments as comparators and based on the inclusion criteria (e.g. the patients in Trial 1 had not received anthracyclines) the resistance patterns are likely to be different, the same docetaxel dosage regimen of 100 mg/m^2, infused over 1 h was used.

A total of 1,092 patients were evaluated; 553 (50.7%) received docetaxel. The differences in trial design make evaluation of pooled data difficult. In Trial 1, paclitaxel produced a significantly faster response compared to the control (doxorubicin). The RR to docetaxel, defined as the proportion of patients with greater than 50% reduction in tumour size, ranged from 30% (Trial 1) to 54% (Trial 4). The RR of docetaxel was superior to doxorubicin (48% vs 33%) in Trial 1; to mitomycin C plus vinblastine (30% vs 12%) in Trial 2 and methotrexate plus 5-fluorouracil (43% vs 21%) in Trial 3. Preliminary results from Trial 4 show no significant differences between docetaxel and 5-fluorouracil and navelbine (54% vs 44%).

The Kaplan Meier curves for Trials 1, 2 and 3 showed a similar median time to progression between docetaxel and comparator in Trial 1; a significantly longer median time to progression for docetaxel in Trial 2 (19 weeks vs 11 weeks, $p = 0.001$); a significantly longer median time to progression for docetaxel in Trial 3 (25.2 weeks vs 12 weeks, $p = 0.001$); in Trial 4, the time to disease progression was 28 weeks for docetaxel compared with 20 weeks for the comparator but no statistical analysis was performed. In terms of overall survival, the Kaplan Meier curves revealed no statistically significant difference between the arms of Trial 1 (15 months vs 14 months). A difference in favour of paclitaxel was observed in Trial 2 (11.4 months vs 8.7 months, $p = 0.01$). No difference was observed in Study 3; however, in this trial, the majority of patients in the comparator group crossed over to docetaxel on progression.

The numbers of patients surviving at 6, 12, 18, 24, 30 and 36 months in each of the arms was compared. Many of the patients in Trial 3 had crossed over to docetaxel therapy on progression compromising survival analysis.

There was no significant difference between arms in Trials 1 or 2 in the numbers of patients surviving. At one year, significantly more patients in the docetaxel arm of Trial 2 had survived (38% vs 25%) but there was no difference in Trial 1 (52% vs 50%). A similar situation was found at 18 months where survival rates were 20% vs 9% in Trial 2 and 27% vs 22% in Trial 1. There were no statistically significant differences between docetaxel and control arms in either trial at 24 months.

In terms of side-effects, significantly more patients in the docetaxel arm of Trial 2 experienced neutropenia (relative risk: 1.49) and a greater proportion of patients suffered serious infections in this trial (relative risk: 10.3) and Trial 3 (relative risk: 4.5). Less than 10% of patients in any docetaxel arm experienced febrile neutropenia. Thrombocytopenia was rare in the docetaxel groups (1–4%). Leucopenia was reported in Trial 3 only, but affected more than 75% of patients receiving docetaxel. In the docetaxel arms, the incidence of gastrointestinal effects was low: nausea, 3–6%; vomiting, 2.5–3%; stomatitis, 5–9%; diarrhoea, 7.5–11%; constipation, 0.5%. Five per cent of patients on docetaxel experienced neurosensory or neuromotor adverse events or peripheral neuropathy. There were no cardiovascular events reported with docetaxel. Alopecia occurred in 74–91% of patients taking docetaxel and asthenia was noted in 12–16%. Severe fluid retention was noted in 3.1%, 4.0% and 2.2% of patients taking docetaxel in Trials 1, 2 and 3, respectively. Cardiotoxicity is always a concern when anthracycline compounds are used. In Trial 1, left ventricular ejection fraction (LVEF) was monitored and 32% of patients receiving doxorubicin experienced an LVEF reduction of greater than 20% compared with 8% given docetaxel. Sixteen per cent of patients taking doxorubicin experienced reductions in LVEF of 40% or more, 4% developed congestive heart failure and 2% of patients subsequently died of this. There were no such incidents with docetaxel.

Summarising, three of the four trials (1, 2 and 3) found docetaxel to be superior to the comparator in terms of response, which ranged from 30% to 54%. The time to disease progression ranged from 19 to 28 weeks; this was significantly longer than the control in two studies (Trials 2 and 3). The overall length of survival ranged from 10.4 to 15 months and this was significantly longer than the control in one study (Trial 2); there were no differences in median survival times in Trials 1, 3 and 4.

Haematological side-effects were relatively frequent and with the exception of thrombocytopenia, more common in the docetaxel arms. Gastrointestinal events were rare. Neurological events were more common with docetaxel, but cardiovascular events were more frequent with the anthracycline containing regimens. Alopecia presented commonly with docetaxel and asthenia was significantly more common with docetaxel in Trials 2 and 3. A small but important minority of patients suffered fluid retention with docetaxel. The two trials that investigated quality of life measured as global health status (Trials 1 and 2) demonstrated no significant advantages of docetaxel over the comparator.

Trials 1, 2 and 3 were high quality trials with median follow-ups of 23, 19 and 11 months, respectively. Over this time, about 67% of patients had died and the data is thus sufficiently mature to permit reliable analysis. In Trial 3 however, many patients crossed from the comparator to docetaxel on progression; in such instances, one is measuring the effects of sequential administration of both regimens. Survival curves were similar, however. Accrual in Trial 4 is ongoing.

All four trials used the same dosage regimen for docetaxel in line with current recommended UK licensed indications ($100\,mg/m^2$) given as a 1-h infusion every three weeks. Superiority to doxorubicin was demonstrated in Trial 1; this is particularly relevant as in the absence of established cardiac disease, this drug is likely to have been first-line therapy.

The results suggest that docetaxel increases the length of progression-free survival of patients previously treated with anthracyclines compared to mitomycin C plus vinblastine (Trial 2) and methotrexate plus 5-fluorouracil (Trial 3). Docetaxel also increased overall survival in Trial 2.

No advantage for docetaxel over doxorubicin could be demonstrated for progression free or overall survival among patients previously treated with alkylating agents; however, such patients may not be eligible for anthracycline therapy due to cardiac disease or sensitivity and docetaxel appears to be an equally effective therapy but without cardiotoxicity.

In a more detailed analysis (Lister-Sharp *et al.*, 2000) make the point that while there are other, published Phase II trials investigating the use of docetaxel in this indication (see above), the majority are non-randomised and of poorer quality than Trials 1–4 and that a review of this data adds little to the conclusions drawn above.

DOCETAXEL COMBINATIONS

Docetaxel has been combined successfully with the anthracyclines and anthracenediones without a significant increase in cardiotoxicity. Response rates of 53–80% have been observed in four small studies, two with doxorubicin (Nabholtz *et al.*, 1999b; Misset *et al.*, 1998); the latter trial also included cyclophosphamide). Docetaxel has also been combined with epirubicin (Kerbrat *et al.*, 1998) and mitoxantrone (Alexopoulos *et al.*, 1998).

Docetaxel/cisplatin combinations have produced response rates of 47% (Bernard *et al.*, 1998) and 60% (Llombart-Cussac *et al.*, 1997) when used second line in anthracycline resistant patients. A response rate of 33% was achieved when docetaxel was used with 5-fluorouracil, second line (Watanabe *et al.*, 1996). Combined with vinorelbine, docetaxel produced response rates of 66% (Fumoleau *et al.*, 1996) and 58% (Escudero *et al.*, 1998). Similarly, combinations with cyclophosphamide (Valero *et al.*, 1997); and gemcitabine (Mavroudis *et al.*, 1998) have also been successful with respective response rates of 69% and 50%; but firm conclusions await the results of larger trials of these combinations.

Summary in breast cancer and the future

The taxanes have proven to be among the most active single agents in the treatment of metastatic breast cancer. Further work is needed to define clearly key features such as optimal dosage regimens and how they fit into the jigsaw of potential combination with other agents. The taxanes remain unlicensed in the UK for adjuvant treatment of early breast cancer or first-line treatment of advanced breast cancer. However, trials in this area are ongoing or have been completed recently and as of February 2001, applications for taxane use in first line and adjuvant settings are imminent.

There seems little justification for doses higher than those currently recommended; because while slightly higher RRs may be achieved, no overall survival benefit is observed. Weekly, as opposed to tri-weekly schedules look promising but have not been compared head to head in Phase III clinical trials. Combination with the anthracyclines is promising, but there are increased risks of cardiotoxicity in some trials. The solution may be to develop analogues with less cardiotoxocity or increase the respite period between administration of the two agents. Combined microtubule blockade with the vinca alkaloids seems logical and has produced good results in small, early trials; confirmation of these early trials is pending. This combination may represent a useful alternative to anthracycline-based combinations in patients where these are contraindicated or poorly tolerated. Miller and Sledge (1999) have postulated that the taxanes may find a valuable place in combined adjuvant therapy where palliation with the taxanes has been demonstrated. They reviewed several promising results from trials of the taxanes as adjuvant chemotherapy, particularly when added to doxorubicin/cyclophosphamide regimens; ongoing trials should illuminate our understanding of the place of the taxanes here. Further Phase III trials are ongoing to provide more information on the advantages of the taxanes in the treatment of metastatic disease either first line in combination with doxorubicin (the EORTC-10961 trial with paclitaxel) or epirubicin (the MRC-97002 trial with paclitaxel; the SWOG-S9520 trial with paclitaxel). Docetaxel/anthracycline combinations are being studied in similar first-line trials; for example, the NCI CAN-NCIC-MA15 trial with epirubicin, and the ECOG-E1196 trial with doxorubicin. The importance of drug sequencing is also being investigated (the NCCT 96-32-52 trial). These studies should yield valuable information on the toxicity and efficacy of taxane/anthracycline combinations.

Approximately 25–40% of breast cancer patients display amplification/over-expression of the HER2 (c-erB-2) cancer gene and preliminary investigations indicate that, as with doxorubicin, patients displaying this had an increased likelihood of complete response following therapy with paclitaxel or docetaxel (Seidman *et al.*, 1996). If this were to be confirmed in larger studies, then determination of HER2 status might be a useful predictor for tailoring chemotherapy to individual tumour characteristics and boosting long-term response to the taxanes.

Ovarian cancer

Ovarian cancer is the fourth most common cause of cancer death in women. Advanced disease is known to respond favourably to platinum-based combination chemotherapy; however up to

30% of patients do not respond and in those that do, many will either relapse or progress within two years of a response with resistant disease. The introduction of paclitaxel has changed this picture to the extent that paclitaxel/platinum combinations are becoming standard therapy for advanced ovarian cancer. The evidence base for docetaxel is much weaker at present (Trimble *et al.*, 1994).

Both taxanes have been used in second-line therapy for platinum-refractory patients; indeed, in pilot studies, paclitaxel produced the highest reported RRs of any second-line agent in this context (McGuire *et al.*, 1989, RR = 30%; Einzig *et al.*, 1992, RR = 20%; Trimble *et al.*, 1993, RR = 18%; Thigpen *et al.*, 1994, RR = 37%). Thus, attention has been focused on its use first-line.

Paclitaxel as first-line therapy

Four phase III trials (Trial 1: McGuire *et al.*, 1995 and 1996a,b,c; Trial 2: Muggia *et al.*, 1997; Trial 3: Piccart *et al.*, 1997; Stuart *et al.*, 1999; Trial 4: Harper, 1999) involving 3,770 patients are reported in the literature in a number of publications and reviewed by Lister-Sharp *et al.* (2000). The main aim of these studies was to investigate the efficacy of paclitaxel in combination with cisplatin, against a variety of controls, all containing platinum analogues, either alone or in combination with cyclophosphamide. The salient features of these trials are summarised in Table 11.4; a variety of paclitaxel dosage regimens were employed. Only Trials 1 and 2 used the same dosage regimen of the combination (paclitaxel 135 mg/m^2 and cisplatin 75 mg/m^2, as a 24-h infusion); Trials 1–3 used cisplatin and Trial 4 used carboplatin. Differences in inclusion criteria, previous surgical debulking, dosage regimens and previous chemotherapy make pooling of the data inadvisable.

Overall RRs (complete plus partial response) for paclitaxel plus cisplatin therapy arms in Trials 1, 2 and 3 were 73%, 46% and 52%, respectively. There was no significant difference with the control in Trials 1 and 3. Cisplatin alone had a superior response to the combination in Trial 2 (74% vs 46%). No data is currently available from Trial 4.

Kaplan Meier curves for each trial allowed calculation of median progression free survival for each trial; for the paclitaxel combination this was 18, 14.1 and 16.5 months for Trials 1, 2 and 3, respectively. Trials 1 and 3 showed significantly greater median length of progression-free survival for the paclitaxel combination compared to the control. Median length of survival for patients treated with the paclitaxel combination was 38, 26 and 35 months for Trials 1, 2 and 3, respectively; these were significantly greater than the control times in Trials 1 and 3. In contrast, patients receiving single agent platinum appeared to fare better than patients receiving the combination in terms of both median progression-free survival and median length of survival, although a detailed statistical analysis is not available.

Adverse events were recorded in differing ways for the four trials, making comparisons difficult. There were notable deviations from the established side-effect profile of palclitaxel. Haematological events were quite common; for example, white cell deficits and neutropenia were reported in 92% of patients receiving the combination and 83% receiving cisplatin alone in Trial 1. Neurosensory and neuromotor events were significantly more common with combination therapy than with controls in Trials 3 and 4. No data is available on quality of life measurements in any of these trials.

Thorough analysis is hampered by only partial publication. Furthermore all trials allowed alternate therapy to be given on disease progression. Sandercock *et al.* (1998) have argued that, if this is not properly validated, it could lead to violation of randomisation. One is, therefore, left to derive only general conclusions. Summarising, about half the patients receiving paclitaxel/platinum combinations responded to therapy (from 46% to 52%). Apart from Trial 2, where

Table 11.4 Summary of randomised Phase III trials of paclitaxel as first-line therapy for advanced ovarian cancer

Trial	Key reference	Disease studied	Previous therapy	Treatment ARM 1 (mg/m²)	Treatment ARM 2 (mg/m²)	Treatment ARM 3 (mg/m²)	Evaluable patients		Treatment response (%)[a]	
							P + C	Control	P + C	Control
Trial 1	McGuire et al., 1996	Epithelial ovarian cancer, Stage III/IV	No prior radiotherapy or chemotherapy.	P, 135, 24 h[b] with cisplatin 75. Six, 21-day cycles. Standard pre-medication	Cylophosphamide, 750. Six, 21-day cycles.		100 Arm 1	116 Arm 2	72 (72) Arm 1	70 (60) Arm 2
Trial 2 1997	Muggia et al., P, 135, 24 h	Epitherlial ovarian cancer, Stage III/IV	No prior radiotherapy or chemotherapy.	P, 200, 24 h. Six, 21-day cycles. Standard	Cisplatin, 100. Six, 21-day cycles. pre-medication.	cisplatin 75. Six, 21-day Antiemetics. Standard pre-medication. Antiemetics.	213 Arm 3 cycles.	200 Arm 2	98 (46) Arm 3	148 (74) Arm 2
Trial 3	Stuart et al., 1999	Epithelial ovarian cancer, Stages IIb–IV	No prior radiotherapy or chemotherapy.	P, 175, 3 h with cisplatin, 75. Up to 9, 21-day cycles. Standard pre-medication.	Cyclophosphamide, 750 with cisplatin, 75. Up to 9, 21-day cycles. Antiemetics.		149 Arm 2	151 Arm 1	77 (52) Arm 2	66 (44) Arm 1
Trial 4 mature	Harper, 1999	Invasive ovarian carcinoma	No prior radiotherapy or chemotherapy.	P, 175, 3 h with carboplatin, 6AUC. Six, 21-day cyles.	Carboplatin, 6AUC. Six, 21-day cycles. Antiemetics.	Cyclophosphamide, 500, doxorubicin, 50, cisplatin, 50. Six, 21-day cycles. Antiemetics.	No mature available		No mature results available	

Notes

P – paclitaxel; C – cisplatin; AUC – area under the time/plasma concentration curve.

a Overall response rate (complete plus partial responses).

b Length of infusion in hours. See text for further details.

cisplatin monotherapy was superior, there were no significant differences between response rates. Trials 1 and 3 found significantly superior progression-free survival rates for the paclitaxel combination. There were no differences in Trials 2 and 4; Trial 2 may well be confounded by crossover and Trial 4 is reported only as an interim analysis of an ongoing study. In terms of median length of overall survival, Trials 1 and 3 together suggest that for one extra patient to survive without progression to one year, six patients would have to be treated with the combination.

Carboplatin is increasingly being considered as a substitute for cisplatin in a range of chemotherapy regimes, because of its apparent equivalent efficacy and better tolerability. Ozols (2000) reported interim results of a randomised comparison of paclitaxel/cisplatin vs paclitaxel/carboplatin in 798 evaluable patients with stage III advanced ovarian cancer. Both treatments had comparable efficacy with almost identical median recurrence-free survival (21.7 vs 22.0 months); survival rate data were not available. A greater proportion of patients receiving carboplatin (51.5%) rather than cisplatin (45.2%) had negative second-look surgery. Traditional platinum-related toxicities were reduced in the carboplatin arm. Two further trials, reported in abstract only, support these results (Aravantinos *et al.*, 1999; Harper *et al.*, 1999).

Docetaxel

Four Phase II trials of docetaxel, $100 \, \text{mg/m}^2$, given in three-week cycles in 206 patients gave RRs of 25–41% (Aapro *et al.*, 1994; Francis *et al.*, 1994; Kavanagh *et al.*, 1994; Cortes and Pazdur, 1995; Piccart *et al.*, 1995; van Oosterom and Schrijvers, 1995). These studies also showed that docetaxel is active in platinum-resistant disease and overall, about a third of patients responded, irrespective of the level of platinum resistance (Francis *et al.*, 1994; Kavanagh *et al.*, 1994). On their own, the results are an insufficient evidence base for recommendations of the use of docetaxel in this indication and, at present, there are no published details of Phase III trials with docetaxel.

Lung cancer

Paclitaxel – non-small cell disease

Hainsworth and Greco (1999) provide an excellent overview of work in this area. Murphy *et al.* (1993) studied the use of first line, high-dose paclitaxel ($200–250 \, \text{mg/m}^2$ as a 24-h infusion) in patients with stage IIIB and IV non-small cell lung cancer. A 24% RR was observed from 25 evaluable patients with a median response duration of 28 weeks and a median survival of 40 weeks (56 weeks in the responders). Chang *et al.* (1993) randomised patients to receive paclitaxel, merbarone or piroxantrone; RR were 21%, 0% and 2%, respectively. The median duration of response to paclitaxel was 6.5 months and the survival at one year was 41.7%, which compared favourably with 21.6% and 22.6% for merbarone and piroxantrone respectively; median survival was 24.1 weeks. Shorter dosage regimens (1-h infusions) gave at least some response in 26% and 23% of untreated and treated lung cancer patients, respectively (Hainsworth *et al.*, 1995).

Bonomi *et al.* (1996) compared a combination of cisplatin plus paclitaxel as a 24-h infusion vs conventional therapy of etoposide plus cisplatin in patients with Stage IV non-small cell lung cancer; both high and low doses of paclitaxel were studied. There was a significantly higher RR with both doses of paclitaxel ($p < 0.001$) compared to standard therapy (32%, 27% and 12%, respectively). Median survival time was shorter on standard therapy (7.7 months) compared with

408   David T. Brown

both low- and high-dose paclitaxel regimens (9.6 and 10 months, respectively). Granulocyte colony stimulating factor (GCSF) was used with the high-dose paclitaxel regimen.

Paclitaxel – small cell disease

Paclitaxel has been studied in combination with carboplatin, in untreated, advanced NSCLC (Langer *et al.*, 1995). The RR was 62%, with a median progression-free survival time of 28 weeks, a median survival time of 53 weeks and a survival rate at 1 year of 54%. Kelly *et al.* (1998) reported a RR of 82% of 23 patients treated with a triplet regime of paclitaxel (175 mg/m^2), cisplatin and etoposide in small cell lung cancer. Also in small cell disease, Hainsworth and Greco (1999) summarise the results of trials with paclitaxel in combination with carboplatin and etoposide. The first involved paclitaxel doses of 135 mg/m^2 and achieved RRs of 93% (14 of 15 patients) in limited stage disease and 65% (15 of 23 patients) in extensive stage disease. Patients with limited stage disease also received radiotherapy. In a second study, using higher doses of paclitaxel (200 mg/m^2) and carboplatin, with the same dose of etoposide, RRs were 98% (40 of 41 patients) and 84% (32 of 38 patients) in limited and extensive stage disease, respectively. Median survival among patients with extensive disease was improved significantly; differences in survival for limited stage disease were not statistically, significantly different. Reck *et al.* (1997) reported similar results – an RR of 89% in 16 patients – using the same combination. One trial (Raefsky *et al.*, 1998) investigated a combination of carboplatin, paclitaxel and topotecan; an RR of 70% was reported in ten patients with refractory, advanced small cell lung cancer.

When used in doublet therapy with cispalatin, paclitaxel 175 mg/m^2 produced good RRs (64% and 89%, respectively) in 50 patients (Faoro *et al.*, 1997) and 44 patients (Nair *et al.*, 1997); all had extensive stage disease. A dose of 200 mg/m^2 combined with carboplatin gave an RR of 59% in 42 patients (Deppermann *et al.*, 1997). One combination with topotecan achieved a 92% RR in a very small patient group (Jett *et al.*, 1997). Combination with vinorelbine produced a response of 32% in 22 patients with advanced, refractory small cell lung cancer (Jaffaiolo *et al.*, 1997).

Small cell lung cancer has also responded to single agent paclitaxel therapy. Doses of 250 mg/m^2 given over 24 h produced RRs of 31% of 36 patients (Ettinger *et al.*, 1995) and 41% in 37 patients (Kirschling *et al.*, 1994). More recently, Smit *et al.* (1998) obtained an RR of 29% of 24 patients with progressive disease when used second line, although the duration of response was truncated in most patients.

Docetaxel – non-small cell disease

Cortes and Pazdur (1995) and van Oosterom and Schriijvers (1995) report early trials where docetaxel (100 mg/m^2) produced RRs of 19–38% in non-small cell lung cancer patients who had received the drug first or second line; activity appears to be dose-related. Lower doses (60–75 mg/m^2) produced 19–25% RRs when used first line in other studies (Miller *et al.*, 1994; Kunitoh *et al.*, 1996).

As reviewed by Langer (2000), docetaxel, either alone or in combination with cisplatin or carboplatin, has shown real promise in advanced non-small cell carcinoma. In all, 572 patients were enrolled in 12 trials of docetaxel as single, first-line agent therapy, with dosages ranging from 60 to 100 mg/m^2. Overall RRs ranged from 17% to 54%, with some evidence of a dose response relationship. There was no obvious effect of dose on median survival, which ranged between 6.3 and 11 months. The principle dose-limiting toxicity was myelosuppression. Severe fluid retention was also noted in approximately 8% of subjects.

Two hundred forty-one patients received docetaxel as second-line (salvage) therapy in non-small cell lung cancer in five Phase II studies (Langer, 2000). The majority of patients had platinum-resistant or platinum-refractory disease; all received docetaxel, $100 \, \text{mg/m}^2$ in a three-week cycle. RRs lay in the range 15–25%, with median survival times of 30–42 weeks. A similar RR of 22% was obtained in a sixth, recently published study (Gridelli *et al.*, 2000).

In the first of two Phase III trials, Fosella *et al.* (1999) compared docetaxel, 100 or $75 \, \text{mg/m}^2$ to vinorelbine/ifosfamide in more than 400 patients. All had received prior platinum treatment. RRs were 10%, 6% and 1%, respectively. Quality of life scores were superior for the high dose docetaxel arm of the trial. In a second study, Shepherd *et al.* (1999) compared the same two doses of docetaxel to best supportive care in a trial involving 200 patients. RRs for docetaxel 75 and $100 \, \text{mg/m}^2$ doses were 6.3% and 5.5%, respectively, and the time to progression was 9.1 and 12.3 weeks, respectively, compared to 5.9 weeks for best supportive care.

Several Phase I/II trials have considered the combination of docetaxel and cisplatin at various doses in advanced stages of this disease (Langer, 2000). Five trials involving 232 patients have yielded RRs of 33–52%; as with other trials, neutropenia was most severe when high doses of both agents were combined. Combinations of docetaxel with carboplatin have yielded preliminary RRs of 36–58% with less pronounced myelosuppression (Schutte, 1997; Capozolli *et al.*, 1998). Docetaxel combinations have achieved notable RRs with the following agents: vinorelbine (RR = 23–51%); gemcitabine (RR = 25–43%); irinotecan (RR = 31%) (Langer, 2000).

Docetaxel – small cell disease

Docetaxel has shown similar activity to paclitaxel in two trials against small cell lung cancer. The RRs in patients who had had previous chemotherapy were 25% (Smyth *et al.*, 1994) and in patients who were therapy naive, 23% (Hesketh *et al.*, 2000). A third trial involving just 12 patients with extensive disease gave a partial response in just one patient (Latreille *et al.*, 1996).

Head and neck cancers

Both taxanes have demonstrated activity in locally recurrent or metastatic squamous cell cancers (Khattab and Urba, 1999).

Paclitaxel

Forastiere *et al.* (1993) reported an early Phase II study of paclitaxel, $250 \, \text{mg/m}^2$ given first line to patents with a variety of qualifying advanced cancers that resulted in a RR of 43%. The authors observed that this was comparable to the response achievable with conventional therapy. A second study by Smith *et al.* (1995) reported an RR of 36% in 17 patients with squamous cell head and neck cancers. A third study (Au *et al.*, 1996) noted a response of 26% in patients with nasopharyngeal carcinoma.

When combined with other agents, paclitaxel has produced modest RRs. For example, when used with carboplatin, the RR was 32% of 49 evaluable patients (Fountzilas *et al.*, 1997); 78% in 28 patients treated with paclitaxel plus cisplatin (Hitt *et al.*, 1997); 38% and 71% in 21 and 14 patients treated with paclitaxel, cisplatin and 5-fluorouracil (Benasso *et al.*, 1997; Hussain *et al.*, 1997); and 57% of 53 patients showed a response with a combination of paclitaxel, ifosfamide and cisplatin (Shin *et al.*, 1998). A trial where paclitaxel was followed by ifosfamide in 13 patients with recurrent or metastatic disease was terminated early due to severe myelosupression (Forastiere and Urba, 1995); although five PRs were observed.

There is good *in vitro* evidence that the taxanes are capable of sensitising cancer cells to radiation (see section on 'Lung cancer'). The limited clinical data available suggests that combination with radiotherapy will prove useful. For example, Conley *et al.* (1997) added paclitaxel to carboplatin with concurrent radiotherapy in 18 patients with stage III and stage IV head and neck cancers and attained an RR of 44%. Chougule *et al.* (1997) reported a similar trial but with a more aggressive paclitaxel dose (60 vs 45 mg/m^2) where an RR of 96% was observed in 26 evaluable patients.

Docetaxel

At a dose of 100 mg/m^2, docetaxel produced RRs of 23–42% in patients with advanced head and neck cancer (Catimel *et al.*, 1994; Dreyfuss *et al.*, 1996). Docetaxel has also provided some success when used in combination (Colevas and Posner, 1998). For example, Schoffski *et al.* (1996) treated 44 patients with paclitaxel and cisplatin and achieved an RR of 41%. The addition of 5-flurouracil to this combination increased the RR to 75% in 21 patients with advanced disease (Janinis *et al.*, 1997). Most recently, Couteau *et al.* (1999) treated 24 patients with metastatic squamous cell carcinoma of the head and neck, with 100 mg/m^2, 1-h infusions of docetaxel in three-week cycles, without prophylactic anti-emetics or G-CSF. PRs were seen in 5 (24%) of 21 evaluable patients with a median duration of 11 weeks. Although the authors described the tolerance as good, Grade IV neutropenia was seen in 79.2% of 24 evaluable patients (but with short duration, median four days); febrile neutropenia was not observed. The incidence of moderate/severe fluid retention was 29.2% with one discontinuation because of this. Other toxicities were: nausea 16.7%; diarrhoea 12.5%; stomatitis 16.7%; vomiting 8.3%; hypersensitivity reaction 8.3%. This trial is of particular interest as it gives an indication of the side-effect profile that is likely to be seen on recommended doses in the absence of precautionary measures.

Other cancers

A summary of the results of trials of the taxanes vs a range of cancers is given in Table 11.5.

The list requires constant update as there is intense activity in a number of areas. For additional details of these trials, the reader is referred to the original references. Summarising, paclitaxel on its own appears to be inactive in colorectal, renal, prostate, pancreatic, gastric and CNS cancer (Rowinsky *et al.*, 1992b; Glantz *et al.*, 1996); but docetaxel has shown modest activity in cancer of the pancreas and stomach; colorectal and renal cancer show no response to docetaxel.

Bajorin (2000) and Vaughn (1999) provide excellent reviews of the efficacy of paclitaxel in combination with platinum compounds, for the treatment of advanced urothelial cancer. Trials have demonstrated comparable efficacy to conventional cisplatin-based therapy but appear to be better tolerated. Paclitaxel is one of the most active single agents available to treat urothelial cancer and the drug may be particularly useful when used with carboplatin in this group of patients, who commonly present with renal impairment, as the combination is not nephrotoxic. Docetaxel also shows activity, but there is much less trial data available at present.

Motzer (2000) reviews the results from trials involving paclitaxel alone and in combination in patients with germ cell tumours. The drug showed modest activity against tumours that were resistant to combination cisplatin therapy and in an interim report, promising activity was noted in two, first-line combination salvage programmes. Combined with ifosfamide and cisplatin, paclitaxel achieved a RR of 71% in 21 patients with relapsed germ cell tumours originating from a testicular primary site. In the second trial, patients with cisplatin-resistant germ cell tumours received paclitaxel plus ifosfamide followed by carboplatin plus etoposide in rapid cyclical succession. The total RR in 37 patients was 62%. These encouraging results suggest potential for first-line therapy with taxane combinations. Interesting results have been obtained by Gill *et al.* (1996) in a small trial of paclitaxel in Kaposi's sarcoma, where a RR of 53% was achieved; this trial should be repeated in a larger number of subjects.

Current clinical guidelines

The reader is referred to National marketing approval labelling for paclitaxel and docetaxel indications and dosage regimens in the country of origin. A summary of current general information applying in the UK is supplied here for comparison. This may change as new evidence is presented to the licensing authority.

Marketing approval indications and dosing

PACLITAXEL

Most of the accumulated clinical trial data concerns a 24-h infusion schedule, which was the first to obtain licensing approval. Shorter schedules, down to 3-h also produce significant antitumour activity; indeed, both 3- and 24-h infusions have shown equivalent activity against recurrent ovarian cancer (Eisenhauer and Vermorken, 1998). However, a much longer schedule (96-h) has been shown to be effective in advanced breast cancer (Wilson *et al.*, 1994) and a very short infusion (1 h) has shown good activity in advanced non-small cell lung cancer (Hainsworth *et al.*, 1995).

For the first-line chemotherapy of ovarian cancer, combination with cisplatin is recommended. Paclitaxel 175 mg/m^2 is infused over 3 h followed by cisplatin at 75 mg/m^2 every three weeks. Alternatively, a dose of 135 mg/m^2 is given as a 24-h infusion followed by cisplatin 75 mg/m^2 with a three-week interval between courses. Alone, as second-line therapy for ovarian and breast carcinoma, the recommended dose of paclitaxel is 175 mg/m^2 administered as a three-hour infusion with a three-week interval between courses. Paclitaxel is also licensed for non-small cell lung cancer, where a dose of 175 mg/m^2, administered over three hours is recommended, followed by cisplatin 80 mg/m^2 again with a three-week interval between courses. With all of these indications, the number of three-week cycles is dependent on individual response and tolerance. Pre-medication is recommended, with dexamethasone 20 mg orally or i.v. at 12 and 6 h prior to paclitaxel; plus diphenhydramine, 50 mg or chlorpheniramine 10 mg i.v., 30–60 min prior to treatment and cimetidine (300 mg) or ranitidine (50 mg) i.v., 30–60 min before treatment, all as prophylaxis against the hypersensitivity reactions that may occur. With adequate pre-medication, the incidence of significant hypersensitivity reactions is quoted as being less than 1% (Taxol UK Summary of Product Characteristics, December 2000).

Paclitaxel is contraindicated in patients with proven severe hypersensitivity, in pregnancy and lactation and where baseline neutrophils are $<1.5 \times 10^9$/L and platelets are $<100 \times 10^9$/L.

Due to the extensive hepatic metabolism, dose modifications should be made in patients with liver impairment. Official guidance is unclear but the drug is best avoided in severe liver disease. Moderate elevations in markers of hepatic damage such as bilirubin and liver enzymes are indications that the dose should be halved (Rowinsky, 1995). No such dosage reductions are required in renal impairment.

DOCETAXEL

The vast majority of trial work has been carried out with a dosage regimen of 100 mg/m^2 as a 1-h i.v. infusion every three weeks (Cortes and Pazdur, 1995; van Oosterom and Schrijvers,

Table 11.5 Summary detail of Phase II clinical trials of the taxanes in a range of cancers

Reference	Cancer type	Taxane/dose mg/m^2/regimen (i.v. infusion hours)	First (1) or second (2) line	Evaluable patients	CR/PR	Response rate (%)
Legha et al., 1990	Melanoma	P, 250, 24 h[b]	1	25	0/3	12
Einzig et al., 1991	Melanoma	P, 250, 24h	1	28	3/1	14
Bedikian et al., 1995	Melanoma	D, 100, 1 h	1	40	1/4	13
Aamdal et al., 1994	Melanoma	D, 100, 1 h	1	30	0/5	17
Einzig et al., 1994	Melanoma	D, 100, 1 h	1	13	1/0	8
Hudes et al., 1997	Prostate	P, 120, 96h, with estramustine	2	32	17[a]	53
Smith and Pienta, 1999	Prostate	P, 135, 3 h with estramustine	2	10	3[a]	30
Smith and Pienta, 1999	Prostate	P, 135, 1h with estramustine and etoposide	2	38	24[a]	63
Kreis et al., 1999	Prostate	D, 70, 1 h with estramustine	1	17	14[a]	82
Ajani et al., 1994	Oesophagus	P, 250, 24h	1	50	1/15	32
Einzig et al., 1995	Upper GI tract	P, 250, 24h	1	22	10/1	50
Sulkes et al., 1994	Stomach	D, 100, 1h	1	33	0/8	24
Einzig et al., 1995	Upper GI tract	D, 100, 1h	1	22	1/2	14
McGuire et al., 1996d	Squamous cervical cancer	P, 170, 24h	1	52	2/7	17
Kudelka et al., 1996a	Squamous cervical cancer	P, 250, 24h	1	22	0/5	23
Kudelka et al., 1996b	Squamous cervical cancer	D, 100, 1h	1	16	0/2	13
Thigpen et al., 1995	Endometrium	P, 240, 24h	1	28	4/6	35
Woo et al., 1996	Endometrium	P, 170, 3 h	2	7	0/3	43
Roth et al., 1994	Urothelium – transitional cell	P, 250, 24h	1	26	7/4	42
Dreicer et al., 1996	Urothelium	P, 175–250, 24h	1/2	9	0/5	56
Bajorin, 2000	Urothelium	P, 135–225, 3–24 h with: (a) cisplatin (b) carboplatin	1	(a) 76 (in 3 trials) (b) 210 (in 7 trials)	17/33 28/55	(a) 65 (b) 41

Reference	Disease	Treatment				
Meluch et al., 1999	Advanced transitional cell bladder cancer	P, 200, 1 h with gemcitabine	1/2	25	2/13	60
Marini et al.,1999	Advanced or metastatic transitional cell bladder cancer	P, 150, 3 h with gemcitabine	2	15	3/5	53
Bajorin et al., 1998; McCaffrey et al., 1999	Advanced transitional cell bladder cancer	P, 200, 3 h with ifosfamide and cisplatin	1	44	10/20	68
Bellmunt et al., 1999	Advanced transitional cell bladder cancer	P, 80, 1 h with ifosfamide and gemcitabine	1	29	6/17	79
Vaishampayan et al., 1999	Advanced transitional cell bladder cancer	P, 200, 3 h with cisplatin and gemcitabine	1	19	5/6	58
Tu et al., 1995	Refractory metastatic bladder cancer	P, 200, 3 h with cisplatin and methotrexate	2	25	0/10	40
Meyers et al., 1998	Advanced or metastatic transitional cell bladder cancer	P, 200, 3 h with cisplatin and methotrexate	1	25	4/10	56
Calvo et al., 1997	(a) Metastatic transitional cell bladder cancer (b) Recurrent disease	P, 100, 3 h with cisplatin and fluorouracil	1	(a) 12 (b) 11	(a) 1/8 (b) 0/5	(c) 75% (b) 45%
De Wit et al., 1998	Advanced transitional cell bladder cancer	D, 100, 1 h	1	29	4/5	31
McCaffrey et al., 1995	Urothelium	D, 100, 1 h	2	20	0/4	20
Christou et al., 1996	Germ cell	P, 170, 24 h	2	18	1/1	11
Bokemeyer et al., 1996	Germ cell	P, 225, 3 h	2	24	1/5	25
Motzer et al., 1994	Germ cell	P, 250, 24 h	2	31	3/5	26
Motzer et al., 1997	Cisplatin resistant, advanced germ cell cancers	P, 250, 24 h with ifosfamide followed by carboplatin with etoposide	2	18	10/0	56

(Continued)

Table 11.5 (Continued)

Reference	Cancer type	Taxane/dose mg/m²/regimen (i.v. infusion hours)	First (1) or second (2) line	Evaluable patients	CR/PR	Response rate (%)
Motzer, 2000	Cisplatin resistant, advanced germ cell cancers	P, 200, 24 h with ifosfamide and cisplatin; rapid cycling	2	37	2/21	62
Bruntsch et al., 1994	Renal cell cancer	D, 100, 1 h	1	27	0/1	4
Mertens et al., 1994	Renal cell cancer	D, 100, 1 h	1	18	0/0	0
Pazdur et al., 1994	Colorectum	D, 100, 1 h	1	19	0/0	0
Sternberg et al., 1994	Colorectum	D, 100, 1 h	1	33	1/2	9
Prados et al., 1996	Glioma	P, 210–140, 3 h	1/2	40	0/4	10
Chamberlain and Komanink, 1995	Glioma	P, 175, 3 h	2	20	0/4	20
Forsyth et al., 1996	Glioma	D, 100, 1 h	1	18	0/0	0
Waltzman et al., 1996	Soft tissue sarcoma	P, 250, 3 h	1/2	27	0/1	4
Van Hoesal et al., 1994	Soft tissue sarcoma	D, 100, 1 h	2	25	0/5	17
Blackstein et al., 1995	Soft tissue sarcoma	D, 100, 1 h	1	28	0/3	11
Wilson et al., 1995	Non-Hodgkin's lymphoma	P, 140, 96 h	2	29	0/5	17
Younes et al., 1995	Non-Hodgkin's lymphoma	P, 200, 3 h	2	53	6/6	23
Younes et al., 1996	Non-Hodgkin's lymphoma	P, 140, 96	2	12	0/0	0
Goss et al., 1996	Non-Hodgkin's lymphoma	P, 175, 3 h	2	15	0/2	13
Dimopoulos et al., 1994	Multiple myeloma	P, 125, 24 h	1	33	0/5	15

Notes
Refer to the text for details of trials in breast, ovary, lung, and head/neck cancer.
CR – Complete response; PR – Partial response; P – paclitaxel; D – docetaxel.
a decrease in prostate specific antigen >50%.
b Length of infusion in hours. See text for further details.

1995) and this is the recommended dosage for monotherapy in breast cancer. As first-line combination therapy, docetaxel 75 mg/m^2 is given in combination with doxorubicin (50 mg/m^2). In non-small cell lung cancer, the recommended dosage is 75 mg/m^2 as a 1-h infusion, every three weeks. Docetaxel has similar contraindications to paclitaxel and as with paclitaxel, hypersensitivity reactions are an annoying and potentially dangerous feature of therapy. Several pre-medication schedules are used as prophylaxis, the most popular being dexamethasone, 8 mg orally, twice daily for three days starting one day prior to chemotherapy with or without histamine antagonists given i.v., 30 min before docetaxel (van Oosterom and Schrijvers, 1995; Cortes and Pazdur, 1995). In the case of mild to severe liver disease, dosage reductions of at least 25% are recommended and liver function tests should be conducted at baseline and before each cycle. In recent years, weekly rather than three-weekly regimens have become of interest (Earhart, 1999) and Phase I and Phase II trials have shown promising results in breast cancer, non-Hodgkin's lymphoma and prostate cancer although at reduced doses (typically 35–40 mg/m^2/week). This use is currently off-label in the UK.

National guidelines

Both taxanes currently available are expensive. In the UK, the National Institute of Clinical Excellence (NICE) reviews the evidence base for clinical effectiveness and provides prescribers with guidance on how new drugs might be used to best overall advantage. In breast cancer, no distinction between the two is made. The NICE has recommended (NICE, 2000) that both should be used for the treatment of advanced breast cancer where initial cytotoxic chemotherapy (including anthracyclines) has failed or is inappropriate. The use of taxanes for adjuvant treatment of early breast cancer for the first-line treatment of advanced breast cancer should be limited to clinical trials. In the case of ovarian cancer, NICE recommends that paclitaxel, in combination with a platinum-containing compound (cisplatin or carboplatin), should be the standard initial therapy for ovarian cancer following surgery. Treatment of recurrent or resistant ovarian cancer with paclitaxel and platinum combination therapy is only recommended in a clinical trial context or if the patient has not previously received this combination. Docetaxel is not licensed for ovarian cancer in the UK. This advice may change as the evidence base expands.

An economic footnote

Cancer chemotherapy is not cheap. Economic evaluations of the taxanes are reviewed in detail elsewhere (Lister-Sharp *et al.*, 2000). The answer one gets rather depends, among other things, on the perspective of the investigator, the context of usage and the uncertainty of efficacy in that usage. Other factors are the cross-national and cross-cultural variations in the importance placed on cancer treatments, the available alternatives and the health burden on national economies. In a publication such as this it is unwise to generalise. The UK National Institute of Clinical Excellence (2000) has calculated that in the setting of the use of taxanes for advanced breast cancer in the year 2000, the extra cost to the UK National Health Service (NHS), compared to conventional treatments, is approximately £4,000 per patient; this translates to between £7,000 and £23,500 per life year gained. Annual use of both drugs to treat anthracycline-resistant, advanced breast cancer is likely to cost the NHS £20 m (assuming 5,000 patients at £4,000 each). This figure represents some 0.05% of the National NHS drug budget.

Similar calculations have been made for the use of taxanes for ovarian cancer (National Institute of Clinical Excellence, 2000). The addition of paclitaxel to standard platinum therapy for ovarian cancer would cost approximately £28 m annually, assuming that 4,000 patients were eligible for treatment per year and that the average cost per patient was £7,000. (4.5 cycles at £1,500 per cycle); this sum presents 0.07% of the total NHS drug budget.

Side-effects

A brief overview of the side-effect profiles of the taxanes currently used in the clinic is included here, because this represents one of the main foci for development of safer taxane derivatives, discussed in the section on 'Future prospects for taxane chemotherapy'.

Early animal work indicated that paclitaxel was most toxic in tissues with rapid cell turnover such as reproductive, haematopoietic, gastrointestinal and lymphocytic systems (National Cancer Institute, 1983). As a timely warning, its preferred vehicle, Cremophor-EL, also produced dyspnoea, vasodilation, lethargy, hypotension and fatalities when given to dogs in large, single doses. These effects sounded a note of caution that was heeded in subsequent clinical trials. Subsequent Phase I and II studies confirmed the effects seen in earlier animal studies.

Hypersensitivity reactions

Hypersensitivity reactions are a feature of both paclitaxel and docetaxel therapy. These can be ameliorated or abolished by suitable premedication with a corticosteroid and histamine antagonists, according to dose regimens given in the dosage guideline section. Before these agents were used, early clinical trials were characterised by a high incidence of dyspnoea, bronchospasm, urticaria and hypotension, with serious reactions occurring some 2–3 min after taxane administration and almost all such reactions occurring within 10 min of the first or second dose. With standard prophylaxis, the incidence of serious hypersensitivity reactions has been reduced to below 3% (Eisenhauer and Vermorken, 1998). The docetaxel formulation contains Tween 80 instead of Cremophor EL as the solubilising agent. This too has been associated with hypersensitivity reactions in as many as 33% of patients in the absence of suitable prophylaxis (Rowinsky and Donehower, 1996). The pattern of onset and recovery after cessation of docetaxel treatment is similar to that seen after paclitaxel. Taxanes should not be administered again to patients who have experienced serious hypersensitivity reactions.

Haematological toxicity

Most early Phase II studies used relatively high doses of paclitaxel (up to 250 mg/m^2) and these were associated with severe neutropenia, often requiring rescue with GCSF (Rowinsky *et al.*, 1992b).

Both paclitaxel and docetaxel suppress white blood cell populations. As with other cytotoxic agents, neutropenia occurs 8–10 days after the onset of therapy, but reverses rapidly after cessation and is not cumulative (Rowinsky *et al.*, 1993b); the severity is dose-related and neutropenia is more likely during longer dosage regimens (Huizing *et al.*, 1993; Gianni *et al.*, 1995a; Cortes and Pazdur, 1995) and to occur sooner in heavily pre-treated patients; infections and fever can result and fatal infections have been recorded. Neutropenia responds to GCSF and should be considered likely at paclitaxel doses of 200–250 mg/m^2. Serious thrombocytopenia and erythrocytopenia are relatively rare with paclitaxel or docetaxel alone.

Neurotoxicity

Neurotoxicity remains a more or less permanent feature of clinical trial reports; while this is usually of secondary importance it is more likely to occur with larger doses and particularly with dose intensification allowed by administration of GCSF to enhance recovery from taxane-induced neutropenia (Sarosy and Reed, 1993) where this side-effect may become dose-limiting.

Peripheral neuropathy is a feature of both paclitaxel and docetaxel therapy that is more likely at higher doses or repeated courses. A symmetrical glove and stocking presentation of numbness or paraesthesia is typical (Rowinsky *et al.*, 1993b; Chaudhry *et al.*, 1994). Mild to moderate neuro-sensory signs are characterised by paraesthesia, dysthesia or pain, which may be burning. Dose reductions are recommended for both agents in cases of severe peripheral neuropathy. The onset of symptoms may be as soon as two to six days after starting therapy. Severe neuropathy is rare at recommended doses, even in patients who have received other neurotoxic agents, such as cisplatin (McGuire *et al.*, 1989). Pre-existing neuropathies can be exacerbated, for example, in patients with diabetic neuropathy, and motor and autonomic dysfunction has been observed. Other neural disorders associated with paclitaxel include: scintillation scotoma (Capri *et al.*, 1994), myalgia and myopathy, particularly at high dose in combination with cisplatin (Rowinsky *et al.*, 1993c; Capri *et al.*, 1994). Myalgia and malaise have been observed with docetaxel.

Cardiotoxicity

Cardiotoxicity has been observed, but its significance is still unclear (Rowinsky *et al.*, 1991b; Arbuck *et al.*, 1993b). Disturbances in heart rhythm, notably bradycardia have been observed (Rowinsky *et al.*, 1993b; Arbuck *et al.*, 1993b); heart block has been noted, but rarely (Arbuck *et al.*, 1993b). These events were all asymptomatic, reversible and observed in trials using continuous cardiac monitoring. It is likely that paclitaxel was the culprit, however, as similar effects have been observed in yew poisonings (Arbuck *et al.*, 1993b). Myocardial ischaemia, atrial arrhythmias and ventricular tachycardia have also been observed during paclitaxel theray, however, a causal relationship has not been established (Rowinsky *et al.*, 1993b; Arbuck *et al.*, 1993b). The incidence of congestive heart failure was observed to be greater in patients taking a combination of rapid-dose (3 h) paclitaxel and doxorubicin than in patients taking anthracyclines alone, suggesting additive toxicity. Caution seems warranted when these two classes of drug are used together. In patients with pre-existing cardiac disorders that predispose to bradyarrhythmias, cardiac monitoring should be routine, even when the taxanes are used alone. Myocardial infarction and heart failure have been seen very rarely in clinical trials, but causality was not established.

Skin reactions

Skin reactions are common with both agents, occurring in 50–75% of patients (Van Oosterom and Schrijvers, 1995; Cortes and Pazdur, 1995; Rowinsky and Donehower, 1996); about three-quarters of these events are reversible within 21 days. Erythematous, pruritic, maculopapular rashes on the hand and forearm are typical; other manifestations include desquamation of the hands and feet, palmar and plantar erythrodysthesia and onchodystrophy. Alopecia and loss of body hair occurs with both agents, especially on repeated dosing.

Hepatotoxicity

Severe liver impairment is a contraindication for both paclitaxel and docetaxel. Patients on maximum recommended doses of docetaxel infrequently record increases in serum AST, ALT, bilirubin and AP greater than 2.5 times the upper limit of normal. Paclitaxel has been associated with five times normal values of AST, AP or bilirubin in 5%, 4% and <1% of patients, respectively, in phase III clinical trials. Hepatic necrosis and encephalopathy have been reported rarely outside of such trials. A counsel of caution should be adopted when the taxanes are used in patients with pre-existing liver disease.

Other side-effects

Gastrointestinal side-effects such as nausea, vomiting, stomatitis and diarrhoea appear to be relatively infrequent (Rowinsky *et al.*, 1992a; Rowinsky *et al.*, 1993b; Rowinsky and Donehower, 1995). Mucositis has been observed in leukaemia patients receiving prolonged (96-h) infusions of paclitaxel (Rowinsky *et al.*, 1989; Wilson *et al.*, 1994). Stomatitis, appears to be more common with docetaxel than with paclitaxel (Rowinsky, 1997). Asthenia may present in approximately half the patients treated with docetaxel and may be severe in about 10%. Arthralgia and myalgia has also been observed. Rowinsky (1997) reported a unique fluid retention syndrome associated with cumulative docetaxel doses greater than $400\,mg/m^2$ and characterised by pleural effusion and ascites that resolved only slowly after stopping therapy. Fluid retention is cumulative in incidence and severity produces peripheral oedema that starts in the lower extremities, becoming more generalised with a weight gain of 3kg or more. Retention may take months rather than weeks to reverse. Pericardial effusion, ascites and lacrymation may also be features. Fluid retention may be minimised by prophylaxis with corticosteroids and histamine antagonists (Rowinsky and Donehower, 1996) and if it occurs, is most successfully managed by early diuretic treatment. The incidence may also be reduced by using single lower doses (60–75 mg/m^2). The cause of this reaction is unclear; the most likely mechanism is increased capillary permeability (Rowinsky, 1997). Fluid retention is not accompanied by oliguria or hypertension.

Future prospects for taxane chemotherapy

Clinical research on taxanes, and the biochemical and pharmacological research that underpin it, are proceeding rapidly. In the next ten years it is likely that the drug development team players mentioned below will have answers to some, if not all of the following questions:

Pharmaceutical chemists and formulators

1   Can existing drugs be reformulated, perhaps as liposomes, or nanoparticals, to obviate the need for the present co-solvent systems and their associated toxicities (Straubinger, 1995)?

2   Is it possible, through a better understanding of taxane pharmacokinetics, to design prodrugs with greater distribution to and concentration within tumour cells and a better safety profile? Many alternatives have been suggested (Straubinger, 1995; Long *et al.*, 1996; Pendri *et al.*, 1996).

3   Has the total synthesis of paclitaxel announced by Nicolau *et al.* (1994) seven years ago, been adequately exploited to provide opportunities for the synthesis of taxane analogues? Has it truly abolished the need to claim the drug from often ecologically sensitive natural resources?

Biochemists and toxicologists

1   How can drug resistance, neurotoxicity, cardiotoxicity and hypersensitivity reactions be overcome?

2   Can we clarify the exact mode(s) of action of the taxanes; for example, the exact microtubule binding sites? Such information may provide a better idea of structure/activity relationships and synthesis of more potent taxane analogues.

3   Do taxanes have useful activities that we have overlooked? For example, paclitaxel appears to be able to activate macrophages in animals (Manthey *et al.*, 1994); this could contribute to its therapeutic effect and might be amplified in an appropriate analogue.

Clinicians

Paclitaxel has been studied the most and has been around the longest. Thus, it is further along the path of clinical development than docetaxel. It is likely that both drugs will have similar activity and thus clinical indications. Both have shown notable activity in advanced carcinomas of the breast, ovary, oesophagus, bladder, head and neck. The evidence to support the use of paclitaxel as second-line therapy for advanced breast cancer is currently weak. There are on-going, multi-centre randomised, phase III controlled trials, one comparing epirubicin and paclitaxel vs epirubicin and cyclophosphamide (ABO1); and another comparing doxorubicin and paclitaxel vs doxorubicin and cyclophosphamide (by the European Organisation for Research and Treatment of Cancer – EORTC). These trials should provide a clearer picture of its use.

As the third International Collaborative Ovarian Neoplasm study (ICON3 – Harper, 1999) trial matures, it may be able to indicate which subgroups of women with ovarian cancer are more likely to benefit from the paclitaxel/carboplatin combination. In addition, when complete and mature, the SCOTROC phase III comparison of docetaxel/carboplatin vs paclitaxel/carboplatin as first-line therapy for ovarian cancer should provide information on the comparative merits of the two taxanes. Early reports of activity against other cancers requires substantiation in further trials before the taxanes can be considered part of any 'standard' regimen.

1   Adequate dose–response studies have yet to be completed for many potentially sensitive tumours.
2   What are the optimal dosage regimens (and combinations!) for the newer indications mentioned in Table 11.5? For example, weekly, rather than three-weekly docetaxel or paclitaxel regimens have proved feasible in a number of cancers (Seidman *et al.*, 1998). The dose-limiting toxicity was asthenia rather than myelosuppression. Weekly administration with radiation therapy is also feasible but the dose-limiting toxicity becomes oesophagitis.
3   For the new indications mentioned above, can we define the likely activity of taxane therapy, alone or in combination, in previously treated patients when the drugs are used second line?
4   Are there new applications for the taxanes? For example, paclitaxel has been recently shown to be active both *in vitro* and *in vivo* against a range of human anaplastic thyroid carcinoma cell lines at concentrations achievable by conventional dosing (Ain *et al.*, 1996). Anaplastic thyroid carcinoma is a rapidly fatal cancer that is resistant to any known chemotherapeutic regimen.
5   For the existing drugs that have a wide spectrum of antitumour activity, often in advanced stage of disease, what potential is there for earlier treatment, where these drugs may have their greatest impact on survival and effecting a cure?
6   There has been recent, extensive preclinical investigation of the taxanes in combination with radiotherapy. The taxanes arrest cells in the radiosensitive G2/M phase of their growth cycle. Such studies provide strong evidence that taxanes can enhance radiation sensitivity by factors ranging from 1.1 to 3.0. Limited *in vivo* experiments with murine mammary and ovarian carcinomas showed that the taxanes can enhance tumour response to radiation by factors between 1.2 and 2.0; normal tissue radio-response was much less affected by the

taxanes (Milas *et al.*, 1999). Overall, preclinical studies have shown that the taxanes can enhance radiosensitivity of tumour cells, potentiate tumour response and increase the therapeutic ratio of radiotherapy. Vogt *et al.* (1997) reported improved pilot study results when paclitaxel was used with radiotherapy in patients with non-small cell lung cancer, inoperable, locally advanced cervical cancer and recurrent bladder cancer. The combination has already shown encouraging activity in at least one Phase I study of non-small cell lung cancer (Choi *et al.*, 1994). Further trials in this area are required.

The demonstrated activity of the taxanes in a range of cancers raises all kinds of possibilities and avenues for future research with these compounds. They have come a long way since the first demonstration of cytotoxicity of paclitaxel in 1996 and it seems they have the potential to travel much further.

References

Aamdal, S., Wolff, I. and Kaplan, S. (1994) Docetaxel (taxotere) in advanced malignant melanoma: a phase II study of the EORTC Early Clinical Trials Group. *European Journal of Cancer* 30, 1061–1064.

Aapro, M. (1998) Docetaxel versus doxorubicin in patients with metastatic breast cancer who have failed alkylating therapy: a preliminary report of a randomised phase III trial. *Seminars in Oncology* 25(5), 19–24.

Aapro, M., Pujade-Lauraine, E. and Lhomme, C. (1994) EORTC Clinical Screening group: phase II study of taxotere in ovarian cancer. *Annals of Oncology* 5(Suppl. 5), 202.

Abrams, J., Vens, D. and Baltz, V. (1995) Paclitaxel activity in heavily pretreated breast cancer: a National Cancer Institute treatment referral centre trial. *Journal of Clinical Oncology* 13, 2056–2065.

Adams, J.D., Flora, K.F., Goldspiel, B.R., Wilson, J.W., Arbuck, S.G. and Finley, R. (1993) Taxol: a history of pharmaceutical development and current pharmaceutical concerns. *Monographs of the National Cancer Institute* 15, 141–147.

Ain, K.B., Tofiq, S. and Taylor, K.D. (1996) Antineoplastic activity of taxol against human anaplastic thyroid caricinoma cell lines *in vitro* and *in vivo*. *Journal of Clinical Endocrinology and Metabolism* 81, 3650–3653.

Ajani, J.A., Ilson, D. and Daugherty, L. (1994) Activity of taxol in patients with squamous cell carcinoma and adenocarcinoma of the esophagus. *Journal of the National Cancer Institute* 86, 1086–1091.

Alabiso, O., Durando, A. and Malossi, A. (1998) Paclitaxel and doxorubicin regimen as first-line therapy in metastatic breast cancer. A Phase I–II study. *Proceedings of the American Society of Clinical Oncology* 17, 125A.

Alexopoulos, A., Kouroussis, C. and Kakolyris, S. (1998) A phase I/II study of docetaxel and mitoxantrone in metastatic breast cancer. *Proceedings of the American Society of Clinical Oncology* 17, 126A.

Aravantinos, G., Fountzilas, G. and Kosmidis, P. (1999) Paclitaxel plus carboplatin versus alternating carboplatin and cisplatin for initial treatment of advanced ovarian cancer (AOC): A Hellenic Co-Operative Group Study. *Proceedings of the American Society of Clinical Oncology* 18, 367A.

Arbuck, S.G. and Blaylock, B.A. (1995) Taxol: clinical results and current issues in development. In: M. Suffness (ed.) *Taxol – Science and Applications*. CRC Press, London, pp. 379–416.

Arbuck, S.G., Christian, M.C., Fisherman, J.S., Cazenave, L.S., Sarosy, G., Suffness, M., Adams, J., Canetta, R., Cole, K.E. and Friedman, M.A. (1993a) The clinical development of taxol. *Monographs of the National Cancer Institute* 15, 11–22.

Arbuck, S.G., Strauss, H., Rowinsky, E., Christian, M., Suffness, M., Adams, J., Oakes, M., McGuire, W., Reed, E., Gibbs, H., Greenfiled, R.A. and Montello, M. (1993b) A reassessment of cardiac toxicity associated with taxol. *Monograph of the National Cancer Institute* 15, 117–123.

Au, E., Ang, P.T. and Chua, E.J. (1993) Paclitaxel in metastatic nasopharyngeal cancer. *Proceedings of the American Society of Clinical Oncology* 15, 332A.

Awada, A., Paridaens, R. and Bruning, P. (1997) Doxorubicin or Taxol as first-line chemotherapy for metastatic breast cancer: results of EORTC-IDBBC/ECSG randomised trial with crossover. *Breast Cancer Research and Treatment* 46, 23.

Bajorin, D.F. (2000) Paclitaxel in the treatment of advanced urothelial cancer. *Oncology* 14(1), 43–57.

Bajorin, D.F., McCaffrey, J.A. and Hilton, S. (1998) Treatment of patients with transitional cell carcinoma of the urothelial tract with ifosfamide, paclitaxel and cisplatin: a Phase II trial. *Journal of Clinical Oncology* 16, 2722–2727.

Baldini, E., Innocenti, F. and Michelotti, A. (1996) Paclitaxel by 3-hour infusion plus bolus epirubicin: a feasibility and pharmacokinetic study in metastatic breast cancer. *Annals of Oncology* 7(Suppl. 5), 21A.

Bedikian, A.Y., Weiss, G.R. and Legha, S.S. (1995) Phase II trial of docetaxel in patients with advanced cutaneous melanoma previously untreated with chemotherapy. *Journal of Clinical Oncology* 13, 1895–1899.

Bellmunt, J., Guillem, V. and Paz-Ares, L. (1999) A Phase II trial of paclitaxel, cisplatibn and gemcitabine in patients with advanced transitional cell carcinoma of the urothelium. *Proceedings of the American Society of Clinical Oncology* 18, 332A.

Belotti, D., Vergani, V. and Drudis, T. (1996) The microtubule-affecting drug paclitaxel has antiangiogenic activity. *Clinical Cancer Research* 2, 1843–1849.

Benasso, M., Numico, G. and Roisso, R. (1997) Chemotherapy for relapsed head and neck cancer: paclitaxel, cisplatin, and 5-fluorouracil in chemotherapy-naive patients. A dose finding study. *Seminars in Oncology* 24, 46.

Bernard, A., Antoine, A. and Grozy, M. (1998) Docetaxel and cisplatin in anthracycline pretreated advanced breast cancer: results of a phase II pilot study. *Proceedings of the American Society of Clinical Oncology* 17, 128A.

Bishop, J.D., Toner, G., Smith, J., Tattersall, M.H., Olver, I.N. and Ackland, S. (1997) A randomised study of paclitaxel versus cyclophosphamide/methotrexate/5-fluorouracil/prednisone in previously untreated patients with advanced breast cancer – preliminary results. *Seminars in Oncology* 24(5), S17–S19.

Bishop, J.D., Toner, G., Tattersall, M.H., Olver, I.N. and Ackland, S. (1999) Initial paclitaxel improves outcome compared with CMFP combination chemotherapy as front-line therapy in untreated metastatic breast cancer. *Journal of Clinical Oncology* 17(8), 2355–2364.

Bissery, M.C., Nohynek, G., Sanderink, G.J. and Lavelle, F. (1995) Docetaxel (Taxotere): a review of clinical and preclinical experience. Part 1: Preclinical experience. *Anticancer Drugs* 6, 339–355.

Blackstein, M., Eisenauer, E.A. and Bramwell, V. (1995) Docetaxel (Taxotere) as first line therapy for metastatic or recurrent soft tissue sarcoma: a Phase II trial of the National Cancer Institute of Canada Clinical Trials Group. *European Journal of Cancer* 31A(Suppl. 5), S177.

Bokemeyer, C., Beyer, J. and Metzner, B. (1996) Phase II study of paclitaxel in patients with relapsed or cisplatin refractory testicular cancer. *Annals of Oncology* 7, 31–34.

Bokkel Huinink, W.W., Eisenhauer, E. and Swenerton, K. (1993) Preliminary evaluation of a multicentre, randomised comparative study of taxol (paclitaxel): dose and infusion length in platinum-treated ovarian cancer. *Cancer Treatment Review* 19(Suppl.C), 79–84.

Bonneterre, J., Roche, H., Monnier, A., Serin, D., Fargeot, P. and Guastalla, J.P. (1997) Taxotere versus 5-fluorouracil plus navelbine as second line chemotherapy in patients with metastatic breast cancer – preliminary results. *Proceedings of the Annual Meeting of the American Society of Clinical Oncology* 16, 564A.

Bonomi, P., Kim, K., Chang, D. and Johnson, D. (1996) Phase III trial comparing etoposide, cisplatin, versus taxol with cisplatin-GCSF versus taxol-cisplatin in advanced non-small cell lung cancer. *Proceedings of the American Society of Clinical Oncology* 15, 382A.

Bruntsch, U., Heinrich, B. and Kaye, S.B. (1994) Docetaxel (Taxotere) in advanced renal cell cancer: a phase II trial of the EORTC Early Clinical Trials Group. *European Journal of Cancer*, 30A, 1064–1067.

Calvo, E., Aramendia, J.M. and Garcia-Foncillas, J. (1997) Paclitaxel combination chemotherapy in untreated and in resistant stage IV transitional cell carcionoma of the urinary bladder: a retrospective study. *Proceedings of the American Society of Clinical Oncology* 16, 330A.

Capozolli, M.J.B., Belani, C.P., Einzig, A., Bonomi, P., Dobbs, T. and Kozak, C. (1998). Multi-institutional phase II trial of docetaxel and carboplatin combination in patients with advanced stage IIIB and IV non-small cell lung cancer. *Proceedings of the American Society of Clinical Oncology* 17, 479A.

Capri, G., Munzone, E. and Tarenzi, E. (1994) Optic nerve disturbances: a new form of paclitaxel neurotoxocity. *Journal of the National Cancer Institute* 86, 1101.

Capri, G., Tarenzi, E. and Bertuzzi, A. (1996) Paclitaxel by 3-hour infusion and bolus epirubicin on day 1 and 8 every three weeks in women with metastatic breast cancer. *Annals of Oncology* 7(Suppl. 5), 16–22.

Carmichael, J., Jones, A. and Hutchinson, T. (1997) A phase II trial of epirubicin plus paclitaxel in metastatic breast cancer. *Seminars in Oncology* 24(5, Suppl. 17), S17-44–S17-47.

Catimel, G., Verweij, J. and Mattijssen, V. (1994) Docetaxel (Taxotere): an active drug in the treatment of patients with advanced squamous cell carcinoma of the head and neck. *Annals of Oncology* 5, 533–537.

Cazap, E., Ventriglia, M. and Rubio, G. (1996) Taxol plus doxorubicin treatment of metastatic breast cancer in ambulatory patients. *Proceedings of the American Society of Clinical Oncology* 15, 146A.

Chamberlain, M.C. and Kormanink, P. (1995) Salvage chemotherapy with paclitaxel for recurrent primary brain tumours. *Journal of Clinical Oncology* 13, 2066–2071.

Chan, S. (1997a) Docetaxel (Taxotere) vs doxorubicin in patients with metastatic breast cancer (MBC) who have failed alkylating therapy. Randomised, multicentre phase III trial. *Proceedings of the American Society of Clinical Oncology* 16, 154A.

Chan, S. (1997b) Docetaxel vs doxorubicin in matastatic breast cancer resistant to alkylating chemotherapy. *Oncology Hunting* 11(8), 19–24.

Chan, S., Friedrichs, K., Noel, D., Duarte, R., Vorobiof, D. and Pinter, T. (1997) A randomized phase III study of Taxotere (T) versus doxorubicin (D) in patients with metastatic breast cancer who have failed on alylating regimens: preliminary results. *Proceedings of the American Society of Clinical Oncology* 16, 540A.

Chan, S., Noel, D., Pinter, T., Van Belle, S., Vorobiof, D. and Duarte, R. (1999) Prospective randomised trial of docetaxel versus doxorubicin in patients with metastatic breast cancer. *Journal of Clinical Oncology* 17(8), 2341–2354.

Chang, A., Kim, K. and Glick, J. (1993) Phase II study of taxol, merbarone and piroxantrone in stage IV non-small cell lung cancer: the ECOG results. *Journal of the National Cancer Institute* 85, 388–394.

Chaudhry, V., Rowinsky, E.K. and Sartorious, S.E. (1994) Peripheral neuropathy from taxol and cisplatin combination chemotherapy: clinical and electrophysiological studies. *Annals of Neurology* 35, 490–497.

Chervinski, D.S., Brecher, M.L. and Hoelcle, M.J. (1993) Cremophor-EL enhances taxol efficacy in a multi-drug resistant C1300 neuroblastoma cell line. *Anticancer Research* 13, 93–98.

Chevallier, B., Fumoleau, P. and Kerbrat, P. (1995) Docetaxel is a major cytotoxic drug for the treatment of advanced breast cancer: a phase II trial of the Clinical Screening Cooperative Group of the European Organisation for Research and Treatment of Cancer. *Journal of Clinical Oncology* 13, 314–322.

Choi, H., Akerley, W. and Safran, H. (1994) Phase I trial of outpatinet weekly paclitaxel and concurrent radiation therapy for advanced non-small cell lung cancer. *Journal of Clinical Oncology* 12, 2682–2686.

Chougule, P., Wanebo, H. and Akerley, W. (1997) Concurrent paclitaxel, carboplatin and radiotherapy in head and neck cancer. A phase II study – preliminary results. *Seminars in Oncology* 24, 57.

Christou, A., Roth, B. and Fox, S. (1996) Phase II trial of paclitaxel in refractory germ cell neoplasms. *Proceedings of the American Society of Clinical Oncology* 15, 249A.

Citardi, M., Rowinsky, E. and Schaefer, K. (1990) Sequence-dependent cytotoxicity between cisplatin and antimicrotubule agents taxol and vincristine. *Proceedings of the American Association of Cancer Research* 31, 2431A.

Colevas, A.D. and Posner, M.R. (1998) Docetaxel in head and neck cancer. *American Journal of Clinical Oncology* 21(5), 482–486.

Conley, B., Jacobs, M. and Suntharalingam, M. (1997) A pilot trial of paclitaxel, carboplatin and concurrent radiotherapy for unresectable squamous cell carcinoma of the head and neck. *Seminars in Oncology* 24, 78.

Conte, P., Michelotti, A. and Baldini, E. (1996) Epirubicin plus paclitaxel, a highly active combination devoid of significant cardiotoxicity in the treatment of metastatic breast cancer. *Proceedings of the American Society of Clinical Oncology* 15, 118A.

Cortes, J.E. and Pazdur, R. (1995) Docetaxel. *Journal of Clinical Oncology* 13(10), 2643–2655.

Couteau, C., Chouaki, N., Leyvraz, S., Oulid-Aissa, D., Lebecq, A. and Domenge, C. (1999) A phase II study of docetaxel in patients with metastatic squamous cell carcinoma of the head and neck. *British Journal of Cancer* 81(3), 457–462.

Croom, E.M. (1995) Taxus for taxol and taxoids. In: M. Suffness (ed.) *Taxol – Science and Applications.* CRC Press, London, pp. 37–70.

Crown, J. (1997) Summay of results of a randomised Phase III trial of docetaxel versus doxorubicin in patients with metastatic breast cancer and prior alkylating agent exposure. *Proceedings of the 20th Annual San Antonio Breast Cancer Conference*, 2–3.

Deppermann, K.M., Kurzeja, A. and Lichey, J. (1997) Paclitaxel and carboplatin in advanced SCLC: a phase II study. *Lung Cancer* 18, 46A.

De Witt, R., Kruit, W.H., Stoter, G., De Boer, M., Kerger, J. and Verwei, J. (1998) Docetaxel (Taxotere) – an active agent in metastatic urothelial cancer: results of a Phase II study in non-chemotherapy pretreated patients. *British Journal of Cancer* 78(10), 1342–1345.

Diaz, J.F. and Andreu, J.M. (1993) Assembly of purified GDP-tubulin into microtubules induced by taxol and toxotere: reversibility, ligand stoichiometry and competition. *Biochemistry* 32, 2747–2755.

Di Constanzo, F., Sdrobolini, A. and Bilancia, D. (1996) Phase I–II trial of mitoxantrone and taxol in advanced breast cancer. *Proceedings of the American Society of Clinical Oncology* 15, 139A.

Dieras, V., Fumoleau, P. and Chevallier, B. (1994) Second EORTC Clinical Screening group Phase II trial of taxotere (docetaxel) as first line chemotherapy in advanced breast cancer. *Proceedings of the American Society of Clinical Oncology* 13, 78A.

Dieras, V., Marty, M., Tubiana, N., Corette, L., Morvan, F. and Serin, D. (1995) Phase 2 randomised study of paclitaxel versus mitomycin in advanced breast cancer. *Seminars in Oncology* 22(Suppl. 8), 33–39.

Dimopoulos, M.A., Arbuck, S. and Huber, M. (1994) Primary therapy of multiple myeloma with paclitaxel (Taxol). *Annals of Oncology* 5, 757–759.

Dombernowsky, P., Boesgaard, M. and Andersen, E. (1997) Doxorubicin plus paclitaxel in advanced breast cancer. *Seminars in Oncology* 24(5, Suppl. 17), S17-15–S17-18.

Dreicer, R., Gustin, D. and See, W.A. (1996) Paclitaxel in advanced urothelial carcinoma: its role in patients with renal insufficiency and as salvage therapy. *Proceedings of the American Society of Clinical Oncology* 15, 242A.

Dreyfuss, A., Clark, J. and Norris, C.N. (1996) Docetaxel (Taxotere): an active drug against squamous cell carcinoma of the head and neck. *Journal of Clinical Oncology* 14, 1627–1678.

Earhart, R.H. (1999) Docetaxel (Taxotere): preclinical and general clinical information. *Seminars in Oncology* 26(5), 8–13.

Einzig, A., Hochster, H. and Wiernik, P.H. (1991) A phase II study of taxol in patients with malignant melanoma. *Investigational New Drugs* 9, 59–64.

Einzig, A., Lipsitz, S. and Wiernik, P.H. (1995a) Phase II trial of taxol in patients with adenocarcinoma of the upper gastrointestinal tract. *Investigational New Drugs* 13, 223–227.

Einzig, A., Lipsitz, S., Wiernik, P.H. and Benson, A.B. (1995b) Phase II trial of taxotere in patients with adenocarcinoma of the upper gastrointestinal tract previously untreated with cytotoxic chemotherapy. *Proceedings of the American Society of Clinical Oncology* 14, 191A.

Einzig, A., Schichter, L.M. and Wadler, S. (1994) Phase II trials of Taxotere (RP56976) in patients with metastatic melanoma previously untreated with cytotoxic chemotherapy. *Proceedings of the American Society of Clinical Oncology* 13, 395A.

Einzig, A., Wiernik, P.H., Sasloff, J. (1992) Phase II study and long term follow up of patients treated with Taxol for advanced adenocarcinoma. *Journal of Clinical Oncology* 10, 1748–1753.

Eisenhauer, E.A. and Vermorken, J.B. (1998) The taxoids: comparative clinical pharmacology and therapeutic potential. *Drugs* 55(1), 5–30.

Ellis, G., Gralow, J. and Pierce, H. (1998) Paclitaxel/vinorelbine chemotherapy with concurrent GCSF for metastatic breast cancer: Phase I-II study in doxorubicin-treated patients. *Proceedings of the American Society of Clinical Oncology* 17, 138A.

Escudero, P., Bueso, P. and Mayordome, J. (1998) Docetaxel plus vinorelbine is an active combination for patients with anthracycline refractory metastatic breast cancer. *Proceedings of the American Society of Clinical Oncology* 17, 139A.

Ettinger, D.S., Finkelstein, D.M., Sarma, R. and Johnson, D.H. (1995) Phase II study of paclitaxel in patients with extensive disease small cell lung cancer. An Eastern Cooperative group study. *Journal of Clinical Oncology* 13, 1430–1435.

Fan, D., Ajani, J.A. and Baker, F.L. (1987) Comparison of antitumour activity of standard and investigational drugs at equivalent granulocyte-macrophage colony-forming cell inhibitory concentrations in the adhesive tumour cell culture system: an *in vitro* method of screening new drugs. *European Journal of Cancer and Clinical Oncology* **23**, 1469–1476.

Faoro, C., Wolf, M.E. and Schroder, M. (1997) Paclitaxel/cisplatin chemotherapy in extensive stage small cell lung cancer. *Lung Cancer* **18**, 51A.

Fisherman, J, Cowan, K. and Noone, M. (1996) Phase I/II study of 72-hour infusional paclitaxel and doxorubicin with granulocyte colony stimulating factor in patients with metastatic breast cancer. *Journal of Clinical Oncology* **14**, 774–782.

Fjallskog, M.L., Frij, L. and Bergh, J. (1993) Is Cremophor EL, solvent for paclitaxel, cytotoxic? *Lancet* **342**, 873–874.

Forastiere, A. (1994) Paclitaxel (Taxol) for the treatment of head and neck cancer. *Seminars in Oncology* **21**(Suppl. 8), 49.

Forastiere, A., Neuberg, D. and Taylor, S.G. (1993) Phase II evaluation of taxol in advanced head and neck cancer. *Monographs of the National Cancer Institute* **15**, 181–184.

Forastiere, A. and Urba, S.G. (1995) Single-agent paclitaxel and paclitaxel plus ifosfamide in the treatment of head and neck cancer. *Seminars in Oncology* **22**(3), 24–27.

Forsyth, P., Caircross, G. and Stewart, D. (1996) Phase II trial of docetaxel in patients with recurrent malignant glioma. *Investigational New Drugs* **14**, 203–206.

Fosella, F., DeVore, R., Kerr, R., Crawford, J. and Natale, R. (1999) Phase III trial of taxotere (100 mg/m^2 or 75 mg/m^2) versus vinorelbine or ifosfamide in non-small cell lung cancer patients previously treated with platinum-based chemotherapy. *Proceedings of the American Society of Clinical Oncology* **18**, 1776A.

Fountzilas, G., Dimopoulos, A. and Bafaloukos, D. (1998) Paclitaxel and carboplatin as first-line chemotherapy in advanced breast cancer. *Proceedings of the American Society of Clinical Oncology* **17**, 141A.

Fountzilas, G., Athanassiadis, A. and Giannakakis, T. (1996) A phase II study of paclitaxel in advanced breast cancer resistant to anthracyclines. *European Journal of Cancer* **32**, 47–51.

Fountzilas, G., Athanassiadis, A. and Samantas, E. (1997) Paclitaxel and carboplatin in recurrent or metastatic head and neck cancer. A phase II study. *Seminars in Oncology* **19**, 28.

Francis, P., Schneider, J. and Hann, L. (1994) Phase II trial of docetaxel in patients with platinum refractory ovarian cancer. *Journal of Clinical Oncology* **12**, 2301–2308.

Frassineti, G., Zoli, W. and Silvestro, L. (1997) Paclitaxel plus doxorubicin in breast cancer: an Italian experience. *Seminars in Oncology* **5**(Suppl. 17), S17-19–S17-25.

Fumoleau, P., Chevallier, B. and Kerbrat, P. (1996a) A multicentre Phase II study of the efficacy and safety of docetaxel as first-line treatment of advanced breast cancer: report of the Clinical Screening Group of the EORTC. *Annals of Oncology* **7**, 165–171.

Fumoleau, P, Delacroix, V. and Perrocheau, G. (1996b) Docetaxel in combination with vinorelbine as first line chemotherapy in patients with metastatic breast cancer. *Proceedings of the American Society of Clinical Oncology* **15**, 142A.

Gardin, G., Pronzato, P. and Tognoni, A. (1998) A Phase II trial of vinorelbine and paclitaxel in patients with metastatic breast cancer who have failed prior anthracycline containing chemotherapy. *Proceedings of the American Society of Clinical Oncology* **17**, 142A.

Gehl, J., Bosegaard, M. and Paaske, T. (1996) Combined doxorubicin and paclitaxel in advanced breast cancer: effective and cardiotoxic. *Annals of Oncology* **7**, 687–693.

Gelmon, K., O'Reilly, S. and Tolcher, A. (1996) Phase I/II trial of bi-weekly paclitaxel and cisplatin in the treatment of metastatic breast cancer. *Journal of Clinical Oncology* **14**, 1185–1191.

Georg, G.I., Boge, T.C., Cheruvallath, Z.S., Clowers, J.S., Harriman, G.C., Hepperle, M. and Park, H. (1995) The medicinal chemistry of taxol. In: M. Suffness (ed.) *Taxol – Science and Applications*. CRC Press, London, pp. 317–378.

Geyer, C., Green, S. and Moinpour, C. (1996) A Phase II trial of paclitaxel in patients with metastatic refractory carcinoma of the breast: a Southwest Oncology Group (SWOG) study. *Proceedings of the American Society of Clinical Oncology* **15**, 107A.

Gianni, L., Domberernowsky, P. and Sledge, G. (1998) Cardiac function following combination therapy with taxol and doxorubicin for advanced breast cancer. *Proceedings of the American Society of Clinical Oncology* 17, 115A.

Gianni, L., Munzone, E. and Capri, G. (1995a) Paclitaxel in metastatic breast cancer: a trial of two doses by a 3-hour infusion in patients with disease recurrent after prior therapy with anthracyclines. *Journal of the National Cancer Institute* 87, 1169–1175.

Gianni, L., Kearns, C. and Gianni, A. (1995b) Non-linear pharmacokinetics and metabolism of paclitaxel and its pharmacokinetic/pharmacodynamic relationships in humans. *Journal of Clinical Oncology* 13, 180–190.

Gianni, L., Munzone, E. and Capri, G. (1995c) Paclitaxel by 3-hour infusion in combination with bolus doxorubicin in women with untreated metastatic breast cancer: high antitumor efficacy and cardiac effects in a dose-finding and sequence-finding study. *Journal of Clinical Oncology* 13, 2688–2699.

Gianni, L., Vigano, L. and Locarelli, A. (1994) Human pharmacokinetic characterisation and *in vitro* study of the interaction between doxorubicin and paclitaxel in patients with breast cancer. *Journal of Clinical Oncology* 15, 1906–1915.

Gill, P.S., Tulpule, A. and Reynolds, T. (1996) Paclitaxel (Taxol) in the treatment of relapsed or refractory AIDS-related Kaposi's sarcoma. *Proceedings of the American Society of Clinical Oncology* 15, 306A.

Glantz, M.J., Choy, H. and Kearns, C.M. (1996) Phase I study of weekly outpatient paclitaxel and concurrent cranial irradiation in adults with astrocytomas. *Journal of Clinical Oncology* 14, 600–609.

Goss, P., Stewart, A. and Couture, F. (1996) A phase II study of paclitaxel in patients with refractory lymphomas. *Proceedings of the American Society of Clinical Oncology* 15, 419A.

Greco, F. and Hainsworth, J. (1997) Paclitaxel with mitoxantrone with or without 5-fluorouracil and high-dose leukovorin in the treatment of metastatic breast cancer. *Seminars in Oncology* 24(5, Suppl. 17), S17-61–S17-64.

Gridelli, C., Frontini, L., Barletta, E., Rossi, A. and Barzelloni, M.L. (2000) Single agent docetaxel plus granulocyte-colony stimulating factor (G-CSF) in previously treated patients with advanced non small cell lung cancer. A phase II study and review of the literature. *Anticancer Research* 20, 1077–1084.

Hainsworth, J.D. and Greco, F.A. (1999) The current role and future prospects of paclitaxel in the treatment of small cell lung cancer. *Seminars in Oncology* 26(1), 60–66.

Hainsworth, J.D., Jones, S. and Mennel, R. (1996) Paclitaxel with mitoxantrone, fluorouracil and high-dose leukovorin in the treatment of metastatic breast cancer: a phase II trial. *Journal of Clinical Oncology* 14, 1611–1616.

Hainsworth, J.D., Thompson, D.S. and Greco, F.A. (1995) Paclitaxel by 1-hour infusion: an active drug in metatastic non-small cell lung cancer. *Journal of Clinical Oncology* 13, 1609–1614.

Harper, P.A. (1999) A randomised comparison of paclitaxel and carboplatin versus a control arm of single agent carboplatin or CAP (cyclophosphamide, doxorubicincisplatin): 2075 patients randomised into the 3rd International Collaborative Ovarian Neoplasm Study. *Proceedings of the American Society of Clinical Oncology* 18, 1375A.

Harris, J.W., Katki, A., Anderson, L.W., Chmurny, G.N., Paukstelis, J.V. and Collins, J.M. (1994) Isolation, structural determination and biological activity of 6-alpha-hydroxytaxol, the principal human metabolite of taxol. *Journal of Medicinal Chemistry* 37, 706–709.

Hesketh, P.J., Crowley, J.J. and Burris, H.A. (2000) Evaluation of docetaxel in previously untreated extensive stage small cell lung cancer: a Southwest Oncology Group phase II trial. *Proceedings of the American Society of Clinical Oncology* 19, 132A.

Hill, B.T., Whelan, R.D.H., Shellard, S.A., McClean, S. and Hosking, L.K. (1992) Differential cytotoxic effects of taxotere in a range of mammalian tumour cell lines *in vitro*. *Annals of Oncology* 3(Suppl. 1), 120.

Hitt, R., Paz-Ares, L. and Hidalgo, M. (1997) Phase I/II study of paclitaxel/cisplatin as a first line therapy for locally advanced head and neck cancer. *Seminars in Oncology* 24, 10.

Holmes, F.A., Newman, E.A. and Madden, T. (1994) Schedule dependent pharmacokinetics in a phase I trial of taxol and doxorubicin as initial chemotherapy for metastatic breast cancer. *Annals of Oncology* 5, 191A.

Holmes, F.A., Walters, R.S., Thierault, R.L., Forman, A.D., Newton, L.K., Raaber, M.N., Buzdar, A.U., Frye, D.K. and Hortobagyi, G.N. (1991) Phase II trial of taxol, an active drug in the treatment of metatastic breast cancer. *Journal of the National Cancer Institute* **83**, 1797–1805.

Holton, R.A., Biediger, R.J. and Boatman, P.D. (1995) Synthesis of taxol and taxotere. In: M. Suffness (ed.) *Taxol – Science and Applications*. CRC Press, London, pp. 97–122.

Hoskins, W.J., Brady, M.F., Kucera, P.R., Partidge, E.E., Look, K.Y., Pearson, D.L. and Davidson, M. (1997) Combination paclitaxel-cisplatin vs cyclophosphamide-cisplatin as primary therapy in patients with suboptimally debulked advanced ovarian cancer. *International Journal of Gynecological Cancer* **1**, 9–13.

Hudes, G.R., Nathan, F. and Khater, C. (1997) Phase II trial of 96-hour paclitaxel plus oral estramustine phosphate in metastatic hormone refractory prostate cancer. *Journal of Clinical Oncology* **15**, 3156–3163.

Hudes, G.R., Obasaju, C. and Chapman, A. (1995) Phase I study of estramustine: preliminary activity in hormone refractory prostate cancer. *Seminars in Oncology* **22**(Suppl. 6), 6–11.

Hudis, C.A., Seidman, A.D. and Crown, J.P.A. (1996) Phase II and pharmacologic study of docetaxel as initial chemotherapy for metastatic breast cancer. *Journal of Clinical Oncology* **14**, 58–65.

Huizing, M.T., Keung, A.C.F. and Rosing, H. (1993) Pharmacokinetics of paclitaxel and metabolites in a randomised comparative study in platinum pre-treated ovarian cancer patients. *Journal of Clinical Oncology* **11**, 2127–2135.

Hussain, M., Salwen, W. and Kucuk, O. (1997) Paclitaxel, cisplatin and 5-fluorouracil in patients with advanced or recurrent squamous cell carcinoma of the head and neck. A preliminary report. *Seminars in Oncology* **24**, 43.

Jaffaioloi, R.V., Faccini, G. and Tortoriello, A. (1997) Phase I study of vinorelbine and paclitaxel in small cell lung cancer. *Cancer Chemotherapy and Pharmacology* **41**, 86–90.

Janinis, J., Papadakou, M. and Pangos, G. (1997) A phase II study of combined chemotherapy with docetaxel, cisplatin and 5-fluorouracil in patients with advanced squamous cell carcinoma of the head and neck and nasophyngeal carcinoma. *Proceedings of the American Society of Clinical Oncology* **14**, 36A.

Jett, J.R., Day, R. and Levitt, M. (1997) Topotecan and paclitaxel in extensive stage small cell lung cancer in patients without prior therapy. *Lung Cancer* **18**, 13A.

Johnson, D.H., Chang, A.Y. and Ettinger, D.S. (1994) Taxol (paclitaxel) in the treatment of lung cancer: the Eastern Cooperative Oncology Group experience. *Annals of Oncology* **5**(Suppl. 6), S45–S50.

Johnson, D., Paul, D. and Hande, K. (1997) Paclitaxel, 5-fluorouracil and folinic acid in metastatic breast cancer: BRE-26, a Phase II trial. *Seminars in Oncology* **24**(Suppl. 1), S22–S25.

Jordan, M.A., Wendell, K. and Garddiner, S. (1996) Mitotic block induced in HeLa cells by low concentrations of paclitaxel (Taxol) results in abnormal mitotic exit and apoptotic cell death. *Cancer Research* **56**, 816–825.

Kaufman, P.A. (1999) Paclitaxel and anthracycline combination chemotherapy for metatstatic breast cancer. *Seminars in Oncology* **26**(3), 39–46.

Kavanagh, J., Kudelka, A. and Gonzales de Leon, C. (1994) Phase II study of docetaxel in patients with epithelial ovarian carcinoma refractory to platinum. *Journal of Clinical Oncology* **2**, 837–842.

Kennedy, M., Zahurak, M. and Donehower, R. (1996) Phase I and pharmacologic study of the sequences of paclitaxel and cyclophosphamide supported by granulocyte colony stimulating factor in women with previously treated metastatic breast cancer. *Journal of Clinical Oncology* **14**, 783–791.

Kelly, K., Wood, M.E. and Bunn, P.A. (1998) A phase I study of cisplatin, etoposide and paclitaxel (PET) in small cell lung cancer. *Lung Cancer* **18**, 28A.

Kerbrat, P., Viens, P. and Roche, H. (1998) Docetaxel in combination with epirubicin as first line chemotherapy of metastatic breast cancer: final results. *Proceedings of the American Society of Clinical Oncology* **17**, 151A.

Khattab, J. and Urba, S.G. (1999) Chemotherapy in head and neck cancer. *Hematology/Oncology Clinics of North America* **13**(4), 753–768.

Kirschling, R.J., Jung, S.H. and Jett, J.R. (1994) A phase II trial of taxol and GCSF in previously untreated patients with extensive stage small cell lung cancer. *Proceedings of the American Society of Clinical Oncology* **13**, 326A.

Klassen, U., Wilke, H. and Muller, C. (1997) Infusional 5-fluorouracil/leucovorin plus paclitaxel and cisplatin in the first-line treatment of metastatic breast cancer. *Proceedings of the American Society of Clinical Oncology* 14, 122A.

Klauber, N., Parangi, S. and Flynn, E. (1997) Inhibition of angiogenesis and breast cancer in mice by the microtubule inhibitors 2-methoxyestradiol and Taxol. *Cancer Research* 57, 81–86.

Kohler, U., Olbricht, S. and Fuechsel, G. (1997) Weekly paclitaxel with epirubicin as second-line therapy of metastatic breast cancer; results of a clinical phase II study. *Seminars in Oncology* 24(5, Suppl. 17), S17-40–S17-43.

Kreis, W., Budman, D.R., Fetten, J., Gonzales, A.L., Barile, B. and Visiguerra, V. (1999) Phase I trial of the combination of daily estramustine phosphate and intermittent docetaxel in patients with metastatic hormone refractory prostate carcinoma. *Annals of Oncology* 10(1), 33–38.

Kudelka, A.P., Winn, R. amd Edawards, C.L. (1996a) Advanced squamous cell cancer of the cervix: a multicentre phase II study of paclitaxel (Taxol) 250 mg/m^2 administered intravenously over 3 h every 21 days with GCSF support. *Proceedings of the American Society of Clinical Oncology* 15, 281A.

Kudelka, A.P., Verschraegen, C.F. and Levy, T. (1996b) Preliminary report of the activity of docetaxel in advanced recurrent squamous cell cancer of the cervix. *Anticancer Drugs* 7, 398–401.

Kunitoh, H., Watanabe, K. and Onoshi, T. (1996) Phase II trials of docetaxel in previously untreated advanced non-small cell lung cancer. *Journal of Clinical Oncology* 14, 1649–1655.

Langer, C.J. (2000) Advanced non-small cell lung carcinoma: the emerging role of docetaxel. *Investigational New Drugs* 18, 17–28.

Langer, C.J., Leighton, J.C. and Comis, R.L. (1995) Paclitaxel and carboplatin in combination in the treatment of advanced non-small cell lung cancer: a phase II toxicity, response and survival analysis. *Journal of Clinical Oncology* 13, 1860–1870.

Latreille, J., Cormier, Y., Martins, H., Gross, G. and Fisher, B. (1996) Phase II study of docetaxel in patients with previously untreated extensive small cell lung cancer. *Investigational New Drugs* 13(4), 343–345.

Lavelle, F., Bissery, M.C., Combeau, C., Riou, J.F., Vrignaud, P. and Andre, S. (1995) Preclinical evaluation of docetaxel (Taxotere). *Seminars in Oncology* 22(2), 3–16.

Legha, S.S., Ring, S. and Papadopoulos, N. (1990) A phase II trial of taxol in metastatic melanoma. *Cancer* 65, 2478–2481.

Leonard, R., O'Brien, M. and Barrett-Lee, P. (1996) A prospective analysis of 390 advanced breast cancer patients treated with taxotere throughout the UK. *Annals of Oncology* 7(Suppl. 5), 19A.

Lister-Sharpe, D., McDonagh, M., Khan, K.S. and Kleijnen, J. (2000) A systematic review of the effectiveness of the taxanes used in the treatment of advanced breast and ovarian cancer. Report Commissioned by the NHS HTA Programme on behalf of the UK National Institute of Clinical Excellence, NHS Centre for Reviews and Dissemination, University of York, 2000. See also www.nice.org.uk

Llombart-Cusac, A., Pielman, M. and Dohollou, N. (1997) Cisplatin-taxotere phase I/II trial in patients with anthracycline-resistant advanced breast cancer. *Proceedings of the American Society of Clinical Oncology* 16, 180A.

Long, B.H., Wasserman, A.J. and Wei, J. (1996) Potential paclitaxel analogs that induce aberrant tubulin polymerisation *in vitro*. *Proceedings of the American Association of Cancer Research* 37, 441A.

Luck, H., Thomssen, C. and du Bois, A. (1997) Phase II study of paclitaxel and epirubicin as first-line therapy in patienst with metastatic breast cancer. *Seminars in Oncology* 24(5, Suppl. 17), S17-35–S17-39.

McCaffrey, J.A., Dodd, P.M. and Hilton, S. (1999) Ifosfamide plus paclitaxel plus cisplatin chemotherapy for patients with unresctable or metatstatic transitional cell carcinoma. *Proceedings of the American Society of Clinical Oncology* 18, 329A.

McCaffrey, J.A., Hilton, S. and Bajorin, D. (1995) Docetaxel in patients with advanced transitional cell cancer who failed cisplatin based chemotherapy. *Proceedings of the American Society of Clinical Oncology* 14, 233A.

McCaskill-Stevens, W., Ansari, R. and Fisher, W. (1996) Phase II study of bi-weekly cisplatin and paclitaxel in the treatment of metastatic breast cancer. *Proceedings of the American Society of Clinical Oncology* 15, 120A.

McGuire, W.P. (1995) In: Thigpen, T., Vance, R.B. and Khansur, T. (1995) The platinum comnpounds and paclitaxel in the management of carcinomas of the endometrium and uterine cervix. *Seminars in Oncology* 22(5), 67–75.

McGuire, W.P., Brady, M.F., Kucera, P.R., Partridge, E.E., Look, K.Y. and Davidson, M. (1995) Taxol and cisplatin improves outcome in advanced ovarian cancer as compared to cytoxan and cisplatin. *Proceedings of the Annual Meeting of the American Society of Clinical Oncology* 14, 771A.

McGuire, W.P., Hoskins, W.J., Brady, M.F., Kucera, P.R., Partridge, E.E. and Look, K.Y. (1996a) Cyclophosphamide and cisplatin compared with paclitaxel and cisplatin in patients with stage III and stage IV ovarian cancer. *New England Journal of Medicine* 334(1), 1–6.

McGuire, W.P., Brady, M.F., Kucera, P.R., Partridge, E.E., Look, K.Y., Clarke, K.Y., Pearson, D.L. and Davidson, M. (1996b) Comparison of combination therapy with paclitaxel and cisplatin versus cyclophosphamide – cisplatin as primary therapy in patients with suboptimal stage III and stage IV ovarian cancer. *International Journal of Gynecological Cancer* 6(5), 2–8.

McGuire, W.P., Hoskins, W.J., Brady, M.F., Kucera, P.R., Partridge, E.E., and Look, K.Y. (1996c) Cyclophosphamide and cisplatin versus paclitaxel and cisplatin: a phase II randomised trial in patients with suboptimal stage III/IV ovarian cancer. *Seminars in Oncology* 23, 40–47.

McGuire, W.P., Blessing, J.A. and Moore, D. (1996d) Paclitaxel has moderate activity in squamous cervix cancer: a Gynecologic Oncology Group study. *Journal of Clinical Oncology* 14, 792–795.

McGuire, W.P., Rowinsky, E.K., Rosenhein, N.B., Grumbine, F.C., Ettiger, D.S., Armstrong, D.K. and Donehower, R.C. (1989) Taxol: a unique antineoplastic agent with significant activity in advanced ovarian epithelial neoplasms. *Annals of Internal Medicine* 111, 273–279.

Mamounas, E., Brown, A. and Fisher, B. (1995) 3-hour, high-dose Taxol infusion in advanced breast cancer. An NSABP pahse II study. *Proceedings of the American Society of Clinical Oncology* 14, 127A.

Manfredi, J.J. and Horwitz, S.B. (1994) Taxol: an antimitotic agent with a new mechanism of action. *Pharmacology and Therapeutics* 25, 83–125.

Manthey, C.L., Perera, P.Y., Salkowsi, C.A. and Vogel, S.N. (1994) Taxol provides a second signal for murine macrophage tumoricidal activity. *Journal of Immunology* 152, 825–828.

Marini, L., Sternberg, C.N. and Sella, A. (1999) A new regimen of gemcitabine and paclitaxel in previously treated patients with advanced transitional cell carcinoma. *Proceedings of the American Society of Clinical Oncology* 18, 346A.

Markman, M. (1991) Taxol: an important new drug in the management of epithelial ovarian cancer. *Yale Journal of Biological Medicine* 64, 583–587.

Martin, M., Garcia Carbonero, I. and Lluch, A. (1998) Paclitaxel plus vinorelbine: an active regimen in metastatic breast cancer patients with prior anthracycline exposure. *Proceedings of the American Society of Clinical Oncology* 17, 158A.

Martin, M., Lluch, A. and Ojeda, B. (1997) Paclitaxel plus doxorubicin in metastatic breast cancer: preliminary analaysis of cardiotoxicity. *Seminars in Oncology* 24(Suppl. 17), S17-26–S17-30.

Mavroudis, D., Kouroussis, C. and Malamos, N. (1998) A phase II trial of docetaxel and gemcitabine as second line treatment in metastatic breast cancer. *Proceedings of the American Society of Clinical Oncology* 17, 158A.

Meluch, A.A., Greco, F.A. and Hurris, H.A. (1999) Gemcitabine and paclitaxel in combination for advanced transitional cell carcinoma of the urothelial tract. *Proceedings of the American Society of Clinical Oncology* 18, 347A.

Mereten, W.C., Eisenhauer, E.A. and Jolivet, J. (1994) Docetaxel in advanced renal carcinoma: a phase II trial of the National Cancer Institute of Canada Clinical Trials group. *Annals of Oncology* 5, 185–187.

Meyers, F.J., Edelman, M.J. and Houston, J. (1998) Phase I/II trial of paclitaxel, carboplatin and methotrexate in advanced transitional cell carcinoma. *Proceedings of the American Society of Clinical Oncology* 17, 317A.

Milas, L., Milas, M.M. and Mason, K.A. (1999) Combination of taxanes with radiation: preclinical studies. *Seminars in Radiation Oncology* 9(2), 12–26.

Miller, V.A., Rigas, J.R. and Kris, M.G. (1994) Phase II trial of docetaxel given at a dose of 75 mg/m^2 with prednisone premedication in non-small cell lung cancer. *Proceedings of the American Society of Clinical Oncology* 13, 364A.

Miller, K.D. and Sledge, G.W. (1999) Taxanes in the treatment of breast cancer: a prodigy comes of age. *Cancer Investigation* 17(2), 121–136.

Millward, M.J., Webster, D. and Rischin, D. (1996) Plasma concentrations of polysorbate 80 (Tween 80) in patients following docetaxel (Taxotere) or etoposide. *Proceedings of the 9th NCI-EORTC Symposium on New Drugs in Cancer Therapy*, 95A.

Misset, J., Dieras, V. and Bozec, L. (1998) Long term follow up of the phase I/II study of docetaxel and doxorubicin as first line chemotherapy of metastatic breast cancer. *Proceedings of the American Society of Clinical Oncology* 17, 160A.

Moliterni, A., Tarenzi, E. and Capri, G. (1997) Pilot study of primary chemotherapy with doxorubicin plus paclitaxel in women with locally advanced or operable breast cancer. *Seminars in Oncology* 24(5, Suppl. 17), S17-10–S17-14.

Motzer, R.J. (1997) Paclitaxel in salvage therapy for germ cell tumours. *Seminars in Oncology* 24(5, Suppl. 15), S15-83–S15-85.

Motzer, R.J. (2000) Paclitaxel (Taxol) combination therapy for resistant germ cell tumours. *Seminars in Oncology* 27(1), 33–35.

Motzer, R.J., Bajorin, D. and Schwartz, L.H. (1994) Phase II trial of paclitaxel shows antitumour activity in patients with previously treated germ cell tumours. *Journal of Clinical Oncology* 12, 2277–2283.

Muggia, F., Brady, P. and Brady, M.F. (1997) Phase III study of cisplatin and or paclitaxel versus the combination in suboptimal stage III and IV epithelial ovarian cancer. *Proceedings of the Annual Meeting of the American Society of Clinical Oncology* 16, 1257A.

Munzone, E., Capri, G. and Demicheli, R. (1993) Activity of taxol by 3-hour infusion in breast cancer patients with clinical resistance to anthracyclines. *European Journal of Cancer* 29, S79.

Murad, A., Guirmaraes, R. and Amorin, W. (1997) Phase II trial of paclitaxel and ifosfamide as a salvage treatment in metastatic breast cancer. *Breast Cancer Research and Treatment* 45, 47–53.

Murphy, W.K., Fossella, F.V. and Winn, R.J. (1993) Phase II study of taxol in patients with untreated non-small cell lung cancer. *Journal of the National Cancer Institute* 85, 384–388.

Nabholtz, J.T. (1997a) Docetaxel, doxorubicin and cyclophosphamide in the treatment of metastatic breast cancer. *Proceedings of the 20th Annual San Antonio Breast Cancer Conference*, 7–9.

Nabholtz, J.T. (1997b) Docetaxel (Taxotere) vs mitomycin C and vinblastine in patients with metastatic breast cancer who have failed on anthracycline-containing regimen. Preliminary evaluation of a randomized phase III study. *Proceedings of the American Society of Clinical Oncology* 16, 148A.

Nabholtz, J.T., Bezwoda, W.R., Melnychuk, D., Deschenes, L., Douma, J., Vandenberg, T.A. and Rapoport, B. (1997a) Docetaxel vs mitomycin plus vinblastine in anthracycline-resistant metastatic breast cancer. *Oncology Huntingt* 11(8), 25–30.

Nabholtz, T.T., Thuerlimann, B. and Bezwoda, W. (1997b) Docetaxel vs mitomycin plus vinblastine in anthracycline-resistant metastatic breast cancer. *Oncology* 11(8, Suppl. 8), 25–30.

Nabholtz, T.T., Thuerlimann, B., Betwoda, W., Merlnychuk, D., Alak, M. and Murawsky, M. (1997c) Taxotere vs mitomycin C plus vinblastine in patients with metastatic breast cancer who have failed an anthracycline containing regimen: preliminary results of a randomized phase III study. *Proceedings of the Annual Meeting of the American Society of Clinical Oncology* 16, 519A.

Nabholtz, J.T., Bezwoda, W.R., Melnychuk, D., Deschenes, L., Douma, J., Vandenberg, T.A. and Rapoport, B. (1999a) Prospective randomized trial of docetaxel versus mitomycin plus vinblastine in patients with metastatic breast cancer progressing despite previous anthracycline-containing chemotherapy. *Journal of Clinical Oncology* 17(5), 1413–1424.

Nabholtz, T.T., Falkson, G., Campos, D., Szanto, J., Martin, M. and Chan, S. (1999b) A phase III trial comparing doxorubicin and docetaxel to doxorubicin and cyclophosphamide as first line chemotherapy for MBC. *Proceedings of the American Society of Clinical Oncology* 18, 127A.

Nabholtz, J.M., Gelmon, K. and Bontenbal, M. (1996) Multicentre, randomised comparative study of two doses of paclitaxel in patients with metastatic breast cancer. *Journal of Clinical Oncology* 14, 1858–1867.

Nair, S., Marschke, R. and Grill, J. (1997) A phase II study of paclitaxel (Taxol) and cisplatin in the treatment of extensive stage small cell lung cancer. *Proceedings of the American Society of Clinical Oncology* 16, 454A.

National Cancer Institute (1983) Clinical Brochure, Taxol (NSC 125973), Division of Cancer Treatment, NCI, Bethesda, pp. 6–12.

National Institute of Clinical Excellence (2000) Guidance on the use of taxanes for breast cancer. www.nice.org.uk.

Neuberg, D., Sledge, G. and Fetting, J. (1997) Changes in quality of life during induction therapy in patients enrolled in a randomised trial of adriamycin, taxol and adriamycin plus taxol. *Proceedings of the American Society of Clinical Oncology* 16, 54A.

Nicholson, B., Paul, D. and Hande, K. (1996) Phase II trial of paclitaxel, 5-fluorouracil and leukovorin in metastatic breast cancer. *Proceedings of the American Society of Clinical Oncology* 15, 102A.

Nicolau, K.C., Yang, Z., Liu, J.J., Ueno, H., Nantermet, P.G., Guy, R.K., Claiborne, C.F., Renaud, J., Couladouros, E.A., Paulvannin, K. and Sornsen, E.J. (1994) Total synthesis of taxol. *Nature*, 367, 630–631.

Ozols, R.F. (2000) Paclitaxel (Taxol)/carboplatin combination chemotherapy in the treatment of advanced ovarian cancer. *Seminars in Oncology* 27(3), 3–7.

Panagos, G. (1997) Treatment of advanced and relapsing breast cancer with a combination of paclitaxel and mitoxantrone. *Seminars in Oncology* 24(1, Suppl. 3), S17–S21.

Parekh, H. and Simpkins, H. (1997) The transport and binding of taxol. *General Pharmacology* 29(2), 167–172.

Paridaens, R., Bruning, P., Klijn, J., Gamucci, T., Biganzoli, L. and Van Vreckam, A. (1997) An EORTC crossover trial comparing single agent Taxol and doxorubicin as first- and second-line chemotherapy in advanced breast cancer. *Proceedings of the American Society of Clinical Oncology* 16, 539A.

Paridaens, R., Bruning, P., Klijn, J., Gamucci, T., Houston, R. and Coleman, J. (2000) Paclitaxel versus doxorubicin as first-line single-agent chemotherapy for metastatic breast cancer: a European Organisation for Research and treatment of Cancer randomised study with cross-over. *Journal of Clinical Oncology* 18(4), 724–733.

Paul, D. and Garrett, A. (1995) A phase II trial of paclitaxel, 5-fluorouracil leucovorin (TFL) in metastatic breast cancer. *Proceedings of the American Society of Clinical Oncology* 14, 140A.

Pazdur, R., Lassere, Y. and Soh, L.T. (1994) Phase II trial of docetaxel (Taxotere) in metastatic colorectal cancer. *Annals of Oncology* 5, 468–470.

Pendri, A., Greenwald, R. and Gilber, C. (1996) Drug delivery systems: water soluble taxol-2-polyethylene glycol ester prodrugs – design and *in vivo* effectiveness. *Proceedings of the American Society of Cancer Research* 37, 376A.

Perez, E., Suman, V. and Krook, J. (1998) Phase II study of paclitaxel plus carboplatin as first-line therapy for women with metastatic breast cancer: a North Central Cancer Treatment Group trial. *Proceedings of the American Society of Clinical Oncology* 17, 165A.

Piccart-Gebhart, M., Bertelson, K. and Stuart, G. (1997) Is cisplatin–paclitaxel the standard first-line treatment in advanced ovarian cancer? The EORTC-COCG, NOCOVA, NCIC CTG and Scottish Intergroup experience. *Proceedings of the American Society of Clinical Oncology* 16, 1394A.

Piccart-Gebhart, M.B., Gamucci, T., Klijn, J., Roy, J.A. and Kusenda, Z. (1996) An ongoing European Organization for Research and Treatment of Cancer crossover trial comparing single-agent paclitaxel and doxorubicin as first- and second-line treatment for advanced breast cancer. *Seminars in Oncology* 25(5), 11–15.

Piccart, J.H., Gore, M. and ten Bokkel, H.W. (1995) Docetaxel: an active new drug for the treatment of advanced ovarian cancer. *Journal of the National Cancer Institute* 87, 686–681.

Pluzzanska, A. and Jassem, J. (1999) Randomised, open label phase III multicentre trial comparing Taxol/doxorubicin vs 5-fluorouracil and cyclophosphamide as first line treatment for patients with metastatic breast cancer. *European Journal of Cancer* (Suppl. 4) (S314), 1260A.

Prados, M.D., Schold, S.C. and Spence, A.M. (1996) Phase II study of paclitaxel in patients with recurrent malignant glioma. *Journal of Clinical Oncology* 14, 2316–2321.

Raefsky, E.L., Hainsorth, J.D. and Burris, H.A. (1998) Phase I trial of topotecan, paclitaxel and carboplatin in patients with advanced refractory malignancies. *Proceedings of the American Society of Clinical Oncology* 17, 498A.

Rao, S., Krauss, N.E. and Heerding, J.M. (1994) 3-(p-azidobenzamido) taxol photolabels the N-terminal 31 amino acids of beta-tubulin. *Journal of Biological Chemistry* **269**, 3132–3134.

Ravdin, P., Burris, H.A. and Cook, G. (1995) Phase II trial of docetaxel in advanced anthracycline-resistant or anthracenedione-resistant breast cancer. *Journal of Clinical Oncology* **13**, 2879–2885.

Ravdin, P.M. and Valero, V. (1995) Review of docetaxel (taxotere), a highly active new agent for the treatment of metastatic breast cancer. *Seminars in Oncllology* **22**(2), 17–21.

Reck, M., Jagos, V. and Kaukel, E. (1997) Chemotherapy of limited stage small cell lung cancer with cisplatin, paclitaxel and oral etoposide: a phase II trial. *Lung Cancer* **18**, 31A.

Reichman, B., Seidman, A. and Crown, J. (1993) Paclitaxel and recombinant human granulocyte colony stimulating factor as initial therapy for metatstatic breast cancer. *Journal of Clinical Oncology* **11**, 1943–1951.

Ries, F., Duhem, C. and Kleiber, K. (1997) Phase I/II clinical trial of epirubicin and paclitaxel followed by GCSF in a 2-week schedule in patients with advanced or metastatic breast cancer. *Seminars in Oncology* **24**(5, Suppl. 17), S17-48–S17-51.

Ringel, I. and Horwitz, S.B. (1991) Studies with RP56976 (Taxotere): a semisynthetic analogue of taxol. *Journal of the National Cancer Institute* **83**, 288–291.

Riondel, J., Jacrot, M. and Picot, F. (1986) Therapeutic response to taxol of six human tumours xenografted into nude mice. *Cancer Chemotherapy and Pharmacology* **17**, 137–142.

Riseberg, D., Cowan, K. and Tolcher, A. (1995) 96-hour paclitaxel without and with r-verapamil in patients with metastatic breast cancer previously treated with 3- or 24-hour paclitaxel. *Proceedings of the American Society of Clinical Oncology* **14**, 180A.

Rose, W.C. (1995) Preclinical tumour activity of taxanes. In: M. Suffness (ed.) *Taxol – Science and Applications.* CRC Press, London, pp. 209–236.

Roth, B, J., Dreicer, R. and Einhorn, L.H. (1994) Significant activity of paclitaxel in advanced transitional-cell carcinoma of the urothelium: a phase II study of the ECOG. *Journal of Clinical Oncology* **12**, 2264–2270.

Rowinsky, E.K. (1995) Pharmacology and metabolism. In: E.K. McGuire and E.K. Rowinsky (eds) *Paclitaxel in Cancer Treatment*, Dekker, New York, pp. 91–120.

Rowinsky, E.K. (1997) The development and clinical utility of the taxane class of antimicrotubule chemotherapy agents. *Annual Review of Medicine*, **48**, 353–374.

Rowinsky, E.K., Burke, P.J. and Karp, J.E. (1989) Phase 1 and pharmacodynamic study of taxol in refractory adult acute leukemia. *Cancer Research* **49**, 4640–4647.

Rowinsky, E.K., Cazenave, L.A. and Donehower, R.C. (1990) Taxol: a novel investigational antimicrotubule agent. *Journal of the National Cancer Institute* **82**, 1247–1254.

Rowinsky, E.K., Citardi, M., Noe, D.A. and Donehower, R.C. (1993a) Sequence dependent cytotoxicity between cisplatin and the antimicrotubule agents taxol and vincristine. *Journal of Cancer Research and Clinical Oncology* **119**, 737–743.

Rowinsky, E.K., Eisenhauer, E.A. and Chaudhry V. (1993b) Clinical toxicities encountered with taxol. *Seminars in Oncology* **20**(Suppl. 3), 1–15.

Rowinsky, E.K., Chaudhry, V., Cornblath, D.T. and Donehower, R.C. (1993c) The neurotoxicity of taxol. *Monographs of the National Cancer Institute* **15**, 107–115.

Rowinsky, E.K. and Donehower, R.C. (1995) Drug therapy: paclitaxel (Taxol). *New England Journal of Medicine* **332**, 1004–1014.

Rowinsky, E.K. and Donehower, R.C. (1996) Antimicrotubule agents. In: Chabner, B and Collins, J. (eds.) *Cancer Chemotherapy*, Lippincott-Raven, Philadelphia, pp. 263–296.

Rowinsky, E.K., Gilbert, M. and McGuire, W.P. (1991a) Sequences of taxol and cisplatin: a phase 1 and pharmacologic study. *Journal of Clinical Oncology* **9**, 1692–1703.

Rowinsky, E.K., McGuire, W.P., Guarnieri, T., Fishermann, J.S., Christian, M.C. and Donehower, R.C. (1991b) Cardiac disturbances during the administration of taxol. *Journal of Clinical Oncology* **9**, 1704–1725.

Rowinsky, E.K., Onetto, N., Canetta, R.M. and Arbuck, S.G. (1992a) Taxol, the first of the taxanes, an important new class of antitumour agents. *Seminars in Oncology* **19**, 646–656.

Rowinsky, E.K., Onetto, N., Canetta, R.M. and Arbuck, S.G. (1992b) Taxol: the prototypic taxane, an important new class of antitumour agents. *Seminars in Oncology* 19, 646–662.

Sandercock, J., Parmar, M. and Torri, V. (1998) First line chemotherapy for advanced ovarian cancer: paclitaxel, cisplatin and the evidence. *British Journal of Cancer* 78(11), 1471–1478.

Sarosy, G. and Reed, E. (1993) Taxol dose intensification and its clinical implications. *Journal of the National Medical Association* 85, 427–430.

Schiff, P.B., Gubits, R., Kashimawo, S. and Geard, C.R. (1995) Paclitaxel with ionizing radiation. In: E.K. McGuire and E.K. Rowinsky (eds) *Paclitaxel in cancer treatment*, Dekker, New York, pp. 81–90.

Schiff, P.B., Taylor, H.L. and Horwitz S.P. (1979) Promotion of microtubule assembly *in vitro* by taxol. *Nature* 277, 665–667.

Schoffski, P., Wanders, J. and Catimel, G. (1996) A promising regimen for the treatment of head and neck cancer: docetaxel and cisplatin. *Proceedings of the Annual Meeting of the European Society of Medical Oncology*, 373A.

Schutte, W. (1997) Phase II trial of docetaxel and carboplatin in the treatment of advanced non-small cell cancer. *European Journal of Cancer* 33, 1230–1234.

Schwartzmann, G., Menke, C. and Caleffi, M. (1996) Phase II trial of taxol, doxorubicin plus GCSF in patients with metastatic breast cancer. *Proceedings of the American Society of Clinical Oncology* 15, 126A.

Seidman, A.D. (1994) Single agent use of Taxol (paclitaxel) in breast cancer. *Annals of Oncology* 5(Suppl. 6), S17–S22.

Seidman, A.D. (1995) The emerging role of paclitaxel in breast cancer therapy. *Clinical Cancer Research* 1, 247–250.

Seidman, A.D., Baselga, J. and Yao, T.Y. (1996) HER-/neu over expression and clinical taxane sensitivity: a multivariate analysis in patienst with metastatic breast cancer. *Proceedings of the American Society of Clinical Oncology* 15, 104A.

Seidman, A.D., Crown, J.P.A. and Reichman, B.S. (1993) Lack of cross-resistance of taxol with anthracycline in the treatment of metastatic breast cancer. *Proceedings of the American Society of Clinical Oncology* 12, 63A.

Seidman, A.D., Hochhauser, D. and Gollub, M. (1996) Ninety-six hour paclitaxel infusion after progression during short taxane exposure: a phase II pharmacokinetic and pharmacodynamic study in metastatic breast cancer. *Journal of Clinical Oncology* 14, 1877–1884.

Seidman, A.D., Hudis, C.A., Albanel, J., Tong, W., Tepler, I. and Currie, V. (1998) Dose–dense therapy with weekly, 1-hour paclitaxel infusions in the treatment of metastatic disease. *Journal of Clinical Oncology* 16(10), 3353–3361.

Seidman, A.D., Reichman, B.S. and Crown, J.P.A. (1995a) Paclitaxel as second and subsequent therapy for metastatic breast cancer: activity independent of prior anthracycline response. *Journal of Clinical Oncology* 13, 1152–1159.

Seidman, A.D., Tiersten, A. and Hudis, C. (1995b) Phase II trial of paclitaxel by 3-hour infusion as initial and salvage chemotherapy for metastatic breast cancer. *Journal of Clinical Oncology* 13, 2575–2581.

Sessa, C., Pagani, O. and Martinelli, G. (1994) Dose finding study of paclitaxel and cyclophosphamide in women with advanced breast cancer. *Seminars in Oncology* 24(Suppl. 17), S17-52–S17-57.

Shepherd, F., Ramlau, R., Mattson, K., Cressot, L., O'Rourke, M. and Vincent, M. (1999) Randomised study of taxotere versus best supportive care in NSCLC patients previously treated with platinum based chemotherapy. *Proceedings of the American Society of Clinical Oncology* 18, 1794A.

Shin, D., Glisson, B. and Khuri, F. (1998) Role of paclitaxel, ifosfamide and cisplatin in patients with recurrent or metastatic squamous cell carcinoma of the head and neck. *Seminars in Oncology* 25, 40.

Sjostrom, J., Mouridsen, H. and Pluzanska, A. (1998) Taxotere versus methotrexate- 5-flurouracil in patienst with advanced anthracycline-resistant breast cancer: preliminary results of a randomized phase III study by the Scandinavian Breast Cancer group. *Proceedings of the American Society of Clinical Oncology* 17, 111A.

Sledge, G.W., Neuberg, D., Ingle, J., Martino, S. and Wood, W. (1997) Phase III trial of doxorubicin vs paclitaxel vs doxorubicin plus paclitaxel as first-line therapy for metastatic breast cancer: an intergroup trial. *Proceedings of the Annual Meeting of the American Society of Clinical Oncology* 16, 1A.

Slichenmeyer, W.J. and Von Hoff, D.D. (1991) Taxol: a new and effective anticancer drug. *Anticancer Drugs* 2, 519–523.

Smit, E.F., Fokkema, E. and Biesma, B. (1998) A phase II study of paclitaxel in heavily pretreated patients with small cell lung cancer. *British Journal of Cancer* 77, 347–351.

Smith, D.C. and Pienta, K.J. (1999) Paclitaxel in the treatment of hormone refractory prostate cancer. *Seminars in Oncology* 26(1), 109–111.

Smith, R., Thornton, D. and Allen, J. (1995) A phase II trial of paclitaxel in squamous cell carcinoma of the head and neck. *Seminars in Oncology* 22, 41.

Smyth, J.F., Smith, I.E. and Sessa, C. (1994) Activity of docetaxel (Taxotere) in small cell lung cancer. *European Journal of Cancer* 30, 1058–1060.

Sonnichsen, D.S. and Relling, M.V. (1994) Clinical pharmacokinetics of paclitaxel. *Clinical Pharmacokinetics* 27(4), 256–269.

Sonnichsen, D.S., Hurwitz, C. and Pratt, C. (1994) Saturable pharmacokinetics and paclitaxel pharmacodynamics in children with solid tumours. *Journal of Clinical Oncology* 12, 532–538.

Sparano, J., Neuberg, D. and Glick, J. (1996) A phase II trial of bi-weekly paclitaxel and cisplatin in patients with advanced breast cacinoma: an Eastern Cooperative Oncology (ECOG) trial (E1194). *Proceedings of the American Society of Clinical Oncology* 15, 114A.

Sparreboom, A., van Tellingen, O., Nooijen, W.J. and Beijnen, J.H. (1998) Preclinical pharmacokinetics of paclitaxel and docetaxel. *Anticancer Drugs* 9, 1–17.

Sternberg, C.N., Sodillo, P.P. and Cheng, E. (1987) Evaluation of new anticancer agents against human pancraetic carcinomas in nude mice. *American Journal of Clinical Oncology* 10, 219–221.

Sternberg, C.N., Ten Bokkel Huinink, W.W. and Smyth, J.F. (1994) Docetaxel (Taxotere), a novel taxoid, in the treatment of advanced colorectal carcinoma: an EORTC Early Clinical Trials Group Study. *British Journal of Cancer* 70: 376–379.

Straubinger, R.M. (1995) Biopharmaceutics of paclitaxel (taxol): formulation, activity and pharmacokinetics. In: M. Suffness (ed.) *Taxol – Science and Applications*. CRC Press, London, pp. 237–258.

Stuart, G., Bertelson, K. and Mangioni, C. (1999) Updated analysis shows a highly significant overall improved survival for cisplatin–paclitaxel as fisrt-line treatment of advanced ovarian cancer: mature results of the EORTC-COCG, NOCOVA, NCIC CTG and Scottish Intergroup trial. *Proceedings of the American Society of Clinical Oncology* 17, 1394A.

Suffness, M. and Wall, M.E. (1995) Discovery and development of taxol. In: M. Suffness (ed.) *Taxol – science and Applications*. CRC Press, London, pp. 3–25.

Sulkes, A., Smyth, J. and Sessa, C. (1994) Docetaxel in advanced gastric cancer: results of a phase II clinical trial. *British Journal of Cancer* 70, 380–384.

Swain, S., Honig, S. and Walton, L. (1995) Phase II trial of paclitaxel (taxol) as first-line chemotherapy for metastatic breast cancer. *Proceedings of the American Society of Clinical Oncology* 14, 132A.

Taguchi, T., Adachi, I. and Enomoto, K. (1994) Docetaxel (RP56976) in advanced or recurrent breast cancer: early and late phase II clinical studies in Japan. *Proceedings of the American Society of Clinical Oncology* 13, 331A.

Ten Bokkel, H.W., Prove, A.M. and Piccart, M. (1994) Phase II trial with docetaxel (Taxotere) in second-line treatment with chemotherapy for advanced breast cancer. *Annals of Oncology* 5, 527–532.

Thigpen, J.T., Blessing, J.A., Ball, H., Hummel, S.J. and Barrett, R.J. (1994) Phase II trail of paclitaxel in patients with progressive ovarian carcinoma after platinum-based chemotherapy. *Journal of Clinical Oncology* 12, 1748–1753.

Thigpen, J.T., Vance, R.B. and Khansur, T. (1995) The platinum compounds and paclitaxel in the management of carcinomas of the endometrium and uterine cervix. *Seminars in Onclogy* 22, 67–75.

Tortoriello, A., Facchini, G. and Caponigro, F. (1996) A phase I/early Phase II trial with vinorelbine plus paclitaxel plus CSFs in advanced breast cancer: preliminary data. *Annals of Oncology* 7(Suppl. 5), 23A.

Trandafir, L., Chafine, A. and Spielman, M. (1996) Efficacy of taxotere in advanced breast cancer patients not eligible for further anthracycline. *Proceedings of the American Society of Clinical Oncology* 15, 105A.

Trimble, E.L., Adams, J.D. and Vena, D. (1993) Taxol in patients with platinum-refractory ovarian cancer. *Proceedings of the American Society of Clinical Oncology* 12, 829A.

Trimble, E.L., Arbuck, S.G. and McGuire, W.P. (1994) Options for primary chemotherapy of epithelial ovarian cancer: taxanes. *Gynecological Oncology* 55, S114–S121.

Trudeau, M.E., Eisenhauer, E.A. and Higgins, B.P. (1996) Docetaxel in patients with metastatic breast cancer: a Phase II study of the National Cancer Institute of Canada – Clinical Trials group. *Journal of Clinical Oncology* 14, 422–428.

Tu, S.M., Hossan, E. and Amato, R. (1995) Paclitaxel, cisplatin and methotrexate combination chemotherapy is active in the treatment of refractory urothelial malignancies. *Journal of Urology* 154, 1719–1722.

Untch, M., Untch, A. and Seven, B. (1994) Comparison of paclitaxel and docetaxel (Taxotere) in gynecologic and breast cancer cell lines with the ATP-cell viability assay. *Anticancer Drugs* 5, 24–30.

Vaishampayan, U., Smith, D. and Redman, B. (1999) Phase II evaluation of carboplatin, paclitaxel and gemcitabine in advanced urothelial cancer. *Proceedings of the American Society of Clinical Oncology* 18, 333A.

Valero, V. (1997) Combination docetaxel/cyclophosphamide in patients with advanced solid tumours. *Oncology* 11(8, Suppl. 8), 34–36.

Valero, V., Holmes, F.A. and Walters, R.S. (1995) Phase II trail of docetaxel: a new highly effective anti-neoplastic agent in the management of patients woth anthracycline-resistant metastatic breast cancer. *Journal of Clinical Oncology* 13, 2886–2894.

Valero, V., Jones, S.E. and Von Hoff, D.D. (1998) A phase II study of docetaxel in patients with paclitaxel-resistant metastatic breast cancer. *Journal of Clinical Oncology* 16, 3362–3368.

Van Hoesel, Q., Verweij, J. and Catimel, G. (1994) Phase II study with docetaxel (Taxotere) in advanced soft tissue sarcoma of the adult. *Annals of Oncology* 5, 539–542.

Van Oosterom, A.T. and Schrijvers, D. (1995) Docetaxel (Taxotere), a review of preclinical and clinical experience. Part II: Clinical experience. *Anticancer Drugs* 6, 356–368.

Vaughn, D.J. (1999) Review and outlook for the role of paclitaxel in urothelial carcinoma. *Seminars in Oncology* 26(1), 117–122.

Vermorken, J., Huizing, M. and Liefting, A. (1993) High dose taxol with GCSF in patients with advanced breast cancer refractory to anthracycline therapy. *European Journal of Cancer* 29, S79.

Vogt, H., Martin, T., Kolotas, C., Schneider, L., Strassman, G. and Zamboglou, N. (1997) Simaltaneous paclitaxel and radiotherapy: initial clinical experience in lung cancer and other malignancies. *Seminars in Oncology* 24(4), S12101—S12105.

Von Hoff, D.D. (1991) In: W.J. Slichenmeyer and D.D. von Hoff (eds), Taxol: a new and effective anti-cancer drug. *Anticancer Drugs* 2, 519–529.

Von Hoff, D.D. (1997) The taxoids: same roots, different drugs. *Seminars in Oncology* 24(4), S13-3–S13-10

Wall, M.E. and Wani, M.C. (1967) Recent progress in plant antitumor agents. Paper M-006, 153rd National Meeting, American Chemical Society, Miami Beech, Florida.

Waltzman, R., Schwartz, G.K. and Shorter, S. (1996) Lack of efficacy of paclitaxel (Taxol) in patients with advanced soft tissue sarcoma. *Proceedings of the American Society of Clinical Oncology* 15, 526A.

Wani, M.C., Taylor, H.C., Wall, M.E., Coggan, P. and McPhail, A.T. (1971) The isolation and structure of taxol, a novel antileukaemic and antitumor agent from *Taxus brevifolia*. *Jornal of the American Chemical Society* 93, 2325–2327.

Wasserheit, C., Frazein, A. and Oratz, R. (1996) Phase II trial of paclitaxel and cisplatin in women with advanced breast cancer: an active regimen with limiting neurotoxicity. *Journal of Clinical Oncology* 14, 1993–1999.

Watanabe, T., Adachi, I. and Sasaki, Y. (1996) A phase I/II study of docetaxel in combination with 5-fluorouracil for advanced or recurrent breast cancer. *Annals of Oncology* 7(Suppl. 5), 21A.

Webster, L., Linsenmeyer, M., Milward, M., Morton, C., Bishop, J. and Woodcock, D. (1993) Measurement of Cremophor-EL following taxol: plasma levels sufficient to reverse drug exclusion mediated by the multidrug-resistant phenotype. *Journal of the National Cancer Institute* 85, 1685–1689.

Weiss, R.B., Donehower, R.C., Wiernik, P.H., Ohnuma, T., Gralla, R.J., Trump, D.L., Baker, J.R., Van Echo, D.A., Von Hoff, D.D. and Leyland-Jones, B. (1990) Hypersensitivity reactions from taxol. *Journal of Clinical Oncology* 8, 1263–1276.

Wiernik, P.H., Schwartz, E.L., Strauman, J.J., Dutcher, J., Lipton, R. and Paietta, E. (1987a) Phase I clinical and pharmacokinetic study of taxol. *Cancer Research* 47, 2487.

Wiernik, P.H., Schwartz, E.L., Einzig, A., Staruman, J.J., Lipton, R.B. and Dutcher, J.P. (1987b) Phase 1 trial of taxol given as a 24-hour infusion every 21 days: responses observed in malignant melanoma. *Journal of Clinical Oncology* 5, 1232–1240.

Wiley, J., Inbrahim, N. and Walters, R. (1998) Vocal chord paralysis secondary to vinorelbine and paclitaxel by simaltaneous 3-hour infusion with GCSF support as frontline therapy for metastatic breast cancer patients. *Proceedings of the American Society of Clinical Oncology* 7, 183A.

Wilson, H.W., Chabner, B.A. and Bryant, G. (1995) Phase II study of paclitaxel in relapsed non-Hodgkin's lymphoma. *Journal of Clinical Oncology* 13, 381–386.

Wilson, W.H., Berg, S. and Bryant, G. (1994) Paclitaxel in doxorubicin-refractory or mitoxantrone-refractory breast cancer: a phase 1/2 trial of a 96-hour infusion. *Journal of Clinical Oncology* 12, 1621–1629.

Witherup, K.M., Look, S.A., Stasko, M.W., Ghiorzi, T.J., Muschik, G.M. and Cragg, G.M. (1990) Semisynthesis of taxol. *Journal of Natural Products* 53, 1249–1254.

Woo, H.L., Swenerton, K.D. and Hoskins, P.J. (1996) Taxol is active in platinum-resistant endometrial adenocarcinoma. *American Journal of Clinical Oncology* 19, 290–291.

Younes, A., Ayoub, J. and Hagemeister, F.B. (1996) No effect of 96-hour paclitaxel infusion in patients with relapsed non-Hodgkin's lymphoma refractory to a 3-hour schedule. *Journal of Clinical Oncology* 14, 543–548.

Younes, A., Sarris, A. and Melnyk, A. (1995) Three-hour paclitaxel infusion in patients with refractory and relapsed non-Hodgkin's lymphoma. *Journal of Clinical Oncology* 13, 583–587.

12   Paclitaxel content of yews in Ireland

Ingrid Hook and Dairine Dempsey

Yews are one of the sacred trees of Ireland, and the sanctity in which they are held is deeply rooted in Irish folklore, literature and language, with many names of places ending in 'ure' (Old Irish *yure* = yew). The trees were plentiful in Ireland until the late medieval times when deforestation by English soldiers began and 'prodigious quantities were destroyed for making coal for the iron-works' (More, 1898). The absence of very ancient yews, comparable to those in Britain (e.g. the 5,000-year old Fortingall yew in Tayside) can therefore be explained. The oldest yew trees in Ireland are thought to be those planted *c.*1440 at Muckross Abbey in Kerry (Chetan and Brueton, 1994), which were still abundant there, and around the Upper Lake Killarney in 1756 (More, 1898).

Since the fourth century BC the yew has been known for its poisonous properties. It was said by Culpeper to 'have no place among medicinal plants'. Withered branches are reported to be more toxic to cattle than fresh and growing ones (Withering, 1776). Sheep and goats will also succumb to the effects of yew clippings with consumption rates of 2.4–0.4 g leaves/kg goat body weight proving lethal (Cönen and Bohrs, 1994). The toxic compounds, taxine alkaloids, were first isolated in Germany in 1856 (Bryan-Brown, 1932 references therein; Henry, 1949). The amounts of taxines present in the Irish yews were published in 1909 (Moss, 1909). Results showed significant variation in yields depending on gender (*Taxus baccata* male – 0.18%; female – 0.12%), variety (female *T. baccata* Fastigiata – 0.61%) and growing conditions, for example, shade (*T. baccata* male – 0.082%), and although chemically interesting, the taxines were found to be too toxic to be of any pharmaceutical interest (Bryan-Brown, 1932).

It was the isolation of paclitaxel (Syn. taxol) from the bark of *Taxus brevifolia* and the discovery of its novel cytotoxic properties (Wani *et al.*, 1971) that rekindled scientific interest in *Taxus*. The identification of paclitaxel in other *Taxus* species at concentrations similar to those found in *T. brevifolia* (Witherup *et al.*, 1990) initiated more detailed investigations of these species. Our interest was to investigate yews in Ireland as to variability of paclitaxel content, and if possible to identify high-yielding trees with a view to possible commercial exploitation as a plantation crop.

Seasonal variation of paclitaxel levels

Samples were collected using our adopted protocol: twigs *c.*20 cm long taken from two height levels (1 and 3 m) and from four sides of each tree. Clippings were pooled, dried at <35° and stored in the absence of light until required for analyses (Griffin and Hook, 1996). Results from our initial investigation indicated the presence of paclitaxel in dried needles of *T. baccata* in mean concentrations of 0.0075%. Maximum levels were recorded in *T. baccata* Fastigiata (Irish or Florencecourt Yew) in April (0.01%). Growth studies also indicated that extention growth of shoots occurred between May and July, producing *c.*15–20 cm of new growth (Griffin and Hook,

1996). A second investigation with *T. baccata* Fastigiata produced a similar result with the highest levels of paclitaxel being recorded between February (0.021%) and April (0.019%) (Hook *et al.*, 1999). Such seasonal variation in paclitaxel content is in agreement with other *Taxus* species (Wheeler *et al.*, 1992; Vance *et al.*, 1994; ElSohly *et al.*, 1997). The latter proposed that for *T. x media* the best time of year to harvest for paclitaxel was approximately one month prior to the emergence of new growth. Other results obtained by us with samples collected from mainland Europe are at variance with the above, and suggest a climatic difference between maritime (Ireland) and continental growth conditions.

Variation of paclitaxel levels in *T. baccata* L.

The Common, English or European Yew, though originally widely distributed throughout the island was already extremely rare in the north-east (Derry) in 1835 (More, 1898). Around that time many wealthy families established gardens and arboreta, stocking them with exotic plant species. An example is Powerscourt, Co. Wicklow. Samples collected from there in April 1996 yielded the results shown in Table 12.1.

Effect of gender

Yews are dioecious. Mean values from the above data indicate female trees tend to yield higher levels of paclitaxel (female = 0.0095%; male = 0.0074%). Two previous studies by us gave similar results: 1995 collections of *T. baccata* gave mean values for female trees of 0.007% ($n = 7$) and for males 0.004% ($n = 3$). A larger collection taken over two days in August 1997 from 25 trees growing at the one site (Bedgebury Pinetum, Kent, UK) gave mean values for female *T. baccata* trees of 0.0098% ($n = 16$) and 0.0077% ($n = 9$) for male trees. Although these results tend to indicate female trees can be more productive, the effect of gender on paclitaxel content of yew trees has yet to be conclusively resolved. In one study by Nemeth-Kiss *et al.* (1996), which provided limited information on the gender of *T. baccata* samples analysed, the results indicated no gender differences.

Effect of variegation

Withering in his book (1776) *A Botanical Arrangement of all the Vegetables Naturally Growing in Great Britain* recognised two varieties of the Common Baccata '(1) leaves broad and shining, and (2) leaves variegated'. Our original research results indicated that variegated needles contain less paclitaxel than dark green needles (Griffin and Hook, 1996). A similar result was obtained by

Table 12.1 Paclitaxel content of *T. baccata* varieties from Powerscourt arboretum

Species/variety	Gender	Paclitaxel (%)
T. baccata cv.	Male	0.0058
T. baccata Aurea	Male	0.0101
T. baccata Dovastonia 1	Male	0.0053
T. baccata Dovastonia 2	Male	0.0074
T. baccata cv.	Male	0.0082
T. baccata Adpressa	Female	0.0078
T. baccata cv.	Female	0.0115
T. baccata Fastigiata	Female	0.0095
T. baccata Fructolutea	Female	0.0093

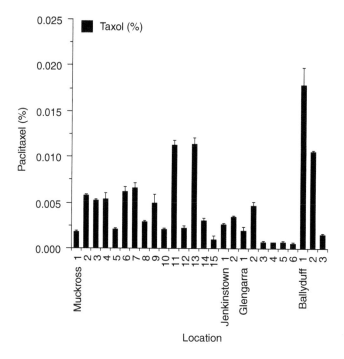

Figure 12.1 Variation in paclitaxel content of European yew trees growing in Ireland.

us, with samples taken from Arboreta in northern France. Examination of the Bedgebury Pinetum data also gave the same result: with mean values for green needles of 0.010% ($n = 13$) and 0.008% ($n = 12$) for variegated varieties.

Country-wide collections

Samples were taken from several trees growing at four southern locations. Samples were taken in May 1996 from mature trees of *T. baccata* growing at Ballyduff, Co. Waterford, Glengarra Wood, Co. Tipperary, Jenkinstown, Co. Waterford and Muckross Park, Co. Kerry.

It is obvious from Figure 12.1 that significant variation exists within each population of trees. The highest paclitaxel-yielding individual tree was found in Ballyduff, Co. Waterford (0.018%).

Variation of paclitaxel levels in *T. baccata* Fastigiata

The Irish yew is a female, columnar, that is fastigiate, variety of *T. baccata* that was originally noticed growing at the Florence Court Estate, Co, Fermanagh in 1780. All columnar yews are derived from cuttings taken off the two original trees. Our samples were collected in May 1996 from mature trees of *T. baccata* Fastigiata growing at Ballinasloe, Co. Galway, Collooney, Co. Sligo, Dundalk, Co. Louth, Fermoy, Co. Cork, New Ross, Co. Wexford and Rathmore, Co. Kerry. Four trees were sampled at each location (Sligo = 3) and all collections were made over a two-day period to minimise seasonal variations.

As expected, considerable variation was found to exist within each sample population. Lowest average paclitaxel yields were found in the most northerly samples (Sligo, 0.0053%), while

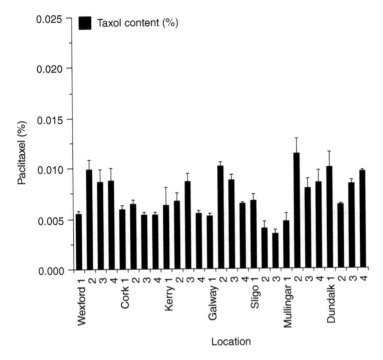

Figure 12.2 Variation in paclitaxel content of Irish yews growing at different locations.

highest mean yields came from samples collected from trees growing in the most easterly locations (Dundalk 0.0086% and Wexford 0.0083%). The average paclitaxel yield for all the samples was 0.0073%, which was in agreement with previous May collections of Irish yew (0.0071% in 1993), (Griffin and Hook, 1996) and 0.0093% in 1996 (Figure 12.2).

Hereditability of paclitaxel content

Rooted cuttings (*c.* two years old) originally taken from a series of *T. baccata* trees previously evaluated for paclitaxel content (see Figure 12.1) were harvested and extracted. Two rooted cuttings from two high yielding, two intermediate yielding and two low yielding trees were selected and analysed for paclitaxel content. Results from this small experiment (Figure 12.3) suggest that the trait for paclitaxel production is inherited, with cuttings taken from high yielding parent trees maintaining that productivity. This results suggest the importance of selecting high yielding cultivars for establishing *Taxus* plantations.

Field cultivation

Small trees (400 × 30–50 cm in height) of *T. baccata* Fastigiata were originally purchased and planted in April 1997 at the sites below. Equal numbers were monitored for biomass yields, clipping frequency and paclitaxel yields. Plants were equally divided between two locations:

1   Aughrim, Co. Wicklow, a warm coastal, sunny site with a rich mineral soil (loam).
2   Tullamore, Co. Offaly, a frost-prone midlands site consisting of peat/mineral soil overlying an alkaline marl (marginal, cut-over peat bog).

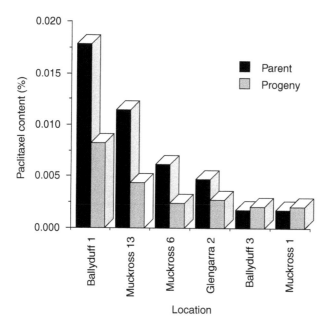

Figure 12.3 Comparison of paclitaxel content in parent trees with progeny cuttings.

Plants were planted 0.5 or 1 metre apart in rows, with a 1 m spacing between rows. Shoots were harvested in the winter by hand clipping. The following were the harvesting regimes:

- annually – once every year
- biennially – once every 2nd year
- triennially – once every 3rd year

From the mean biomass results (g/tree; $n = 25$) shown below it was obvious that the Aughrim site was more productive giving a total fresh weight for 3 × annual clippings of 332.8 g/tree, in contrast to 203 g produced by the less fertile Tullamore site (Figure 12.4).

The results also showed that annual clipping for three years of the same trees yielded more biomass than one triennial clipping, that is, 332.8 g vs 280.3 g at Aughrim and 203 g vs 127.5 g at Tullamore. Results from the mean height measurements showed that, as expected, plants grew better in Co. Wicklow than in Co. Offaly. However at both sites annual height increases in these young trees were small, confirming the slow growing nature of Irish yews (Table 12.2 and Figure 12.5).

Analyses of soil samples taken from the two sites were carried out by the Irish Forestry Board (Coillte) and are shown in Table 12.3. The very high levels of calcium and magnesium at the Tullamore site are obvious. They can be explained by the fact that as the upper layer of peat was removed the exposed sub-soil is a heavy shell marl, which is usually high in free calcium. This high level causes problems for growing commercial conifers and has been shown above to reduce the growth of *Taxus*. This environmental stress has however increased the yield of paclitaxel.

Results of the field trial obtained with small established trees of Irish yew proved that they could grow on nutritionally deficient marginal land, that is, Tullamore site. Not only did the trees grow sufficiently to give an annual harvest of biomass, but also their paclitaxel content was higher than in trees grown at more productive sites, that is, Aughrim. To overcome the shortage of biomass, trees could be planted more densely. The soil analyses indicated very high levels of calcium in the sample taken from Tullamore, which confirmed the importance of calcium in paclitaxel production. *Taxus* is traditionally known to grow well on limestone soils (Elwes and Henry, 1906).

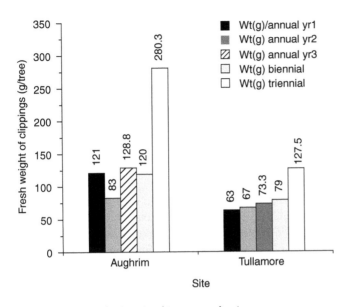

Figure 12.4 Taxus field trials – biomass production.

Table 12.2 Paclitaxel content of *T. baccata* 'Fastigiata' trees growing in Aughrim and Tullamore field sites

Site	Paclitaxel (mg/kg) annual end of yr 1	Paclitaxel (mg/kg) annual end of yr 2	Paclitaxel (mg/kg) biennial end of yr 2
Aughrim	95	205	211
Tullamore	123	188	226

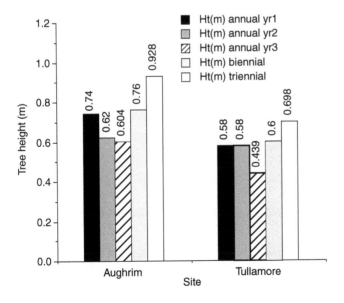

Figure 12.5 Taxus field trials – height.

Table 12.3 Soil analyses of samples taken from Aughrim and Tullamore (units for P, K, Ca and Mg are μg/g soil)

Site	pH	Free lime	Phosphorous	Potassium	Calcium	Magnesium
Aughrim	5.2	none	26	104	322	45
Tullamore	5.1	none	7	89	11,633	362

Acknowledgements

For financial support to the following:

European Community: AIR 3 - CT 94-1979 (DG 12SSMA) 'Selection, cultivation and harvesting of Yew: silviculture and biotechnological production as alternatives in the development of tax-anes with anticancer activity'. *Enterprise Ireland and the Irish Forestry Board* (Coillte Teoranta) for Research Project HE/1998/236 'Yew (*Taxus baccata*)-optimization of its field culture and establishment as a plantation crop'. Also David Thompson and John Fennessy, Coillte Teoranta Research Centres, for yew collections and field work.

References

Bryan-Brown, T. (1932) The pharmacological actions of taxine. *Quar. J. Pharm. Pharmacol.*, **V**(1): 205–219.

Chetan, A. and Brueton, D. (1994) *The Sacred Yew*. Ch. 13. Arkana Penguin Books Middlesex, England.

Coenen, M. and Bohrs, F. (1994) Fatal yew poisoning in goats as a result of ingestion of foliage from garden prunings. *Deutsche Tieratzliche Wochenschrift*, **101**: 364–367.

ElSohly, H.N., Croom, Jr, E.M., Kopycki, W.J., Joshi, A.S. and McChesney, J.D. (1997) Diurnal and seasonal effects on the taxane content of the clippings of certain *Taxus* cultivars. *Phytochem. Anal.*, **8**: 124–129.

Elwes, H.J. and Henry, A.H. (1906) In *The Trees of Great Britain and Ireland*, Vol. 1. S.R. Publishers Ltd., London. pp. 98–126.

Griffin, G. and Hook, I. (1996) Taxol content of Irish yews. *Planta Med.*, **62**: 370–372.

Henry, T. (1949) *The Plant Alkaloids*, 4th edn, Churchill, London. 769 pp.

Hook, I., Poupat, C., Ahond, A., Guenard, D., Gueritte, F., Adeline, M.-T., Wang, X.-P., Dempsey, D., Breuillet, S. and Potier, P. (1999) Seasonal variation of neutral and basic taxoid contents in shoots of European Yew (*Taxus baccata*). *Phytochemistry*, **52**: 1041–1045.

More, A.G. (1898) *Cybele Hibernica. Geographical distribution of plants in Ireland*. Edward Ponsonby, Dublin, pp. 330–331.

Moss, R. (1909) The taxine in Irish yew. *Taxus baccata* var *Fastigiata*. *The Proc. Royal Dub. Soc. XII* (NS) No. 10: 92–96.

Nemeth-Kiss, V., Forgacs, E., Cserhati, T. and Schmudt, G. (1996) Taxol content of various *Taxus* species in Hungary. *J. Pharm. Biomed. Anal.*, **14**: 997–1001.

Vance, N.C., Kelsey, R.G. and Sabin, T.E. (1994) Seasonal and tissue variation in taxane concentrations of *Taxus brevifolia*. *Phytochemistry*, **36**: 1241–1244.

Wani, M.C., Taylor, H.L., Wall, M.E., Coggan, P. and McPhail, A.T. (1971) Plant antitumour agents, IV. The isolation and structure of taxol, a novel antileukaemic and antitumour agent from *Taxus brevifolia*. *J. Am. Chem. Soc.*, **93**: 2325–2327.

Wheeler, N., Jech, K., Masters, S., Brobst, S.W., Alvarado, B.A., Hoover, A.J. and Snader, K.M. (1992) Effects of genetic, epigenetic and environmental factors on taxol content in *Taxus brevifolia* and related species. *J. Nat. Prod.*, **55**: 432–440.

Withering, W. (1776) *A Botanical Arrangement of all the Vegetables*, M. Swinney, Birmingham, 620 pp.

Witherup, K.M., Look, S.M., Stasko, M.W., Ghiorzi, T.J. and Muschik, G.M. (1990) *Taxus* Sp. needles contain amounts of taxol comparable to the bark of *Taxus brevifolia*: Analysis and isolation. *J. Nat. Prod.*, **53**: 1249–1255.

Name index

Subject index

Milton Keynes UK
Ingram Content Group UK Ltd.
UKHW052023071024
449327UK00027B/2399